Chemical Reactivity Theory

A Density Functional View

Chemical Reactivity Theory

A Density Functional View

Edited by
Pratim Kumar Chattaraj

CRC Press
Taylor & Francis Group
Boca Raton London New York

CRC Press is an imprint of the
Taylor & Francis Group, an **informa** business

CRC Press
Taylor & Francis Group
6000 Broken Sound Parkway NW, Suite 300
Boca Raton, FL 33487-2742

First issued in paperback 2020

© 2009 by Taylor & Francis Group, LLC
CRC Press is an imprint of Taylor & Francis Group, an Informa business

No claim to original U.S. Government works

ISBN-13: 978-0-367-57736-0 (pbk)
ISBN-13: 978-1-4200-6543-5 (hbk)

Library of Congress Cataloging-in-Publication Data

Chattaraj, Pratim Kumar.
    Chemical reactivity theory : a density functional view / Pratim Kumar Chattaraj.
        p. cm.
    Includes bibliographical references and index.
    ISBN 978-1-4200-6543-5 (hardcover : alk. paper)
        1. Density functionals. 2. Electron distribution. 3. Chemical reaction, Conditions and laws of. I. Title.

QD462.6.D45C43 2009
541'.394--dc22   2008054117

Visit the Taylor & Francis Web site at
http://www.taylorandfrancis.com

and the CRC Press Web site at
http://www.crcpress.com

Contents

Preface

Chemical reactions take place due to the redistribution of electron density among the reacting partners. Focusing on changes in electron density, which accompany the breaking and forming of chemical bonds, instead of the changes in the wave function accompanying them, allows us to use the "classical" three-dimensional language. Conceptual density functional theory (DFT) quantifies the possible responses of the system to various changes in density. Popular concepts like electronegativity, hardness, and electrophilicity, which explain a large number of diverse types of reactions in a systematic fashion, are grounded in conceptual DFT.

The aim of this book is to introduce various aspects of DFT and their connections to a chemical reactivity theory at a broadly accessible level. To this end, 34 chapters have been written by 65 eminent scientists from 13 different countries. Although the book is designed for readers with little or no prior knowledge of the subject, the breadth of the book and the expertise of the authors ensure that even experienced scientists will benefit from its contents.

The book comprises chapters on bonding, interactions, reactivity, dynamics, toxicity, and aromaticity as well as fundamental aspects of DFT. Several chapters are minireviews of the key global and local reactivity descriptors and their variations under different perturbations.

I am grateful to all the authors and the reviewers who cooperated with me to ensure the publication of the book on time. It is a great pleasure to express my gratitude to my teachers, Professors S.C. Rakshit, B.M. Deb, and R.G. Parr, for kindly introducing me to the fascinating field of quantum mechanics as applied to many-electron systems. I would especially like to thank Professor Paul Ayers, Lance Wobus, David Fausel, and Santanab Giri. Finally, I must express my gratitude to my wife Samhita and my daughter Saparya for their wholehearted support.

Pratim Kumar Chattaraj

Editor

Pratim Kumar Chattaraj joined the faculty of IIT Kharagpur after obtaining his BSc and MSc from Burdwan University and his PhD from the Indian Institute of Technology (IIT) Bombay. He is now a professor and the head of the department of chemistry and also the convener of the Center for Theoretical Studies at IIT Kharagpur. He was a postdoctoral research associate at the University of North Carolina at Chapel Hill and has served as a visiting professor to several other universities throughout the world. Apart from teaching, Professor Chattaraj is involved in research on density functional theory, the theory of chemical reactivity, ab initio calculations, quantum trajectories, and nonlinear dynamics. He has been invited to deliver special lectures at several international conferences and to contribute chapters to many edited volumes. Professor Chattaraj is a member of the editorial board of the *Journal of Molecular Structure* (*Theochem*) and the *Journal of Chemical Sciences*, among others. He is a council member of the Chemical Research Society of India and a fellow of the Indian Academy of Sciences (Bangalore, India) and the Indian National Science Academy (New Delhi, India).

Contributors

Paul W. Ayers
Department of Chemistry
McMaster University
Hamilton, Ontario, Canada

P. Balanarayan
Department of Chemistry
University of Pune
Pune, Maharashtra, India

Libero J. Bartolotti
Department of Chemistry
East Carolina University
Greenville, North Carolina

Subhash C. Basak
Center for Water and Environment
Natural Resources Research Institute
University of Minnesota Duluth
Duluth, Minnesota

Josep Maria Bofill
Departament de Química Orgánica
Universitat de Barcelona i Parc Cientific
 de Barcelona
Barcelona, Spain

Alexander I. Boldyrev
Department of Chemistry and
 Biochemistry
Utah State University
Logan, Utah

Patrick Bultinck
Department of Inorganic and Physical
 Chemistry
Ghent University
Ghent, Belgium

Ramon Carbó-Dorca
Department of Inorganic and Physical
 Chemistry
Ghent University
Ghent, Belgium

and

Department of Chemistry
Institute of Computational Chemistry
University of Girona
Girona, Spain

Andrés Cedillo
Departamento de Química
Universidad Autónoma Metropolitana-
 Iztapalapa
Mexico City, México

Asit K. Chandra
Department of Chemistry
North-Eastern Hill University
Shillong, Meghalaya, India

P. K. Chattaraj
Department of Chemistry and Center
 for Theoretical Studies
Indian Institute of Technology
Kharagpur, West Bengal, India

Abhijit Chatterjee
Material Science
Accelrys KK
Tokyo, Japan

Sofie Van Damme
Department of Inorganic and Physical
 Chemistry
Ghent University
Ghent, Belgium

Susmita De
Department of Inorganic and Physical
  Chemistry
Indian Institute of Science
Bangalore, Karnataka, India

B. M. Deb
Indian Institute of Science Education
  and Research
Kolkata, West Bengal, India

P. Fuentealba
Departamento de Física
Facultad de Ciencias
Universidad de Chile
Santiago, Chile

Shridhar R. Gadre
Department of Chemistry
University of Pune
Pune, Maharashtra, India

Marcelo Galván
Departamento de Química
División de Ciencias Básicas
  e Ingeniería
Universidad Autónoma Metropolitana-
  Iztapalapa
Mexico City, México

Jorge Garza
Departamento de Química
División de Ciencias Básicas
  e Ingeniería
Universidad Autónoma Metropolitana-
  Iztapalapa
Iztapalapa, México

José L. Gázquez
Departamento de Química
División de Ciencias Básicas
  e Ingeniería
Universidad Autónoma Metropolitana-
  Iztapalapa
Mexico City, México

Paul Geerlings
Eenheid Algemene Chemie
Faculty of Sciences
Free University of Brussels
Brussels, Belgium

Swapan K. Ghosh
Theoretical Chemistry Section
Bhabha Atomic Research Centre
Mumbai, Maharashtra, India

Xavier Giménez
Departament de Química Física
Universitat de Barcelona i Parc Cientific
  de Barcelona
Barcelona, Spain

S. Giri
Department of Chemistry and Center
  for Theoretical Studies
Indian Institute of Technology
Kharagpur, West Bengal, India

D. Guerra
Department of Chemistry
Universidad Técnica Federico
  Santa María
Valparaíso, Chile

Brian D. Gute
Center for Water and Environment
Natural Resources Research Institute
University of Minnesota Duluth
Duluth, Minnesota

Soledad Gutiérrez-Oliva
Laboratorio de Quimica Teórica
  Computacional
Facultad de Quimica
Pontificia Universidad Católica
  de Chile
Santiago, Chile

Manoj K. Harbola
Department of Physics
Indian Institute of Technology
Kanpur, Uttar Pradesh, India

Eluvathingal D. Jemmis
Indian Institute of Science Education
  and Research
CET Campus
Tiruvananthapuram, Tamil Nadu, India

Rahul Kar
Physical Chemistry Division
National Chemical Laboratory
Pune, Maharashtra, India

P. Kolandaivel
Department of Physics
Bharathiar University
Coimbatore, Tamil Nadu,
  India

György Lendvay
Institute of Structural Chemistry
Chemical Research Center
Hungarian Academy of Sciences
Budapest, Hungary

M. Levy
Department of Chemistry
Duke University
Durham, North Carolina

Shubin Liu
Research Computing Center
Renaissance Computing Institute
University of North Carolina
Chapel Hill, North Carolina

Eduard Matito
The Lundbeck Foundation Center
  for Theoretical Chemistry
Department of Chemistry
University of Aarhus
Aarhus, Denmark

Denise Mills
Center for Water and Environment
Natural Resources Research Institute
University of Minnesota Duluth
Duluth, Minnesota

Salvador Miret-Artés
Department of Atomic, Molecular,
  and Cluster Physics
Instituto de Física Fundamental
Consejo Superior de Investigaciones
  Científicas
Madrid, Spain

Jane S. Murray
Department of Chemistry
University of New Orleans
New Orleans, Louisiana

and

Department of Chemistry
Cleveland State University
Cleveland, Ohio

Á. Nagy
Department of Theoretical Physics
University of Debrecen
Debrecen, Hungary

Roman F. Nalewajski
Department of Theoretical Chemistry
Jagiellonian University
Cracow, Poland

Ramanathan Natarajan
Department of Chemical Engineering
Lakehead University
Thunder Bay, Ontario, Canada

Minh Tho Nguyen
Department of Chemistry
Mathematical Modeling and
  Computational Science Center
University of Leuven
Leuven, Belgium

Sourav Pal
Physical Chemistry Division
National Chemical Laboratory
Pune, Maharashtra, India

Robert G. Parr
Department of Chemistry
University of North Carolina
Chapel Hill, North Carolina

Ralph G. Pearson
Chemistry and Biochemistry
  Department
University of California
Santa Barbara, California

Jordi Poater
Institut de Química Computacional
  and Departament de Química
Universitat de Girona
Girona, Spain

Peter Politzer
Department of Chemistry
University of New Orleans
New Orleans, Louisiana

and

Department of Chemistry
Cleveland State University
Cleveland, Ohio

Paul Popelier
Manchester Interdisciplinary Biocenter
University of Manchester
Manchester, United Kingdom

Frank De Proft
Eenheid Algemene Chemie
Vrije Universiteit Brussel
Brussels, Belgium

D. R. Roy
Department of Chemistry and Center
  for Theoretical Studies
Indian Institute of Technology
Kharagpur, West Bengal, India

Ángel S. Sanz
Department of Atomic, Molecular,
  and Cluster Physics
Instituto de Física Fundamental
Consejo Superior de Investigaciones
  Científicas
Madrid, Spain

G. Narahari Sastry
Molecular Modeling Group
Indian Institute of Chemical Technology
Hyderabad, Andhra Pradesh, India

A. Savin
Laboratoire de Chimie Theorique
CNRS et Universite Pierre et Marie
  Curie Paris VI
Paris, France

Paul von Ragué Schleyer
Department of Chemistry and Center
  for Computational Chemistry
University of Georgia
Athens, Georgia

K. D. Sen
School of Chemistry
University of Hyderabad
Hyderabad, Andhra Pradesh, India

Patrick Senet
Institut Carnot de Bourgogne
UMR 5209 CNRS-Université
  de Bourgogne
Dijon, France

Alina P. Sergeeva
Department of Chemistry and
  Biochemistry
Utah State University
Logan, Utah

Miquel Solà
Institut de Química Computacional
  and Departament de Química
Universitat de Girona
Girona, Spain

V. Subramanian
Chemical Laboratory
Central Leather Research Institute
Chennai, Tamil Nadu, India

Alejandro Toro-Labbé
Laboratorio de Quimica Teórica
  Computacional
Facultad de Quimica
Pontificia Universidad Católica de Chile
Santiago, Chile

David J. Tozer
Department of Chemistry
University of Durham
Durham, United Kingdom

Rubicelia Vargas
Departamento de Química
División de Ciencias Básicas
  e Ingeniería
Universidad Autónoma Metropolitana-
  Iztapalapa
Mexico City, México

P. Venuvanalingam
School of Chemistry
Bharathidasan University
Tiruchirapalli, Tamil Nadu, India

Weitao Yang
Department of Chemistry
Duke University
Durham, North Carolina

Dmitry Yu. Zubarev
Department of Chemistry and
  Biochemistry
Utah State University
Logan, Utah

1 How I Came about Working in Conceptual DFT

Robert G. Parr

CONTENTS

When Pratim Chattaraj asked me to provide some kind of "foreword" to this book, my first reaction was "no," since everyone knows that the past is not so important in science and that one person's recollections often are faulty. What we had here was just a long-laboring quantum chemist with a rigorous training in classical Gibbsian chemical thermodynamics, always enchanted with the chemical potential. So when quantum chemistry was suddenly confronted with the density functional theory (DFT), I was ready and happy to plunge into work with DFT, the chemical potential again taking a central role. To say a little more, below is reproduced a short piece which I provided for a 2003 Springer book entitled *Walter Kohn* (two photographs which were in the original are omitted). What this contains is the story of how DFT came into chemistry proper, broadening computational chemistry and enlightening chemical concepts both old and new. Concepts are what this volume is mainly about: conceptual DFT.

1.1   BONDING OF QUANTUM PHYSICS WITH QUANTUM CHEMISTRY*

The bond that developed between quantum physics and quantum chemistry, that led to the award of a big chemistry prize to the physicist Walter Kohn in 1998, developed not without trial. Here I give an account of it. An element in this bond has been a friendship between Walter Kohn and me. My having reached 80 first, he has already kindly spoken of this [1]. Now it is my turn.

In the 20s and early 30s there was a flush of successes in establishing the ability of quantum mechanics to describe the simplest molecules accurately: the Born-Oppenheimer approximation, the nature of chemical bonding, and the fundamentals

---

* I thank Springer for allowing me to reproduce this article.

of molecular spectroscopy. But then the quantitative theory of molecular structure, which we call quantum chemistry, was stymied, by the difficulty of solving the Schrödinger equation for molecules. The senior chemical physicists of the 30s pronounced the problem unsolvable. But the younger theoreticians in the period coming out of WWII thought otherwise. Clearly one could make substantial progress toward the goal of complete solution, because the equation to solve was known and had a simple universal structure.

The boundary conditions too were known. It would not be as easy as handling an infinite periodic solid, but a number of us set to work. The special demand of chemistry was to quantify very small molecular changes. Successes came slowly, but with the development of computers and a lot of careful, clever work, by the 90s the quantitative problem was essentially solved. The emergent hero of the chemical community was John Pople, whose systematic strategy and timely method developments were decisive. The methods of what is termed "ab initio" quantum chemistry became available and used everywhere.

Over the years the quantum chemists did a lot more than gradually improve their ability to calculate wavefunctions and energies from Schrödinger's equation. All the while they have served molecular spectroscopy, physical inorganic chemistry, and physical organic chemistry. Relevant for the present story was the development by Per-Olov Löwdin in 1955 of the density matrix reduction of the Schrödinger equation, especially the identification and mathematical physics of natural spin orbitals and their occupation numbers. The hope was, although hope floundered, that the Schrödinger problem could be resolved in terms of the first- and second-order density matrices. Foundering came because of the difficulty of incorporating the Pauli principle.

Beginning way back in the 20s, Thomas and Fermi had put forward a theory using just the diagonal element of the first-order density matrix, the electron density itself. This so-called statistical theory totally failed for chemistry because it could not account for the existence of molecules. Nevertheless, in 1968, after years of doing wonders with various free-electron-like descriptions of molecular electron distributions, the physicist John Platt wrote [2] "We must find an equation for, or a way of computing directly, total electron density." [This was very soon after Hohenberg and Kohn, but Platt certainly was not aware of HK; by that time he had left physics.]

From the end of the 40s, I was a happy participant in most of these things, ab initio and the rest, although from about 1972 I became pretty much an observer. We plunged into density-functional theory.

DFT soon intoxicated me. There were the magnificent Hohenberg-Kohn and Kohn-Sham 1964–65 papers. The Xalpha method of John Slater was popular in those days, but it was not sufficient for the high accuracy needed. And I was much taken by the work of Walter Kohn, whom I had known since 1951. There were many things to do: Improve upon the LDA to reach the accuracy needed for chemical applications. Shift the emphasis on fixed, very large electron number toward variable, small number, since that most concerns chemistry. Enlarge the language to include chemical as well as solid-state concepts. Introduce into DFT, as appropriate, some of the theoretical advances already made within quantum chemistry. All of these things subsequently came about. The methods and concepts of DFT became available and used throughout the chemical community.

I had been on the faculty of Carnegie Institute of Technology for a couple of years when Walter Kohn arrived in 1951. I was aware from the beginning of the strength of physics at Carnegie, especially solid state physics. Fred Seitz was the Head when I arrived, and several other solid state experts also were there. I bought Seitz's great book for $6.38 and browsed in it, noting in particular the fine description of the Hartree-Fock method (but not finding any treatment of the invariance to unitary transformation of orbitals that is so important for understanding the equivalence of localized and non-localized descriptions of molecules). I enjoyed pleasant interactions with a number of the physicists. Soon after Kohn arrived, I had two physicist postdocs, Tadashi Arai from Japan and Fausto Fumi from Italy, who became acquainted with him. On the thesis examination committee of Walter's graduate student Sy Vosko, I learned that it was okay to use trial wavefunctions with discontinuous derivatives. I was pleased to attend an evening party at the Kohns, and I was disappointed when Walter left Carnegie for elsewhere.

I do not recall when I first heard of the Hohenberg-Kohn-Sham papers, but I do know that the quantum chemistry community at first paid little attention to them. In June of 1966 Lu Sham spoke about DFT at a Gordon Conference. But in those days, there was more discussion about another prescription that had been on the scene since 1951, the Slater Xa method. The Xa method was a well-defined, substantial improvement over the Thomas-Fermi method, a sensible approximation to exact Kohn-Sham. Debate over Xa went on for a number of years. Slater may never have recognized DFT as the major contribution to physics that it was. [When I asked John Connolly five or six years ago how he thought Slater had viewed DFT, he replied that he felt that Slater regarded it as "obvious."]

Walter Kohn's appearance at the Boulder Theoretical Chemistry Conference of 1975 was memorable. On June 24, he presented a formal talk, in which he outlined DFT to the assemblage of skeptical chemists. There were many sharp questions and a shortage of time, so the chair of the conference decided to schedule a special session for the afternoon of June 26. With quite a crowd for an informal extra session like this, Walter held forth on his proof. In his hand, he held a reprint of the HK paper, from which he quietly read as he slowly proceeded: ".... and now we say ...". The audience sobered down quickly. It was a triumph. The interest of quantum chemists in DFT began to grow at about this time.

Our group began contributing to DFT in the 1970s. In some of our first work, my graduate student Robert Donnelly generalized the original idea to functionals of the first-order density matrix. In 1977, I described the central result at Walter Kohn's luncheon seminar in San Diego: All natural orbitals with nonzero occupation numbers have the same chemical potential. Discussing this with Walter at the blackboard afterwards, I remember his saying "This must be correct." [Walter himself recently recalled this incident.] First-order density matrix functional theory is receiving fresh attention nowadays.

As we ourselves kept plugging along, the quantum chemical community largely was negative about DFT, even antagonistic. Their "house journal" *International Journal of Quantum Chemistry*, in 1980, published a pointed criticism of it [3]: "There seems to be a misguided belief that a one-particle density can determine the exact N-body ground state." In 1982, Mel Levy and John Perdew replied with a letter that was both incisive and eloquent [4]: "The belief is definitely not misguided." Yet, in the same issue of *IJQC*, the editors called for further discussion of the "controversial" subject [5]. It was going to be awhile before quantum chemists were convinced.

Over the period 1979–1982, Mel Levy supplied a major advance with introduction and careful discussion of the constrained search formulation of density functional theory. This greatly heightened confidence in the theory (and it still does!). Then in 1983 came Elliott Lieb's masterly detailed analysis, which validated DFT as rigorous mathematical physics. [Once in the 70s I asked Barry Simon, the mathematician who with Lieb had done a famous rigorous analysis of the Thomas-Fermi theory, what his opinion was of DFT. "It may be good physics," he said, "but it is not good mathematics"]. Lieb's paper signaled the end of the period of doubt about DFT. The space for further development was now wide open and the interest of chemists began to accelerate.

What computational chemists wanted above all else was calculational methods for molecules, and the LDA just was not enough. The need for more accurate exchange-correlation functionals was met in the 80s, with an accuracy that has proved quite good enough for the times. The Nobel award in 1998, one may point out, was specifically designated to be a prize for computational chemistry. Well, good, and immensely deserved in my opinion. I note, however, that there is another whole side of DFT which has concerned and still concerns many of us, the "conceptual" side. This side is rich in potential, and it is not without accomplishment. The concepts of DFT neatly tie into older chemical reasoning, and they are useful for discussing molecules in course of reaction as well as for molecules in isolation. Where solid-state physics has Fermi energy, chemical potential, band gap, density of states, and local density of states, quantum chemistry has ionization potential, electron affinity, hardness, softness, and local softness. Much more too. DFT is a single language that covers atoms, molecules, clusters, surfaces, and solids.

Walter Kohn has been a great help to many scientists over many years, an expert consultant and helpmate and a fine, unobtrusive, even-handed host of good meetings in lovely places. We thank him. In recent years I have discussed with him (among other things), circulant orbitals, the monotonic density theorem, and the information theory point of view on what constitutes an atom in a molecule, the latter during a stolen few minutes in a Stockholm hotel in December of 1999 [6]. Walter may or may not "like" chemistry [7], and he claims not to have studied chemistry in the university. But what does one call a great teacher of chemical principles? I would say, CHEMIST, full caps.

REFERENCES

1. W. Kohn, in *Reviews of Modern Quantum Chemistry*. K.D. Sen (Ed.). Vols. I, II, World Scientific, Singapore, 2002, pp. v–vii.
2. John R. Platt, letter to RGP, dated October 23, 1968.
3. *International Journal of Quantum Chemistry* 18, 1029, 1980.
4. M. Levy and J.P. Perdew, *International Journal of Quantum Chemistry* 21, 511, 1982.
5. *International Journal of Quantum Chemistry* 21, 357, 1982.

6. RGP was delighted to be Walter Kohn's guest. See P. Hohenberg, in *Walter Kohn: Personal Stories and Anecdotes Told by Friends and Collaborators*, M. Scheffler and P. Weinberger (Eds.), Springer, Berlin, 2003, p. 120.
7. E. Eliel, in *Walter Kohn: Personal Stories and Anecdotes Told by Friends and Collaborators*, M. Scheffler and P. Weinberger (Eds.), Springer, Berlin, 2003, p. 79.

2 Chemical Reactivity Concepts in Density Functional Theory

José L. Gázquez

CONTENTS

2.1 INTRODUCTION

The description of chemical reactivity implies, among other aspects, the study of the way in which a molecule responds to the attack of different types of reagents. In order to establish this response, one usually adopts the electronic structure of the molecule in its isolated state as the reference point and considers the effects of an attacking reagent on this state. This procedure leads to the description of what we may call the inherent chemical reactivity of a molecule.

The inherent chemical reactivity of a great variety of molecules has been described over the years through different concepts and principles. To mention some, the concept of electronegativity, together with the electronegativity equalization principle, has been used to qualitatively establish the distribution of electronic charge between the different atoms in a molecule, or the direction of the flow of charge when two species interact. The concepts of hardness and softness, together with the hard and soft acids and bases principle, have been used to explain the vast world of Lewis acid–base chemistry qualitatively.

Undoubtedly, from a theoretical viewpoint, from the wave function approach to the description of chemical processes, molecular orbitals have been amply used for a basically qualitative, but at the same time conceptually simple and general, interpretation of the inherent chemical reactivity. Particularly, the frontier orbital theory has been used to understand fundamental aspects of a great variety of organic and inorganic reactions. In order to achieve greater accuracy in the description of chemical processes through a wave function language, one needs to include correlation effects at the expense of losing the conceptual simplicity of the orbital picture.

In the last three decades, density functional theory (DFT) has been extensively used to generate what may be considered as a general approach to the description of chemical reactivity [1–5]. The concepts that emerge from this theory are response functions expressed basically in terms of derivatives of the total energy and of the electronic density with respect to the number of electrons and to the external potential. As such, they correspond to conceptually simple, but at the same time, chemically meaningful quantities.

Unlike the wave function description, in which increasing accuracy implies a greater complexity in the interpretation, the density functional approach maintains its simplicity, since the derivatives can be evaluated as accurately as possible, but their chemical meaning remains the same.

The objective of the present chapter is to analyze the chemical reactivity criteria that emerge from DFT. Thus, in Section 2.2, we present the extension of DFT to noninteger number of electrons. In Section 2.3, we discuss the behavior of the energy as a function of the number of electrons in order to link the concept of electronegativity with the negative of the chemical potential of DFT, and to identify the concept of hardness. In Section 2.4, we examine the derivatives of the electronic density with respect to the number of electrons to present the concepts of the Fukui function and the dual descriptor, together with their main properties. In Section 2.5, we derive the expressions for all these derivatives in the Kohn–Sham (KS) and the Hartree–Fock (HF) methods to establish their interpretation in an orbital language. Finally, in Section 2.6, we give some concluding remarks.

2.2 DENSITY FUNCTIONAL THEORY FOR NONINTEGER NUMBER OF ELECTRONS

In DFT, the ground-state energy of an atom or a molecule is written in terms of the electronic density $\rho(\mathbf{r})$, and the external potential $v(\mathbf{r})$, in the form [1,6]

$$E[\rho] = F[\rho] + \int d\mathbf{r}\rho(\mathbf{r})v(\mathbf{r}), \qquad (2.1)$$

where $F[\rho]$ is the universal Hohenberg–Kohn functional,

$$F[\rho] = T[\rho] + V_{ee}[\rho], \qquad (2.2)$$

where
    $T[\rho]$ represents the electronic kinetic energy functional
    $V_{ee}[\rho]$ the electron–electron interaction energy functional

The minimization of the total energy, subject to the condition that the total number of electrons N is fixed,

$$N = \int d\mathbf{r} \rho(\mathbf{r}), \tag{2.3}$$

leads to an Euler–Lagrange equation of the form

$$\mu = \left(\frac{\delta E}{\delta \rho(\mathbf{r})} \right)_v = v(\mathbf{r}) + \frac{\delta F}{\delta \rho(\mathbf{r})}, \tag{2.4}$$

where μ, the Lagrange multiplier, is the chemical potential. The solution of this equation leads to the ground-state density, from which one can determine the ground-state energy.

The external potential [1] is responsible for keeping the electrons confined to a region of space. For the case of an isolated molecule, the external potential is the potential generated by its nuclei. When one considers the interaction between a molecule and another species, then the external potential is the one generated by the nuclei of both species, and it acts on all the electrons. However, when they are very far apart from each other, since the electrons of both species are localized in, basically, separated regions, then the external potential of each species may be assumed to be the one generated by its own nuclei, and by the nuclei and the electrons of the other species.

The starting point of the DFT of chemical reactivity is the identification of the concept of electronegativity, χ, with the chemical potential, through the relationship [7]

$$\mu = \left(\frac{\partial E}{\partial N} \right)_v = -\chi, \tag{2.5}$$

where the second equality corresponds to the definition of electronegativity given by Iczkowski and Margrave [8] as a generalization of the definition of Mulliken [9], and the first equality, established by Parr et al. [7], links the chemical potential of DFT with the derivative of the energy with respect to the number of electrons.

The chemical potential of DFT measures the escaping tendency of the electrons from a system [1]. That is, electrons flow from the regions with higher chemical potential to the regions with lower chemical potential, up to the point in which μ becomes constant throughout the space. Thus, the chemical potential of DFT is equivalent to the negative of the concept of electronegativity, and the principle of electronegativity equalization [10,11] follows readily from this identification [1,7,12,13].

The next step is the identification of the concept of chemical hardness, η, with the second derivative of the energy with respect to the number of electrons, formulated by Parr and Pearson [14]

$$\eta = \left(\frac{\partial^2 E}{\partial N^2} \right)_v = \left(\frac{\partial \mu}{\partial N} \right)_v. \tag{2.6}$$

The original definition was established as $1/2$ of the second derivative; however, the one given in Equation 2.6 is more convenient and it has become the most common one in the recent literature.

The identification given by Equation 2.6 is consistent with the hard and soft acids and bases principle [12,15–21] established originally by Pearson [22] to explain many aspects of Lewis acid–base chemistry.

It is important to mention that when one considers the derivative of a quantity at constant external potential, it means that the changes in the quantity are analyzed for a fixed position of the nuclei. These types of changes are known as vertical differences, precisely because the nuclei positions are not allowed to relax to a new position associated with a lower total energy.

Since the definitions given by Equations 2.5 and 2.6 require the energy to be well defined for noninteger number of electrons and to be differentiable with respect to N, let us present first the extension of DFT to open systems that can exchange electrons with its environment, developed by Perdew, Parr, Levy, and Balduz (PPLB) [23].

An open system with a fluctuating number of particles is described by an ensemble or statistical mixture of pure states and the fractional electron number may arise as a time average. Thus, let N_0 be an integer electron number, and N be an average in the interval $N_0 < N < N_0 + 1$. In this case Γ is an ensemble or statistical mixture of the N_0-electron pure state with wave function Ψ_{N_0} and probability $1 - \tau$, and the $(N_0 + 1)$-electron pure state with wave function Ψ_{N_0+1} and probability τ, where $0 < \tau < 1$.

Now, PPLB showed, through a constrained search of the minimum energy over all ensembles Γ yielding a given density and then over all densities integrating to a given electron number, that the ground-state density and the ground-state energy for the N electrons subject to the external potential $v(\mathbf{r})$, are

$$\rho_N(\mathbf{r}) = \rho_{N_0+\tau}(\mathbf{r}) = (1 - \tau)\rho_{N_0}(\mathbf{r}) + \tau\rho_{N_0+1}(\mathbf{r}), \qquad (2.7)$$

and

$$E_N = E_{N_0+\tau} = (1 - \tau)E_{N_0} + \tau E_{N_0+1}, \qquad (2.8)$$

where $\rho_{N_0}(\mathbf{r})$ and E_{N_0} represent the ground-state density and the ground-state energy of the N_0-electron system, while $\rho_{N_0+1}(\mathbf{r})$ and E_{N_0+1} correspond to the $(N_0 + 1)$-electron system, in both cases subject to the external potential $v(\mathbf{r})$.

Integrating Equation 2.7 over the whole space leads to

$$N = (1 - \tau)N_0 + \tau(N_0 + 1) = N_0 + \tau. \qquad (2.9)$$

The relationship expressed in Equation 2.8 indicates that the total energy as a function of the number of electrons is given by a series of straight lines connecting the ground-state energies of the systems with integer number of electrons (see Figure 2.1). Thus, since the energy is a continuous function of the number of electrons, the differentiation with respect to electron number is justified. However, it is clear from Equation 2.8 that the first derivative will present discontinuities.

Finally, it is important to mention that the joined straight line structure has been confirmed through arguments based on the size consistency of the energy without invoking the grand canonical ensemble [24,25].

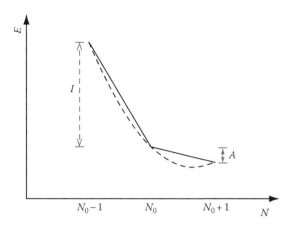

FIGURE 2.1 Plot of the total energy as a function of the number of electrons. The solid straight lines correspond to Equation 2.8, while the dashed curve corresponds to Equation 2.18.

2.3 DERIVATIVES OF THE ENERGY WITH RESPECT TO THE NUMBER OF ELECTRONS

Let us first rewrite Equation 2.8 in the form

$$\frac{E_{N_0+\tau} - E_{N_0}}{\tau} = E_{N_0+1} - E_{N_0} = -A, \tag{2.10}$$

and the equivalent of Equation 2.8 for the interval between $N_0 - 1$ and N_0 in the form

$$\frac{E_{N_0} - E_{N_0-\tau}}{\tau} = E_{N_0} - E_{N_0-1} = -I, \tag{2.11}$$

where I and A are the vertical first ionization potential and the vertical electron affinity, respectively. Note that the energy difference in the right-hand side of Equations 2.10 and 2.11 is independent of τ. Thus, if one takes the limit $\tau \to 0$ in Equations 2.10 and 2.11, then [23,26,27]

$$\lim_{\tau \to 0} \frac{E_{N_0+\tau} - E_{N_0}}{\tau} = \left(\frac{\partial E}{\partial N}\right)_v^+ = \mu^+ = E_{N_0+1} - E_{N_0} = -A, \tag{2.12}$$

and

$$\lim_{\tau \to 0} \frac{E_{N_0} - E_{N_0-\tau}}{\tau} = \left(\frac{\partial E}{\partial N}\right)_v^- = \mu^- = E_{N_0} - E_{N_0-1} = -I, \tag{2.13}$$

where Equation 2.5 has been used.

A very important consequence of the behavior of the energy as a function of the number of electrons is that the left (μ^-) and right (μ^+) first derivatives are not equal. From a chemical perspective, this differentiation with respect to charge addition or charge removal is important, because, in general, one could expect that the chemical species will respond differently to these two processes, and therefore the left derivative μ^- (response to charge removal) could be different from the right derivative μ^+ (response to charge addition).

Note that the arithmetic average of the one-sided derivatives

$$\left(\frac{\partial E}{\partial N}\right)_v^0 = \mu^0 = \frac{1}{2}(\mu^- + \mu^+) = -\frac{1}{2}(I + A), \tag{2.14}$$

corresponds to the definition of electronegativity given by Mulliken [9].

According to the results obtained for the first derivative, Equations 2.12 and 2.13, the second derivative, i.e., the hardness, is zero when evaluated from the left or from the right, and it is not defined for integer number of electrons. However, Ayers [25] has shown that if one makes use of the Heaviside step function

$$\Theta(x) = \begin{cases} 0, & x < 0, \\ 1, & x > 0, \end{cases} \tag{2.15}$$

the chemical potential can be written in the form

$$\mu^{(N_0+x)} = \mu^- + \Theta(x)(\mu^+ - \mu^-), \quad -1 \le x \le 1. \tag{2.16}$$

Thus, since the derivative of the Heaviside function is the Dirac delta function, $(d\Theta(x)/dx) = \delta(x)$, taking the derivative in Equation 2.16 with respect to x, one finds, using Equation 2.6, that the hardness is given by

$$\eta^{(N_0+x)} = (\mu^+ - \mu^-)\delta(x) = (I - A)\delta(x), \quad -1 \le x \le 1, \tag{2.17}$$

where Equations 2.12 and 2.13 have been used.

Up to this point, we have worked with the nonsmooth expression for the energy that results from the ensemble extension to fractional electron numbers, and that has been confirmed through the size consistency of the energy. However, in order to incorporate the second-order effects associated with the charge transfer processes, the most common approach has been to make use of a smooth quadratic interpolation [14]. That is, with the two definitions given in Equations 2.5 and 2.6, the energy change ΔE due to the electron transfer ΔN, when the external potential $v(\mathbf{r})$ is kept fixed, may be approximated through a second order Taylor series expansion of the energy as a function of the number of electrons

$$\Delta E = \mu\Delta N + \frac{1}{2}\eta(\Delta N)^2. \tag{2.18}$$

The values of the two derivatives, μ and η, at the reference point N_0, may be approximated through this smooth quadratic interpolation between the points $E(N_0 - 1)$, $E(N_0)$ and $E(N_0 + 1)$, when combined with the two conditions, $E(N_0 - 1) - E(N_0) = I$ and $E(N_0) - E(N_0 + 1) = A$ (see Figure 2.1). This procedure leads to the well-known finite differences approximations

$$\mu = -\chi \approx -\frac{I+A}{2}, \tag{2.19}$$

and

$$\eta \approx I - A. \tag{2.20}$$

A remarkable fact is that Equation 2.19 is the same as Equation 2.14, and Equation 2.20 is the same as Equation 2.17 if one ignores the Dirac delta function, although Equations 2.14 and 2.17 result from the ensemble approach, while Equations 2.19 and 2.20 result from the smooth quadratic interpolation. Thus, the expressions given by Equations 2.19 and 2.20 are fundamental to evaluate the chemical potential (electronegativity) and the chemical hardness.

When the experimental values of I and A are known, one can determine through these expressions the values of μ and η. Since for atoms and molecules, the trends shown by these values of μ and η are, in general, in line with those provided by several empirical scales constructed intuitively by chemists, the identification of these global DFT descriptors with their associated chemical concepts is strengthened. In other words, the quantity $(I+A)/2$ shows, in general, the same behavior as that of the electronegativity concept, while the quantity $(I-A)$ shows, also in general, the same behavior as that of the chemical hardness concept.

Finally, it is important to mention that in the case of Equation 2.18 the energy and its derivatives are continuous functions of the number of electrons around N_0, so that a single value for μ and a single value for η are used to describe charge transfer processes. That is, the advantage of this procedure is to have well-defined first and second derivatives and the disadvantage is that they are the same for charge addition and for charge removal processes.

On the other hand, in the case of Equation 2.8, in spite of the mathematical difficulties associated with the discontinuities, one has the advantage of being able to differentiate the response of the system to charge donation from that corresponding to charge acceptance.

2.4 DERIVATIVES OF THE ELECTRONIC DENSITY WITH RESPECT TO THE NUMBER OF ELECTRONS

Let us consider now the response functions that arise when a chemical system is perturbed through changes in the external potential. These quantities are very important in the description of a chemical event, because for the early stages of the interaction, when the species are far apart from each other, the change in the external potential of one of them, at some point $\mathbf{r}$, is the potential generated by the nuclei and electrons of the other one at that point, and vice versa. As a consequence, the functional derivatives of the energy, the chemical potential, and the chemical hardness, with respect to the external potential, provide information on the response of a molecule to the presence of a reagent.

For the energy, one can make use of first-order perturbation theory for a non-degenerate state. Thus, if the state Ψ_k^0 is perturbed to the state $\Psi_k = \Psi_k^0 + \Psi_k^1$ by the one-electron perturbation $\Delta V = \sum_i \delta v(\mathbf{r}_i)$, the energy change to first order is [1]

$$\delta E = E_k^1 = \int dx^N \Psi_k^0 \Delta V \Psi_k^0 = \int d\mathbf{r} \rho(\mathbf{r}) \delta v(\mathbf{r}), \qquad (2.21)$$

so that, according to the definition of functional derivative, one has that

$$\left(\frac{\delta E}{\delta v(\mathbf{r})} \right)_N = \rho(\mathbf{r}). \qquad (2.22)$$

For the chemical potential [28] and the chemical hardness [29,30] one finds that

$$\left(\frac{\delta \mu}{\delta v(\mathbf{r})} \right)_N = \left(\frac{\partial \rho(\mathbf{r})}{\partial N} \right)_v = f(\mathbf{r}), \qquad (2.23)$$

and

$$\left(\frac{\delta \eta}{\delta v(\mathbf{r})} \right)_N = \left(\frac{\partial^2 \rho(\mathbf{r})}{\partial N^2} \right)_v = \Delta f(\mathbf{r}), \qquad (2.24)$$

through the use of Equations 2.5, 2.6, and 2.22. The function $f(\mathbf{r})$ is known as the Fukui function, and the function $\Delta f(\mathbf{r})$ is known as the dual descriptor. From Equation 2.3 one has

$$\int d\mathbf{r} f(\mathbf{r}) = 1, \qquad (2.25)$$

and

$$\int d\mathbf{r} \Delta f(\mathbf{r}) = 0. \qquad (2.26)$$

It is important to mention that the chemical potential and the hardness, μ and η, are global-type response functions that characterize the molecule as a whole, while the electronic density $\rho(\mathbf{r})$, the Fukui function $f(\mathbf{r})$, and the dual descriptor $\Delta f(\mathbf{r})$ are local-type response functions whose values depend upon the position within the molecule.

Now, let us rewrite Equation 2.7 in the form

$$\frac{\rho_{N_0+\tau}(\mathbf{r}) - \rho_{N_0}(\mathbf{r})}{\tau} = \rho_{N_0+1}(\mathbf{r}) - \rho_{N_0}(\mathbf{r}), \qquad (2.27)$$

and the equivalent of Equation 2.7 for the interval between $N_0 - 1$ and N_0 in the form

$$\frac{\rho_{N_0}(\mathbf{r}) - \rho_{N_0-\tau}(\mathbf{r})}{\tau} = \rho_{N_0}(\mathbf{r}) - \rho_{N_0-1}(\mathbf{r}). \tag{2.28}$$

Note that again, as in the case of the energy, the densities' difference in the right-hand side of Equations 2.27 and 2.28 is independent of τ. Thus, if one takes the limit $\tau \to 0$ in Equations 2.27 and 2.28, then [31,32]

$$\lim_{\tau \to 0} \frac{\rho_{N_0+\tau}(\mathbf{r}) - \rho_{N_0}(\mathbf{r})}{\tau} = \left(\frac{\partial \rho(\mathbf{r})}{\partial N}\right)_v^+ = f^+(\mathbf{r}) = \rho_{N_0+1}(\mathbf{r}) - \rho_{N_0}(\mathbf{r}), \tag{2.29}$$

and

$$\lim_{\tau \to 0} \frac{\rho_{N_0}(\mathbf{r}) - \rho_{N_0-\tau}(\mathbf{r})}{\tau} = \left(\frac{\partial \rho(\mathbf{r})}{\partial N}\right)_v^- = f^-(\mathbf{r}) = \rho_{N_0}(\mathbf{r}) - \rho_{N_0-1}(\mathbf{r}), \tag{2.30}$$

where Equation 2.23 has been used. The relationships given by Equations 2.29 and 2.30 are notable in the sense that for the exact DFT, the finite differences lead to the exact Fukui functions.

The Fukui function $f^-(\mathbf{r})$ corresponds to the case in which the system donates charge, because it is interacting with an electrophilic reagent, while the Fukui function $f^+(\mathbf{r})$ corresponds to the case in which the system accepts charge, because it is interacting with a nucleophilic reagent. In the case in which the system is interacting with a neutral (or radical) reactant, the arithmetic average of the one-sided derivatives seems to be a good approximation [28], that is,

$$f^0(\mathbf{r}) = \left(\frac{\partial \rho(\mathbf{r})}{\partial N}\right)_v^0 = \frac{1}{2}(f^-(\mathbf{r}) + f^+(\mathbf{r})) = \frac{1}{2}(\rho_{N_0+1}(\mathbf{r}) - \rho_{N_0-1}(\mathbf{r})). \tag{2.31}$$

In Equations 2.29 through 2.31, $\rho_{N_0-1}(\mathbf{r})$, $\rho_{N_0}(\mathbf{r})$, and $\rho_{N_0+1}(\mathbf{r})$ are the electronic densities of the $N_0 - 1$-, N_0-, and $N_0 + 1$-electron systems, calculated for the external potential of the ground-state of the N_0-electron system.

A remarkable fact about the interpretation of the Fukui function [31] is that the most stable way to distribute the additional charge ΔN in a molecule is given by the product $\Delta N f^+(\mathbf{r})$, while the most stable way to remove the charge ΔN from a molecule is given by the product $\Delta N f^-(\mathbf{r})$. This means that a molecule accepts charge at the regions where $f^+(\mathbf{r})$ is large and it donates charge from the regions where $f^-(\mathbf{r})$ is large. Thus, the Fukui functions provide information about the site reactivity within a molecule.

A common simplification of the Fukui function is to condense its values to individual atoms in the molecule [33]. That is, through the use of a particular population analysis, one can determine the number of electrons associated with every atom in the molecule. The condensed Fukui functions is then determined through a finite differences approach, so that for the kth atom in the molecule A one has that

$$f_{Ak}^- = N_{Ak}(N_0) - N_{Ak}(N_0 - 1), \quad \text{for electrophilic attack,} \tag{2.32}$$

$$f_{Ak}^+ = N_{Ak}(N_0 + 1) - N_{Ak}(N_0), \quad \text{for nucleophilic attack,} \tag{2.33}$$

and

$$\begin{aligned} f_{Ak}^0 &= \frac{1}{2} \left(f_{Ak}^- + f_{Ak}^+ \right) \\ &= \frac{1}{2} (N_{Ak}(N_0 + 1) - N_{Ak}(N_0 - 1)), \quad \text{for neutral (or radical) attack.} \end{aligned} \tag{2.34}$$

In these relationships $N_{Ak}(N_0 - 1)$, $N_{Ak}(N_0)$, and $N_{Ak}(N_0 + 1)$ are the number of electrons associated with the kth atom in the molecule A, when the total number of electrons in the molecule is $N_0 - 1$, N_0, and $N_0 + 1$ electrons, respectively. The calculation of the $N_0 - 1$- and the $N_0 + 1$-electron system is done at the ground-state geometry of the N_0-electron system.

For the second derivative of the electronic density with respect to the number of electrons, the dual descriptor, one can proceed as in the case of the energy. That is, the Fukui function using the Heaviside function [25] is written as

$$f^{(N_0+x)}(\mathbf{r}) = f^-(\mathbf{r}) + \Theta(x)(f^+(\mathbf{r}) - f^-(\mathbf{r})), \quad -1 \leq x \leq 1. \tag{2.35}$$

Then, taking the derivative in Equation 2.35 with respect to x, one finds, using Equation 2.24, that the dual descriptor is given by

$$\Delta f^{(N_0+x)}(\mathbf{r}) = (f^+(\mathbf{r}) - f^-(\mathbf{r}))\delta(x), \quad -1 \leq x \leq 1. \tag{2.36}$$

Now, as in the case of the energy, up to this point, we have worked with the nonsmooth expression for the electronic density. However, in order to incorporate the second-order effects associated with the charge transfer processes, one can make use of a smooth quadratic interpolation. That is, with the two definitions given in Equations 2.23 and 2.24, the electronic density change $\Delta\rho(\mathbf{r})$ due to the electron transfer ΔN, when the external potential $v(\mathbf{r})$ is kept fixed, may be approximated through a second-order Taylor series expansion of the electronic density as a function of the number of electrons,

$$\Delta\rho(\mathbf{r}) = f^0(\mathbf{r})\Delta N + \frac{1}{2}\Delta f(\mathbf{r})(\Delta N)^2. \tag{2.37}$$

The values of the two derivatives, $f^0(\mathbf{r})$ and $\Delta f(\mathbf{r})$, at the reference point N_0, may be approximated through this smooth quadratic interpolation between the points $\rho_{N_0-1}(\mathbf{r})$, $\rho_{N_0}(\mathbf{r})$, and $\rho_{N_{0+1}}(\mathbf{r})$, when combined with the two conditions,

$f^-(\mathbf{r}) = \rho_{N_0}(\mathbf{r}) - \rho_{N_0-1}(\mathbf{r})$ and $f^+(\mathbf{r}) = \rho_{N_0+1}(\mathbf{r}) - \rho_{N_0}(\mathbf{r})$. This procedure leads to Equation 2.31 for $f^0(\mathbf{r})$ and to

$$\Delta f(\mathbf{r}) = f^+(\mathbf{r}) - f^-(\mathbf{r}). \tag{2.38}$$

Through a similar procedure, one finds that for the condensed dual descriptor

$$\Delta f_{Ak} = f_{Ak}^+ - f_{Ak}^-. \tag{2.39}$$

From the interpretation given to the Fukui function, one can note that the sign of the dual descriptor is very important to characterize the reactivity of a site within a molecule toward a nucleophilic or an electrophilic attack [29,30]. That is, if $\Delta f(\mathbf{r}) > 0$, then the site is favored for a nucleophilic attack, whereas if $\Delta f(\mathbf{r}) < 0$, then the site may be favored for an electrophilic attack.

2.5   DERIVATIVES IN AN ORBITAL LANGUAGE

Up to this point, we have established general expressions for the derivatives of the energy and density with respect to the number of electrons. These general expressions require basically the knowledge of the total energy and the electronic density of the reference system and its corresponding cation and anion. Consequently, one can make use of any molecular electronic structure method to calculate these quantities, from which one can determine the DFT reactivity criteria. However, it is important to analyze these quantities through an orbital language in order to establish their relationship with orbital concepts.

Thus, let us consider first the KS approach [34] in which the spin–orbitals $\psi_i(\mathbf{r})$ are self-consistent solutions of the equations

$$\left[-\frac{1}{2}\nabla^2 + v_S(\mathbf{r}) \right]\psi_i(\mathbf{r}) = \varepsilon_i\psi_i(\mathbf{r}), \tag{2.40}$$

with

$$v_S(\mathbf{r}) = v(\mathbf{r}) + \frac{\delta J[\rho]}{\delta\rho(\mathbf{r})} + \frac{\delta E_{XC}[\rho]}{\delta\rho(\mathbf{r})}, \tag{2.41}$$

where
   $J[\rho] = \frac{1}{2}\iint d\mathbf{r}\,d\mathbf{r}'\rho(\mathbf{r})\rho(\mathbf{r}')/|\mathbf{r}-\mathbf{r}'|$ is the classical Coulomb energy
   $E_{XC}[\rho]$ is the exchange-correlation energy functional

The electronic density is given by

$$\rho(\mathbf{r}) = \sum_i n_i|\psi_i(\mathbf{r})|^2, \tag{2.42}$$

where the occupation numbers n_i equals 1 for all the spin–orbitals up to N_0, and equals 0 for all the spin–orbitals above N_0.

PPLB showed that for a nondegenerate ground state, if n_i equals 1 for all the spin–orbitals below N_0, equals 0 for all the spin–orbitals above N_0, and equals τ for the spin–orbital corresponding to N_0, then [23,26,27]

$$\mu^- = \varepsilon_{N_0}(N_0) = -I_{N_0} = E(N_0) - E(N_0 - 1), \quad N_0 - 1 < N < N_0, \quad (2.43)$$

while for the case corresponding to a fractional occupation τ in the $N_0 + 1$ spin–orbital,

$$\mu^+ = \varepsilon_{N_0+1}(N_0 + 1) = -A_{N_0} = E(N_0 + 1) - E(N_0), \quad N_0 < N < N_0 + 1, \quad (2.44)$$

where Equations 2.12 and 2.13 have been used. Thus, from Equations 2.43 and 2.44 one has

$$I_{N_0} - A_{N_0} = \mu^+ - \mu^- = \varepsilon_{N_0+1}(N_0 + 1) - \varepsilon_{N_0}(N_0) \quad (2.45)$$

or, adding and subtracting $\varepsilon_{N_0+1}(N_0)$,

$$I_{N_0} - A_{N_0} = \varepsilon_{N_0+1}(N_0) - \varepsilon_{N_0}(N_0) + \Delta_{XC}, \quad (2.46)$$

where Δ_{XC} is the discontinuity of the exact exchange correlation potential,

$$\Delta_{XC} = \varepsilon_{N_0+1}(N_0 + 1) - \varepsilon_{N_0+1}(N_0). \quad (2.47)$$

Since $\varepsilon_{N_0+1}(N_0)$ and $\varepsilon_{N_0}(N_0)$ are the eigenvalues of the lowest unoccupied molecular spin–orbital (LUMO), ε_L, and the highest occupied molecular spin–orbital (HOMO), ε_H, respectively, it is clear that their difference cannot be identified directly with the chemical hardness, expressed as $I_{N_0} - A_{N_0}$, because it must be corrected by the discontinuity Δ_{XC}. Besides, common approximations to the exchange-correlation potential [1,35] such as the local density approximation (LDA) or the generalized gradient approximation (GGA), which are continuum functionals and do no exhibit the derivative discontinuity, approximately average over it in the energetically important regions where electrons are concentrated. However, they fail to do so asymptotically [36–38]. This behavior leads to a good description of the electronic density and the total energy, but a poor description of the KS eigenvalue spectrum.

Thus, in these cases, the calculation of the chemical potential (electronegativity) and the chemical hardness through the values of I and A is accurate when it is done through energy differences, but it is poorly described when it is done through the KS eigenvalues.

Nevertheless, it is important to mention that from a qualitative viewpoint, the approximations

$$\mu = -\chi \approx \frac{1}{2}(\varepsilon_L + \varepsilon_H), \quad (2.48)$$

and

$$\eta \approx \varepsilon_{\mathrm{L}} - \varepsilon_{\mathrm{H}}, \tag{2.49}$$

are conceptually very important and useful to understand many aspects of chemical reactivity.

Now, the Fukui function is closely related to the frontier orbitals. This can be seen from Equations 2.29 and 2.30, together with Equation 2.42, because if one determines the electron densities of the $N_0 - 1$- and the $N_0 + 1$-electron systems with the orbitals set corresponding to the N_0-electron system, then

$$f^-(\mathbf{r}) \approx \rho_{\mathrm{H}}(\mathbf{r}) \quad \text{and} \quad f^+(\mathbf{r}) \approx \rho_{\mathrm{L}}(\mathbf{r}), \tag{2.50}$$

and

$$\Delta f(\mathbf{r}) \approx \rho_{\mathrm{L}}(\mathbf{r}) - \rho_{\mathrm{H}}(\mathbf{r}), \tag{2.51}$$

where $\rho_{\mathrm{H}}(\mathbf{r})$ and $\rho_{\mathrm{L}}(\mathbf{r})$ are the densities of the highest occupied and lowest unoccupied molecular orbitals, respectively. However, one can see that the Fukui function, in contrast with frontier orbital theory, includes the orbital relaxation effects associated with electron addition or removal and the electron correlation effects. In some cases, these effects are very important. Nevertheless, again, from a qualitative viewpoint, the approximations given by Equations 2.50 and 2.51, together with the condensed version of these expressions for a Mulliken-like population analysis [39], are conceptually very important and useful to understand many aspects of chemical reactivity.

Finally, it is important to mention that in the case of the HF method, the calculation of the chemical potential and the hardness, through energy differences, to determine I and A, leads, in general, to a worst description than in the KS approach, because the correlation energy is rather important, particularly for the description of the anions. However, the HF frontier eigenvalues provide, in general, a better description of μ and η, through Equations 2.48 and 2.49, because they lie closer to the values of $-I$ and $-A$ than the LDA- or GGA-KS values, as established by Koopmans' theorem.

On the other hand, the calculation of the Fukui function with the HF frontier orbitals is, in general, qualitatively very similar to the one obtained through KS orbitals. However, there may be cases where the absence of correlation effects in HF may lead to large differences with respect to the KS description.

2.6 CONCLUDING REMARKS

In the preceding sections, we have analyzed the derivatives of the energy and of the density with respect to the number of electrons. The former is identified with the concepts of chemical potential (electronegativity) and hardness and measure the global response of a chemical species to changes in the number of electrons since they are independent of the position. The latter give rise to the concepts of the Fukui function and the dual descriptor and measure the local response of the chemical potential and the hardness to changes in the external potential, since they depend on the position.

The usual way to make use of these concepts to describe the inherent chemical reactivity of a molecule is, in general, through a second-order Taylor series expansions of the energy as a function of the number of electrons N and the external potential $v(\mathbf{r})$, around the reference state that corresponds to that of the isolated species [40]. Thus, on one hand, there will be terms associated with the chemical potential and the chemical hardness that will give information on the global behavior of the chemical species as a whole. On the other hand, there will be terms corresponding to integrals over the whole space involving the Fukui function or the dual descriptor. In this case, the values of these integrals will have a strong dependence on the overlap between these local reactivity descriptors and the change in the external potential (the potential generated by the nuclei and electrons of the reagent). Consequently, the Fukui function and the dual descriptor provide information on site selectivity [29,30,41–43].

Although we have concentrated in this chapter on the derivatives of the energy and density, there are other chemically meaningful concepts that can be derived from the ones presented here [44–46]. Among these, the chemical softness, the inverse of the chemical hardness, and the local softness [47,48] have proven to be quite useful to explain intermolecular reactivity trends.

Also, it is interesting to note that in the smooth quadratic interpolation, the curve of the total energy as a function of the number of electrons shows a minimum for some value of N beyond N_0 (see Figure 2.1). This point has been associated by Parr et al. [49] with the electrophilicity index that measures the energy change of an electrophile when it becomes saturated with electrons. Together with this global quantity, the philicity concept of Chattaraj et al. [50,51] has been extensively used to study a wide variety of different chemical reactivity problems.

As already mentioned, through DFT, it has been possible to explain the electronegativity equalization principle [1,7,10–13] and the hard and soft acids and bases principle [12,15–22] and, additionally, it has also been possible to introduce new ones like the maximum hardness principle [52,53] and the local hard and soft acids and bases principle [20,54–56].

In conclusion, the reactivity concepts that emerge from DFT provide a conceptually simple, but at the same time, chemically meaningful framework to explain the behavior of a wide variety of systems. In this chapter, we have analyzed some of the fundamental aspects required to understand its basis.

ACKNOWLEDGMENTS

I wish to thank Marcelo Galván and Alberto Vela for their important comments on the manuscript and appreciate the support from Conacyt grant C01-39621.

REFERENCES

1. Parr, R. G. and Yang, W., *Density Functional Theory of Atoms and Molecules*, Oxford University Press, New York, 1989.
2. Chermette, H., *J. Comput. Chem.* 1999, *20*, 129–154.
3. Geerlings, P., De Proft, F., and Langenaeker, W., *Chem. Rev.* 2003, *103*, 1793–1873.
4. Nalewajski, R. F., *Adv. Quantum Chem.* 2003, *43*, 119–184.
5. Chattaraj, P. K., Sarkar, U., and Roy, D. R., *Chem. Rev.* 2006, *106*, 2065–2091.
6. Hohenberg, P. and Kohn, W., *Phys. Rev. B* 1964, *136*, B864–B871.
7. Parr, R. G., Donnelly, R. A., Levy, M., and Palke, W. E., *J. Chem. Phys.* 1978, *68*, 3801–3807.
8. Iczkowski, R. and Margrave, J. L., *J. Am. Chem. Soc.* 1961, *83*, 3547–3551.
9. Mulliken, R. S., *J. Chem. Phys.* 1934, *2*, 782–793.
10. Sanderson, R. T., *Science* 1951, *114*, 670–672.
11. Sanderson, R. T., *Chemical Bonds and Bond Energy*, Academic Press, New York, 1971.
12. Cedillo, A., Chattaraj, P. K., and Parr, R. G., *Int. J. Quantum Chem.* 2000, *77*, 403–407.
13. Ayers, P. W., *Theor. Chem. Acc.* 2007, *118*, 371–381.
14. Parr, R. G. and Pearson, R. G., *J. Am. Chem. Soc.* 1983, *105*, 7512–7516.
15. Chattaraj, P. K., Lee, H., and Parr, R. G., *J. Am. Chem. Soc.* 1991, *113*, 1855–1856.
16. Gázquez, J. L., *J. Phys. Chem. A* 1997, *101*, 4657–4659.
17. Ayers, P. W., *J. Chem. Phys.* 2005, *122*, 141102.
18. Chattaraj, P. K. and Ayers, P. W., *J. Chem. Phys.* 2005, *123*, 086101.
19. Ayers, P. W., Parr, R. G., and Pearson, R. G., *J. Chem. Phys.* 2006, *124*, 194107.
20. Ayers, P. W., *Faraday Discuss.* 2007, *135*, 161–190.
21. Chattaraj, P. K., Ayers, P. W., and Melin, J., *Phys. Chem. Chem. Phys.* 2007, *9*, 3853–3856.
22. Pearson, R. G., *J. Am. Chem. Soc.* 1963, *85*, 3533–3539.
23. Perdew, J. P., Parr, R. G., Levy, M., and Balduz, J. L., *Phys. Rev. Lett.* 1982, *49*, 1691–1694.
24. Yang, W. T., Zhang, Y. K., and Ayers, P. W., *Phys. Rev. Lett.* 2000, *84*, 5172–5175.
25. Ayers, P. W., *J. Math. Chem.* 2008, *43*, 285–303.
26. Perdew, J. P. and Levy, M., *Phys. Rev. Lett.* 1983, *51*, 1884–1887.
27. Perdew, J. P. and Levy, M., *Phys. Rev. B* 1997, *56*, 16021–16028.
28. Parr, R. G. and Yang, W. T., *J. Am. Chem. Soc.* 1984, *106*, 4049–4050.
29. Morell, C., Grand, A., and Toro-Labbe, A., *J. Phys. Chem. A* 2005, *109*, 205–212.
30. Morell, C., Grand, A., and Toro-Labbe, A., *Chem. Phys. Lett.* 2006, *425*, 342–346.
31. Ayers, P. W. and Parr, R. G., *J. Am. Chem. Soc.* 2000, *122*, 2010–2018.
32. Ayers, P. W. and Levy, M., *Theor. Chem. Acc.* 2000, *103*, 353–360.
33. Yang, W. and Mortier, W. J., *J. Am. Chem. Soc.* 1986, *108*, 5708–5711.
34. Kohn, W. and Sham, L. J., *Phys. Rev.* 1965, *140*, 1133–1138.
35. Perdew, J. P. and Kurth, S., In *A Primer in Density Functional Theory*, Fiolhais, C., Nogueira, F., Marques, M. A. L., Eds., Springer, Berlin, 2003, p. 1.
36. Casida, M. E., Casida, K. C., and Salahub, D. R., *Int. J. Quantum Chem.* 1998, *70*, 933–941.
37. Tozer, D. J. and Handy, N. C., *J. Chem. Phys.* 1998, *109*, 10180–10189.
38. Casida, M. E. and Salahub, D. R., *J. Chem. Phys.* 2000, *113*, 8918–8935.
39. Contreras, R. R., Fuentealba, P., Galvan, M., and Perez, P., *Chem. Phys. Lett.* 1999, *304*, 405–413.
40. Ayers, P. W., Anderson, J. S. M., and Bartolotti, L. J., *Int. J. Quantum Chem.* 2005, *101*, 520–534.
41. Berkowitz, M., *J. Am. Chem. Soc.* 1987, *109*, 4823–4825.

42. De Proft, F., Ayers, P. W., Fias, S., and Geerlings, P., *J. Chem. Phys.* 2006, *125*, 214101.
43. Ayers, P. W., Morell, C., De Proft, F., and Geerlings, P., *Chem. Eur. J.* 2007, *13*, 8240–8247.
44. Nalewajski, R. F. and Parr, R. G., *J. Chem. Phys.* 1982, *77*, 399–407.
45. Nalewajski, R. F., *J. Chem. Phys.* 1983, *78*, 6112–6120.
46. Fuentealba, P. and Parr, R. G., *J. Chem. Phys.* 1991, *94*, 5559–5564.
47. Yang, W. T. and Parr, R. G., *Proc. Natl. Acad. Sci. U.S.A.* 1985, *82*, 6723–6726.
48. Berkowitz, M. and Parr, R. G., *J. Chem. Phys.* 1988, *88*, 2554–2557.
49. Parr, R. G., Von Szentpaly, L., and Liu, S.B., *J. Am. Chem. Soc.* 1999, *121*, 1922–1924.
50. Chattaraj, P.K., Maiti, B., and Sarkar, U., *J. Phys. Chem. A* 2003, *107*, 4973–4975.
51. Gázquez, J. L., Cedillo, A., and Vela, A., *J. Phys. Chem. A* 2007, *111*, 1966–1970.
52. Pearson, R. G., *J. Chem. Educ.* 1987, *64*, 561–567.
53. Parr, R. G. and Chattaraj, P. K., *J. Am. Chem. Soc.* 1991, *113*, 1854–1855.
54. Gázquez, J. L. and Méndez, F., *J. Phys. Chem.* 1994, *98*, 4591–4593.
55. Méndez, F. and Gázquez, J. L., *J. Am. Chem. Soc.* 1994, *116*, 9298–9301.
56. Anderson, J. S. M., Melin, J., and Ayers, P. W., *J. Chem. Theory Comput.* 2007, *3*, 358–374.

3 Quantum Chemistry of Bonding and Interactions

P. Kolandaivel, P. Venuvanalingam, and G. Narahari Sastry

CONTENTS

The concept of valence and bonding are central to chemistry and helps to understand molecular structure and reactivity in a systematic way. The principles that govern the formation of molecules from atoms and intermolecular interactions are of paramount interest and have attracted much attention. Initially it was believed that certain types of chemical species were joined together by means of chemical affinity to form a chemical bond. A clear concept of a chemical bond emerged only after Lewis introduced the concept of electron pair bond in his landmark paper in 1916 [1] and his ideas continue to dominate a chemist's perception toward chemical bonding and molecular structure. The theory of chemical bonding has received considerable attention ever since and several theories have been put forward to understand the nature of atoms and how atoms come together to form molecules, followed by the Lewis theory of paired electron bond [2–9]. Linus Pauling, in his highly influential book *The Nature of the Chemical Bond*, has orchestrated rules for the shared electron-bond on the basis of electron paired bond on which the valence bond (VB) theory was built [2]. This chapter aims to introduce chemical bonding at the molecular and supramolecular levels starting from the Lewis concept. Early ideas are presented as a historical note and the development of chemical bonding models after the advent of quantum mechanics are described in a logical way. The importance and computation of electron correlation are discussed and then various kinds of noncovalent interactions are outlined quoting a few standard examples. This will form a basis on how the early concepts, which are mostly wave function based, describe bonding and reactivity in greater detail. This would naturally be a prelude to understand reactivity from density functional theory (DFT) point of view.

3.1 QUANTUM MECHANICAL TREATMENT OF CHEMICAL BONDING

A rigorous mathematical formalism of chemical bonding is possible only through the quantum mechanical treatment of molecules. However, obtaining analytical solutions for the Schrödinger wave equation is not possible even for the simplest systems with more than one electron and as a result attempts have been made to obtain approximate solutions; a series of approximations have been introduced. As a first step, the Born–Oppenheimer approximation has been invoked, which allows us to treat the electronic and nuclear motions separately. In solving the electronic part, mainly two formalisms, VB and molecular orbital (MO), have been in use and they are described below. Both are wave function-based methods. The wave function Ψ is the fundamental descriptor in quantum mechanics but it is not physically measurable. The squared value of the wave function $|\Psi|^2 d\tau$ represents probability of finding an electron in the volume element $d\tau$.

Heilter and London made the first quantum mechanical treatment of a chemical bond in 1927 [10] and their ideas have laid the foundations for the general theory of chemical bonding known as VB theory, with seminal contributions from London and Pauling. In contrast to the two-electron wave functions in the VB theory, a mathematically elegant formalism based on the one-electron wave function, MO theory, has been introduced and developed parallelly by Lennard-Jones in 1929, which was refined and applied to a large number of systems by Mulliken and Hund. VB and MO theories have become two alternatives to explain chemical bonding. While VB theory [11] is chemically intuitive and is primarily responsible for the understanding of chemical concepts based on lone pairs and bond pairs, MO theory [12] has been highly successful in predicting spectroscopic properties and more importantly the MO wave functions are orthogonal and follow the group theoretical principles. All these lead to the emergence of MO theory as the method of choice over VB theory during the second half of the twentieth century and the position of MO has further consolidated upon the arrival of computers and quantum mechanical programs. With increasing computer power, the computationally attractive orthogonal MO formalism has become much more tractable on medium-sized molecules, while the progress in VB is restricted due to the technical bottlenecks in the implementation of the formalism.

VB and MO theories can be applied to simple molecular systems as follows. According to VB theory, if φ_a and φ_b are wave functions of independent systems a and b then the total wave function ψ and total energy E are written as follows:

$$\psi = \varphi_a \varphi_b \tag{3.1}$$

$$E = E_a + E_b \tag{3.2}$$

If ψ_1, ψ_2, ψ_3, ... are acceptable wave functions for the system, then according to the principle of superposition the true wave function Ψ could be expanded as the linear combination of them:

$$\Psi = c_1\psi_1 + c_2\psi_2 + c_3\psi_3 + \cdots \tag{3.3}$$

For example, for the hydrogen molecule,

$$\psi_1 = 1s_a(1)1s_b(2) \tag{3.4}$$

where
  a and b denote the atoms
  1 and 2 represent the electrons

Particle indistinguishability allows that the product $\psi_2 = 1s_a(2)\,1s_b(1)$ is also equally acceptable.

Therefore, the true wave function can be written as

$$\psi = c_1 1s_a(1)1s_b(2) + c_2 1s_a(2)1s_b(1) \tag{3.5}$$

In hydrogen molecule, because of symmetry the component wave functions $1s_a(1)$ $1s_b(2)$ and $1s_a(2)$ $1s_b(1)$ contribute with equal weight:

$$\psi = c_1 1s_a(1)1s_b(2) + c_2 1s_a(2)1s_b(1) \tag{3.6}$$

$$c_1^2 = c_2^2 \quad \text{and} \quad c_1 = \pm c_2 \tag{3.7}$$

$$\psi_s = 1s_a(1)1s_b(2) + 1s_a(2)1s_b(1) \tag{3.8}$$

$$\psi_a = 1s_a(1)1s_b(2) - 1s_a(2)1s_b(1) \tag{3.9}$$

while
  ψ_s represents a bonding state
  ψ_a corresponds to the antibonding or repulsive state (Figure 3.1)

The curve ψ_1 exhibits a minimum but the stabilization is not significant. It should be noted that when electrons are allowed to interchange, there is a substantial stabilization, which comes from exchange interaction as a consequence of particle indistinguishability. Bonding energy explained using ψ_s is still substantially above the experimental value and this indicates that the wave function should be improved further. One way to improve the wave function is considering the admixture of other electron configurations, such as ionic structures

$$\psi_3 = 1s_a(1)1s_a(2) \cdots H_A^- H_B^+ \tag{3.10}$$

$$\psi_4 = 1s_b(1)1s_b(2) \cdots H_A^+ H_B^- \tag{3.11}$$

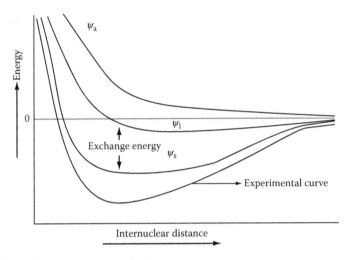

FIGURE 3.1   VB potential energy for H_2 molecule.

Again owing to the symmetry, ψ_3 and ψ_4 are virtually identical and thus are equally probable. Now the total wave function including ionic contribution is written as

$$\psi_s = \psi_{covalent} + \lambda\psi_{ionic} \tag{3.12}$$

where

$$\psi_{covalent} = 1s_a(1)1s_b(2) + 1s_a(2)1s_b(1) \tag{3.13}$$

$$\psi_{ionic} = 1s_a(1)1s_a(2) + 1s_b(1)1s_b(2) \tag{3.14}$$

and λ is a measure of contribution of ionic structures. Obviously, inclusion of ionic structures has led to further stabilization of the bonding state, which is known as the resonance stabilization energy (Figure 3.1).

The total wave function is a product of orbital (space) and spin wave functions. In a two-electron system like the H_2 molecule, the possible spin wave functions are $\alpha(1)\alpha(2)$, $\beta(1)\beta(2)$, $\alpha(1)\beta(2)$, and $\alpha(2)\beta(1)$. The first two spin states represent the parallel spin and the last two antiparallel spins. Particle indistinguishability forces the linear combination $1/\sqrt{2}(\alpha(1)\beta(2) \pm \alpha(2)\beta(1))$. Here the functions $\alpha(1)\alpha(2), \beta(1)\beta(2)$, and $1/\sqrt{2}(\alpha(1)\beta(2) + \alpha(2)\beta(1))$ represent the triplet state and $1/\sqrt{2}(\alpha(1)\beta(2) - \alpha(2)\beta(1))$ represent the singlet state. An H_2 molecule is a closed shell system and will have a singlet multiplicity in the ground state. Further improvement on the VB wave function has been done by allowing the orbital exponent to change when atoms approach during the bond formation. Though the VB theory successfully explained bonding in molecules, it has some drawbacks. A major bottleneck is that the localized VB wave functions are not orthonormal, thus leading to complicated equations. The impact of VB declined greatly as the MO theory introduced by Lennard-Jones in 1929 became much popular. It is quite interesting to see that the practitioners of the alternative formalisms MO and VB are more at loggerheads than in harmony [13].

The MO theory differs greatly from the VB approach and the basic MO theory is an extension of the atomic structure theory to molecular regime. MOs are delocalized over the nuclear framework and have led to equations, which are computationally tractable. At the heart of the MO approach lies the linear combination of atomic orbitals (LCAO) formalism

$$\psi_i = \sum_i a_i \varphi_i \tag{3.15}$$

Suppose atoms a and b form a molecule and their atomic orbitals are φ_a and φ_b, respectively, then

$$\psi_{MO} = c_a \varphi_b + c_b \varphi_b \tag{3.16}$$

The combining orbitals should lie closer in energy and have same symmetry with optimal overlap in space. The positive overlap leads to bonding, negative overlap to antibonding, and zero overlap to nonbonding situations. The MOs are arranged in ascending order of energy and electrons are fed into their following Pauli and Aufbau principles. For example, the MOs of hydrogen molecule can be written as

$$\psi_{MO} = c_a 1s_a + c_b 1s_b \tag{3.17}$$

Because of the symmetry $c_a^2 = c_b^2; c_a = \pm c_b$

$$\psi_b = 1s_a + 1s_b \tag{3.18}$$

$$\psi_a = 1s_a - 1s_b \tag{3.19}$$

where ψ_b and ψ_a represent bonding and antibonding states. They are also referred to as $1s\sigma_g$ and $1s\sigma_u$ states. The ground state electron configuration of the hydrogen molecule is $1s\sigma_g^2$. The ground state wave function of the hydrogen molecule in Slater determinant form is as follows:

$$\psi = \frac{1}{\sqrt{2}} \begin{vmatrix} 1s\sigma_g(1) & 1\bar{s}\sigma_g(1) \\ 1s\sigma_g(2) & 1\bar{s}\sigma_g(2) \end{vmatrix} \tag{3.20}$$

$$\psi_{MO} = 1s\sigma_g(1) 1s\sigma_g(2) \frac{1}{\sqrt{2}} (\alpha_1 \beta_2 - \beta_1 \alpha_2) \tag{3.21}$$

and this corresponds to the singlet state of H_2 where the spin part of the wave function is antisymmetric with respect to electronic interchange and the space part is a symmetric combination of $1s_a$ and $1s_b$. The contribution of covalent and ionic terms to the wave function can be understood by expanding the space part of its wave function

$$\begin{aligned} \psi_s &= \{1s_a(1) + 1s_b(1)\}\{1s_a(2) + 1s_b(2)\} \\ &= \{1s_a(1)1s_b(2) + 1s_a(2)1s_b(1)\} + \{1s_a(1)1s_a(2) + 1s_b(1)1s_b(2)\} \\ &= \psi_{covalent} + \psi_{ionic} \end{aligned} \tag{3.22}$$

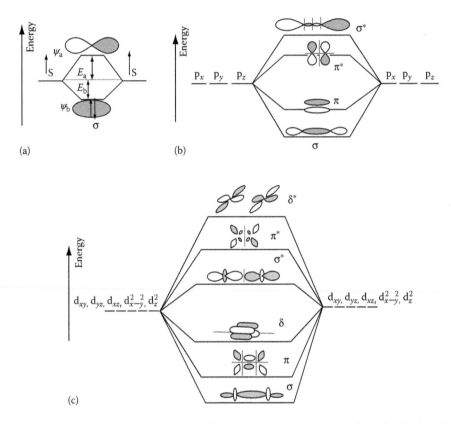

FIGURE 3.2 MO diagram: Formation of chemical bonds due to overlap of orbitals. (a) σ-bonding: A result of overlapping s-orbitals. (b) σ- and π-bonding: A result of overlapping p-orbitals. (c) σ-, π-, and δ-bonding: A result of overlapping d-orbitals.

The method of generating MOs from LCAOs has been proved to be quite useful in establishing the relationship between the atoms and molecules from an electronic structure point of view. The ways in which s, p, and d orbitals interact with each other to form MOs of various symmetry kinds such as σ, π, and δ are depicted in Figure 3.2a through 3.2c, respectively.

It is also possible that orbitals of different kinds on the two atomic centers such as s-p_z, p_z-d_z^2, d_{xz}-p_x, etc. can combine to generate the MO for the diatomic molecules. As the one-center atomic orbitals are not orthogonal in molecules, for the depiction of electronic structure, the concept of hybridization is quite useful.

MO wave functions in the above form give equal importance to covalent and ionic structures, which is unrealistic in homonuclear diatomic molecules like H_2. This should be contrasted with ψ_{VB}, which in its simple form neglects the ionic contributions. Both ψ_{VB} and ψ_{MO} are inadequate in their simplest forms; while in the VB theory the electron correlation is overemphasized, simple MO theory totally neglects it giving equal importance to covalent and ionic structures. Therefore neither of them is able to predict binding energies closer to experiment. The MO theory could be improved through configuration interaction (CI). Excitation of one or more electrons could be done from occupied levels to unoccupied ones and several singly, doubly, and multiply occupied "configurations" can be generated. MO wave functions could be improved by incorporating Slater determinants corresponding to such excited configurations. In the hydrogen molecule case, doubly excited configurations have proper symmetry to interact with ground state configuration and stabilize the bonding state. Physically, inclusion of such electronic configurations, which belong to the exited state leads, by populating antibonding orbitals, which results in the depletion of electron density in the internuclear region. Therefore, although VB and MO theories appear to represent two extremes in treating electron correlation, improvements in their procedures lead to convergence and binding energies very close to experiment.

Due to the simplicity and the ability to explain the spectroscopic and excited state properties, the MO theory in addition to easy adaptability for modern computers has gained tremendous popularity among chemists. The concept of directed valence, based on the principle of maximum overlap and valence shell electron pair repulsion theory (VSEPR), has successfully explained the molecular geometries and bonding in polyatomic molecules.

MO theories for polyatomic molecules could be broadly classified into two, based on the rigor. They are electron-independent or non self-consistent field (SCF) and electron-dependent or SCF methods. Electron-independent theories do not consider electron–electron interaction explicitly and they include Hückel [12,14] and the extended Hückel theory [15]. While the former considers only π basis, the latter takes into account all valence basis. They involve many approximations and parameterizations. Approximations are mathematical neglect and parameterization means replacing certain integrals through parameters derived from experiment. SCF methods iteratively solve many-electron Schrödinger wave equations in matrix form based on Hartree–Fock (HF) theory. They are either called ab initio method when the Fock matrix is constructed from first principles and semiempirical when certain approximations are invoked and parameters introduced. In recent years, density-based methods are gaining popularity due to the considerable time advantage and conceptual simplicity. The following chapters deal with this subject in sufficient detail.

3.2 ELECTRON CORRELATION

The HF method does not consider the instantaneous electrostatic interactions, but it treats the interaction of one electron with the average field due to the other electrons. Thus the correlated motions of electrons are neglected and as a consequence HF energies are always higher than the exact energy of the system. The difference between the exact and HF energies is defined as the correlation energy. It may be noted that exchange correlation, which says that two electrons of same spin cannot occupy a single orbital, is already included in the HF theory. Inclusion of electron correlation is necessary for the reliable description of structure and properties of molecules, and therefore the development of post-HF methods have been of paramount importance. The following procedures are available to improve the HF-SCF theory, which includes the correlation energy: (1) CI, (2) many-body perturbation theory (MBPT), and (3) coupled cluster (CC) theory.

$$\text{Slater determinant} = D = \frac{1}{\sqrt{n!}} \begin{vmatrix} u_1(1) & u_2(1) & \ldots & \ldots & u_n(1) \\ u_1(2) & \ldots & \ldots & \ldots & u_n(2) \\ u_1(3) & \ldots & \ldots & \ldots & \ldots \\ \ldots & \ldots & \ldots & \ldots & \ldots \\ u_1(n) & u_2(n) & u_3(n) & \ldots & u_n(n) \end{vmatrix}$$

FIGURE 3.3 Slater determinant. The spin–orbital u is the product of orbital (θ) and spin $(\sigma = \alpha/\beta)$ functions.

It is possible to divide electron correlation as dynamic and nondynamic correlations. Dynamic correlation is associated with instant correlation between electrons occupying the same spatial orbitals and the nondynamic correlation is associated with the electrons avoiding each other by occupying different spatial orbitals. Thus, the ground state electronic wave function cannot be described with a single Slater determinant (Figure 3.3) and multiconfiguration self-consistent field (MCSCF) procedures are necessary to include dynamic electron correlation.

However, in a large number of closed shell molecules, a single Slater determinant describes the ground state wave function fairly accurately. Even in such cases inclusion of excited state configuration results in substantial lowering of total electronic energy, and this is referred to as nondynamic electron correlation.

3.2.1 Configuration Interaction

HF method determines the energetically best determinantal trial wave function (ϕ_0) and this would be improved further by including more "configurations." Let φ_0 be represented as

$$\phi_0 = |\phi_1 \phi_2 \cdots \phi_a \phi_b \cdots \phi_n| \tag{3.23}$$

where ϕ_a, ϕ_b are occupied spin orbitals. Excited configuration could be generated by promoting electron from occupied orbitals to virtual orbitals as follows:

$$\phi_a^p = |\phi_1 \phi_2 \cdots \phi_p \phi_b \cdots \phi_n| \tag{3.24}$$

$$\phi_{ab}^{pq} = |\phi_1 \phi_2 \cdots \phi_p \phi_q \cdots \phi_n| \tag{3.25}$$

where ϕ_a^p and ϕ_{ab}^{pq} represent singly, doubly excited configurations, respectively. In a similar way, any multiply excited configuration could be generated and used. Here the multideterminantel wave function could be written as

$$\psi_{CI} = a_o \varphi_{HF} + \sum_S a_S \varphi_S + \sum_D a_D \varphi_D + \sum_T a_T \varphi_T + \cdots + \sum_{i=0} a_i \varphi_i \tag{3.26}$$

S, D, and T stand for singly, doubly, and triply excited states relative to the HF configuration.

3.3 MULTICONFIGURATION SELF-CONSISTENT FIELD

In the multiconfiguration SCF, the MOs are used for constructing the determinants and the coefficients were optimized by the variational principle. The MCSCF optimization is iterative like SCF the procedure, where the iterations required for achieving convergence tend to increase with the number of configurations included. The major problem with MCSCF method is selecting the configurations that are necessary to include for the property of interest. One of the most popular approaches in this theory is the complete active space self-consistent field method (CASSCF). The selection of configuration is performed by partitioning the MOs into active and inactive spaces. The active MOs will have some of the highly occupied and some of the lowest unoccupied MOs from an RHF calculation. However, the highly stable orbitals as well as the very high lying virtual orbitals, which are not part of the active space, are referred to as inactive orbitals. Thus, the inactive orbitals are either doubly occupied or completely unoccupied in all the excitations that are considered.

3.4 MANY-BODY PERTURBATION THEORY

The perturbation method is a unique method to determine the correlation energy of the system. Here the Hamiltonian operator consists of two parts, H_0 and H', where H_0 is the unperturbed Hamiltonian and H' is the perturbation term. The perturbation method always gives corrections to the solutions to various orders. The Hamiltonian for the perturbed system is

$$H = H_0 + \lambda H' \tag{3.27}$$

where λ is a parameter determining the strength of the perturbation.

$$H_0\phi_i = E_i\phi_i \quad i = 0, 1, 2, \ldots, \infty \tag{3.28}$$

The solution of the unperturbed Hamiltonian operator forms a complete orthonormal set. The perturbed Schrödinger equation is given by

$$H\psi = W\psi \tag{3.29}$$

If $\lambda = 0$, then $H = H_0$, $\psi = \phi_0$, and $W = E_0$. As the perturbation is increased from zero to a finite value, the energy and wave function also change continuously and can be written as a Taylor expansion in the power of the perturbation parameter λ:

$$W = \lambda^0 W_0 + \lambda^1 W_1 + \lambda^2 W_2 + \lambda^3 W_3 + \cdots \tag{3.30}$$

$$\psi = \lambda^0 \psi_0 + \lambda^1 \psi_1 + \lambda^2 \psi_2 + \lambda^3 \psi_3 + \cdots \tag{3.31}$$

$\psi_1, \psi_2, \ldots$, and $W_1, W_2, \ldots$, are first- and second-order corrections to the wave function and energies.

3.5   COUPLED CLUSTER THEORY

The CC method was developed for the system of interacting particles. The basic equation for this theory is

$$\psi = e^{\mathrm{T}}\varphi_0 \tag{3.32}$$

where
   ψ is the ground state molecular electronic wave function
   ϕ_0 is the normalized ground state HF wave function
   e^{T} is defined by the Taylor series as

$$e^{\mathrm{T}} \equiv 1 + \hat{T} + \frac{\hat{T}^2}{2!} + \frac{\hat{T}^3}{3!} + \cdots = \sum_{k=0}^{\infty} \frac{\hat{T}^k}{k!} \tag{3.33}$$

The cluster operator T is defined as

$$T = T_1 + T_2 + T_3 + \cdots + T_n \tag{3.34}$$

where
   n is the total number of electrons
   various T_s are the excitation operators

The one-particle excitation operator $\hat{T}_1$ and the two-particle excitation operator $\hat{T}_2$ are defined by

$$T_1\phi_0 = \sum_{i}^{\mathrm{occ}} \sum_{a}^{\mathrm{vir}} t_i^a \phi_i^a \tag{3.35}$$

$$T_1\phi_0 = \sum_{i<j}^{\mathrm{occ}} \sum_{a<b}^{\mathrm{vir}} t_{ij}^{ab} \phi_{ij}^{ab} \tag{3.36}$$

where
   ϕ_i^a is a singly excited Slater determinant
   t_i^a is a numerical coefficient
   ϕ_{ij}^{ab} is a Slater determinant with the occupied spin-orbitals
   t_{ij}^{ab} is a numerical coefficient

The trial wave functions of a Schrödinger equation are expressed as determinant of the HF orbitals. This will give coupled nonlinear equations. The amplitudes were solved usually by some iteration techniques so the cc energy is computed as

$$\langle \phi_0 | H | e^{\mathrm{T}} \phi_0 \rangle = E_{\mathrm{CC}} \tag{3.37}$$

The cc correlation energy is determined by the single and double amplitudes and the two-electron MO integrals. However, the recent progress in computational methods is largely influenced by DFT. One primary advantage of DFT over conventional HF-SCF procedures is that the former includes electron correlation fairly adequately (some times too much) at a fractional cost compared to a typical post-SCF (CI, MBPT, or CC) calculation.

3.6 INTERMOLECULAR INTERACTIONS

Having understood the formation of molecules from atoms through chemical bonds, it is important to understand the molecular aggregation and their relationship to materials. Conventionally, intermolecular interactions, also known as noncovalent or nonbonded interactions, are weak and it is assumed that the change in the geometric and electronic structures of the individual components is minimum during the complex formation. Obviously, hydrogen bonding is the most important and elaborately studied nonbonded interactions [16,17]. Figure 3.4 depicts a couple of examples of dimers, which are bound through nonbonded interactions. Such interactions play a pivotal role in determining the structure, stability, and dynamics of biological systems such as proteins and DNA and thus the accurate theoretical description of these interactions is of critical importance [18–20].

Noncovalent interactions operate at larger internuclear distances of several angstroms. The formation of a covalent bond requires overlapping of partially occupied orbitals of interacting atoms, which share a pair of electrons. In noncovalent interactions, no overlapping is necessary because the attraction comes from the electrical properties of the building blocks. Noncovalent or van der Waals interactions were first recognized by J. D. van der Waals in the nineteenth century. Their role in nature has been unraveled only during the past three to four decades.

The noncovalent interactions or van der Waals forces involved in supramolecular entities may be a combination of several interactions, e.g., ion-pairing, hydrophobic, hydrogen bonding, cation–π, π–π interactions, etc. They comprise interactions between permanent multipoles, between a permanent multipole and an induced multipole, and between a time-variable multipole and an induced multipole.

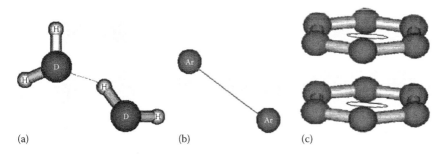

(a)   (b)   (c)

FIGURE 3.4 (See color insert following page 302.)   The prototypical noncovalent interactions between (a) water dimer, (b) Ar dimer, and (c) benzene dimer.

The stabilizing energy of noncovalent complexes is generally said to consist of the following energy contributions: electrostatic (or coulombic), induction, charge transfer (CT), and dispersion. These terms are basically attractive terms. The repulsive contribution, which is called exchange repulsion, prevents the subsystems from drawing too close. The term induction refers to general ability of charged molecules to polarize neighboring species, and dispersion (London) interaction results from the interactions between fluctuating multipoles. In charge transfer (CT) interactions, the electron flow from the donor to the acceptor is indicated. The term van der Waals (*v*dW) forces is frequently used to describe dispersion and exchange repulsion contributions, but sometimes other long-range contributions are also included in the definition. The *v*dW energy is calculated by the following equation:

$$E_{vdW}(r) = \varepsilon\left[(r_0/r)^{12} - 2(r_0/r)^6\right] \tag{3.38}$$

where
    r is the current distance between atoms
    r_0 is the equilibrium distance
    ε is the minimum energy

All of these interactions involve a host and a guest as well as their surroundings like solvation, crystal lattice, and gas phase. Electrostatic interactions are the driving force behind the ion pairing (ion–ion, ion–dipole, dipole–dipole, etc.) interactions, which are undeniably important in natural and supramolecular systems. The electrostatic interaction energy E is given by

$$E = \frac{q_1 q_2}{\varepsilon r} \tag{3.39}$$

where
    q_1, q_2 are the charges
    r the distance between the two
    ε is the dielectric coefficient

Ion–ion interactions are between two oppositely charged particles and because of the electrostatic force between them, they stick together and a considerable amount of energy will be required to separate them. They form an ion pair, a new particle, which has a positively charged area and a negatively charged area. There are fairly strong interactions between these ion pairs and free ions, so that these clusters tend to grow, and they will eventually fall out of the gas phase as a liquid or solid (depending on the temperature). The presence of these interactions plays a crucial role in the transportation of ions through ion-channels in biochemical pathways. An ion–dipole force is an attractive force that results from the electrostatic attraction between an ion and a neutral molecule that has a dipole. They are most commonly found in solutions, and especially important for solutions of ionic compounds in polar liquids. Various methods have been put forward to understand the role of electrostatics in a variety of molecules.

Charges are heavily delocalized in organic ions, which complicate the theoretical analysis of ion pairing. Between neutral polar molecules the electrostatic contributions comes mostly from dipole–dipole interactions. Perhaps van der Waals interactions are the most important class of dipole–dipole interactions where one or both molecules do not have a permanent dipole. These interactions are valid for any two atoms that come into close contact with each other, and are called van der Waals interactions. Another very important noncovalent interaction is the hydrophobic interaction. As the term hydrophobic suggests, this interaction is an effective interaction between two nonpolar molecules that tend to avoid water and, as a result, prefer to cluster around each other.

While the qualitative understanding of the nonbonded interactions is well established, quantitative evaluation of their interaction strength and other properties are very challenging. As the very presence of several biological systems and supramolecular assemblies are due to the presence of the nonbonded interactions, clearly the quest to unravel their structure and energetics are of great importance. Virtually all quantum chemical calculations computing the interaction energy between two noncovalently bound molecules is based on ab initio MO theory. One noteworthy development is to employ methods, which delineate the composition of various components, such as electrostatic, polarization, and other terms, to the total interaction energy, a procedure known as energy decomposition analysis. It started with an interesting paper by Morokuma in 1971 [21] and later several ingenious ways in which the contributions of various components to the total interaction energy is calculated were developed. In recent years, there is a heightened activity in gauging the effect of basis set superposition error [22], electron correlation, and cooperativity [23] (nonadditivity of the interaction strengths) among various nonbonded interactions. In addition to the energy decomposition analysis, the natural bond order (NBO) analysis [24] and the evaluation of the electron density distributions through atoms in molecules (AIM) theory [25] are also employed to probe into the nature of the noncovalent interactions. Several studies are reported in the literature, where the hydrogen bonding and van der Waals interactions are characterized using the AIM theory.

Ab initio MO based methods have played a pivotal role in qualitatively and quantitatively understanding covalent and noncovalent interactions in chemistry. These methods form the basis for modeling chemical and biological reactivity. The exponential raise, starting from early 1990s, the application of DFT-based calculations made an enormous impact on theoretical and computational quantum chemistry. Following this, there is a renewed interest in devising and exploring conceptual DFT-based protocols in understanding the structure and reactivity of molecules, which will be described in detail in the forthcoming chapters [26,27].

ACKNOWLEDGMENT

The authors thank G. Praveena, A. Abiram, N. Santhanamoorthi, P. Deepa, and G. Gayatri for their technical support and help in the preparation of the manuscript.

REFERENCES

1. Lewis, G. N. 1916. The atom and the molecule. *J. Am. Chem. Soc.* 38: 762.
2. Pauling, L. 1960. *The Nature of the Chemical Bond*, 3rd ed. Cornell University Press, Ithaca, NY.
3. Malrieu, J. P., Gauthéry, N., Calzad, C. J., and Angeli, C. 2007. Bond electron pair: Its relevance and analysis from the quantum chemistry point of view. *J Comput. Chem.* 28: 35–50.
4. Shaik, S. 2007. The Lewis legacy: The chemical bond—A territory and heartland of chemistry. *J. Comput. Chem.* 28: 51–61.
5. Boys, S. F. 1966. In *Quantum Theory of Atoms, Molecules and the Solid State*, Löwdin, P. O. (Ed.). Academic, New York.
6. Robb, M. A. and Bernardi, F. 1989. In *New Theoretical Concepts for Undertaking Organic Reactions*, Bertran, J., Csizmadia, I. G. (Eds.). Kluwer, Dordrecht.
7. Frenking, G. and Krapp, A. 2007. Unicorns in the world of chemical bonding models. *J. Comput. Chem.* 28: 15–24.
8. Bitter, T., Ruedenberg, K., and Schwarzl, W. H. E. 2007. Toward a physical understanding of electron-sharing two-center bonds. I. General aspects. *J. Comput. Chem.* 28: 411–422.
9. Albright, T. A. Burdett, J. K., and Whangbo, M. H. 1985. *Orbital Interactions In Chemistry*. Wiley, New York.
10. Heitler, W. and London, F. 1927. Interaction of neutral atoms and homopolar binding according to the quantum mechanics. *Z. Phys.* 44: 455–472.
11. Shaik, S. and Hiberty, P. C. 2007. *A Chemist's Guide to Valence Bond Theory*. Wiley Interscience, New York.
12. Strietvieser Jr. A. 1961. *Molecular Orbital Theory for Organic Chemists*. John Wiley, New York.
13. Hoffmann, R., Hiberty, P. C., and Shaik, S. 2003. Conversation on VB vs. MO theory: A never ending rivalry? *Acc. Chem. Res.* 36: 750–756.
14. Hückel, E. The Hückel theory. (a) 1930. *Z. Phys.* 60: 423; (b) 1931. *Z. Phys.* 70: 204.
15. Hoffmann, R. 1963. An extended Hückel theory I. Hydrocarbons. *J. Chem. Phys.* 39: 1397.
16. Černÿ, J. and Hobza, P. 2007. Non-covalent interactions in biomacromolecules. *Phys. Chem. Chem. Phys.* 9: 5291–5303.
17. Alabugin, I. V. and Manoharan, M. 2007. Rehybridization as a general mechanism for maximizing chemical and supramolecular bonding and a driving force for chemical reactions. *J. Comp. Chem.* 28: 373–390.
18. Desiraju, G. R. and Steiner, T. 1999. *The Weak Hydrogen Bond In Structural Chemistry and Biology*. Oxford University Press, Oxford.
19. Grabowski, S. 2006. *Hydrogen Bonding—New Insights*. Springer, Berlin.
20. Hobza, P. and Zahradnik, R. 1980. *Weak Intermolecular Interactions in Chemistry and Biology*. Elsevier, Amsterdam.
21. Morokuma, K. 1971. Molecular orbital studies of hydrogen bonds. III. $C = O \cdots H - O$ hydrogen bond in $H_2CO \cdots H_2O$ and $H_2CO \cdots 2H_2O$. *J. Chem. Phys.* 55: 1236.
22. Boys, S. B. and Bernardi, F. 1970. The calculation of small molecular interactions by the differences of separate total energies. Some procedures with reduced errors. *Mol. Phys.* 19: 553.
23. Reddy, A. S., Vijay, D., Sastry, G. M., and Sastry, G. N. 2006. From subtle to substantial: Role of metal ions on π–π interactions, *J. Phys. Chem. B.* 110: 2479.
24. Weinhold, F. and Landis, C. R. 2005. *Valency and Bonding: A Natural Bond Orbital Donor-Acceptor Perspective*. Cambridge University Press, Cambridge, UK.

25. Bader, R. F. W. 1990. *Atoms in Molecules. A Quantum Theory*. Oxford University Press, Oxford.
26. Geerlings, P. de Proft, F., and Langenaekar, W. 2003. Conceptual density functional theory. *Chem. Rev.* 103: 1793.
27. Chattaraj, P. K., Sarkar, U. and Roy, D. R. 2006. Electrophilicity index. *Chem. Rev.* 106: 2065.

4 Concepts in Electron Density

B. M. Deb

CONTENTS

4.1 INTRODUCTION [1–3]

During the last five decades, an alternative way of looking at the quantum theory of atoms, molecules, and solids in terms of the electron density in three-dimensional (3D) space, rather than the many-electron wave function in the multidimensional configuration space, has gained wide acceptance. The reasons for such popularity of the density-based quantum mechanics are the following:

1. Electron density is a fundamental variable and can be determined experimentally.
2. Considerable conceptual advantages are gained through the building of various transparent, interpretative models of structure, properties (including reactivity), and dynamics in terms of the single-particle density. Being simple but rigorous, such conceptual models ought to replace popular, ad hoc models in chemistry.
3. Simplicity and accuracy in computation, especially with large molecules for which other ab initio quantum chemical methods currently in vogue require computational labor of at least one order of magnitude greater for delivering results of comparable accuracy.

The three main approaches based on the single-particle density are the density functional theory (DFT), quantum fluid dynamics (QFD), and studying the properties of a system through local quantities in 3D space. In this chapter, we present simple discussions on certain conceptual and methodological aspects of the single-particle density; for details, the reader may consult the references listed at the end of this chapter.

4.2 WHAT IS A SINGLE-PARTICLE DENSITY AND WHY IS IT IMPORTANT? [1–5]

To answer these questions, let us first consider the normalized wave function for a system of N electrons, given by $\psi(x_1, x_2, \ldots, x_N)$, where x_i denotes the set of space and spin coordinates for the ith electron, i.e., $x_i \equiv (r_i, s_i)$, r_i being the position vector in 3D space and s_i the spin variable. ψ is postulated to contain all information about the system. One can define a single-particle reduced density matrix (RDM) and a two-particle RDM as follows:

$$\rho_1\left(x_1 | x_1'\right) = N \int \psi(x_1, x_2, \ldots, x_N)\psi^*\left(x_1', x_2, \ldots, x_N\right) dx_2 \ldots dx_N \qquad (4.1)$$

$$\Gamma_2\left(x_1, x_2 | x_1', x_2'\right) = (1/2)N(N-1) \int \psi(x_1, x_2, \ldots, x_N)\psi^*\left(x_1', x_2', \ldots, x_N\right) dx_3 \ldots dx_N$$

$$(4.2)$$

In Equations 4.1 and 4.2, the numbers before the integral signs occur due to the indistinguishability of electrons and electron pairs, respectively. The single-particle density $\rho(x)$ is defined as the diagonal element of the single-particle density matrix $\rho_1(x_1 | x_1')$, viz.,

$$\rho(x) = N \int \psi(x, x_2, \ldots, x_N)\psi^*(x, x_2, \ldots, x_N) dx_2 \ldots dx_N \qquad (4.3)$$

The spin-averaged single-particle density is given by

$$\rho(r) = \int \rho(x) ds \qquad (4.4)$$

where $\rho(r)$ has three important properties, viz.,

$$\rho(r) \geq 0, \quad \text{for all } r; \quad \int \rho(r)\mathrm{d}r = N; \quad \int \left|\mathrm{grad}\,\rho(r)^{1/2}\right|^2 \mathrm{d}r < \infty \qquad (4.5)$$

The second property in Equation 4.5 normalizes $\rho(r)$ to the total number of electrons in the system by integrating over the whole 3D space. Note that in atomic units (used throughout this chapter unless otherwise mentioned), the number density $\rho(r)$ becomes the electronic charge density, thereby paving the way to various useful, interpretative approaches as described below.

The nonrelativistic quantum mechanics of many-electron systems (atoms, molecules, nanomaterials, and condensed matter) can be formulated entirely in terms of the two-particle RDM (for Hartree–Fock systems or single-determinantal wave functions, the single-particle RDM will suffice because the two-particle RDM can be written in terms of the single-particle RDM), bypassing the many-particle wave function. Thus, the two-electron RDM contains all information about a system and the many-electron wave function, involving $3N$ space and N spin variables, is not necessary. This is undoubtedly a great simplification. However, since the wave function is being bypassed, a question arises: How does one know whether a given one-electron or two-electron RDM corresponds to an antisymmetric wave function? This is the well-known N-representability problem, which has been solved for the one-electron RDM but is unfortunately intractable for the two-electron RDM. In other words, the promising RDM approach cannot bypass the wave function approach.

However, it is indeed fortunate that the N-representability problem for the electron density $\rho(r)$ greatly simplifies itself. In fact, the necessary and sufficient conditions that a given $\rho(r)$ be N-representable are actually given by Equation 4.5 above. Nevertheless, question remains: Can the single-particle density contain all information about a many-electron system, at least in its ground state? An affirmative answer to this question can be given from Kato's cusp condition for a nuclear site in the ground state of any atom, molecule, or solid, viz.,

$$\partial \rho(r)/\partial r\big|_{r=0} = -2Z\rho(r = 0) \qquad (4.6)$$

where Z is the nuclear charge at the site of the cusp. Equation 4.6 has a profound significance. It says that the electron density $\rho(r)$ contains all information about the system in the ground state as follows: Let $\rho(r)$ alone be given for an unknown system in its ground state. One can, in principle, calculate the slope of the density at many points in 3D space and thereby hit upon all the cusps given by Equation 4.6. Thus, all the nuclei in the system and the total number of electrons become known, from which the Schrödinger equation for the system can be written and all information about the system can, in principle, be obtained. In other words, the single-particle density contains all information about the system, at least in the ground state. This conclusion forms the core of modern DFT (see Section 4.5).

4.3 SOME FUNDAMENTAL AND INTERPRETATIVE PROPERTIES OF ELECTRON DENSITY

4.3.1 ASYMPTOTIC BEHAVIOR [1]

While Equation 4.6 describes the short-range behavior of $\rho(r)$, the long-range behavior is given by the asymptotic relation

$$\rho(r) \sim \exp\left[-2(2I)^{1/2}r\right], \quad \text{for } r \to \infty \tag{4.7}$$

where I is the first ionization potential of the system.

4.3.2 MONOTONIC VARIATION AND SHELL STRUCTURE IN ATOMS [2,5,6]

The electron density $\rho(r)$ in the ground state of any atom falls off monotonically from the nuclear site until it vanishes asymptotically at infinity. For any atom in the ground state, a plot of the radial probability density $4\pi r^2 \rho(r)$ against r (or, more clearly against $r^{1/2}$) reveals the atomic shell structure in two ways: (1) The number of maxima equals the number of shells and (2) the locations of the minima indicate the approximate regions where the preceding shell ends and the next one begins. The changes in sign of $\nabla^2 \rho(r)$—where ∇^2 is the Laplacian—also indicates atomic shell structure. For a pair of atoms involved in the formation of a chemical bond, $\rho(r)$ has a saddle-point between the two nuclei; the saddle-point is a minimum along the bond direction and a maximum perpendicular to the bond direction. Therefore, from either of the two nuclear sites or cusps, the density falls off monotonically toward the saddle-point.

4.3.3 ELECTROSTATIC HELLMANN–FEYNMAN THEOREM [7,8]

Assuming the validity of the Born–Oppenheimer approximation, the electrostatic Hellmann–Feynman (H–F) theorem expresses the force F_A on a nucleus A, of charge Z_A, in a molecule or solid, as

$$F_A = Z_A \sum_{B \neq A} Z_B R_{AB}/R_{AB}^3 - Z_A \int \left(\rho(r)r_A/r_A^3\right)dr \tag{4.8}$$

where
    R_{AB} is the distance between the nuclei A and B
    r_A is the position vector from A

The first and second terms on the right-hand side of Equation 4.8 represent the nuclear–nuclear repulsive force and electron–nuclear attractive force, respectively. The great simplicity and visuality of this force concept in chemistry, involving $\rho(r)$, have been of enormous advantage in obtaining detailed qualitative and quantitative insights into the nature of chemical binding, molecular geometry, chemical reactivity as well as other properties.

4.3.4 ELECTROSTATIC POTENTIAL [9]

Since force is the negative gradient of the corresponding potential, one can also define an electrostatic potential (ESP) at a point r in the space around a molecule or solid, in terms of $\rho(r)$, as

$$V(r) = \sum_A Z_A/|\mathbf{R}_A - r| - \int (\rho(r')/|r - r'|)dr' \tag{4.9}$$

Equation 4.9 has been extensively applied to study the mechanisms of electrophilic (e.g., protonation) reactions, drug–nucleic acid interactions, receptor-site selectivities of pain blockers as well as various other kinds of biological activities of molecules in relation to their structure. Indeed, the ESP has been hailed as the "most significant discovery in quantum biochemistry in the last three decades." The ESP also occurs in density-based theories of electronic structure and dynamics of atoms, molecules, and solids. Note, however, that Equation 4.9 appears to imply that $\rho(r)$ of the system remains unchanged due to the approach of a unit positive charge; in this sense, the interaction energy calculated from $V(r)$ is correct only to first order in perturbation theory. However, this is not a serious limitation since using the correct $\rho(r)$ in Equation 4.9 will improve the results.

4.3.5 ATOMS IN MOLECULES [2,6]

Chemists have long been intrigued by the question, "Does an atom in a molecule somehow preserve its identity?" An answer to this question comes from studies on the topological properties of $\rho(r)$ and grad $\rho(r)$. It has been shown that the entire space of a molecule can be partitioned into "atomic" subspaces by following the trajectories of grad $\rho(r)$ in 3D space. These subspaces themselves extend to infinity and obey a subspace virial theorem $(2\langle T\rangle + \langle V\rangle = 0)$. The subspaces are bounded by surfaces of zero flux in the gradient vectors of $\rho(r)$, i.e., for all points on such a surface,

$$\text{grad}\,\rho(r) \cdot \mathbf{n}(r) = 0 \tag{4.10}$$

where $\mathbf{n}(r)$ is a unit vector normal to the surface at r. Both $\rho(r)$ and grad $\rho(r)$ vanish at infinity.

4.3.6 PROPERTY DENSITIES IN 3D SPACE [2,3]

For a system, one can define a property density function (PDF) $p(r)$ in 3D space such that the corresponding property P is given by

$$P = \int p(r)dr \tag{4.11}$$

Examples of $p(r)$ are energy density, charge density, current density (see Section 4.6), difference density (difference between a final density and an initial density), electric moment density, magnetic moment density, local reactivity functions (see Section 4.5.2), force density, etc. Note that, for ensuring the stability of matter, the net force density must vanish everywhere in space. The concept of a PDF has generated many significant developments in interpretative quantum chemistry.

From the preceding discussion, it is quite clear that $\rho(r)$ is indeed a fundamental quantum mechanical entity of no less significance than the wave function and that $\rho(r)$ generates numerous attractive and transparent models of chemical behavior. How does one calculate $\rho(r)$? One way would of course be to calculate it from the normalized occupied orbital densities, viz.,

$$\rho(r) = \sum_i n_i |\varphi_i(r)|^2 \tag{4.12}$$

where $\varphi_i(r)$ is an orbital with occupation number n_i. For a usual atomic or molecular orbital, n_i equals 0, 1, or 2, whereas for a natural orbital—in principle, the most accurate orbital description of a many-electron system, since it is based on an accurate single-particle density matrix—one has $0 \leq n_i \leq 1$. However, Equation 4.12 still ties the single-particle density to the apron strings of the wave function and it is therefore interesting to enquire whether $\rho(r)$ can be *directly* calculated bypassing the wave function formalism. As discussed in Section 4.5, this is achieved by modern DFT, at least for the ground state.

4.4 THOMAS–FERMI STATISTICAL MODEL AND ITS MODIFICATIONS [2,4,5]

The Thomas–Fermi (TF) model (1927) for a homogeneous electron gas provides the underpinnings of modern DFT. In the following discussion, it will be shown that the model generates several useful concepts, relates the electron density to the potential, and gives a universal differential equation for the direct calculation of electron density. The two main assumptions of the TF model are as follows:

1. The electrons in an atom or any many-electron system move under an effective potential. The electronic distribution results from feeding two electrons into a volume h^3 of the 6D phase space; this is in accord with Pauli exclusion principle.
2. The effective potential is determined by nuclear charges and this electronic distribution.

In order to calculate the electronic energy, consider the 3D space to be divided into small cubic cells, each cell of length l, volume $\Delta V = l^3$, containing ΔN electrons. The system is at 0 K, the cells are independent, and the electrons move independently of one another. The familiar expression for the energy of a particle in a cubical box (with potential energy taken as zero) is

$$\varepsilon(n_x, n_y, n_z) = (h^2/8ml^2)\left(n_x^2 + n_y^2 + n_z^2\right) = (h^2/8ml^2)R^2 \qquad (4.13)$$

where

n_x, n_y, $n_z = 1, 2, 3, \ldots$, etc.

R is the radius of a sphere described by the quantum numbers

For high quantum numbers, i.e., high R, the number of distinct energy levels with energy not greater than ε is given by (note that only the first octant of the sphere of quantum numbers, without the origin, is to be taken into account)

$$\Phi(\varepsilon) = (1/8)(4\pi R^3/3) = (\pi/6)(8ml^2\varepsilon/h^2)^{3/2} \qquad (4.14)$$

Define a density of states $g(\varepsilon)$ at energy ε, so that the number of states between ε and $\varepsilon + \delta\varepsilon$ is

$$g(\varepsilon)\delta\varepsilon = \Phi(\varepsilon + \delta\varepsilon) - \Phi(\varepsilon) = (\pi/4)(8ml^2/h^2)^{3/2}\varepsilon^{1/2}\delta\varepsilon + O\left((\delta\varepsilon)^2\right) \qquad (4.15)$$

To calculate the total energy of the cell with ΔN electrons, one needs $f(\varepsilon)$, the probability for the state with energy ε to be occupied by an electron. This is given by Fermi–Dirac statistics as

$$f(\varepsilon) = 1/(1 + \exp[\beta(\varepsilon - \varepsilon_F)]); \quad \beta = 1/kT \qquad (4.16)$$

where ε_F is the highest energy (Fermi energy) of an occupied state. At $T = 0$ K, $f(\varepsilon)$ becomes unity for $\varepsilon < \varepsilon_F$ and vanishes for $\varepsilon > \varepsilon_F$, since $\beta \to \infty$; in other words, all states with $\varepsilon < \varepsilon_F$ are occupied and all states with $\varepsilon > \varepsilon_F$ are unoccupied. Therefore, the total energy (kinetic energy only) of the electrons in this cell is (see assumption 1 above)

$$\Delta\varepsilon = 2 \int \varepsilon f(\varepsilon)g(\varepsilon)d\varepsilon = 4\pi(2m/h^2)^{3/2}l^3 \int \varepsilon^{3/2}d\varepsilon$$
$$= (8\pi/5)(2m/h^2)^{3/2}l^3\varepsilon_F^{5/2} \qquad (4.17)$$

the integration limits being 0 to ε_F. Using the same integration limits, one has

$$\Delta N = 2 \int f(\varepsilon)g(\varepsilon)d\varepsilon = (8\pi/3)(2m/h^2)^{3/2}l^3\varepsilon_F^{3/2} \qquad (4.18)$$

Dividing Equation 4.17 by Equation 4.18 and simplifying by using Equation 4.18, one obtains

$$\Delta\varepsilon = (3/5)\Delta N\varepsilon_F = (3h^2/10m)(3/8\pi)^{2/3}l^3(\Delta N/l^3)^{5/3} \qquad (4.19)$$

Replace $(\Delta N/l^3)$ by ρ, the finite density of the homogeneous electron gas. Taking $\Delta V \to 0$, ρ can be locally replaced by $\rho(r)$. Using atomic units and summing the contributions from all cells, the total TF kinetic energy becomes a functional of the electron density (see Ref. [4] for an account of functional calculus), viz.,

$$T_{TF}[\rho] = c_{TF} \int \rho^{5/3}(r)dr \qquad (4.20)$$

where the TF constant $c_{TF} = (3/10)(3\pi^2)^{2/3} = 2.8712$. Equation 4.20 is called the local density approximation (LDA) to the kinetic energy of a many-electron system. The total electronic energy of an atom (of nuclear charge Z), neglecting two-electron quantum effects such as exchange and correlation but including a classical Coulomb repulsion term, can now be written as

$$E_{TF}[\rho] = c_{TF} \int \rho^{5/3}(r)dr - Z \int (\rho(r)/r)dr + 1/2 \iint (\rho(r)\rho(r')/|r - r'|)drdr' \quad (4.21)$$

with $\int \rho(r)dr = N$, the total number of electrons. The term $(-Z/r)$ is called the external (electron–nuclear attraction) potential (see Section 4.5.2) for the electron gas in an atom. For obtaining a differential equation for the direct determination of electron density, we now perform a constrained variation whereby the density is always kept normalized to the total number of electrons, as

$$\delta\left[E_{TF}[\rho] - \mu_{TF}\left(\int \rho(r)dr - N\right)\right] = 0 \qquad (4.22)$$

where the Lagrange multiplier μ_{TF} can be identified as the TF chemical potential (note its dimensional equivalence with the thermodynamic chemical potential). It will be seen in Section 4.3.4 that the concept of chemical potential plays a fundamental role in DFT.

By functional differentiation, Equation 4.22 leads us to the Euler–Lagrange deterministic equation for the electron density, viz.,

$$\mu_{TF} = \delta E_{TF}[\rho]/\delta\rho = (5/3)c_{TF}\rho^{2/3}(r) - \varphi(r) \qquad (4.23)$$

where the ESP is

$$\varphi(r) = Z/r - \int (\rho(r')/|r - r'|)dr' \qquad (4.24)$$

As mentioned before, the ESP has been a quantity of great significance in quantum biochemistry. Using Poisson's equation of classical electrostatics, as applied to an atom, one can write

$$\nabla^2\varphi(r) = 4\pi\rho(r) - 4\pi Z\delta(r) \qquad (4.25)$$

where $\delta(r)$ is the Dirac delta function. Using the substitutions

$$x = \alpha r; \quad \alpha = 1.1295Z^{1/3}, \quad \text{and} \quad \chi(r) = (r/Z)\varphi(r) \qquad (4.26)$$

where vectors are replaced by scalars and $\chi(r)$ is a dimensionless function, one eventually obtains the TF universal differential equation,

$$d^2\chi(x)/dx^2 = x^{-1/2}\chi^{3/2}(x); \quad \chi(0) = 1, \quad \chi(\infty) = 0 \tag{4.27}$$

One can improve upon the TF model by incorporating two-electron effects into $E_{TF}[\rho]$ as the approximate, local Dirac exchange energy functional (c_X is the Dirac exchange constant)

$$E_X[\rho] = -c_X \int \rho^{4/3}(r)dr; \quad c_X = (3/4\pi)(3\pi^2)^{1/3} = 0.7386 \tag{4.28}$$

with the exchange potential proportional to $\rho^{1/3}(r)$ and the approximate, local Wigner correlation energy functional

$$E_c[\rho] = -\int \rho(r)v_c[\rho]dr \tag{4.29}$$

where the correlation potential $v_c[\rho]$ is given by

$$v_c[\rho] = \left(a + b\rho^{-1/3}(r)\right) \Big/ \left(a + (3b/4)\rho^{-1/3}(r)\right)^2; \quad a = 9.810, \quad b = 28.583 \tag{4.30}$$

The interesting implication of Equations 4.28 and 4.29 is that the single-particle density can incorporate two-particle effects such as exchange and correlation. In modern DFT, the three density functionals $E_{TF}[\rho]$, $E_X[\rho]$, and $E_C[\rho]$ are sometimes employed as nonlocal functionals, involving gradients of $\rho(r)$. The universal value of the ratio c_{TF}/c_X has been shown to correlate nicely with variations in atomic radii, ionic radii, van der Waals radii, Wigner–Seitz radii, atomic polarizability, London dispersion coefficient, lanthanide contraction, etc. among the rows and columns of the Periodic Table. The Thomas–Fermi–Dirac (TFD) atomic energy can also be expanded in terms of $Z^{1/3}$, the first term involving $Z^{7/3}$.

The TFD model does not reveal atomic shell structure as well as the correct short-range and long-range behavior of the electron density. Furthermore, it does not show any chemical binding, thereby wrongly implying that no molecule or solid can exist. This problem occurs due to replacing (see Equation 4.20) the particle-in-a-box wave functions locally by plane waves. An explicit correction to the TFD kinetic energy, involving deviation from plane waves, is provided by the Weizsäcker inhomogeneity correction

$$T_W[\rho] = (1/8)\int \left((grad\,\rho)^2/\rho\right)dr \tag{4.31}$$

This provides partial chemical binding. Note that $T_{TF}[\rho]$ and $T_W[\rho]$ constitute the first two terms in a gradient expansion of the kinetic energy.

The conclusion that it may be possible to formulate the quantum mechanics of many-electron systems solely in terms of the single-particle density was put on a firm foundation by the two Hohenberg–Kohn theorems (1964), which are stated below, without proof.

4.5 SIMPLE ACCOUNT OF TIME-INDEPENDENT DENSITY FUNCTIONAL THEORY

4.5.1 HOHENBERG–KOHN THEOREMS FOR THE GROUND STATE OF A MANY-ELECTRON SYSTEM [4,5,10–12]

Theorem 1. *The external potential $v(r)$ is determined, within a trivial additive constant, by the electron density $\rho(r)$.* (The implication of this existence theorem is that $\rho(r)$ determines the wave function and therefore all electronic properties in the ground state; see also Equation 4.6.)

Theorem 2. *For a trial density $\tilde{\rho}(r)$, satisfying the N-representability conditions, the trial ground-state energy E_v satisfies the relation*

$$E_0 \leq E_v[\tilde{\rho}] = \int \tilde{\rho}(r)v(r)dr + F[\tilde{\rho}] \tag{4.32}$$

where
  E_0 is the true ground-state energy
  the universal functional $F[\rho]$ is a sum of the electronic kinetic energy $T[\rho]$, classical Coulomb repulsion energy $J[\rho]$, and quantum exchange-correlation energy $E_{XC}[\rho]$ functionals

The implication of this theorem is that it gives a prescription for the variational determination of the ground-state electron density, since the latter minimizes the energy.

4.5.2 HOHENBERG–KOHN–SHAM EQUATIONS FOR DETERMINING DENSITY [4,5,10–12]

One can write the total electronic energy of the system as

$$E[\rho] = \int \rho(r)v(r)dr + T[\rho] + J[\rho] + E_{XC}[\rho] \tag{4.33}$$

Since the ground-state electron density minimizes the energy, subject to the normalization constraint, $\int \rho(r)dr - N = 0$, the Euler–Lagrange equation (see Equation 4.23) becomes

$$\mu = \delta E[\rho]/\delta\rho = v(r) + \delta T[\rho]/\delta\rho + \delta J[\rho]/\delta\rho + \delta E_{XC}[\rho]/\delta\rho \tag{4.34}$$

where the chemical potential μ (see Section 4.4) has been shown to be the zero-temperature limit of the chemical potential defined for the finite-temperature grand canonical ensemble in statistical mechanics. Of the four terms on the right-hand side of Equation 4.34, only the functionals $T[\rho]$ and $E_{XC}[\rho]$ are unknown.

The problem of $T[\rho]$ is cleverly dealt with by mapping the interacting many-electron system on to a system of "noninteracting electrons." For a determinantal wave function of a system of N "noninteracting electrons," each electron occupying a normalized orbital $\psi_i(r)$, the Hamiltonian is given by

$$H_s = \sum_{i=1}^{N} (-\nabla_i^2 2) + \sum_{i=1}^{N} v_{si}(r) \tag{4.35}$$

where
∇_i^2 is the Laplacian operator for the ith electron
v_{si} is the ith electron–nuclear attraction term

Therefore, the density and the kinetic energy can be written as

$$\rho(r) = \sum_{i=1}^{N} |\psi_i(r)|^2 \tag{4.36}$$

$$T_s[\rho] = \sum_{i=1}^{N} \langle \psi_i | -\nabla_i^2/2 | \psi_i \rangle \tag{4.37}$$

One now replaces the interacting $T[\rho]$ in Equation 4.34 by the noninteracting $T_s[\rho]$. This means that any kinetic energy missing as a result of this replacement must be included in $E_{XC}[\rho]$; clearly, $T[\rho] > T_s[\rho]$. Equation 4.34 now becomes

$$\mu = v_{eff}(r) + \delta T_s[\rho]/\delta\rho \tag{4.38}$$

where the effective potential $v_{eff}(r)$ is given by

$$v_{eff}(r) = v(r) + \delta J[\rho]/\delta\rho + \delta E_{XC}[\rho]/\delta\rho$$
$$= v(r) + \int (\rho(r')/|r - r'|)dr' + v_{XC}(r) \tag{4.39}$$

where $v_{XC}(r)$ is the exchange-correlation potential. Now, for N noninteracting electrons, one can obviously write the Schrödinger equation

$$\left[-\nabla^2/2 + v_s(r)\right]\varphi_i(r) = \lambda_i\varphi_i(r); \quad i = 1, 2, \ldots, N \tag{4.40}$$

where λ_i is the energy eigenvalue for the orbital $\varphi_i(r)$. In view of the above mapping of an interacting electron system into a "noninteracting electron" system, it is now possible to write the Hohenberg–Kohn–Sham (HKS) "Schrödinger-like" equations for the interacting electron system as

$$\left[-\nabla^2/2 + v_{eff}(r)\right]\psi_i(r) = \varepsilon_i\psi_i(r); \quad i = 1, 2, \ldots, N \tag{4.41}$$

where ε_i is the energy eigenvalue for the DFT orbital ψ_i (r). In Equation 4.41, $v_{\text{eff}}(r)$ itself depends on the solutions of Equation 4.41, which is therefore a nonlinear Schrödinger equation with potentially interesting mathematical properties, some of which remain to be explored. Equation 4.41 can be solved iteratively and self-consistently as follows: (1) Assume a trial set of orbitals $\{\psi_i$ $(r)\}$, $i = 1, 2, \ldots, N$, and an accurate, though approximate, form of $E_{\text{XC}}[\rho]$; (2) calculate $\rho(r)$, $v_{\text{XC}}(r)$, and $v_{\text{eff}}(r)$; (3) solve for new sets $\{\psi_i(r)\}$ and $\{\varepsilon_i\}$; (4) repeat (2) and (3) iteratively until the final $\{\varepsilon_i\}$ agrees with the $\{\varepsilon_i\}$ in the previous iteration within a prescribed limit of tolerance. This process yields the self-consistent DFT orbitals and their energies. The computations are economical in time, with the labor being comparable to that of the Hartree method although the results are significantly more accurate than the Hartree–Fock method. This is the main reason behind the enormous popularity and wide applicability of DFT in dealing with atoms, molecules, nanosystems, and condensed matter in general, involving chemical, physical, biological, and geological phenomena.

Some explanations concerning Equation 4.41 are necessary. First, note that all electron interactions are included in $v_{\text{eff}}(r)$, which vanishes as $r \to \infty$. Second, unlike the Koopmans' theorem in Hartree–Fock theory, the DFT orbital energies $\{\varepsilon_i\}$ do not have a simple interpretation. One can, however, show that $\varepsilon_i = (\partial E/\partial n_i)$, where n_i is the occupation number of the ith orbital, as well as $\varepsilon_{\text{max}} = -I$, the ionization potential. Most interesting, however, is the chemical potential μ, which yields a deterministic equation for the electron density through $\mu = \delta E/\delta \rho$. Using simple arguments, it has been shown that $\mu \approx -(I+A)/2 = -\kappa_M$, where κ_M is Mulliken electronegativity, thus validating the concept of electronegativity quantum mechanically. Therefore, electronegativity plays a basic variational role in DFT, similar to the role played by energy in wave function–based variation theory. This again emphasizes that DFT ought to be a fundamental theory of chemistry. Note that the principle of "electronegativity equalization" may serve as an important guideline for interactions (e.g., bond-making, bond-breaking, and chemical reactivity in general) between atomic/molecular species with differing electronegativities.

Several other quantities of chemical significance have been defined in terms of the chemical potential, viz.,

$$\text{Hardness, } \eta = (1/2)(\partial\mu/\partial N)_v = (1/2)(\partial^2 E/\partial N^2)_v \approx (I-A)/2; \quad \eta \geq 0 \quad (4.42)$$

$$\text{Softness, } S = 1/2\eta = (\partial N/\partial\mu)_v \quad (4.43)$$

$$\text{Fukui function (a local reactivity index), } f(r) = (\partial\rho/\partial N)_v$$

$$= (\partial\mu/\partial v)_N; \int f(r)dr = 1 \quad (4.44)$$

$$\text{Local softness, } s(r) = (\partial\rho/\partial\mu)_v; \ S = \int s(r)dr; \ f(r) = s(r)/S \quad (4.45)$$

$$\text{Local hardness, } \eta(r) = (1/2N)\iint\{\delta^2 F[\rho]/(\delta\rho(r)\delta\rho(r'))\}\rho(r')dr';$$

$$\eta = \int f(r)\eta(r)dr \quad (4.46)$$

Both the local functions $f(r)$ and $s(r)$ contain useful information about relative activities of different sites in a molecule or solid.

At this point, it is necessary to say a few words about the v-representability of the electron density. An electron density is said to be v-representable if it is associated with the antisymmetric wave function of the ground state, corresponding to an external potential $v(r)$, which may or may not be a Coulomb potential. Not all densities are v-representable. Furthermore, the necessary and sufficient conditions for the v-representability of an electron density are unknown. Fortunately, since the N-representability (see Section 4.2) of the electron density is a weaker condition than v-representability, one needs to formulate DFT only in terms of N-representable densities without unduly worrying about v-representability.

4.5.3 EXCHANGE-CORRELATION HOLE [4,5,13]

As indicated in Equation 4.21, the interelectronic Coulomb repulsion energy functional $J[\rho]$ is written as the classical expression

$$J[\rho] = 1/2 \iint (\rho(r)\rho(r')/|r - r'|)\mathrm{d}r\mathrm{d}r' \tag{4.47}$$

Since the Coulomb, exchange, and correlation energies are all consequences of the interelectronic $1/r_{12}$ operator in the Hamiltonian, one can define the exchange energy functional $E_X[\rho]$ in the same manner as

$$E_X[\rho] = 1/2 \iint (\rho(r)\rho_X(r, r')/|r - r'|)\mathrm{d}r\mathrm{d}r' \tag{4.48}$$

where the exchange density matrix $\rho_X(r, r')$ has the property

$$\int \rho_X(r, r')\mathrm{d}r' = -1, \quad \text{for all } r \tag{4.49}$$

Equation 4.49 defines the exchange or Fermi hole. It is as if an electron of a given spin "digs" a hole around itself in space in order to exclude another electron of the same spin from coming near it (Pauli exclusion principle). The integrated hole charge is unity, i.e., there is exactly one electron inside the hole. Likewise, the correlation energy functional can be defined as

$$E_C[\rho] = 1/2 \iint (\rho(r)\rho_C(r, r')/|r - r'|)\mathrm{d}r\mathrm{d}r' \tag{4.50}$$

where

$$\int \rho_C(r, r')\mathrm{d}r' = 0, \quad \text{for all } r \tag{4.51}$$

indicating that the correlation "hole" is overall neutral. Assuming that $E_{XC}[\rho] = E_X[\rho] + E_C[\rho]$, one can write an exchange-correlation density matrix $\rho_{XC}(r, r') = \rho_X(r, r') + \rho_C(r, r')$, thereby defining an exchange-correlation hole such that

$$\int \rho_{XC}(r, r')dr' = -1, \quad \text{for all } r \tag{4.52}$$

4.6 QUANTUM FLUID DYNAMICS: TIME-DEPENDENCE OF SINGLE-PARTICLE DENSITY [2,3,14,15]

Since the beginning of quantum mechanics, there have been significant attempts to explain quantum phenomena based on familiar "classical" concepts. An earlier attempt was by Madelung who transformed the one-particle, time-dependent Schrödinger equation (TDSE) into two fluid dynamical equations of classical appearance, viz., a continuity equation signifying the absence of any source or sink, as well as a Euler-type equation of motion. The continuity of the density over the whole space imparts to the quantum or Schrödinger fluid more fluid-like character than even a classical fluid, thereby raising the possibility of a "classical" description of quantum mechanics through the fluid dynamical viewpoint. The Madelung transformation for a single-particle wave function is defined in the following.

One may write the time-dependent wave function in the polar form, viz.,

$$\Psi(r, t) = R(r, t) \exp[iS(r, t)/\hbar] \tag{4.53}$$

Then, the TDSE can be reformulated in terms of two fluid-dynamical equations:

$$\text{Continuity equation: } \partial\rho/\partial t + \text{div}(\rho v) = 0 \tag{4.54}$$

$$\text{Euler-type equation of motion: } m\rho dv/dt = -\rho\text{grad}(V + V_{qu}) \tag{4.55}$$

where

$$\rho = R^2; \, v = (1/m)\text{grad}\,S; \, j = \rho v \tag{4.56}$$

$$dv/dt = \partial v/\partial t + (v \cdot \text{grad})v \tag{4.57}$$

and

$$V_{qu} = (-\hbar^2/2m)\nabla^2 R/R \tag{4.58}$$

In Equation 4.56, the real quantities ρ, v, and j are the charge density, velocity field, and current density, respectively. The above equations provide the basis for the fluid dynamical approach to quantum mechanics. In this approach, the time evolution of a quantum system in any state can be completely interpreted in terms of a continuous, flowing fluid of charge density $\rho(r, t)$ and the current density $j(r, t)$, subjected to forces arising from not only the classical potential $V(r, t)$ but also from an additional potential $V_{qu}(r, t)$, called the quantum or Bohm potential; the latter arises from the kinetic energy and depends on the density as well as its gradients. The current density vanishes in the ground state but not in a general excited state. This is the reason why while the ground-state density contains all information about the system (HK Theorem 1), the same is not true for the pure-state density of a general excited state because then the current density will also be involved. Obviously, in general, the complex-valued wave function can be replaced by two real-valued functions (ρ and j) but not by just one of them, except in special cases.

The above fluid dynamical analogy to quantum mechanics has been extended to many-electron systems. Subsequently, this has provided the foundations for the developments of TD DFT and excited-state DFT, two areas which had remained unaccessed for many years. However, these developments are outside the scope of the present chapter.

4.7 CONCLUSION

Five decades of extensive studies on the statics and dynamics of the single-particle density in atoms, molecules, clusters, and solids have established that an alternative quantum mechanics of many-electron systems can be constructed solely in terms of the single-particle density, thus bypassing the wave function for most practical purposes. The ability of $\rho(r)$ to yield transparent and deep insights into problems concerning binding, structure, properties and dynamics covering physical, chemical, biological, and even geological phenomena has been astounding. Undoubtedly, such developments would continue in the years to come.

REFERENCES

1. Smith, V. H., Jr. and Absar, I. *Israel J. Chem.* 1977, *16*, 87.
2. Bamzai, A. S. and Deb, B. M. *Rev. Mod. Phys.* 1981, *53*, 95, 593.
3. Ghosh, S. K. and Deb, B. M. *Phys. Rep.* 1982, *52*, 1.
4. Parr, R. G. and Yang, W. *Density Functional Theory of Atoms and Molecules*, Oxford: Oxford University Press, 1989.
5. March, N. H. *Electron Density Theory of Atoms and Molecules*, London: Academic Press, 1992.
6. Bader, R. F. W. *Atoms in Molecules*, Oxford: Clarendon Press, 2003.
7. Deb, B. M. *Rev. Mod. Phys.* 1973, *45*, 22.
8. Deb, B. M. (Ed). *The Force Concept in Chemistry*, New York: Van Nostrand-Reinhold, 1981.
9. Murray J. S. and Sen, K. (Eds.), *Molecular Electrostatic Potentials: Concepts and Applications*, Amsterdam: Elsevier, 1996.
10. Kryachko, E. S. and Ludena, E. V. *Energy Density Functional Theory of Many-Electron Systems*, Dordrecht: Kluwer Academic, 1990.
11. Kohn, W., Becke, A. D., and Parr, R. G. *J. Phys. Chem.* 1996, *100*, 12974.
12. Kohn, W. *Rev. Mod. Phys.* 1999, *71*, 1253.
13. McWeeny, R. and Sutcliffe, B. T. *Methods of Molecular Quantum Mechanics*, Chap. 4, London: Academic Press, 1978.
14. Deb, B. M. and Ghosh, S. K. *The Single-Particle Density in Physics and Chemistry*, N. H. March and B. M. Deb (Eds.), London: Academic Press, 1987.
15. Wyatt, R. E. *Quantum Dynamics with Trajectories: Introduction to Quantum Hydrodynamics*, New York: Springer, 2005.

5 Atoms and Molecules: A Momentum Space Perspective

Shridhar R. Gadre and P. Balanarayan

CONTENTS

The discussion of atoms and molecules by physicists and chemists until the 1920s, was limited to position space. Exploration of these systems in momentum space began with the pioneering work [1] of Pauling and Podolsky in 1929, in which they applied a Fourier–Dirac transformation [2], as given by Jordan in 1927, to the hydrogenic orbitals. The aim of this was to obtain the wave function in momentum space and thereby the probability of an electron having momentum in a given range. This was related to the experimental Compton line shapes giving electron momentum densities (EMDs) for an atomic system [3]. It was shown by Pauling and Podolsky [1] that the orbitals in position space transformed to momentum space yield the familiar associated Legendre functions $P_l^m(\cos\theta)$ multiplied by $e^{\pm im\phi}$. In momentum space, the radial part was described by a Gegenbauer polynomial instead of the Laguerre polynomial in position space [1].

Figure 5.1 illustrates the Fourier transform (FT) of a simple function, viz., a Gaussian. The relatively sharp Gaussian function with the exponent $\alpha = 1$ depicted in Figure 5.1a, yields a diffuse Gaussian (in dotted line) in momentum space. A flat Gaussian function in position space with $\alpha = 0.1$, transforms to a sharp one (cf. Figure 5.1b). Connected by an FT, the wave functions in position and momentum

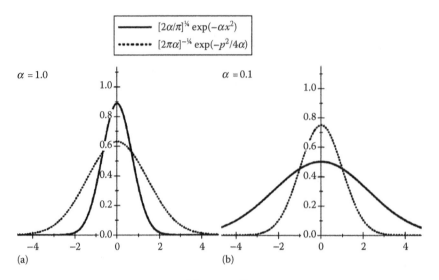

FIGURE 5.1 FT of a Gaussian $(2\alpha/\pi)^{1/4}\exp(-\alpha x^2)$ for two different values: (a) $\alpha = 1.0$ and (b) $\alpha = 0.1$.

spaces bear a reciprocal relation, i.e., a narrow wave function in position space leads to a broad one in momentum space and vice versa.

The atomic and molecular wave functions are usually described by a linear combination of either Gaussian-type orbitals (GTO) or Slater-type orbitals (STO). These expressions need to be multiplied by a center dependent factor $\exp(-ip\cdot A)$. Further the STOs in momentum space need to be multiplied by $Y_{lm}(\theta_p,\phi_p)$. Examining the expressions [4], one notices the Gaussian nature of the GTOs even after the FT. The STOs are significantly altered on FT. From the expressions in Table 5.1, STOs are seen to exhibit a decay $\approx p^{-4}$, which is the decay of the slowest 1s

TABLE 5.1

Expressions for GTO in Cartesian Momentum Coordinates and STO in Spherical Polar Momentum Space Coordinates

GTO	STO
(s) $\dfrac{1}{(2\pi\alpha)^{3/4}}\exp(-p^2/4\alpha)$	(1s) $\left(\dfrac{2\alpha}{\pi}\right)^{1/2}\dfrac{8\pi\alpha^{5/2}}{(\alpha^2+p^2)^2}$
(p$_x$) $\dfrac{1}{(2\pi)^{3/4}\alpha^{5/4}}ip_x\exp(-p^2/4\alpha)$	(2s) $\left(\dfrac{2\alpha}{3\pi}\right)^{1/2}\dfrac{4\alpha^2(3\alpha^2-p^2)}{(\alpha^2+p^2)^3}$
(d$_z^2$) $\dfrac{2^{1/4}}{\sqrt{3}(\alpha\pi)^{1/4}}\left(1-\dfrac{p_z^2}{2\alpha}\right)\exp(-p^2/4\alpha)$	(2p) $\left(\dfrac{2\alpha}{3\pi}\right)^{1/2}\dfrac{16p\alpha^3}{(\alpha^2+p^2)^3}$
(d$_{xy}$) $\dfrac{-1}{(2\pi)^{3/4}\alpha^{7/4}}p_xp_y\exp(-p^2/4\alpha)$	

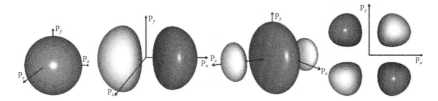

FIGURE 5.2 The s, p_x, d_z^2, d_{xy} Gaussian functions with exponent $\alpha = 1.0$ in momentum space. For p_x, the function is imaginary. The isosurfaces for ± 0.01 a.u. are given.

function. Isosurfaces of the GTOs in momentum space are plotted in Figure 5.2, which are qualitatively similar in shape to their position space counterparts.

The many-particle momentum space wave function, $\Phi(\mathbf{p}_1, \mathbf{p}_2, \mathbf{p}_3, \ldots, \mathbf{p}_N)$ is obtained as the $3N$ dimensional FT of the corresponding position space counterpart viz., $\Psi(\mathbf{r}_1, \mathbf{r}_2, \mathbf{r}_3, \ldots, \mathbf{r}_N)$ {only the spatial parts are explicitly denoted}:

$$\Phi(\mathbf{p}_1, \mathbf{p}_2, \ldots, \mathbf{p}_N) = (2\pi)^{-3N/2} \int \Psi(\mathbf{r}_1, \mathbf{r}_2, \mathbf{r}_3, \ldots, \mathbf{r}_N)$$
$$\exp[-i(\mathbf{p}_1 \cdot \mathbf{r}_1 + \cdots + \mathbf{p}_N \cdot \mathbf{r}_N)] d^3 r_1 \cdots d^3 r_N \tag{5.1}$$

The EMD, $\gamma(\mathbf{p})$ can be extracted from $\Phi(\mathbf{p}_1, \mathbf{p}_2, \ldots, \mathbf{p}_N)$ on squaring, followed by a suitable integration over all but one momentum coordinate. Here $\mathbf{y}$ depends both on $\mathbf{p}$ and the spin:

$$\gamma(\mathbf{p}) = N \sum_\sigma \int |\Phi(\mathbf{y}, \mathbf{y}_2, \ldots, \mathbf{y}_N)|^2 d^3 p_2 \cdots d^3 p_N \tag{5.2}$$

5.1 ATOMIC MOMENTUM DENSITIES

In spite of their familiarity, there still remains a lot to be said about the nature of atomic charge densities in position space. Atomic densities have a fairly simple but intriguing structure. A spherical symmetry for the atomic density profiles is one that is always chosen out of convenience. The ground state atomic charge densities possess a finite maximum at the nuclear position. The spherically averaged charge density $\rho(r)$ satisfies the Kato's cusp condition [5] at the nuclear position, i.e., $d\rho/dr|_{r=0} = -2Z\rho(0)$, where Z is the nuclear charge. Thus the charge density for an atom resembles $\exp(-2Zr)$ near the nucleus and $\exp(-2\sqrt{2I}r)$ (I is the ionization potential) at the asymptotes [6]. There exists (a yet unproven) postulate termed as the monotonic density postulate [7] which states that $d\rho/dr < 0$ for all r, for ground states of atoms.

The spherically symmetric atomic momentum densities, in contrast, exhibit monotonic as well as nonmonotonic behavior even in their ground states. Further, it was

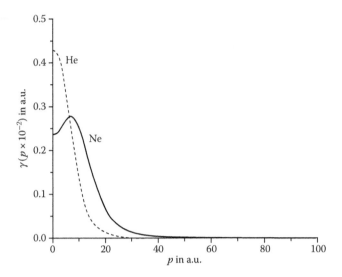

FIGURE 5.3 Monotonic and nonmonotonic atomic momentum densities for He (dotted) and Ne are portrayed.

shown by Smith and Benesch [8] that they exhibit an asymptotic decay $\sim p^{-8}$. Two examples of ground state atomic densities for He and Ne are depicted in Figure 5.3. In contrast to the monotonic momentum density of He, the momentum density of Ne has a minimum at $p=0$ and a maximum at $p=0.9$ a.u. The behavior of the spherically averaged ground state atomic densities has been studied in detail [9]. A nonmonotonic behavior has been noticed by Thakkar [10a] for carbon, nitrogen, oxygen, fluorine, neon, and argon atoms. The carbon, nitrogen, oxygen, and fluorine atoms were found to show atmost three maxima in their spherically averaged momentum densities. Later, Koga et al. [10b] classified the 103 elements of the periodic table into three groups with atoms being: (1) monotonic (2) nonmonotonic, and (3) those having two maxima, one at $p=0$ and the other at $p>0$.

At this point, one may wonder why there is an interest in the atomic momentum densities and their nature and what sort of information does one derive from them. In a system in which all orientations are equally probable, the full three-dimensional (3D) momentum density is not experimentally measurable, but its spherical average is. The moments of the atomic momentum density distributions are of experimental significance. The moments and the spherically averaged momentum densities are defined in the equations below.

$$I(p) = (1/4\pi) \int \gamma(p)p^2 \sin \theta_p \, d\theta_p \, d\phi_p \tag{5.3}$$

$$\langle p^n \rangle = 4\pi \int I(p)p^n dp, \quad \text{for } -2 \leq n \leq 4 \tag{5.4}$$

These moments have a physical significance with $n = 2$, giving a measure of the kinetic energy of the atom, $n = -1$ being proportional to the peak heights of experimentally measurable Compton profiles, $n = 1$ giving measures of shielding in nuclear magnetic resonance, and $n = 4$ being equivalent to the Breit–Pauli relativistic correction.

5.2 MOLECULAR ELECTRON MOMENTUM DENSITIES

The chemistry of momentum densities becomes more interesting when one goes over to molecules. There is an inherent stabilization when atoms bond together to form molecules. One of the first chemical interpretations of the EMD was presented by Coulson [11]. The simplest valence bond (VB) and molecular orbital (MO) wave functions for H_2 molecule given in the following equations were used for this purpose:

$$\Psi_{MO} = \frac{1}{\sqrt{2(1 + S^2)}} [\varphi_a(\mathbf{r}_1) + \varphi_b(\mathbf{r}_1)] \tag{5.5}$$

$$\Psi_{VB} = \frac{1}{\sqrt{2(1 + S)}} [\varphi_a(\mathbf{r}_1)\varphi_b(\mathbf{r}_2) + \varphi_a(\mathbf{r}_2)\varphi_b(\mathbf{r}_1)] \tag{5.6}$$

Here φ_a is the 1s atomic Slater functions $(\alpha^{3/2}/\pi^{1/2}) \exp(-\alpha|\mathbf{r} - \mathbf{r}_a|)$ with the atom being centered at position vectors $\mathbf{r}_a$. The overlap between these functions is given by S. After an FT and integrating over momentum coordinates of one particle, the EMD of H_2 molecule within VB and MO theory are derived as

$$|\gamma(\mathbf{p})|^2 = \frac{1 + \cos[\mathbf{p} \cdot (\mathbf{r}_a - \mathbf{r}_b)]}{1 + S} |A(\mathbf{p})|^2 \tag{5.7}$$

$$|\gamma(\mathbf{p})|^2 = \frac{1 + S\cos[\mathbf{p} \cdot (\mathbf{r}_a - \mathbf{r}_b)]}{1 + S^2} |A(\mathbf{p})|^2 \tag{5.8}$$

where

$(\mathbf{r}_a - \mathbf{r}_b)$ is the bond vector $A(\mathbf{p})$ is the FT of the 1s atomic Slater function.

It is interesting to note that the factor $\cos[\mathbf{p} \cdot (\mathbf{r}_a - \mathbf{r}_b)]$ induces a maximality in a momentum direction perpendicular to $(\mathbf{r}_a - \mathbf{r}_b)$, i.e., bonding direction because for $\mathbf{p} \cdot (\mathbf{r}_a - \mathbf{r}_b) = \pi/2$, it becomes zero and for $\mathbf{p} \cdot (\mathbf{r}_a - \mathbf{r}_b) = 0$, it attains a maximum. Coulson termed the factor $1 + \cos[\mathbf{p} \cdot (\mathbf{r}_a - \mathbf{r}_b)]$ as a diffraction factor in analogy with intensity of diffraction patterns. The maximal nature of the EMD perpendicular to the position space bonding directions was termed as the bond directionality principle. The maximal directions in the EMD and the corresponding position space density for H_2 molecule are illustrated in Figure 5.4. The coordinate space charge density shows the cusp maxima at the nuclear positions, even evidencing the bonding, where the density is a minimum on a line connecting the two nuclei and a maximum in comparison with all directions perpendicular to it. Analyzing the isosurfaces of the corresponding momentum density, it is seen that the contours have a larger extent

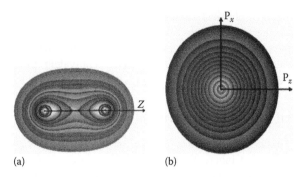

(a)   (b)

FIGURE 5.4 (See color insert following page 302.) Coordinate and momentum space charge densities of H_2 molecule illustrating the bond directionality principle. Isosurfaces from 0.04 to 0.01 a.u. are plotted for the coordinate space charge density (a). Isosurfaces from 1.0 to 0.01 a.u. are plotted for the momentum space charge density (b).

along the P_x direction and a smaller extent along the P_z axis, which is perpendicular to the X-axis in the coordinate space.

The next step for Coulson was towards polyatomic systems of ethylene, butadiene, and naphthalene [11]. He concluded that, "Whenever there is a certain direction in a molecule such that several bonds lie exactly or close along it, then the momentum is likely to have a small component rather than a large one in this direction. In the case of consecutive or conjugated bonds, this tendency is considerably enhanced due to the high mobility of certain electrons. High mobility in any direction usually gives rise to low values of momentum in this direction." Thus was the graphical generalization of the bond directionality principle which would be again referred to in the latter part of this chapter. Several studies on molecular EMDs and Compton profiles were reported by Epstein and Lipscomb [12] in 1970s, in which localized MOs were employed.

The EMD is closely related to intensities obtained from Compton scattering experiments, in which the obtained distribution depends on the incident wavelength and the scattering angle. The intensity of the scattered radiation is proportional to the theoretically obtained Compton profile given by the equation

$$J(q) = \frac{1}{2} \int\limits_{q}^{\infty} \frac{I(p)\mathrm{d}p}{p} \tag{5.9}$$

where $I(p)$ is the spherically averaged EMD given by Equation 5.3. These relationships provided a method for estimating the electronic energy of a molecular system from the Compton line shapes via the virial theorem. By virial theorem, the electronic energy of Coulombic systems is the negative of the kinetic energy and can be obtained from the Compton profiles as [13],

$$E_{\mathrm{mol}} = -\frac{1}{2} \int\limits_{0}^{\infty} p^2 I(p)\mathrm{d}p = \int\limits_{0}^{\infty} p^3 \left(\frac{\mathrm{d}J}{\mathrm{d}q}\right)_{q=p} \mathrm{d}p = -3 \int p^2 J(p)\mathrm{d}p \tag{5.10}$$

In chemistry, several properties such as enthalpy of formation, dipole moments, etc., are analyzed for molecules on the basis of an additivity approximation. The same was applied to Compton profiles by Eisenberger and Marra [14], who measured the Compton profiles of hydrocarbons and extracted bond Compton profiles by a least squares fitting. This also enabled an approximate evaluation of the energy of these systems from the virial theorem.

All these and other related studies gave rise to some qualitative and semiquantitative principles as a guide to interpret molecular EMDs. These principles apart from the previously described FT one, are as follows:

1. Virial theorem: Lowering the energy of system either by an improved wave function or by a chemical change such as bonding, leads to a shift of momentum density from regions of lower to higher momentum.
2. Bond oscillation principle: In the spirit of the bond directionality principle advocated by Coulson, Kaijser and Smith [15] proposed that the momentum distributions and Compton profiles associated with the bond will exhibit oscillations.
3. Hybrid orbital principle: Increased p character in an sp^n-type hybrid orbital results in increased density at higher momentum.

The above qualitative principles are useful for chemical interpretation of molecular EMDs. The first step for a quantitative analysis is to study the nature of EMDs and characteristics with respect to symmetry and topography.

5.3 SYMMETRY IN MOMENTUM SPACE

The 3D profiles of momentum densities on examination, initially, baffle a chemist who is very familiar with coordinate space densities, because of the absence of a nuclei-centric structure. Symmetry is manifested in chemistry via molecular geometries as revealed by nuclear structure. However, EMDs have an additional characteristic of being inversion symmetric, viz., $\gamma(\mathbf{p}) = \gamma(-\mathbf{p})$. This inversion symmetry arises [15–17] due to the FT involved in obtaining the momentum space wave function, which is a precursor to the momentum density. Examining the definition of momentum densities as given by the Equations 5.1 and 5.2 and taking a complex conjugate of the integral, the following expression is derived:

$$\gamma(\mathbf{p}) = \gamma(\mathbf{p})^* = \left[(2\pi)^{-3} \int e^{-i\mathbf{p}\cdot(\mathbf{r}-\mathbf{r}')} \Gamma(\mathbf{r}|\mathbf{r}') d^3r d^3r' \right]^*$$

$$= (2\pi)^{-3} \int e^{i\mathbf{p}\cdot(\mathbf{r}-\mathbf{r}')} \Gamma(\mathbf{r}|\mathbf{r}') d^3r d^3r' = \gamma(-\mathbf{p}) \qquad (5.11)$$

Here $\Gamma(\mathbf{r}|\mathbf{r}')$ signifies what is termed as the first-order reduced density matrix, which is defined as

$$\Gamma(\mathbf{r}|\mathbf{r}') = N \int \Psi(\mathbf{r}, \mathbf{r}_2, \mathbf{r}_3, \ldots, \mathbf{r}_N) \Psi^*(\mathbf{r}', \mathbf{r}_2, \mathbf{r}_3, \ldots, \mathbf{r}_N) d^3r_2 d^3r_3 \ldots d^3r_N \qquad (5.12)$$

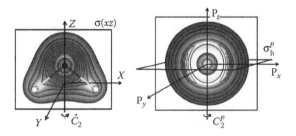

FIGURE 5.5 Symmetry enhancement from coordinate to momentum space: C_{2v} to D_{2h} for water molecule.

TABLE 5.2

Direct Product Table for C_{2v} Point Group in Coordinate Space with i to Give the D_{2h} Point Group in Momentum Space

$C_{2v} \rightarrow D_{2h}$	E	$C_2(z)$	$\sigma(xz)$	$\sigma(yz)$	i	$\sigma(xy)$	$C_2(x)$	$C_2(y)$
E	E	$C_2(z)$	$\sigma(xz)$	$\sigma(yz)$	i	$\sigma(xy)$	$C_2(x)$	$C_2(y)$
$C_2(z)$	$C_2(z)$	E	$\sigma(yz)$	$\sigma(xz)$	$\sigma(xy)$	i	$C_2(y)$	$C_2(x)$
$\sigma(xz)$	$\sigma(xz)$	$\sigma(yz)$	E	$C_2(z)$	$C_2(y)$	$C_2(x)$	$\sigma(xy)$	i
$\sigma(yz)$	$\sigma(yz)$	$\sigma(xz)$	$C_2(z)$	E	$C_2(x)$	$C_2(y)$	i	$\sigma(xy)$
i	i	$\sigma(xy)$	$C_2(y)$	$C_2(x)$	E	$C_2(z)$	$\sigma(yz)$	$\sigma(xz)$
$\sigma(xy)$	$\sigma(xy)$	i	$C_2(x)$	$C_2(y)$	$C_2(z)$	E	$\sigma(xz)$	$\sigma(yz)$
$C_2(x)$	$C_2(x)$	$C_2(y)$	$\sigma(xy)$	i	$\sigma(yz)$	$\sigma(xz)$	E	$C_2(z)$
$C_2(y)$	$C_2(y)$	$C_2(x)$	i	$\sigma(xy)$	$\sigma(xz)$	$\sigma(yz)$	$C_2(z)$	E

A visual manifestation of this inversion symmetry is given in Figure 5.5, for the momentum density of water molecule. The position space symmetry of water molecule as per the familiar group theoretical terms, is given by the point group C_{2v}, whose symmetry operations have been given in the figure. This point group is purely due to the geometry of the water molecule. In momentum space, the nuclei are absent and hence the symmetry is given by the profiles of the 3D EMD. For the EMD, it can be seen that it is the maximal nature of the profiles that dictates the symmetry. The effect of the inversion property is quite evident in the profile which converts it to a D_{2h} point group [17]. Obtaining theoretically, the point groups in momentum space involves a direct product of the group in position space with the inversion operator i. This is worked out in Table 5.2 for the point group C_{2v}. The point group in position and momentum spaces remains the same when the group already contains the inversion operator i.

5.4   TOPOGRAPHY OF ELECTRON MOMENTUM DENSITIES

The next step toward a systematic analysis of the structure of EMDs is an examination of its topography (Greek, *topos*, place; *graphia*, writing) [18]. Topography is described in terms of isosurfaces, critical points (CPs), gradient paths, etc. Bader [19] has pioneered the use of topographical concepts in chemistry, with major emphasis on the nature of the coordinate space charge density. The CP of a 3D function is defined as a point at which all its first-order partial derivatives vanish. The nature of the CP of a 3D function $f(x_1, x_2, x_3)$ is then determined by the Hessian matrix at the CP, viz., $H_{ij} = \partial^2 f / \partial x_i \partial x_j |_{CP}$ [17]. The number of nonzero eigenvalues of the Hessian matrix is termed as its rank R which is equal to the order of the largest square submatrix whose determinant is nonzero. If all the eigenvalues of the Hessian matrix are nonzero, then the CP is termed to be nondegenerate. To determine whether the CP is a minimum, maximum or a saddle, a quantity defined as the signature, σ, is used. The signature of a CP is equal to the algebraic sum of the signs of the eigenvalues. For a 3D function, there can be four types of nondegenerate CPs. The notation used for them will be of the form (R, σ) and they are

1. $(3, +3)$, a true minimum in all directions
2. $(3, +1)$, saddle which is a maximum at least in one direction
3. $(3, -1)$, saddle which is a minimum at least in one direction
4. $(3, -3)$, a true maximum in all directions

The topography of EMD was extensively mapped and analyzed by Kulkarni and Gadre [20], for a variety of molecules. The inversion symmetric nature of the scalar field requires the point $\mathbf{p} = \mathbf{0}$ to be a CP. The nondegenerate CP at $\mathbf{p} = \mathbf{0}$ is unique in its nature as determined by the value of EMD and its signature. The CPs occurring elsewhere follow a hierarchy as shown in Table 5.3, with the signature increasing with respect to the CP at $\mathbf{p} = \mathbf{0}$. The CP at $\mathbf{p} = \mathbf{0}$ is thus a harbinger of the topography of the complete scalar field. When a $(3, +3)$ is at $\mathbf{p} = \mathbf{0}$, all the other types of CPs $((3, +1), (3, -1)$ and $(3, -3))$ are found elsewhere. For a $(3, +1)$ at $\mathbf{p} = \mathbf{0}$, $(3, -1)$, and $(3, -3)$ CPs and for a $(3, -1)$ CP at $\mathbf{p} = \mathbf{0}$ only $(3, -3)$ CPs occur at other $\mathbf{p}$. A true maximum at $\mathbf{p} = \mathbf{0}$ is a unique CP with no other CP occurring elsewhere. The question "why is there such a structure for EMDs?" is unanswered and needs further examination and attention.

TABLE 5.3

Depiction of the Hierarchy Principle of CPs in EMD

CP at $\mathbf{p} = \mathbf{0}$	CP Found Elsewhere
$(3, +3)$	$(3, +1), (3, -1), (3, -3)$
$(3, +1)$	$(3, -1), (3, -3)$
$(3, -1)$	$(3, -3)$
$(3, -3)$	No other CP found

Source: Kulkarni, S.A., Gadre, S.R., and Pathak, R.K., *Phy. Rev. A.*, 45, 4399, 1992; Balanarayan, P. and Gadre, S.R., *J. Chem. Phy.*, 122, 164108, 2005.

5.5 CHEMISTRY IN MOMENTUM SPACE

Although the absence of nuclei-centric structure makes direct chemical interpretation difficult, the EMD does have some other advantages. For instance, it is related to energy via the virial theorem stated previously and carries the valence information around $\mathbf{p}=\mathbf{0}$. The entire nature of EMD topography is fixed by that at $\mathbf{p}=\mathbf{0}$, as described by the hierarchy principle.

Analyzing the chemical reactions in momentum space, armed with just the information around $\mathbf{p}=\mathbf{0}$ has been one of the attractive prospects of momentum space [20]. This information could be in the form of $\gamma(0)$ or the entire set of eigenvalues at 0. The information around $\mathbf{p}=\mathbf{0}$ indeed refers to the "valence region," which is of supreme importance for understanding the chemical reactions. For a corresponding study in position space, a search for various bond-, ring-, and cage-type CPs has to be carried out in three dimensions. On the other hand, in the momentum space, the information around a *single point* could be used as a practical tool for probing the chemical reactions [20]. The nature of the EMD for the isomerization reaction of HCN to HNC is portrayed in Figure 5.6. The snapshots of the momentum densities at various points on the reaction path are plotted. The close connection with energetics is apparent in momentum space, with the nature of the CP at $\mathbf{p}=\mathbf{0}$ changing around the transition state and the EMD values at zero momentum closely following the energetics.

Yet another chemical concept that has been related to EMDs is strain [21]. The classical definition of molecular strain originated in the work of Baeyer. In modern

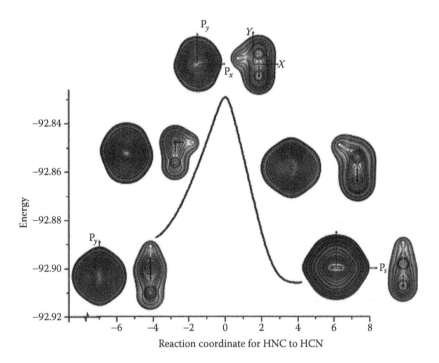

FIGURE 5.6 (See color insert following page 302.) Momentum and coordinate space charge density profiles for the reaction path from HNC to HCN.

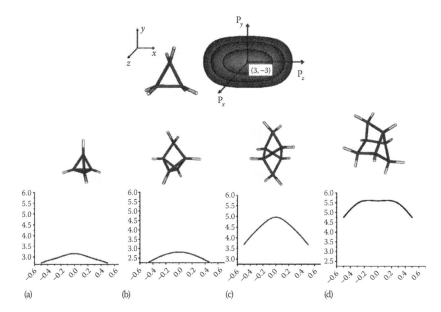

FIGURE 5.7 Momentum density isosurfaces (in a.u.) of cyclopropane and (a) momentum density of tetrahedrane along the p_z axis. Parts b, c, and d show the variation of the density for the same axis for a single, two, and three CH_2 insertions, respectively.

terms, carbon atoms whose bond angles deviate substantially from standard bond angles 109.5°, 120°, and 180°, respectively, for sp^3, sp^2, and sp hybrid orbitals are said to be strained. Interestingly, the "bent banana bond" model of cyclopropane was introduced by Coulson and Moffitt [22], whose preliminary findings triggered the interest in molecular EMDs. The topographical features of EMD provide a measure of molecular strain through angular deviations and bond ellipticities as put forth by Bader and coworkers [19]. As noted earlier, EMDs have an interpretative problem in this respect, as they lack an atom-centric visual description of molecular structure. However, EMDs are intimately connected to kinetic energy densities, which in turn can be closely linked with strain, as it is heuristically expected.

The effects of ring strain are manifested in the momentum densities as: (1) The momentum density at $\mathbf{p}=\mathbf{0}$ is higher in magnitude for the strained system as compared to its unstrained isomeric counterparts, with the distribution tending to a maximum at low momenta. (2) A partitioning of the total distribution into atoms reveals that the carbons accumulate more density in comparison with the unstrained reference, and the hydrogens are more positive. (3) Within the strained systems, the carbons become less negative with increasing strain. (4) The spherically averaged values also seem to indicate crowding of the density around $\mathbf{p}=\mathbf{0}$, although there is a loss of information on spherical averaging.

Particular instances in which the absence of nuclear structure for EMD has turned out to be a blessing in disguise are those in quantitative structure–activity relationships (QSAR) [24–27] and molecular similarity. One of the most popular similarity measures based on coordinate space charge densities is known as the Carbó index [24–27], which is a correlation measure between two charge densities. An important step in its evaluation involving translation and rotation of one of the species, is to find a particular alignment that maximizes this correlation measure and hence does involve some computation. However, such a problem of alignment does not arise at all in momentum space due to the absence of a nuclei-centric structure. The only operation involved is rotation around any fixed axis. This advantage for EMDs [24–27] has made it a good "descriptor" in several QSAR studies such as similarity measures for Hammond's postulate and prediction of water–octanol partition coefficients.

The nature of bonding in solids can also be probed via experimental Compton profile measurements. Philips and Weiss [28a] had investigated the bonding in LiF by analyzing the Compton profiles. They noticed that the momentum density of LiF was more similar to superposition of the momentum densities of Li^+ and F^-, rather than a superposition of the momentum densities of Li^0 and F^0. Thus it was possible to conclude about the degree of ionicity involved in the bonding of LiF. There have been several experiments on solids, metals, etc., involving the measurement of momentum densities. A discussion of these is beyond the scope of this chapter. The reader is referred to a comprehensive book giving these details [28b].

5.6 MOMENTUM DENSITY MEASUREMENTS: SPECTROSCOPIC TECHNIQUES

Several experimental techniques such as Compton scattering, positron annihilation, angular correlation, etc., are used for measuring momentum densities. One of the most popular techniques involved in measuring momentum densities is termed as electron momentum spectroscopy (EMS) [29]. This involves directing an electron beam at the surface of the metal under study. Hence EMS techniques fall under what is classified as coincidence spectroscopy.

The experiment uses a beam of electrons of known kinetic energy E_0 and known momentum $\mathbf{k}_0$. The beam is incident on a target consisting of atoms, molecules, or a solid film. The incident electron knocks a second electron out of the target both of which are detected. Their kinetic energies and momenta, E_f and $\mathbf{k}_f$ for the faster one and E_s and $\mathbf{k}_s$ for the slower one, are observed. The binding energy ε and momentum $\mathbf{q}$ of the target electron are given by $\varepsilon = E_0 - E_s - E_f$ and $\mathbf{q} = \mathbf{k}_s - \mathbf{k}_f - \mathbf{k}_0$. In the independent-particle model, the probability that one measures a certain binding energy–momentum combination is proportional to the absolute square of the momentum space orbital of the target electron, which hence gives the momentum density of the orbital, via a fitting procedure of theoretically obtained orbital momentum densities.

These experiments have been applied to a variety of molecules, the predominant ones being strained organic molecules such as cubane, norbornane, cyclopropane, etc. [30–33]. Recently, van der Waals complexes of formic acid have been subjected to EMS spectroscopy [34].

5.7 INTERCONNECTIONS BETWEEN POSITION AND MOMENTUM SPACES

The Schrödinger equation in momentum space for a single particle system is obtained on taking an FT of its position space counterpart in the form

$$\frac{p^2}{2}\phi(\mathbf{p}) + \int V(\mathbf{p} - \mathbf{p}')\phi(\mathbf{p}')\mathrm{d}^3 p' = E\phi(\mathbf{p}) \tag{5.13}$$

where $V(\mathbf{p}) = (-2\pi)^{-3/2} \int V(\mathbf{r})\exp(-i\mathbf{p}\cdot\mathbf{r})\mathrm{d}^3 r$. The multiplicative potential energy operator in position space thus leads to a convolution procedure involving FT. The kinetic energy operator has become local and multiplicative. Attempts at solving the momentum space Schrödinger equation for the multielectron He atom were pioneered by McWeeny and Coulson [35a] who presented a solution for the H_2^+ molecular ion. However, due to difficulties in solving integral equations accurately, no significant progress seems to have been made in the literature, although there have been a few attempts in that direction [35b,c].

The position space density study has immensely benefitted from the development of density functional theory (DFT) [36], wherein the energy is evaluated solely from the knowledge of the 3D function, i.e., the position space charge density. Such a development for momentum densities has not yet been achieved, the difficulty being in the handling of integrals of the Coulombic kind. The Hohenberg–Kohn theorem, which is the fundamental basis of DFT emerges as an existence theorem in momentum space [37]. The advantage in momentum space is again the local multiplicative nature of the kinetic energy functional. Invoking quasiclassical consideration, Pathak and Gadre [38] had formulated an approximate DFT for atoms in momentum space, which yielded quick estimates for gross properties in momentum space.

A standing problem related to momentum space densities is a one-to-one mapping with coordinate space densities, which to date has not yet been solved. Given just the position space charge density of a molecule what would be its corresponding momentum density and vice versa? A direct answer to such a question would certainly aid in chemical interpretation. The FT of the coordinate space charge density leads to the scattering factor $f(\mathbf{k})$, which is measured in an X-ray crystallographic experiment. The inverse FT of the momentum density also does not yield the position space charge density. Hence, there have been several approximate relations found connecting the position and momentum space. A simple and approximate semiclassical procedure has been outlined in the literature [39]. In this study, the spherically averaged momentum densities and thereby the Compton profiles were estimated only from the knowledge of the coordinate space charge density. The momentum p is evaluated from the coordinate space charge density $\rho(\mathbf{r})$ as $p = (3/10)[3\pi^2 \rho(\mathbf{r})]^{1/3}$ and the approximate spherically averaged momentum density for this p is related to the volume enclosed by the isosurface of $\rho(\mathbf{r})$ at that momentum value [39].

Yet another indirect connection between momentum and coordinate space charge densities is derived via a quantity called the Shannon information entropy

[40]. The information entropy of a probability distribution is defined as $S[p_i] = -\Sigma\, p_i$ $\ln p_i$, where $\{p_i\}$ forms the set of probabilities of a distribution. For continuous probability distributions such as momentum densities, the information entropy is given by $S[\gamma] = -\int \gamma(\mathbf{p}) \ln \gamma(\mathbf{p})\, d^3 p$, with an analogous definition in position space [41]. The information entropy is a quantitative measure of the uncertainty in a distribution with lower information entropy giving lesser uncertainty, which implies a higher probability for a particular event. The FT relationship between position and momentum space and the inverse behavior in both the spaces hence connect the information entropies of the distributions as well. An interesting uncertainty type relation has been given by Białynicki-Birula and Mycielski [42] in terms of the information entropies as $-\langle \ln|\Psi|^2 \rangle - \langle \ln|\Phi|^2 \rangle \geq n(1 + \ln \pi)$. Gadre and Bendale [43] derived a lower bound to the sum of information entropies in position and momentum spaces, viz., $S[\rho] + S[\gamma] \geq 3N(1 + \ln \pi) - 2N \ln N$. This led to postulating a new maximum entropy principle [43] in dual spaces, which has been used in a variety of applications.

5.8 CONCLUDING REMARKS

Starting from the FT of hydrogenic orbitals to EMDs of molecules and solids, the work on momentum densities has progressed through several phases, dealing with interpretation, spectroscopic measurements, and practical aspects of energy evaluation. As seen earlier, the Schrödinger equation is not easily solvable in momentum space. Rigorous as well as pragmatic interconnections between position and momentum densities need to be built. For a chemist and physicist, the momentum density still remains a mystery with the absence of nuclear structure, making it difficult to interpret. In spite of this, the very absence of nuclear structure has made EMDs amicable to similarity studies in QSAR. The local nature of kinetic energy and the availability of valence information around $\mathbf{p} = \mathbf{0}$ has proved to be useful in a practical sense.

To derive some chemistry out of the momentum distribution, a bridge between the electron densities in position and momentum spaces needs to be built. The studies in momentum space are therefore still a breeding ground for further conceptual and developmental work.

ACKNOWLEDGMENTS

The members of the Theoretical Chemistry Group at the Department of Chemistry, University of Pune, who were instrumental in developing the property evaluation package INDPROP [44], which has been used to evaluate the EMDs portrayed in the figures, are gratefully acknowledged. The graphics in the figures were constructed using the property visualization package [45], UNIVIS-2000. One of the authors, P.B. thanks C-DAC (Center for Development of Advanced Computing), Pune, for a research fellowship. S.R.G. is grateful to the Department of Science and Technology for the award of J. C. Bose fellowship.

REFERENCES

1. Podolsky, B. and Pauling, L. *Phys. Rev.* 1929, *34*, 109.
2. Jordan, P. *Zeit. f. Phys.* 1927, *40*, 809.
3. Du Mond, J. W. M. *Phys. Rev.* 1929, *33*, 643.
4. Kaijser, P. and Lindner, P. *Phil. Mag.* 1975, *31*, 871.
5. Kato, T. *Commun. Pure Appl. Math.* 1957, *10*, 151.
6. Morrell, M. M., Parr, R. G., and Levy, M. *J Chem. Phys.* 1975, *62*, 549.
7. Weinstein, H., Politzer, P., and Serebrenik, S., *Theor. Chim. Acta (Berl.)* 1975, *38*, 159.
8. Benesch, R. and Smith, V. H. Jr. in *Wave Mechanics—The First Fifty Years*, eds. W. C. Price, S. S. Chissick, and T. Ravensdale (Butterworths, London, 1973), pp. 357–377.
9. Gadre, S. R., Pathak, R. K., and Chakravorty, S. *J. Chem. Phys.* 1983, *78*, 4581 and references therein.
10. (a) Thakkar, A. J. *J. Chem. Phys.* 1982, *76*, 746; (b) Koga, T., Matsuyama, H., Inomata, H., Romera, E., Dehesa, J. S., and Thakkar, A. J. *J. Chem. Phys.* 1998, *109*, 1601.
11. (a) Coulson, C. A. *Proc. Camb. Phil. Soc.* 1941, *37*, 55; (b) Coulson, C. A. *Proc. Camb. Phil. Soc.* 1941, *37*, 74.
12. Epstein, I. R. and Lipscomb, W. N. *J. Chem. Phys.* 1970, *53*, 4418; Epstein, I. R. *J. Chem. Phys.* 1970, *53*, 4425. See also Epstein, I. R. and Tanner, A. C. in *Compton Scattering*, ed. B. Williams (McGraw-Hill, New York, 1977).
13. Coulson, C. A. *Mol. Phys.* 1973, *26*, 507.
14. Eisenberger, P. and Marra, W. C. *Phys. Rev. Lett.* 1971, *27*, 1413.
15. Kaijser, P. and Smith, V. H. Jr. in *Quantum Science Methods and Structure*, eds. J. L. Calais, O. Gosciniski, J. Linderberg, and Y. Öhrn (Plenum, New York, 1976), p. 417.
16. Thakkar, A. J., Simas, A. M., and Smith, V. H. Jr. *J. Chem. Phys.* 1984, *81*, 2953.
17. Defranceschi, M. and Berthier, G. *J. Phys. France* 1990, *51*, 2791; (b) Gadre, S. R., Limaye, A. C., and Kulkarni, S. A. *J. Chem. Phys.* 1991, *94*, 8040.
18. Gadre, S. R. in *Computational Chemistry: Reviews of Current Trends*, ed. J. Lesczynski (World Scientific, Singapore, 2000), Vol. 4, pp. 1–53.
19. Bader, R. F. W. *Atoms in Molecules: A Quantum Theory* (Clarendon, Oxford, 1990).
20. Kulkarni, S. A., Gadre, S. R., and Pathak, R. K. *Phys. Rev. A* 1992, *45*, 4399.
21. Kulkarni, S. A. and Gadre, S. R. *J. Am. Chem. Soc.* 1993, *115*, 7434.
22. (a) Coulson, C. A. and Moffitt, W. E. *Philos. Mag.* 1949, *40*, 1; (b) Coulson, C. A. and Googwin, T. H. *J. Chem. Soc.* 1963, *2851*, 3161.
23. Balanarayan, P. and Gadre, S. R. *J. Am. Chem. Soc.* 2006, *128*, 10702.
24. Amat, L., Carbó-Dorca, R., Cooper, D. L., and Allan, N. L. *Chem. Phys. Lett.* 2003, *267*, 207.
25. Amat, L., Carbó-Dorca, R., Cooper, D. L., Allan, N. L., and Ponec, R. *Mol. Phys.* 2003, *101*, 3159.
26. Al-Fahemi, J. H., Cooper, D. L., and Allan, N. L. *J. Mol. Str. (Theochem)* 2005, *727*, 57.
27. Amat, L., Carbó-Dorca, R., Cooper, D. L., Allan, N. L., and Ponec, R. *Mol. Phys.* 2003, *101*, 3159.
28. (a) Philips, W. C. and Weiss, R. J. *Phys. Rev.* 1968, *171*, 790; (b) For a comprehensive review of basic theory, experimental methods and measurements on solids, see *X-ray Compton Scattering*, eds. M. J. Cooper, P. E. Mijnarends, M. Shiotani, N. Sakai and A. Bansil, (Oxford University Press, Oxford, 2004).
29. Vos, M. and McCarthy, E. *Am. J. Phys.* 1997, *65*, 544.
30. See Adcock, W., Brunger, M. J., Clark, C. I., McCarthy, I. E., Michalewicz, M. T., von Niessen, W., Weigold, E., and Winkler, D. A. *J. Am. Chem. Soc.* 1997, *119*, 2896 and references therein.
31. Adcock, W., Brunger, M. J., McCarthy, I. E., Michalewicz, M. T., von Niessen, W., Wang, F., Weigold, E., and Winkler, D. A. *J. Am. Chem. Soc.* 2000, *122*, 3892.

32. Knippenberg, S., Nixon, K. L., Brunger, M. J., Maddern, T., Campbell, L., Trout, N., Wang, F., Newell, W. R., Deleuze, M. S., Francois, J. -P., and Winkler, D. A. *J. Chem. Phys.* 2004, *121*, 10525.

33. Adcock, W., Brunger, M. J., Michalewicz, M. T., and Winkler, D. A. *Aus. J. Phys.* 1998, *51*, 707.

34. Nixon, K. L., Lawrance, W. D., Jones, D. B., Euripides, P., Saha, S., Wang, F., and Brunger, M. J. *Chem. Phys. Lett.* 2008, *451*, 18.

35. (a) McWeeny, R. and Coulson, C. A. *Proc. Phys. Soc. (Lond.)* 1949, *62*, 509; (b) Navaza, J. and Tsoucaris, G. *Phys. Rev. A* 1981, *24*, 683; (c) See Koga, T. *J. Chem. Phys.* 1985, *83*, 2328 and other related works.

36. Parr, R. G. and Yang, W. in *Density Functional Theory of Atoms and Molecules* (Oxford, New York, 1989).

37. Henderson, G. A. *Phys. Rev. A* 1981, *23*, 19.

38. Pathak, R. K., Panat, P.V., and Gadre, S. R. *J. Chem. Phys.* 1981, *74*, 5925.

39. Gadre, S. R., Gejji, S. P., and Pathak, R. K. *Phys. Rev. A* 1982, 26, 3073.

40. Shannon, C. E. *Bell System Tech. J.* 1948, *27*, 379.

41. Gadre, S. R., Sears, S. B., Chakravorty, S. J., and Bendale, R. D. *Phys. Rev.* 1985, *32*, 2602.

42. Białynicki-Birula, I. and Mycielski, J. *Commun. Math. Phys.* 1975, *44*, 129.

43. (a) Gadre, S. R. and Bendale, R. D. *Phys. Rev. A* 1987, *36*, 1932; (b) Gadre, S. R. and Bendale, R. D. *Int. J. Quant. Chem.* 1985, *28*, 311.

44. INDPROP, The molecular properties calculation package developed at the Theoretical Chemistry Group, Department of Chemistry, University of Pune, Pune, India. See also Bapat, S. V., Shirsat, R. N., and Gadre, S. R. *Chem. Phys. Lett.* 1992, *200*, 373. See also Balanarayan, P. and Gadre, S. R. *J. Chem. Phys.* 2003, *115*, 5037.

45. UNIVIS-2000, The molecular properties visualization package, developed at the Theoretical Chemistry Group, Department of Chemistry, University of Pune, Pune, India; http://chem.unipune.ernet.in/univis.html. See also Limaye, A. C. and Gadre, S. R. *Curr. Sci. (India)* 2001, *80*, 1296.

6 Time-Dependent Density Functional Theory of Many-Electron Systems

Swapan K. Ghosh

CONTENTS

6.1 INTRODUCTION

The study of behavior of many-electron systems such as atoms, molecules, and solids under the action of time-dependent (TD) external fields, which includes interaction with radiation, has been an important area of research. In the linear response regime, where one considers the external field to cause a small perturbation to the initial ground state of the system, one can obtain many important physical quantities such as polarizabilities, dielectric functions, excitation energies, photo-absorption spectra, van der Waals coefficients, etc. In many situations, for example, in the case of interaction of many-electron systems with strong laser field, however, it is necessary to go beyond linear response for investigation of the properties. Since a full theoretical description based on accurate solution of TD Schrodinger equation is not yet within the reach of computational capabilities, new methods which can efficiently handle the TD many-electron correlations need to be explored, and time-dependent density functional theory (TDDFT) is one such valuable approach.

One of the important concepts that has led to tremendous advantages for describing the properties of atoms, molecules, and solids is the concept of single-particle electron density [1]. It provides a conceptually simple and computationally economic route to the description of many-particle systems within a single-particle framework through the well-known theoretical approach known as density functional theory (DFT) [2–4], which uses the single-particle electron density as a basic variable. While there has been conspicuous success of DFT for the ground state of many-electron systems, the progress in the area of excited states and TD situations has been rather less spectacular.

The precursor of DFT is the well-known Thomas–Fermi (TF) theory [5], which was proposed as a thermodynamic and electrostatic model of the electron cloud in atoms, soon after the Schrodinger equation-based formulation of quantum mechanics was proposed. A corresponding TD version of the TF theory followed soon through the work of Bloch [6]. The static TF theory, although introduced initially in a somewhat intuitive manner, was later upgraded to the status of a rigorous DFT when it was derived as its approximate version with a local density form for the kinetic energy density functional and neglecting the exchange correlation (XC) contribution altogether. The TD theory of Bloch is also essentially a phenomeno-logical and approximate version of a more general TDDFT. Both TF and Bloch theory have been strongly based on an underlying classical picture of the electron cloud. The desire to have a classical interpretation of the strange world of quantum mechanics was so strong that soon after the Schrodinger equation was proposed, its hydrodynamic analog was derived by Madelung [7] and the resulting framework of quantum fluid dynamics (QFD) [8,9] played a major role in the development of TDDFT in later years. A major aspect in which TDDFT differs from DFT is the roping of the additional density variable, i.e., the current density in addition to the electron density.

DFT was formally born through the pioneering work of Hohenberg and Kohn (HK) [3], which demonstrated the uniqueness of the density to potential mapping for the ground state of many-electron systems. A formal TDDFT was born much later mainly through the works of Peuckert [10], Bartolotti [11], and Deb and Ghosh [12] on oscillating time dependence and subsequently, Runge and Gross [13] paved the way to a generalized TDDFT for an arbitrary scalar potential, which was later extended to TD electric and magnetic fields with arbitrary time dependence by Ghosh and Dhara [14]. Besides the formal foundations of DFT and TDDFT, for practical application, a suitable form of the XC energy density functional is also needed. In absence of an exact form for this quantity, several approximate forms are in use and the situation is more critical for TD potentials where the development has been rather less exhaustive. For excellent recent reviews, see Refs. [15–19].

In what follows, we present in this short review, the basic formalism of TDDFT of many-electron systems (1) for periodic TD scalar potentials, and also (2) for arbitrary TD electric and magnetic fields in a generalized manner. Practical schemes within the framework of quantum hydrodynamical approach as well as the orbital-based TD single-particle Schrodinger-like equations are presented. Also discussed is the linear response formalism within the framework of TDDFT along with a few miscellaneous aspects.

6.2 DENSITY FUNCTIONAL THEORY OF MANY-ELECTRON SYSTEMS FOR TIME-INDEPENDENT AND PERIODIC TIME-DEPENDENT POTENTIALS

The DFT, as already stated, was formally born with the pioneering works of Hohenberg and Kohn [3] who proved a theorem demonstrating a unique mapping between the electron density of a many-electron system and the external potential that characterizes the system. In the Hohenberg–Kohn–Sham version of DFT, the energy of an N-electron system characterized by an external potential $v_N(\mathbf{r})$ (say, due to the nuclei) is expressed as a functional of the single-particle electron density $\rho(\mathbf{r})$, and is given by

$$E[\rho] = T_s[\rho] + \int d\mathbf{r} v_N(\mathbf{r})\rho(\mathbf{r}) + U_{\text{int}}[\rho] + E_{\text{xc}}[\rho] \tag{6.1}$$

where $T_s[\rho]$ is the kinetic energy of a fictitious noninteracting system of same density, given by

$$T_s[\rho] = -\frac{\hbar^2}{2m} \sum_k < \psi_k^*(\mathbf{r})|\nabla^2|\psi_k(\mathbf{r}) >, \tag{6.2}$$

$U_{\text{int}}[\rho]$ is the classical Coulomb energy and $E_{\text{xc}}[\rho]$ is the XC energy. It is the functional form of this XC functional, which is usually approximated in absence of an exact expression. The one-electron orbitals $\{\psi_k(\mathbf{r})\}$ are obtained through self-consistent solution of the Kohn–Sham equations

$$\left\{ -\frac{\hbar^2}{2m}\nabla^2 + v_{\text{eff}}(\mathbf{r}) \right\}\psi_k(\mathbf{r}) = \varepsilon_k\psi_k(\mathbf{r}), \tag{6.3}$$

where the effective potential is given by

$$v_{\text{eff}}(\mathbf{r}) = v_N(\mathbf{r}) + v_{\text{SCF}}(\mathbf{r}); \quad v_{\text{SCF}}(\mathbf{r}) = v_{\text{COUL}}(\mathbf{r}) + v_{\text{XC}}(\mathbf{r});$$

$$v_{\text{COUL}} = \left(\frac{\delta U_{\text{int}}}{\delta\rho}\right); \quad v_{\text{XC}} = \left(\frac{\delta E_{\text{xc}}}{\delta\rho}\right). \tag{6.4}$$

Although originally proved for the ground state, the scope of the theorem was later extended to include excited states, and oscillating TD potentials. For systems subjected to external potential $v_{\text{ext}}(\mathbf{r}, t)$ with periodic time dependence, Deb and Ghosh [12] proved a HK-like theorem by considering the Hamiltonian for the steady state, i.e., $(H(t) - i\hbar(\partial/\partial t))$ and following a HK-like procedure using the minimal property of the quasienergy quantity defined as a time average of its expectation value over a period. They derived the TD analog of the Kohn–Sham equation, given by

$$\left\{ -\frac{\hbar^2}{2m}\nabla^2 + v_N(\mathbf{r}) + v_{\text{ext}}(\mathbf{r}, t) + v_{\text{SCF}}(\mathbf{r}, t) \right\}\psi_k(\mathbf{r}, t) = \left(i\hbar\frac{\partial}{\partial t} \right)\psi_k(\mathbf{r}, t) \tag{6.5}$$

to calculate the electron density and the current density using the relations

$$\rho(\mathbf{r}, t) = \sum_k \psi_k^*(\mathbf{r}, t)\psi_k(\mathbf{r}, t); \tag{6.6a}$$

$$\mathbf{j}(\mathbf{r}, t) = -\frac{i\hbar}{2m} \sum_k \left[\psi_k^*(\mathbf{r}, t)\nabla\psi_k(\mathbf{r}, t) - \psi_k(\mathbf{r}, t)\nabla\psi_k^*(\mathbf{r}, t) \right]. \tag{6.6b}$$

The effective potential has a form similar to that for the time-independent case with the density argument replaced by the TD density variable. Also some of the functionals may be current density dependent.

6.3 FORMAL FOUNDATIONS OF DENSITY FUNCTIONAL THEORY FOR TIME-DEPENDENT ELECTRIC AND MAGNETIC FIELDS

As already mentioned, the scope of the original HK theorem was extended to include excited states, oscillating TD potentials, and also the case of arbitrary time dependence. The elegant work of Runge and Gross [13] on the TDDFT corresponding to arbitrary TD scalar potentials [20] has been extended [14] to aribitrary TD scalar as well as TD vector potentials, thus enabling it to be applicable to the study of interaction with electromagnetic radiation, magnetic field, etc. We present here the formalism for TD electric and magnetic field, so that the other cases can be derived as special cases of this generalized situation.

We consider an N-electron system where the electrons experience the mutual Coulomb interaction along with an external potential $v_N(\mathbf{r})$ due to the nuclei. The system is then subjected to an additional TD scalar potential $\phi(\mathbf{r}, t)$ and a TD vector potential $A(\mathbf{r}, t)$. The many-electron wavefunction $\psi(\mathbf{r}_1, \mathbf{r}_2, \ldots, \mathbf{r}_N; t)$ of the system evolves according to the TD Schrodinger equation

$$\left(i\hbar\frac{\partial}{\partial t} - \hat{H} \right)\psi = 0, \tag{6.7}$$

where the many-body Hamiltonian $\hat{H}$ can be written, using the Coulomb gauge $\nabla \cdot \mathbf{A} = 0$, as

$$\hat{H} = \hat{H}_0 - e\sum_k \phi(\mathbf{r}_k, t) - i\left(\frac{e\hbar}{mc}\right)\sum_k A(\mathbf{r}_k, t) \cdot \nabla_k + \left(\frac{e^2}{2mc^2}\right)\sum_k A^2(\mathbf{r}_k, t) \tag{6.8a}$$

with

$$\hat{H}_0 = -\left(\frac{\hbar^2}{2m}\right)\sum_k \nabla_k^2 + \sum_k v_N(\mathbf{r}_k) + e^2\sum_{k<l}\frac{1}{|\mathbf{r}_k - \mathbf{r}_l|}, \tag{6.8b}$$

where the spin-dependent terms are omitted and other symbols have their usual significance.

The single-particle electron density $\rho(\mathbf{r}, t)$ and the current density $\mathbf{j}(\mathbf{r}, t)$, which are defined as the expectation values,

$$\rho(\mathbf{r}, t) = \langle \psi | \hat{\rho} | \psi \rangle; \quad \mathbf{j}(\mathbf{r}, t) = \langle \psi | \hat{\mathbf{j}} | \psi \rangle, \tag{6.9}$$

of the density operator $\hat{\rho}$ and the current density operator $\hat{\mathbf{j}}$ given, respectively, by

$$\hat{\rho} = \sum_k \delta(\mathbf{r} - \mathbf{r}_k) \tag{6.10a}$$

and

$$\hat{\mathbf{j}} = \hat{\mathbf{j}}_0 + \frac{e}{mc} \sum_k \mathbf{A}(\mathbf{r}_k, t)\, \delta(\mathbf{r} - \mathbf{r}_k);$$

$$\hat{\mathbf{j}}_0 = -\frac{i\hbar}{2m} \sum_k [\nabla_k \delta(\mathbf{r} - \mathbf{r}_k) + \delta(\mathbf{r} - \mathbf{r}_k)\nabla_k], \tag{6.10b}$$

are the two basic variables of the TDDFT for such systems.

Following the approach of Runge and Gross [13], Ghosh and Dhara [14] proved that if the two potentials $\phi(\mathbf{r}, t)$ and $A(\mathbf{r}, t)$ can be expanded into Taylor series with respect to time around $t = t_0$, both the potentials are uniquely (apart from only an additive TD function) determined by the current density $\mathbf{j}(\mathbf{r}, t)$ of the system.

Let us first consider two scalar potentials $\phi(\mathbf{r}, t)$ and $\phi'(\mathbf{r}, t)$, which differ by more than a mere TD function, and two vector potentials $\mathbf{A}(\mathbf{r}, t)$ and $\mathbf{A}'(\mathbf{r}, t)$, which are also different. Their Taylor expansions demand that there must exist at least one time derivative of the potentials differing from zero, in the case of scalar as well as vector potentials. In other words, there must exist a minimal nonnegative integer, say k for the scalar potential and l for the vector potential, such that

$$\left(\frac{\partial^n}{\partial t^n}\right)(\phi(\mathbf{r}, t) - \phi'(\mathbf{r}, t))\bigg|_{t=t_0} \quad \begin{aligned} &\neq 0, \, n = k \\ &= 0, \, 0 \leq n < k \end{aligned} \tag{6.11a}$$

$$\left(\frac{\partial^n}{\partial t^n}\right)(\mathbf{A}(\mathbf{r}, t) - \mathbf{A}'(\mathbf{r}, t))\bigg|_{t=t_0} \quad \begin{aligned} &\neq 0, \, n = l \\ &= 0, \, 0 \leq n < l. \end{aligned} \tag{6.11b}$$

Let the two densities corresponding to the two potential sets $\{\phi(\mathbf{r}, t), \mathbf{A}(\mathbf{r}, t)\}$ and $\{\phi'(\mathbf{r}, t), \mathbf{A}'(\mathbf{r}, t)\}$ be denoted by $\rho(\mathbf{r}, t)$ and $\rho'(\mathbf{r}, t)$ and the two current densities be $\mathbf{j}(\mathbf{r}, t)$ and $\mathbf{j}'(\mathbf{r}, t)$, respectively. We assume, without any loss of generality, the evolution to be from a fixed initial state $\psi(t_0)$, i.e., at $t = t_0$, $\rho(t_0) = \rho'(t_0)$, $\mathbf{j}(t_0) = \mathbf{j}'(t_0)$, and also $\phi(t_0) = \phi'(t_0)$; $\mathbf{A}(t_0) = \mathbf{A}'(t_0)$.

The time derivative of the current density can easily be obtained from the general equation of motion for an arbitrary operator $\hat{\Omega}(t)$, given by

$$i\hbar \left(\frac{d}{dt}\right) \langle \psi(t) | \hat{\Omega}(t) | \psi(t) \rangle = \left\langle \psi(t) \left| i\hbar \left(\frac{\partial}{\partial t}\right)\hat{\Omega}(t) + \left[\hat{\Omega}(t), \hat{H}(t)\right] \right| \psi(t) \right\rangle \tag{6.12}$$

by specializing to the current density operator $\hat{\mathbf{j}}(t)$ given by Equation 6.10b, thus leading to

$$
i\hbar\left(\frac{d}{dt}\right)\mathbf{j}(\mathbf{r},t) = \left\langle \psi(t)\left|i\hbar\left(\frac{\partial}{\partial t}\right)\hat{\mathbf{j}}(t) + \left[\hat{\mathbf{j}}(t),\hat{H}(t)\right]\right|\psi(t)\right\rangle \tag{6.13a}
$$

$$
= \left\langle \psi(t)\left|i\hbar\left(\frac{e}{mc}\right)\sum_k\left(\frac{\partial}{\partial t}\right)A_k(t)\delta(\mathbf{r}-\mathbf{r}_k) + \left[\hat{\mathbf{j}}(t),\hat{H}(t)\right]\right|\psi(t)\right\rangle. \tag{6.13b}
$$

By applying the equation of motion repeatedly, say for n times and making use of the initial conditions at $t=t_0$, one obtains considering both $\mathbf{j}$ and $\mathbf{j}'$ after some algebra, the result

$$
\left(i\hbar\frac{\partial}{\partial t}\right)^n[\mathbf{j}(\mathbf{r},t) - \mathbf{j}'(\mathbf{r},t)]\bigg|_{t=t_0}
$$

$$
= \left(\frac{e}{mc}\right)\rho(\mathbf{r},t_0)\left(i\hbar\frac{\partial}{\partial t}\right)^n[\mathbf{A}(\mathbf{r},t) - \mathbf{A}'(\mathbf{r},t)]\bigg|_{t=t_0}
$$

$$
+ i\hbar\left(\frac{e}{m}\right)\rho(\mathbf{r},t_0)\nabla\left\{\left(i\hbar\frac{\partial}{\partial t}\right)^{n-1}[\phi(\mathbf{r},t) - \phi'(\mathbf{r},t)]\bigg|_{t=t_0}\right\} \tag{6.14}
$$

where all the lower derivatives of the differences of the potential terms have been assumed to vanish. Equation 6.14 is the key equation on which the proof that follows is based.

Ghosh and Dhara [14] considered three possibilities for the relative values of n, k, and l, i.e., the cases (1) $l > k+1$, (2) $l < k+1$, and (3) $l = k+1$. In the first case, i.e., for the choice $n = k+1 < l$, clearly the first term on the right-hand side of Equation 6.14 vanishes but the second term is nonzero according to Equation 6.11a, thus implying that the left-hand side of Equation 6.14 is nonzero. Hence it is clear that there exists at least one (nth with $n = k+1$) derivative of the difference of the two current densities $\mathbf{j}(\mathbf{r},t)$ and $\mathbf{j}'(\mathbf{r},t)$ appearing in their Taylor expansion, which is nonzero. It implies that they will differ infinitesimally after $t=t_0$. In the second case, for the choice $n = l < k+1$, the second term on the right-hand side of Equation 6.14 vanishes but the first term remains as nonzero and hence once again, at least one (nth with $n = l$) derivative of the two current densities $\mathbf{j}(\mathbf{r},t)$ and $\mathbf{j}'(\mathbf{r},t)$ appearing in their Taylor expansion is different. This situation also clearly implies that the two current densities will differ infinitesimally after $t=t_0$. Analogous conclusion has been reached for the third case as well. For the first case, it has also been shown by coupling Equation 6.14 with the continuity equation that at least one ($k+2$)th derivative of the two electron densities $\rho(\mathbf{r},t)$ and $\rho'(\mathbf{r},t)$ is different, implying that the two densities will differ infinitesimally after $t=t_0$.

Thus, it is clear that the two sets of TD potentials $\{\phi(\mathbf{r},t), \mathbf{A}(\mathbf{r},t)\}$ and $\{\phi'(\mathbf{r},t), \mathbf{A}'(\mathbf{r},t)\}$ cannot lead to the same current density $\mathbf{j}(\mathbf{r},t)$ even if only any one of them is different. Thus, the current density $\mathbf{j}(\mathbf{r},t)$ fixes both scalar and vector potentials uniquely (apart from arbitrary constant or gauge transformation). The current density $\mathbf{j}(\mathbf{r}, t)$, of course, determines the electron density uniquely as is obvious from the continuity equation which relates the two. The energy and other quantities for the TD problems can, therefore, be treated as functionals of the current density or more conveniently as functionals of both $\mathbf{j}(\mathbf{r}, t)$ and $\rho(\mathbf{r}, t)$ unlike the stationary ground state where electron density alone suffices as a basic variable.

6.4 PRACTICAL SCHEMES FOR THE CALCULATION OF DENSITY AND CURRENT DENSITY VARIABLES IN TDDFT

It has been shown that for practical calculation of the density quantities $\rho(\mathbf{r}, t)$ and $\mathbf{j}(\mathbf{r}, t)$, one can have several schemes of which we discuss only two. In the first scheme, one has to solve the hydrodynamical equations, i.e., the continuity equation

$$\left(\frac{\partial}{\partial t}\right)\rho(\mathbf{r}, t) + \nabla . \mathbf{j}(\mathbf{r}, t) = 0 \tag{6.15}$$

and the Euler equation

$$\left(\frac{\partial}{\partial t}\right)\mathbf{j}(\mathbf{r}, t) = P_{\{v\}}[\rho(\mathbf{r}, t), \mathbf{j}(\mathbf{r}, t)] \tag{6.16}$$

where the vector $P_{\{v\}}[\rho(\mathbf{r}, t), \mathbf{j}(\mathbf{r}, t)]$ is a functional of the two densities $\rho(\mathbf{r}, t)$ and $\mathbf{j}(\mathbf{r}, t)$ for specified external potential denoted by $\{v\}$, and can be written as

$$P_{\{v\}}[\rho(\mathbf{r}, t), \mathbf{j}(\mathbf{r}, t)] = \frac{1}{i\hbar}\langle\psi(t)|[\hat{\mathbf{j}}_0, \hat{H}_0]|\psi(t)\rangle$$
$$- \frac{1}{m}\rho(\mathbf{r}, t)[\nabla v_N(\mathbf{r}) + e\mathbf{E}(\mathbf{r}, t)] - \frac{e}{mc}\mathbf{j}(\mathbf{r}, t) \times \mathbf{B}(\mathbf{r}, t). \tag{6.17}$$

Here $\mathbf{E}(\mathbf{r}, t)$ and $\mathbf{B}(\mathbf{r}, t)$ denote the TD electric and magnetic fields defined by the scalar and vector potentials as

$$\mathbf{E}(\mathbf{r}, t) = -\nabla\phi(\mathbf{r}, t) - \frac{1}{c}\frac{\partial}{\partial t}\mathbf{A}(\mathbf{r}, t); \quad \mathbf{B}(\mathbf{r}, t) = \nabla \times \mathbf{A}(\mathbf{r}, t) \tag{6.18}$$

and the operators $\hat{\mathbf{j}}_0$ and $\hat{H}_0$ are defined in Equations 6.10b and 6.8b, respectively. In Equation 6.17, the last two terms on the right-hand side denote the classical forces due to various external potentials, while the first term corresponds to forces of quantum origin and includes the XC contributions, which are to be expressed as density functionals of $\rho(\mathbf{r}, t)$ and $\mathbf{j}(\mathbf{r}, t)$ so that the set of hydrodynamical equations (Equations 6.15 through 6.18) can be used for the direct calculation of these density quantities.

While the hydrodynamical scheme mentioned above involves the density quantities directly, an alternative second scheme based on their orbital partitioning along the lines of the Kohn–Sham [4] version of time-independent DFT has been derived by Ghosh and Dhara [14]. In this scheme, one obtains the exact densities $\rho(\mathbf{r}, t)$ and $\mathbf{j}(\mathbf{r}, t)$ from the TD orbitals $\psi_k(\mathbf{r}, t)$ obtained by solving the effective one-particle TD Schrodinger-like equations given by

$$\left\{ \frac{1}{2m} \left[-i\hbar\nabla + \frac{e}{c} \mathbf{A}_{\text{eff}}(\mathbf{r}, t) \right]^2 + v_{\text{eff}}(\mathbf{r}, t) \right\} \psi_k(\mathbf{r}, t) = \left(i\hbar \frac{\partial}{\partial t} \right) \psi_k(\mathbf{r}, t) \qquad (6.19)$$

and using the relations

$$\rho(\mathbf{r}, t) = \sum_k \psi_k^*(\mathbf{r}, t) \psi_k(\mathbf{r}, t) \qquad (6.20a)$$

and

$$\mathbf{j}(\mathbf{r}, t) = -\frac{i\hbar}{2m} \sum_k \left[\psi_k^*(\mathbf{r}, t) \nabla \psi_k(\mathbf{r}, t) - \psi_k(\mathbf{r}, t) \nabla \psi_k^*(\mathbf{r}, t) \right]$$
$$+ \frac{e}{mc} \rho(\mathbf{r}, t) \mathbf{A}_{\text{eff}}(\mathbf{r}, t). \qquad (6.20b)$$

In Equation 6.19, which can be called TD Kohn–Sham-type equation, the effective scalar and vector potentials $v_{\text{eff}}(\mathbf{r}, t)$ and $\mathbf{A}_{\text{eff}}(\mathbf{r}, t)$, respectively, consist of contributions from the external potentials augmented by internal contributions determined by the density variables and can be expressed as

$$v_{\text{eff}}(\mathbf{r}, t) = v_N(\mathbf{r}) - e\phi(\mathbf{r}, t) + \frac{c}{e} \left(\frac{\delta U_{\text{int}}}{\delta \rho} \right) + \frac{c}{e} \left(\frac{\delta E_{\text{xc}}}{\delta \rho} \right) + \left(\frac{e^2}{2mc^2} \right) \left(A_{\text{eff}}^2 - A^2 \right) \qquad (6.21a)$$

$$\mathbf{A}_{\text{eff}}(\mathbf{r}, t) = \mathbf{A}(\mathbf{r}, t) + \frac{c}{e} \left(\frac{\delta U_{\text{int}}}{\delta \mathbf{j}} \right) + \frac{c}{e} \left(\frac{\delta E_{\text{xc}}}{\delta \mathbf{j}} \right). \qquad (6.21b)$$

The derivation of these expressions involves lengthy algebra details which can be found in Ghosh and Dhara [14]. Here, the internal energy $U_{\text{int}}[\rho, \mathbf{j}]$ is basically the classical Coulomb energy, while the term $E_{\text{xc}}[\rho, \mathbf{j}]$ denotes the well-known XC energy density functional. With a suitable chosen form for $E_{\text{xc}}[\rho, \mathbf{j}]$, Equations 6.19 through 6.21 have to be solved self-consistently for the density and the current density.

The picture that emerges here is that the density and current density of the actual system of interacting electrons characterized by the given external potentials are obtainable by calculating the same for a system of noninteracting particles moving in the field of effective scalar and vector potentials $v_{\text{eff}}(\mathbf{r}, t)$ and $\mathbf{A}_{\text{eff}}(\mathbf{r}, t)$, respectively. It is interesting to note that, in a recent work, Vignale [21] has provided a generalization of this aspect proving that the TD density and current density of a many-electron system evolving under the action of fixed external potentials from an initial quantum state can be reproduced in another many-particle system with a different two-particle interaction subjected to other suitable external potentials and starting from an initial state which yields the same density and current as the original state.

6.5   LINEAR RESPONSE WITHIN TDDFT

From the discussion so far, it is clear that the mapping to a system of noninteracting particles under the action of suitable effective potentials provides an efficient means for the calculation of the density and current density variables of the actual system of interacting electrons. The question that often arises is whether there are effective ways to obtain other properties of the interacting system from the calculation of the noninteracting model system. Examples of such properties are the one-particle reduced density matrix, response functions, etc. An excellent overview of response theory within TDDFT has been provided by Casida [15] and also more recently by van Leeuwen [17]. A recent formulation of density matrix-based TD density functional response theory has been provided by Furche [22].

Here, we consider the response theory which has been successful for many investigations. For simplicity, we consider the N-electron system to be initially in the ground state, which is subjected to an external TD electric field. The density change $\delta\rho(\mathbf{r}, t)$ induced by an external perturbation $\delta v_{\text{ext}}(\mathbf{r}, t)$ can be written in the response theory framework, in terms of quantities in frequency domain as

$$\delta\rho(\mathbf{r}, \omega) = \int d\mathbf{r}'\chi(\mathbf{r}, \mathbf{r}'; \omega)\delta v_{\text{ext}}(\mathbf{r}', \omega) \tag{6.22a}$$

which can also be rewritten in the form of perturbation in the effective potential $\delta v_{\text{eff}}(\mathbf{r}', t)$ for an equivalent noninteracting system within the Kohn–Sham theory as

$$\delta\rho(\mathbf{r}, \omega) = \int d\mathbf{r}'\chi_0(\mathbf{r}, \mathbf{r}'; \omega)\delta v_{\text{eff}}(\mathbf{r}', \omega). \tag{6.22b}$$

Here, the frequency-dependent response functions $\chi(\mathbf{r}, \mathbf{r}'; \omega)$ and $\chi_0(\mathbf{r}, \mathbf{r}'; \omega)$ correspond, respectively, to the actual interacting system and the equivalent Kohn–Sham noninteracting system. Using the expression of the effective potential, one can write

$$\delta v_{\text{eff}}(\mathbf{r}, \omega) = \delta v_{\text{ext}}(\mathbf{r}, \omega) + \delta v_{\text{SCF}}(\mathbf{r}, \omega) \tag{6.23a}$$

$$\delta v_{\text{SCF}}(\mathbf{r}, \omega) = \int d\mathbf{r}' \left[\frac{1}{|\mathbf{r} - \mathbf{r}'|} + f_{\text{XC}}(\mathbf{r}, \mathbf{r}', \omega) \right] \delta\rho(\mathbf{r}', \omega), \tag{6.23b}$$

where $f_{\text{XC}}(\mathbf{r}, \mathbf{r}', \omega)$ is defined in the time domain as

$$f_{\text{XC}}(\mathbf{r}, \mathbf{r}', t - t') = \frac{\delta v_{\text{XC}}(\mathbf{r}, t)}{\delta\rho(\mathbf{r}', t')}. \tag{6.24}$$

Using Equation 6.22a for $\delta\rho(\mathbf{r}', \omega)$ in Equation 6.23 and using Equation 6.22b, one obtains the integral relation involving the two response kernels, viz.,

$$\chi(\mathbf{r}, \mathbf{r}'; \omega) = \chi_0(\mathbf{r}, \mathbf{r}'; \omega) + \int d\mathbf{r}'' \int d\mathbf{r}''' \chi_0(\mathbf{r}, \mathbf{r}''; \omega)$$

$$\times \left[\frac{1}{|\mathbf{r}'' - \mathbf{r}'''|} + f_{XC}(\mathbf{r}'', \mathbf{r}'''; \omega) \right] \chi(\mathbf{r}''', \mathbf{r}'; \omega). \qquad (6.25)$$

Thus, the response kernel for the interacting system can be obtained from that of the noninteracting system if one has a suitable functional form for the XC energy density functional for TD systems. The standard form for the kernel $\chi_0(\mathbf{r}, \mathbf{r}''; \omega)$ for the noninteracting system, expressed in terms of the Kohn Sham orbitals $\psi_k(\mathbf{r})$, their energy eigenvalues ε_k, and the occupation numbers n_k, is given [17,19] by

$$\chi_0(\mathbf{r}, \mathbf{r}'; \omega) = \lim_{\eta \to 0+} \sum_{k,l} (n_k - n_l) \frac{\psi_k(\mathbf{r}) \psi_l^*(\mathbf{r}) \psi_k^*(\mathbf{r}') \psi_l(\mathbf{r}')}{\omega - (\varepsilon_k - \varepsilon_l) + i\eta}. \qquad (6.26)$$

While the present discussion of the response property is limited to scalar potential only, an analogous description is also possible for the vector potential and a similar equation can be derived.

6.6 CONCLUDING REMARKS

In this short review, a brief overview of the underlying principles of TDDFT has been presented. The formal aspects for TDDFT in the presence of scalar potentials with periodic time dependence as well as TD electric and magnetic fields with arbitrary time dependence are discussed. This formalism is suitable for treatment of interaction with radiation in atomic and molecular systems. The Kohn–Sham-like TD equations are derived, and it is shown that the basic picture of the original Kohn–Sham theory in terms of a fictitious system of noninteracting particles is retained and a suitable expression for the effective potential is derived.

Although TDDFT is considered to be a well-established tool for the investigation of dynamical properties of molecular systems, development of better and more accurate XC functionals of density and current density is still an ongoing process. Spin polarization has been neglected in the present discussion, which is, however, important particularly in view of the many recent developments in the areas of magnetism and spintronics. While only a few chosen aspects have been covered in this chapter to provide a glimpse of the basic formalism, there have been many new developments in this exciting area of research in recent years.

ACKNOWLEDGMENTS

I am indebted to Professor B.M. Deb for introducing me to the subject of DFT. It is a pleasure to thank Dr A.K. Dhara for many fruitful collaborations. I am also thankful to Professor P.K. Chattaraj for inviting and encouraging me to write this chapter.

REFERENCES

1. N.H. March and B.M. Deb (Eds.), *The Single-Particle Density in Physics and Chemistry*, Academic Press, New York, 1987.
2. R.G. Parr and W. Yang, *Density Functional Theory of Atoms and Molecules*, Oxford University Press, New York, 1989.
3. P. Hohenberg and W. Kohn, *Phys. Rev.* 136, B864, 1964.
4. W. Kohn and L.J. Sham, *Phys. Rev.* 140, A1133, 1965.
5. L.H. Thomas, *Proc. Camb. Phil. Soc.* 23, 542, 1926; E. Fermi, *Z. Phys.* 48, 73, 1928.
6. F. Bloch, *Z. Phys.* 81, 363, 1933.
7. E. Madelung, *Z. Phys.* 40, 332, 1926.
8. S.K. Ghosh and B.M. Deb, *Phys. Rep.* 92, 1, 1982.
9. P.K. Chattaraj, S. Sengupta, and A. Poddar, *Int. J. Quant. Chem.*, 69, 279, 1998.
10. V. Peuckert, *J. Phys.* C11, 4945, 1978.
11. L.J. Bartolotti, *Phys. Rev.* A24, 1661, 1981.
12. B.M. Deb and S.K. Ghosh, *J. Chem. Phys.* 77, 342, 1982.
13. E. Runge and E.K.U. Gross, *Phys. Rev. Lett.* 52, 997, 1984.
14. S.K. Ghosh and A.K. Dhara, *Phys. Rev.* A 38, 1149, 1988.
15. M.E. Casida, in *Recent Advances in Density Functional Methods*, Part I, D.P. Chong (Ed.), World Scientific, Singapore 1995, p. 155.
16. E.K.U. Gross and W. Kohn, *Adv. Quant. Chem.* 21, 255, 1990.
17. R. van Leeuwen, *Int. J. Mod. Phys.* B15, 1969, 2001.
18. J. Schirmer and A. Dreuw, *Phys. Rev.* A75, 22513, 2007.
19. N.T. Maitra, K. Burke, H. Appel, E.K.U. Gross, and R. van Leeuwen, in *Reviews in Modern Quantum Chemistry: A Celebration of the Contributions of Robert Parr*, K.D. Sen (Ed.), World Scientific, Singapore, 2002, p. 1186.
20. A.K. Dhara and S.K. Ghosh, *Phys. Rev.* A35, 442, 1987.
21. G. Vignale, *Phys. Rev.* B70, 20102, 2004.
22. F. Furche, *J. Chem. Phys.* 114, 5982, 2001.

7 Exchange-Correlation Potential of Kohn–Sham Theory: A Physical Perspective

Manoj K. Harbola

CONTENTS

7.1 INTRODUCTION

Among various theories of electronic structure, density functional theory (DFT) [1,2] has been the most successful one. This is because of its richness of concepts and at the same time simplicity of its implementation. The new concept that the theory introduces is that the ground-state density of an electronic system contains all the information about the Hamiltonian and therefore all the properties of the system. Further, the theory introduces a variational principle in terms of the ground-state density that leads to an equation to determine this density. Consider the expectation value $\langle H \rangle$ of the Hamiltonian (atomic units are used)

$$H = \sum_i \left(-\frac{1}{2}\nabla_i^2 + v_{\text{ext}}(\vec{r}_i) \right) + \frac{1}{2}\sum_{\substack{i,j \\ i \neq j}} \frac{1}{|\vec{r}_i - \vec{r}_j|} \qquad (7.1)$$

of a system of N electrons. In the expression above, $v_{\text{ext}}(\vec{r})$ is the potential with which the electrons are moving in. For example, in an atom $v_{\text{ext}}(\vec{r}) = -(Z/r)$, in a molecule $v_{\text{ext}}(\vec{r}) = \sum_i Z_i/|\vec{r} - \vec{R}_i|$, where Z_i indicates the nuclear charge on the ith atom of the molecule and $\vec{R}_i$ its position, and $v_{\text{ext}}(\vec{r}) = \frac{1}{2}kr^2$ if electrons are moving in a harmonic potential. While in conventional theory the expectation value $\langle H \rangle$ is a functional $E[\Psi]$ of the wave function Ψ, in DFT it is [3] a functional $E[\rho]$ of the ground-state density ρ. The density is given in terms of the wave function as

$$\rho(\vec{r}) = N \int |\Psi(\vec{r}, \vec{r}_2, \vec{r}_3 \ldots \vec{r}_N)|^2 d\vec{r}_2 d\vec{r}_3 \ldots d\vec{r}_N \qquad (7.2)$$

The equation satisfied by the wave function Ψ, the Schrödinger equation, is obtained by minimizing the functional $E[\Psi]$ with respect to Ψ, with the energy of the system appearing as a Lagrange multiplier to ensure the normalization of the wave function. Similarly in DFT, the equation for the density is obtained by minimizing the functional $E[\rho]$ with respect to the density ρ and leads to the Euler equation

$$\frac{\delta E[\rho]}{\delta \rho(\vec{r})} = \mu \qquad (7.3)$$

where μ appears as a Lagrange multiplier to ensure that the density integrates to the correct number of electrons N. The physical interpretation of μ as the chemical potential and its derivatives has been discussed in other chapters of this book. For the purposes of this chapter, we note that the chemical potential of an electronic system equals the negative of its ionization energy when an electron is removed from it and negative of its electron affinity when an electron is added to it [4].

Equation 7.3 can alternatively be written [5] in terms of single particle orbitals as

$$\left(-\frac{1}{2}\nabla^2 + v_{\text{ext}}(\vec{r}) + \int \frac{\rho(\vec{r}')}{|\vec{r} - \vec{r}'|} d\vec{r}' + v_{\text{xc}}(\vec{r}) \right) \phi_i = \varepsilon_i \phi_i \qquad (7.4)$$

where $v_{\text{xc}}(\vec{r})$ is known as the exchange-correlation potential. This equation is the famous Kohn–Sham equation of DFT. Its solutions $\{\phi_i\}$ are called the Kohn–Sham orbitals. By construction, the connection of this equation with Equation 7.3 is through the following two relations: (1) density $\rho(\vec{r})$ is given in terms of the single particle orbitals as

$$\rho(\vec{r}) = \sum_i |\phi_i(\vec{r})|^2 \qquad (7.5)$$

and (2) the chemical potential μ is equal to the eigenvalue ε_{max} of the highest occupied orbital, i.e.,

$$\mu = \varepsilon_{\text{max}} \qquad (7.6)$$

Coupled with the fact that the chemical potential equals the ionization energy of a system, Equation 7.6 implies that the eigenvalue ε_{max} will be equal [4] to the negative of the ionization potential I of a system, i.e.,

$$\varepsilon_{max} = -I \tag{7.7}$$

This is known as the ionization potential theorem. Equation 7.7 between ε_{max} and ionization potential I can also be obtained alternatively by looking at the asymptotic behavior of the density of a many-electron system. For atoms and molecules, the asymptotic decay of the density is given as [6–11]

$$\rho(r \rightarrow \infty) \sim \exp\left(-2\sqrt{2I}r\right) \tag{7.8}$$

On the other hand, asymptotic density of the corresponding Kohn–Sham system is determined completely by the highest occupied orbital and is given as

$$\rho(r \rightarrow \infty) \sim \exp\left(-2\sqrt{-2\varepsilon_{max}}r\right) \tag{7.9}$$

A comparison of Equations 7.8 and 7.9 leads to Equation 7.7.

As an example of these ideas, we plot in Figure 7.1 the exact Kohn–Sham potential for the neon atom. The potential has been obtained by applying the Zhao–Parr (ZP) method [12], which generates the exact Kohn–Sham potential for a given density to a highly accurate density of neon [13]. The corresponding

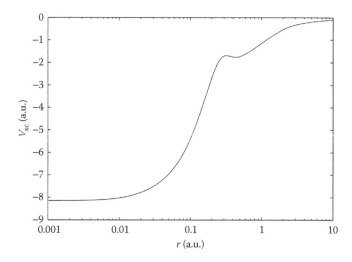

FIGURE 7.1 Exchange-correlation potential V_{xc} (in atomic units) for neon as a function of distance r (in atomic units) from the nucleus. The potential is obtained from the ground-state density by employing the ZP method.

ε_{max} equals -0.773 a.u. that has essentially the same magnitude as the experimental ionization potential of neon, which is [14] 0.793 a.u.

The total energy of the system in terms of the Kohn–Sham orbitals is given as

$$E[\rho] = \sum_i \left\langle \phi_i \left| -\frac{1}{2}\nabla^2 \right| \phi_i \right\rangle + \int v_{\text{ext}}(\vec{r})\rho(\vec{r})d\vec{r} + \frac{1}{2}\int\int \frac{\rho(\vec{r})\rho(\vec{r}')}{|\vec{r}-\vec{r}'|}d\vec{r}d\vec{r}' + E_{\text{xc}}[\rho]$$

$$= \sum_i \varepsilon_i - \frac{1}{2}\int\int \frac{\rho(\vec{r})\rho(\vec{r}')}{|\vec{r}-\vec{r}'|}d\vec{r}d\vec{r}' - \int v_{\text{xc}}(\vec{r})\rho(\vec{r})d\vec{r} + E_{\text{xc}}[\rho] \qquad (7.10)$$

In the equation above, the functional $E_{\text{xc}}[\rho]$ is the exchange-correlation (XC) energy functional and is the sum of the conventional quantum mechanical XC energy

$$E_{\text{xc}}^{\text{QM}}[\rho] = \left\langle \Psi \left| \frac{1}{2}\sum_{\substack{i,j \\ i\neq j}} \frac{1}{|\vec{r}_i-\vec{r}_j|} \right| \Psi \right\rangle - \frac{1}{2}\int\int \frac{\rho(\vec{r})\rho(\vec{r}')}{|\vec{r}-\vec{r}'|}d\vec{r}d\vec{r}' \qquad (7.11)$$

and the difference

$$T_{\text{c}} = \left\langle \Psi \left| \sum_i -\frac{1}{2}\nabla_i^2 \right| \Psi \right\rangle - \sum_i \left\langle \phi_i \left| -\frac{1}{2}\nabla^2 \right| \phi_i \right\rangle \qquad (7.12)$$

between the kinetic energy of the true system and that of the Kohn–Sham system. Thus the XC energy in Kohn–Sham theory is given as

$$E_{\text{xc}}[\rho] = \left\langle \Psi \left| \frac{1}{2}\sum_{\substack{i,j \\ i\neq j}} \frac{1}{|\vec{r}_i-\vec{r}_j|} \right| \Psi \right\rangle - \frac{1}{2}\int\int \frac{\rho(\vec{r})\rho(\vec{r}')}{|\vec{r}-\vec{r}'|}d\vec{r}d\vec{r}' + T_{\text{c}} \qquad (7.13)$$

We note that T_{c} is small in magnitude; in the above example of neon its value is 0.34 a.u. in comparison to the total kinetic energy, which is 128.93 a.u. and the quantum mechanical XC energy, which is 12.79 a.u. (see Ref. [15] for results for a number of atoms including neon).

In the Kohn–Sham equation above, the Coulomb potential and the XC potential are obtained from their energy counterparts by taking the functional derivative of the latter with respect to the density. Thus

$$\int \frac{\rho(\vec{r}')}{|\vec{r}-\vec{r}'|}d\vec{r}' = \frac{\delta}{\delta\rho(\vec{r})}\left[\frac{1}{2}\int\int \frac{\rho(\vec{r})\rho(\vec{r}')}{|\vec{r}-\vec{r}'|}d\vec{r}d\vec{r}'\right] \qquad (7.14)$$

and

$$v_{\text{xc}}(\vec{r}) = \frac{\delta E_{\text{xc}}[\rho]}{\delta\rho(\vec{r})} \qquad (7.15)$$

However, while the Coulomb potential above is easy to understand—it is the electrostatic potential produced by the charge distribution $\rho(\vec{r})$—physical meaning of the XC potential is not clear. The focus of this chapter is in providing this understanding. For this, we begin with the definition of the exchange-correlation energy in terms of the corresponding XC hole.

7.2 EXCHANGE-CORRELATION ENERGY AND EXCHANGE-CORRELATION HOLE

We start with a physical understanding of the conventional XC energy in terms of XC hole. To see this, we write the conventional XC energy in Equation 7.11 as

$$
E_{xc}^{QM}[\rho] = \frac{1}{2} \int \int \frac{\rho(\vec{r})}{|\vec{r} - \vec{r}'|}
$$
$$
\times \left\{ \frac{N(N-1)}{\rho(\vec{r})} \int |\Psi(\vec{r}, \vec{r}', \vec{r}_3 \ldots \vec{r}_N)|^2 d\vec{r}_3 \ldots d\vec{r}_N - \rho(\vec{r}') \right\} d\vec{r} d\vec{r}' \tag{7.16}
$$

after performing some algebraic manipulations and using the fact that $|\Psi(\vec{r}_1, \vec{r}_2, \ldots)|^2$ is symmetric with respect to an interchange of $\vec{r}_i$ and $\vec{r}_j$. We identify the expression in curly brackets in Equation 7.16 as the XC hole

$$
\rho_{xc}^{QM}(\vec{r}, \vec{r}') = \left\{ \frac{N(N-1)}{\rho(\vec{r})} \int |\Psi(\vec{r}, \vec{r}', \vec{r}_3 \ldots \vec{r}_N)|^2 d\vec{r}_3 \ldots d\vec{r}_N - \rho(\vec{r}') \right\} \tag{7.17}
$$

and write the XC energy as

$$
E_{xc}^{QM}[\rho] = \frac{1}{2} \int \int \frac{\rho(\vec{r})\rho_{xc}^{QM}(\vec{r}, \vec{r}')}{|\vec{r} - \vec{r}'|} d\vec{r} d\vec{r}' \tag{7.18}
$$

Thus the XC energy is the energy of interaction between the electrons and a charge distribution represented by $\rho_{xc}^{QM}(\vec{r}, \vec{r}')$. The question we wish to answer now is whether the expression in Equation 7.18 is just the rewriting of the XC energy in a different way or does it have a physical interpretation. We now show that it indeed has a physical interpretation: the term represents the deficit in the density of electrons at $\vec{r}'$ when an electron is at $\vec{r}$.

Since the probability density of finding an electron at $\vec{r}$ is $\rho(\vec{r})/N$, one expects the probability density $P_2(\vec{r}, \vec{r}')$ that one electron is at $\vec{r}$ and another at $\vec{r}'$, would be given by multiplying the probability density $\rho(\vec{r})/N$ that an electron is found at $\vec{r}$ and the probability density that another electron (from the $N-1$ left) is found at $\vec{r}'$. Normally one would calculate the latter by subtracting from density $\rho(\vec{r}')$, the average density of one electron $\rho(\vec{r}')/N$ and dividing the resulting expression by $(N-1)$. Thus,

$$P_2(\vec{r}, \vec{r}') = \frac{\rho(\vec{r})}{N} \frac{1}{(N-1)} \left\{ \rho(\vec{r}') - \frac{\rho(\vec{r}')}{N} \right\} \qquad (7.19a)$$

$$= \frac{\rho(\vec{r})}{N} \frac{\rho(\vec{r}')}{N} \qquad (7.19b)$$

which is nothing but the product of the two probability densities. This, however, would be true only if all the electrons were moving independent of each other and therefore an electron does not affect the motion of any other electron. On the other hand, this is not the case in an interacting electron system. Thus this probability is modified. The probability density that an electron is at $\vec{r}$ and another at $\vec{r}'$ is given by $\int |\Psi(\vec{r}, \vec{r}', \vec{r}_3, \ldots, \vec{r}_N)|^2 \, d\vec{r}_3 \ldots d\vec{r}_N$. We now rewrite it, after some manipulations, in the form given by Equation 7.19a as

$$P_2(\vec{r}, \vec{r}') = \frac{\rho(\vec{r})}{N} \frac{1}{(N-1)}$$
$$\times \left[\rho(\vec{r}') + \left\{ \frac{N(N-1)}{\rho(\vec{r})} \int |\Psi(\vec{r}, \vec{r}', \vec{r}_3, \ldots, \vec{r}_N)|^2 d\vec{r}_3 \ldots d\vec{r}_N - \rho(\vec{r}') \right\} \right] \quad (7.20)$$

Further, if the wave function depends also on the electron spins, spin variables over all electrons should also be integrated; we will see this below, in the calculation of exchange hole. The expression in the curly brackets above is exactly the XC hole $\rho_{xc}^{QM}(\vec{r}, \vec{r}')$ defined in Equation 7.17. A comparison with Equation 7.19a shows that adding the hole to the density is similar to subtracting the density of one electron $\rho(\vec{r}')/N$ from it. The hole thus represents a deficit of one electron from the density. This is easily verified by integrating $\rho_{xc}^{QM}(\vec{r}, \vec{r}')$ over the volume $d\vec{r}'$, which gives a value of -1. However, the structure of the hole is not simple and this is because of the motion of different electrons correlated due to the Pauli exclusion principle and the Coulomb interaction between them. Finally we note that the product $\rho(\vec{r})\rho_{xc}^{QM}(\vec{r}, \vec{r}')$ is symmetric with respect to an exchange in the variables $\vec{r}$ and $\vec{r}'$.

The physical picture that emerges out of the exercise above is that the electron–electron interaction energy

$$\left\langle \frac{1}{2} \sum_{\substack{i,j \\ i \neq j}} \frac{1}{|\vec{r}_i - \vec{r}_j|} \right\rangle$$

is not simply the Coulomb energy of a charge distribution given by $\rho(\vec{r})$ corrected only for the electronic charge being finite. If that were the case, this energy would be obtained by subtracting the self energy of N electrons from the Coulomb energy of charge distribution $\rho(\vec{r})$ and will be given as

$$\left(1 - \frac{1}{N} \right) \frac{1}{2} \int \int \frac{\rho(\vec{r})\rho(\vec{r}')}{|\vec{r} - \vec{r}'|} d\vec{r} d\vec{r}' \qquad (7.21)$$

Rather, the energy is given by the expectation value

$$\left\langle \frac{1}{2} \sum_{\substack{i,j \\ i \neq j}} \frac{1}{|\vec{r}_i - \vec{r}_j|} \right\rangle = \frac{1}{2} \int \int \frac{\rho(\vec{r})\rho(\vec{r}')}{|\vec{r} - \vec{r}'|} d\vec{r} d\vec{r}' + E_{xc}^{QM} \tag{7.22}$$

The XC energy represents the correction to the Coulomb energy for the self-energy of an electron in a many-electron system. The latter is due to both the direct self-energy of the electron as well as the redistribution of electronic density around each electron because of the Pauli exclusion principle and the Coulomb interaction. As an example, we now discuss the case of Fermi hole and the exchange energy in Hartree–Fock (HF) theory [16]. For brevity, we restrict ourselves to closed-shell cases.

7.2.1 EXCHANGE ENERGY AND EXCHANGE HOLE

For a closed-shell system, the wave function in HF theory is given as a Slater determinant

$$\Psi_{HF} = \frac{1}{\sqrt{N!}} \begin{vmatrix} \chi_1(x_1) & \chi_1(x_2) & \cdots & \chi_1(x_N) \\ \chi_2(x_1) & \chi_2(x_2) & \cdots & \chi_2(x_N) \\ \cdots & \cdots & \cdots & \cdots \\ \cdots & \cdots & \cdots & \cdots \\ \chi_N(x_1) & \chi_N(x_2) & \cdots & \chi_N(x_N) \end{vmatrix} \tag{7.23}$$

where $x = (\vec{r}, \sigma)$ represents the space and spin variables of an electron and $\chi(x) = \phi_\sigma(\vec{r})\alpha(\sigma)$ or $\chi(x) = \phi_\sigma(\vec{r})\beta(\sigma)$ are the spin orbitals. Here we have explicitly taken into account the possibility that the space orbitals may have a dependence on the spin. The wave function can alternatively be written as

$$\Psi_{HF} = \frac{1}{\sqrt{N!}} \sum_P (-1)^P P(\chi_1(x_1)\chi_2(x_2)\ldots\chi_N(x_N)) \tag{7.24}$$

where P represents a permutation of the electron variables $(x_1, x_2, \ldots, x_N)$. In this case, the Fermi–Coulomb hole is known as the Fermi hole $\rho_x(\vec{r}, \vec{r}')$ because HF wave function has only the Pauli exclusion principle and not the Coulomb correlations built into the wave function. Substituting Equation 7.24 in Equation 7.17 for the hole results in

$$\rho_x(\vec{r}, \vec{r}') = \frac{N(N-1)}{N!} \frac{1}{\rho(\vec{r})} \int \sum_{P,P'} (-1)^P (-1)^{P'} P(\chi_1(\vec{r}, \sigma_1)\chi_2(\vec{r}', \sigma_2)\ldots\chi_N(x_N))$$

$$\times P'(\chi_1(\vec{r}, \sigma_1)\chi_2(\vec{r}', \sigma_2)\ldots\chi_N(x_N))d\sigma_1 d\sigma_2 dx_3 dx_4 \ldots dx_N - \rho(\vec{r}) \tag{7.25}$$

and describes the deficit in the density of the electron at $\vec{r}'$ if there is an electron at $\vec{r}$, irrespective of their spins. That is why their spins (σ_1, σ_2) have been integrated over (integration over spin implies a sum over them). To evaluate the expression in Equation 7.25, we write the permutation over the variables $(\vec{r}, \sigma_1)$ and $(\vec{r}', \sigma_2)$ explicitly so that one of the permutations is

$$P_{ij}(\chi_1(\vec{r}, \sigma_1)\chi_2(\vec{r}', \sigma_2)\ldots\chi_N(x_N)) = \chi_i(\vec{r}, \sigma_1)\chi_j(\vec{r}', \sigma_2)P_{N-2}$$
$$\times ((N-2) \text{ spin orbitals}) \tag{7.26}$$

Here P_{N-2} represents the permutations over the rest $(N-2)$ orbitals, and i and j are two occupied spin orbitals. Thus the sum over permutations is $\sum_P = \sum_{\substack{i,j \\ j\neq i}} \sum_{P_{N-2}}$. For a given permutation P_{ij}, there will be two permutations contributing from P' that have rest of the orbitals in the same order, P'_{ij} with a positive sign and P'_{ji} with a negative sign. Thus the expression under the integral sign in Equation 7.25 can be written as

$$\left\{ \int \sum_{\substack{i,j \\ j\neq i}} \chi_i^*(\vec{r}, \sigma_1)\chi_j^*(\vec{r}', \sigma_2)(\chi_i(\vec{r}, \sigma_1)\chi_j(\vec{r}', \sigma_2) - \chi_i(\vec{r}', \sigma_2)\chi_j(\vec{r}, \sigma_1))d\sigma_1 d\sigma_2 \right\}$$

$$\times \int \sum_{P_{N-2}, P'_{N-2}} (-1)^{P_{N-2}}(-1)^{P'_{N-2}} P_{N-2}((N-2) \text{ orbitals})P'_{N-2}((N-2) \text{ orbitals}) \, dx_3 dx_4 \ldots dx_N$$

$$\tag{7.27}$$

The second integral above is a standard integral in the HF theory and gives $(N-2)!$ In the first integral, we can remove the restriction $(j \neq i)$ in the summation because that term cancels. Second, the integration over the spin variables forces the spin of the ith and jth orbitals to be the same in the second term inside the curly brackets. Taking all these facts into account, the Fermi hole comes out to be

$$\rho_x(\vec{r}, \vec{r}') = -\frac{1}{\rho(\vec{r})} \sum_{\substack{i,j \\ \sigma_j = \sigma_i}} \phi_{i,\sigma_i}^*(\vec{r})\phi_{j,\sigma_j}^*(\vec{r}')\phi_{i,\sigma_i}(\vec{r}')\phi_{j,\sigma_j}(\vec{r}) \tag{7.28}$$

Thus in HF theory, the deficit in density around an electron at $\vec{r}$ arises solely from electrons of the same spin. The corresponding exchange energy is given as

$$E_x = -\frac{1}{2} \sum_{\substack{i,j \\ \sigma_j = \sigma_i}} \int \frac{\phi_{i,\sigma_i}^*(\vec{r})\phi_{j,\sigma_j}^*(\vec{r}')\phi_{i,\sigma_i}(\vec{r}')\phi_{j,\sigma_j}(\vec{r})}{|\vec{r} - \vec{r}'|} d\vec{r} d\vec{r}' \tag{7.29}$$

Note again that the product $\rho(\vec{r})\rho_x(\vec{r}, \vec{r}')$ is also symmetric with respect to an interchange of $\vec{r}$ and $\vec{r}'$.

In the above discussion, we have shown that the conventional exchange-correlation energy of a many-electron system can be thought of as the energy of interaction between the electrons and the Fermi–Coulomb hole. In Kohn–Sham DFT too, the XC energy can be expressed [17] in exactly the same manner, except that the Fermi–Coulomb hole is going to be slightly different from that given above because it also includes the effects of the difference in the kinetic energy given by Equation 7.10 above. Thus

$$E_{xc}^{DFT}[\rho] = \frac{1}{2} \int \int \frac{\rho(\vec{r})\rho_{xc}^{DFT}(\vec{r}, \vec{r}')}{|\vec{r} - \vec{r}'|} d\vec{r}d\vec{r}' \tag{7.30}$$

The expression for the XC hole of Kohn–Sham theory is derived by a coupling constant integration over the quantum mechanical expression. The exchange hole and the exchange energy, on the other hand, are calculated by employing the Kohn–Sham orbitals in the expressions of Equations 7.28 and 7.29. We note that the XC hole of DFT also satisfies the symmetry that the product $\rho(\vec{r})\rho_{xc}^{DFT}(\vec{r}, \vec{r}')$ remains unchanged if $\vec{r}$ and $\vec{r}'$ are interchanged in it.

7.3   EXCHANGE-CORRELATION POTENTIAL FROM THE FERMI–COULOMB HOLE

Having understood the physical meaning of the XC energy in terms of the corresponding hole, the next step would be to apply the same meaning to the XC potential of Equation 7.4 and calculate it as the electrostatic potential arising from the XC hole. Thus one is tempted to write the XC potential as

$$v_{xc}(\vec{r}) = \int \frac{\rho_{xc}^{DFT}(\vec{r}, \vec{r}')}{|\vec{r} - \vec{r}'|} d\vec{r}' \tag{7.31}$$

Can this be true? Let us examine it in the case of exchange potential because it can be calculated in terms of orbitals.

7.3.1   EXCHANGE POTENTIAL

The exchange potential of Equation 7.31 is called the Slater potential [12], because it was Slater who had proposed [18] that the nonlocal exchange potential of HF theory can be replaced by the potential

$$v_x^{Slater}(\vec{r}) = \int \frac{\rho_x(\vec{r}, \vec{r}')}{|\vec{r} - \vec{r}'|} d\vec{r}' \tag{7.32}$$

so that the solution of HF equations is simplified. In Figure 7.2 we compare the exact exchange potential obtained by applying the ZP method to the HF density (obtained from the analytical HF orbitals [19]) and self-consistently determined Slater potential for the neon atom. It is seen that the Slater potential overestimates the exact exchange potential over the entire atom. In the outer regions, however, both the potentials go as $-(1/r)$. What could be the reason for this overestimate? We now answer this question.

The key to understanding the difference between the Slater potential and the exact exchange potential lies in the explicit dependence of the Fermi hole $\rho_x(\vec{r}, \vec{r}')$ on

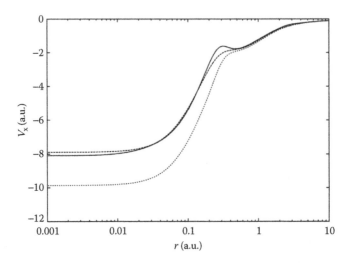

FIGURE 7.2 Different exchange potentials V_x (in atomic units) for neon, as functions of distance r (in atomic units) from the nucleus. The solid line indicates the potential obtained from the HF density by employing the ZP method, long dashes (——) the W_x potential of Equation 7.31 and short dashes (----) the Slater potential of Equation 7.29.

position $\vec{r}$ of the electron. Because of this dependence, the Fermi hole changes [16,20] as an electron moves in a many-electron system. In such a situation, calculating the electrostatic potential as given by Equation 7.32 is not correct. Rather the potential should be calculated [20] as the work done in moving an electron in the electric field

$$\vec{F}_x(\vec{r}) = \int \frac{\rho_x(\vec{r}, \vec{r}')}{|\vec{r} - \vec{r}'|^3}(\vec{r} - \vec{r}')d\vec{r}' \tag{7.33}$$

of the Fermi hole. Note that if the Fermi hole did not depend on the position of the electron, the field in Equation 7.33 would have been equal to the gradient of the Slater potential of Equation 7.32; it is not precisely because $\rho_x(\vec{r}, \vec{r}')$ depends on position $\vec{r}$. We draw a parallel of this proposal with a textbook example. Consider the surface charge density induced on the surface of a grounded conductor because of a charge in front of it. The charge density depends on the position of the charge in front. In this case too, the image potential is not the electrostatic potential as calculated using the standard formula. However, if the work done in moving the charge in the electric field of the induced charge is calculated, it is indeed the correct way of calculating the potential and gives the image potential. Let us, therefore, write the exchange potential as

$$W_x(\vec{r}) = -\int_{\infty}^{\vec{r}} \vec{F}_x(\vec{r}') \cdot d\vec{l}' \tag{7.34}$$

At this point one question must be answered: Is the potential calculated in the manner above path independent [21]? Equivalently, is the field given by Equation 7.33 curl-free? For one-dimensional cases and within the central field approximation for atoms, it is. For other systems, there is a small solenoidal component [21,22] and we will see later that it arises from the difference in the kinetic energy of the true system and the corresponding Kohn–Sham system (in this case the HF system and its Kohn–Sham counterpart). For the time being, we explore whether the physics of calculating the potential in the manner prescribed above is correct in the cases where the curl of the field vanishes.

Plotted in Figure 7.2 is also the self-consistently determined exchange potential W_x for neon. As is evident from the figure, the potential is highly accurate. This indicates the correctness of the physics invoked to calculate the potential as the work done in moving an electron in the field created by its Fermi–Coulomb hole. Notice that we have calculated the potential directly from the hole rather than taking the functional derivative of the exchange energy functional, as is done in Equation 7.15. We now connect the exchange potential of Equation 7.34, potential and the exchange energy given by Equation 7.29, mathematically.

The relationship between the exchange potential of DFT and the corresponding energy functional is established through the virial theorem. The two are related via the following relationship derived by Levy and Perdew [23]

$$E_x = -\int \rho(\vec{r})\vec{r} \cdot \vec{\nabla} v_x(\vec{r})d\vec{r} \qquad (7.35)$$

Since the gradient of the potential is the field given by Equation 7.33, its substitution in Equation 7.35 gives

$$E_x = -\sum_{\substack{i,j \\ \sigma_j=\sigma_i}} \int \frac{\phi^*_{i,\sigma_i}(\vec{r})\phi^*_{j,\sigma_j}(\vec{r}')\phi_{i,\sigma_i}(\vec{r}')\phi_{j,\sigma_j}(\vec{r})}{|\vec{r}-\vec{r}'|^3}\vec{r} \cdot (\vec{r}-\vec{r}')d\vec{r}d\vec{r}' \qquad (7.36)$$

Now interchanging $\vec{r}$ and $\vec{r}'$ in Equation 7.36 does not affect the value of the integral above. Thus the exchange energy can be written as the sum of expression of Equation 7.36 and that obtained by interchanging $\vec{r}$ and $\vec{r}'$ and dividing the sum by 2. This immediately leads to the exchange energy expression of Equation 7.29. Thus we see that the exchange potential proposed satisfies the virial theorem sum rule that connects the local potential of Kohn–Sham theory to the corresponding energy functional. The next question that we ask is if self-consistent solutions of Equation 7.4 can be obtained with the proposed exchange potential. The answer is in the affirmative and we discuss the results next.

7.3.1.1 Self-Consistent Solutions for the Ground States

Given in Table 7.1 are the results [24] of the total energy of some atoms obtained by solving the Kohn–Sham equation self-consistently with the exchange potential W_x within the central field approximation. The energy is obtained from Equation 7.10

TABLE 7.1

Negative of the Energies (Atomic Units) of Some Atoms Calculated with the Potential W_x Equation 7.34

Atom	Configuration	W_x	HF
He	$1s^2(^1S)$	2.862	2.862
Be	$[He]2s^2(^1S)$	14.571	14.573
C	$[He]2s^22p^2(^3P)$	37.685	37.689
O	$[He]2s^22p^4(^3P)$	74.805	74.809
Ne	$[He]2s^22p^6(^1S)$	128.542	128.547
Al	$[Ne]3s^23p^1(^2P)$	241.868	241.877
Cl	$[Ne]3s^23p^5(^2P)$	459.472	459.482
Ar	$[Ne]3s^23p^6(^1S)$	526.804	526.818
Ca	$[Ar]4s^2(^1S)$	676.743	676.758
Kr	$[Ar]3d^{10}4s^24p^6(^1S)$	2752.030	2752.055

Note: Comparison is made with the corresponding HF energies. Note that the W_x energies are slightly above those of HF.

by substituting the exchange energy of Equation 7.29 for the exchange-correlation energy. As such, the results are compared with those of HF theory [25]. As is evident from the numbers presented, the local exchange potential W_x gives the energies which are very close to their HF counterparts. In fact, the difference is in parts per million (ppm). It will be discussed later why this potential leads to slightly higher energies.

In Table 7.2, we show the eigen energies corresponding to the highest occupied orbital of the atoms in Table 7.1. Again it is seen that these eigen energies are very close to their HF counterparts as well as the negative of their ionization energies.

A severe test that a potential can be put to is to see whether it can give self-consistent solutions for the negative ions. The potential proposed above gives the solutions [26] for the negative ions also with energies close to their HF energies. The energies for the negative ions of hydrogen, lithium, fluorine, and chlorine are shown in Table 7.3. In Table 7.4, the eigen energies for the highest occupied orbital are shown and compared with the corresponding HF eigenvalues and the experimental electron affinity [27] of the neutral atoms (ionization energy of the negative ions). The accuracy of the numbers obtained is self-evident.

7.3.1.2 Excited States

An advantage of obtaining the exchange potential W_x through physical arguments is that unlike its Kohn–Sham counterpart, it is equally valid for the excited states [20]. Thus the densities and energies of excited states can also be obtained by solving Equation 7.4 in the excited-state configuration and by employing potential W_x for

TABLE 7.2

Negative of the Highest Occupied Orbital Eigen Energies (Atomic Units) of Some Atoms Calculated with the Potential W_x Equation 7.34

Atom	W_x	HF	Experimental Ionization Potential
He	0.918	0.918	0.904
Be	0.313	0.310	0.343
C	0.409	0.434	0.414
O	0.625	0.632	0.500
Ne	0.857	0.850	0.793
Al	0.203	0.210	0.220
Cl	0.503	0.507	0.477
Ar	0.589	0.591	0.579
Ca	0.201	0.196	0.250
Kr	0.518	0.524	0.515

Note: Comparison is made with the corresponding HF energies and experimental ionization potential.

the XC potential. The excitation energy is then calculated as the difference in the excited-state energy and the ground-state energy. A large number of calculations for the excited states of atoms have been performed [28] employing the potential W_x alone or by employing the sum of W_x and a local correlation potential. All these calculations give highly accurate results for the excitation energies of the systems studied. As a demonstration, we discuss the case of excited states of negative ions. As pointed out above, negative ions pose a real challenge for a theory to be tested.

TABLE 7.3

Negative of the Energies (Atomic Units) of Some Negative Ions Calculated with the Potential W_x Equation 7.34

Ion	W_x	HF
H$^-$	0.488	0.488
Li$^-$	7.427	7.428
F$^-$	99.455	99.459
Cl$^-$	459.565	459.576

Note: Comparison is made with the corresponding HF energies.

TABLE 7.4

Negative of the Highest Occupied Orbital Eigen Energies (Atomic Units) of Some Negative Ions Calculated with the Potential W_x Equation 7.34

Ion	W_x	HF	Experimental Electron Affinity [20]
H^-	0.046	0.046	0.028
Li^-	0.015	0.015	0.023
F^-	0.178	0.181	0.125
Cl^-	0.144	0.150	0.133

Note: Comparison is made with the corresponding HF energies and experimental electron affinity of the corresponding neutral atoms.

In Table 7.5, we display the results [29] for energies of some excited states of the negative ions Li^- and Be^- of lithium and beryllium. The calculations have been performed both at the exchange-only level by employing the potential W_x alone, and also at the XC level by adding the Lee–Yang–Parr (LYP) [30] correlation potential to W_x. The exchange-only results are compared with those of HF theory [31] and the XC results with other accurate calculations [32]. As these results show, the accuracy of the potential W_x in obtaining the excited-state energies is the same as that for the ground-state energies. In Table 7.6, we show the transition wavelengths corresponding to two transitions in these ions as calculated [29] using the ($W_x + LYP$) potential. A comparison with the experimental numbers [32,33] given in the table shows the results obtained to be highly accurate.

TABLE 7.5

Negative of the Energies (Atomic Units) of Some Excited States of Li^- and Be^- along with the HF and the Fully Correlated Energies

Ion	State	W_x	HF	$W_x + LYP$	Literature
Li^-	$1s2s2p^2(^5P)$	5.364	5.364	5.393	5.383–5.387
	$1s2p^3(^5S)$	5.222	5.223	5.261	5.254–5.256
	$1s2s2p3p(^5P)$	5.329	—	5.368	5.368
Be^-	$[He]2s2p^2(^4P)$	14.508	14.509	14.581	14.571–14.578
	$[He]2p^3(^4S)$	14.327	14.328	14.408	14.400–14.406
	$1s2s2p^3(^6S)$	10.428	10.429	10.476	10.462–10.471

Note: The present results at exchange-only level are given in the third column (W_x) and those with correlation included in the fifth column ($W_x + LYP$).

TABLE 7.6

Transition Wavelengths (in nm) for Transitions in Li⁻ and Be⁻ as Calculated Employing the ($W_x + $ LYP) Potential

Transition	Wavelength ($W_x + $ LYP)	Experiment
Li⁻ $1s2s2p^2(^5P) \rightarrow 1s2p^3(^5S)$	345.96	349.07, 349.0
Be⁻ $[He]2s2p^2(^4P) \rightarrow [He]2p^3(^4S)$	264.14	265.301, 265.318, 265.331

Note: Comparison is made with the corresponding experimental numbers.

As is evident from the above, both the physics invoked to derive the potential of Equation 7.31 and the numerical results presented show that W_x gives an accurate exchange potential for the excited states. When the proposal was initially made, there was no mathematical proof of the existence of a Kohn–Sham equation for excited states. It is only during the past few years that DFT of excited states [34–37], akin to its ground-state counterpart, is being developed.

An important aspect of calculations with potential W_x is that the transition energy of a single-electron excitation is well estimated [38] by the eigen energy difference obtained in a ground-state calculation, of the orbitals involved in transition. The same is not the case in HF theory. This is because in HF theory the eigen energy of the unoccupied orbital corresponds to that of an $(N + 1)$ electron system, whereas in calculations with a local potential this is not the case. Thus the eigen energy differences as obtained in a calculation performed with W_x potential can be taken to give a decent estimate of the energy difference between an excited state and the ground state. For example, experimental transition energies [39] in Na atom for transitions $3^2S \rightarrow 4^2S$, $3^2S \rightarrow 5^2S$, $3^2S \rightarrow 3^2P$, and $3^2S \rightarrow 4^2P$ are 0.117, 0.151, 0.077, and 0.138 a.u., respectively. The corresponding W_x eigenvalue differences are [31] 0.115, 0.153, 0.079, and 0.141 a.u.

Having shown the correctness of physics in calculating the exchange potential as the work done in moving an electron in the electric field of its Fermi hole, we next discuss if the XC potential could also be obtained in the same way from the Fermi–Coulomb hole. We will see that in calculating the XC potential in the prescribed manner, the kinetic energy contribution to it, as indicated by Equation 7.13 is left out. Following that we discuss the work of Holas and March [40] who proved that the XC potential can indeed be thought of as the work done in moving an electron in a field, which is the sum of the field of its Fermi–Coulomb hole and a very small field arising from the difference as given by Equation 7.12 in the kinetic energy of an interacting and a noninteracting system, and in the process deriving an expression for the latter. Further, we will see that it is precisely this difference in the kinetic energy that is responsible [41] for the slight difference in the energies of atoms obtained by employing the W_x potential and those of HF theory.

7.3.2 EXCHANGE-CORRELATION POTENTIAL

If we wish to obtain the XC potential of Equation 7.15 as the work done in moving an electron in the field of its Fermi–Coulomb hole $\rho_{xc}(\vec{r}, \vec{r}')$, we first calculate the field as

$$\vec{F}_{xc}^{DFT}(\vec{r}) = \int \frac{\rho_{xc}^{DFT}(\vec{r}, \vec{r}')}{|\vec{r} - \vec{r}'|^3} (\vec{r} - \vec{r}')d\vec{r}' \tag{7.37}$$

and then obtain the potential

$$W_{xc}(\vec{r}) = -\int_{\infty}^{\vec{r}} \vec{F}_{xc}^{DFT}(\vec{r}') \cdot d\vec{l}' \tag{7.38}$$

by doing a line integration. Let us first see if the potential derived in this manner satisfies the Levy–Perdew relationship [23], similar to the one in Equation 7.35, for the XC potential. The relationship for the XC potential is given as

$$E_{xc}^{DFT} = -\int \rho(\vec{r})\vec{r} \cdot \vec{\nabla} v_{xc}(\vec{r})d\vec{r} + T_c \tag{7.39}$$

where T_c is given by Equation 7.12. Now if we substitute the expression for $\vec{F}_{xc}^{DFT}(\vec{r})$ of Equation 7.37 for the gradient $\vec{\nabla} v_{xc}(\vec{r})$ of the XC potential in Equation 7.39 and use the symmetry of the product $\rho(\vec{r})\rho_{xc}^{DFT}(\vec{r}, \vec{r}')$ with respect to an interchange of $\vec{r}$ and $\vec{r}'$, we get

$$T_c = 0 \tag{7.40}$$

which is not correct [42]. Thus the definition (Equation 7.38) for the exchange-correlation potential misses out on the kinetic energy component of the XC energy and potential of DFT. In other words, the definition (Equation 7.38) represents the potential only due to the quantum mechanical exchange-correlation hole $\rho_{xc}^{QM}(\vec{r}, \vec{r}')$. In the original work, it was therefore proposed [43] that this component has to be added separately and would represent the functional derivative $\delta T_c[\rho]/\delta\rho(\vec{r})$, i.e., the kinetic energy component of the DFT exchange-correlation potential. How this component is obtained from the many-particle wave function became clear only after Holas and March [40] derived the differential virial theorem and on the basis of it, obtained an expression for the XC potential as a line integral. This, in turn, also gives a mathematical derivation starting from the many-electron Schrödinger equation of our physical interpretation of the XC potential. Next we discuss the derivation by Holas and March.

7.4 DIFFERENTIAL VIRIAL THEOREM AND EXCHANGE-CORRELATION POTENTIAL

We now look at how the ideas developed so far on the basis of purely physical arguments, can be derived starting from the many-body Schrödinger equation. This is done through the differential form of the virial theorem. Consider the Hamiltonian of Equation 7.1 and its eigenfunction $\Psi(\vec{r}_1, \vec{r}_2, \ldots, \vec{r}_N)$ for a bound state. Then the differential virial theorem relates the gradient $\vec{\nabla}v_{xc}(\vec{r})$ of the external potential to the wave function through the following relationship:

$$-\vec{\nabla}v_{\text{ext}}(\vec{r}) = -\frac{1}{4\rho(\vec{r})}\vec{\nabla}\nabla^2\rho(\vec{r}) + \frac{\vec{z}(\vec{r})}{\rho(\vec{r})} - \int \frac{(\vec{r}-\vec{r}')}{|\vec{r}-\vec{r}'|^3}\left(\rho(\vec{r}') + \rho_{\text{xc}}^{\text{QM}}(\vec{r},\vec{r}')\right)d\vec{r}' \quad (7.41)$$

where $\rho_{\text{xc}}^{\text{QM}}(\vec{r},\vec{r}')$ is the XC hole as given by Equation 7.17, and vector $\vec{z}(\vec{r})$ is related to the kinetic energy tensor defined next. The kinetic energy tensor

$$t_{\alpha\beta}(\vec{r}) = \frac{1}{4}\int \left(\begin{matrix} \frac{\partial}{\partial r_{1\alpha}}\Psi^*(\vec{r}_1, \vec{r}_2, \vec{r}_3 \ldots \vec{r}_N)\frac{\partial}{\partial r_{1\beta}}\Psi(\vec{r}_1, \vec{r}_2, \vec{r}_3 \ldots \vec{r}_N) \\ + \frac{\partial}{\partial r_{1\beta}}\Psi^*(\vec{r}_1, \vec{r}_2, \vec{r}_3 \ldots \vec{r}_N)\frac{\partial}{\partial r_{1\alpha}}\Psi(\vec{r}_1, \vec{r}_2, \vec{r}_3 \ldots \vec{r}_N) \end{matrix} \right) d\vec{r}_2 d\vec{r}_3 \ldots d\vec{r}_N \bigg|_{\vec{r}_1=\vec{r}_2=\vec{r}} \quad (7.42)$$

where $\alpha, \beta = 1, 2, 3$ stand for the Cartesian components. The component z_α of the vector $\vec{z}(\vec{r})$ is related to the kinetic energy tensor through the relationship

$$z_\alpha = 2\sum_\beta \frac{\partial}{\partial r_\beta}t_{\alpha\beta}(\vec{r}) \quad (7.43)$$

The quantity $t_{\alpha\beta}(\vec{r})$ is known as the kinetic energy tensor because the total kinetic energy T is related to it through $\vec{z}(\vec{r})$ as

$$2T = \int \rho(\vec{r})\vec{r} \cdot \vec{z}(\vec{r})d\vec{r} \quad (7.44)$$

The reason why the relationship in Equation 7.41 is called the differential virial theorem is because if we take the dot product of both sides with vector $\vec{r}$, multiply both sides by $\rho(\vec{r})$, and then integrate over the entire volume, it gives

$$2T = \left\langle \Psi \left| \sum_i \vec{r}_i \cdot \vec{\nabla}_i \left(v_{\text{ext}}(\vec{r}_i) + \sum_{\substack{j \\ j\neq i}} \frac{1}{|\vec{r}_i - \vec{r}_j|} \right) \right| \Psi \right\rangle \quad (7.45)$$

which is the global virial theorem.

Our aim here is to apply the differential virial theorem to get an expression for the Kohn–Sham XC potential. To this end, we assume that a noninteracting system giving the same density as that of the interacting system exists. This system satisfies Equation 7.4, i.e., the Kohn–Sham equation. Since the total potential term of Kohn–Sham equation is the external potential for the noninteracting system, application of the differential virial relationship of Equation 7.41 to this system gives

$$-\vec{\nabla}v_{ext}(\vec{r}) + \int \frac{(\vec{r} - \vec{r}')}{|\vec{r} - \vec{r}'|^3} \rho(\vec{r}')d\vec{r}' - \vec{\nabla}v_{xc}(\vec{r}) = -\frac{1}{4\rho(\vec{r})}\vec{\nabla}\nabla^2\rho(\vec{r}) + \frac{\vec{z}_S(\vec{r})}{\rho(\vec{r})} \qquad (7.46)$$

Here the vector $\vec{z}_S(\vec{r})$ is constructed from the kinetic energy tensor obtained by employing the solutions of the Kohn–Sham equation in Equation 7.42. Thus it is, in general, different from the vector $\vec{z}(\vec{r})$. A comparison of Equations 7.41 and 7.46 gives

$$\vec{\nabla}v_{xc}(\vec{r}) = \frac{\vec{z}(\vec{r}) - \vec{z}_S(\vec{r}')}{\rho(\vec{r})} - \int \frac{(\vec{r} - \vec{r}')}{|\vec{r} - \vec{r}'|^3}\rho_{xc}^{QM}(\vec{r}, \vec{r}')d\vec{r}' \qquad (7.47)$$

Performing a line integral of the vector field given on the right-hand side of the equation above leads to the exact XC potential of a system. We now compare this expression with that of Equation 7.38 that calculates the exchange-correlation potential as the work done in moving an electron in the field of its XC hole.

The exact expression of Equation 7.47 also calculates the XC potential as the work done in moving an electron in a field that is a sum of the electric field due to its Fermi–Coulomb hole and a non-Coulombic field related to the difference in the kinetic energy tensor of the interacting and Kohn–Sham systems. It is easily verified that the potential given by Equation 7.47 satisfies the Levy–Perdew relationship of Equation 7.39. Further, the curl of the expression on its right-hand side vanishes because it represents the gradient of a scalar function. As noted in the beginning, the value of the difference in the kinetic energies of the interacting and Kohn–Sham systems is very small. Thus, although numerically the term $(\vec{z}(\vec{r}) - \vec{z}_S(\vec{r}')/\rho(\vec{r}))$ represents a very small correction to the potential of Equation 7.38, it is significant for important qualitative reasons.

As noted above, the curl of the expression on the right-hand side of Equation 7.47 vanishes. However, it does not mean that the Coulombic and non-Coulombic components—the former is the electric field produced by the Fermi–Coulomb hole and the latter arises from the kinetic energy tensor—of this field also have vanishing curl. Thus the potential W_{xc} of Equation 7.38 may sometimes be path dependent [21].

7.4.1 DIFFERENTIAL VIRIAL THEOREM AND HARTREE–FOCK THEORY

Now we discuss the differential virial theorem for HF theory and the corresponding Kohn–Sham system. The Kohn–Sham system in this case is constructed [41] to reproduce the HF density. As such, the equation describing the differential virial theorem in HF theory is

$$-\vec{\nabla}v_{\text{ext}}(\vec{r}) = -\frac{1}{4\rho_{\text{HF}}(\vec{r})}\vec{\nabla}\nabla^2\rho_{\text{HF}}(\vec{r}) + \frac{\vec{z}_{\text{HF}}(\vec{r})}{\rho_{\text{HF}}(\vec{r})} - \int\frac{(\vec{r}-\vec{r}')}{|\vec{r}-\vec{r}'|^3}$$
$$\times\ (\rho_{\text{HF}}(\vec{r}') + \rho_x(\vec{r}, \vec{r}'))d\vec{r}' \qquad (7.48)$$

where the subscript "HF" implies that all the quantities in the equation above have been calculated by employing HF orbitals. Similarly, the XC hole $\rho_{\text{xc}}(\vec{r}, \vec{r}')$ is now replaced by the Fermi hole $\rho_x(\vec{r}, \vec{r}')$ calculated by using HF orbitals in Equation 7.28. The local exchange potential $v_x(\vec{r})$ whose orbitals give the same density then has the gradient

$$\vec{\nabla}v_x(\vec{r}) = \frac{\vec{z}_{\text{HF}}(\vec{r}) - \vec{z}_S(\vec{r}')}{\rho_{\text{HF}}(\vec{r})} - \int\frac{(\vec{r}-\vec{r}')}{|\vec{r}-\vec{r}'|^3}\rho_x(\vec{r}, \vec{r}')d\vec{r}' \qquad (7.49)$$

Thus for the exchange potential also there is a non-Coulombic contribution arising from the difference in the kinetic energy of the HF theory and the corresponding Kohn–Sham theory, although both the theories are based on single-particle orbitals. This is expected as the Kohn–Sham system is constructed to reproduce the HF density only and it does not mean that the corresponding kinetic energy will also be equal to the HF kinetic energy. It is because of this difference that the exchange potential W_x of Equation 7.34, calculated from the Fermi hole alone, may also be path dependent. However, as the results presented in Sections 7.3.1.1 and 7.3.1.2 show, this difference is not very significant numerically.

7.5 CONCLUDING REMARKS

In this chapter, we have discussed how the XC potential of Kohn–Sham theory can be interpreted in a physical way. Before this work, the potential was thought of purely in mathematical terms as the functional derivative of the XC energy functional; within the exchange-only theory, the local exchange potential was generated [44] numerically by looking for a local potential whose orbitals minimize the HF expression for the energy. Our work therefore provides an alternate way of understanding the exchange and XC potential. Further, the potential is absolutely general and can be applied to the ground as well as excited states. In fact, when the interpretation was proposed, there was no other theory that could be applied to perform excited-state calculations with a local potential.

Mathematical derivation of the potential as the work done in moving an electron in a field is provided through differential virial theorem. The theorem proves that the XC potential of Kohn–Sham theory is indeed the work done in moving an electron in the field of its Fermi–Coulomb hole plus a non-Coulombic field related to the difference in the kinetic energies of the interacting and the corresponding Kohn–Sham system. This way of looking at Kohn–Sham theory has also been given the name *Quantal Density-Functional Theory*, and many results derived and understood (as an example see Refs. [45,46]) on the basis of the physical interpretation are discussed in an eponymous book [47].

ACKNOWLEDGMENT

I thank Professor Pratim Chattaraj for inviting me to write a chapter in this book.

REFERENCES

1. Parr, R.G. and Yang, W. *Density Functional Theory of Atoms and Molecules.* Oxford University Press, Oxford, 1989.
2. Dreizler, R.M. and Gross, E.K.U. *Density Functional Theory*, Springer, Berlin, 1990.
3. Hohenberg, P. and Kohn, W. *Phys. Rev.* 1964, 136, B864.
4. Perdew, J.P., Parr, R.G., Levy, M., and Balduz, J.L. *Phys. Rev. Lett.* 1982, 49, 1691.
5. Kohn, W. and Sham, L.J. *Phys. Rev.* 1965, 140, A1133.
6. Morrel, M.M., Parr, R.G., and Levy, M. *J. Chem. Phys.* 1975, 62, 549.
7. Katriel, J. and Davidson, E.R. *Proc. Natl. Acad. Sci. USA* 1980, 77, 4403.
8. Levy, M., Perdew, J.P., and Sahni, V. *Phys. Rev. A* 1984, 30, 2745.
9. Almbladh, C.O. and von Barth, U. *Phys. Rev. B* 1985, 31, 3231.
10. Tal, Y. *Phys. Rev. A* 1978, 18, 1781.
11. Shamim, Md. and Harbola, M.K. *Chem. Phys. Rev. A.*
12. Zhao, Q. and Parr, R.G. *J. Chem. Phys.* 1993, 98, 543.
13. Bunge, V. and Esquivel, R.O. *Phys. Rev. A* 1986, 34, 853.
14. Striganov, A.R. and Sventitskii, N.S. *Tables of Spectral Lines of Neutral and Ionized Atoms*, IFI/Plenum, New York, 1968.
15. Morrison, R.C. and Zhao, Q. *Phys. Rev. A* 1995, 51, 1980.
16. Harbola, M.K. and Sahni, V. *J. Chem. Edu.* 1993, 70, 920.
17. See Kohn, W. and Vashishta, P. in *Theory of Inhomogeneous Electron Gas*, Lundquist, S. and March N.H. (Eds.), Plenum, New York, 1983.
18. Slater, J.C. *Phys. Rev.* 1951, 81, 385.
19. Clementi, E. and Roetti, C. *Atomic Data and Nuclear Data Tables,* 1974, 14, 177.
20. Harbola, M.K. and Sahni, V. *Phys. Rev. Lett.* 1989, 62, 489; Sahni, V. and Harbola, M.K. *Int. J. Quant. Chem. S.* 1990, 24, 569.
21. Ou-Yang, H. and Levy, M. *Phys. Rev. A* 1990, 41, 4038.
22. Harbola, M.K., Slamet, M., and Sahni, V. *Phys. Lett. A* 1991, 157, 60; Slamet, M., Sahni, V., and Harbola, M.K. *Phys. Rev. A* 1994, 49, 809.
23. Levy, M. and Perdew, J.P. *Phys. Rev. A* 1985, 32, 2010.
24. Sahni, V., Li, Y., and Harbola, M.K. *Phys. Rev. A* 1992, 45, 1434 (1992).
25. Fischer, C.F. *The Hartree–Fock Method for Atoms*, Wiley, New York, 1977.
26. Sen, K.D. and Harbola, M.K. *Chem. Phys. Lett.* 1991, 178, 347.
27. Hotop, H. and Lineberger, W.C. *J. Phys. Chem. Ref. Data* 1975, 4, 539.
28. Singh, R. and Deb, B.M. *Phys. Rep.* 1999, 311, 47.
29. Roy, A.K. and Jalbout, A.F. *Chem. Phys. Lett.* 2007, 445, 355.
30. Lee, C., Yang, W., and Parr, R.G. *Phys. Rev. B* 1988, 37, 785.
31. Gálvez, F.J., Buendiá, E., and Sarsa, A. *Eur. Phys. J D* 2006, 40, 161; Beck, D.R., Nicolaides, C.A., and Aspromallis, G. *Phys. Rev. A* 1981, 24, 3252.
32. Berry, H.G., Bromander J., and Buchta, R. *Nucl. Instrum. Meth.* 1970, 90, 269; Mannervik, S., Astner, G., and Kisielinski, M. *J. Phys. B* 1980, 13, L441.

33. Gaardsted, J.O. and Andersen, T. *J. Phys. B* 1989, 22, L57; Kristensen, P., Petrunin, V.V., Andersen, H.H., and Andersen, T. *Phys. Rev. A* 1995, 52, R2508; Andersen, H.H., Balling, P., Petrunin, V.V., and Andersen, T. *J. Phys. B* 1996, 29, L415.
34. Görling, A. *Phys. Rev. A* 1999, 59, 3359.
35. Levy M. and Nagy, Á. *Phys. Rev. Lett.* 1999, 83, 4361; Nagy, Á. and Levy, M. *Phys. Rev. A* 2001, 63, 052502.
36. Harbola, M.K. *Phys. Rev. A* 2004, 69, 042512.
37. Samal, P. and Harbola, M.K. *J. Phys. B* 2005, 38, 3765; Samal, P. and Harbola, M.K. *J. Phys. B* 2006, 39, 4065.
38. Sen, K.D. *Chem. Phys. Lett.* 1992, 188, 510.
39. Bashkin, S. and Stoner, J.O. *Atomic Energy Levels and Grotarian Diagrams*, Vol. 1, North-Holland, Amsterdam, 1975.
40. Holas, A. and March, N.H. *Phys. Rev. A* 1995, 55, 2040.
41. Sahni, V. *Phys. Rev. A* 1997, 55, 1846.
42. Nagy, Á. *Phys. Rev. Lett.* 1990, 65, 2608.
43. Harbola, M.K. and Sahni, V. *Phys. Rev. Lett.* 1990, 65, 2609; Harbola, M.K. Ph.D. Thesis, City University of New York, 1989.
44. Sharp, R.T. and Horton, G.K. *Phys. Rev.* 1953, 90, 3876; Talman, J.D. and Shadwick, W.F. *Phys. Rev. A* 1976, 14, 36.
45. Harbola, M.K. *Phys. Rev. A* 1998, 57, 4253.
46. Qian, Z. and Sahni, V. *Phys. Lett. A* 1998, 248, 393.
47. Sahni, V. *Quantal Density Functional Theory*, Springer, Berlin, 2004.

8 Time-Dependent Density Functional Theory from a Bohmian Perspective

Ángel S. Sanz, Xavier Giménez, Josep Maria Bofill, and Salvador Miret-Artés

CONTENTS

8.1 INTRODUCTION

Since the early days of quantum mechanics, the wave function theory has proven to be very successful in describing many different quantum processes and phenomena. However, in many problems of quantum chemistry and solid-state physics, where the dimensionality of the systems studied is relatively high, ab initio calculations of the structure of atoms, molecules, clusters, and crystals, and their interactions are very often prohibitive. Hence, alternative formulations based on the direct use of the probability density, gathered under what is generally known as the density matrix theory [1], were also developed since the very beginning of the new mechanics. The independent electron approximation or Thomas–Fermi model, and the Hartree and Hartree–Fock approaches are former statistical models developed in that direction [2]. These models can be considered direct predecessors of the more recent density functional theory (DFT) [3], whose principles were established by Hohenberg,

Kohn, and Sham [4,5] in the mid-1960s. According to this theory, the fundamental physical information about a many-body system is provided by single-particle densities in a three-dimensional space, which are obtained variationally within a time-independent framework. When compared with other previous formalisms, DFT presents two clear advantages: (1) it is able to treat many-body problems in a sufficiently accurate way and (2) it is computationally simple. This explains why it is one of the most widely used theories to deal with electronic structure—the electronic ground-state energy as a function of the position of the atomic nuclei determines molecular structures and solids, providing at the same time the forces acting on the atomic nuclei when they are not at their equilibrium positions. At present, DFT is used routinely to solve many problems in gas phase and condensed matter. Furthermore, it has made possible the development of accurate molecular dynamics schemes in which the forces are evaluated quantum mechanically "on the fly." Nonetheless, DFT is a fundamental tool provided the systems studied are relatively large; for small systems, standard methods based on the use of the wave function render quite accurate results [6]. Moreover, it is also worth stressing that all practical applications of DFT rely on essentially uncontrolled approximations [7] (e.g., the local density approximation [4,5], the local spin-density approximation, or generalized gradient approximations [8]), and therefore the validity of DFT is conditioned to its capability to provide fairly good values of the experimental data.

As mentioned above, standard DFT is commonly applied to determine ground states in time-independent problems. Hence, reactive and nonreactive scatterings as well as atoms and molecules in laser fields have been out of the reach of the corresponding methodology. Nevertheless, though it is less known than the standard DFT, a very interesting work in this direction can also be found in the literature [9–15], where DFT is combined with quantum hydrodynamics (or quantum fluid dynamics [QFD]) (QFD-DFT) in order to obtain a quantum theory of many-electron systems. In this case, the many-electron wave function is replaced by single-particle charge and current densities. The formal grounds of QFD-DFT rely on a set of hydrodynamical equations [10–12]. It has the advantage of dealing with dynamical processes evolving in time in terms of single-particle time-dependent (TD) equations, as derived by different authors [14]. Apart from QFD-DFT, there are other TD-DFT approaches based on similar grounds, such as the Floquet DFT [16,17] or the quantal DFT [18]. Furthermore, we would like to note that TD-DFT does not necessarily require to pass through a QFD or QFD-like formulation in order to be applied [19]. As happens with standard DFT, TD-DFT can also be started directly from the many-body TD Schrödinger equation, the density being determined from solving a set of TD Schrödinger equations for single, noninteracting particles [12].

Although trajectories are not computed in QFD-DFT, it is clear that there is a strong connection between this approach and the trajectory or hydrodynamical picture of quantum mechanics [20], independently developed by Madelung [21], de Broglie [22], and Bohm [23], which is also known as Bohmian mechanics. From the same hydrodynamical equations, information not only about the system configuration (DFT calculations) but also about its dynamics (quantum trajectories) is possible to obtain. This fact is better understood when the so-called quantum potential is considered, as it allows to associate the probability density (calculated from DFT) with the quantum trajectories. Note that this potential is determined by the curvature of the probability density and, at the same time, it governs the behavior displayed by the quantum trajectories. Because of the interplay between probability density and quantum potential, the latter conveys fundamental physical information: it transmits the nonseparability contained in the probability density (or, equivalently, the wave function) to the particle dynamics. This property, on the other hand, is connected with the inherent nonlocality of quantum mechanics [24], i.e., two distant parts of an entangled or nonfactorizable system will keep a strong correlation due to coherence exhibited by its quantum evolution.

The purpose of this chapter is to show and discuss the connection between TD-DFT and Bohmian mechanics, as well as the sources of lack of accuracy in DFT, in general, regarding the problem of correlations within the Bohmian framework or, in other words, of entanglement. In order to be self-contained, a brief account of how DFT tackles the many-body problem with spin is given in Section 8.2. A short and simple introduction to TD-DFT and its quantum hydrodynamical version (QFD-DFT) is presented in Section 8.3. The problem of the many-body wave function in Bohmian mechanics, as well as the fundamental grounds of this theory, are described and discussed in Section 8.4. This chapter is concluded with a short final discussion in Section 8.5.

8.2 MANY-BODY PROBLEM IN STANDARD DFT

There are many different physical and chemical systems of interest, which are characterized by a relatively large number of degrees of freedom. However, in most of the cases, the many-body problem can be reduced to calculations related to a sort of inhomogeneous gas, i.e., a set of interacting point-like particles which evolve quantum mechanically under the action of a certain effective potential field. This is the typical DFT scenario, with an ensemble of N electrons in a nuclear or external potential representing the system of interest. DFT thus tries to provide an alternative approach to the exact, nonrelativistic N-electron wave function $\Psi(\mathbf{r}_1 s_1, \ldots, \mathbf{r}_N s_N)$, which satisfies the time-independent Schrödinger equation and where $\mathbf{r}_N$ and s_N are the space and spin coordinates, respectively. Because the methodology based on DFT is easy and computationally efficient in its implementation, this theory is still enjoying an ever-increasing popularity within the physics and chemistry communities involved in many-body calculations.

To understand the main idea behind DFT, consider the following. In the absence of magnetic fields, the many-electron Hamiltonian does not act on the electronic spin coordinates, and the antisymmetry and spin restrictions are directly imposed on the wave function $\Psi(\mathbf{r}_1 s_1, \ldots, \mathbf{r}_N s_N)$. Within the Born–Oppenheimer approximation, the energy of an N-electron system with a fixed M-nuclei geometry $\mathbf{R}$ takes the following form in atomic units:

$$E = -\frac{1}{2} \int\limits_{\mathbf{r}_1=\mathbf{r}_1'} [\nabla \cdot \nabla^T \gamma_1(\mathbf{r}_1; \mathbf{r}_1')] d\mathbf{r}_1 + \int v_{\text{ext}}(\mathbf{R}, \mathbf{r}_1)\gamma_1(\mathbf{r}_1)d\mathbf{r}_1$$

$$+ \int \frac{\gamma_2(\mathbf{r}_1, \mathbf{r}_2)}{r_{12}} d\mathbf{r}_1 d\mathbf{r}_2, \tag{8.1}$$

where $\gamma_1(\mathbf{r}_1)$ and $\gamma_2(\mathbf{r}_1, \mathbf{r}_2)$ are the diagonal elements of $\gamma_1(\mathbf{r}_1; \mathbf{r}_1')$ and $\gamma_2(\mathbf{r}_1, \mathbf{r}_2; \mathbf{r}_1', \mathbf{r}_2')$, respectively, which represent the one-electron (or one-particle) density and the electron–electron (or two-particle) correlation function, commonly used in DFT and electronic structure theory. In principle, it might seen that all the information about the system, necessary to evaluate the energy, is contained in $\gamma_1(\mathbf{r}_1)$ and $\gamma_2(\mathbf{r}_1; \mathbf{r}_2)$, and therefore one could forget about manipulating the wave function. However, in order to avoid unphysical results in the evaluation of the energy, it is still necessary to compute the wave function $\Psi(\mathbf{r}_1 s_1, \dots, \mathbf{r}_N s_N)$ that generates the correct $\gamma_1(\mathbf{r}_1)$ and $\gamma_2(\mathbf{r}_1; \mathbf{r}_2)$ densities. Equation 8.1 is the starting point of DFT, which aims to replace both $\gamma_1(\mathbf{r}_1; \mathbf{r}_1')$ and $\gamma_2(\mathbf{r}_1, \mathbf{r}_2)$ by $\rho(\mathbf{r})$. If we are only interested in the system ground state, the Hohenberg–Kohn theorems state that the exact ground-state total energy of any many-electron system is given by a universal, unknown functional of the one-electron density. However, only the second term of Equation 8.1 is an explicit functional of $\rho(\mathbf{r})$. The first term corresponds to the kinetic energy, which is a functional of the complete one-electron density function $\gamma_1(\mathbf{r}_1; \mathbf{r}_1')$. For N-electron systems, the most important contribution to the electron–electron term comes from the classical electrostatic self-energy of the charge interaction, which is an explicit functional of the diagonal one-electron function. The remaining contribution to the electron–electron term is still unknown. These two terms are a functional of the one-electron density, namely the "exchange-correlation" functional. Thus, it is possible to define a universal functional, which is derivable from the one-electron density itself and with no reference to the external potential $v_{\text{ext}}(\mathbf{R}, \mathbf{r})$. According to McWeeny [25], we can reformulate the DFT by ensuring not only that a variational procedure leads to $\rho(\mathbf{r})$—which is derivable from a wave function $\Psi(\mathbf{r}_1 s_1, \dots, \mathbf{r}_N s_N)$ (the so-called N-representability problem)—but also the wave function belongs to the totally irreducible representation of the spin permutation group A. From a mathematical point of view, the above proposition can be expressed (in atomic units) as

$$E = \min_{\substack{\rho \to \gamma_1 \text{ derived from } \Psi \in A}} \left\{ -\frac{1}{2} \int\limits_{\mathbf{r}_1=\mathbf{r}_1'} [\nabla \cdot \nabla^T \gamma_1(\mathbf{r}_1; \mathbf{r}_1')] d\mathbf{r}_1 + \int v_{\text{ext}}(\mathbf{R}, \mathbf{r}_1)\gamma_1(\mathbf{r}_1)d\mathbf{r}_1 \right.$$

$$\left. + \frac{1}{2} \int \frac{\gamma_1(\mathbf{r}_1)(L - \hat{P}_{12})\gamma_1(\mathbf{r}_2; \mathbf{r}_2')}{r_{12}} d\mathbf{r}_1 d\mathbf{r}_2 + \min_{\substack{\gamma_2 \text{ derived from } \Psi \in A}} E_{\text{corr}}[\gamma_2(\mathbf{r}_1, \mathbf{r}_2)] \right\}. \tag{8.2}$$

This equation shows the relationship between the one-electron function, $\gamma_1(\mathbf{r}_1; \mathbf{r}_1')$, and the main part of the energy functional—the rest of the functional, which is the electron–electron repulsion, depends on $\gamma_2(\mathbf{r}_1, \mathbf{r}_2)$. The last term is also a functional of the one-electron density. In the new reformulation of DFT, the methodology is almost universally based on the Kohn–Sham (KS) approach and only differs in a particular way to model the unknown "exchange-correlation" term.

8.3 TIME-DEPENDENT DENSITY FUNCTIONAL THEORY

An extension of standard DFT is its TD version. This generalization is necessary when dealing with intrinsic TD phenomena. In addition, it preserves the appealing flavor of the classical approach to the theory of motion.

The rigorous foundation of the TD-DFT was started with the works by Bartolotti [10] and Deb and Ghosh [11]. However, the proofs of the fundamental theorems were provided by Runge and Gross [12]. One of those theorems corresponds to a Hohenberg–Kohn-like theorem for the TD Schrödinger equation. The starting point for the derivation of the TD KS equations is the variational principle for the quantum mechanical action (throughout this section, atomic units are also used):

$$S[\Psi] = \int_{t_0}^{t_1} \left\langle \Psi(t) \left| \left[i \frac{\partial}{\partial t} - \hat{H}(t) \right] \right| \Psi(t) \right\rangle dt. \tag{8.3}$$

This variational principle is not based on the total energy because in TD systems, the total energy is not conserved. The so-called Runge–Gross theorem then states that there exists a one-to-one mapping between the external potential (in general, TD), $v_{\text{ext}}(\mathbf{r}, t)$, and the electronic density, $\rho(\mathbf{r}, t)$, for many-body systems evolving from a fixed initial state, $\Psi(t_0)$. Runge and Gross thus open the possibility of deriving the TD version of the KS equations. This procedure yields the TD Schrödinger equation for the KS electrons described by the orbitals $\phi_k(\mathbf{r}, t)$,

$$i \frac{\partial \phi_k(\mathbf{r}, t)}{\partial t} = H_{\text{KS}}(\mathbf{r}, t) \phi_k(\mathbf{r}, t), \tag{8.4}$$

where the KS Hamiltonian is

$$H_{\text{KS}}(\mathbf{r}, t) = -\frac{1}{2} \nabla^2 + v_{\text{KS}}[\rho(\mathbf{r}, t)], \tag{8.5}$$

with a TD-KS effective potential, usually given by the sum of three terms which account for external, classical electrostatic, and exchange interactions. The latter is the source of all nontrivial, nonlocal, strongly correlated many-body effects.

By construction, the exact TD density of the interacting system can then be calculated from a set of noninteracting, single-particle orbitals fulfilling the TD-KS Equation 8.4 and reads

$$\rho(\mathbf{r}, t) = \sum_{k=1}^{N} |\phi_k(\mathbf{r}, t)|^2. \tag{8.6}$$

Further analysis from the minimum action principle shows that the exchange (xc) potential is then the functional derivative of that quantity in terms of the density:

$$v_{xc}(\mathbf{r}, t) = \frac{\delta S_{xc}}{\delta \rho(\mathbf{r}, t)},\tag{8.7}$$

where S_{xc} includes all nontrivial many-body parts of the action. The above equations provide the starting ground for further derivations of the theory. Thus, in addition to the TD-KS scheme, other variants have been proposed over the years, which include the TD spin-DFT, the TD current-DFT, the TD linear response DFT, and the basis-set DFT [26]. Each method has its range of applicability, but discussing them is out of the scope of this chapter.

Here we focus on yet another implementation, the single-particle hydrodynamic approach or QFD-DFT, which provides a natural link between DFT and Bohmian trajectories. The corresponding derivation is based on the realization that the density, $\rho(\mathbf{r}, t)$, and the current density, $\mathbf{j}(\mathbf{r}, t)$ satisfy a coupled set of "classical fluid," Navier–Stokes equations:

$$\frac{\partial \rho(\mathbf{r}, t)}{\partial t} = -\nabla \mathbf{j}(\mathbf{r}, t),\tag{8.8}$$

$$\frac{\partial \mathbf{j}(\mathbf{r}, t)}{\partial t} = \mathbf{P}[\rho](\mathbf{r}, t),\tag{8.9}$$

with

$$\mathbf{P}[\rho](\mathbf{r}, t) = -i\langle \Psi[\rho](t)|[\mathbf{j}(\mathbf{r}), H(t)]|\Psi[\rho](t)\rangle,\tag{8.10}$$

being a functional of the density and with initial conditions $\rho(\mathbf{r}, t_0)$ and $\mathbf{j}(\mathbf{r}, t_0)$.

One can finally show that the above coupled equations translate into one single-particle nonlinear differential equation for the hydrodynamical wave function $\Phi(\mathbf{r}, t) = \rho(\mathbf{r}, t)^{1/2} e^{iS(\mathbf{r},t)}$ in terms of potential energy functionals:

$$\left(-\frac{1}{2}\nabla^2 + v_{\text{eff}}[\rho]\right)\Phi(\mathbf{r}, t) = i\frac{\partial \Phi(\mathbf{r}, t)}{\partial t},\tag{8.11}$$

with $v_{\text{eff}}[\rho]$ given by

$$v_{\text{eff}}[\rho] = \frac{\delta E_{\text{el−el}}}{\delta \rho} + \frac{\delta E_{\text{nu−el}}}{\delta \rho} + \frac{\delta E_{xc}}{\delta \rho} + \frac{\delta T_{\text{corr}}}{\delta \rho} + \frac{\delta E_{\text{ext}}}{\delta \rho}\tag{8.12}$$

where $\mathbf{j}(\mathbf{r}, t) = \rho(\mathbf{r}, t)\mathbf{v}(\mathbf{r}, t)$, with $\nabla S(\mathbf{r}, t) = \mathbf{v}(\mathbf{r}, t)$. For many-particle systems, this is still an open problem (see Section 8.4.2 for a new discussion). In Equation 8.12, each term corresponds, respectively, to the interelectronic repulsion energy, the Coulomb nuclear–electron attraction energy, the exchange and correlation energy, the non-classical correction term to Weizsäcker's kinetic energy, and the electron–external field interaction energy functionals. A judicious choice in the form of the above functionals yields surprisingly good results for selected applications.

As a simple mathematical approach to QFD-DFT, let us consider that the N-electron system is described by the TD orbitals $\phi_k(\mathbf{r}, t)$ when there is an external periodic, TD potential, for which we want to obtain the (TD) density $\rho(\mathbf{r}, t)$. These orbitals can be expressed in polar form:

$$\phi_k(\mathbf{r}, t) = R_k(\mathbf{r}, t)e^{iS_k(\mathbf{r},t)}, \tag{8.13}$$

where the amplitudes $R_k(\mathbf{r}, t)$ and phases $S_k(\mathbf{r}, t)$ are real functions of space and time, and the former are subjected to the normalization condition:

$$\iint_t R_k(\mathbf{r}, t)R_l(\mathbf{r}, t)d\mathbf{r} = \delta_{kl}, \tag{8.14}$$

where $\int_t$ denotes the time-averaged integration over one period of time. The kinetic energy associated with this (noninteracting) N-electron system reads [10] as

$$T_s[\{R_k, S_k\}]_t = -\frac{1}{2}\sum_{k=1}^{N}\iint_t R_k(\mathbf{r}, t)\left[\nabla^2 R_k(\mathbf{r}, t)\right] - R_k(\mathbf{r}, t)\left[\nabla S_k(\mathbf{r}, t)^2\right]d\mathbf{r}. \tag{8.15}$$

Similar to the time-independent case, we also assume the constraint that the sum of the squares of the R_k gives the exact density $\rho(\mathbf{r}, t)$, i.e.,

$$\sum_{k=1}^{N} R_k^2(\mathbf{r}, t) = \rho(\mathbf{r}, t). \tag{8.16}$$

Moreover, we introduce an additional constraint: the conservation of the number of particles,

$$\sum_{k=1}^{N}\frac{\partial R_k^2}{\partial t}\left(=\frac{\partial \rho}{\partial t}\right) = -\nabla \cdot \mathbf{j}, \tag{8.17}$$

where $\mathbf{j}$ is the single-particle quantum density current vector. After minimizing Equation 8.15 with respect to the R_k (which is subject to the previous constraints), we reach the Euler–Lagrange equation:

$$-\frac{1}{2}\nabla^2 R_k + v_{\text{eff}}R_k = \varepsilon_k R_k. \tag{8.18}$$

where $v_{\text{eff}}(\mathbf{r}, t)$ and $\varepsilon_k(\mathbf{r}, t)$ are the Lagrange multipliers associated with the constraint defined in Equation 8.16 and the conservation of the number of particles given by Equations 8.14 and 8.17, respectively. Moreover, $\varepsilon_k(\mathbf{r}, t)$ can be split up as a sum of two terms:

$$\varepsilon_k(\mathbf{r}, t) = \varepsilon_k^{(0)} + \varepsilon_k^{(1)}(\mathbf{r}, t). \tag{8.19}$$

The quantity $\varepsilon_k^{(0)}$ is a result of the normalization constraint, while $\varepsilon_k^{(1)}$ are the Lagrange multipliers associated with the charge-current conservation defined by Equation 8.17. On the other hand, if Equation 8.18 is divided by R_k we can reexpress the corresponding equation as

$$Q_k(\mathbf{r}, t) + v_{\text{eff}}(\mathbf{r}, t) = \varepsilon_k(\mathbf{r}, t) \tag{8.20}$$

where Q_k is the so-called quantum potential associated with the state ϕ_k,

$$Q_k(\mathbf{r}, t) = -\frac{1}{2} \frac{\nabla^2 R_k}{R_k}. \tag{8.21}$$

Next, we minimize $T_s[\{R_k, S_k\}]_t$ with respect to S_k to be subjected to the constraint

$$\frac{\partial S_k}{\partial t} = -\varepsilon_k(\mathbf{r}, t). \tag{8.22}$$

The resulting Euler–Lagrange equation is given by

$$\frac{\partial R_k^2}{\partial t} + \nabla \cdot \left(R_k^2 \nabla S_k \right) = 0. \tag{8.23}$$

The coupled equations, Equations 8.18 and 8.23, provide a means for determining the exact TD density of the system of interest. We note that, at the solution point, the current vector is given by

$$\mathbf{j}(\mathbf{r}, t) = \sum_{k=1}^{N} R_k^2(\mathbf{r}, t) \nabla S_k(\mathbf{r}, t). \tag{8.24}$$

Note that, in the limit that the time dependence is turned off, the TD-DFT approach correctly reduces to the usual time-independent DFT one, as ∇S_k vanishes, Equations 8.17, 8.22, and 8.23 are identically satisfied, and Equation 8.15 will reduce to the time-independent kinetic energy of an N-electron system.

8.4 BOHMIAN MECHANICS: A TRAJECTORY PICTURE OF QUANTUM MECHANICS

8.4.1 SINGLE-PARTICLE TRAJECTORIES

Apart from the operational, wave or action-based pictures of quantum mechanics provided by Heisenberg, Schrödinger, or Feynman, respectively, there is an additional, fully trajectory-based picture: Bohmian mechanics [20,23]. Within this picture, the standard quantum formalism is understood in terms of trajectories defined by very specific motion rules. Although this formulation was independently formulated by Bohm, it gathers two former conceptual ideas: (1) the QFD picture proposed by Madelung and (2) the pilot role assigned to the wave function, proposed by de Broglie. In this way, the time evolution or dynamics of the system is described as an ideal quantum fluid with no viscosity; the evolution of this flow of identical particles is "guided" by the wave function.

The Bohmian formalism follows from the Schrödinger one in the position representation after considering a change of variables, from the complex wave function field (Ψ, Ψ^*) to the real fields (ρ, S) according to the transformation relation:

$$\Psi(\mathbf{r}, t) = R(\mathbf{r}, t)e^{iS(\mathbf{r}, t)/\hbar}, \tag{8.25}$$

with $\rho = R^2$. Substituting this relation into the TD Schrödinger equation for a single particle of mass m,

$$i\hbar \frac{\partial \Psi(\mathbf{r}, t)}{\partial t} = \left[-\frac{\hbar^2}{2m}\nabla^2 + V(\mathbf{r}) \right] \Psi(\mathbf{r}, t), \tag{8.26}$$

and then separating the real and imaginary parts from the resulting expression, two real coupled equations are obtained:

$$\frac{\partial \rho}{\partial t} + \nabla \cdot \left(\rho \frac{\nabla S}{m} \right) = 0, \tag{8.27a}$$

$$\frac{\partial S}{\partial t} + \frac{(\nabla S)^2}{2m} + V_{\text{eff}} = 0, \tag{8.27b}$$

where

$$V_{\text{eff}} = V + Q = V - \frac{\hbar^2}{2m}\frac{\nabla^2 R}{R} = V - \frac{\hbar^2}{4m}\left[\frac{\nabla^2 \rho}{\rho} - \frac{1}{2}\left(\frac{\nabla \rho}{\rho} \right)^2 \right] \tag{8.28}$$

is an effective potential resulting from the sum of the "classical" contribution V and the so-called quantum potential Q which depends on the quantum state via ρ, or equivalently on the instantaneous curvature of the wave function via R. Note that in the case $V = v_{\text{eff}}$ and Ψ given as in Section 8.3, one gets $V_{\text{eff}} = \varepsilon_k$ or $V_{\text{eff}} = \varepsilon_k(\mathbf{r}, t)$ depending on whether we are considering the time-independent or the TD case, respectively. The action of the whole ensemble through the wave function on the particle motion can be seen as a dynamical manifestation of quantum nonlocality. Equation 8.27a is the continuity equation for the particle flow (or the probability density, from a conventional viewpoint) and Equation 8.27b is a generalized (quantum) Hamilton–Jacobi equation. As in classical mechanics, the characteristics or solutions S of Equation 8.27b define the particle velocity field,

$$\mathbf{v} = \frac{\nabla S}{m},\qquad(8.29)$$

from which the quantum trajectories are known. Uncertainty arises from the unpredictability in determining the particle initial conditions—distributed according to $\rho(\mathbf{r}, t = 0)$ [20]—but not from the impossibility to know the actual (quantum) trajectory pursued during its evolution.

An alternative way to obtain the quantum trajectories is by formulating the Bohmian mechanics as a Newtonian-like theory. Then, Equation 8.29 gives rise to a generalized Newton's second law:

$$m\frac{d\mathbf{v}}{dt} = -\nabla V_{\text{eff}}.\qquad(8.30)$$

This formulation results very insightful; according to Equation 8.30, particles move under the action of an effective force $-\nabla V_{\text{eff}}$, i.e., the nonlocal action of the quantum potential here is seen as the effect of a (nonlocal) quantum force. From a computational viewpoint, these formulation results are very interesting in connection to quantum hydrodynamics [21,27]. Thus, Equations 8.27 can be reexpressed in terms of a continuity equation and a generalized Euler equation. As happens with classical fluids, here also two important concepts that come into play: the quantum pressure and the quantum vortices [28], which occur at nodal regions where the velocity field is rotational.

Since TD-DFT is applied to scattering problems in its QFD version, two important consequences of the nonlocal nature of the quantum potential are worth stressing in this regard. First, relevant quantum effects can be observed in regions where the classical interaction potential V becomes negligible, and more important, where $\rho(\mathbf{r}, t) \approx 0$. This happens because quantum particles respond to the "shape" of $\mathbf{\Psi}$, but not to its "intensity," $\rho(\mathbf{r}, t)$. Notice that Q is scale-invariant under the multiplication of $\rho(\mathbf{r}, t)$ by a real constant. Second, quantum-mechanically the concept of asymptotic or free motion only holds locally. Following the classical definition for this motional regime,

$$m\frac{d\mathbf{v}}{dt} \approx 0,\qquad(8.31)$$

which means in Bohmian mechanics that $\nabla V_{\text{eff}} \approx 0$, i.e., the local curvature of the wave function has to be zero (apart from the classical-like requirement that $V \approx 0$). In scattering experiments, this condition is satisfied along the directions specified by the diffraction channels [29]; in between, although $V \approx 0$, particles are still subjected to strong quantum forces.

8.4.2 BOHMIAN TRAJECTORIES DESCRIBING MANY-BODY SYSTEMS

In the case of a many-body problem, the Bohmian mechanics for an N-body dynamics follows from the one for a single system, but replacing Equation 8.25 by

$$\Psi(\mathbf{r}_1, \mathbf{r}_2, \ldots, \mathbf{r}_N; t) = R(\mathbf{r}_1, \mathbf{r}_2, \ldots, \mathbf{r}_N; t)e^{iS(\mathbf{r}_1, \mathbf{r}_2, \ldots, \mathbf{r}_N; t)/\hbar}, \tag{8.32}$$

with $\rho(\mathbf{r}_1, \mathbf{r}_2, \ldots, \mathbf{r}_N; t) = R^2(\mathbf{r}_1, \mathbf{r}_2, \ldots, \mathbf{r}_N; t)$. If we are interested in the density of a single particle, we need to "trace" over the remaining $N - 1$ degrees of freedom in the corresponding density matrix (see Section 8.4.3). On the other hand, in order to know the specific trajectory pursued by the particle associated with the kth degree of freedom, we have to integrate the equation of motion

$$\mathbf{v}_k = \frac{\nabla_k S}{m}, \tag{8.33}$$

where $\nabla_k = \partial/\partial \mathbf{r}_k$. The velocity field is irrotational in nature except at nodal regions. Obviously, there will be as many equations of motion as degrees of freedom. Note that as each degree of freedom represents a particle that is interacting with the remaining $N - 1$ particles in the ensemble, the corresponding trajectory will be strongly influenced by the evolution of those other $N - 1$ particles. This entanglement is patent through the quantum potential, which is given here as

$$Q = -\frac{\hbar^2}{2m} \sum_{k=1}^{N} \frac{\nabla_k^2 R}{R}, \tag{8.34}$$

where $Q = Q(\mathbf{r}_1, \mathbf{r}_2, \ldots, \mathbf{r}_N; t)$, is nonseparable and therefore, strongly nonlocal. Note that this nonlocality arises from correlation among different degrees of freedom, which is different from the nonlocality that appears when considering symmetry properties of the wave function, not described by the Schrödinger equation but by quantum statistics. In this sense, we can speak about two types of entanglement: symmetry and dynamics. The general N-body wave function (Equation 8.32) is entangled in both aspects.

Now, if the many-body (electron) problem can be arranged in such a way that the many-body, nonseparable wave function is expressed in terms of a separable wave function, which depends on N single-particle wave functions (Hartree approximation), i.e.,

$$\Psi(\mathbf{r}_1, \mathbf{r}_2, \ldots, \mathbf{r}_N; t) = \prod_{k=1}^{N} \psi_k(\mathbf{r}_k; t) = \prod_{k=1}^{N} R_k(r_k; t)e^{iS_k(r_k; t)/\hbar}, \tag{8.35}$$

then, in terms of trajectories, we find a set of uncoupled equations of motion,

$$\mathbf{v}_k = \frac{\nabla_k S_k}{m}, \tag{8.36}$$

which will only depend implicitly (through v_{eff}) on the other particles. Note that the factorization of the wave function implies that the quantum potential becomes a separable function of the N particle coordinates and time,

$$Q = \frac{\hbar^2}{2m} \sum_{k=1}^{N} \frac{\nabla_k^2 R_k}{R_k} = \sum_{k=1}^{N} Q_k, \qquad (8.37)$$

where each Q_k measures the local curvature of the wave function associated with the ith orbital associated to the corresponding particle. Therefore, each degree of freedom can be studied separately from the rest (with the exception that we have to take into account the mean field created by the remaining $N - 1$ particles). Factorizability implies physical independence, statistical independence, or in other words, that particles obey Maxwell–Boltzmann statistics (they are distinguishable) and the associate wave function is, therefore, not entangled.

In TD-DFT, the wave function is antisymmetrized and therefore, nonfactorizable or entangled. However, as said above, it is not entangled from a dynamical point of view because the quantum forces originated from a nonseparable quantum potential as in Equation 8.34 are not taken into account.

8.4.3 REDUCED QUANTUM TRAJECTORY APPROACH

In Section 8.4.2, we considered the problem of the reduced dynamics from a standard DFT approach, i.e., in terms of single-particle wave functions from which the (single-particle) probability density is obtained. However, one could also use an alternative description which arises from the field of decoherence. Here, in order to extract useful information about the system of interest, one usually computes its associated reduced density matrix by tracing the total density matrix $\hat{\rho}_t$ (the subscript t here indicates time-dependence), over the environment degrees of freedom. In the configuration representation and for an environment constituted by N particles, the system reduced density matrix is obtained after integrating $\hat{\rho}_t \equiv |\Psi\rangle_{t}{}_{t}\langle\Psi|$ over the $3N$ environment degrees of freedom, $\{\mathbf{r}_k\}_{k=1}^{N}$,

$$\tilde{p}(\mathbf{r}, \mathbf{r}'; t) = \int \langle \mathbf{r}, \mathbf{r}_1, \mathbf{r}_2, \ldots, \mathbf{r}_N | \Psi(t)\rangle \langle \Psi(t) | \mathbf{r}', \mathbf{r}_1, \mathbf{r}_2, \ldots, \mathbf{r}_N \rangle d\mathbf{r}_1 d\mathbf{r}_2 \ldots d\mathbf{r}_N. \quad (8.38)$$

The system (reduced) quantum density current can be derived from this expression, being

$$\tilde{j}(\mathbf{r}, t) \equiv \frac{\hbar}{m} \text{Im}[\nabla_r \tilde{p}(\mathbf{r}, \mathbf{r}'; t)]|_{\mathbf{r}'=\mathbf{r}}, \qquad (8.39)$$

which satisfies the continuity equation

$$\dot{\tilde{\rho}} + \nabla \tilde{\mathbf{j}} = 0. \qquad (8.40)$$

where $\tilde{\rho}$ is the diagonal element (i.e., $\tilde{\rho} \equiv \tilde{\rho}(\mathbf{r}, \mathbf{r}; t)$) of the reduced density matrix. Taking into account Equations 8.39 and 8.40, now we define the velocity field $\dot{\mathbf{r}}$ associated with the (reduced) system dynamics as

$$\tilde{\mathbf{j}} = \tilde{\rho}\dot{\mathbf{r}}, \qquad (8.41)$$

which is analogous to the Bohmian velocity field. Now, from Equation 8.41, we define a new class of quantum trajectories as the solutions to the equation of motion

$$\dot{\mathbf{r}} \equiv \frac{\hbar}{m} \frac{\mathrm{Im}[\nabla_{\mathbf{r}} \tilde{\rho}(\mathbf{r}, \mathbf{r}'; t)]}{\mathrm{Re}[\tilde{\rho}(\mathbf{r}, \mathbf{r}'; t)]} \bigg|_{\mathbf{r}'=\mathbf{r}} \tag{8.42}$$

These new trajectories are the so-called reduced quantum trajectories [30], which are only explicitly related to the system reduced density matrix. The dynamics described by Equation 8.42 leads to the correct intensity (time evolution of which is described by Equation 8.40) when the statistics of a large number of particles are considered. Moreover, Equation 8.42 reduces to the well-known expression for the velocity field in Bohmian mechanics, when there is no interaction with the environment.

8.5 FINAL DISCUSSION AND CONCLUSIONS

Nowadays the success of DFT and TD-DFT is out of question in both the physics and chemistry communities. The numerical results obtained are most of cases in good agreement to those from experimental and other theoretical methods with a relative small computational effort. However, in this chapter, our goal is to present the TD-DFT from a Bohmian perspective and to analyze, from a conceptual level, some of the aspects which are deeply rooted in DFT.

Working with a system of fermions, where the total wave function has to be antisymmetrized with respect to two-particle interchanges, it gives rise to the appearance of new quantum forces from the quantum potential, which are not described by the DFT Hamiltonian. The DFT wave function will then be nonfactorizable and therefore, entangled from a symmetry point of view but not from a dynamical point of view. In this sense, as mentioned above, the effective potential V_{eff} plays a fundamental role not only in the nonlocality of the theory, but in the so-called invertibility problem of the one-to-one mapping up to an additive TD function between the density and v_{eff}. In our opinion, the central theorems of TD-DFT should be written in terms of V_{eff} instead of v_{eff}, as the quantum potential is also state-dependent and a functional of the density. An infinite set of possible quantum potentials can be associated with the same physical situation and Schrödinger equation and therefore, the invertibility should be questioned. Moreover, for scattering problems, when v_{eff} is negligible in the asymptotic region, the quantum potential can still be active and the time propagation should be extended much farther in order to obtain a good numerical convergence.

In Bohmian mechanics, the way the full problem is tackled in order to obtain operational formulas can determine dramatically the final solution due to the context-dependence of this theory. More specifically, developing a Bohmian description within the many-body framework and then focusing on a particle is not equivalent to directly starting from the reduced density matrix or from the one-particle TD-DFT equation. Being well aware of the severe computational problems coming from the first and second approaches, we are still tempted to claim that those are the most natural ways to deal with a many-body problem in a Bohmian context.

ACKNOWLEDGMENTS

This work was supported by the Ministerio de Ciencia e Innovación (Spain) under Projects FIS2007-62006 and CTQ2005-01117/BQU and Generalitat de Catalunya under Projects 2005SGR-00111 and 2005SGR-00175. A.S. Sanz acknowledges the Consejo Superior de Investigaciones Cientificas for a JAE-Doc Contract.

REFERENCES

1. For an excellent account on DMT see, for example: K. Blum, *Density Matrix Theory and Applications* (Plenum Press, New York, 1981).
2. A. Szabo and N. S. Ostlund, *Modern Quantum Chemistry* (Dover Publications, New York, 1996).
3. W. Koch and M. C. Holthausen, *A Chemist's Guide to Density Functional Theory* (Wiley-VCH, Weinheim, 2002).
4. P. Hohenberg and W. Kohn, *Phys. Rev. B* **136**, 864, 1964.
5. W. Kohn and L. J. Sham, *Phys. Rev. A* **140**, 1133, 1965.
6. R. McWeeny and B. T. Sutcliffe, *Methods of Molecular Quantum Mechanics* (Academic Press, London, 1969).
7. For example, see the *Proceedings of the VIth International Conference on the Applications of Density Functional Theory*, Paris, France, 29 Aug.–1 Sept. 1995 [*Int. J. Quantum Chem.* **61**, 181 (1997)].
8. X. Hua, X. Chen, and W. A. Goddard III, *Phys. Rev. B* **55**, 16103, 1997; D. C. Langreth and J. P. Perdew, *Phys. Rev. B* **21**, 5469, 1980; J. P. Perdew and Y. Wang, *Phys. Rev. B* **33**, 8800, 1986.
9. F. Bloch, *Z. Physik* **81**, 363, 1933.
10. L. J. Bartolotti, *Phys. Rev. A* **24**, 1661, 1981; L. J. Bartolotti, *Phys. Rev. A* **26**, 2243, 1982.
11. B. M. Deb and S. K. Ghosh, *J. Chem. Phys.* **77**, 342, 1982; B. M. Deb and S. K. Ghosh, in *The Single-Particle Density in Physics and Chemistry*, eds. N. H. March and B. M. Deb (Academic Press, London, 1987, p. 219).
12. E. Runge and E. K. U. Gross, *Phys. Rev. Lett.* **52**, 997, 1984.
13. B. M. Deb and P. K. Chattaraj, *Phys. Rev. A* **39**, 1696, 1989.
14. G. P. Lawes and N. H. March, *Phys. Scr.* **21**, 402, 1980; M. Levy, J. P. Pardew, and V. Sahni, *Phys. Rev. A* **30**, 2745, 1984; N. H. March, *Phys. Lett. A* **113**, 476, 1986.
15. M. McClendon, *Phys. Rev. A* **38**, 5851, 1988.
16. N. T. Maitra and K. Burke, *Chem. Phys. Lett.* **359**, 237, 2002; N. T. Maitra and K. Burke, *Chem. Phys. Lett.* **441**, 167, 2007.
17. P. Samal and M. K. Harbola, *Chem. Phys. Lett.* **433**, 204, 2006.
18. V. Sahni, *Quantal Density Functional Theory* (Springer, Berlin, 2004).
19. S. Botti, A. Schindlmayr, R. Del Sole, and L. Reining, *Rep. Prog. Phys.* **70**, 357, 2007.
20. P. R. Holland, *The Quantum Theory of Motion* (Cambridge University Press, Cambridge, 1993).
21. E. Madelung, *Z. Physik* **40**, 332, 1926.
22. L. de Broglie, *Compt. Rend.* **184**, 273, 1927.
23. D. Bohm, *Phys. Rev.* **85**, 166, 180, 1952; **89**, 458, 1953.
24. J. S. Bell, *Rev. Mod. Phys.* **38**, 447, 1966.
25. R. McWeeny, *Philos. Mag. B* **69**, 727, 1994; F. Illas, I. de P. R. Moreira, J. M. Bofill, and M. Filatov, *Phys. Rev. B* **70**, 132414, 2004; F. Illas, I. de P. R. Moreira, J. M. Bofill, and M. Filatov, *Theor. Chem. Acc.* **116**, 587, 2006.

26. M. A. L. Marques and E. K. U. Gross, *Annu. Rev. Phys. Chem.* **55**, 427, 2004.
27. R. E. Wyatt, *Quantum Dynamics with Trajectories* (Springer, Berlin, 2005).
28. A. S. Sanz, F. Borondo, and S. Miret-Artés, *Phys. Rev. B* **69**, 115413, 2004.
29. A. S. Sanz, F. Borondo, and S. Miret-Artés, *Phys. Rev. B* **61**, 7743, 2000.
30. A. S. Sanz and F. Borondo, *Eur. Phys. J.D* **44**, 319, 2007.

9 Time-Independent Theories for a Single Excited State

Á. Nagy, M. Levy, and Paul W. Ayers

CONTENTS

9.1   INTRODUCTION

Density functional theory was originally formalized for the ground state [1]. It is valid for the lowest energy state in each symmetry class [2,3]. To calculate excitation energies, Slater [4] introduced the transition state method, which proved to be a reasonably good one to calculate excitation energies.

Density functional theory was first generalized for excited states by Theophilou [5]. The density functional variational principle for excited states was studied by Perdew and Levy [6] and Lieb [7]. Formalisms for excited states were also provided by Fritsche [8] and English et al. [9]. The subspace theory of Theophilou was enlarged into the theory of unequally weighted ensembles of excited states by Gross et al. [10]. The relativistic generalization of this formalism was also done [11]. A theory of excited states was presented utilizing Görling–Levy perturbation theory [12,13]. Kohn [14] proposed a quasilocal density approximation and excitation energies of He atom were calculated using this method [10]. Excitation energies of several atoms [15–18] were determined using the ensemble theory, and several ground-state approximate functionals were tested [19]. The coordinate scaling for the density matrix of ensembles was explored [20]. The adiabatic connection formula was extended to the ensemble exchange-correlation energy and a simple local ensemble exchange potential was presented [21]. The subspace density and pair density at the coincidence of the first excited state for two harmonically interacting electrons with antiparallel spins under isotropic harmonic confinement were calculated [22]. (For reviews of excited-state theories, see Refs. [18,23].)

Unfortunately, the exchange-correlation part of the ensemble Kohn–Sham potential is not known exactly. The optimized potential method (OPM) and its approximations turned to be very successful in treating the exchange exactly in the ground-state theory [24–28], were generalized for ensembles of excited states. The first generalization was based on the ensemble Hartree–Fock method [29]. Later, a ghost-interaction correction to this scheme was proposed [30]. Then another more appropriate OPM was developed [31]. The combination of this method, the self-interaction-free Perdew–Zunger approximation, and ghost- and self-interaction corrected (GSIC) ensemble Kohn–Sham potential was constructed and applied to determine ensemble energies and excitation energies [32–34]. The virial theorem was also derived in the ensemble theory [35]. Based on the ensemble theory, a relationship between excitation energy and hardness was derived, the concept of the ensemble Kohn–Sham hardness was introduced, and it was proposed that the first excitation energy can substitute for the hardness as a reactivity index [36].

Two theories for a single excited state [37–40] are the focus of this chapter. A nonvariational theory [37,38] based on Kato's theorem is reviewed in Section 9.2. Sections 9.3 and 9.4 summarize the variational density functional theory of a single excited state [39,40]. Section 9.5 presents some application to atoms and molecules. Section 9.6 is devoted to discussion.

There are other noteworthy single excited-state theories. Görling developed a stationary principle for excited states in density functional theory [41]. A formalism based on the integral and differential virial theorems of quantum mechanics was proposed by Sahni and coworkers for excited state densities [42]. The local scaling approach of Ludena and Kryachko has also been generalized to excited states [43].

An alternative theory is the popular time-dependent density functional theory [44], in which transition energies are obtained from the poles of dynamic linear response properties. There are several excellent reviews on time-dependent density functional theory. See, for instance, Ref. [45].

9.2   NONVARIATIONAL THEORY FOR A SINGLE EXCITED STATE

According to the Hohenberg–Kohn theorem of the density functional theory, the ground-state electron density determines all molecular properties. E. Bright Wilson [46] noticed that Kato's theorem [47,48] provides an explicit procedure for constructing the Hamiltonian of a Coulomb system from the electron density:

$$Z_\beta = \frac{1}{2\bar{n}(r)} \frac{\partial \bar{n}(r)}{\partial r}\bigg|_{r=R_\beta}. \tag{9.1}$$

Here $\bar{n}$ denotes the angular average of the density n and the right-hand side is evaluated at the position of nucleus β. From Equation 9.1, the cusps of the density tell us where the nuclei are (R_β) and what the atomic numbers Z_β are. The integral of the density gives us the number of electrons:

$$N = \int n(\mathbf{r}) d\mathbf{r}. \tag{9.2}$$

Consequently, from the density the Hamiltonian can be readily obtained, and then every property of the system can be determined by solving the Schrödinger equation to obtain the wave function. One has to emphasize, however, that this argument holds only for Coulomb systems. By contrast, the density functional theory formulated by Hohenberg and Kohn is valid for any external potential.

Kato's theorem is valid not only for the ground state but also for the excited states. Consequently, if the density n_i of the i-th excited state is known, the Hamiltonian $\hat{H}$ is also known in principle and its eigenvalue problem

$$\hat{H}\Psi_k = E_k\Psi_k \quad (k = 0, 1, \ldots, i, \ldots) \tag{9.3}$$

can be solved, where

$$\hat{H} = \hat{T} + \hat{V} + \hat{V}_{ee}. \tag{9.4}$$

$$\hat{T} = \sum_{j=1}^{N} \left(-\frac{1}{2}\nabla_j^2\right), \tag{9.5}$$

$$\hat{V}_{ee} = \sum_{k=1}^{N-1} \sum_{j=k+1}^{N} \frac{1}{|\mathbf{r}_k - \mathbf{r}_j|} \tag{9.6}$$

and

$$\hat{V} = \sum_{k=1}^{N} \sum_{J=1}^{M} \frac{-Z_J}{|r_k - R_J|} \tag{9.7}$$

are the kinetic energy, the electron–electron energy, and the electron–nuclear energy operators, respectively.

There are certain special cases, however, where Equation 9.1 does not provide the atomic number. The simplest example is the 2p orbital of the hydrogen atom, where the density

$$n_{2p}(r) = cr^2 e^{-Zr} \tag{9.8}$$

and the derivative of the density are zero at the nucleus. Though Kato's theorem (Equation 9.1) is valid in this case too, it does not give us the desired information, that is, the atomic number. Similar cases occur in other highly excited atoms, ions, or molecules, for which the spherical average of the derivative of the wave function is zero at the nucleus, that is where we have no s-electrons.

Pack and Brown [49] derived cusp relations for the wave functions of these systems. We derived the corresponding cusp relations for the density [50,51]. Let us define

$$\eta_l(\mathbf{r}) = \frac{n(\mathbf{r})}{r^{2l}},$$ (9.9)

where l is the smallest integer for which η_l is not zero at the nucleus. The new cusp relations for the density are

$$\left.\frac{\partial \bar{\eta}^l(\mathbf{r})}{\partial r}\right|_{r=0} = -\frac{2Z}{l+1}\eta^l(0).$$ (9.10)

For the example of a one-electron atom in the 2p state, Equation 9.9 leads to

$$\eta_{2p}(r) = \frac{n_{2p}}{r^{2l}} = ce^{-Zr}$$ (9.11)

and the new cusp relation has the form:

$$-2Z\eta_{2p}(0) = 2\eta'_{2p}(0).$$ (9.12)

So we can again readily obtain the atomic number from the electron density. Other useful cusp relations have also been derived [52,53]. There are several other works concerning the cusp of the density [54–63].

Next, using the concept [2,64] of adiabatic connection, Kohn–Sham-like equations can be derived. We suppose the existence of a continuous path between the interacting and the noninteracting systems. The density n_i of the ith electron state is the same along the path.

$$\hat{H}^\alpha_i \Psi^\alpha_k = E^\alpha_k \Psi^\alpha_k,$$ (9.13)

where

$$\hat{H}^\alpha_i = \hat{T} + \alpha \hat{V}_{ee} + \hat{V}^\alpha_i.$$ (9.14)

The subscript i denotes that the density of the given excited state is supposed to be the same for any value of the coupling constant α. $\alpha = 1$ corresponds to the fully interacting case, while $\alpha = 0$ gives the noninteracting system:

$$\hat{H}^0_i \Psi^0_k = E^0_k \Psi^0_k.$$ (9.15)

For $\alpha = 1$, the Hamiltonian $\hat{H}^\alpha_i$ is independent of i. For any other value of α, the "adiabatic" Hamiltonian depends on i and we have different Hamiltonians for different excited states. Thus the noninteracting Hamiltonian ($\alpha = 0$) is different for different excited states. If there are several "external" potentials $V^{\alpha=0}$ leading to the same density n_i, we select that potential for which the one-particle density matrix is closest to the interacting one-particle density matrix.

To solve the Kohn–Sham-like equation (Equation 9.15), one has to find an approximation to the potential of the noninteracting system. The OPM [25] can be generalized for a single excited state also. It was shown [37] that because the energy is stationary at the true wave function, the energy is stationary at the true potential. This is the consequence of the well-known fact that when the energy is considered to be a functional of the wave function, the only stationary points of $E[\Psi]$ are those associated with the eigenvalues/eigenvectors of the Hamiltonian

$$\frac{\delta E}{\delta \Psi_k} = 0 \quad (k = 1, \ldots, i, \ldots). \tag{9.16}$$

From the density of a given excited state n_i, one can obtain the Hamiltonian, the eigenvalues and eigenfunctions, and (through adiabatic connection) the noninteracting effective potential $V_i^{\alpha=0}$. The solution of equations of the noninteracting system then leads to the density n_i. Thus, we can consider the total energy to be a functional of the noninteracting effective potential:

$$E[\Psi_i] = E\big[\Psi_i\big[V_i^0\big]\big]. \tag{9.17}$$

Making use of Equation 9.16, we obtain

$$\frac{\delta E}{\delta V_i^0} = \int \frac{\delta E}{\delta \Psi_i} \frac{\delta \Psi_i}{\delta V_i^0} + c.c. = 0. \tag{9.18}$$

So an optimized effective potential can be found for the given excited state. The Knieger-Li-Iafrate (KLI) approximation to the optimized effective potential can also be derived [37].

Exchange identities utilizing the principle of adiabatic connection and coordinate scaling and a generalized Koopmans' theorem were derived and the excited-state effective potential was constructed [65]. The differential virial theorem was also derived for a single excited state [66].

9.3 VARIATIONAL THEORY FOR A NONDEGENERATE SINGLE EXCITED STATE

The theory discussed in the Section 9.2 is a nonvariational one. It presumes that the interacting excited-state density is known. Two of the authors showed, accordingly, that there exists a variational excited-state density functional theory that generates the interacting excited-state density as well as the corresponding energy [39]. Consider first the nondegenerate case. The functionals in this variational theory are universal bifunctionals. That is, they are functionals of not only the given excited-state trial density n, but of the ground-state density n_0 as well.

The variational principle is written in the form of a constrained search

$$E_i = \min_n \ \min_{\substack{\Psi \to n \\ \Psi \perp \Psi_0, \ldots, \Psi_{i-1}}} \ \langle \Psi | \hat{H} | \Psi \rangle. \tag{9.19}$$

The minimization process is done in two steps. The first minimization is over all wave functions that are orthogonal to the first $i - 1$ states of $\hat{H}$ and simultaneously gives the trial the density, n. The second minimization is over the set of all N-electron trial densities. We can write this procedure as

$$E_i = \min_n \left\{ \int v(\mathbf{r})n(\mathbf{r})d\mathbf{r} + F_i[n, n_0] \right\} = \int v(\mathbf{r})n_i(\mathbf{r})d\mathbf{r} + F_i[n_i, n_0], \qquad (9.20)$$

where the universal functional $F_i[n, n_0]$ is defined as

$$F_i[n, n_0] = \min_{\substack{\Psi \to n \\ \Psi \perp \Psi_0, \dots, \Psi_{i-1}}} \langle \Psi | \hat{T} + \hat{V}_{ee} | \Psi \rangle = \langle \Psi[n, n_0] | \hat{T} + \hat{V}_{ee} | \Psi[n, n_0] \rangle. \qquad (9.21)$$

In Equation 9.21, Ψ yields n and is orthogonal to the first $i-1$ state of the Hamiltonian for which n_0 is the ground-state density. Here, this Hamiltonian is the $\hat{H}$ in Equation 9.19. Note that instead of the ground-state electron density n_0, we could use the external potential v or any ground-state Kohn–Sham orbital, etc. Thus we could use $F_i[n, v]$. The extension to degenerate states is studied in Section 9.4.

Now define the noninteracting kinetic energy $T_{s,i}[n, n_0]$ by

$$T_{s,i}[n, n_0] = \min_{\substack{\Phi \to n \\ \Phi \perp \Phi_0, \dots, \Phi_{m-1}}} \langle \Phi | \hat{T} | \Phi \rangle = \langle \Phi[n, n_0] \hat{T} | \Phi[n, n_0] \rangle, \qquad (9.22)$$

where each Φ is orthogonal to the first $m-1$ (with $m \geq 1$) states of that noninteracting Hamiltonian whose ground state resembles n_0 closest, say in a least squares sense, and for which n_i is the mth state density. With

$$w_i([n_i, n_0]; \mathbf{r}) = -\left. \frac{\delta T_{s,i}[n, n_0]}{\delta n} \right|_{n=n_i} \qquad (9.23)$$

we then have the minimum principle

$$T_{s,i}[n_i, n_0] + \int w_i([n_i, n_0]; \mathbf{r})n_i(\mathbf{r})d\mathbf{r}$$
$$= \min_n \left\{ T_{s,i}[n, n_0] + \int w_i([n_i, n_0]; \mathbf{r})n(\mathbf{r})d\mathbf{r} \right\}. \qquad (9.24)$$

The Kohn–Sham equations take the form

$$\left[-\frac{1}{2}\nabla^2 + w_i([n_i, n_0]; \mathbf{r}) \right] \phi_j^i(\mathbf{r}) = \varepsilon_j^i \phi_j^i(\mathbf{r}), \qquad (9.25)$$

where the orbitals are occupied as necessary, so that

$$n_i = \sum_{k=1} \lambda_k^i |\phi_k^i|^2. \qquad (9.26)$$

The occupation numbers λ_k^i will be 0, 1, or 2 for a nondegenerate system. Since n_i is an excited-state density of a noninteracting system whose potential is w_i, at least one of the λ_k^i will be zero. As in the usual Kohn–Sham scheme, w_i is obtained by first approximating it with a starting guess for n_i in the Kohn–Sham potential, and then the Kohn–Sham equations are solved in a self-consistent manner. The total excited-state energy is

$$E_i = \int v(\mathbf{r}) n_i(\mathbf{r}) d(\mathbf{r}) - \frac{1}{2} \sum_j \lambda_j^i \left\langle \phi_j^i \middle| \nabla^2 \middle| \phi_j^i \right\rangle + G_i[n_i, n_0], \tag{9.27}$$

where

$$G_i[n, n_0] = F_i[n, n_0] - T_{s,i}[n, n_0] \tag{9.28}$$

is the sum of the Coulomb and exchange-correlation energy. For practical calculations, G must be approximated. The OPM and the KLI approximation mentioned above can also be generalized to approximate G and the Kohn–Sham potential w. Consider partitioning $G_i[n, n_0]$ as

$$G_i[n, n_0] = Q_i[n, n_0] + E_c^i[n, n_0], \tag{9.29}$$

where
Q_i is the Coulomb plus exchange component
E_c^i is the correlation component of G_i

A crucial constraint for approximating Q_i and $\delta Q_i / \delta n$ is [39]

$$\left\langle \Phi[n_i, n_0] \middle| \hat{V}_{ee} \middle| \Phi[n_i, n_0] \right\rangle - \left\langle \Phi^{N-1}[n_i, n_0] \middle| \hat{V}_{ee} \middle| \Phi^{N-1}[n_i, n_0] \right\rangle$$
$$= \int d(\mathbf{r}) \left[n_i(\mathbf{r}) - n_i^{N-1}(\mathbf{r}) \right] \frac{\delta Q_i[n, n_0]}{\delta n} \Bigg|_{n=n_i}, \tag{9.30}$$

where Φ^{N-1} is the ground state of the noninteracting Hamiltonian with potential $w_i([n_i, n_0]; \mathbf{r})$ (but with $N-1$ electrons), and n_i^{N-1} is the density of Φ^{N-1}. Also, it is understood that both w_i and $\delta Q_i / \delta n$ vanish as $|r| \to \infty$. Equation 9.30 is analogous to the ground-state exchange-only Koopmans' relation that has been previously obtained for finite systems [27,67]. Moreover, the highest occupied orbital energy must equal the exact excited-state ionization energy, unless prevented by symmetry [68]. Other useful constraints can also be derived [39,69].

(By the way, through ensemble theory with unequal weights, Ref. [68] identifies an effective potential derivative discontinuity that links physical excitation energies to excited Kohn–Sham orbital energies from a ground-state calculation.)

9.4 VARIATIONAL THEORY FOR A DEGENERATE SINGLE EXCITED STATE

Now, we turn to the degenerate case. Consider the solutions of the Schrödinger equation

$$\hat{H} |\Psi_\gamma^i\rangle = E^i |\Psi_\gamma^i\rangle \quad (\gamma = 1, 2, \ldots, g_i), \tag{9.31}$$

where g_i is the degeneracy of the ith excited state. For the sake of simplicity, only one index is used to denote the symmetry both in spin and spatial coordinates. Instead of treating one wave function, the subspace S_i spanned by a set of wave functions Ψ_γ^i will be considered.

We define the density matrix in subspace S_i as

$$\hat{D}^i = \sum_{\gamma=1}^{g_i} \eta_\gamma^i \left| \Psi_\gamma^i \right\rangle \left\langle \Psi_\gamma^i \right|, \tag{9.32}$$

where the weighting factors η_γ^i satisfy the conditions

$$1 = \sum_{\gamma=1}^{g_i} \eta_\gamma^i \tag{9.33}$$

and

$$\eta_\gamma^i \geq 0. \tag{9.34}$$

In principle, any set of weighting factors η_γ^i satisfying the above conditions in Equations 9.33 and 9.34 can be used. Now, we define the subspace density as

$$n_i = N \sum_{\gamma=1}^{g_i} \eta_\gamma^i \int \left| \Psi_\gamma^i \right|^2 ds_1 dx_2 \ldots dx_N, \tag{9.35}$$

where x stands for a space–spin coordinate. The weighting factors η_γ^i should satisfy the conditions Equation 9.33 and Equation 9.34. If all η_γ^i are equal, the density has the property of transforming according to the totally symmetric irreducible representation [70,71]. (For instance, for atoms the subspace density will be spherically symmetric.) But, it is possible to select other values for the weighting factors η_γ^i. Equal weighting factors have the advantage that the subspace density has the symmetry of the external potential.

Now, we define the functional

$$F[n_i, n_0] = \underset{S \to n_i}{\text{Min}} \sum_{\gamma=1}^{g_i} \eta_\gamma \langle \Psi_\gamma | \hat{T} + \hat{V}_{ee} | \Psi_\gamma \rangle. \tag{9.36}$$

It can be rewritten applying the density matrix

$$F[n_i, n_0] = \underset{S \to n_i}{\text{min}} \, \text{tr}\{\hat{D}(\hat{T} + \hat{V}_{ee})\}, \tag{9.37}$$

where n_i and n_0 are arbitrary densities. n_0 is a ground-subspace density, while n_i is the trial excited subspace density we are considering. All the subspaces corresponding to the first $i - 1$ states of that Hamiltonian for which n_0 is the ground-state subspace density are supposed to be orthogonal to the subspace considered. Thus the total energy of the ith excited state has the form

$$
\begin{aligned}
E_i &= \min_{S_i} \sum_{\gamma=1}^{g_i} \eta_\gamma \left\langle \Psi_\gamma^i \middle| \hat{H}_v \middle| \Psi_\gamma^i \right\rangle \\
&= \min_{n_i} \left\{ \min_{S_i \to n} \sum_{\gamma=1}^{g_i} \eta_\gamma \left\langle \Psi_\gamma^i \middle| \hat{H} \middle| \Psi_\gamma^i \right\rangle \right\} \\
&= \min_{n_i} \left\{ F[n_i, n_0] + \int n_i(\mathbf{r}) v(\mathbf{r}) d\mathbf{r} \right\}.
\end{aligned}
\tag{9.38}
$$

E_i can be rewritten with the density matrix:

$$
E_i = \min_{S_i} tr\{\hat{D}^i \hat{H}\} = \min_{n_i} \left\{ \min_{S_i \to n_i} tr\{\hat{D}^i \hat{H}\} \right\}.
\tag{9.39}
$$

The noninteracting Kohn–Sham system is defined by adiabatic connection,

$$
\hat{H}^{i,\alpha} = \hat{T} + \alpha \hat{V}_{ee} + \sum_{k=1}^{N} v_\alpha^i(\mathbf{r}_k).
\tag{9.40}
$$

$v_\alpha^i(\mathbf{r}; [n_i, n_0])$ is defined so that the subspace density n_i (a) remains independent of α and (b) the ground state of $\hat{H}^{i,\alpha}$ is closest to n_0 in a least squares sense. The noninteracting Kohn–Sham Hamiltonian is obtained for $\alpha = 0$:

$$
\hat{H}_w^i = \hat{H}^{i,\alpha=0} = \hat{T} + \sum_{j=1}^{N} w^i([n_i, n_0]; \mathbf{r}_j).
\tag{9.41}
$$

Both the noninteracting Hamiltonian $\hat{H}_w^i$ and the Kohn–Sham-like potential $w^i([n_i, n_0]; \mathbf{r}) = v_0^i(\mathbf{r})$ depend on i; they are different for different excited states. The Kohn–Sham-like equations read

$$
\hat{H}^{i,0} \middle| \Psi_\gamma^{i,0} \rangle = E^{i,0} \middle| \Psi_\gamma^{i,0} \rangle \quad (\gamma = 1, 2, \ldots, g_i),
\tag{9.42}
$$

where the noninteracting density matrix can be constructed from the wave functions $\Psi_\gamma^{i,0}$ as

$$
\hat{D}_s^i = \sum_{\gamma=1}^{g_i} \eta_\gamma \middle| \Psi_\gamma^{i,0} \rangle \langle \Psi_\gamma^{i,0} \middle|,
\tag{9.43}
$$

while the noninteracting kinetic energy takes the form

$$
T_{s,i}[n_i, n_0] = tr\{\hat{D}_s^i \hat{T}\}.
\tag{9.44}
$$

$T_{s,i}$ can also be given variationally as

$$T_{s,i}[n_i, n_0] = \underset{S_i \to n_i}{\text{Min}} \, tr\{\hat{D}^i \hat{T}\} = tr\{\hat{D}^i_s[n_i, n_0] \hat{T}\}, \tag{9.45}$$

where each S_i is orthogonal to all subspaces corresponding to the first $m-1$ states of $\hat{H}^i_w$ and n_i is the subspace density of the mth excited state of $\hat{H}^i_w$. $D^i_s[n_i, n_0]$ is the noninteracting excited-state density matrix of $\hat{H}^i_w$ whose subspace density is n_i.

Minimizing the noninteracting kinetic energy

$$T_{s,i}[n_i, n_0] + \int n_i(\mathbf{r}) w^i([n_i, n_0]; \mathbf{r}) d\mathbf{r} = \underset{n_i}{\text{Min}} \left\{ T_s[n_i, n_0] + \int n_i(\mathbf{r}) w^i([n_i, n_0]; \mathbf{r}) d\mathbf{r} \right\} \tag{9.46}$$

leads to the Euler equation

$$w^i([n_i, n_0]; \mathbf{r}) + \left. \frac{\delta T_{s,i}[n_i, n_0]}{\delta n} \right|_{n=n_i} = \mu_i, \tag{9.47}$$

where μ_i is a Lagrange parameter. The Kohn–Sham potential takes the form

$$w^i(\mathbf{r}) = v(\mathbf{r}) + \left. \frac{\delta G[n_i, n_0]}{\delta n} \right|_{n=n_i}, \tag{9.48}$$

where the functional $G[n_i, n_0]$ is defined by the partition

$$F[n_i, n_0] = T_{s,i}[n_i, n_0] + G[n_i, n_0]. \tag{9.49}$$

A further partition of $G[n_i, n_0]$ gives

$$G[n_i, n_0] = J[n_i] + E_x[n_i, n_0] + E_c[n_i, n_0] \tag{9.50}$$

J, E_x, and E_c are the Coulomb, exchange, and the correlation components of G:

$$J[n_i] + E_x[n_i, n_0] = tr\{\hat{D}^i \hat{V}_{ee}\}, \tag{9.51}$$

$$E_c[n_i, n_0] = tr\{\hat{D}^i \hat{V}_{ee}\} - tr\{\hat{D}^i_s \hat{V}_{ee}\}. \tag{9.52}$$

Therefore, the Kohn–Sham potential has the form

$$w^i(\mathbf{r}) = v(\mathbf{r}) + v^i_J(\mathbf{r}) + v^i_{xc}(\mathbf{r}), \tag{9.53}$$

where $v(\mathbf{r})$, v^i_J, and $v^i_{xc}(\mathbf{r})$ are the external, Coulomb, and the exchange-correlation potentials, respectively.

The Kohn–Sham equations can be obtained from the minimalization of the noninteracting kinetic energy after expressing it with one-electron orbitals. Because $\Psi^{i,0}_\gamma$ is generally a linear combination of several Slater determinants, the form of the

Kohn–Sham equations is rather complicated for an arbitrarily selected set of weighting factors, and has to be derived separately for every different case of interest. For a spherically symmetric external potential and equal weighting factors, however, the Kohn–Sham equations have a very simple form, as shown in Ref. [72]. In this case the noninteracting kinetic energy is given by

$$T_{s,i} = \sum_{j=1}^{N} \lambda_j^i \int P_j^i \left[-\frac{1}{2} \left(P_j^i \right)'' + \frac{l_j^i (l_j^i + 1)}{2r^2} P_j^i \right] dr, \tag{9.54}$$

where P_j^i and λ_j^i are the radial wave functions and the occupation numbers corresponding to the given configuration, respectively. In Equation 9.54, $(P_j^i)''$ denotes the second derivative of P_j^i with respect to r. The radial subspace density

$$\varrho_i = \sum_{j=1}^{N} \lambda_j^i \left(P_j^i \right)^2 \tag{9.55}$$

in this particular case is spherically symmetric. The minimization of the noninteracting kinetic energy leads to the radial Kohn–Sham equations

$$-\frac{1}{2} \left(P_j^i \right)'' + \frac{l_j(l_j + 1)}{2r^2} P_j^i + w^i P_j^i = \varepsilon_j^i P_j^i. \tag{9.56}$$

9.5 APPLICATION TO ATOMS AND MOLECULES

To perform excited-state calculations, one has to approximate the exchange-correlation potential. Local self-interaction-free approximate exchange-correlation potentials have been proposed for this purpose [73]. We can try to construct these functionals as orbital-dependent functionals. There are different exchange-correlation functionals for the different excited states, and we suppose that the difference between the excited-state functionals can be adequately modeled through the occupation numbers (i.e., the electron configuration). Both the OPM and the KLI methods have been generalized for degenerate excited states [37,40].

Table 9.1 presents excitation energies for a few atoms and ions. Calculations were performed with the generalized KLI approximation [69,74]. For comparison, experimental data and the results obtained with the local-spin-density (LSD) exchange-correlation potential [75] are shown. The KLI method contains only the exchange.

The inclusion of correlation in OPM and KLI methods is straightforward in principle. One needs a correlation functional as a functional of the orbitals and then the method of derivation and calculation is exactly the same. Orbital-based correlation functionals already exist for the ground state (e.g., the functionals of Becke [76] and Mori–Sánchez et al. [77]). Unfortunately, we do not have excited-state correlation functionals. In the existing approximating functionals, exchange and correlation are treated together and if we change only the exchange part (into KLI

TABLE 9.1

Calculated and Experimental Excitation Energies (in Ry)

Atom	Transition	KLI	LSD	Exp.
Li	$2s\ ^2S \rightarrow 2p\ ^2P$	0.135	0.136	0.136
Na	$3s\ ^2S \rightarrow 3p\ ^2P$	0.144	0.164	0.154
K	$4s\ ^2S \rightarrow 4p\ ^2P$	0.103	0.124	0.118
Ne$^+$	$2s^2 2p^5\ ^2S \rightarrow 2s2p^6\ ^2S$	2.166	1.671	1.978
C	$2s^2 2p^2\ ^3P \rightarrow 2s\ 2p^3\ ^3D$	0.588	0.591	0.584
Si	$3s^2 3p^2\ ^3P \rightarrow 3s\ 3p^3\ ^3D$	0.469	0.486	0.441
O	$2s^2 2p^4\ ^3P \rightarrow 2s\ 2p^5\ ^3P$	1.252	1.058	1.151
F$^+$	$2s^2 2p^4\ ^3P \rightarrow 2s\ 2p^5\ ^3P$	1.597	1.363	1.505
O$^+$	$2s^2 2p^3\ ^4S \rightarrow 2s\ 2p^4\ ^4P$	1.106	1.082	1.094
B	$2s^2 2p\ ^2P \rightarrow 2s\ 2p^2\ ^2D$	0.433	0.411	0.436

or OPM), the balance between the exchange and correlation is ruined and we might receive worse results than in the exchange-only case. To find an appropriate approximate correlation functional for excited states will be the subject of future research.

In a recent paper, Glushkov and Levy [78] have presented an OPM algorithm that takes into account the necessary orthogonality constraints to lower states. One has to solve the problem

$$\hat{P}(\hat{H} - E)\hat{P}|\Psi_i\rangle = 0, \qquad \Psi_i\rangle = \hat{P}|\Psi_i\rangle, \quad i = 1, 2, \ldots, m, \tag{9.57}$$

$$\langle u_s|\Psi_i\rangle = 0 \quad s = 1, 2, \ldots q < m \tag{9.58}$$

on a finite dimentional subspace M, $\dim(M) = m$ with the associated projector P. For the lowest excited state, the Kohn–Sham equations have the form

$$\hat{P}_1\left[-\frac{1}{2}\nabla^2 + V_{\text{eff}}^\sigma\right]\hat{P}_1|\phi_{1,i}^{\text{sigma}}\rangle = \varepsilon_{1,i}^\sigma|\phi_{1,i}^\sigma\rangle, \quad \sigma = \alpha, \beta, \tag{9.59}$$

subjected to the constraints

$$\hat{P}_n^\alpha|\phi_{1,j}^\alpha\rangle = 0 \quad j = 1, 2, \ldots, n^\alpha, \tag{9.60}$$

where

$$\hat{P}_n^\alpha = |\phi_{0,n}^\alpha\rangle\langle\phi_{0,n}^\alpha| \tag{9.61}$$

and $\hat{P}_1$ is the orthoprojector defined from the basis set chosen for the excited state under consideration. OPM equations were derived using a variational principle with orthogonality constraints. The Kohn–Sham potential is expressed in a parametrized

TABLE 9.2
Excited State Energies (in Ry) of the HeH
Molecule at $R = 1.5$ bohrs

State	OPM	HF	CI
A $^2\Sigma^+$	−6.1296	−6.1346	−6.2254
C $^2\Sigma^+$	−6.0298	−6.0308	−6.1116
D $^2\Sigma^+$	−5.9688	−5.9698	−6.0600

form as a direct mapping of the external potential [79,80]. The second column of Table 9.2 presents the exchange-only OPM excited-state energies for the HeH molecule. For comparison, the Hartree–Fock and configuration interaction (CI) results are also shown.

9.6 DISCUSSION

We would like to emphasize that the generalizations of the Hohenberg–Kohn theorems to excited states reviewed in Sections 9.3 and 9.4 are different from the ground-state Hohenberg–Kohn theorems. The universal variational functionals for the kinetic and electron–electron repulsion energies in this excited-state variational theory are bifunctionals. That is, they are functionals of either the trial excited-state density and the ground-state density or of either the trial excited-state density and the external potential of interest. The standard Hohenberg–Kohn theorems for a single excited-state density do not exist [81–83]. Indeed, in recent studies, Gaudoin and Burke [82] and Sahni et al. [83] have presented examples of the nonuniquess of the potential. That is, they presented cases where a given excited-state density corresponds to several different "Kohn–Sham" potentials. Samal et al. [84], on the other hand, have argued that in the Levy–Nagy theory, the criterion based on the ground-state density of the Kohn–Sham potential may fix the density-to-potential map uniquely. In another recent paper, Samal and Harbola [85] have proposed a different criterion. Based on numerical examples, they recommend a criterion based on the kinetic energy instead of the ground-state density.

Two approaches to the excited-state problem have been the focus of this chapter. The nonvariational one, based on Kato's theorem, is pleasing in that it does not require a bifunctional, but it presumes that the excited-state density is known. On the other hand, the bifunctional approach is appealing in that it actually generates the desired excited-state density, which results in the generation of more known constraints on the universal functional for approximation purposes.

ACKNOWLEDGMENTS

Á. Nagy acknowledges OTKA No. (1) T67923 grant. P.W. Ayers, thanks NSERC, the Canada Research Chairs, and Sharcnet for support.

REFERENCES

1. P. Hohenberg and W. Kohn, *Phys. Rev.* **136**, B864, 1864.
2. O. Gunnarsson and B.I. Lundqvist, *Phys. Rev. B* **13**, 4274, 1976; O. Gunnarsson, M. Jonson, and B.I. Lundqvist, *Phys. Rev. B* **20**, 3136, 1979.
3. U. von Barth, *Phys. Rev. A* **20**, 1693, 1979.
4. J.C. Slater, *Quantum Theory of Molecules and Solids*, vol. 4. (McGraw-Hill, New York, 1974).
5. A.K. Theophilou, *J. Phys. C* **12**, 5419, 1978.
6. J.P. Perdew and M. Levy, *Phys. Rev. B* **31**, 6264, 1985; M. Levy and J.P. Perdew, in *Density Functional Methods in Physics*, Eds. R.M. Dreizler and J. da Providencia, NATO ASI Series B (Plenum, New York, 1985), vol. 123. p. 11.
7. E.H. Lieb, in *Density Functional Methods in Physics*, Eds. R.M. Dreizler and J. da Providencia, NATO ASI Series B (Plenum, New York, 1985) vol. 123, p. 31.
8. L. Fritsche, *Phys. Rev. B* **33**, 3976, 1986; *Int. J. Quantum. Chem.* **21**,15, 1987.
9. H. English, H. Fieseler, and A. Haufe, *Phys. Rev. A* **37**, 4570, 1988.
10. E.K.U. Gross, L.N. Oliveira, and W. Kohn, *Phys. Rev. A* **37**, 2805, 2809, 2821, 1988.
11. Á. Nagy, *Phys. Rev. A* **49**, 3074, 1994.
12. A. Görling, *Phys. Rev. A* **54**, 3912, 1996.
13. A. Görling and M. Levy, *Int. J. Quantum. Chem. S.* 29, 93, 1995; *Phys. Rev. B* **47**, 13105, 1993.
14. W. Kohn *Phys. Rev. A* **34**, 737, 1986.
15. Á. Nagy, *Phys. Rev. A* **42**, 4388, 1990.
16. Á. Nagy, *J. Phys. B* **24**, 4691, 1991.
17. Á. Nagy *Int. J. Quantum. Chem. S.* **29**, 297, 1995.
18. Á. Nagy, *Adv. Quant. Chem.* **29**, 159, 1997.
19. Á. Nagy and I. Andrejkovics, *J. Phys. B* **27**, 233, 1994; *Chem. Phys. Lett.* **296**, 489, 1998.
20. Á. Nagy, *Int. J. Quantum. Chem.* **56**, 225, 1995.
21. Á. Nagy, *J. Phys. B* **29**, 389, 1996.
22. Á. Nagy, I.A. Howard, N.H. March, and Zs. Jánosfalvi, *Phys. Lett. A* **335**, 347, 2005.
23. R. Singh and B.D. Deb, *Phys. Rep.* **311**, 47, 1999.
24. R.T. Sharp and G.K. Horton, *Phys. Rev.* **30**, 317, 1953.
25. J.D. Talman and W.F. Shadwick, *Phys. Rev. A* **14**, 36, 1976.
26. K. Aashamar, T.M. Luke, and J.D. Talman, *At. Data Nucl. Data Tables* **22**, 443, 1978.
27. J.B. Krieger, Y. Li, and G.J. Iafrate, *Phys. Rev. A* **45**, 101, 1992; *Phys. Rev. A* **46**, 5453, 1992.
28. Á. Nagy, *Phys. Rev. A* **55**, 3465, 1997.
29. Á. Nagy, *Int. J. Quantum. Chem.* **69**, 247, 1998.
30. N.I. Gidopoulos, P. Papakonstantinou, and E.K.U. Gross, *Phy. Rev. Lett.* **88**, 033003, 2001.
31. Á. Nagy, *J. Phys. B* **34**, 2363, 2001.
32. F. Tasnádi and Á. Nagy, *Int. J. Quantum Chem.* **92**, 234, 2003.
33. F. Tasnádi and Á. Nagy, *J. Phys. B* **36**, 4073, 2003.
34. F. Tasnádi and Á. Nagy, *J. Chem. Phys.* **119**, 4141, 2003.
35. Á. Nagy, *Acta Phys. et Chim. Debr.* **34–35**, 99, 2002.
36. Á. Nagy, *J. Chem. Sci.* (formerly *Proc. Indian Acad. Sci., Chem. Sci.*) **117**, 437 2005.
37. Á. Nagy, *Int. J. Quantum. Chem.* **70**, 681, 1998.
38. Á. Nagy, in *Electron Correlations and Materials Properties*, Eds. A. Gonis, N. Kioussis, and M. Ciftan (Kluwer, New York, 1999), p. 451.
39. M. Levy and Á. Nagy, *Phys. Rev. Lett.* **83**, 4631, 1999.
40. Á. Nagy and M. Levy, *Phys. Rev. A* **63**, 2502, 2001.
41. A. Görling, *Phys. Rev. A* **59**, 3359, 1999; *Phys. Rev. Lett.* **85**, 4229, 2000.

42. V. Sahni, L. Massa, R. Singh, and M. Slamet, *Phys. Rev. Lett.* **87**, 113002, 2001; V. Sahni and X.Y. Pan, *Phys. Rev. Lett.* **90**, 123001, 2003.
43. E.S. Kryachko, E.V. Ludena, and T. Koga, *J. Math. Chem.* **11**, 325, 1992.
44. E.K.U. Gross, J.F. Dobson, and M. Petersilka, in *Density Functional Theory (Topics in Current Chemistry)*, Ed. R. Nalewajski (Springer-Verlag, Heidelberg, 1996), vol. 181, p. 81.
45. M.A.L. Marques, C.A. Ullrich, F. Nogueira, A. Rubio, K. Burke, and E.K.U. Gross, *Time-Dependent Density Functional Theory* (Springer-Verlag, Heidelberg, 2006), *Series Lecture Notes in Physics*, vol. 706 and references therein.
46. N.C. Handy, in *Quantum Mechanical Simulation Methods for Studying Biological Systems*, Eds. D. Bicout and M. Field (Springer-Verlag, Heidelberg,1996), p. 1.
47. T. Kato, *Commun. Pure Appl. Math.* **10**, 151, 1957.
48. E. Steiner, *J. Chem. Phys.* **39**, 2365, 1963; N.H. March, *Self-Consistent Fields in Atoms* (Pergamon, Oxford, 1975).
49. R.T. Pack and W.B. Brown, *J. Chem. Phys.* **45**, 556, 1966.
50. Á. Nagy and K.D. Sen, *J. Phys. B* **33**, 1745, 2000.
51. P.W. Ayers, *Proc. Natl. Acad. Sci.* **97**, 1959, 2000.
52. Á. Nagy and K.D. Sen, *Chem. Phys. Lett.* **332**, 154, 2000.
53. Á. Nagy and K.D. Sen, *J. Chem. Phys.* **115**, 6300, 2001.
54. F.J. Gálvez, J. Porras, J.C. Angulo, and J.S. Dehesa, *J. Phys. B.* **21**, L271, 1988.
55. J.C. Angulo, J.S. Dehesa, and F.J. Gálvez, *Phys. Rev. A* **42**, 641, 1990; erratum *Phys. Rev. A* **43**, 4069, 1991.
56. J.C. Angulo and J.S. Dehesa, *Phys. Rev. A* **44**, 1516, 1991.
57. F.J. Gálvez and J. Porras, *Phys. Rev. A* **44**, 144, 1991.
58. J. Porras and F.J. Gálvez, *Phys. Rev. A* **46**, 105, 1992.
59. R.O. Esquivel, J. Chen, M.J. Stott, R.P. Sagar, and V.H. Smith, Jr., *Phys. Rev. A* **47**, 936, 1993.
60. R.O. Esquivel, R.P. Sagar, V.H. Smith, Jr., J. Chen, and M.J. Stott, *Phys. Rev. A* **47**, 4735, 1993.
61. J.S. Dehesa, T. Koga, and E. Romera, *Phys. Rev. A* **49**, 4255, 1994.
62. T. Koga, *Theor. Chim. Acta* **95**, 113 1997; T. Koga and H. Matsuyama, *Theor. Chim. Acta* **98**, 129, 1997
63. J.C. Angulo, T. Koga, E. Romera, and J.S. Dehesa, *THEOCHEM* **501–502**, 177, 2000.
64. D.C. Langreth and J.P. Perdew, *Phys. Rev. B* **15**, 2884, 1977; J. Harris and R.O. Jones, *J. Phys. F* **4**,1170, 1974; J. Harris, *Phys. Rev. A* **29**,1648, 1984.
65. Á. Nagy, in *New Trends in Quantum Systems in Chemistry and Physics (Progress in Theoretical Chemistry and Physics, Vol. 6)*, Eds. J. Maruani, C. Minot, R. McWeeny Y.G. Smeyers, and S. Wilson, (Kluwer, Dordrecht, 2001), vol. 1, p. 13; Á. Nagy, *Adv. Quant. Chem.* **39**, 35, 2001.
66. Á. Nagy, in *Recent Advances in Computational Chemistry*, vol. 1; *Recent Advances in the Density Functional Methods*, Eds. V. Barone, A. Bencini, and P. Fantucci (World Scientific, Singapore, 2002), Part III, p. 247.
67. M. Levy and A. Görling, *Phys. Rev. A* **53**, 3140, 1996.
68. M. Levy, *Phys. Rev. A* **52**, 4313, 1995.
69. Á. Nagy, *Int. J. Quantum Chem.* **99**, 256, 2004.
70. A. Görling, *Phys. Rev. A* **47**, 2783, 1993.
71. Á. Nagy, *Phys. Rev. A* **57**, 1672 1998.
72. Á. Nagy, *J. Phys. B* **32**, 2841, 1999.
73. F. Tasnádi and Á. Nagy, *Chem. Phys. Lett.* **366**, 496, 2002.
74. Á. Nagy, *Adv. Quant. Chem.* **42**, 363, 2003.
75. M.K. Harbola, *Phys. Rev. A* **65**, 052504, 2002.
76. A.D. Becke, *J. Chem. Phys.* **122**, 064101, 2005.

77. P. Mori-Sanchez, A.J. Cohen, and W. Yang, *J. Chem. Phys.* **124**, 091102, 2006.
78. V.N. Glushkov and M. Levy, *J. Chem. Phys.* **126**, 174106, 2007.
79. A.K. Theophilou and V.N. Glushkov, *Int. J. Quantum Chem.* **104**, 538, 2005; *J. Chem. Phys.* **124**, 034105, 2006.
80. S. Thanos and A.K. Theophilou, *J. Chem. Phys.* **124**, 204109, 2006.
81. M. Levy and J.P. Perdew, in *Density Functional Methods in Physics*, Eds. R.M. Dreizler and J. da Providencia (Plenum, New York, 1985).
82. R. Gaudoin and K. Burke, *Phys. Rev. Lett.* **93**, 173001, 2004.
83. V. Sahni, M. Slamet, and X.Y. Pan, *J. Chem. Phys.* **126**, 204106, 2007, 2003.
84. P. Samal, M.K. Harbola, and A. Holas, *Chem. Phys. Lett.* **419**, 217, 2006.
85. P. Samal and M.K. Harbola, *J. Phys. B* **38**, 3765 2005; **39**, 4065, 2006.

10 Spin-Polarized Density Functional Theory: Chemical Reactivity

Rubicelia Vargas and Marcelo Galván

CONTENTS

10.1 INTRODUCTION

Taking into account the spin degree of freedom of electrons is important to develop theories and methods robust enough to handle open shell systems and magnetic properties. In addition, nowadays manipulate spin transfer is a reality due to the development of spintronics. However, the relativistic nature of spin imposes a strong restriction to build theories with feasible wide applicability in electronic structure studies. Such restriction is present in wave function and density functional approaches. In both worlds, most of the applications of spin-dependent methods are done in the nonrelativistic spin-polarized limit. For the wave function approach, the limit is quite easy to achieve at least in a conceptual form: solving the Schrö-dinger equation with a fully flexible wave function constructed as an infinite linear combination of Slater determinants. The spin variable is introduced in the form of spin orbitals. On the density functional theory (DFT) side, the development of spin-dependent theories was introduced since the very beginning [1]. In this chapter, a brief description of the DFT relativistic method is outlined and the most widely used nonrelativistic limit is presented. Also, the chemical reactivity analysis that emerged from such limit is summarized in the light of several practical applications. In order to emphasize the physical meaning of the reactivity parameters, the discussion of them is done in the context of their formal definition rather than in the environment of their working equations within the spin-polarized Kohn–Sham method.

10.2 SPIN-POLARIZED DENSITY FUNCTIONAL THEORY

The Schrödinger formulation of quantum mechanics does not include the spin as an explicit variable in the Hamiltonian. This limitation is partially eliminated by the introduction of the spin variable in the wave function. This is done by using spin orbitals as building blocks for the Slater determinants. In this way, each element of the Slater determinant is a simple product of a spatial and a spin function; by construction the antisymmetry property required for the fermion nature of the electrons is satisfied. In the wave function approach, the spin is properly taken into account by the Dirac equation, but its complexity has avoided the possibility of using it for the systematic study of polyatomic systems. Accordingly, one may say that the spin is treated in the Schrödinger formulation in a nonrelativistic framework. On the side of DFT, there is a similar distinction in the approaches that address the treatment of spin variable. The original Hohenberg–Kohn–Sham formalism was introduced in the context of a nonrelativistic treatment of the spin variable [1,2]. Indeed, the spin was only mentioned in a short discussion of the spin susceptibility of a uniform electron gas. The first step in the direction of an appropriate treatment of the spin was done, almost simultaneously, by Von Barth and Hedin [3], Pant and Rajagopal [4], and by Rajagopal and Callaway [5]; they introduced the nonrelativistic limit for spin-polarized systems in a way that prevails until now as the most widely used spin-polarized DFT method. On the other hand, Ragajopal and Callaway provided the first demonstration that the Hohenberg–Kohn theorems can be extended to the case of many-electron systems characterized in terms of quantum electrodynamics [5]. It was until the late 1970s that the seminal works by Rajagopal [6] and by MacDonald and Vosko [7] settled down the relativistic formalism for a spin-polarized system. In the next section, a brief summary of this formalism is presented following the notation and approach presented in more detail in the Dreizler and Gross book [8].

10.2.1 RELATIVISTIC DENSITY FUNCTIONAL THEORY

In the relativistic version of DFT, the ground state energy of the system is a unique functional of the fermion four-current density, $J_\mu^A(r)$:

$$E_A\left[J_\mu^A\right] = F_A\left[J_\mu^A\right] + \int dr J_\mu^A(r) A_{ext}^\mu(r). \tag{10.1}$$

In this equation, A_{ext}^μ is a given external four-potential and $J_\mu^A(x)$ is the four-current density. As the treatment is for stationary states, the external four-potential and the fermion four-current are independent of time.

The variational principle can be stated as

$$E_A\left[J_\mu^A\right] \le E_A\left[\left(J_\mu^A\right)'\right],$$
(10.2)

in which $(J_\mu^A)'$ is any nonequivalent A_μ-representable four-current density.

It is also possible to extend the Kohn–Sham formalism by defining an energy term T_S that includes the kinetic energy of the noninteracting system, and the total rest mass of the electron [8]:

$$T_S = \frac{1}{2}\int dr \left\{ \sum_{\varepsilon_n < \varepsilon_f} \bar{\psi}_n(x)[-i\boldsymbol{\gamma}\cdot\nabla + \boldsymbol{m}]\psi_n(x) - \sum_{\varepsilon_n > \varepsilon_f} \bar{\psi}_n(x)[-i\boldsymbol{\gamma}\cdot\nabla + \boldsymbol{m}]\psi_n(x) \right\}.$$
(10.3)

The equation above is written using the units $\hbar = c = 1$. The quantity $\boldsymbol{\gamma}$ is a vector of Dirac matrices, $\boldsymbol{m}$ is the electron mass multiplied by a Dirac matrix. $\psi_n(x)$ is a spinor dependent on the space and time coordinates, x, and $\bar{\psi}_n(x)$ is its corresponding adjoint. ε_n is an eigenvalue and ε_f is the Fermi level. With this definition the energy functional is

$$E[J^\mu] = T_S[J^\mu] + \int dr J_\mu(x) A_{\text{ext}}^\mu(r) + \frac{1}{2}\int dr dr' \frac{J^\mu(x)J_\mu(x')}{|r - r'|} + E_{XC}[J^\mu].$$
(10.4)

It is important to notice that the Coulomb-like term (the third term in the equation) is written in terms of the four-current, and the exchange and correlation energy is given by

$$E_{XC}[J^\mu] = F[J^\mu] - T_S[J^\mu] - \frac{1}{2}\int dr dr' \frac{J^\mu(x)J_\mu(x')}{|r - r'|}.$$
(10.5)

In Equations 10.1, 10.4, and 10.5, the standard implicit sum on the index μ is assumed. The relativistic Kohn–Sham equations are obtained by minimization of the Equation 10.4 with respect to the orbitals. In contrast to the nonrelativistic case, the variational procedure gives rise to an infinite set of coupled equations (see the summation restrictions in Equation 10.3) that have to be solved in a self-consistent manner:

$$[\boldsymbol{\gamma}\cdot(-i\nabla - A_{\text{eff}}(r)) + m + \gamma_0 v_{\text{eff}}(r)]\psi_n(r) = \varepsilon_n \gamma_0 \psi_n(r);$$
(10.6)

in these equations

$$v_{\text{eff}}(r) = -e\left\{ A_{\text{ext}}^0(r) + \int dr' \frac{J^0(r')}{|r - r'|} + \frac{\delta E_{XC}[J^\mu]}{\delta J^0(r)} \right\}$$
(10.7)

and

$$A_{\text{eff}}(r) = -e\left\{A_{\text{ext}}(r) + \int dr' \frac{J(r')}{|r - r'|} + \frac{\delta E_{\text{XC}}[J^\mu]}{\delta J(r)}\right\}. \tag{10.8}$$

Solving this iterative process gives rise to a set of orbitals to construct the ground state four-current, $J^\mu(x)$, including vacuum polarization corrections due to the external field as well as the field mediating the interaction between the electrons. As the charge density in the nonrelativistic case, the four-current has the form of a reference noninteracting N-electron system, J_μ^A

$$J^\mu(x) \equiv J_S^\mu(x) = -e \sum_{-m < \varepsilon_n < \varepsilon_f} \bar{\psi}_n(x)\gamma^\mu\psi_n(x) + J_S^\mu(x)_{\text{vac}}. \tag{10.9}$$

The summation starts in $-mc^2$, and the energy in the current units is just m. To simplify the iterative process, a standard approximation neglects the vacuum polarization effects giving a simpler structure of the four-current [5,7]:

$$J_S^\mu(r) = -e \sum_{-m < \varepsilon_n < \varepsilon_f} \bar{\psi}_n(x)\gamma^\mu\psi_n(x). \tag{10.10}$$

The general relativistic Hohenberg–Kohn–Sham formalism, outlined above, contains the spin degrees of freedom in a complete form. Consequently, the spin and kinetic motion effects are not separable. Indeed, they are contained in the external potential term as one can see if such term is written using the orbital current density as [9]

$$\langle H_{\text{ext}} \rangle = \int dr \left\{ \hat{J}_0(r)A_{\text{ext}}^0(r) - \hat{j}_{\text{orb}}(r) \cdot A_{\text{ext}}(r) + \frac{\mu_B}{2} \hat{\bar{\psi}}(x)\sigma_{ij}\hat{\psi}(x)F_{\text{ext}}^{ji}(r) \right\}. \tag{10.11}$$

In this expression the orbital current density is

$$j_{\text{orb}}(r) = -\frac{e}{2m} \left\{ \begin{array}{l} \hat{\bar{\psi}}(x)\left[(-i\nabla - e\hat{A}(x) - eA_{\text{ext}}(r))\hat{\psi}(x)\right] \\ -\left[(-i\nabla - e\hat{A}(x) - eA_{\text{ext}}(r))\hat{\bar{\psi}}(x)\right]\hat{\psi}(x) \end{array} \right\}. \tag{10.12}$$

It is important to notice that in Equation 10.11, only the sum of the three terms is Lorentz invariant. The first term corresponds to the interaction of the charge density with the external Coulomb potential and the last term can be written in the form

$$-\int dr\, m(r) \cdot B_{\text{ext}}(r), \tag{10.13}$$

where $m(r)$ is the magnetization and $B_{\text{ext}}(r)$ a external magnetic field. As only the combination of the three terms in Equation 10.11 is covariant, any explicit approximations in terms of the charge density $\rho(r)$ and the magnetization $m(r)$ alone would depend on the Lorentz frame chosen. That is the case for the nonrelativistic scheme described below.

10.2.2 SPIN-POLARIZED NONRELATIVISTIC LIMIT

As in the case of the Schrödinger approach in which spin is introduced by giving a specific form to the wave function, the spin dependence in the Hohenberg–Kohn–Sham formalism in a nonrelativistic framework is introduced by imposing some form of restrictions to the functional. Namely, the total energy can be written as [3,5]

$$E_{v,B}[\rho, m] = F[\rho, m] + \int dr[v(r)\rho(r) - B(r) \cdot m(r)], \qquad (10.14)$$

where $F[\rho, m]$ is a universal functional of the electron density $\rho(r)$ and of the magnetization $m(r)$; $v(r)$ is the external potential related to $A_{ext}^0(r)$ in Equation 10.7.

This functional satisfy a variational principle [3,5] $E_{v_0,B_0}[\rho_0, m_0] \leq E_{v_0,B_0}[\rho, m]$. $E_{v_0,B_0}[\rho_0, m_0]$ denote the ground state energy with density $\rho_0(r)$, and magnetization $m_0(r)$ of a particular system characterized by the external fields $(v_0(r), B_0(r))$. One of the main differences between the spin-restricted and spin-polarized cases is that the one-to-one relation between the external potential and the density cannot be extrapolated to the set of quantities $(v_0(r), B_0(r))$ and $(\rho_0(r), m_0(r))$ [3].

If the external magnetic field $B(r)$, and $m(r)$ have only a nonvanishing z-component, $B(r) = (0, 0, B(r))$ and $m(r) = (0, 0, m(r))$, the universal functional $F[\rho, m]$ may then be considered as a functional of the spin densities $\rho_S(r)$ and $\rho(r)$, $F[\rho_S(r), \rho(r)]$, because the spin density is proportional to the z-component of the magnetization: $m(r) = -\mu_B \rho_S(r)$; μ_B is the electron Bohr magneton. It is of worth mentioning that it is possible to define two spin densities that are the diagonal elements of the density matrix introduced by von Barth and Hedin [3]. These correspond to the spin-up (alpha) electrons density $\rho_\uparrow(r)$, and the spin-down (beta) electrons density $\rho_\downarrow(r)$. In terms of these quantities, the electron and spin densities can be written as

$$\rho(r) = \rho_\uparrow(r) + \rho_\downarrow(r) \quad \text{and} \quad \rho_S(r) = \rho_\uparrow(r) - \rho_\downarrow(r). \qquad (10.15)$$

Then, it is clear that the functional of the energy could be written in terms of any of the two sets of variables $E[\rho, \rho_S]$ or $E[\rho_\uparrow, \rho_\downarrow]$. Consequently, the nonrelativistic spin-polarized DFT can be developed in both sets of variables $\{\rho, \rho_S\}$ or $\{\rho_\uparrow, \rho_\downarrow\}$. In the next section, the set of variables $\{\rho, \rho_S\}$ will be used in the discussion of the reactivity parameters.

The impact of Equation 10.14 in the treatment of spin-polarized systems is in two directions: on the one hand, it allows the inclusion of external magnetic field effects in the description of N-electron systems; on the other hand, when the limit of zero magnetic field is imposed, the formalism becomes useful for N-electron systems having spin-polarized ground state in the absence of a external magnetic field. This last feature has been widely exploited in practical applications within the framework of the Kohn–Sham method.

10.3 CHEMICAL REACTIVITY

The ionization potential and electron affinity are some of the first concepts introduced in chemistry courses to understand chemical reactivity. These quantities measure the energy changes when the system loses or gains electrons. However, when this happens, the system also suffers changes in the paired or unpaired electron number, because the number of electrons N is given by $N = N_\uparrow + N_\downarrow$, where $N_\uparrow$ are the number of electrons with spin-up or α, and $N_\downarrow$ are the number of electrons with spin-down or β. The examples above and many other physical situations indicate that to analyze the chemical reactivity from a more general point of view, a spin-polarized theory is necessary. The spin-polarized density functional theory (SP-DFT) distinguishes between the changes produced by charge transfer, it means changes in N and those produced by the redistribution of the electronic density, i.e., changes in the spin number, $N_S = N_\uparrow + N_\downarrow$.

As mentioned in the previous section, the SP-DFT may be developed in both set of variables $\{\rho_\uparrow, \rho_\downarrow\}$ or $\{\rho, \rho_S\}$. Ghosh and Ghanty [10] used the first set of variables and generated some reactivity indexes within this approach, however, we are presenting here the SP-DFT using the $\{\rho, \rho_S\}$ set that has received more attention since it was presented in 1988 by Galván et al. [11].

If we consider a system in the presence of a magnetic field $B(r)$ in the z-direction, the energy functional of Equation 10.14 can be written as

$$E[\rho, \rho_S, v, B] = F[\rho, \rho_S] - \int v(r)\rho(r)dr - \mu_B \int B(r)\rho_S(r)dr, \qquad (10.16)$$

where $F[\rho, \rho_S]$ is a universal functional of ρ and ρ_S, independent of $v(r)$ and $B(r)$.

One may consider that ρ and ρ_S are independent functions, this implies that the changes in ρ are decoupled from changes in ρ_S. Thus, the minimization of the energy functional of Equation 10.16 can be done with respect to both the variables using the Lagrange multiplier technique.

To assure the correct value of N and N_S, the restrictions to minimize Equation 10.16 are $\int \rho(r)dr = N$ and $\int \rho_S(r)dr = N_S$. In this way, $\{\delta E - \mu_N N\} = 0$ and $\{\delta E - \mu_S N_S\} = 0$ must be solved, where μ_N and μ_S are the Lagrange multipliers associated to the restrictions. Thus, there are two Euler–Lagrange equations for the minimization of Equation 10.16:

$$\mu_N = \left(\frac{\delta E}{\delta \rho(r)}\right)_{\rho_S} = v(r) + \left(\frac{\delta F}{\delta \rho(r)}\right)_{\rho_S} \qquad (10.17)$$

and

$$\mu_S = \left(\frac{\delta E}{\delta \rho_S(r)}\right)_{\rho} = -\mu_B B(r) + \left(\frac{\delta F}{\delta \rho_S(r)}\right)_{\rho}. \qquad (10.18)$$

It is important to notice that the above functional derivatives are performed along specific paths in the $\{\rho, \rho_S\}$ space. Namely, those paths in which $\rho_\uparrow$ and $\rho_\downarrow$ change according to Equation 10.15, to satisfy the condition imposed. By solving Equations 10.17 and 10.18 simultaneously, the Lagrange multipliers and the charge and spin densities are obtained. The total energy is computed by substituting ρ and ρ_S in Equation 10.16. It is important to emphasize that two Lagrange multipliers appeared due to the fact that N and N_S are kept fixed in the minimization, and this is equivalent to fix $N_\uparrow$ and $N_\downarrow$. It may be possible to keep just N fixed during the minimization, thus, only μ_N as in Equation 10.17 would remain and μ_S in Equation 10.18 would be zero.

It can be proved that

$$\mu_N = \left(\frac{\partial E}{\partial N}\right)_{N_S, v(r), B(r)} \tag{10.19}$$

and

$$\mu_S = \left(\frac{\partial E}{\partial N_S}\right)_{N, v(r), B(r)}. \tag{10.20}$$

Equations 10.19 and 10.20 are known as global reactivity indexes within the SP-DFT, because they are constant over the entire space in the molecule. Equation 10.19 resembles the electronic chemical potential found in the spin-restricted DFT, it also measures the energy changes when the electron number in the system varies, and it is called chemical potential as well. However, it is important to note that the derivative in Equation 10.19 is carried out at N_S fixed, this is a different path and we cannot expect that μ of the spin-restricted case and μ_N of the SP-DFT must be equal. Actually, it has been shown for atoms that μ_N does not correspond to the negative of the electronegativity while μ does [12].

Due to the fact that N and N_S are fixed during the minimization, μ_S in Equation 10.20 has an equivalent role to that of μ_N in the mathematical structure of the theory. Thus, μ_S is related to the tendency of the system to change the spin polarization or multiplicity, for this reason this quantity is called the spin potential. In Figure 10.1, a schematic plot for the total energy as a function of N_S for an open-shell atom in the absence of a magnetic field is depicted, there are some important features to observe in this curve. There is a discontinuous first derivative around the ground state multiplicity, it implies that for an open shell system there is one derivative toward higher multiplicities (μ_S^+) and a different one toward lower multiplicities (μ_S^-) and the average (μ_S^0) between both. The spin transfer in a closed shell system only exists toward higher multiplicities. In this way, the values of the slope of the curve E vs. N_S may be used to predict the direction of spin transfer when the system interacts. Some applications of spin potential will be discussed later in this chapter.

There is another important feature to note in the curve of Figure 10.1, the second derivative of the energy with respect to N_S is discontinuous at the ground state multiplicity and must be negative in both directions, due to the fact that both branches in the plot have negative curvatures. This second derivative, as in the

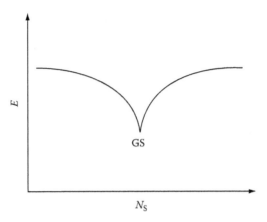

FIGURE 10.1 Schematic representation of the energy as a function of spin number.

spin-restricted DFT case, is related to the system hardness and it is called the spin hardness (η_{SS}),

$$\eta_{SS} = \left(\frac{\partial^2 E}{\partial N_S^2}\right)_{N,v(r),B(r)} = \left(\frac{\partial \mu_S}{\partial N_S}\right)_{N,v(r),B(r)}. \tag{10.21}$$

The nature of the negative of spin hardness was first established by Ortiz [13].

The chemical and spin potentials according to Equations 10.19 and 10.20 can be seen as a function of N, N_S, $v(r)$, and $B(r)$, then one can write the variations of μ_N and μ_S,

$$\delta\mu_N = \eta_{NN}dN + \eta_{NS}dN_S + \int f_{NN}(r)\delta v(r)dr + \mu_B \int f_{SN}(r)\delta B(r)dr \tag{10.22}$$

and

$$\delta\mu_S = \eta_{SN}dN + \eta_{SS}dN_S + \int f_{NS}(r)\delta v(r)dr + \mu_B \int f_{SS}(r)\delta B(r)dr. \tag{10.23}$$

Let us to make a parenthesis in these equations, to analyze the physics behind them. Equations 10.22 and 10.23 must be interpreted in terms of chemical reactivity, for this we must understand first what $\delta v(r)$ and $\delta B(r)$ are. Keeping in mind the meaning of external potential, its changes $\delta v(r)$ may be produced by changes in the geometry of the reacting molecule, but also due to the potential generated by the species that surround it. Similarly, the change in the magnetic field $\delta B(r)$ can also be analyzed as changes produced by the magnetic field generated by the species that surround the reacting molecule, provoking a magnetic interaction. Thus, the response of the chemical system interacting or reacting with another may be understood in general through Equations 10.22 and 10.23, by considering that the changes in the external potential and magnetic field arise from the presence of the surrounding species. As the spin and chemical potentials must be constant over the entire space, if two different species interact to form a new one, two equalization processes must be carried out: the chemical potential equalization and the spin potential equalization. The first one will be achieved by the change in the electron number, that is, charge transfer. For the second, a "spin transfer" must be observed, and it actually corresponds to redistributions in the spin-up and spin-down densities that imply changes in the regional values of N_S, that is, multiplicity changes. Thus, in terms of spin and chemical potentials, there may be cases in which the charge transfer drives the process, because the chemical potential between two species may be very different. Or, there may be another in which the species may have very large differences in their spin potentials, and the spin redistributions drive the process. Or may be both potentials have big differences, then spin and charge transfer would be important. In this sense, all the quantities that appear in Equations 10.22 and 10.23 are reactivity indexes that can give us information about the changes that are carried out when two chemical species interact. We may distinguish in Equations 10.22 and 10.23, the global and local reactivity indexes. The global hardnesses, as the generalized hardnesses

$$\eta_{NS} \equiv \left(\frac{\partial \mu_N}{\partial N_S}\right)_{N,v(r),B(r)} = \left(\frac{\partial \mu_S}{\partial N}\right)_{N_S,v(r),B(r)} \equiv \eta_{SN}, \tag{10.24}$$

$$\eta_{NN} \equiv \left(\frac{\partial \mu_N}{\partial N}\right)_{N_S,v(r),B(r)}, \tag{10.25}$$

and the spin hardness η_{SS} which was defined before in Equation 10.21, are constant over the entire space as the spin and chemical potentials. The local reactivity indexes, appearing in Equations 10.22 and 10.23, are generalizations of Fukui function for the spin-polarized case, so they are called generalized Fukui functions, and they are defined as

$$f_{NN} \equiv \left(\frac{\partial \rho(r)}{\partial N}\right)_{N_S,v(r),B(r)} = \left(\frac{\delta \mu_N}{\delta v(r)}\right)_{N_S,N,B(r)}, \tag{10.26}$$

$$f_{NS} \equiv \left(\frac{\partial \rho(r)}{\partial N_S}\right)_{N,v(r),B(r)} = \left(\frac{\delta \mu_S}{\delta v(r)}\right)_{N_S,N,B(r)}, \tag{10.27}$$

$$f_{SN} \equiv \left(\frac{\partial \rho_S(r)}{\partial N}\right)_{N_S,v(r),B(r)} = -\frac{1}{\mu_B}\left(\frac{\delta \mu_N}{\delta B(r)}\right)_{N_S,N,v(r)}, \tag{10.28}$$

and, finally

$$f_{SS} \equiv \left(\frac{\partial \rho_S(r)}{\partial N_S}\right)_{N,v(r),B(r)} = -\frac{1}{\mu_B}\left(\frac{\delta \mu_S}{\delta B(r)}\right)_{N_S,N,v(r)}. \tag{10.29}$$

These generalized Fukui functions are local reactivity indexes, as they depend on the coordinates r where they are determined, in other words their values vary from one point to another within the molecule. It is clear that these quantities can be used to know how the charge or (and) spin densities respond when there are charge or (and) spin transfer to the reacting molecule.

The relevance of the reactivity indexes that emerge in SP-DFT may be clarified if we turn to the Parr and Yang postulate [14]; it establishes that the preferred direction of a reaction will be such that it provokes the maximum initial chemical potential response of a reactant. In SP-DFT, the chemical potential response ($|\delta\mu_N|$) not only depends on the response of charge density as in the spin-restricted case, but also on $f_{SN}(r)$ which, according to Equation 10.27, takes into account the response of the spin density due to the local charge transfer. It may be the cases of reactants where this could be more important, thus the preferred reactive sites will be identified with large changes in $\rho_S(r)$. In a different situation, the preferred direction of a reaction may be governed by the initial maximum change in the spin potential ($|\delta\mu_S|$), in this case the reactive sites will be those where $f_{NS}(r)$ or $f_{SS}(r)$ are large.

In this way, the SP-DFT using $\{\rho, \rho_S\}$ as variables set provides global and local reactivity indexes that give us the possibility to study processes that involve changes in the number of electrons, multiplicity (changes in the spin number), or both. Some examples are discussed in Section 10.4.

10.4   APPLICATIONS OF GLOBAL SP-DFT REACTIVITY INDEXES

In the $\{N, N_S\}$ representation, the total energy is a function of N, N_S and a functional of $v(r)$ and $B(r)$. Thus, the energy functional can be expanded in a Taylor series around a reference ground state $\left(N^0, N_S^0, v^0(r)\right)$ in the absence of a magnetic field, as [15]

$$
\begin{aligned}
\Delta E \cong{}& E[N, N_S, v(r)] - E\left[N^0, N_S^0, v^0(r)\right] \\
={}& \mu_N^0 \Delta N + \mu_S^0 \Delta N_S + \int \rho^0(r)\Delta v(r)dr + \eta_{NN}^0 (\Delta N)^2 + \frac{1}{2}\eta_{SS}^0 (\Delta N_S)^2 \\
&+ \eta_{NS}^0 (\Delta N)(\Delta N_S) + \Delta N \int f_{NN}^0(r)\Delta v(r)dr \\
&+ \Delta N_S \int f_{NS}^0(r)\Delta v(r)dr + \frac{1}{2}\int\int \chi^0(r, r')\Delta v(r)\Delta v(r')drdr'.
\end{aligned}
\tag{10.30}
$$

In this equation ΔN, ΔN_S, and $\Delta v(r)$ are the changes with respect to each variable in the expansion; $\chi^0(r, r')$ is the linear response function at the reference ground state, and the other quantities have been already defined in the previous section; the upper index 0 indicates that all reactivity indexes are evaluated at the reference state.

The capability of the Equation 10.30 to describe the energy changes has been applied on process at fixed N; it means at a constant number of electrons, in this case, one may analyze changes in the multiplicity without charge transfer. Thus, Equation 10.30 takes the form,

$$\Delta E \cong E\big[N^0, N_S, v(r)\big] - E\big[N^0, N_S^0, v^0(r)\big]$$

$$= \mu_S^0 \Delta N_S + \int \rho^0(r)\Delta v(r)dr + \frac{1}{2}\eta_{SS}^0(\Delta N_S)^2$$

$$+ \Delta N_S \int f_{NS}^0(r)\Delta v(r)dr + \frac{1}{2}\int\int \chi^0(r, r')\Delta v(r)\Delta v(r')drdr'. \qquad (10.31)$$

The changes in Equation 10.31 can be understood by splitting them into different paths (Figure 10.2). Following path I, one may analyze changes where N_S is changing at fixed external potential (vertical multiplicity energy changes), for example in atoms or molecules with a fixed molecular geometry, then Equation 10.31 becomes

$$\Delta E_v \cong \mu_S^0 \Delta N_S + \frac{1}{2}\eta_{SS}^0(\Delta N_S)^2, \qquad (10.32)$$

where the lower index v refers to a vertical change in the energy. It is important to point out that in Equation 10.32 just global reactivity indexes appear and they are evaluated at the ground state.

Using the path II, we may analyze a process when the multiplicity does not change but the external potential does, that is, the relaxation of the geometry would provoke the system going from one stationary state in the potential energy surface to another. Now, Equation 10.31 is reduced to

$$\Delta E_{N_S} \cong \int \rho^0(r)\Delta v(r)dr + \frac{1}{2}\int\int \chi^0(r, r')\Delta v(r)\Delta v(r')drdr'. \qquad (10.33)$$

The second term in Equation 10.33 implies to evaluate a nonlocal quantity, $\chi(r, r')$; the linear response function depends on two different points within the molecule.

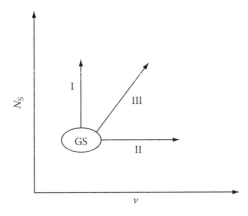

FIGURE 10.2 Schematic representation of the different process that involve changes in the spin number and external potential.

The path III in Figure 10.2 represents the changes in the energy when the multiplicity and external potential change (relaxation on the geometry is permitted); it means adiabatic multiplicity energy changes. This trajectory may be analyzed with the full Equation 10.30.

10.4.1 Vertical Energy Changes

In this section some practical applications of path I are described. As the first example, let us analyze the case of pairing energy, defined as the energy required to produce the spin arrangement of the low spin complex [16,17]; this quantity is useful to predict the preference of a metallic ion to produce high or low spin complexes. For metallic ions, the pairing energy can be related with changes to lower multiplicities without changing the external potential, in this sense Equation 10.32 has been used to analyze this process.

Remembering the previous discussion about Figure 10.1, it is important to distinguish the direction of the process, in the case of pairing energy the direction is toward lower multiplicities and ΔN_S is equal to -2. Thus Equation 10.32 must be expressed as

$$\Delta E_v \cong -2\mu_S^- + 2\eta_{SS}^-. \tag{10.34}$$

It is important to keep in mind that all global quantities appearing in the equations above can be evaluated within the spin-polarized Kohn–Sham method. All the working equations to calculate them depend on the eigenvalues of the frontier Kohn–Sham spin-down and up orbitals. The expressions can be consulted in Refs. [12,18].

By an explicit calculation using frontier orbital eigenvalues, it is shown that μ_S^- follows the trend of pairing energy for cations of metallic atoms, indicating that the spin potential, in this case, measures the tendency of a metal cation to form high or low spin complexes [12]. We also showed that including the spin hardness, as it is described in Equation 10.34, helps to get better quantitative agreement with metal cation pairing energy.

As another example of the applications of Equation 10.32, let us analyze singlet–triplet energy gaps. On the contrary to pairing energy, in this case one has to analyze the energy change from a lower multiplicity state toward a higher multiplicity state. For singlet–triplet energy changes $\Delta N_S = 2$, as for the initial singlet state $N_S = 0$ and for the final triplet state $N_S = 2$. Thus Equation 10.32 becomes

$$\Delta E_v \cong 2\mu_S^{(s)+} + 2\eta_{SS}^{(s)+}. \tag{10.35}$$

It is important to note that the derivatives of the energy (spin potential and spin hardness) must be evaluated to higher multiplicities at fixed external potential, both at the singlet (S) ground state. The Equation 10.35 was applied to study vertical singlet–triplet energy differences, it means without relaxation on the geometry, on halocarbenes, carbenes, silylenes, germylenes, and stannylenes [12,19,20]. It was found that the first order of Equation 10.35, $2\mu_S^{(S)+}$, follows the trend of singlet–triplet vertical energy, exhibiting the spin potential meaning as the tendency of a system to change its multiplicity. In order to include the second-order contribution on Equation 10.35, $\eta_{SS}^{(s)+}$ must be evaluated, it implies a second derivative. However, by a finite difference scheme,

$$\eta_{SS}^{(S)+} \cong \frac{\mu_S^{(T)-} - \mu_S^{(S)+}}{2}, \qquad (10.36)$$

where $\mu_S^{(T)-}$ is the spin potential of the triplet toward lower multiplicity. Note that for the estimation of derivatives up to the second order, two different states are necessary. When the second order is included in Equation 10.35, the singlet–triplet vertical energy is quantitatively described. It was also found that the contribution of the relaxation of the geometry to obtain the singlet–triplet adiabatic energy gaps is constant as this correlates linearly with the vertical singlet–triplet energy gaps. Whether the correlation observed for this class of molecules could be valid for other systems or is just valid for this particular case, is a question waiting to be solved. Such kind of correlations would represent an important simplification of Equation 10.31.

10.4.2 SPIN-PHILICITY AND SPIN-DONICITY

Starting from Equation 10.32, Perez et al. [21] defined spin-donicity and spin-philicity. Following a variational calculation, as used by Parr et al. [22] to define the electrophilic power, they obtain, for the maximum change in energy (ΔE_{max}) when the system modifies its spin number from N_S to $N_S + \Delta N_S$ in a "reservoir of spins,"

$$\Delta E_{max,v} = -\frac{\left(\mu_S^0\right)^2}{2\eta_{SS}^0}. \qquad (10.37)$$

In the direction of increasing spin multiplicity ($\Delta N_S > 0$), using this energy difference the spin-philicity ω_S^+ can be defined as

$$\omega_S^+ \equiv \frac{\left(\mu_S^+\right)^2}{2\eta_{SS}^0}. \qquad (10.38)$$

To obtain results according to chemical intuition, Olah et al. proposed the following convention [19]: a large negative number for the spin-philicity index is obtained when the energy change between the higher and lower spin states is large, and a small negative spin-philicity index when the energy difference is small. It was found that molecules with large negative spin-philicity are good spin catalysts.

In the direction of decreasing spin multiplicity ($\Delta N_S < 0$), the spin-donicity can be defined as

$$\omega_S^- \equiv \frac{\left(\mu_S^-\right)^2}{2\eta_{SS}^0}. \qquad (10.39)$$

A large spin-donicity number is expected when the energy difference is smaller between the triplet and singlet states. Thus, spin-philicity and spin-donicity are also a measure of the energy differences between singlet and triplet states, furthermore, it has been demonstrated the applicability of these reactivity indexes in the prediction of the spin transfer observed in the spin-catalysis phenomenon [21]. Equivalent quantities at fixed N_S ($\omega_{NN}^{\pm}$) have been defined [20].

Spin potential, spin hardness, spin-donicity, and spin-philicity indexes have also been applied successfully to other specific problems [19,23,24].

10.4.3 LOCAL REACTIVITY INDEXES IN SP-DFT

Up to this point, just applications of global reactivity indexes have been analyzed. However, as we can see in Equation 10.30, besides the charge density, generalized Fukui functions ($f_{NN}^0(r)$ and $f_{NS}^0(r)$) come to light and must be considered. When there are changes in the spin number without charge transfer (Equation 10.31), just the charge density and the $f_{NS}^0(r)$ as local reactivity indexes are involved. Some efforts have been taken to apply $f_{NS}^0(r)$. According to Equation 10.27, $f_{NS}^0(r)$ gives us information about the changes in the charge density when there is a "spin transfer," in other words, multiplicity changes. It is well known that the first derivative of the energy with respect to the electron number has discontinuities, furthermore it has been shown that the curve of the energy with respect to N_S, also presents this behavior [12]. These discontinuities are mapped on the charge density, thus, its derivatives must be evaluated in the direction where the changes are carried on. There is a study of $f_{NS}^-(r)$ in atoms, from $Z = 5$ to $Z = 9$, where the valence shell is systematically populated and the unpaired electrons are diminishing [25]. Figure 10.3 shows an example for $Z = 9$, that illustrates this series of general behavior of this generalized Fukui function for this

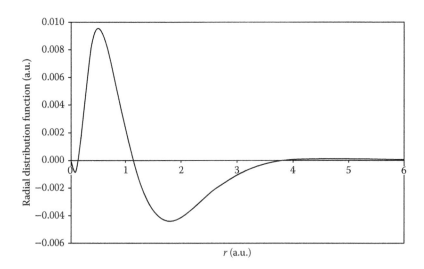

FIGURE 10.3 Radial distribution function of the f_{NS}^- for fluorine atom ($Z = 9$) obtained by numerical derivation of atomic density.

series of atoms. In the series studied, as the nuclear charge increases, the peak in the inner region is closer to the nucleus; the size of that peak depends on the number of unpaired electrons and is smaller at $Z = 9$. To understand this behavior, an approach based on the $\{\rho_\uparrow, \rho_\downarrow\}$ set was applied; [25] the derivatives of $\rho_\uparrow(r)$ and $\rho_\downarrow(r)$ with respect to N_S were analyzed. It was found that the redistribution of $\rho_\uparrow(r)$ toward the nucleus is more pronounced for nitrogen and decays as the number of unpaired electrons decreases, indicating a correlation of the $\rho_\uparrow(r)$ contraction with the number of unpaired electrons. This tendency is in agreement with the idea that the stability of half-filled shells is a consequence of the charge redistributions induced by Fermi correlation.

There is no more research on the analysis or applications of SP-DFT generalized Fukui functions, per se. Instead, condensed-to-atoms SP-DFT Fukui function schemes have been developed and applied to different chemical reactivity problems. In these schemes, the information of the Fukui functions is condensed on an atomic position. In addition, the Fukui function $f_{SS}^{\pm}(r)$ is related with the extension of global to local spin-donicity and spin-philicity, defined as [20]

$$\omega_S^{\pm}(r) = \omega_S^{\pm} f_{SS}^{\pm}(r), \tag{10.40}$$

where instead of $f_{SS}^{\pm}(r)$ the condensed-to-atoms Fukui function models are used and applied to the analysis of chemical reactivity [23,26].

The SP-DFT has been shown to be useful in the better understanding of chemical reactivity, however there is still work to be done. The usefulness of the reactivity indexes in the $\{\rho_\uparrow, \rho_\downarrow\}$ representation has not been received much attention but it is worth to explore them in more detail. Along this line, the new experiments where it is able to separate spin-up and spin-down electrons may be an open field in the applications of the theory with this variable set. Another issue to develop in this context is to define response functions of the system associated to first and second derivatives of the energy functional defined by Equation 10.1. But the challenge in this case would be to find the physical meaning of such quantities rather than build the mathematical framework because this is due to the linear dependence on the four-current and external potential.

ACKNOWLEDGMENTS

The authors thank CONACYT for financial support through projects 49057-F and 60116.

REFERENCES

1. Kohn, W. and Sham, L.J. 1965. Self-consistent equations including exchange and correlation effects. *Phys. Rev.* 140: A1133–A1138.
2. Hohenberg, P. and Kohn, W. 1964. Inhomogeneus electron gas. *Phys. Rev.* 136: B864–B871.
3. Von Barth, U. and Hedin, L. 1972. A local exchange-correlation potential for the spin polarized case: I. *J. Phys. C: Solid State Phys.* 5: 1629–1642.

4. Pant, M.M. and Rajagopal, A.K. 1972. Theory of inhomogeneous magnetic electron gas. *Solid State Comm.* 10: 1157–1160.

5. Rajagopal, A.K. and Callaway, J. 1973. Inhomogeneous electron gas. *Phys. Rev. B.* 7: 1912–1919.

6. Rajagopal, A.K. 1978. Inhomogeneous relativistic electron gas. *J. Phys. C: Solid State Phys.* 11: L943–L948.

7. MacDonald, A.H. and Vosko, S.H. 1979. *J. Phys. C: Solid State Phys.* 12: 2977–2990.

8. Dreizler, R.M. and Gross, E.K.U. 1990. *Density Functional Theory: An approach to the Quantum Many-Body Problem.* Springer-Verlag, New York.

9. Baym, G. 1964. *Lectures on Quantum Mechanics.* Benjamín, New York. p. 552.

10. (a) Ghanty, T.K. and Ghosh, S.K. 1992. Electronegativity, hardness, and chemical-binding in simple molecular-systems. *Inorg. Chem.* 31: 1951–1955. (b) Ghanty, T.K. and Ghosh, S.K. 1994. A frontier orbital density-functional approach to polarizability, hardness, electronegativity, and covalent radius of atomic systems. *J. Am. Chem. Soc.* 116: 8801–8802. (c) Ghanty, T.K. and Ghosh, S.K. 1991. Electronegativity and covalent binding in homonuclear diatomic-molecules. *J. Phys. Chem.* 95: 6512–6514. (d) Ghanty, T.K. and Ghosh, S.K. 1994. A new simple density-functional approach to chemical binding. *J. Phys. Chem.* 98: 1840–1843. (e) Ghanty, T.K. and Ghosh, S.K. 1994. Simple density-functional approach to polarizability, hardness, and covalent radius of atomic systems. *J. Phys. Chem.* 98: 9197–9201. (f) Ghanty, T.K. and Ghosh, S.K. 1994. Spin-polarized generalization of the concepts of electronegativity and hardness and the description of chemical-binding. *J. Am. Chem. Soc.* 116: 3943–3948. (g) Ghosh, S.K. 1994. Electronegativity, hardness, and a semiempirical density-functional theory of chemical-binding. *Int. J. Quantum Chem.* 49: 239–251.

11. Galván, M., Vela, A., and Gázquez, J.L. 1988. Chemical reactivity in spin-polarized density functional theory. *J. Phys. Chem.* 92: 6470–6474.

12. Galván, M. and Vargas, R. 1992. Spin potential in Kohn–Sham theory. *J. Phys. Chem.* 96: 1625–1630.

13. Ortiz, E. 1990. Reactividad Química y Funcionales de la Densidad. Ph. D. Thesis. Universidad Autónoma Metropolitana Iztapalapa: México.

14. Parr, R.G. and Yang, W. 1984. Density functional approach to the frontier-electron theory of chemical reactivity. *J. Am. Chem. Soc.* 106: 4049–4050.

15. Vargas, R., Galván, M., and Vela, A. 1998. Singlet–triplet gaps and spin potentials. *J. Phys. Chem.* 102: 3134–3140.

16. Griffith, J.S. 1956. On the stabilities of transition metal complexes. I. Theory of the energies. *J. Inorg. Nucl. Chem.* 2: 1–10.

17. Griffith, J.S. 1956. On the stabilities of transition metal complexes. II. Magnetic and thermodynamic properties. *J. Inorg. Nucl. Chem.* 2: 229–236.

18. Vargas, R., Cedillo, A., Garza, J., and Galván, M. 2002. Reactivity criteria in spin-polarized density functional theory. In *Reviews of Modern Quantum Chemistry*, ed. K.D. Sen, pp. 936–965. World Scientific Publishing, Singapore.

19. Olah, J., De Proft, F., Veszpremi, T., and Geerlings, P. 2004. Spin-philicity and spin-donicity of substituted carbenes, silylenes, germylenes, and stannylenes. *J. Phys. Chem. A* 108: 490–499.

20. Chamorro, E., Pérez, P., De Proft, F., and Geerlings, P. 2006. Philicity indices within the spin-polarized density-functional theory framework. *J. Chem. Phys.* 124: 044105–044105–7.

21. Perez, P., Andres, J., Safont, V.S., Tapia, O., and Contreras, R. 2002. Spin-philicity and spin-donicity as auxiliary concepts to quantify spin-catalysis phenomena. *J. Phys. Chem. A* 106: 5353–5357.

22. Parr, R.G., Szentpály, L.V., and Liu, S. 1999. Electrophilicity index. *J. Am. Chem. Soc.* 121: 1922–1924.

23. Guerra, D., Andrés, J., Chamorro, E., and Pérez, P. 2007. Understanding the chemical reactivity of phenylhalocarbene systems: an analysis based on the spin-polarized density functional theory. *Theor. Chem. Acc.* 118: 325–335.
24. Oláh, J., Veszprémi, T., and Nguyen, M.T. 2005. Spin-philicity and spin-donicity of simple nitrenes and phosphinidenes. *Chem. Phys. Lett.* 401: 337–341.
25. Vargas, R. and Galván, M. 1996. On the stability of half-filled shells. *J. Phys. Chem.* 100: 14651–14654.
26. (a) Chamorro, E. and Pérez, P. 2005. Condensed-to-atoms electronic Fukui functions within the framework of spin-polarized density-functional theory. *J. Chem. Phys.* 123: 114107-1–114107-9. (b) Pintér, B., De Proft, F., Van Speybroeck, V., Hemelsoet, K., Waroquier, M., Chamorro, E., Veszprémi, T. and Geerlings, P. Spin-polarized conceptual density functional theory study of the regioselectivity in ring closures of radicals. 2007. *J. Org. Chem.* 72: 348–356. (c) De Proft, F., Fias, S., Van Alsenoy, C. and Geerlings, P. 2005. Spin-polarized conceptual density functional theory study of the regioselectivity in the [2 + 2] photocycloaddition of enones to substituted alkenes *J. Phys. Chem. A* 109: 6335–6343. (d) Guerra, D., Castillo, R., Andrés, J., Fuentealba, P., Aizman, A. and Contreras, R. 2006. Homofugality: A new reactivity index describing the leaving group ability in homolytic substitution reactions. *Chem. Phys. Lett.* 424: 437–442. (e) Guerra, D., Fuentealba, P., Aizman, A. and Contreras, R. 2007. β-Scission of thioimidoyl radicals (R$_1$-N-C=S-R$_2$): A theoretical scale of radical leaving group ability. *Chem. Phys. Lett.* 443: 383–388.

11 Hardness of Closed Systems

Ralph G. Pearson

CONTENTS

11.1 INTRODUCTION

Normally, density functional theory (DFT) is used to study large chemical systems, as systems with only a few electrons and nuclei have already been solved by traditional methods. Nevertheless, there are reasons for looking at simple systems in terms of DFT. Being simple, it is much easier to visualize and comprehend any results. Also, we already have a wealth of information available on these systems from earlier work.

The original premise of Hohenberg and Kohn has led to new methods of calculating energies, very useful for large systems [1]. We shall not go into this important area, but instead concentrate on another aspect of DFT: the new chemical concepts that have arisen, mainly from the work of Parr and his coworkers [2].

Two quantities derived from DFT are the electronic chemical potential μ and the chemical hardness η [2]. The definitions of these quantities are

$$\mu = \left(\frac{\partial E}{\partial N}\right)_v \tag{11.1}$$

$$\eta = \left(\frac{\partial \mu}{\partial N}\right)_v \tag{11.2}$$

where
  E is the total energy
  v is the potential due to the nuclei, held fixed in position
  N is the number of electrons

The chemical potential is constant everywhere if the system is in equilibrium. The hardness, however, is a function of position. Equations 11.1 and 11.2 also imply

$$\mu = (\delta E/\delta \rho)_v \qquad (11.3)$$

$$\eta = (\delta \mu/\delta \rho)_v \qquad (11.4)$$

so that changes in ρ will change both μ and η. That is, they have a functional dependence on the electron density ρ.

If a wave function is an approximate one, as is usually the case, μ will not be constant everywhere and electron density should move from regions where μ is too positive to regions where μ is too negative. This will equalize μ and also change the assumed ρ to a better one. The hardness acts as a resistance to changes in ρ due to differences in μ. The energy will decrease from E^0, the value calculated for the approximate wave function, to E, a better one.

Using Equations 11.1 and 11.2, the energy can be expanded about E^0 as a power series in ΔN. From this, the total change in energy can be calculated:

$$\Delta E = E - E^0 = (\mu - \bar{\mu})\Delta N + \frac{\eta}{2}(\Delta N)^2 \qquad (11.5)$$

where ΔN is the weighted sum over all space, of the small changes in electron density for each volume element. If not weighted, this sum would be zero. We are not changing N, the total number of electrons, but rearranging them.

The weighting factor is $(\mu - \bar{\mu}) = \Delta\mu$, where $\bar{\mu}$ is the average value of the chemical potential of the approximate wave function, ψ^0. When $\Delta\mu$ is positive, ΔN is negative; when $\Delta\mu$ is negative, ΔN is positive. In Equation 11.5, the first term is negative and is energy-lowering due to electron density shifting to or from μ to $\bar{\mu}$. The second term is positive and is due to any energy increase that accompanies the overall changes in μ. Actually the electron density at those points in space where $\mu = \bar{\mu}$ does not change. Electron density transferred there is balanced by density transferred out. The average value $\bar{\mu}$ is simply a convenient marker.

The virial theorem tells us that

$$\Delta E = \frac{\Delta V}{2} = -\Delta T \qquad (11.6)$$

so that the energy can only be lowered by lowering the potential energy. The kinetic energy must increase half as much as ΔV is lowered. There are also the characteristics of Equation 11.5, if the first term is all potential energy, then the second term all kinetic energy.

The minimum energy of Equation 11.5 is reached when the fraction of an electron transferred is given by

$$\Delta N = -\frac{\Delta\mu}{\eta} \qquad (11.7)$$

and the energy lowering is a maximum,

$$\Delta E = \frac{\Delta N \Delta \mu}{2} = -\frac{(\Delta \mu)^2}{2\eta} \qquad (11.8)$$

Since the approximate μ is a function of position, Equations 11.7 and 11.8 apply to volume elements ΔV, and must be summed over space:

$$\Delta E = -\int \frac{(\mu - \bar{\mu})^2}{2\eta} \rho dV \qquad (11.9)$$

Equation 11.9 certainly implies that the local hardness should be used. It is known that this depends only on the functional dependence of the kinetic energy and electron repulsion terms upon the value of ρ [3]. However, it is difficult to calculate local values. In spite of this uncertainty, Equation 11.9 or its equivalent has often been used to calculate the interaction between two chemical systems [4].

In atomic theory or molecular orbital theory, the chemical potential is related to the orbital energy [5,6]. In the case of one or two electrons in the same orbital, the local orbital energy ε is equal to the local chemical potential:

$$\varepsilon = \mu = v_n + v_e - \frac{\nabla^2 \Phi}{2\Phi} \qquad (11.10)$$

where
   v_n is the potential due to the nuclei
   v_e the potential of the other electrons, and the last term is the kinetic energy
   Φ is an orbital

Confusion can arise because, in this chapter, both the chemical potential and the hardness will be treated as variables dependent on position. The original articles introducing the chemical hardness gave a useful operational definition [7].

$$\eta = (I - A) \qquad (11.11)$$

where
   I is ionization potential
   A the electron affinity of the system [8]

Another approximate equation is

$$\eta = (\varepsilon_{LUMO} - \varepsilon_{HOMO}) \qquad (11.12)$$

where the hardness is the energy gap between the highest occupied molecular orbitals (HOMO) and the lowest unoccupied molecular orbitals (LUMO) [8]. In addition, there are a number of other approximate equations [9]*. The corresponding operational definition for the electronic potential is

$$-\mu = \frac{1}{2}(I + A) \qquad (11.13)$$

---

* A factor of 1/2 from Ref. [7] has been dropped.

for the chemical potential [10]. In the latter case, $-\mu$ has been called the absolute electronegativity [11].

These are constant numbers obtained from experiments or independent calculations. They are properties of the atom or the molecule and have been useful in predicting the chemical behavior of the system. In addition to its use as a scale of electronegativity, the chemical potential is also a measure of the intrinsic strength of generalized acids and bases [12].

The hardness measures the stability of the system. A hard molecule resists changes within itself, or in reaction with others. As a result, a molecule will arrange itself to be as hard as possible, the principle of maximum hardness. This usually is interpreted as the placing of the nuclei.

The variable hardness in this work is the local hardness as given by the basic theory [2]. The electronic chemical potential in this work is a property if a given molecule (arrangement of nuclei) is also of the approximate wave function used to describe it. This does not represent an equilibrium system. The variation of the chemical potential is a consequence.

The average value of the orbital energy can be calculated from the assumed wave function. If the wave function were exact, the orbital energy would indeed be constant, as required. For the approximate Ψ^0, the average orbital energy is equal to the ionization potential I, according to Koopmans' theorem. In Equation 11.13, the electron affinity A has dropped out.

This is reasonable as Equations 11.11 and 11.13 apply to open systems where electrons can be gained from or lost to systems external to the one of interest. Quantities such as A or $\varepsilon_{\text{LUMO}}$ have much relevance.

In this chapter, we are trying to improve the energy of an approximate wave function by rearranging the electron density corresponding to it. We are dealing with closed systems in which electrons can only be lost. The hardness of a closed system is quite different from that of an open system. The latter is useful for predicting chemical behavior. The former is useful for predicting the stable structure of chemical systems and the energies of such structures. It is the hardness to which the principle of maximum hardness applies [13,14].

11.2 HARDNESS

Basic theory does give recipes for finding both the local hardness and the average, or global hardness [2,15]. Unfortunately, they are very difficult to use. Let us examine an equation for the local hardness $\eta(r)$,

$$\eta(r) = \int \frac{\delta^2 F \rho^1 dV^1}{\delta\rho\delta\rho^1} \tag{11.14}$$

where F is the local value of the kinetic energy and electron-repulsion energy of the system [22]. The equation may be interpreted as the change in F due to a change in ρ, as influenced by the value of ρ that already exists. Obviously this is why v_n is not included in F. The local hardness is the average value of a quantity called the hardness kernel. The global hardness, in turn, is the average value of the local hardness. The interpretation of Equation 11.14 is that $\eta(r)$ is the change in F caused by a change in ρ, as influenced by the value of ρ that existed before the change.

The examples to be considered are the ground states of He and the related two-electron ions from H to Ne^{8+}. In all cases, the single-zeta function of Kellner will be the approximate wave function used [16]. This function is $\Psi^0 = Ne^{-\alpha_0 r}$. The local chemical potential is given by

$$\mu^0 = \varepsilon^0 - \frac{2}{r} + \frac{1}{r_{12}} - \frac{\alpha_0{}^2}{2} + \frac{\alpha_0}{r} \tag{11.15}$$

The terms in Equation 11.15 are v_n, v_e, and t, in that order, for He, a_0 is 1.6875. Now consider the two-electron ions from H^- to Ne^{8+}. All values of α_0 are given by the simple relation, $\alpha_0 = Z - 0.3125$, where Z is the nuclear charge. Therefore, α_0 will vary from 0.6875 to 9.6875 in going from H^- to Ne^{8+}. Effects scaling with either α_0 or $\alpha_0{}^2$ should be distinguishable. All the properties of the two-electron systems are the same, except for the changes due to α_0. We also know that v_e scales as α_0 and t scales as $\alpha_0{}^2$.

Table 11.1 shows the values of $E^0 = -\alpha_0{}^2$, E_{HF} the Hartree–Fock energy, and E_0, the nearly exact values for the series. In spite of the large range of α_0 values, the values of $E - E_{HF}$ and $E - E_0$ are almost constant. They range from 0.0550 to 0.0591 for $E - E_0$. If H^- is omitted, the ranges are even smaller. It is important that the corrections needed are small, showing that the single-zeta wave function is quite good. Equation 11.8, which we are testing, is based on Equation 11.5. This is obviously an approximation, valid for small ΔN. This is only possible if the starting wave function is a good one.

TABLE 11.1

Various Energies for the Two-Electron Ions

Ion	$-E^0$	$-E_{HF}$[a]	$-E_0$[b]	$-E_c$
H^-	0.4727	0.4877	0.5278	0.5263
(He)	2.8477	2.8616	2.9037	2.9013
Li^+	7.2227	7.2365	7.2799	7.2763
Be^{2+}	13.5977	3.6113	13.6556	13.6513
B^{3+}	21.9727	21.9862	22.0310	22.0263
C^{4+}	32.3477	32.3612	32.4063	32.4013
N^{5+}	44.7227	44.7362	44.7815	44.7763
O^{6+}	59.0977	59.1111	59.1566	59.1513
F^{7+}	75.4727	75.4861	75.5317	75.5267
Ne^{8+}	93.8477	93.8611	93.9068	93.9013

[a] C.C.J. Roothaan and A.W. Weiss, *Rev. Mod. Phys.*, 186, 32, 1960.

[b] C.L. Pekeris, *Phys. Rev.* 1699, 112, 1958.

In addition to the difference between t and v_e in their dependence on α_0, there is another important difference between them. The repulsive potential felt by an electron of, say, α-spin, depends on the existing density of β-spin. But the kinetic energy depends on the existing density of α-spin. Any change in the density of α-spin will cause a change in the kinetic energy. Since the hardness acts as a resistance to change, the increase in kinetic energy that occurs (Equation 11.6) must be more important than the change in v_e, which decreases. This leads to a result for $\eta(r)$ as

$$\eta(r) = t_1 + t_2 = \frac{\alpha_0}{r_1} + \frac{\alpha_0}{r_2} \tag{11.16}$$

Averaging over the two-electron density function ρ leads to the global value $\eta = 2\alpha_0^2$. The constant part of the kinetic energy does not change with changes in ρ. Also, it has a negative value.

There are two ways of expressing v_e. One is to find the average value of the interelectron potential by integrating over the coordinates of one electron. This gives v_e as a function of the position of the second electron:

$$v_e = \frac{1}{r} - \frac{e^{-2\alpha_0 r}}{r}(1 + \alpha_0 r) \tag{11.17}$$

The overall average can now be found by integrating over the coordinates of the second electron.

This gives the well-known result $\langle v_e \rangle = (5\alpha_0/8)$. The second way is to use the instantaneous value, $v_e = (1/r_{12})$. The average value can now be found by using the Hylleraas coordinates $u = r_{12}$, $s = (r_1 + r_2)$, and $t = (r_1 - r_2)$ [17]. This also gives $\langle v_e \rangle = (5\alpha_0/8) = \langle {}^1/u \rangle$.

But Equation 11.9 also requires finding $\langle {}^1/u^2 \rangle$. The Hylleraas coordinates give a value of $(2\alpha_0^2/3)$. In an earlier work [5], the potential of Equation 11.17 was simply squared leading to an average value of $\langle {}^1/u^2 \rangle$ equal to $(\alpha_0^2/2)$. Using this result and using the global value of $\eta = 2\alpha_0^2$, we find

$$-\Delta E = \frac{\langle (\mu - \bar{\mu})^2 \rangle}{2\eta} = \frac{0.0548\alpha_0^2}{4\alpha_0^2} = 0.0137 \tag{11.18}$$

This correction is constant for all values of α_0 and is very close to the average value of $(E - E_{HF})$ for all the systems, excluding H^-, of value 0.0136. The independence of the correction on α_0 is important, as it shows that the hardness must indeed depend only on the kinetic energy, as was assumed.

It is also reasonable that using the average value of the interelectronic potential should correct the energy only to the Hartree–Fock level. There is no correlation between the two electrons in Ψ^0, and Equation 11.18 introduces none. However, the instantaneous value of $\langle {}^1/r_{12} \rangle$ does introduce correlation. Note that the correlation is in μ and not in Ψ^0, which is unchanged.

Using the instantaneous value of $(1/u)$ and the correct value of $\langle 1/u^2 \rangle$, we can recalculate $\langle (\mu - \bar{\mu})^2 \rangle$, which is $0.2956\alpha_0^2$, much greater than the $0.0548\alpha_0^2$ of Equation 11.18. We will also find ΔE greater than before, showing correlation. But if we use $\eta = 2\alpha_0^2$, as before, we find $\Delta E = -0.0739$ a.u., whereas we only need a correction of about -0.058 a.u. The average hardness needed must be larger. Adding v_e to η almost works for He but is inadequate for Ne^{8+}, because $\langle v_e \rangle$ depends on α_0, not on α_0^2. The hardness must be increased in a way that keeps it dependent on α_0^2.

Fortunately the local hardness $\eta(r)$ is simple enough so that we can solve Equation 11.9 directly. This gives ΔE directly and the result of -0.0586 a.u. is valid for all the systems. Adding this correction to E^0 gives the corrected energies E_c, shown in Table 11.1. The closeness of E_c to E_0 is better than could have been expected, even with H^- included.

The average error is -0.005 a.u., which is 3.15 kcal/mol. Clearly this is near the range of a chemically significant accuracy. Of course, if we now use the local hardness and the v_e of Equation 11.17, we no longer get the excellent agreement that the average hardness gave. The calculated ΔE is only about -0.006 a.u. The good results when the average hardness is used to correct the average potential must be regarded as largely fortuitous. But there may be some connection between using average values in both cases.

11.3   SUMMARY

It appears that the simple single-zeta function contains all the information needed to calculate a nearly exact energy for the ground energy of all two-electron atomic systems. The calculations are very simple. We need expressions for the local chemical potential μ^0, and the local hardness $\eta(r)$. These are properties of ψ^0 only. The correction to the energy occurs by electron density moving to a lower energy chemical potential. If the instantaneous value of the interelectronic potential is used, the movement is correlated. Density moves from small values of r_{12} to larger values. This causes a substantial lowering of $\langle v_e \rangle$.

It is remarkable that the energy scale given by the chemical potential of an approximate wave function can lead to an energy close to that of the exact wave function. The implications are, of course, very great. But it is by no means certain that these results for a two-electron, single-orbital system, can be generalized.

The conclusion that the local hardness is given entirely by the variable parts of the kinetic energy is very logical. It is the kinetic energy increase which limits the distribution of electron density in all systems with fixed nuclei. Since the equilibrium state of atoms and molecules is characterized by minimum energy, they will also be marked by maximum kinetic energy because of the virial theorem. This will put them in agreement with the principles of maximum hardness, for which much evidence exists.

The proof of the equation for local hardness is as much heuristic as mathematical. The constant value for $(E_0 - E^0)$ over a large range of α_0 values is well established. Since $(\mu - \bar{\mu})^2$ varies as α_0^2, the hardness must also vary as α_0^2. The validity of Equations 11.5 and 11.8 is supported by calculations showing that

ΔE does equal $\langle (\Delta N \Delta \mu/2) \rangle$, as predicted by Equation 11.8 [18]. This strengthens Equation 11.9. The accuracy of ΔE is probably limited by Equation 11.5, which is based on an expansion in which ΔN is small.

In earlier work, it was thought that $(\mu - \bar{\mu})$ was an approximation for $(\mu - \mu_0)$, where μ_0 is the exact chemical potential. This is incorrect, as there is no assurance that the calculation will lead to the exact value of the energy, or the exact ρ_0. The advantage of using $\bar{\mu}$ is that it will always lower the energy and reduce the variance of μ, which are the desired results. Also it is an easily calculated property. Finally, if ρ_{HF} is calculated independently, using Ψ_{HF}, it is found that $(\rho_{HF} - \rho^0)$ is zero when $(\mu - \bar{\mu}) = 0$. That is, there is no driving force for density transfer at these points in space [5]. This is not true for $\mu = \mu_0$, or any other value for the chemical potential.

Extension of this method for correcting the energies of approximate wave functions to systems containing more electrons and orbitals would be very useful. But difficulties quickly arise. The interelectronic effects become complicated because of exchange and correlation. More importantly, in DFT, it is only the highest occupied orbital whose energy is equal to the electronic chemical potential. This potential is valid for the total electron density.

However, in studies of the atoms from lithium to neon, good results have been obtained by simply assuming that the orbital energies of lower-lying orbitals are equal to "orbital" electronic potentials [5,19]. The calculations were made using a global hardness equal or nearly equal to that derived in this work. The average electron potential of Equation 11.17 was used. The energy corrections were made only to the Hartree–Fock level, as expected. Similar calculations have been made for H_2^+ and H_2 [5].

REFERENCES

1. Hohenberg, P.; Kohn, W. *Phys. Rev.* 1964, 136B, 864.
2. See Yang, W.; Parr R.G. *Density Functional Theory of Atoms and Molecules*, Oxford University Press, New York, 1989, for a full discussion.
3. Berkowitz, M.; Parr, R.G. *J. Chem. Phys.* 1988, 88, 2554.
4. For examples see Pearson, R.G. *Chemical Hardness*, John Wiley-VCH, Weinheim, 1997.
5. Pearson, R.G.; Palke, W.E. *Int. J. Quantum Chem.* 1990, 37, 103.
6. Alonso, J.A.; March, N.H. *J. Chem. Phys.* 1970, 53, 451.
7. Parr, R.G.; Pearson, R.G. *J. Am. Chem. Soc.* 1983, 105, 7512.
8. Pearson, R.G., *Proc. Natl. Acad. Sci. U.S.A.* 1986, 83, 8440.
9. Harbola, M.K.; Parr, R.G.; Lee, C. *J. Chem. Phys.* 1991, 94, 6055. Torrent-Succarat, M.; Duran, M.; Sola, M. *J. Phys. Chem. A.* 2002, 106, 4632.
10. Parr, R.G.; Donnelly, R.A.; Levy, M.; Palke, W.E. *J. Chem. Phys.* 1978, 68, 3801.
11. Mulliken, R.S. *J. Chem. Phys.* 1934, 2, 782.
12. Ayers, P.W.; Parr, R.G.; Pearson, R.G. *J. Chem. Phys.* 2006, 124, 194, 107.
13. Pearson, R.G. *Acc. Chem. Res.* 1993, 26, 250.
14. Parr, R.G.; Zhou, Z. *Acc. Chem. Res.* 1993, 26, 256.
15. Chattaraj, P.K.; Cedillo, A.; Parr, R.G. *J. Chem. Phys.* 1995, 103, 7645.
16. Kellner, G.W., *Z. f. Physik.* 1927, 44, 91, 110.
17. Hylleraas, E. *Adv. Quantum Chem.* 1964, 1, 1.
18. Pearson, R.G. *Int. J. Quantum Chem.* in press.
19. Pearson, R.G. *Int. J. Quantum Chem.* 2002, 86, 273.

12 Fukui Function and Local Softness as Reactivity Descriptors

Asit K. Chandra and Minh Tho Nguyen

CONTENTS

12.1 INTRODUCTION

In density functional theory (DFT), electron density is the key quantity determining the properties of a molecular system. Electron density is always positive and its value constitutes a fundamental descriptor. However, chemical reactivity of a molecule cannot be described by its electron density alone, because the course of a reaction is rather determined by its response toward different perturbations caused by an approaching reagent. Sensitivities of an electron density toward structural modifications and its responses to changes in external potential and conditions are actually more important in reflecting the reactivity of the corresponding system, than its absolute values. Several global and local reactivity indices have thus been derived within the framework of DFT that are basically the measures of molecular system's responses. As discussed in various chapters of this book, these global and local reactivity indices, such as chemical potential (μ), hardness (η), and Fukui function, are defined as the first or second derivative of electronic energy and electron density.

The global parameters help understanding the behavior of a system and lead to applicable and useful principles such as the principle of maximum hardness (MHP) [1]. In this chapter, however, our main focus is to introduce the working formula of local reactivity parameters, their actual computations, and practical ways of application to different types of organic reactions. In this process, we mention briefly some of the relevant global reactivity parameters and their calculations as well just to have continuity in the subject matter.

12.2   GLOBAL PARAMETERS

The chemical potential (μ), electronegativity (χ), hardness (η), and softness (S) are defined as follows [2]:

$$\mu = \left(\frac{\partial E}{\partial N}\right)_{v(r)} = -\chi \tag{12.1}$$

$$\eta = \frac{1}{2}\left(\frac{\partial \mu}{\partial N}\right)_{v(r)} = \frac{1}{2}\left(\frac{\partial^2 E}{\partial N^2}\right)_{v(r)} \tag{12.2}$$

$$S = \frac{1}{2\eta} \tag{12.3}$$

where
    E is the energy
    N the number of electrons
    v the external potential of the molecular system under consideration

Evaluation of μ and η faces a practical difficulty due to the discontinuity of the energy E with respect to the variation of N [3]. The implication of this discontinuity in conceptual DFT has also been pointed out recently [4]. One generally makes a finite difference approximation to calculate these quantities, and Scheme 12.1

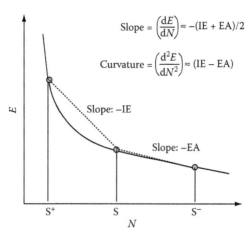

$$\text{Slope} = \left(\frac{dE}{dN}\right) \approx -(\text{IE} + \text{EA})/2$$

$$\text{Curvature} = \left(\frac{d^2E}{dN^2}\right) \approx (\text{IE} - \text{EA})$$

Slope: $-$IE

Slope: $-$EA

SCHEME 12.1

provides a practical way for evaluating them. Accordingly, for a substrate S, the energies of S, S^+, and S^- are determined, all at the optimized structure of S to keep the external potential constant, and the first vertical ionization energy (IE) and electron affinity (EA) can be evaluated from the respective energy differences.

Once the energies of the neutral (S), cationic (S^+), and anionic (S^-) systems are known, chemical potential and hardness (or softness) values can easily be estimated from the following formulas:

$$\mu = -(\text{IE} + \text{EA})/2 = (E_{N+1} - E_{N-1})/2 \tag{12.4}$$

$$\eta = (\text{IE} - \text{EA})/2 = (E_{N-1} + E_{N+1} - 2E_N)/2 \tag{12.5}$$

In the simplest frozen orbital approach, both IE and EA values can be approximated as the negative of the highest occupied molecular orbital (HOMO) and lowest unoccupied molecular orbital (LUMO) energies, respectively, following the Koopmans' theorem. A better way is to calculate the energies of the system and its cationic and anionic counterparts separately and then estimate μ and η from Equations 12.4 and 12.5, respectively.

The hardness (or softness) is a global molecular property and provides information about the general behavior of a reactive molecule. Theoretical justification of the well-known hard and soft acids and bases (HSAB) principle [5] has also been given using chemical potential and softness [6]. Further explanations about the basis of HSAB rule have been given from the minimum energy principle and MHP. It has been shown that the fundamental driving force behind the HSAB principle is electron transfer [7]. When the HSAB principle was proposed, the size, charge, and dipole polarizability (α) of systems were used for a qualitative hard and soft classifications. There have been, in fact, numerous analytical and numerical evidences that softness has a closer link with those properties, especially with polarizability [8,9] and can thus be used for the hard–soft classification of molecules in a quantitative way.

12.3 LOCAL REACTIVITY DESCRIPTORS

In order to understand the detailed reaction mechanism such as the regio-selectivity, apart from the global properties, local reactivity parameters are necessary for differentiating the reactive behavior of atoms forming a molecule. The Fukui function [10] (f) and local softness [11] (s) are two of the most commonly used local reactivity parameters.

The Fukui function is primarily associated with the response of the density function of a system to a change in number of electrons (N) under the constraint of a constant external potential [$v(r)$]. To probe the more global reactivity, indicators in the grand canonical ensemble are often obtained by replacing derivatives with respect to N, by derivatives with respect to the chemical potential μ. As a consequence, in the grand canonical ensemble, the local softness $s(r)$ replaces the Fukui function $f(r)$. Both quantities are thus mutually related and can be written as follows:

$$f(r) = \left(\frac{\partial \rho(r)}{\partial N}\right)_{v(r)} = \left(\frac{\delta \mu}{\delta v(r)}\right)_N \tag{12.6}$$

$$s(r) = \left(\frac{\partial \rho(r)}{\partial \mu}\right)_{v(r)} = \left(\frac{\partial \rho(r)}{\partial N}\right)_v \left(\frac{\partial N}{\partial \mu}\right)_v = S \cdot f(r) \qquad (12.7)$$

Accordingly, Fukui function also represents the response of the chemical potential of a system to a change in external potential. As the chemical potential is a measure of the intrinsic acidic or base strength, and the local softness incorporates the global reactivity, both parameters provide us with a pair of indices to probe for example, the specific sites of interaction between two reagents. Generally, it is demonstrated that the larger the value of the Fukui function, the greater the reactivity of the corresponding site.

Once again, due to the discontinuity of the electron density with respect to N, finite difference approximation leads to three types of Fukui function for a system, namely (1) $f^+(r)$ for nucleophilic attack measured by the electron density change following addition of an electron, (2) $f^-(r)$ for electrophilic attack measured by the electron density change upon removal of an electron, and (3) $f^0(r)$ for radical attack approximated as the average of both previous terms. They are defined as follows:

$$f^+(r) = \rho_{N+1}(r) - \rho_N(r)$$
$$f^-(r) = \rho_N(r) - \rho_{N-1}(r) \qquad (12.8)$$
$$f^0(r) = \frac{1}{2}[\rho_{N+1}(r) - \rho_{N-1}(r)]$$

Using one-electron orbital picture, Fukui functions can be approximately defined as

$$f^+(r) = \rho_{\text{LUMO}}(r)$$
$$f^-(r) = \rho_{\text{HOMO}}(r) \qquad (12.9)$$
$$f^0(r) = \frac{1}{2}[\rho_{\text{LUMO}}(r) - \rho_{\text{HOMO}}(r)]$$

These relations highlight the fact that the formalism of DFT-based chemical reactivity built by Parr and coworkers, captures the essence of the pre DFT formulation of reactivity under frontier molecular orbital theory (FMO). Berkowitz showed that similar to FMO, DFT could also explain the orientation or stereoselectivity of a reaction [12]. In addition, DFT-based reactivity parameters are augmented by more global terms expressed in the softness.

For studying reactivity at the atomic level, however, a more convenient way of calculating the $f(r)$ functions at atomic resolution is used. The condensed-to-atom Fukui functions for an atom k in a molecule are expressed as [13]

$$f_k^+ = q_k(N+1) - q_k(N)$$
$$f_k^- = q_k(N) - q_k(N-1) \qquad (12.10)$$
$$f_k^0 = \frac{1}{2}[q_k(N+1) - q_k(N-1)]$$

where q_k is the electronic population of atom k in the molecule under consideration.

Fukui functions and other response properties can also be derived from the one-electron Kohn–Sham orbitals of the unperturbed system [14]. Following Equation 12.9, Fukui functions can be connected and estimated within the molecular orbital picture as well. Under frozen orbital approximation (FOA of Fukui) and neglecting the second-order variations in the electron density, the Fukui function can be approximated as follows [15]:

$$f^\alpha(r) \approx \left| \phi_f^\alpha(r) \right|^2 \tag{12.11}$$

where $\phi_f^\alpha(r)$ is a particular FMO chosen depending upon the value of $\alpha = +$ or $-$. Expanding the FMO in terms of the atomic basis functions, an orbital (say μ) component of the Fukui function can be defined as

$$f_\mu^\alpha = |c_{\mu\alpha}|^2 + c_{\mu\alpha} \sum_{\nu \neq \varpi} c_{\nu\alpha} S_{\mu\nu} \tag{12.12}$$

where
$c_{\mu\alpha}$ is the expansion coefficient for orbital α
$S_{\mu\nu}$ is the overlap integral between basis functions $\chi_\mu(\mathbf{r})$ and $\chi_\nu(\mathbf{r})$

Similar to condensed Fukui function defined in terms of atomic charge, one can also define condensed-to-atom (k) Fukui function, by summing over contribution of all the basis functions centered at k:

$$f_k^\alpha = \sum_{\mu \in k} f_\mu^\alpha \tag{12.13}$$

This provides us with an avenue for the direct evaluation of Fukui function without considering the cationic and anionic systems. However, this approach is not generally accepted due to many inherent limitations, and Fukui functions are evaluated from finite difference formula (Equation 12.10) using atomic charges. Once the Fukui function is evaluated following a particular scheme, condensed-to-atom softness can easily be evaluated from the relation (following Equation 12.7)

$$s_k^\alpha = S \cdot f_k^\alpha \quad (\alpha = +, - \text{ or } 0) \tag{12.14}$$

Since all charge schemes obtained from partition of the total electron density are arbitrary (do not come from first-principles), evaluation of Fukui functions from Equation 12.10 suffers from the same deficiency. The value of the Fukui function inherently depends upon the charge partitioning scheme and also on the quantum chemical method and one-electron basis set used for electronic structure calculations. Atomic net charges obtained from the natural population analysis (NPA) [16] and electrostatic potential driven charges are perhaps the good choice for calculating Fukui functions. Recently, it has been advocated that Hirshfelder population analysis (HPA)-based atomic charges may be a better choice for Fukui function calculations,

TABLE 12.1

Total Energies (E_T in a.u.) and Net Atomic Charges (q) on Carbon and Oxygen Atom of H_2CO and Its Corresponding Cationic (H_2CO^+) and Anionic (H_2CO^-) Species as Obtained from the B3LYP/6-311G(d,p) Calculations at the Optimized Structure of H_2CO

	H_2CO	H_2CO^+	H_2CO^-
E_T	-114.53634	-114.13970	-114.47139
q_c	0.29	0.33	-0.26
q_0	-0.48	0.06	-0.75

especially to avoid seemingly unrealistic negative Fukui function [17], and plausible explanations have also been given to show why partitioning scheme based on HPA would be a better choice [18]. Another interesting aspect about these parameters is that although their genesis is in DFT, they can however be calculated using any electronic structure theory, either ab initio MO theory or DFT.

Let us consider an example for the calculation of reactivity parameters of formaldehyde (H_2CO) from DFT calculations (Table 12.1).

Accordingly, we can derive the different parameters as

- Hardness: $\eta = (-114.13970 - 114.47139 + 2 \times 114.53634) = 0.2307$ a.u. $= 6.28$ eV
- Condensed-to-atom Fukui functions for nucleophilic and electrophilic attacks on H_2CO estimated from Equation 12.10:

$$f_c^+ = 6.26 - 5.71 = 0.55 \quad \text{and} \quad f_0^+ = 8.75 - 8.48 = 0.27$$

$$f_c^- = 5.71 - 5.67 = 0.04 \quad \text{and} \quad f_0^- = 8.48 - 7.94 = 0.54$$

Condensed-to-atom softness values for nucleophilic and electrophilic attacks can easily be estimated by multiplying the respective FF values by global softness value.

12.4 SOME APPLICATIONS

12.4.1 APPLICABILITIES

The most useful and important application of Fukui function and local softness resides in the interpretation and thereby, prediction of reaction mechanism, especially in the site selectivity or regioselectivity. Since long FMO theory has generally been used to probe the regioselective nature of a reaction, in particular of organic compounds, but the DFT-based local reactivity parameters have emerged as useful alternative tool for rationalizing, interpreting, and predicting diverse aspects of chemical bonding and molecular mechanism. Let us first outline the general procedures for applying local reactivity parameters.

In general, an application of Fukui function is based on the following consideration: "Of the two different sites with generally similar dispositions for reacting with a given reagent, the reagent prefers the one which is associated with the maximum response of the system's chemical potential. Thus, the greater the Fukui function value the higher should be the reactivity" [10].

The next consideration is the HSAB principle formulated at a local level. Let us consider the interaction energy between two chemical species A and B, in which one is electrophilic and the other nucleophilic. From a global point of view and neglecting the effect of change in external potential of A and B, the change in grand canonical potential can be expressed as [7a]

$$\Delta\Omega_A = -\frac{1}{2}\frac{(\mu_B - \mu_A)^2}{(S_A + S_B)^2}S_B^2 S_A \tag{12.15}$$

$$\Delta\Omega_B = -\frac{1}{2}\frac{(\mu_B - \mu_A)^2}{(S_A + S_B)^2}S_A^2 S_B \tag{12.16}$$

It can be shown that grand potential of all the atoms in A and B becomes minimum, when A and B have an approximately equal global softness. Extending the idea to the atomic level, when two molecules A and B approach to each other to form a new molecule AB, then the change in grand potential for each atom in A (say i) and B (say k) can be written as [19]

$$\Delta\Omega_{Ai} = -\frac{1}{2}\frac{(\mu_B - \mu_A)^2}{(S_A + S_B)^2}S_B^2 S_A f_{Ai} \tag{12.17}$$

$$\Delta\Omega_{Bk} = -\frac{1}{2}\frac{(\mu_B - \mu_A)^2}{(S_A + S_B)^2}S_A^2 S_B f_{Bk} \tag{12.18}$$

If the interaction between A and B occurs through the ith atom of A and the kth atom of B, then the most favorable situation that arises from the minimization of $\Delta\Omega_{Ai}$ and $\Delta\Omega_{Bk}$ leads to $s_{Ai} = s_{Bk}$. Hence, the interaction between A and B is favored when it takes place between those atoms whose softnesses are approximately equal. This is essentially the local HSAB principle.

It has been shown [20] from the energy perturbation analysis that for hard reaction, the site having a minimal Fukui function is the most reactive site, whereas the site having the maximal Fukui function is preferred for soft reaction. This differs slightly from the original proposition of Parr and Yang, but considering the fact that hard reactions are generally dominated by long-range electrostatic interactions, such a trend is expected. In fact, recently it was shown from the deprotonation study of 1,2-dialkylpyridinium ions, that reactivity for hard–hard interactions can be explained only from minimal Fukui function criteria [21]. Of course, this conjecture is yet to be tested for a wide variety of systems. Meanwhile, it has been demonstrated taking protonation of hydroxylamine as a test case that the hard–hard interactions are actually charge controlled and thus charge is the better descriptor, whereas Fukui function is the ideal descriptor for soft–soft interactions [22]. Moreover, for strong polarizing reagent, or for reaction leading to strong charge reorganization (such as protonation), this reactivity parameter may not be applicable, because they are based on isolated molecular properties. Nevertheless, in most of the cases in organic reaction, both hard and soft reactions can be rationalized by the softness matching criteria originated from local HSAB principle.

12.4.2 INTRAMOLECULAR REACTIVITY

Evaluation of the only appropriate Fukui function is required for investigating an intramolecular reaction, as local softness is merely scaling of Fukui function (as shown in Equation 12.7), and does not alter the intramolecular reactivity trend. For this type, one needs to evaluate the proper Fukui functions (f^+ or f^-) for the different potential sites of the substrate. For example, the Fukui function values for the C and O atoms of H_2CO, shown above, predicts that O atom should be the preferred site for an electrophilic attack, whereas C atom will be open to a nucleophilic attack. Atomic Fukui function for electrophilic attack (f_C^-) for the ring carbon atoms has been used to study the directing ability of substituents in electrophilic substitution reaction of monosubstituted benzene [23]. In some cases, it was shown that relative electrophilicity (f^+/f^-) or nucleophilicity (f^-/f^+) indices provide better intramolecular reactivity trend [23]. For example, basicity of substituted anilines could be explained successfully using relative nucleophilicity index (f^-/f^+) [23]. Note however that these parameters are not able to differentiate the preferred site of protonation in benzene derivatives, determined from the absolute proton affinities [24].

The Fukui function for nucleophilic attack (f^+) should be considered for nucleophilic addition reaction. Experimentally observed higher reactivity of the β-position for nucleophilic attack on different α,β-unsaturated aldehydes and ketones could be explained from the higher value of f^+ for the β-position than α-position [25].

12.4.3 INTERMOLECULAR REACTIVITY

To study the intermolecular reactivity of two partners, the general procedures to be considered are

1. Classification of the two reactants as electrophilic or nucleophilic
2. Calculation of the appropriate local reactivity parameters (values for nucleophilic attack for the electrophilic reactant and vice versa)
3. Application of the local HSAB principle to determine the preferred site of attack

Each of the reactants (A and B) can be classified as electrophilic or nucleophilic by evaluating the energy cost for an electron transfer from A to B, described by the

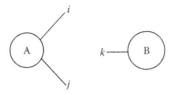

SCHEME 12.2

term $IE_A - EA_B$, or vice versa. The process associated with the lower energy cost should be the preferred one in determining the reactant character.

12.4.3.1 [2 + 1] Addition

In this simplest type of addition reaction, a single interaction site at one of the partners can react with two possible sites of the other partner (Scheme 12.2). In general, two modes of attack ($k \rightarrow i$ or $k \rightarrow j$) will have different reaction barrier introducing different kinetics and regioselectivity.

The pathway characterized with the lower energy barrier is expected to be the preferred reaction channel, especially when the addition leads to the same product. Following the local HSAB principle, one has to look at the softness matching criteria, and the minimum of $|s_{Ai} - s_{Bk}|$ and $|s_{Aj} - s_{Bk}|$ will determine the preferred site of attack.

Example 1

Addition of isocyanide to dipolarophiles [26]. Energy cost analysis from the terms $IE_{X=Y} - EA_{HNC}$ and vice versa, shows that HNC acts as a nucleophile in these addition reactions. Thus s^- for the C atom of HNC and s^+ for the X and Y atoms should be considered for determining the preferred reaction site (Scheme 12.3).

It is clear from the Δs values in Table 12.2 that the site associated with the lower Δs value, implying a better satisfaction of the local HSAB principle, is the preferred site of attack for this type of reaction [26].

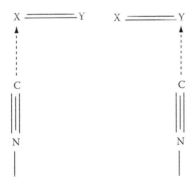

SCHEME 12.3

TABLE 12.2

Global and Atomic Softness Values of HNC and Various Dipolarophiles Calculated from B3LYP/6-31G(d,p) Method

Molecule	S	Atom[a]	s^+	Δs
HN≡C	1.694	C	1.88^b	
$H_2C{=}SiH_2$	2.733	C	−0.31	2.19
		Si*	2.31	0.43
$H_2C{=}NH$	2.049	C*	1.00	0.88
		N	0.56	1.32
$H_2C{=}O$	2.121	C*	1.30	0.58
		O	0.45	1.43
$H_2Si{=}O$	2.481	Si*	2.34	0.46
		O	0.10	1.78
HP=NH	2.751	P*	2.11	0.23
		N	0.48	1.40

Note: Δs measures the difference between the softness of electrophilic attack of the carbon atom of HNC and softness for nucleophilic attack for the X or Y atom of dipolarophile. Values are given in a.u.

[a] Preferred site of attack as determined by barrier height calculations is shown by asterisk (*).

[b] Correspond to s^- value.

12.4.3.2 Cycloaddition Reactions

Perhaps the most successful application of Fukui function and local softness is in the elucidation of the region-selective behavior of different types of pericyclic reactions including the 1,3-dipolar cycloadditions (13DC), Diels–Alder reactions, etc. These reactions can be represented as shown in Scheme 12.4. Considering the concerted approach of the two reactants A and B, there are two possible modes of addition as shown in Pathway-I and Pathway-II.

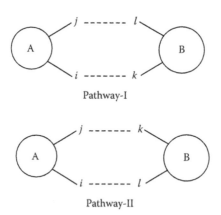

Pathway-I

Pathway-II

SCHEME 12.4

These two paths are normally associated with different barrier heights introducing, thus, a regio-selectivity in the cycloadditive process. The path associated with the lower energy barrier should be preferred, and the corresponding cycloadduct will be dominant. Now, direct application of HSAB at the local level is not possible here, because it has to be satisfied for both the termini simultaneously. A softness matching criteria, thus, needs to be defined for the multisite interaction that measures the extent of the fulfillment of local HSAB principle. A quantity (Δs) can, thus, be defined to measure the softness matching criteria for the two paths in a least square sense, and the minimum value of this quantity should be preferable [27]:

$$\Delta s_{ij}^{kl} = \left(s_i^- - s_k^+\right)^2 + \left(s_j^- - s_l^+\right)^2$$
$$\Delta s_{ij}^{lk} = \left(s_i^- - s_l^+\right)^2 + \left(s_j^- - s_k^+\right)^2$$

(12.19)

The smaller value of these two quantities should indicate the lower energy TS and thereby, the preferred mode of cycloaddition:

$$\min \ \Delta_{ij}^{kl}(TS) \rightarrow TS^*$$

The regiochemistry of several types of 13DC reactions have successfully been explained using this softness matching critera [28,29]. Interestingly, it was also observed that the lower energy transition structure (TS) has larger hardness and lower polarizability compared to the other TS of addition [28,29]. Let us consider a typical 13DC whose main characteristics are shown in Example 2 and Scheme 12.5.

Example 2

Cycloaddition reaction of phenylazide with styrene and phenylacetylene (Scheme 12.5) [29a]. Two possible cycloaddition products are shown, but in both the cases

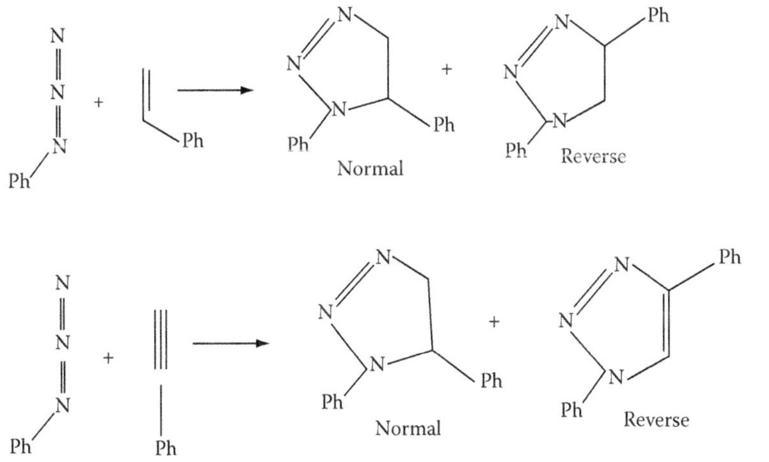

SCHEME 12.5

the first one (normal) is known to be the major product. The 1,3-dipole phenylazide acts as a nucleophile in these reactions. The Δs values were estimated for both possible orientations for additions to various dipolarophiles, from the appropriate atomic softness values calculated from the B3LYP/6-31G(d,p) method. The Δs values for the two products of phenylazide and styrene reaction are 0.11 and 0.42, respectively, whereas the Δs values are 0.17 and 0.46 for the two products of phenylazide and phenylacetylene reaction. Thus in both the cases, the approach that satisfies local HSAB principle to a larger extent, is found to be the preferred product.

Applications of local HSAB principle have been used for the determination of the softer regions in Si clusters by using Ga as probe atom [30a], or the site for H-atom adsorption on Si clusters. In the latter case, the isomer predicted by the Fukui function was found but it is not always the most stable one. The use of the reactivity indices is only valid when the adsorption process does not induce strong deformation of the cluster [30b].

Local HSAB principle can also be used to calculate the relative homolytic bond dissociation energies (BDE). For the homolytic dissociation of *para*-substituted phenols:

$$X-C_6H_4-OH \rightarrow X-C_6H_4O + H$$

the negative of interaction energy (ΔE_{int}) can be estimated from the expression [31]

$$-\Delta E_{int} = -\frac{(\mu_{xC_6H_4O} - \mu_H)^2}{2(\eta_H + \eta_O)} - \frac{\lambda(\eta_H \eta_O)}{2(\eta_H + \eta_O)} \qquad (12.20)$$

where

μ_x and η_x are the chemical potential and hardness for the species x where
$X = O, H$; O stands for XC_6H_4O
λ is an adjustable parameter related to the effective number of electrons

The ΔE_{int} value provides the corresponding BDE. It was shown that good relative BDE could be obtained from a suitable choice of value for the parameter λ.

The use of a dual descriptor defined in terms of the variation of hardness with respect to the external potential, and it is written as the difference between nucleophilic and electrophilic Fukui functions, Equation 12.21, can also be used as an alternative to rationalize the site reactivity [32]:

$$\Delta f(r) = f^+(r) - f^-(r) \qquad (12.21)$$

12.4.3.3 Radical Reactions

Addition of radicals to a different unsaturated substrate is an important class of organic reactions. To understand its regiochemistry, one needs to examine the condensed Fukui function (f^0) or atomic softness (s^0) for radical attack of the different potential sites within the reactant substrate. We consider now a simple problem summarized in Example 3.

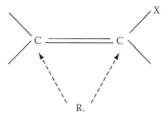

SCHEME 12.6

Example 3

The regiochemistry of methyl radical attack (CH_3) to a series of substituted ethylenes (Scheme 12.6) [33]. Generally, the radical attack occurs at the less substituted end of of the olefins. It has been found that while there is no correlation between the global softness (S) for radicals and the barrier heights for radical addition, the barrier tends to decrease with the increase in electronegativity of the radicals.

The f^0 value calculated at the B3LYP/6-31G(d,p) level was found to be consistently larger for the less substituted carbon atom than the substituted bearing carbon atom.

Radical addition takes place at the carbon atom with greater atomic softness value for radical attack, in agreement with the numerous experimental observations. This proposition works well as long as potential sites of radical attack have similar atoms. There is a certain correlation between the s^0 values and barrier heights when only the additions to carbon atom are taken into account. In contrast, the f^0 values fail to explain the observed regiochemistry of radical addition to heteroatom double bond (C=X), or to heteronuclear ring compounds [34]. In fact, the f^0 or s^0 values are consistently greater for oxygen atom than the carbon atom in aldehydes, but the radical attack takes place at the carbon atom. The main reason behind this failure lies in the inherent deficiency in the definition of the Fukui function for radical attack. The present definition $f^0 = \frac{1}{2}(f^+ + f^-)$ considers the radical as amphoteric and it has no tendency to gain or lose electron. Most importantly, this definition implies the same site selectivity for the addition of all radicals to a particular substrate, which is simply unrealistic.

One possible solution of this problem is to differentiate a radical first as electrophilic or nucleophilic with respect to its partner, depending upon its tendency to gain or lose electron. Then the relevant atomic Fukui function (f^+ or f^-) or softness (s^+ or s^-) should be used. Using this approach, regiochemistry of radical addition to heteratom C=X double bond (aldehydes, nitrones, imines, etc.) and heteronuclear ring compounds (such as uracil, thymine, furan, pyridine, etc.) could be explained [34]. A more rigorous approach will be to define the Fukui function for radical attack in such a way that it takes care of the inherent nature of a radical and thus differentiates one radical from the other.

12.4.3.4   Radical Abstraction Reactions

DFT-based descriptors can be applied for studying the most potential site for hydrogen abstraction reaction of a radical from a substrate. For H-abstraction processes from a series of polycyclic aromatic hydrocarbons (PAHs), local softness is found to be well suited to predict the preferred H-abstraction site [35]. A qualitative agreement is also observed between local softness differences (between softness of H atom and the central atom of the abstracting radical) and energy barriers at 0 K [36]. It was also observed for the H-abstraction reaction of propene with a series of radicals (such as CH_3, CF_3, C_2H, C_2H_3, C_2H_5, OH) that the barrier height for H-abstraction decreases with the increase in electronegativity of the radical, and a linear relation is observed between them [37]. The more electronegative is the attacking radical, the lower is the energy barrier for H-abstraction.

12.5 FURTHER COMMENTS AND OUTLOOK

We discussed mainly some of the possible applications of Fukui function and local softness in this chapter, and described some practical protocols one needs to follow when applying these parameters to a particular problem. We have avoided the deeper but related discussion about the theoretical development for DFT-based descriptors in recent years. Fukui function and chemical hardness can rigorously be defined through the fundamental variational principle of DFT [37,38]. In this section, we wish to briefly mention some related reactivity concepts, known as electrophilicity index (W), spin-philicity, and spin-donicity.

The electrophilicity of a system is defined as [39]

$$\omega = \frac{\mu^2}{2\eta} \tag{12.22}$$

The name stems from the fact that the above relation resembles the equation of power ($= V^2/R$) in classical electricity. It, therefore, represents the electrophilic power of a chemical species.

Following Equation 12.14, the corresponding condensed-to-atom philicity index can be expressed as [40]

$$\omega_k^\alpha = \omega \cdot f_k^\alpha \quad (\alpha = +, -, \text{ or } 0) \tag{12.23}$$

These indices have been used to study the reactivity for a series of chlorobenzenes and a good correlation is observed, for example, between W and toxicity of chlorobenzene [41]. For a detail discussion of this concept and its applications, we refer the readers to a recent review [41,42]. For studying intramolecular reactivity, these philicity indices and local softness contain the same information as obtained from the Fukui functions, because they simply scale the Fukui functions. In some cases the "relative electrophilicity" and "relative nucleophilicity" may be used although they provide similar trends as $s(\vec{r})$ and $\omega(\vec{r})$ in most cases [43]. In the same vein, the spin-donicity and spin-philicity, which refer to the philicity of open-shell systems [44], could also be utilized to unravel the reactivity of high-spin species, such as the carbenes, nitrenes, and phosphinidenes [45].

In summary, the DFT-based reactivity descriptors are conceptually simple and easy to evaluate. They are useful for studying reactivity, especially in probing the regiochemistry of different types of chemical reactions. As these parameters incorporate the essential features of frontier orbital theory, they are expected to have, at least, similar performance, and thus provide an alternative to the latter. In both approaches, one needs to be careful about their applicability to a particular problem and their limitations. The DFT-based reactivity descriptors are good prediction tools, especially for soft–soft interaction where electronic factor dominates the course of a reaction.

REFERENCES

1. Pearson, R.G. *Acc. Chem. Res.* 1993, *26*, 250.
2. Parr, R.G. and Pearson, R.G. *J. Am. Chem. Soc.* 1983, *105*, 7512.
3. (a) Perdew, J.P., Parr, R.G., Levy, M., and Balduz, Jr. J.L. *Phys. Rev. Lett.* 1982, *49*, 1691; (b) Zhang, Y. and Yang, W. *Theor. Chem. Acc.* 2000, *103*, 346.
4. Ayers, P.W. *Math. Chem.* 2008, *43*, 285.
5. Pearson, R.G. *J. Am. Chem. Soc.* 1963, *85*, 3533.
6. Pearson, R.G. *J. Chem. Edu.* 1987, *64*, 561.
7. (a) Chattaraj, P.K., Lee, H., and Parr, R.G. *J. Am. Chem. Soc.* 1991, *113*, 1855; (b) Chattaraj, P.K. and Ayers, P.W. *J. Chem. Phys.* 2005, *123*, 086101; (c) Ayers, P.W. *J. Chem. Phys.* 2005, *122*, 141102; (d) Ayers, P.W., Parr, R.G., and Pearson, R.G. *J. Chem. Phys.* 2006, *124*, 194107.
8. (a) Ghanty, T.K. and Ghosh, S.K. *J. Phys. Chem.* 1993, *97*, 4951; (b) Roy, R., Chandra, A.K., and Pal, S. *J. Phys. Chem.* 1994, *98*, 10447; (c) Simon-Manso, Y. and Fuentealba, P. *J. Phys. Chem. A* 1998, *102*, 2029.
9. Ayers, P.W. *Faraday Discuss.* 2007, *135*, 161.
10. Parr, R.G. and Yang, W. *J. Am. Chem. Soc.* 1984, *106*, 4049.
11. Yang, W. and Parr, R.G. *Proc. Natl. Acad. Sci.* 1985, *82*, 6723.
12. Berkowitz, M. *J. Am. Chem. Soc.* 1987, *109*, 4823.
13. Yang, W. and Mortier, W.J. *J. Am. Chem. Soc.* 1986, *108*, 5708.
14. (a) Cohen, M.H., Ganduglia-Pirovano, M.V., and Kudmovsky, J. *J. Chem. Phys.* 1994, *101*, 8988; (b) Senet, P. *J. Chem. Phys.* 1997, *107*, 2516.
15. (a) Contreras, R.R., Fuentealba, P., Galvan, M., and Perez, P. *Chem. Phys. Lett.* 1999, *304*, 405; (b) Ayers, P.W., De Proft, F., Borgoo, A., and Geerlings, P. *J. Chem. Phys.* 2007, *126*, 224107; (c) Bultinck, P., Fias, S., Van Alsenoy, C., Ayers, P.W., and Carbo-Borca, R. *J. Chem. Phys.* 2008, *7127*, 034102.
16. (a) Reed, A.E. and Weinhold, F. *J. Chem. Phys.* 1983, *78*, 4066; (b) Reed, A.E., Curtiss, L.A., and Weinhold, F. *Chem. Rev.* 1988, *88*, 899.
17. (a) Roy, R.K., Hirao, K., and Pal, S. *J. Chem. Phys.* 2000, *113*, 1372; (b) Roy, R.K., Krishnamurthy, S., Hirao, K., and Pal, S. *J. Chem. Phys.* 2001, *115*, 2901.
18. (a) Ayers, P.W., Morrison, R.C., and Roy, R.K. *J. Chem. Phys.* 2002, *116*, 8731; (b) Bultinck, P. and Carbo-Dorca, R. *J. Math. Chem.* 2003, *34*, 67; (c) Bultine, P., Carbo-Dorca, R., and Langenaeker, W. *J. Chem. Phys.* 2003, *118*, 4349.
19. Gazquez, J.L. and Mendez, F. *J. Phys. Chem.* 1994, *98*, 4591.
20. Li, Y. and Evans, J.N. *J. Am. Chem. Soc.* 1995, *117*, 7756.
21. Gupta, N., Garg, R., Shah, K.K., Tanwar, A., and Pal, S. *J. Phys. Chem. A* 2007, *111*, 8823.
22. Melin, J., Aparicio, F., Subramanian, V., Galvan, M., and Chattaraj, P.K. *J. Phys. Chem. A* 2004, *108*, 2487.
23. (a) Roy, R.K., Krishnamurthy, S., Geerlings, P., and Pal. S. *J. Phys. Chem. A* 1998, *102*, 3746; (b) Roy, R.K., De Proft, F., and Geerlings, P. *J. Phys. Chem. A* 1998, *102*, 7035.

24. Tishchenko, O., Pham-Tran, N.N., Kryachko, E.S., and Nguyen, M.T. *J. Phys. Chem. A* 2001, *105*, 7035.

25. Langenaeker, W., Demel, K., and Geerlings, P. *J. Mol. Struct. (Theochem)* 1992, *259*, 317.

26. (a) Chandra, A.K., Geerlings, P., and Nguyen, M.T. *J. Org. Chem.* 1997, *62*, 6417; (b) Nguyen, T.L., Le, N.T., De Proft, F., Chandra, A.K., Geerlings, P., and Nguyen, M.T. *J. Am. Chem. Soc.* 1999, *121*, 5992; (c) Nguyen, T.L., De Proft, F., Nguyen, M.T., and Geerlings, P. *J. Org. Chem.* 2001, *66*, 4316.

27. (a) Damoun, S., Van de Woude, Mendez, F., and Geerlings, P. *J. Phys. Chem. A* 1997, *101*, 886; (b) Sengupta, D., Chandra, A.K., and Nguyen, M.T. *J. Org. Chem.* 1997, *62*, 6404.

28. (a) Chandra, A.K. and Nguyen, M.T. *J. Phys. Chem. A* 1998, *102*, 6181; (b) Nguyen, M.T., Nguyen, T.L., Chandra, A.K., De Proft, F., Uchimaru, T., Nguyen, M.T., and Geerlings, P. *J. Org. Chem.* 2001, *66*, 6096; (c) Nguyen, T.L., Dao, V.L., De Proft, F., Nguyen, M.T., and Geerlings, P. *J. Phys. Org. Chem.* 2003, *16*, 615.

29. (a) Chandra, A.K. and Nguyen, M.T. *J. Comput. Chem.* 1998, *19*, 195; (b) Nguyen, M.T., Chandra, A.K., Sakai, S., and Morakuma, K. *J. Org. Chem.* 1999, *64*, 65.

30. (a) Galvan, M., Dal Pino, A., and Joannopolous, J.D. *Phys. Rev. Lett.* 1993, *70*, 21; (b) Tiznadom, W., Ona, O.B., Bazterra, V.E., Caputo, M.C., Facelli, J.C., Ferraro, M.B., and Fuentealba, P. *J. Chem. Phys.* 2005, *123*, 214302.

31. Romero M. de L. and Mendez, F. *J. Phys. Chem. A* 2003, *107*, 5874.

32. Morell, C., Grand, A., and Toro-Labbe, A. *J. Phys. Chem. A* 2005, *109*, 205.

33. Chandra, A.K. and Nguyen, M.T. *J. Chem. Soc. Perkin Trans. 2* 1997, 1415.

34. Chandra, A.K. and Nguyen, M.T. *Faraday Discussions* 2007, *135*, 191.

35. Hemelsoet, K., Speybroeck, V.V., Marin, G.B., De Proft, F., Geerlings, P., and Waroquier, M. *J. Phys. Chem. A* 2004, *108*, 7281.

36. Hemelsoet, K., Speybroeck, V.V., and Waroquier, M. *Chem. Phys. Lett.* 2007, *444*, 17.

37. Nguyen, H.M.T., Peeters, J., Nguyen, M.T., and Chandra, A.K. *J. Phys. Chem. A* 2004, 108, 484.

38. Chattaraj, P.K., Cedillo, A., and Parr, R.G. *J. Chem. Phys.* 1995, *103*, 7645; (b) Ayers, P.W. and Parr, R.G. *J. Am. Chem. Soc.* 2000, *122*, 2010; (c) Ayers, P.W. and Parr, R.G. *J. Am. Chem. Soc.* 2001, *123*, 2007; (d) Ayers, P.W. and Levy, M. *Theoret. Chem. Acc.* 2000, *103, 353.*

39. Parr, R.G., von Szentpaly, L., and Liu, S. *J. Am. Chem. Soc.* 1999, *121*, 1922.

40. Chattaraj, P.K., Maiti, B., and Sarkar, U. *J. Phys. Chem. A* 2003, *107*, 4973.

41. Padmanabhan, J., Parthasarathi, R., Subramanian, V., and Chattaraj, P.K. *J. Phys. Chem. A* 2005, *109*, 11043.

42. Chattaraj, P.K., Sarkar, U., and Roy, D.R. *Chem. Rev.* 2006, *106*, 2065.

43. (a) Roy, D.R., Parthasarathi, R., Padmanabhan, J., Sarkar, U., Subramanian, V., and Chattaraj, P.K. *J. Phys. Chem. A* 2006, *110*, 1084; (b) Bagaria, P. and Roy, R.K. *J. Phys. Chem. A* 2008, *112*, 97.

44. Perez, P., Andre, J., Safont, V.S., Tapia, O., and Contreras, R. *J. Phys. Chem. A* 2002, *106*, 5353.

45. Olah, J., Veszpremi, T., and Nguyen, M.T. *Chem. Phys. Lett.* 2005, *401*, 337.

FURTHER READING

1. Parr, R.G. and Yang, W. *Density Functional Theory of Atoms and Molecules*, Oxford University Press, New York, 1989.

2. Pearson, R.G. *Chemical Hardness*. Wiley-VCH, Weinheim, Germany, 1997.

3. Geerlings, P., De Proft, F. and Langenaeker, W. Conceptual density functional theory, *Chem. Rev.* 2003, *103*, 1793.

13 Electrophilicity

Shubin Liu

CONTENTS

13.1 INTRODUCTION

Electrophilicity [1] is the property of being electrophilic and a measure of the the relative reactivity of an electrophile. An electrophile is a reagent attracted to electrons that participates in a chemical reaction by accepting electrons to form a bond with the nucleophile. Because electrophiles accept electrons, they are Lewis acids [2] according to the general acid–base theory of Brönsted and Lowry [3,4]. Most electrophiles are positively charged, having an atom which carries a partial positive charge, or does not have an octet of electrons. Qualitatively, as Lewis acidity is measured by relative equilibrium constants, electrophilicity is measured by relative rate constants for reactions of different electrophilic reagents toward a common substrate (usually involving attack at a carbon atom). Closely related to electrophilicity is the concept of nucleophilicity, which is the property of being nucleophilic, the relative reactivity of a nucleophile. A nucleophile is a reagent that forms a chemical bond to its reaction partner (an electrophile) by donating bonding electrons. Because nucleophiles donate electrons, they are by definition Lewis bases. All molecules or ions with a free pair of electrons can act as nucleophiles, although anions are more potent than neutral reagents.

It is generally believed that it was Ingold [1] in the early 1930s who proposed the first global electrophilicity scale to describe electron-deficient (electrophile) and electron-rich (nucleophile) species based on the valence electron theory of Lewis. Much has been accomplished since then. One of the widely used electrophilicity scales derived from experimental data was proposed by Mayr et al. [5–12]:

$$\log k = s(E + N), \qquad (13.1)$$

where
   k is the equilibrium constant involving the electrophile and nucleophile
   E and N are, respectively, the electrophilicity and nucleophilicity parameters
   s is a nucleophile-specific constant

The second well-known electrophilicity or nucleophilicity scale was by Legon and Millen [13,14]. In this scale, the assigned intrinsic nucleophilicity is derived from the intermolecular stretching force constant k, recorded from the rotational and infrared (IR) spectra of the dimer B ... HX formed by the nucleophile B and a series of HX species (for X halogens) and other neutral electrophiles. The nucleophilicity number in this case is obtained from the empirical relation

$$k = cNE, \qquad (13.2)$$

where
   c is a constant
   N is the nucleophilicity value of B
   E is the electrophilicity value of HX

The implementation of this model is as follows: a nucleophilicity number $n = 10$ for H_2O and an electrophilicity number $E = 10$ for HF were assigned as references.
   In addition to the above prescriptions, many other quantities such as solution phase ionization potentials (IPs) [15], nuclear magnetic resonance (NMR) chemical shifts and IR absorption frequencies [16–18], charge decompositions [19], lowest unoccupied molecular orbital (LUMO) energies [20–23], IPs [24], redox potentials [25], high-performance liquid chromatography (HPLC) [26], solid-state syntheses [27], K_e values [28], isoelectrophilic windows [29], and the harmonic oscillator models of the aromaticity (HOMA) index [30], have been proposed in the literature to understand the electrophilic and nucleophilic characteristics of chemical systems.

13.2   THEORY

Can the concept of electrophilicity be generally formularized on a solid theoretical ground? In 1999, prompted by an earlier proposal by Maynard et al. [31], the concept of electrophilicity index was quantitatively introduced by Parr et al. [32] as the stabilization energy when atoms or molecules in their ground states acquire additional electronic charge from the environment. The question to address is to what extent partial electron transfer contributes to the lowering of the total binding energy by maximal flow of electrons.
   Consider an electrophile immersed in an idealized zero-temperature free electron sea of zero chemical potential, which could be an approximation to its binding environment in a protein, a DNA coil, or a surface. It will become saturated with electrons, to the point that its chemical potential increases to zero, thereby becoming

equal to the chemical potential of the sea. To the second order, the energy change ΔE due to the electron transfer ΔN satisfies the formula [33]

$$\Delta E = \mu \Delta N + 1/2 \eta (\Delta N)^2, \tag{13.3}$$

where μ and η are the chemical potential (negative of the electronegativity) and chemical hardness, respectively defined by

$$\mu = \left(\frac{\partial E}{\partial N}\right)_v, \tag{13.4}$$

and

$$\eta = \left(\frac{\partial^2 E}{\partial N^2}\right)_v, \tag{13.5}$$

with $v(\mathbf{r})$ as the external potential of the electrophile. According to Mulliken [34–38], using a finite difference method, working equations for the calculation of μ and η may be given as

$$\mu = -\chi = -\frac{1}{2}(I + A) \tag{13.6}$$

and

$$\eta = I - A, \tag{13.7}$$

where I and A are the first IP and electron affinity (EA), respectively. According to, the Koopmans' theorem for closed-shell molecules, based on the finite difference approach, I and A can be expressed in terms of the highest occupied molecular orbital (HOMO) energy, ε_{HOMO}, and the LUMO energy, ε_{LUMO}, respectively, $I \approx -\varepsilon_{HOMO}$; $A \approx -\varepsilon_{LUMO}$.

If the electron sea provides enough electrons, the electrophile will become saturated with electrons according to Equation 13.3

$$\frac{d\Delta E}{d\Delta N} = 0, \tag{13.8}$$

leading to the maximum amount of electron charge (see examples in Table 13.1)

$$\Delta N_{max} = -\frac{\mu}{\eta}, \tag{13.9}$$

and the total energy decrease

$$\Delta E_{min} = -\frac{\mu^2}{2\eta}. \tag{13.10}$$

TABLE 13.1

Ionization Potential (IP), Electron Affinity (EA), Maximal Charge Acceptance ΔN_{max}, and Electrophilicity Index ω in the Ground State for the First and Second Row Atoms (Units in eV)

	IP	EA	ΔN_{max}	ω
H	13.60	0.75	0.56	2.01
Li	5.39	0.62	0.63	0.95
B	8.30	0.28	0.54	1.15
C	11.26	1.26	0.63	1.96
N	14.53	0.07	0.50	1.84
O	13.62	1.46	0.62	2.34
F	17.42	3.40	0.74	3.86
Na	5.14	0.55	0.62	0.88
Al	5.99	0.44	0.58	0.93
Si	8.15	1.39	0.71	1.68
P	10.49	0.75	0.58	1.62
S	10.36	2.08	0.75	2.34
Cl	12.97	3.61	0.89	3.67

Notice that since $\eta > 0$, we always have $\Delta E < 0$, i.e., the charge transfer process is energetically favorable. We proposed the new density functional theory (DFT) reactivity index, electrophilicity index ω as

$$\omega \equiv \frac{\mu^2}{2\eta} \qquad (13.11)$$

as the measure of electrophilicity of an electrophile. The reason that ω can be viewed as a measure of the electrophilicity power is because it is analogous to the classical electrostatics power, V^2/R, and μ and η serve the purpose of potential (V) and resistance (R), respectively.

Is the electrophilicity index in Equation 13.11 consistent with the experimental electrophilicity scale proposed by Mayr et al. in Equations 13.1 and 13.2? To test the reliability of the theoretical electrophilicity scale, Pérez et al. [39] selected a series of diazonium ions whose global electrophilicity pattern was evaluated earlier by Mayr et al. [5–12]. The series included the benzenediazonium ion and a series of substituted derivatives containing a wide variety of electron-withdrawing and electron-releasing groups in the ortho- and para-positions. The strong correlation between the E value in Equation 13.1 and the ω value in Equation 13.11, is shown in Figure 13.1, suggesting that the global electrophilicity pattern of a series of diazonium ions, as described by the global electrophilicity index introduced by Equation 13.11, compares fairly well with the kinetic scale of electrophilicity proposed by Mayr et al. in Equation 13.1.

Also, since EA is a quantity that measures the capability of an electrophile to accept an electron, is EA related to ω? It is anticipated that ω should be related to EA,

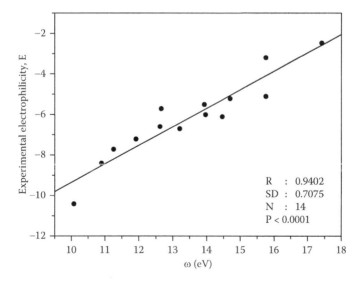

FIGURE 13.1   Correlation between experimental electrophilicity (E) and theoretical electrophilicity (ω) of a series of benzene diazonium ion and its derivatives containing a large variety of electron-releasing and electron-withdrawing groups in the *ortho*- and *para*- positions. (Reprinted from Pérez, P., *J. Org. Chem.*, 68, 5886, 2003; Pérez, P., Aizman, A., and Contreras, R., *J. Phys. Chem. A*, 106, 3964, 2002. With permission.)

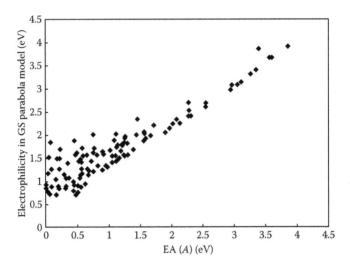

FIGURE 13.2   Correlation between electrophilicity index and EA of 54 neutral atoms and 55 simple molecules in the ground-state parabola model. (Reprinted from Parr, R.G., Szentpaly, L.V., and Liu, S.B., *J. Am. Chem. Soc.*, 121, 1922, 1999. With permission.)

because both ω and EA measure the capability of an agent to accept some number of electrons. However, EA reflects the capability of accepting only one electron from the environment, whereas the electrophilicity index ω measures the energy lowering of a ligand due to maximal electron flow between donor and acceptor. The electron flows may be either less or more than one. Figure 13.2 gives ω versus EA for 54 neutral atoms and 55 simple molecules in the ground state, showing that EA is somehow related to ω but does not correlate well with it. Further examinations indicate that the outliers in the Figure 13.2 are those whose ΔN_{max} values from Equation 13.9 are less than 1.

13.3   EXTENSIONS

Since its inception, the concept of electrophilicity index as the theoretical measure of the electrophilic power of an electrophile has attracted considerable interests in the literature. There has been a comprehensive review in *Chemical Reviews* by Chattaraj et al. [40] specifically on this topic. Only a few recent extensions and developments according to my personal flavor are outlined here. For more information and comprehensive review of the subject, refer to Ref. [40].

13.3.1   Nucleophilicity

The pair of electrophilicity and nucleophilicity comes together in chemistry textbooks. Just as the former is formally defined in Equation 13.11, is there a similar, straightforward formalization for the latter? It turns out that it is not the case. One of the reasons may lie in the theoretical difficulty in dealing with local hardness [41,42], a quantity that is intrinsically related to nucleophilicity. Another reason stems from the more intriguing role that the electrophilic partner plays. Since different electro-philic molecules can accept different amount of charge, a nucleophile can be a good donor for one electrophile but a bad one for another, leading to the difficulty, if not impossible, to define a universal scale of nucleophilicity for a nucleophile.

Recently, Jaramillo et al. [43] introduced a nucleophilicity scale, depending on the electrophilic partner, and suggested that the nucleophilicity index can be written as

$$\omega^- = \frac{1}{2}\left(\frac{\mu_A - \mu_B}{\eta_A + \eta_B}\right)^2 \eta_A, \qquad (13.12)$$

assuming that A is the nucleophile and B is the electrophile. Hence, the correspond-ing nucleophilicity scale is of a relative nature, in contrast to the absolute nature of the electrophilicity scale. They considered the process with the grand canonical ensemble, where the natural variables are the chemical potential and the external potential [44]

$$\Omega = \Omega[\mu, v(\mathbf{r})]. \qquad (13.13)$$

Assuming that the external potential is fixed and is in light of Equation 13.3, one has

$$\Delta\Omega = -N\Delta\mu - 1/2\, S(\Delta\mu)^2. \qquad (13.14)$$

Minimizing $\Delta\Omega$ in Equation 13.14 with respect to $\Delta\mu$, one has

$$\Delta\mu = -N\eta \quad \text{and} \quad \Delta\Omega = -1/2N^2\eta. \qquad (13.15)$$

Making use of the following relation by Parr et al. [33] and Malone [45],

$$N = \frac{\mu_A - \mu_B}{\eta_A + \eta_B}, \qquad (13.16)$$

one then obtains Equation 13.12. Correlating Equation 13.12 with the experimentally defined nucleophilicity scale in Equation 13.2 showed strong statistical significance (see Figure 13.3 as an example).

De Proft and coworkers [46] recently examined the relationship between Equa-tions 13.11 and 13.12 for a group of radicals (Figure 13.4) and found that the global electrophilicity index and the nucleophilicity index for 35 radicals correlate well, but for some weak electrophiles and nucleophiles and the hydroxyl radical, which possess very large values of the chemical hardness, only intermediate to large values of the electronic chemical potential are encountered. For 15 radicals, a comparison between the classifications obtained with the global electrophilicity index and Principal Com-ponent analysis (PCA) was made. The agreement is astonishingly good, considering that the theoretical electrophilicity scale is absolute and free from input of reaction data (neither experimental nor theoretical).

Other proposals on the theoretical quantification of nucleophilicity are available in the literature. For instance, Chattaraj et al. [47] suggested a multiplicative inverse of

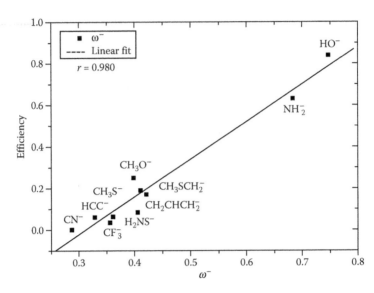

FIGURE 13.3 Comparison between calculated nucleophilicity (ω^-, in eV) and experimental efficiency (nucleophilicity measure) for anions in the $X^- + CH_3Cl$ reaction. (Reprinted from Jaramillo, P., Perez, P., Contreras, R., Tiznado, W., and Fuentealba, P., *J. Phys. Chem. A*, 110, 8181, 2006. With permission.)

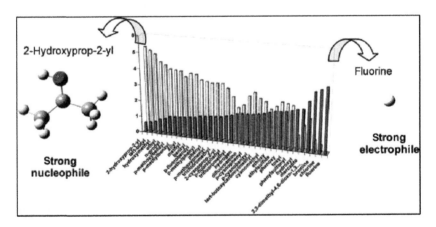

FIGURE 13.4 Electrophilicity versus nucleophilicity (value of ω^- times 10). (Reprinted from De Vleeschouwer, F., Van Speybroeck, V., Waroquier, M., Geerlings, P., and De Proft, F., *Org. Lett.*, 9, 2721, 2007. With permission.)

the electrophilicity index ($1/\omega$), as well as an additive inverse ($1 - \omega$). Cedillo et al. [48] proposed a nucleophilicity index employing the electrostatic potential with a test charge q at r_0, $\omega^- = -\frac{1}{2}|\varphi(r_0)|^2/\langle\chi\rangle_{r_0}$, where $\varphi(r_0)$ is the electrostatic potential at r_0, $\chi(r, r')$ is the first-order static density response function, and the quantity $\langle\chi\rangle_{r_0}$ is defined by

$$\langle \chi \rangle_{r_0} \equiv \iint \frac{\chi(r, r')}{|r - r_0||r' - r_0|} \, dr dr'.$$

Figure 13.5 shows its correlation with the experimental scale [13].

13.3.2 LOCAL EXTENSIONS

To describe the electrophilic character of a reactive site within a molecule, a local electrophilicity index $\omega(\mathbf{r})$ has been proposed [49,50]:

$$\omega(\mathbf{r}) = \omega f^+(\mathbf{r}) \tag{13.17}$$

with $f^+(\mathbf{r})$ the Fukui function for nucleophilic attack. For the computation of $f^+(\mathbf{r})$ the finite differences approximation condensed to atoms using electronic population analyses through, for example, the natural population analysis (NPA) can be used. For the analysis of electrophile–nucleophile interactions, $\omega(\mathbf{r})$ has been found to be a better reactivity descriptor than the corresponding Fukui function, because the local electrophilicity index is a product of a global index, ω, and a local index, $f^+(\mathbf{r})$ [51].

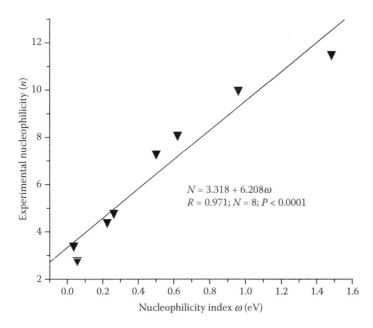

FIGURE 13.5 Comparison between experimental gas-phase nucleophilicity n in Equation 13.2 and the theoretical nucleophilicity index ω^- evaluated at the B3LYP/6-311(d,p) level of theory for the series of neutral nucleophiles that have been fully investigated using the experimental spectroscopic scale given in Ref. [10]. R is the regression coefficient, N the number of points included in the regression, and P the probability that the observed correlation was randomly obtained. (Reprinted from Cedillo, A., Contreras, R., Galván, M., Aizman, A., Andrés, J., and Safont, V.S., *J. Phys. Chem. A*, 111, 2442, 2007. With permission.)

A generalized version of the above local quantity termed as philicity has been made by Chattaraj et al. [52], who extended the local quantity through the resolution of the identity associated with the normalization of the Fukui function as

$$\omega = \omega \int f(\mathbf{r})d\mathbf{r} = \int \omega f(\mathbf{r})d\mathbf{r} = \int \omega(\mathbf{r})d\mathbf{r} \qquad (13.18)$$

where

$$\omega(\mathbf{r}) = \omega f(\mathbf{r}). \qquad (13.19)$$

Notice that from $\omega(\mathbf{r})$, one can also obtain local softness, global softness, and hardness when chemical potential is given. The corresponding condensed-to-atom variants may be defined as

$$\omega_k^\alpha = \omega f_k^\alpha; \quad \text{with } \alpha = +, -, 0 \qquad (13.20)$$

13.3.3   MINIMUM ELECTROPHILICITY PRINCIPLE

Is there a minimum electrophilicity principle (MEP) in chemical processes, analogous to the maximum hardness principle (MHP) [53–55]? Noorizadeh [56,57] and Chattaraj [58] separately addressed the issue and concluded that "the natural direction of a chemical reaction is toward a state of minimum electrophilicity." In a recent work, Noorizadeh [57] investigated the regio and stereoselectivity of a few Paternó–Büchi reactions and found that although in most cases MHP successfully predict the major oxetane products of these reactions, but in all of the considered reactions, with no exception, the main products have the lesser electrophilicity values than the minor isomers, and therefore MEP correctly predicts the most stable stereoisomer of the reaction, suggesting that not only MEP is able to predict correctly the regioselectivity during a photocycloaddition reaction, but it also successfully predicts the major stereoisomer for the reaction. It was claimed that at equilibrium, a chemical system attempts to arrange its electronic structure to generate species with the lesser electrophilicity so that more stable isomers correspond to lesser electrophilicity values.

13.3.4   OTHER DEVELOPMENTS

Two new reactivity indices related to electrophilicity and nucleophilicity, electrofugality and nucleofugality, have recently been introduced by Ayers et al. [59–61]. Electrofugality ΔE_e is defined as

$$\Delta E_e = \omega + I, \qquad (13.21)$$

and ΔE_n as

$$\Delta E_n = \omega - A, \qquad (13.22)$$

assessing the quality of electron-fleeing and electron-accepting, leaving groups, respectively, in terms of the leaving group's ionization potential I and EA A. This definition allows one to measure the inherent quality of a leaving group, as opposed to the effects due to a particular reaction environment (e.g., molecule- and solvent-specific effects).

Extension of Equation 13.11 to spin-polarized DFT has recently been made, from which two new concepts were proposed: spin-philicity and spin-donicity [62,63], where the system of interest was considered to be embedded in a zero-potential sea of spins, emphasizing that the spin properties of electrons have to be considered in the treatment. A good linear correlation has been revealed between the energy difference estimated by the sum of the spin potentials and the vertical triplet energy gaps independent of the ground state of the molecule and the atomic number of the central atom. The spin-philicity values can predict the singlet–triplet gaps to a satisfactory accuracy [62,63].

13.4 RECENT APPLICATIONS

Besides the applications of the electrophilicity index mentioned in the review article [40], following recent applications and developments have been observed, including relationship between basicity and nucleophilicity [64], 3D-quantitative structure activity analysis [65], Quantitative Structure-Toxicity Relationship (QSTR) [66], redox potential [67,68], Woodward–Hoffmann rules [69], Michael-type reactions [70], S_{N2} reactions [71], multiphilic descriptions [72], etc. Molecular systems include silylenes [73], heterocyclohexanones [74], pyrido-di-indoles [65], bipyridine [75], aromatic and heterocyclic sulfonamides [76], substituted nitrenes and phosphinidenes [77], first-row transition metal ions [67], triruthenium ring core structures [78], benzhydryl derivatives [79], multivalent superatoms [80], nitrobenzodifuroxan [70], dialkylpyridinium ions [81], dioxins [82], arsenosugars and thioarsenicals [83], dynamic properties of clusters and nanostructures [84], porphyrin compounds [85–87], and so on.

13.5 CONCLUDING REMARKS

The purpose of this chapter is to give a general, pedagogical introduction about how electrophilicity and related chemical concepts can theoretically be quantified together with brief outlines of some recent developments in the literature. The formulation in Equation 13.11 has been showing tremendous interpretive power and predictive potential in providing insights about structure, properties, stability, reactivity, bonding, toxicity, and dynamics of many-electron systems in ground and excited states for systems from almost every arena of chemistry. The field is still undergoing vivid and ever-expanding developments as witnessed by the growing number of published papers each year pertaining to the title topic.

Looking ahead, I am optimistic that we will see continued growth of our knowledge about this and other conceptual DFT-based reactivity and selectivity descriptors as well as broadening applications in understanding a diverse class of biophysicochemical properties and processes.

REFERENCES

1. Ingold, C. K. *J. Chem. Soc.* 1933, 1120; Ingold, C. K. *Chem. Rev.* 1934, *15*, 225.
2. Lewis, G. N. *Valence and the Structure of Atoms and Molecules*; The Chemical Catalog Co.: 923.
3. Brönsted, J. N. *Recl. TraV. Chim. Pays-Bas* 1923, *42*, 718.
4. Lowry, T. M. *Chem. Ind. (London)* 1923, *42*, 43.
5. Mayr, H. and Patz, M. *Angew. Chem., Int. Ed. Engl.* 1994, *33*, 938.
6. Roth, M. and Mayr, H. *Angew. Chem., Int. Ed. Engl.* 1995, *34*, 2250.
7. Mayr, H., Bug, T., Gotta, M. F., Hering, N., Irrgang, B., Janker, B., Kempf, B., Loos, R., Ofial, A. R., Remennikov, G., and Schimmel, H. *J. Am. Chem. Soc.* 2001, *123*, 9500.
8. Mayr, H., Lang, G., and Ofial, A. R. *J. Am. Chem. Soc.* 2002, *124*, 4076.
9. Bug, T., Hartnagel, M., Schlierf, C., and Mayr, H. *Chem. Eur. J.* 2003, *9*, 4068.
10. Lemek, T. and Mayr, H. *J. Org. Chem.* 2003, *68*, 6880.
11. Tokuyasu, T. and Mayr, H. *Eur. J. Org. Chem.* 2004, 2791.
12. Minegishi, S., Loos, R., Kobayashi, S., and Mayr, H. *J. Am. Chem. Soc.* 2005, *127*, 2641.
13. Legon, A. C. and Millen, D. J. *J. Am. Chem. Soc.* 1987, *109*, 356.
14. Legon, A. C. *Angew. Chem., Int. Ed.* 1999, *38*, 2686.
15. Contreras, R., Andres, J., Safont, V. S., Campodonico, P., and Santos, J. G. *J. Phys. Chem. A* 2003, *107*, 5588. See also: Meneses, L., Tiznado, W., Contreras, R., and Fuentealba, P. *Chem. Phys. Lett.* 2004, *383*, 181; Campodonico, P. R., Aizman, A., and Contreras, R. *Chem. Phys. Lett.* 2006, *422*, 204.
16. Neuvonen, H., Neuvonen, K., Koch, A., Kleinpeter, E., and Pasanen, P. *J. Org. Chem.* 2002, *67*, 995.
17. Neuvonen, H. and Neuvonen, K. *J. Chem. Soc., Perkin Trans.* 1999, 2, 1497.
18. Epstein, D. M. and Meyerstein, D. *Inorg. Chem. Commun.* 2001, *4*, 705.
19. Deubel, D. V., Frenking, G., Senn, H. M., and Sundermeyer, J. *Chem. Commun.* 2000, *24*, 2469.
20. Cronin, M. T. D., Manga, N., Seward, J. R., Sinks, G. D., and Schultz, T. W. *Chem. Res. Toxicol.* 2001, *14*, 1498.
21. Ren, S. *Toxicol. Lett.* 2003, *144*, 313. See also: Pasha, F. A., Srivastava, H. K., and Singh, P. P. *Bioorg. Med. Chem.* 2005, *13*, 6823.
22. Ren, S. and Schultz, T. W. *Toxicol. Lett.* 2002, *129*, 151.
23. Lippa, K. A. and Roberts, A. L. *Environ. Sci. Technol.* 2002, *36*, 2008.
24. Saethre, L. J., Thomas, T. D., and Svensson, S. *J. Chem. Soc., Perkin Trans.* 1997, 2, 749.
25. Topol, I. A., McGrath, C., Chertova, E., Dasenbrock, C., Lacourse, W. R., Eissenstat, M. A., Burt, S. K., Henderson, L. E., and Casas-Finet, J. R. *Protein Sci.* 2001, *10*, 1434.
26. Morris, S. J., Thurston, D. E., and Nevell, T. G. *J. Antibiot.* 1990, *43*, 1286.
27. Dronskowski, R. and Hoffmann, R. *Adv. Mater.* 1992, *4*, 514.
28. Benigni, R., Cotta-Ramusino, M., Andreoli, C., and Giuliani, A. *Carcinogenesis* 1992, *13*, 547.
29. Mekenyan, O. G. and Veith, G. D. *SAR QSAR Environ. Res.* 1993, *1*, 335.
30. Mrozek, A., Karolak-Wojciechowska, J., Amiel, P., and Barbe, J. *THEOCHEM* 2000, *524*, 159.
31. Maynard, A. T., Huang, M., Rice, W. G., and Corel, D. G. *Proc. Natl. Acad. Sci. U.S.A.* 1998, *95*, 11578.
32. Parr, R. G., Szentpaly, L. V., and Liu, S. B. *J. Am. Chem. Soc.* 1999, *121*, 1922.
33. Parr, R. G. and Yang, W. *Density Functional Theory of Atoms and Molecules*, Oxford University Press, Oxford, 1989.
34. Mulliken, R. S. *J. Chem. Phys.* 1934, *2*, 782.
35. Iczkowski, R. P. and Margrave, J. L. *J. Am. Chem. Soc.* 1961, *83*, 3547.

36. Sen, K. D. and Jørgensen, C. K. *Electronegativity, Structure and Bonding*, Springer, Berlin, 1987; Sen, K. D. *Chemical Hardness Structure and Bonding*, Springer, Berlin, 1993.
37. Pearson, R. G. *Chemical Hardness: Applications from Molecules to Solids*, Wiley-VCH, Weinheim, 1997.
38. Geerlings, P., De Proft, F., and Langenaeker, W. *Chem. Rev.* 2003, *103*, 1793.
39. Pérez, P. *J. Org. Chem.* 2003, *68*, 5886; Pérez, P., Aizman, A., and Contreras, R. *J. Phys. Chem. A* 2002, *106*, 3964.
40. Chattaraj, P. K., Sarkar, U., and Roy, D. R. *Chem. Rev.* 2006, *106*, 2065.
41. Meneses, L., Tiznado, W., Contreras, R., and Fuentealba, P. *Chem. Phys. Lett.* 2004, *383*, 181.
42. Harbola, M., Chattaraj, P. K., and Parr, R. G. *Isr. J. Chem.* 1991, *31*, 395.
43. Jaramillo, P., Perez, P., Contreras, R., Tiznado, W., and Fuentealba, P. *J. Phys. Chem. A* 2006, *110*, 8181.
44. Liu, S. B. and Parr, R. G. *J. Chem. Phys.* 1997, *106*, 5578.
45. Malone, J. G. *J. Chem. Phys.* 1933, *1*, 197.
46. De Vleeschouwer, F., Van Speybroeck, V., Waroquier, M., Geerlings, P., and De Proft, F. *Org. Lett.* 2007, *9*, 2721.
47. Chattaraj, P. K. and Maiti, B. *J. Phys. Chem. A* 2001, *105*, 169.
48. Cedillo, A., Contreras, R., Galván, M., Aizman, A., Andrés, J., and Safont, V. S. *J. Phys. Chem. A* 2007, 111, 2442.
49. Perez, P., Toro-Labbe, A., Aizman, A., and Contreras, R. *J. Org. Chem.* 2002, *67*, 4747.
50. Chamorro, E., Chattaraj, P. K., and Fuentealba, P. *J. Phys. Chem. A* 2003, *107*, 7068.
51. Roy, R. K., Usha, V., Paulovic, J., and Hirao, K. *J. Phys. Chem. A* 2005, *109*, 4601.
52. Chattaraj, P. K., Maiti, B., and Sarkar, U. *J. Phys. Chem. A* 2003, *107*, 4973.
53. Pearson, R. G. and Chattaraj, P. K. *J. Am. Chem. Soc.* 1991, *113*, 1854.
54. Chattaraj, P. K., Liu, G. H., and Parr, R. G. *Chem. Phys. Lett.* 1995, *237*, 171.
55. Chattaraj, P. K. *Proc. Indian Natl. Sci. Acad. Part A* 1996, *62*, 1133.
56. Noorizadeh, S. *Chin. J. Chem.* 2007, *25*, 1439.
57. Noorizadeh, S. *J. Phys. Org. Chem.* 2007, *20*, 514.
58. Chattaraj, P. K. *Ind. J. Phys.* 2007, *81*, 871.
59. Ayers, P. W., Anderson, J. S. M., Rodriguez, J. I., and Jawed, Z. *Phys. Chem. Chem. Phys.* 2005, *7*, 1918.
60. Ayers, P. W., Anderson, J. S. M., and Bartolotti, L. J. *Int. J. Quantum Chem.* 2005, *101*, 520.
61. Roos, G., Loverix, S., Brosens, E., Van Belle, K., Wyns, L., Geerlings, P., and Messens, J. *Chem. BioChem.* 2006, *7*, 981.
62. Perez, P., Andres, J., Safont, V. S., Tapia, O., and Contreras, R. *J. Phys. Chem. A* 2002, *106*, 5353.
63. Oláh, J., De Proft, F., Veszprémi, T., and Geerlings, P. *J. Phys. Chem. A* 2004, *108*, 490.
64. Jaramillo, P., Perez, P., and Fuentealba, P. *J. Phys. Org. Chem. A* 2007, *20*, 1050.
65. Pasha, F. A., Chung, H. W., Kang, S. B., and Cho, S. J. *Int. J. Quantum Chem.* 2007, *108*, 391.
66. Chattaraj, P. K., Roy, D. R., Giri, S., Mukherjee, S., Subramanian, V., Parthasarathi, R., Bultinck, P., and Van Damme, S. *J. Chem. Sci.* 2007, *119*, 475.
67. Moens, J., Roos, G., Jaque, P., De Proft, F., and Geerlings, P. *Chem. Euro. J.* 2007, *13*, 9331.
68. Moens, J., Geerlings, P., and Roos, G. *Chem. Eur. J.* 2007, *13*, 8174.
69. Ayers, P. W., Morell, C., De Proft, F., and Geerlings, P. *Chem. Eur. J.* 2007, *13*, 8240.
70. Lakhdar, S., Goumont, R., Berionni, G., Boubaker, T., Kurbatov, S., and Terrier, F. *Chem. Eur. J.* 2007, *13*, 8317.
71. Arnaut, L. G. and Formosinho, S. J. *Chem. Eur. J.* 2007, *13*, 8018.

72. Padmanabhan, J., Parthasarathi, R., Elango, M., Subramanian, V., Krishnamoorthy, B. S., Gutierrez-Oliva, S., Toro-Labbé, A., Roy, D. R., and Chattaraj, P. K. *J. Phys. Chem. A* 2007, *111*, 9130.

73. Olah, J., Veszpremi, T., De Proft, F., and Geerlings, P. *J. Phys.Chem. A* 2007, *111*, 10815.

74. Lira, A. L., Zolotukhin, M., Fomina, L., and Fomine, S. *J. Phys.Chem. A* 2007, *111*, 13606.

75. Kulkarni, B. S., Tanwar, A., and Pal, S. *J. Chem. Sci.* 2007, *119*, 489.

76. Eroglu, E. and Turkmen, H. *J. Mol. Graph. Mod.* 2007, *26*, 701.

77. Rincon, E., Perez, P., and Chamorro, E. *Chem. Phys. Lett.* 2007, *448*, 273.

78. Tsipis, A. C., Kefalidis, C. E., and Tsipis, C. A. *J. Am. Chem. Soc.* 2007, *129*, 13905.

79. Campodonico, P. R., Perez, C., Aliaga, M., Gazitua, M., and Contreras, R. *Chem. Phys. Lett.* 2007, *447*, 375.

80. Chattaraj, P. K. and Giri, S. *J. Phys. Chem. A* 2007, *111*, 11116.

81. Gupta, N., Garg, R., Shah, K. K., Tanwar, A., and Pal, S. *J. Phys. Chem. A* 2007, *111*, 8823.

82. Gu, C. G., Jiang, X., Ju, X. H., Yang, X. L., and Yu, G. F. *SAR QSAR Environ. Res.* 2007, *18*, 603.

83. Regmi, R., Milne, B. F., and Feldmann, J. *Anal. Bioanal. Chem.* 2007, *388*, 775.

84. Liu, S. B. *J. Chem. Sci.* 2005, *117*, 477; Zhong, A. G., Rong, C. Y., and Liu, S. B. *J. Phys. Chem. A* 2007, *111*, 3132.

85. Rong, C., Lian, S., Yin, D., Shen, B., Zhong, A. G., Bartolotti, L., and Liu, S. B. *J. Chem. Phys.* 2006, *125*, 174102.

86. Rong, C., Lian, S., Yin, D., Zhong, A., Zhang, R. Q., and Liu, S. B. *Chem. Phys. Lett.* 2007, *434*, 149.

87. Huang, Y., Zhong, A. G., Rong, C. Y., Xiao, X. M., and Liu, S. B. *J. Phys. Chem. A* 2008, *112*, 305.

14 Application of Density Functional Theory in Organometallic Complexes: A Case Study of Cp$_2$M Fragment (M = Ti, Zr) in C—C Coupling and Decoupling Reactions

Susmita De and Eluvathingal D. Jemmis

CONTENTS

14.1 INTRODUCTION

Our attempt in this chapter is to demonstrate the application of density functional theory (DFT) to real-life problems in transition metal organometallic chemistry through examples. Organometallic chemistry is an area where use of DFT to predict the structure, bonding, and reactivity has become complementary to experimental studies. A major part of the organometallic chemistry can be viewed profitably as resulting from the replacement of small groups in the organic structures by the transition metal fragments. The transformation of the organic moiety as a result of the attachment of a metal fragment is so delicate and specific that seemingly similar fragments make large changes in the system. We have selected the reactions of the biscyclopentadienyl titanium (Cp_2Ti) and biscyclopentadienyl zirconium (Cp_2Zr) complexes as an example to demonstrate the dramatic changes brought in by these metallocenes. The in situ generated metallocenes (Cp_2Ti, Cp_2Zr), with d^2 valence electron count, have been playing a pivotal role in the stoichiometric and catalytic reactions [1]. In their many different forms, these metallocenes are extensively used as catalysts in olefin polymerization [2]. Their importance is attributed to their specific catalytic activity for the generation of stereoregular and stereospecefic polyolefins. Cp_2Ti and Cp_2Zr are also used in the synthesis of several precursors for the organo-metallic chemical vapor deposition of ceramic thin films [3]. Recently, Chirik and coworkers reported the formation of a tetramethylated cyclopentadienyl zirconium complex with side-on bound dinitrogen. This dinitrogen complex on hydrogenation gives ammonia [4a]. Pentamethylated cyclopentadienyl derivative of zirconium complex under similar reaction conditions forms end-on-bound dinitrogen complex and further reaction of this complex with H_2 does not yield ammonia [4]. On the contrary, a similar reaction with dinitrogen complexes of Cp_2Ti does not show any activity toward the addition of H_2 across the Ti—N bonds [4d]. DFT calculations performed by Morokuma and coworkers explain these experimental observations [5]. The metallocenes (Cp_2Ti, Cp_2Zr) play an important role in the C—C coupling and cleavage reactions of unsaturated molecules such as alkynes, olefins, acetylides, and vinylides [1,6,7]. The difference in the reactivity of Cp_2Ti and Cp_2Zr complexes toward C—C coupling and cleavage reaction is dramatic. The systematic study of Rosenthal [7a,b], Erker [7c–e], and others [7f–o] reveals that, from similar starting materials, titanocene forms the C—C coupled structure **1** (the two central carbon C2 and C3 are connected by a bond), whereas zirconocene favors the structure **2**, where the coupling between the two central carbon atoms C2 and C3 is absent (Scheme 14.1) [7].

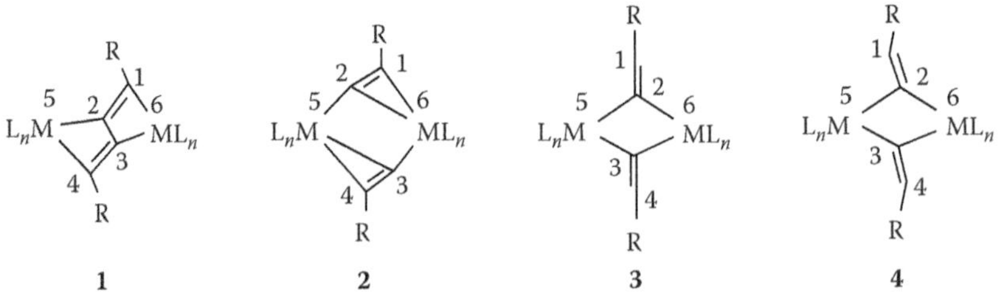

SCHEME 14.1 Schematic representation of the structural variations observed for the acetylide bridging in the bimetallic complexes of various transition and nontransition metals. Structures **1** and **2** are reported for M = Ti, Zr, and structures **3** and **4** are reported for M = Li, Be, Al, Ga, In, Cu, Ag, Er, and Sm. (Reproduced from Pravan Kumar, P.N.V. and Jemmis, E.D., *J. Am. Chem. Soc.*, 110, 125, 1988. With permission.)

Our interest in this started with the structural variations available for the complexes where two acetylide ligands bridge between two metal centers (Scheme 14.1). These structural variations are observed (**1**, **2**, **3**, and **4**) as a function of metal. The main group bimetallic complexes of Li, Na, Be, and Al prefer structures **3** and **4** (Scheme 14.1) and transition metals prefer structures **1** and **2** [7,8]. Even though acetylinic π bonds of the bridged acetylides in **4** are slightly bent, they are not strongly involved in the interaction with the metal centers. This is evident from the C1—C2 distances of the acetylide units, which are almost the same in structures **3** and **4** [8]. The situation changes considerably in the presence of transition metal fragments Cp_2Ti and Cp_2Zr, where either π bridging structure **2** or C—C coupled structure **1** is preferred over structures **3** and **4**. Interestingly, Ti gives the C—C coupled structure **1** as a final product, whereas Zr gives the C—C cleaved structure **2** as the final product [7]. The formation of C—C coupled structure for Ti and C—C decoupled structure for Zr led to the theoretical study of the reactivity of Cp_2M (M = Ti, Zr) fragments. Rosenthal et al. (Scheme 14.2a) [9] and Lang et al. (Scheme 14.2b) [7m] independently suggested the plausible reaction mechanism for this C—C coupling and decoupling reactions exhibited by Cp_2M (M = Ti, Zr)

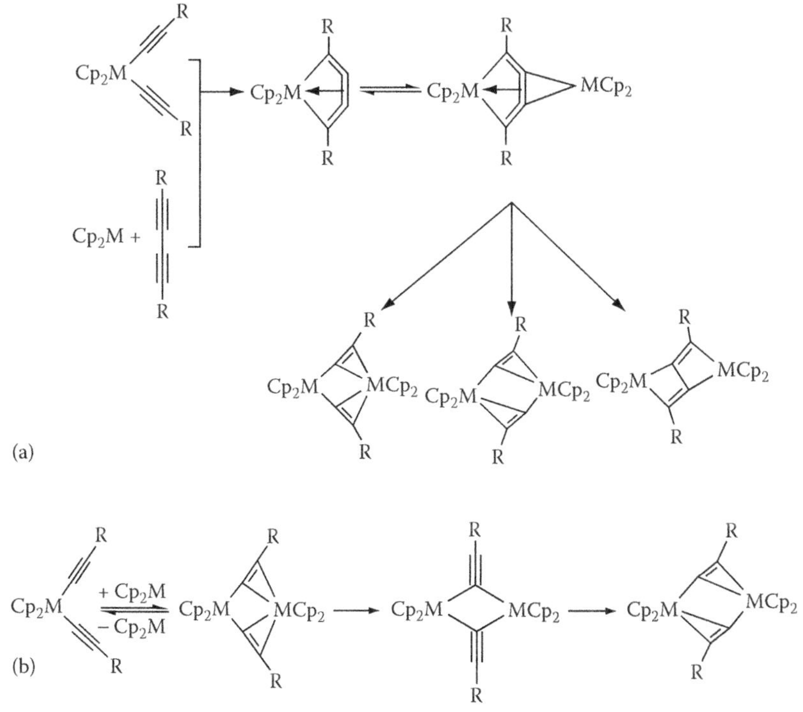

SCHEME 14.2 (a) Mechanistic scheme proposed by Rosenthal and coworkers for the formation of various complexes of Ti and Zr acetylides. (b) Reaction steps proposed by Lang and coworkers for the formation of the C—C decoupled complex of Ti and Zr. (Reproduced from Jemmis, E.D. and Giju, K.T., *J. Am. Chem. Soc.*, 120, 6952, 1998. With permission.)

fragments. The mechanism proposed by Lang included a paramagnetic complex of type **3**, which was also suggested by Royo et al. [71]. This paramagnetic dimer **3** was found to isomerize to the diamagnetic structures **1** and **2** depending on the solvent and temperature. According to Rosenthal, exchange of acetylides is possible at some stage of the reaction, which leads to butadiyne, that is different from what present in the beginning of the reaction [10]. Therefore, this reaction can be considered as an alkane metathesis reaction (Equation 14.1).

$$R-C\equiv C-C\equiv C-R + R'-C\equiv C-C\equiv C-R'$$

$$\xrightarrow[\text{M=Ti, Zr}]{\text{Cp}_2\text{M}} 2\ R-C\equiv C-C\equiv C-R' \qquad (14.1)$$

Over the years, we have developed detailed understanding of the reactions of the Cp_2M fragment to form several metallacycles and further reaction of those metallacycles [11]. The metallacycles are involved in important reactions such as the synthesis of carbocyclic and heterocyclic compounds and they have unusual ability to stabilize highly reactive organic entities [1,6]. We begin our discussion with a description of the nature of the frontier orbitals of bent metallocenes, followed by their reactions with organic ligands to form unusual metallacycles, especially the metallacyclocumulenes. We will particularly concentrate on the structure, bonding, and reactivity of the key intermediate of the C—C coupling and decoupling reactions, the metallacyclocumulene (**5**). The organic counterpart of the metallacyclocumulene is not stable due to high ring strain [11e]. The possible isomeric forms of metallacyclocumulene (**5**) are metallacyclopropene (**6**) and metal bisacetylide (**7**) (Scheme 14.3) [11d]. The preference for these isomers varies with the metal in the Cp_2M (M = Ti, Zr) fragment. Zr prefers the metallacyclocumulene (**5**) complexes and Ti exists in a dynamic equilibrium between the metallacyclocumulene (**5**) and metallacyclopropene (**6**) [7e,9b]. Our DFT calculations explain these observations [11f]. Lastly, we describe the different mechanistic steps involved in the C—C coupling and cleavage reactions of Ti and Zr complexes using DFT studies.

We have used hybrid HF-DFT method, B3LYP [12,13] for optimization of the molecules using Gaussian program package [14]. This method is based on Becke's three-parameter functionals including Hartree–Fock exchange contribution with a nonlocal correction for the exchange potential proposed by Becke [13a,b] together with the nonlocal correction for the correlation energy provided by Lee et al. [13c].

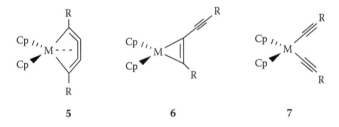

      5   6   7

SCHEME 14.3 Schematic representation of different isomers of $Cp_2M(C_4R_2)$, M = Ti, Zr; metallacyclocumulene (**5**), metalacyclopropene (**6**), and metal bis-acetylide (**7**) complexes.

We used the LANL2DZ basis set with the effective core potentials of Hay and Wadt [15]. Frequency calculations were carried out at the same level of theory to characterize the nature of the optimized structures. A better understanding of the structures is obtained from the fragment molecular orbital calculations [16]. We have obtained reasonable agreement with experiment in terms of relative energetics and structural parameters using this level of theory. The accuracy of the energy values and geometrical parameters, calculated by B3LYP functional, is well documented [17a,b]. The geometrical parameters in the transition metal complexes are reported to be close to the experimental values [17a]. However, the errors in the energy values are usually in the range of 3–5 kcal/mol [17a]. Due to computational limitations, we have replaced the actual ligands on metal and butadiyne with smaller substituents. These replacements will be mentioned in the appropriate contexts during the course of the discussion. The Kohn–Sham DFT has become one of the most popular and widely used tool in the electronic structure calculation due to its modest computational cost, which makes it applicable to large systems as compared with correlated wave function theory [17a–f]. Nevertheless, one should be aware of certain limitations of the DFT. The reliability of this method is limited by the functionals used. The main challenge in a DFT calculation at this level is the quality of the nonclassical exchange and correlation interactions between electrons. Moreover, there are cases where DFT fails [17g–l]. It is inaccurate for interactions which are dominated by correlation energy such as van der Waals attractions, noncovalent interactions, and π–π stacking. Photochemistry, strongly correlated systems, physisorption, and polymers are poorly described in standard DFT. The major drawback in using these DFT methods is that they cannot be systematically improved unlike the wave function-based methods. As a result, the performance testing of the different density functionals is necessary before it is applied to a specific research problem. Several attempts to get better functionals than the popular B3LYP during the last few years have been fruitful [18]. Many groups are actively participating in the design of new functionals to overcome the above mentioned limitations of B3LYP. At present, several new and improved functionals are available in the literature [18]. Some of the newly developed functionals are B2PLYP [18a], B2PLYP-D [18b], B3PW91 [18c], mPW1PW91, mPWPW91 [18d], PBEPBE [18e], M05-2X [18f], M05, PW6B95, PWB6K, and MPWB1K [18g]. However, when we started these studies, B3LYP was the best available functional for the calculation involving transition metal organometallic complexes. The calculations done using B3LYP are retained in this discussion, so that all calculations need not be done again. We have checked several crucial steps with more recent functionals and they give similar results [11h]. It is clear that the study of a new problem must be undertaken after ascertaining the applicability of the functional for that specific system. Other chapters in this book provide several guidelines for this.

14.2   BENT METALLOCENES: Cp$_2$M (M = Ti, Zr)

Unlike ferrocene, group-4 metallocenes are 14-electron species and hence, very reactive [19]. These are generated in situ, for immediate reaction [9b,20] and have bent geometry (Scheme 14.4). The electron deficiency forces them to accept

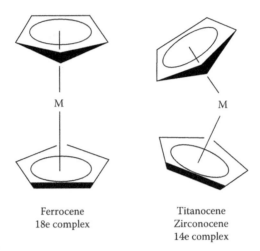

Ferrocene
18e complex

Titanocene
Zirconocene
14e complex

SCHEME 14.4 Schematic representation of ferrocene and bent metallocene (Titanocene and Zirconocene).

electrons from other ligands. The bent metallocenes Cp_2M (M = Ti, Zr) have three in-plane valence molecular orbitals and two electrons (d^2 system) available for bonding with other ligands (Figure 14.1a) [19]. The similarity of the frontier orbitals of Cp_2M (M = Ti, Zr) fragment, with the frontier orbitals of the carbene, can be easily understood (Figure 14.1b) [6b]. Carbene has two in-plane molecular orbitals, whereas Cp_2M fragment has three in-plane molecular orbitals. The highly directed and contracted molecular orbitals available in carbene lead to the typical organic

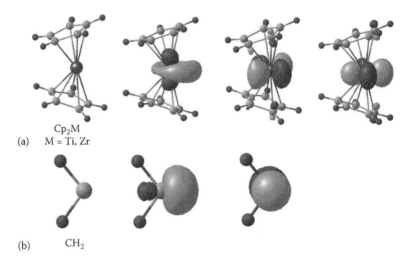

(a) Cp_2M
     M = Ti, Zr

(b) CH_2

FIGURE 14.1 (See color insert following page 302.) Frontier molecular orbitals of (a) bent metallocenes (Cp_2M; M = Ti, Zr) and (b) carbene.

reactions of carbene. Carbene cannot stabilize the strained organic fragments to the extent a Cp_2M fragment can do with its more diffuse orbitals that are available for bonding. Moreover, the three frontier molecular orbitals of Cp_2M fragment can rehybridize to form suitably oriented molecular orbitals for the interaction with the ligand. Therefore, the replacement of CH_2 fragment in organic molecules by Cp_2M (M = Ti, Zr) fragment brings dramatic changes in the structure, bonding, and reactivity of the organometallic moiety as compared with the organic molecule. One such example is obtained by replacing CH_2 fragment in highly strained cyclopentatriene molecule by Cp_2M (M = Ti, Zr) fragment to generate metallacyclocumulenes (**5**).

14.3 STRUCTURE AND BONDING IN METALLACYCLOCUMULENE

These bent metallocenes on reaction with unsaturated molecules such as alkynes, olefins, acetylides, and vinyls form several metallacycles [9,10]. One rather unusual five-membered metallacycle is metallacyclocumulene (**5**). It has been synthesized in several ways. In 1994, the first stable metallacyclocumulene ($Cp_2Zr(\eta^4\text{-}t\text{-}BuC_4\text{-}t\text{-}Bu)$) was synthesized by Rosenthal et al. and was obtained by the reaction of Cp_2Zr (pyridine)($\eta^2\text{-}Me_3SiC_2SiMe_3$) with $t\text{-}BuC\equiv C-C\equiv C\text{-}t\text{-}Bu$ [21]. A few years later, the titanacyclocumulenes $Cp_2Ti(\eta^4\text{-}RC_4R)$, R = t-Bu and Ph, were synthesized from $Cp_2Ti(\eta^2\text{-}Me_3SiC_2SiMe_3)$ and $RC\equiv C-C\equiv CR$ [22]. Later on several derivatives of both titanacyclocumulenes and zirconacyclocumulenes have been synthesized in many different ways (Scheme 14.5) [6h,6i,23–25].

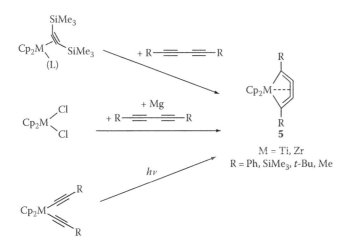

SCHEME 14.5   Different synthetic routes for the preparation of metallacyclocumulene (**5**). Note: The perception of strain energy in the cyclocumulene is so high that initial attempts at publishing the experimental single crystal data of the metallacyclocumulene (**5**) met with strong resistance from referees (personal comments from Prof. U. Rosenthal). Characterization of the structure computationally helped in gaining acceptance of the structure.

TABLE 14.1

Geometrical Parameters of the Metallacyclocumulene Complexes (5), $Cp_2Zr(C_4H_2)$ and $Cp_2Ti(C_4H_2)$, Calculated at B3LYP/LANL2DZ Level of Theory

Molecule	Bond Length (Å)				Bond Angle (°)	
	C1—C2	C2—C3	M—C1	M—C2	C1—C2—C3	M—C1—C2
$Cp_2Zr(C_4H_2)$	1.316	1.342	2.353	2.373	146.6	74.7
	(1.280)	(1.310)	(2.357)	(2.303)	(150.0)	—
$Cp_2Ti(C_4H_2)$	1.302	1.344	2.242	2.259	145.1	73.9
	(1.277)	(1.338)	(2.252)	(2.210)	(147.6)	—

Note: Experimental values are given in parenthesis [25].

Source: Reproduced from Bach, M.A., Parameswaran, P., Jemmis, E.D., Rosenthal, U. *Organometallics*, 26, 2149, 2007.

This complex seems unusual in the beginning, as the organic counterpart of metallacyclocumulene (**5**), i.e., cyclopentatriene (C_5H_4), is highly unstable due to the nonlinear C=C=C=C bond and the resulting ring strain [11e,26]. On the basis of x-ray structural data and IR spectra, there is substantial interaction between the metal and the middle C=C bond of the metallacyclocumulene [6h,i,21–25]. The calculations on metallacyclocumulene (**5**) show that the metal atom and the four carbon atoms of the metallacyclocumulene ring are coplanar [11d,e,g]. All the M—C bonds are within the bonding range, and the middle M—C bonds are marginally longer than the end ones (Table 14.1 and Figure 14.2a). The computed geometrical parameters of the metallacyclocumulenes are not far from the experimental structures (Table 14.1). The C2—C3 bond lengths of the metallacyclocumulenes $Cp_2Zr(C_4H_2)$ and $Cp_2Ti(C_4H_2)$ are of comparable magnitude, indicating a similarity in the electronic structure as well.

The computed C—C and M—C bond indices and populations reveal the cumulenic nature of **5** [11d,e,g]. The electronic structure of metallacyclocumulene is best analyzed from a fragment molecular orbital approach [16]. The metal in the Cp_2M fragment is in the formal oxidation state of +2 with two valence electrons. The three frontier orbitals of Cp_2M are in the MC_4 plane. The in-plane frontier orbitals of the HCCCCH fragment are formed from the two in-plane p-orbitals on the two middle carbon atoms (C2 and C3) and the sp hybrid orbitals on the end carbon atoms, C1 and C4. These form four linear combinations, similar to the orbitals of butadiene. The lowest two orbitals among these are filled. The next MO, the lowest unoccupied molecular orbital (LUMO) of the C_4H_2 fragment, corresponds to the in-plane equivalent of the LUMO of the butadiene π-orbitals and is bonding between C2 and C3. The strongest stabilizing interaction is between this LUMO of C_4H_2 fragment and the highest occupied molecular orbital (HOMO) of Cp_2M fragment (Figures 14.2b and 14.3). This interaction stabilizes the C2—C3 bond (HOMO, Figure 14.2b). This is in contrast with the familiar Dewar–Chatt–Duncanson model of metal to π^* backbonding, which would have lengthened the C2—C3 bond [27]. This effect is tempered by the two π MOs

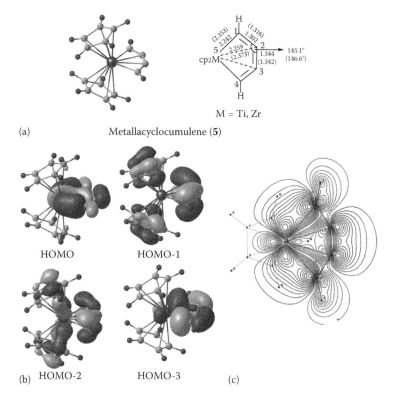

(a)   Metallacyclocumulene (**5**)

(b) HOMO-2   HOMO-3   (c)

FIGURE 14.2 (See color insert following page 302.)   (a) Structure and important geometrical parameters of metallacyclocumulene (**5**). (b) molecular orbitals of metallacyclocumulene (**5**). The values in normal font correspond to M = Ti and the values in parenthesis correspond to M = Zr. (c) A contour diagram of the HOMO of metallacyclocumulene (**5**) showing the in-plane interaction of metal and the ligand. (Reproduced from Bach, M.A., Parameswaran, P., Jemmis, E.D., Rosenthal, U., *Organametallics*, 26, 2149, 2007; Jemmis, E.D., Phukan, A.K., Jiao, H., and Rosenthal, U., *Organometallics*, 22, 4958, 2003. With permission.)

perpendicular to the MC_4 plane, typical of butadiene. Another stabilizing interaction is the donation of electrons from the HOMO of C_4H_2 to the empty d-orbital of the metal atom (HOMO-3, Figures 14.2b and 14.3). The contour plot of the HOMO of **5** indicates significant interaction between the central carbon atoms with metal and in-plane delocalization of electrons in the ring (Figure 14.2c). The π delocalization in the plane of the ring (HOMO and HOMO-3) and in the plane perpendicular to the ring (HOMO-1 and HOMO-2) indicates the bis-homoaromatic nature of the metallacyclocumulene. Homo- and bis-homoaromaticity are usually observed in charged species, but neutral homoaromatic and bis-homoaromatic systems are very rare. We have done nuclear independent chemical shift (NICS) calculations to understand the cyclic delocalization of electrons or aromaticity [29]. This calculation indicates possible interactions between the metal and the carbon atoms in the metallacyclocumulenes (**5**). A very strong aromatic stabilization is found in the metallacyclocumulene with both M = Zr and Ti, as indicated by their

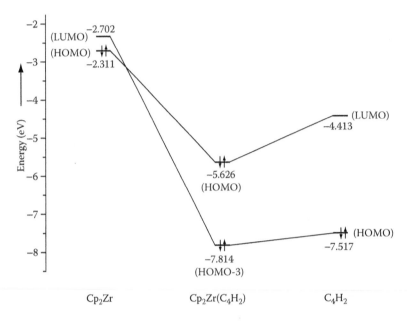

FIGURE 14.3 Interaction diagram between the cumulene (C_4H_2) and Cp_2Zr fragment in metallacyclocumulene complex (**5**, $Cp_2Zr(C_4H_2)$) obtained using ADF2007.01 program package [28]. The other less important interactions are omitted for clarity. A similar interaction diagram is obtained for $Cp_2Ti(C_4H_2)$ complex.

large NICS(0) (−34.4 and −36.2 for M = Zr and Ti, respectively) and NICS(1) (−15.0 and −16.7 for M = Zr and Ti, respectively) values [11e]. These are due to the 3c–2e bonding in the plane of the five-membered ring and π-bond perpendicular to the plane of the five-membered ring. Thus, one might conclude that they are neutral bis-homoaromatic.

The interaction between the central carbon atoms with metal and in-plane delocalization of electrons in the ring reduce the strain in the five-membered metallacyclocumulene as compared to corresponding organic counterpart, cyclopentatriene [11e]. The ring strains were calculated by successive hydrogenation energies from the unsaturated cyclocumulene (C_5H_4, $Cp_2Ti(C_4H_2)$ and $Cp_2Zr(C_4H_2)$) to the saturated cyclopentane (C_5H_{10}, $Cp_2Ti(C_4H_8)$ and $Cp_2Zr(C_4H_8)$) (Scheme 14.6). As

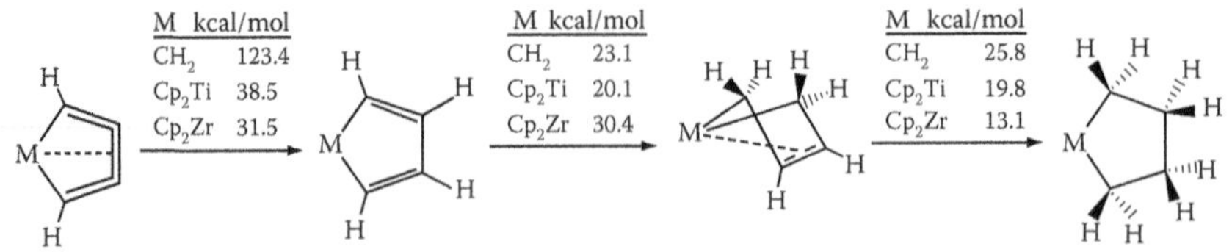

SCHEME 14.6 Successive hydrogenation energies (in kcal/mol) of the unsaturated cyclocumulenes C_5H_4, $Cp_2Ti(C_4H_2)$ and $Cp_2Zr(C_4H_2)$ to form the saturated cyclopentanes C_5H_{10}, $Cp_2Ti(C_4H_8)$ and $Cp_2Zr(C_4H_8)$ at the B3LYP/LANL2DZ level of theory.

expected, the C—C cumulenic bond in cyclopentatriene (C_5H_4) has a very high hydrogenation energy (123.4 kcal/mol), which is five times the value (23.1 kcal/mol) of cyclopentadiene (C_5H_6). This is due to the enhanced strain in the cyclopentatriene, and it is not surprising that this molecule is still elusive [26]. The small hydrogenation energies for metallacyclocumulene (38.5 kcal/mol for M = Ti and 31.5 kcal/mol for M = Zr) indicate that the metal fragment nearly eliminates the strain energy present in the cyclopentatriene molecule [11e]. These reduced hydrogenation energies can be ascribed to very strong stabilizing interaction between the metal center and the cumulene. Another component responsible for the reduction in strain is the longer M—C bonds as opposed to the corresponding C—C bonds in cyclopentatriene. The C1—C2—C3 angle of 114.7° in the cumulinic part of the cyclopentatriene is far away from the ideal linearity anticipated for cumulenes. The complexation with metal increases this angle in metallacyclocumulene to 145.1° for M = Ti and 146.6° for M = Zr (Figure 14.2a) and helps in reducing the strain in five-membered ring. However, this angle is still far away from linearity.

After observing the low value of the strain energy of the metallacyclocumulenes, it is not surprising that these are similar in energy to their less strained isomers such as metallacyclopropenes (**6**) and metal bis-acetylides (**7**). It is interesting to note that the most stable isomer among the organic counterpart of the molecules **5**, **6**, and **7** is the least strained acyclic species **7**; the cyclocumulene derivative **5** is higher in energy by 51.6 kcal/mol and the cyclopropene derivative **6** is higher in energy by 16.7 kcal/mol. The introduction of Cp_2Ti and Cp_2Zr metal fragment brings a remarkable change in the stability of these molecules and such a clear spread of energy between the three structures vanishes [11d]. Calculations indicate that the metallacyclocumulene derivative **5** and the metal bis-acetylide derivative **7** are almost comparable in energy for both Ti and Zr [11d]. The difference between Ti and Zr is shown in the relative energies of the metallacyclopropene derivative **6**. The titanacyclopropene is lower in energy (8.1 kcal/mol) than the other two isomers, while the zirconacyclopropene is higher in energy (4.7 kcal/mol). It is clearly seen from the experiments so far that the Cp_2Ti fragment prefers a metalla-cyclopropene structure [6b,21]. On the other hand, similar experiments with Zr gives structures which may be derived from **5** [6b,24]. Either of these isomers, metal bis-acetylides (**7**) or metallacyclocumulene (**5**), can react further with another Cp_2M fragment to form the bimetallic complexes **1** or **2**. The detailed energetics for these set of reactions are discussed in the next section.

14.4   MECHANISM OF C—C COUPLING AND DECOUPLING REACTIONS OF Cp₂M (M = Ti, Zr)

The possible mechanism for the formation of C—C coupled (**1**) and decoupled (**2**) bimetallic complexes from Cp_2M has been investigated [11b,c]. We followed a stepwise procedure to arrive at a model that was practical and at the same time realistic. In the first stage, the substituted cyclopentadienyls were replaced by Cp and the substituents on acetylides and butadiynes were replaced by H. The relative energies showed that, the C—C coupled structure **1** for M = Ti when L = Cp and R = H is more stable than **2** by 3 kcal/mol, while **2** is calculated to be 14.8 kcal/mol lower in energy than **1** for M = Zr when L = Cp and R = H (Figure 14.4) [11b,c]. This follows the experimental trend. However, the systems with cyclopentadienyl ligand on the metal fragment were too large to perform computation. We had substituted the ligand on Ti and Zr by Cl and calculated the relative energies for the structures **1** and **2** for M = Ti and Zr. Even though, the C—C coupled structure was unfavorable for both Ti and Zr, the C—C coupled structure **1** for Zr complex is higher in energy as compared to the C—C coupled structure for Ti complex. For a detailed study of the mechanistic details, even these systems were too large. Thus while several calculations were carried out with this model, further simplification was sought by replacing the Cl by H. As Figure 14.4 indicates, this did not change the energetics substantially. However, the replacement of Cp by Cl or H has a larger effect on the energetics of the Ti complexes and the energetics of the Zr complexes are less affected by the substitution. It should be noted that, if these studies have been carried out in recent times, the real-life ligands could have been used. Selected

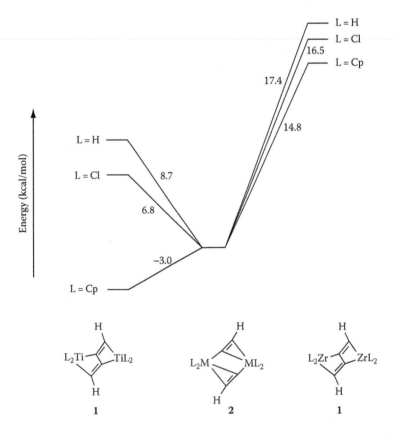

FIGURE 14.4 Comparison of relative energies (in kcal/mol) of the complexes **1** and **2** with M = Ti, Zr and L = H, Cl, and Cp calculated at the B3LYP/LANL2DZ level of theory. (Reproduced from Jemmis, E.D. and Giju, K.T., *J. Am. Chem. Soc.*, 120, 6952, 1998. With permission.)

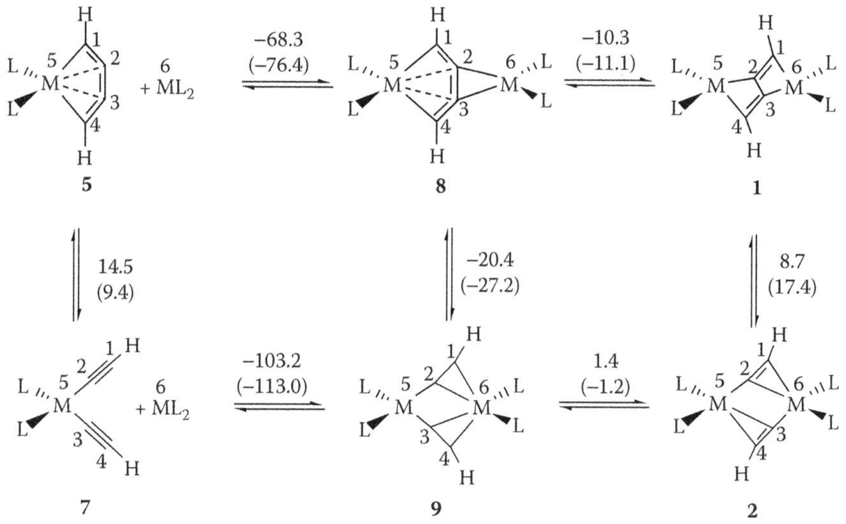

SCHEME 14.7 Reaction energies (in kcal/mol) at the B3LYP/LANL2DZ level of theory between various intermediates of the reaction of Ti and Zr complexes where L = H. The energetics with L = Cl also gives similar trend in the order of the relative energies [11b,c]. The values in normal font correspond to M = Ti and values in parenthesis correspond to M = Zr. (Reproduced from Jemmis, E.D. and Giju, K.T., *J. Am. Chem. Soc.*, 120, 6952, 1998. With permission.)

computations on the structures with Cp ligands indicate that the general conclusions of this study remain the same.

On the basis of the analysis of various steps of the reactions and previous suggestions (Scheme 14.2a and b), we had proposed a reaction mechanism, which accounts for most of the experimental observations in the C—C coupling and decoupling reaction (Scheme 14.7). This mechanistic proposal was based on the structures of homo- or heterobinuclear transition metal complexes isolated previously [3,7,9]. All the intermediates in this reaction scheme, except the complex **8** for homo-bimetallic complexes with M = Ti, Zr, and **1** for M = Zr are reported experimentally for both Ti and Zr complexes. Thus, the chemistry of Ti and Zr provides many similarities along with dramatic contrasts. However, the mechanism proposed by us from DFT calculations is found to be different from the suggested mechanisms (Scheme 14.2) [7m,9]. In the bimetallic complexes **8** and **1**, the central carbon atoms C2 and C3, are bonded to each other. These structures can thus be considered as C—C coupled structures. On the other hand, the structures **9** and **2** do not have a bond between the central carbon atoms (C2 and C3), and hence these structures can be considered as C—C decoupled structures. If we consider the reaction pathway starting from structure **5** to structure **2**, then it is obvious that the central C—C bond in **5** is cleaved during the reaction. Similarly, the reaction pathway from structure **7** to **1** indicates a C—C coupling reaction.

Scheme 14.7 gives the detailed energetics for M = Ti, Zr and L = H of all the possible reaction steps starting from the monometallic complexes **5** and **7**. The order

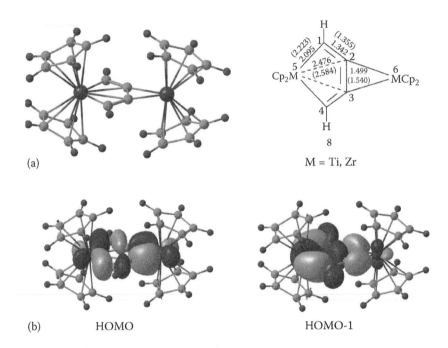

(a)   M = Ti, Zr

(b)   HOMO   HOMO-1

FIGURE 14.5 (See color insert following page 302.) (a) The structure and important geometrical parameters of the bimetallic complex **8**. The values in the normal font correspond to M = Ti and the values in parenthesis correspond to M = Zr. (b) Molecular orbitals of the complex **8** showing the in-plane delocalization.

of the relative energies obtained for L = Cl is not very different from what is obtained for L = H [11b,c]. The first step is the formation of the bimetallic complex **8** from metallacyclocumulene (**5**). This process is exothermic by 68.3 kcal/mol for M = Ti and L = H (Scheme 14.7). The major change observed here is the elongation of the C2—C3 bond in complex **8** as compared to complex **5** (Figures 14.2a and 14.5a). From the earlier bonding analysis of metallacyclocumulene (**5**) it is clear that the perpendicular π-orbitals are delocalized (HOMO-1 and HOMO-2, Figure 14.2b) and hence less available for further interaction with the second metal. There is only the in-plane molecular orbital (HOMO, Figure 14.2b) available in metallacyclocumulene for bonding to another Cp$_2$M fragment [11d,e,g,h]. The HOMO of the metallacyclocumulene donates electrons to the empty d-orbital of the second Cp$_2$M fragment to form the complex **8** (HOMO-1, Figure 14.5b). This reduces the bonding interaction between C2 and C3. At the same time, back donation from the filled d-orbital of Cp$_2$M fragment to the LUMO of metallacyclocumulene also takes place (HOMO, Figure 14.5b). Both these interactions lead to an increase in the antibonding interaction between C2 and C3, which results in the elongation of C2—C3 bond in structure **8** as compared to that in the structure **5** [11d,e,g]. The striking feature of the structure **8** is the coexistence of two adjacent planar tetra-coordinated carbon atoms C2 and C3 [11g]. While there have been several examples of structures with single planar tetra-coordinated carbon atom, structures with two adjacent planar tetra coordinated carbon atoms are not common [9b,10c,30].

The metal in metallacyclocumulene (5) is in the formal +4 oxidation state and retains the +4 oxidation state while forming the bimetallic complex 8. The longer C2—C3 bond distance in 8 justifies the description of C_4 as a buta-1,3-diene tetraanion making both metals formally +4. On the other hand, the bis-acetylide complex 7 (metal is in +4 oxidation state) reacts similarly with ML_2 and forms a complex 9. This can be viewed as formed by the donation of the π-MO in the metal bis-acetylide (7) complex to another ML_2 fragment. The formation of the redox complex 9, in which one metal center formally is in the +4 and the other is in +2 oxidation state, is more exothermic (103.2 kcal/mol for M = Ti and L = H) than the formation of 8 from 5. Thermodynamically, 9 is favored over 8 partly because of the absence of stabilization due to π-coordination towards the metal center in the latter complex. Since the in-plane π bond of C2—C3 is already delocalized with the metal (M5) in metallacyclocumulene (5), the metal–π interaction in complex 8 (interaction of M6 with C2—C3) does not have optimum orbital hybridizations for the best overlap (Figure 14.5b) [11g,h]. The complex 9 can rearrange to 2 by the inversion of one of the C=C π-bond and similarly complex 8 can isomerize to 1 through the movement of one C=C π-bond. The relative energy difference between 9 and 2 is 1.4 kcal/mol for M = Ti and −1.2 kcal/mol for M = Zr. Rosenthal et al. had suggested the rearrangement of 8 to 1 as a possible step in the mechanism (Scheme 14.2a). This is also calculated to be thermodynamically favorable for Ti and Zr (Scheme 14.7). The structural difference between 1 and 2 is not very large and 2 can form 1 by a C—C bond formation between the two central carbon atoms, C2 and C3.

These energetics points out that the thermodynamic preferences of ground-state structures are not sufficient for deciding the most feasible mechanistic steps. The potential energy diagram for the overall reaction mechanism including various transition structures are shown in Figure 14.6 for Ti and Zr complexes. Comparison of barrier heights of different reaction steps helps to predict the favorable pathway. The conversion of the complex 8 to 1 suggested in the experimental Scheme 14.2a by Rosenthal et al. is calculated to have the highest barrier (34.1 kcal/mol for M = Ti and 33.4 kcal/mol for M = Zr) on the potential energy surface. However, the M–M axis in 8 is orthogonal to the middle C—C bond, which is already highly stretched for a conjugated carbon chain, making its activation feasible. The same conversion can be achieved through intermediates 9 and 2. The barriers for the isomerization reaction 8 to 9 are calculated to be practically nil (0.9 kcal/mol for M = Ti and 0.3 kcal/mol for M = Zr). The next step 9 to 2 also have reasonably low barrier heights of 13.4 kcal/mol for M = Ti and 14.3 kcal/mol for M = Zr, justifying the dynamic processes observed in solution [7,9]. The final step, involving the C—C bond formation from 2 to 1, has barriers 10.4 kcal/mol and 18.1 kcal/mol for Ti and Zr complexes, respectively. This is the largest difference calculated between energetics of the Ti and Zr complexes in these sets of reactions. Thus, the analysis of different mechanistic pathways outlined in Scheme 14.7 suggests the possibility of the reaction to proceed via 5-8-9-2-1 rather than any other pathways. Details of the transition structures give hints about the reason for high or low barriers. An interesting case is the transformations of 8 to 1 and 9 to 2 (Scheme 14.8). Both involve shifting of a

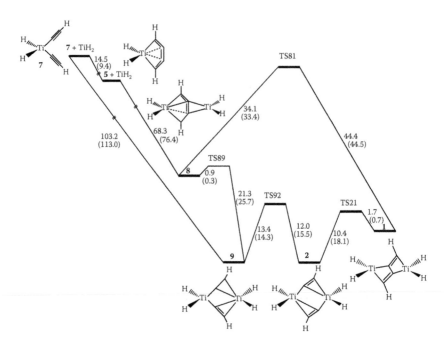

FIGURE 14.6 Potential energy surface at the B3LYP/LANL2DZ level of theory for the mechanism of C—C coupling and decoupling reactions given in Scheme 14.7. The energy values are in kcal/mol. The energetics with L = Cl also gives similar trend in the order of the relative energies [11b,c]. The values in normal font correspond to M = Ti and values in parenthesis correspond to M = Zr. (Reproduced from Jemmis, E.D. and Giju, K.T., *J. Am. Chem. Soc.*, 120, 6952, 1998. With permission.)

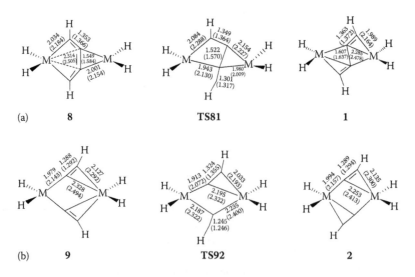

SCHEME 14.8 Important bond lengths of the complexes and transition states for M = Ti, Zr at the B3LYP/LANl2DZ level of theory; (a) **8**, **TS81** and **1**; (b) **9**, **TS92**, and **2**. The values in normal font correspond to M = Ti and values in parenthesis correspond to M = Zr. (Reproduced from Jemmis, E.D. and Giju, K.T., *J. Am. Chem. Soc.*, 120, 6952, 1998. With permission.)

π-bonded C_2 unit from one metal to another. However, the former transformation has a very high barrier compared to the latter one. This is because the middle C—C bond of the C_4H_2 unit in **8** does not allow much room for adjustment in the transition state (**TS81**) while transition state connecting **9** and **2** (**TS92**) has enough feasibility to retain reasonable geometry around the metal even in the transition state structure. Among all the mechanistic steps in Scheme 14.7, the major step in distinguishing Ti from Zr in C—C coupling is the formation of **1** from **2**.

The experiments suggest a delicate thermodynamic balance between the complexes **2** and **1** for the two metals; Zr prefers structure **2** and Ti prefers structure **1** [7,9]. This forced us to study the mechanism of this step of the reaction in greater detail [11b,c]. The formal oxidation states +3 can be assigned to the metals in **2** and +4 to the metals in **1**. Hence, complex **2** has two 17-electron metal centers (when L = Cp) and complex **1** has two 16-electron metal centers (when L = Cp). Therefore, the formation of **1** from **2** may be described as an unusual oxidative C—C coupling (change of oxidation state from +3 to +4 with concomitant C—C coupling). At this stage, it was tempting to conclude that Ti, a first-row transition metal, can accommodate both 16 and 17 electron counts (for L = Cp), but Zr may not be able to do the same. As a result, the formation of C—C decoupled structure **2** is only observed for M = Zr and Ti forms both the complexes **1** and **2**. However, this need not be true, because the substituents on the 1,3-butadiyne may also change the thermodynamic stabilities substantially. The preference for the structure **2** shown by Ti complexes with R = Si(CH$_3$)$_3$ [31] suggests a complementary strategy for stabilizing the C—C coupled structures **1** for Zr. The highly electron-withdrawing substituents such as fluoride on the acetylide bridging group might help in making structure **1** more competitive to **2**. Accordingly, the energetics for the transformation of **2** to **1** were calculated using R = H, F, and CN on the 1,3-butadiyne and L = H (Figure 14.7).

The change in the thermodynamic stability in going from R = H to R = F was as expected. While **1** (R = H) is 17.4 kcal/mol less stable than **2**, the complex **1** (R = F) is 3.9 kcal/mol more stable than **2** for M = Zr (Figure 14.7) [11b,c]. Model studies indicate that Ti prefers C—C coupled product **1** over Zr even when R = H and this preference becomes more prominent with R = F. The C—C coupled structure **1** is more preferred for Zr complexes than the uncoupled structure **2**, when R = F (Figure 14.7). Thus, we suggest that fluoride substituents on 1,3-butadiyne would force the reaction into the side of structure **1** for M = Zr. These trends can be used in designing appropriate ligands for the experimental synthesis of C—C coupled Zr complex. The thermodynamic preference for the structure **2** with M = Zr and R = CN did not reverse, and the coupled product **1** is less favorable by 9.1 kcal/mol than the decoupled product **2** (Figure 14.7). This indicates that the CN substituent is not sufficiently electron accepting to enforce the shift in the equilibrium. Further calculations on the model complexes **1** and **2** for both the metals, where L = Cp and R = H, were carried out to verify the trend obtained with the simplistic model complexes (L = H and R = H). The change in the thermodynamic preference from R = H to R = F for M = Zr retains with L = Cp. The relative energy difference of 14.8 kcal/mol (17.4 kcal/mol with L = H) between the C—C decoupled structure **2**

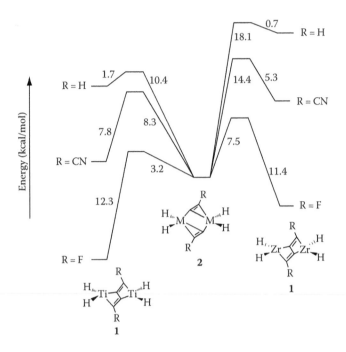

FIGURE 14.7 Potential energy diagram at the B3LYP/LANL2DZ level of theory for the isomerization of **2** to **1** with M = Ti, Zr and R = H, CN, and F. The relative energy values are in kcal/mol. (Reproduced from Jemmis, E.D. and Giju, K.T., *J. Am. Chem. Soc.*, 120, 6952, 1998. With permission.)

and C—C coupled structure **1** with R = H was brought down to −10.6 kcal/mol (−3.9 kcal/mol with L = H) with R = F when Cp is used for M = Zr.

The difference in the energetics of **1** and **2** for Ti and Zr requires an explanation. The difference in the ionic radii between Ti and Zr suggests a possible justification. The larger size of Zr leads to the longer C1—C2 bond and consequently the C1 and C2 are almost equidistant from M5 in structure **2**, for Cp$_2$Zr. This in turn results in a longer C2—C3 bond (3.003 Å, with L = Cp) in **2**. This facilitates π-bonding to M6, while maintaining σ interaction with M5. In the isostructural Ti complex **2**, the corresponding C2—C3 bond length is 2.985 Å (L = Cp), which is further along the path to C—C bond formation. We found that the fluoride substitution also helps in reducing the C2—C3 distance in the Zr complex **2** from 3.003 Å when R = H to 2.985 Å (R = F, L = Cp). The relative energies of the transition state structures indicate that C—C coupling is considerably more favorable for Ti than for Zr (Figure 14.7). The analysis of the transition state structure for the transformation of **2** to **1**, indicates that the movement of the central carbon atoms required for M = Ti is less compared to Zr. During the process of C—C bond formation, with L = H, the C2—C3 distance varies from 2.786 Å (in **2**) → 1.964 Å (in transition state **TS21**) → 1.607Å (in **1**) for Ti complexes. A similar variation in Zr complexes, with L = H, is 2.986Å (in **2**) → 1.944Å (in transition state **TS21**) → 1.637Å (in **1**).

Comparison of the x-ray crystal structure data of similar Ti and Zr complexes **2** shows that the Zr–Zr distance (3.522 Å) is shorter compared to the Ti–Ti distance (3.550 Å). This is interesting for the following reason. The conventional electron counting leads to +3 oxidation state and d^1 electron count for the metals in **2**. The covalent radius of Zr (1.45 Å) is larger than that of Ti (1.32 Å). Despite this, the Zr—Zr distance is shorter than that of the Ti–Ti distance. The charge analysis with L = Cp and R = H for M = Ti and Zr indicates a major electrostatic metal ligand interaction for M = Zr in comparison to M = Ti [11c]. This might lead to a possible antiferomagnetic interaction mediated through the bridging acetylides to the shortening of Zr–Zr distance compared to Ti–Ti distance in **2**. In conclusion, the unusual C—C coupling observed in dimeric titanium complexes and lacking in the corresponding zirconium complexes is a consequence of thermodynamic energy differences.

14.5   CONCLUSION

It is clear from the above discussion that the theoretical calculations using DFT contribute enormously to the understanding of the structure, bonding, and reactivity of the organometallic complexes. It also gives such insight into the structure, bonding, and reactivity, which otherwise could not have been obtained from experiments. Major advances have been made in analyzing the nature of the electron-deficient metallocenes, Cp_2M (M = Ti, Zr). We have also studied successfully the formation of the unusual organometallic complex, the metallacyclocumulene, from these electron-deficient metallocenes. Besides, the reactivity of metallacyclocumulenes toward another Cp_2M fragment is also analyzed using DFT. Theoretical studies could answer several crucial questions in this area of the reaction of Cp_2M fragment with unsaturated organic molecule such as acetylide and butadiyne. It could satisfactorily explain the stability of the unusual metallacyclocumulene complex from the bonding studies, which indicates predominant interaction between ligand and metal as the prime cause for the stability. A thorough mechanistic study at the DFT level explains that the difference in the reactivity of M = Ti and M = Zr to form C—C coupled and C—C decoupled products is governed by the thermodynamic energy difference. Our experience clearly indicates that chemical understanding of great importance can be obtained by considering all aspects of a set of chemical structures and reactions rather than a single chemical transformation.

ACKNOWLEDGMENTS

The authors acknowledge Council of Scientific and Industrial Research, New Delhi (fellowship to S. D.) and Department of Science and Technology (J. C. Bose Fellowship) for financial assistance.

REFERENCES

1. Togni, A., Halterman, R. L. (Eds.), *Metallocenes: Synthesis, Reactivity, Applications*, Wiley-VCH, Weinheim, 1998, Vols. 1 and 2 and references therein.
2. (a) Alt, H. G. and Köppl, A. *Chem. Rev.* 2000, *100*, 1205. (b) Resconi, L., Cavallo, L., Fait, A., and Piemontesi, F. *Chem. Rev.* 2000, *100*, 1253. (c) Kaminsky, W. *J. Chem.*

Soc., Dalton Trans. 1998, 1413. (d) Bochmann, M. *J. Chem. Soc., Dalton Trans.* 1996, 255. (e) Brintzinger, H. H., Fischer, D., Mulhaupt, R., Rieger, B., and Waymouth, R. M. *Angew. Chem., Int. Ed. Engl.* 1995, *34*, 1143.

3. (a) Choukroun, R., Donnadieu, B., Zhao, J.-S., Cassoux, P., Lepetit, C., and Silvi, B. *Organometallics* 2000, *19*, 1901. (b) Choukroun, R. and Cassoux, P. *Acc. Chem. Res.* 1999, *32*, 494. (c) Choukroun, R., Donnadieu, B., Zhao, J., Cassoux, P., Lepetit, C., and Silvi, B. *Organometallics* 2002, *19*, 1901. (d) Danjoy, C., Zhao, J. S., Donnadieu, B., Legros, J.-P., Valade, L., Choukroun, R., Zwick, A., and Cassoux, P. *Chem. Eur. J.* 1998, *4*, 1100.

4. (a) Pool, J. A., Lobkovsky, E., and Chirik, P. J. *Nature* 2004, *427*, 527. (b) Bernskoetter, W. H., Olmos, A. V., Pool, J. A., Lobkovsky, E., and Chirik, P. J. *J. Am. Chem. Soc.* 2006, *128*, 10696. (c) Chirik, P. J. *Dalton Trans.* 2007, 16. (d) Hanna, T. E., Bernskoetter, W. H., Bouwkamp, M. W., Lobkovsky, E., and Chirik, P. J. *Organometallics* 2007, *26*, 2431. (e) Pool, J. A., Bernskoetter, W. H., and Chirik, P. J. *J. Am. Chem. Soc.* 2004, *126*, 14326. (f) Bernskoetter, W. H., Lobkovsky, E., and Chirik, P. J. *Angew. Chem. Int. Ed.* 2007, *46*, 2858.

5. (a) Bobadova-Parvanova, P., Wang, Q., Quinonero-Santiago, D., Morokuma, K., and Musaev, D. G. *J. Am. Chem. Soc.* 2006, *128*, 11391. (b) Bobadova-Parvanova, P., Wang, Q., Morokuma, K., and Musaev, D. G. *Angew. Chem. Int. Ed.* 2005, *44*, 7101. (c) Musaev, D. G., Bobadova-Parvanova, P., and Morokuma, K. *Inorg. Chem.* 2007, *46*, 2709.

6. (a) Sato, F., Urabe, H., and Okamoto, S. *Chem. Rev.* 2000, *100*, 2835. (b) Rosenthal, U., Pellny, P.-M., Kirchbauer, F. G., and Burlakov, V. V. *Acc. Chem. Res.* 2000, *33*, 119. (c) Ohff, A., Pulst, S., Peulecke, N., Arndt, P., Burlakov, V. V., and Rosenthal, U. *Synlett* 1996, 111. (d) Negishi, E. and Takahashi, T. *Acc. Chem. Res.* 1994, *27*, 124. (e) Buchwald, S. L. and Nielsen, R. B. *Chem. Rev.* 1988, *88*, 1047. (f) Miura, K., Funatsu, M., Saito, H., Ito, H., and Hosomi, A. *Tetrahedron Lett.* 1996, *37*, 9059. (g) Rosenthal, U. and Burlakov, V. V. In *Titanium and Zirconium in Organic Synthesis*, Marek, I. (Ed.), Wiley-VCH, Weinheim, 2002, p. 355. (h) Rosenthal, U., Burlakov, V. V., Arndt, P., Baumann, W., and Spannenberg, A. *Organometallics* 2003, *22*, 884. (i) Rosenthal, U. *Angew. Chem., Int. Ed.* 2004, *43*, 3882.

7. (a) Rosenthal, U. and Görls, H. *J. Organomet. Chem.* 1992, *439*, C36. (b) Rosenthal, U. Ohff, A., Tillack, A., Baumann, W., and Görls, H. *J. Organomet. Chem.* 1994, *468*, C4. (c) Erker, G., Frömberg, W., Mynott, R., Gabor, B., and Krüger, C. *Angew. Chem., Int. Ed. Engl.* 1986, *25*, 463. (d) Erker, G. *Angew. Chem., Int. Ed. Engl.* 1989, *28*, 397. (e) Erker, G., Frömberg, W., Benn, R., Mynott, R., Angermund, D., and Kruger, C. *Organometallics* 1989, *8*, 911. (f) Teuben, J. H., de Liefde Meijer, H. J. *J. Organomet. Chem.* 1969, *17*, 87. (g) Sekutowski, D. G. and Stucky, G. D. *J. Am. Chem. Soc.* 1976, *98*, 1376. (h) Wood, G. L., Knobler, C. B., and Hawthorne, M. F. *Inorg. Chem.* 1989, *28*, 382. (i) Cuenca, T., Gómez, R., Gómez-Sal, P., Rodriguez, G. M., and Royo, P. *Organometallics* 1992, *11*, 1229. (j) Varga, V., Mach, K., Hiller, J., Thewalt, U., Sedmera, P., and Polasek, M. *Organometallics* 1995, *14*, 1410. (k) Metzler, N. and Nöth, H. *J. Organomet. Chem.* 1993, *454*, C5. (l) Cano, A., Cuenca, T., Galakhov, M., Rodríguez, G. M., Royo, P., Cardin, C. J., and Convery, M. A. *J. Organomet. Chem.* 1995, *493*, 17. (m) Lang, H., Blau, S., Nuber, B., and Zsolnai, L. *Organometallics* 1995, *14*, 3216. (n) Heshmatpour, F., Wocadlo, S., Massa, W., and Dehnicke, K. *Acta Crystallogr.* 1995, *C51*, 2225. (o) Hsu, D. P., Davis, W. M., and Buchwald, S. L. *J. Am. Chem. Soc.* 1993, *115*, 10394.

8. (a) Schubert, B. and Weiss, E. *Chem. Ber.* 1983, *116*, 3212. (b) Bell, N.A., Nowell, I. W., Coates, G. E., and Shearer, H. M. M. *J. Organomet. Chem.* 1984, *273*, 179. (c) Morosin, B., and Howatson, J. *J. Organomet. Chem.* 1971, *29*, 7. (d) Stucky, G. D., McPherson, A. M., Rhine, W. E., Eisch, J. J., and Consi-dine, J. L. *J. Am. Chem. Soc.* 1974, *96*, 1941. (e) Almenningen, A., Fernholt, L., and Haaland, A. *J. Organomet. Chem.* 1978, *155*,

245. (f) Tecle, B., Ilsley, W. H., and Oliver, J. P. *Inorg. Chem.* 1981, *20*, 2335. (g) Fries, W., Schwarz, W., Hansen, H. D., and Weidlein, J. *J. Organomet. Chem.* 1978, *159*, 373. (h) Atwood, J. L., Hunter, W. E., Wyada, A. L., and Evans, W. J. *Inorg. Chem.* 1981, *20*, 4115. (i) Evans, W. J., Bloom, I., Hunter, W. E., and Atwood, J. L. *Orgunometallics* 1983, *2*, 709. (j) Corfield, P. W. R. and Shearer, H. M. M. *Acta Crystallogr.* 1966, *21*, 957. (k) Corfield, P. W. R. and Shearer, H. M. M. *Acta Crystallogr.* 1966, *20*, 502.

9. (a) Pulst, S., Arndt, P., Heller, B., Baumann, W., Kempe, R., and Rosenthal, U. *Angew. Chem., Int. Ed. Engl.* 1996, *35*, 1112. (b) Pellny, P.-M., Peulecke, N., Burlakov, V. V., Tillack, A., Baumann, W., Spannenberg, A., Kempe, R., and Rosenthal, U. *Angew. Chem., Int. Ed. Engl.* 1997, *36*, 2615.

10. (a) Rosenthal, U., Ohff, A., Baumann, W., Kempe, R., Tillack, A., and Burlakov, V. V. *Organometallics* 1994, *13*, 2903. (b) Rosenthal, U., Pulst, S., Ohff, A., Tillack, A., Baumann, W., Kempe, R., and Burlakov, V. V. *Organometallics* 1995, *14*, 2961. (c) Pulst, S., Kirchbauer, F. G., Heller, B., Baumann, W., and Rosenthal, U. *Angew. Chem. Int. Ed. Engl.* 1998, *37*, 1925.

11. (a) Pavan Kumar, P. N. V. and Jemmis, E. D. *J. Am. Chem. Soc.* 1988, *110*, 125. (b) Jemmis, E. D. and Giju, K. T. *Angew. Chem., Int. Ed.* 1997, *36*, 606. (c) Jemmis, E. D. and Giju, K. T. *J. Am. Chem. Soc.* 1998, *120*, 6952. (d) Jemmis, E. D., Phukan, A. K., and Giju, K. T. *Organometallics* 2002, *21*, 2254. (e) Jemmis, E. D., Phukan, A. K., Jiao, H., and Rosenthal, U. *Organometallics* 2003, *22*, 4958. (f) Burlakov, V. V., Arndt, P., Baumann, W., Spannenberg, A., Rosenthal, U., Parameswaran, P., and Jemmis, E. D. *Chem. Commun.* 2004, 2074. (g) Jemmis, E. D., Parameswaran, P., and Phukan, A. K. *Mol. Phys.* 2005, *103*, 897. (h) Bach, M. A., Parameswaran, P., Jemmis, E. D., and Rosenthal, U. *Organometallics* 2007, *26*, 2149.

12. Hehre, W., Radom, L., Schleyer, P. v. R., and Pople, J. A. *Ab Initio Molecular Orbital Theory*, Wiley, New York, 1986.

13. (a) Becke, A. D. *J. Chem. Phys.* 1993, *98*, 5648. (b) Becke, A. D. *Phys. Rev. A* 1988, *38*, 3098. (c) Lee, C., Yang, W., and Parr, R. G. *Phys. Rev. B* 1988, *37*, 785.

14. Frisch, M. J., Pople, J. A. et al. *Gaussian 94*, Revision D.2, Gaussian, Inc. Pittsburgh, PA, 1995. (b) Frisch, M. J., Pople, J. A. et al. *Gaussian 98*, Revision A.7, Gaussian, Inc. Pittsburgh, PA, 1998. (c) Frisch, M. J., Pople, J. A. et al. *Gaussian 03*, Revision C.02, Gaussian, Inc. Wallingford, CT, 2004.

15. (a) Hay, P. J. and Wadt, W. R. *J. Chem. Phys.* 1985, *82*, 270. (b) Wadt, W. R. and Hay, P. J. *J. Chem. Phys.* 1985, *82*, 284. (c) Hay, P. J. and Wadt, W. R. *J. Chem. Phys.* 1985, *82*, 299.

16. (a) Hoffmann, R. *J. Chem. Phys.* 1963, *39*, 1397. (b) Hoffmann, R. and Lipscomb, W. N. *J. Chem. Phys.* 1962, *36*, 2179. (c) Fujimoto, H. and Hoffmann, R. *J. Phys. Chem.* 1974, *78*, 1167.

17. (a) Siegbahn, P. E. M. *J. Biol. Inorg. Chem.* 2006, *11*, 695. (b) Grimme, S. *J. Comput. Chem.* 2006, *27*, 1787. (c) Mattsson, A. E. *Science* 2002, *298*, 759. (d) Hirva, P., Haukka, M., Jakonen, M., and Moreno, M. A. *J. Mol. Model.* 2008, *14*, 171. (e) Grimme, S. *Angew. Chem. Int. Ed. Engl.* 2006, *45*, 4460. (f) Carbonniere, P. and Barone, V. *Chem. Phys. Lett.* 2004, *399*, 226. (g) Check, C. E. and Gilbert, T. M. *J. Org. Chem.* 2005, *70*, 9828. (h) Kurth, S., Perdew, J. P., and Blaha, P. *Int. J. Quant. Chem.* 1999, *75*, 889. (i) Wodrich, M. D., Corminboeuf, C., and Schleyer, P.v.R. *Org. Lett.* 2006, *8*, 3631. (j) Schreiner, P. R., Fokin, A. A., Pascal, Jr., R. A., and Meijere, A. *Org. Lett.* 2006, *8*, 3635. (k) Wodrich, M. D., Corminboeuf, C., Schreiner, P. R., Fokin, A. A., and Schleyer, P.v.R. *Org. Lett.* 2007, *9*, 1851. (l) Zhao, Y., and Truhlar, D. G. *J. Chem. Theory Comput.* 2006, *2*, 1009.

18. (a) Grimme, S. *J. Chem. Phys.* 2006, *124*, 034108. (b) Schwabe, T. and Grimme, S. *Phys. Chem. Chem. Phys.*, 2007, *9*, 3397. (c) Becke, A. D. *J. Chem. Phys.* 1993, *98*, 5648. (d) Adamo, C. and Barone, V. *J. Chem. Phys.* 1998, *108*, 664. (e) Perdew, J. P., Burke, K.,

and Ernzerhof, M. *Phys. Rev. Lett.* 1996, *77*, 3865. (f) Zhao, Y., Donald G., and Truhlar, D. G. *Org. Lett.* 2006, *8*, 5753. (g) Zhao, Y., Schultz, N. E., and Truhlar, D. G. *J. Chem. Theory Comput.* 2006, *2*, 364.

19. Lauher, J. W. and Hoffmann, R. *J. Am. Chem. Soc.* 1976, *98*, 1729.

20. (a) Negishi, E. and Takahashi, T. *Aldrichim. Acta* 1985, *18*, 31. (b) Negishi, E. *Acc. Chem. Res.* 1987, *20*, 65. (c) Negishi, E. and Takahashi, T. *Synthesis* 1988, 1.

21. Rosenthal, U., Ohff, A., Baumann, W., Kempe, R., Tillack, A., and Burlakov, V. V. *Angew. Chem. Int. Ed. Engl.* 1994, *33*, 1605.

22. Rosenthal, U., Burlakov, V. V., Arndt, P., Baumann, W., Spannenberg, A., Shur, V. B. *Eur. J. Inorg. Chem.* 2004, 4739.

23. Erker, G., Venne-Duncker, S., Kehr, G., Kleigrewe, N., Fröhlich, R., Mück-Lichtenfeld, C., and Grimme, S. *Organometallics* 2004, *23*, 4391.

24. Pellny, P.-M., Kirchbauer, F. G., Burlakov, V. V., Baumann, W., Spannenberg, A., and Rosenthal, U. *J. Am. Chem. Soc.* 1999, *121*, 8313.

25. Rosenthal, U., Burlakov, V. V., Arndt, P., Baumann, W., and Spannenberg, A. *Organometallics* 2005, *24*, 456.

26. (a) Krebs, A. and Wilke, J. *Top. Curr. Chem.* 1983, *109*, 189. (b) Gleiter, R. and Merger, R. In *Modern Acetylene Chemistry*, Stang, P. J., Diederich, F. (Eds.), Wiley-VCH, Weinheim, 1995, p. 284. (c) Jones, W. M. and Klosin, J. *Adv. Organomet. Chem.* 1998, *42*, 147.

27. (a) Dewar, M. J. S., *Bull. Soc. Chim. Fr.* 1951, C71-C79. (b) Chatt, J., and Duncanson, L. A. *J. Chem. Soc.* 1953, 2939.

28. (a) *ADF*2007.01, SCM, Theoretical Chemistry, Vrije Universiteit, Amsterdam, The Netherlands, http://www.scm.com. (b) Velde, G., Bickelhaupt, F. M., Gisbergen, S. J. A. v., Fonseca-Guerra, C., Baerends, E. J., Snijders, J. G., and Ziegler, T. *J. Comput. Chem.* 2001, *22*, 931. (c) Fonseca-Guerra, C., Snijders, J. G., Velde, G., and Baerends, E. J. *Theor. Chem. Acc.* 1998, *99*, 391.

29. (a) Schleyer, P.v.R., Maerker, C., Dransfeld, A., Jiao, H., and Hommes, N. J. R.v. E. *J. Am. Chem. Soc.* 1996, *118*, 6317. (b) Schleyer, P.v.R., Jiao, H., Hommes, N. J. R. v. E., Malkin, G. V., and Malkina, O. L. *J. Am. Chem. Soc.* 1997, *119*, 12669. (c) Schleyer, P. v. R., Manoharan, M., Wang, Z.-X., Kiran, B., Jiao, H., Puchta, R., and Hommes, N. J. R. v. E. *Org. Lett.* 2001, *3*, 2465.

30. (a) Seidel, W., Kreisel, G. and Mennenga, H. Z. *Chem.* 1976, *16*, 492. (b) Cotton, F. A. and Millar, F. *J. Am. Chem. Soc.* 1977, *99*, 7886. (c) Erker, G. *Comments Inorg. Chem.* 1992, *13*, 111. (d) Albrecht, M., Erker, G., and Kruger, C. *Synlett.* 1993, 441. (e) Jones, W. M. and Klosin, J. *Adv. Organomet. Chem.* 1998, *42*, 147. (f) Rottger, D. and Erker, G. *Angew. Chem. Int. Ed. Engl.* 1997, *36*, 812. (g) Erker, G. and Rottger, D. *Angew. Chem. Int. Ed. Engl.* 1993, *32*, 1623. (h) Rottger, D., Erker, G., Frohlich, R., Grehl, M., Silverio, S. J., Hyla-Kryspin, I., and Gleiter, R. *J. Am. Chem. Soc.* 1995, *117*, 10503. (i) Rottger, D., Erker, G., Frohlich, R., Kotila, S. *Chem. Ber.* 1996, *129*, 1.

31. (a) Niihara, K. *J. Ceram. Soc. Jpn.* 1991, *99*, 974. (b) Riedel, R., Kleebe, H.-J., Schönfelder, H., and Aldinger, F. *Nature* 1995, *374*, 526.

15 Atoms in Molecules and Population Analysis

Patrick Bultinck and Paul Popelier

CONTENTS

15.1 INTRODUCTION

"The underlying physical laws necessary for the mathematical theory of a large part of physics and the whole of chemistry are thus completely known...". This assertion stated in 1929 by P. A. M. Dirac [1] sets a clear goal for theoretical chemists. Put simply, every chemist, and the theoretical chemists probably first, should strive to frame chemistry within the context of quantum mechanics. Unfortunately, after nearly 80 years, we are far from having achieved this aspiration. Chemistry is still described by a wealth of concepts that are neither rooted in quantum mechanics nor directly derived from it [2]. Two reasons immediately come to mind. Firstly, chemistry has a long history whereas quantum mechanics is a relatively young discipline. Recasting an older science with its own traditional views within the context of a newer one is hard and often meets opposition. Secondly, in the distant past, chemists had already developed their own parlance, which continues to suit their needs very well till now. A good example is the chemical bond. The chemical bond was already mentioned by Frankland in 1866 [3], long before the Heitler–London wave function for H_2 [4] was derived, and thus much before one could put it in a quantum mechanical framework.

It would have been easier to put vague and intuitive concepts on a firmer quantum mechanical footing if they had not been so successful. Indeed, one can only admire the amount of interpretation and prediction accomplished by chemists using their toolbox of concepts. Some of these "toolbox users" may be distrustful of theoretical chemists trying to fix something that, according to these users, is not broken [5]. On the other hand, chemistry does need the input of quantum mechanics for a different reason. Chemists want to fundamentally understand why a chemical bond between the two same atoms is so similar in different molecules and why molecules in a homologous series behave in such a similar way, for example.

Current computational resources enable highly accurate predictions of molecular structures and properties. As a result, the theoretical chemist is appreciated within a broader chemical community as being able to contribute to the ever growing field of chemistry. This mutual understanding is partially due to all involved speaking the same (often conceptual) chemical language. Suddenly abandoning this language may quickly turn the theoretical chemist into the outsider he once was in the chemical community. But every theoretician should remember Dirac's remark mentioned above and speak the language of quantum mechanics.

The debate on several chemical concepts between those following a more "intuitive" path and those following a "physically rigorous" path remains lively. The present chapter deals exactly with such a concept: the atom in the molecule (AIM). Some consider it a product of the mind, a noumenon [6], others accept only a strict quantum mechanical definition. The dust does not seem to have settled yet as far as this argument is concerned. In order to give the reader an idea of the discussions arising from confronting different AIM methods, emphasis is put on describing different AIM techniques. Some thoughts on the deeper roots of the AIM will be shared at the end. This chapter will also introduce some key ideas on population analysis. Nevertheless, we limit the coverage of population analyses because the concept of an AIM is wider than a mere atomic charge.

15.2 ATOMS IN MOLECULES VERSUS POPULATION ANALYSIS

Although the title of this chapter reads Atoms in Molecules and Population Analysis, it should be clear from the beginning that the two topics in the title need to be differentiated. A population analysis is a computational technique to obtain atomic charges. An intuitively plausible population analysis should quantitatively recover the qualitative consequences of electronegativity, where more electronegative atoms "draw" more electrons to themselves than less electronegative ones. Any sound approach that yields a definition of an AIM should allow obtaining an AIM population and thus an AIM charge. The reverse is not always possible. To show this we discuss the ubiquitous electrostatic potential (ESP)-derived charges [7–10].

The electron density $\rho(\mathbf{r})$ and the molecular M-nuclear framework $\{Z_A, \mathbf{R}_A\}$ together generate an EPS $V(\mathbf{r}_i)$ in all points of space $\mathbf{r}_i$:

$$V(\mathbf{r}_i) = \sum_A^M \frac{Z_A}{|\mathbf{r}_i - \mathbf{R}_A|} - \int \frac{\rho(\mathbf{r})}{|\mathbf{r}_i - \mathbf{r}|} d\mathbf{r} \qquad (15.1)$$

There is much interest in the ESP because it plays a major role in molecular recognition. Repeated evaluation of the ESP is very costly, especially for large molecules during simulations. Therefore finding an alternative way to compute ab initio quality ESPs at much lower cost is important. Methods that simplify

Equation 15.1 have been derived. The key idea is to replace the continuous density function and the nuclear framework with the following much simpler expression (for a review, see Ref. [9]):

$$V'(\mathbf{r}_i) = \sum_{A}^{M} \frac{Q_A}{|\mathbf{r}_i - \mathbf{R}_A|} \tag{15.2}$$

The introduction of point charges in Equation 15.2, usually placed at nuclear positions, is a popular simplification of Equation 15.1. Naturally these charges need to be fitted in order to attain the best possible agreement between $V(\mathbf{r}_i)$ and $V'(\mathbf{r}_i)$. The atomic charges Q_A are optimized under the constraint of adding up to the molecular charge, and often also under the constraint of reproducing some molecular multipole moments. The input for the optimization consists usually of a (very) large set of data points $\{\mathbf{r}_i, V(\mathbf{r}_i)\}$, where the $V(\mathbf{r}_i)$ values were computed from Equation 15.1. The points $\{\mathbf{r}_i\}$ are chosen on some predetermined molecular surface so that the procedure effectively reproduces the ESP in regions around the molecule where interactions are assumed to take place. This simple description of ESP-derived charges shows how a population analysis is performed without any notion of an AIM. In fact, one can easily think of adding more sites to position charges, for instance, in the middle of each bond. Nevertheless, point charges continue to dominate applications both in biochemistry and material science in spite of their well-documented deficiencies. The literature on determination of point charges is relatively large and highlights a surprising lack of mathematical understanding of the fitting procedures used to determine them. A thorough and informed analysis of this problem [11] demonstrated that the fitting procedure is underdetermined. In other words, there are too many point charges (i.e., degrees of freedom) to reproduce the exact ESP V as best as possible. This leads to point charge values being erratically assigned and further reduction of their chemical meaning. Overall, it is more desirable to model intermolecular interactions by means of multipole moments [12–14].

15.3 BASIC REQUIREMENTS OF AN AIM METHOD

If population analysis is not synonymous with the concept of an AIM, it becomes necessary to introduce a proper set of requirements before one can speak of an AIM. An AIM is a quantum object and as such has an electron density of its own. This atomic electron density must obviously be positive definite and the sum of these atomic densities must equal the molecular density. Each atomic density $\rho_A(\mathbf{r})$ can be obtained from the molecular density $\rho(\mathbf{r})$ in the following way:

$$\rho_A(\mathbf{r}) = \hat{w}_A \rho(\mathbf{r}) \tag{15.3}$$

where $\hat{w}_A$ is a positive definite operator. As the atomic densities sum to the molecular density,

$$\sum_{A} \rho_A(\mathbf{r}) = \rho(\mathbf{r}) \tag{15.4}$$

it follows that

$$\sum_A \hat{w}_A = 1 \tag{15.5}$$

All the different AIM methods that will be discussed below basically use this same approach but quite different in the nature of $\hat{w}_A$. Chronologically, we will discuss the Mulliken AIM, the Hirshfeld AIM, and the Bader AIM. This last approach will henceforth be called quantum chemical topology (QCT)*. There are more AIM methods, but most of them can be easily understood by the three selected emblematic approaches.

15.4 MULLIKEN APPROACH

Before introducing the form of the Mulliken operator, $\hat{w}_A^{\text{Mull}}$, it is appropriate to return to the concepts of early days of quantum mechanics. Heitler and London wrote down the singlet wave function for H_2 in terms of the hydrogen 1s atomic orbitals on both hydrogen atoms A and B [4]:

$$\Psi = N(1s_A(1)1s_B(2) + 1s_B(1)1s_A(2))(\alpha(1)\beta(2) - \alpha(2)\beta(1)) \tag{15.6}$$

From this wave function, one sees how even in the early beginning of molecular quantum mechanics, atomic orbitals were used to construct molecular wave functions. This explains why one of the first AIM definitions relied on atomic orbitals. Nowadays, molecular ab initio calculations are usually carried out using basis sets consisting of basis functions that mimic atomic orbitals. Expanding the electron density in the set of natural orbitals and introducing the basis function expansion leads to [15]

$$\rho(\mathbf{r}) = \sum_{\nu\mu} D_{\nu\mu}|\nu\rangle\langle\mu| \tag{15.7}$$

---

* In the literature of the 1980s and 1990s, the acronym AIM was uniquely used to refer to the "quantum theory of atoms in molecules" pioneered by the Bader group. To differentiate this approach from others the acronym, QTAIM was introduced later. Extensive work by the Bader group provided QTAIM with a rigorous quantum mechanical foundation by proving that the topological condition of zero-flux serves as the boundary condition for the application of Schwinger's principle of stationary action in the definition of an open system. One of us has proposed the name quantum chemical topology (QCT) to better capture the essential and unique features of QTAIM. Secondly, QCT facilitates future developments, generalizations, and applications. The acronym AIM (or QTAIM) is actually too narrow because, strictly speaking, it only makes sense as a term if one analyses the electron density topologically. Only then does one recover an "atom in a molecule." A topological analysis of the Laplacian of the electron density (which is part of QTAIM) or the topology of the electron localization function (ELF), for example, does not yield "atoms in molecules." However, they can both be put under the umbrella of QCT since they share the central topological idea of partitioning space by means of a gradient vector field. Also, returning to the electron density, one could use the topological analysis to recover molecules in van der Waals complexes or condensed matter. Again, as a name, AIM does not describe this result.

where we use Dirac's bra-ket notation for the basis functions $|\nu\rangle$ and $\langle\mu|$. The symbol $D_{\nu\mu}$ denotes an element of the charge and bond order matrix [15]. Greek letters refer to the basis functions used in the calculation. In order to establish the Mulliken AIM, the required operator can be written as [16]

$$\hat{w}_A^{\text{Mull}} = \sum_{\sigma\in A}\sum_\lambda S_{\sigma\lambda}^{(-1)}|\sigma\rangle\langle\lambda| \tag{15.8}$$

where $S_{\sigma\lambda}^{(-1)}$ is the element $\sigma\lambda$ of the inverse of the overlap matrix. Letting this operator act on Equation 15.7 one finds that

$$\begin{aligned}
\rho_A^{\text{Mull}}(\mathbf{r}) &= \hat{w}_A^{\text{Mull}}\rho(\mathbf{r}) \\
&= \sum_{\nu\mu}\sum_{\sigma\in A}\sum_\lambda D_{\nu\mu}S_{\sigma\lambda}^{(-1)}|\sigma\rangle\langle\lambda|\nu\rangle\langle\mu| \\
&= \sum_{\nu\mu}\sum_{\sigma\in A}\sum_\lambda D_{\nu\mu}S_{\sigma\lambda}^{(-1)}S_{\lambda\nu}|\sigma\rangle\langle\mu| \\
&= \sum_{\nu\mu}\sum_{\sigma\in A} D_{\nu\mu}\delta_{\sigma\nu}|\sigma\rangle\langle\mu| \\
&= \sum_{\sigma\in A}\sum_\mu D_{\sigma\mu}|\sigma\rangle\langle\mu|
\end{aligned} \tag{15.9}$$

The well-known expression embodying the Mulliken population analysis [17–20] then follows after integration over all space,

$$\int \rho_A^{\text{Mull}}(\mathbf{r})d\mathbf{r} = \sum_{\sigma\in A}(\mathbf{DS})_{\sigma\sigma} \tag{15.10}$$

The operator $\hat{w}_A^{\text{Mull}}$ has an interesting characteristic, namely that

$$\begin{aligned}
\hat{w}_A^{\text{Mull}}\hat{w}_B^{\text{Mull}} &= \sum_{\sigma\in A}\sum_\lambda\sum_{\nu\in B}\sum_\mu S_{\sigma\lambda}^{(-1)}S_{\lambda\nu}S_{\nu\mu}^{(-1)}|\sigma\rangle\langle\mu| \\
&= \sum_{\sigma\in A}\sum_{\nu\in B}\sum_\mu \delta_{\sigma\nu}S_{\nu\mu}^{(-1)}|\sigma\rangle\langle\mu| \\
&= 0 \qquad \text{if } A\neq B \\
&= \hat{w}_A^{\text{Mull}} \quad \text{if } A = B
\end{aligned} \tag{15.11}$$

This means that the operators are mutually exclusive and that the operator is idempotent. Nevertheless, in three-dimensional Cartesian space the atoms do overlap, often even to a large extent. So they have no boundaries.

It is clear that the Mulliken operator works in the Hilbert space of the basis functions, which has repercussions on the way the electron density is assigned to the nuclei. The basis functions are allocated to the atomic nuclei they are centered on; the decision as to which portion of the electron density belongs to which nucleus rests on the centering itself. The decision is encapsulated in expressions such as "$\sigma \in A$" appearing in Equations 15.8 through 15.11. This allocation is a direct consequence of an approach that invokes atom-centered basis functions. Such an allocation scheme will fail when basis functions centered on other locations were included in the basis set, or if the basis functions have no center, as in plane waves. The Mulliken approach is thus intertwined with linear combination of atomic orbitals (LCAO) theory. Secondly, if diffuse basis functions are included in the basis set, the Mulliken population analysis may become "unstable." Large fluctuations in the atomic charges may appear upon changing the number of basis functions in the basis set. Diffuse functions decline slowly while moving away from the nucleus. Hence they contribute to a relatively large part of the electron density that is remote from this nucleus. However, the contribution of the diffuse functions to the electron density is still allocated to the nucleus. This leads to spurious results [21].

15.5   HIRSHFELD APPROACH

This approach, also often called the stockholder scheme, was introduced in 1977 by Hirshfeld [22]. The central idea of the Hirshfeld method originates in x-ray crystallography. It proposes to divide the electron density among the atoms in a molecule, guided by a promolecular density. More precisely, once a molecular geometry is known, a promolecular density $\rho^0(\mathbf{r})$ is composed by simply summing the density of each atom A (denoted $\rho_A^0(\mathbf{r})$) in an isolated state:

$$\rho^0(\mathbf{r}) = \sum_A \rho_A^0(\mathbf{r}) \tag{15.12}$$

where the sum runs over all constituent atoms. The idea of using a promolecule to distinguish AIM was not new in 1977 since it had been proposed earlier by Daudel and coworkers [23–25].

At each point in space, the share of the atom is calculated as

$$w_A^H(\mathbf{r}) = \frac{\rho_A^0(\mathbf{r})}{\sum_A \rho_A^0(\mathbf{r})} \tag{15.13}$$

This share is used as the Hirshfeld operator. The assumption behind the Hirshfeld AIM is that this same weight operator can be used to divide the electron density of the molecule via

$$\rho_A(\mathbf{r}) = w_A^H(\mathbf{r})\rho(\mathbf{r}) = \frac{\rho_A^0(\mathbf{r})}{\sum_A \rho_A^0(\mathbf{r})}\rho(\mathbf{r}) \tag{15.14}$$

This so-called Hirshfeld scheme is particularly popular within the so-called conceptual density functional theory (DFT) [26,27]. The weighting function, which identifies the AIM as one that is most similar to the isolated atom [28], has been shown to be directly derivable from information entropy [6,29–33]. Here again, the atoms do not have sharp boundaries and extend to infinity although a relatively sharp drop in the weight of an atom may be expected in the neighborhood of other atoms. As such, it is also reminiscent of other techniques such as the fuzzy atoms of Mayer and Salvador [34] or Becke's scheme [35].

Hirshfeld atomic populations were found to be mildly dependent on the basis set [36,37], thus remedying one of the important problems of the Mulliken technique. However, new problems are introduced. First, a promolecular density is a nonphysical concept since it does not comply with the Pauli principle. Several other questionable aspects of the Hirshfeld technique have been revealed as well. Firstly, the promolecule is not defined uniquely. Davidson and Chakravorty [38], Bader and Matta [39], Matta and Bader [40], and Bultinck et al. [41] have questioned the apparently conventional choice of neutral ground state isolated atom densities as building blocks for the promolecular density. A different promolecular density results in a different set of atomic charges. Davidson and Chakravorty [38] showed this for the case of N_2, where the atomic charges differed significantly whether one took it as a promolecule N^+ and N^- or as two neutral nitrogen atoms. This is clearly undesirable and Bader and Matta [39] and Matta and Bader [40] raised the question as to what would be the proper Li density to use in the promolecule for LiF: Li^+ or neutral Li? Bultinck and coworkers [37,41,42] investigated this matter and found that using $Li^0 + F^0$ as promolecule resulted in charges of $+0.57$ and -0.57, respectively. Using the combination, $Li^+ + F^-$ as promolecule yielded a charge of ± 0.98, and a charge of ± 0.30 was found when using $Li^- + F^+$ as promolecule. The only constant is that Li is assigned the positive charge. The reported charge fluctuations are dramatic, which corroborates the criticism by Bader and Matta [39] and Matta and Bader [40]. A second concern is that, for covalently bonded systems the Hirshfeld atomic charges are virtually zero. For instance, in the case of H_2O the atomic charges are 0.16 (H) and -0.32 (O). Compared with any other type of population analysis, these values are rather small. Even for cases where one expects significant contributions from charge transfer between atoms, the charges remain very small. Relying on the proof by Ayers that the Hirshfeld, AIM is the one that keeps the AIM as similar as possible to the reference isolated atom [28], this may indicate that indeed the AIM also wants to keep its charge as close as possible to that of the reference atom, i.e., zero. This problem becomes even larger for charged systems.

A third concern follows from the information theory background provided by Parr, Nalewajski, and coworkers [6,29–32]. In order to use the Kullback–Liebler formula for missing information for the AIM, an essential criterion is that the AIM density and the density of the isolated atom used in the promolecule must normalize to the same number [6]. This is almost never the case when using neutral atoms as a reference. All these concerns led Bultinck and coworkers [37,41,42] to develop the iterative Hirshfeld method, denoted Hirshfeld-I. This method, which actually coincides with the suggestion of Davidson and Chakravorty [38] to find a self-consistent Hirshfeld method, proceeds in the following way. First, a regular Hirshfeld population analysis is carried out, resulting in a first set of atomic populations. Then a new promolecule is constructed using atomic densities that normalize to the populations produced in the previous step. Using that promolecule, the Hirshfeld analysis is again carried out. This procedure is repeated until eventually the populations that result from the ith iteration are the same as those that were used in the promolecule in the same iteration. Once self-consistency is obtained, one can proceed with the extraction of Hirshfeld-I AIMs and use them for further analysis.

The results [41,42] of implementing this scheme reveal that the atomic charges tend to grow much bigger, roughly by a factor of 3. Even charges below -1 or above $+1$ do appear now. In many studies, Bader's AIM charges (detailed explanation in the following sections) were criticized for being too large. However, the charges of the self-consistent Hirshfeld scheme seem to grow quite large as well. Secondly, the Hirshfeld-I scheme also makes charged molecules tractable without using promolecular densities that normalize differently from the molecular density. Finally, the Kullback–Liebler formula is strictly valid in assessing information loss [43]. Also, the basis set dependence has been shown to be very small [37]. Returning to LiF, the Hirshfeld-I procedure yields AIM charges of ± 0.932, which agree well with those of an ionic bond. LiH is another molecule for which Hirshfeld and similar approaches are problematic according to Bader and coworkers [39,40]. The regular Hirshfeld method yields charges of ± 0.43, which are indeed very low for this kind of species. On the other hand, using the Hirshfeld-I scheme, the charges become ± 0.930. As is the case for the Bader method [39], the Hirshfeld-I charges of LiH and LiF have become very similar.

Introducing self-consistency in the Hirshfeld-I scheme removes most of the arbitrariness in choosing a promolecule. It was also proven that the corresponding AIM populations [37] are independent from the starting point of the iterative process. Still, one arbitrary decision remains. The states of the isolated atoms used in constructing the promolecule are chosen arbitrarily as the ground state. In principle, this problem can also be solved in the spirit of information theory. One could initiate the Hirshfeld-I procedure for every possible combination of states of every atom, carry it out until convergence, and then compute the information loss. Routine application of this procedure is of course impossible. On the other hand, Rousseau et al. [45] showed that, in the regular Hirshfeld scheme, changing the states of the neutral reference atoms has little impact on the final AIM populations.

15.6 BADER APPROACH

The original "atoms in molecules" method [45–47] as developed by the Bader group is based on the topology of the electron density ρ and that of the Laplacian of ρ. Instead of an in-depth discussion of the topological features of the electron density we just introduce a few essential characteristics here. A key concept is the gradient vector field, which is a collection of gradient paths. Here, a gradient path is a trajectory of steepest ascent in the electron density. The topology of the electron density is best revealed through its gradient vector field. At the so-called critical points, the gradient of the electron density vanishes. The critical points are best classified in terms of their rank and signature. At a critical point, the Hessian of ρ is computed and its eigenvalues are obtained. The rank is the number of nonzero eigenvalues and the signature is the sum of the signs of the Hessian eigenvalues. For example, a maximum in ρ has three negative eigenvalues, and hence its signature is $-3 = (-1) + (-1) + (-1)$. The rank is usually 3, which gives rise to four possible signatures, namely -3, -1, $+1$, and $+3$.

The nature of a critical point is denoted as (rank, signature). For example, a minimum in ρ is designated by $(3, +3)$ and a maximum by $(3, -3)$. The two remaining types of critical points, $(3, -1)$ and $(3, +1)$ are saddle points, called bond critical point and ring critical point, respectively.

Gradient paths can be classified by means of the type of critical points (i.e., their signature) that they connect. This was achieved exhaustively for the first time in 2003 [48]. In the gradient vector field of ρ, the vast majority of gradient paths terminate at $(3, -3)$ critical points, which (approximately) coincide with nuclear positions. A collection of gradient paths that terminates at a given nucleus is called an atomic basin. These basins are mutually exclusive and thus partition three-dimensional space into disjoint (i.e., nonoverlapping) domains. The basin together with the nucleus inside it is then defined as the AIM within the context of QCT. Figure 15.1 shows a few examples of QCT atoms, using a new algorithm based on finite elements [49]. It is clear that a water dimer, which is a van der Waals complex, can also be partitioned into QCT atoms, in the same manner as a single molecule. The electron density's topology does not distinguish intramolecular interactions from intermolecular interactions. In principle, a QCT atom can be completely bounded by topological surfaces called interatomic surfaces provided there are enough neighboring atoms. The hydrogen atom in the middle of the water dimer, for example, is bounded by an interatomic surface on the left and one on the right. Only a very small edge of a nontopological surface bounds it at the top and at the bottom. This nontopological surface is an envelope of constant electron density, typically set at $\rho = 0.001$ a.u. For most atoms in the systems shown, such an envelope is vital to bound an atom visually. Secondly, an atom also needs to be bounded in order to have a finite volume for a numerical integration [50–52] to be possible over the atomic basin. The picture showing cyclopropane marks the bond critical points as purple points. These carbon–carbon bond critical points are at the center of the interatomic surface that separates the two carbons.

Till now, the AIM in QCT comes from an entirely topological origin. The single most important step forward in the theory was the realization that a quantum mechanical AIM coincides exactly with this topological atom [53]. The definition of an interatomic surface is given by

$$\nabla\rho(\mathbf{r}) \cdot \mathbf{n}(\mathbf{r}) = 0 \quad \text{for all } \mathbf{r} \in S \tag{15.15}$$

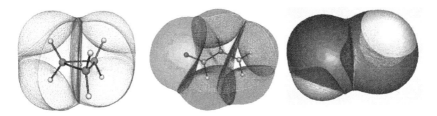

FIGURE 15.1 (See color insert following page 302.) Examples of the QCT partitioning of the electron density. (left) All atoms in cyclopropane (except for the front methylene group); (middle) acrolein; and (right) a water dimer (global minimum).

This equation means that the normal to the surface S, $\mathbf{n}(\mathbf{r})$, is orthogonal to the gradient of the electron density. In other words, the surface is parallel to $\nabla\rho$, or rephrased again, the surface consists of gradient paths. The interatomic surface is a bundle of gradient paths that terminate at the bond critical point at the center of the surface.

It can be shown that Equation 15.15 means no less than the QCT AIM itself is a quantum mechanical object within the global quantum object. A common misunderstanding is that the AIM in this case becomes a closed system. This is incorrect. The QCT AIM should be seen as an open system [54], free to exchange electronic charge, for instance.

In order to frame the QCT method within the general expression (3), we need to give an expression for the weight operator $\hat{w}_A^{QCT}$. From the above discussion, it is clear that this operator will depend on $\mathbf{r}$ and that it is a binary operator, so its value is either 0 or 1. For a given atom A, the operator vanishes at every point in space, except within the basin of A, where it is equal to one. This way all atomic basins are indeed mutually exclusive.

Figure 15.1 also shows molecular graphs, which is a collection of special gradient paths that embody chemical bonds. Two such gradient paths originate from a given bond critical point and when traced in opposite directions, each terminate at a nucleus. This pair of gradient paths is called an atomic interaction line. In an equilibrium geometry, they are known as bond paths. A debate has emerged about whether bond paths can always be identified with a chemical bond. Unfortunately, this debate remained circular until the careful work of the Oviedo group and Gatti in 2007. They recently published an important paper [55], in which the exchange-correlation energy between atoms, $V_{xc}(A,B)$, is invoked to cut the vicious circle that fueled the bond path debate. The authors show that bond paths are "privileged exchange channels" and highlight a remarkable set of observations for classical test systems. It is clear that the competition between various $V_{xc}(A,B)$ terms is expressed by the presence or absence of bond paths. One should note that QCT not only proposes an AIM but also recovers bonding.

15.7 AIM PROPERTIES AND COMPARISON OF AIM METHODS

The properties of a quantum mechanical system such as an AIM are readily calculated from any method as long as they involve an operator acting on the electron density, e.g., for the case of the dipole moment. The problem would seem to become harder for other properties, although the introduction of property densities allows us to generally introduce AIM expectation values [45]. The expectation value of a property A for atom α in the Hirshfeld and QCT methods can be written as

$$A_\alpha = \int w_\alpha(\mathbf{r})d\mathbf{r} \int d\tau' \frac{N}{2}\left[\Psi^*\hat{A}\Psi + \left(\hat{A}\Psi\right)^*\Psi\right] \tag{15.16}$$

where $d\tau'$ denotes integration over all coordinates of all electrons in the system, except one. For QCT, this means that computing the average value of an atomic property warrants integration over the atomic basin only. In the case of both the Hirshfeld schemes (original and I), one integrates over all space but weights the result with the appropriate $w_A^H(\mathbf{r})$. In the case of the Mulliken method, the expressions are analogous, except that manipulations are carried out in the Hilbert space of the basis functions.

Many studies have focused on the comparison of AIM methods. Most concentrate on comparing the results from the AIM populations. Therefore, no such comparison will be repeated here. The issue to be addressed here is of a more fundamental nature. Although all the three methods addressed here have been and continue to be used widely, none of them is without its own problems. These may be computational in nature, such as the basis set dependence of the Mulliken method, the high CPU cost of QCT, or the arbitrariness of the promolecule in the Hirshfeld method. The most fundamental criticism is that in the Mulliken and Hirshfeld methods, the AIM is introduced whereas the AIM can be deduced from the QCT analysis of the electron density. In particular, in the Hirshfeld method, one composes a promolecule based on the atoms that constitute the molecule, thus from the beginning the concept of an atom is introduced. In the Mulliken approach, one needs to rely on the use of atom-centered basis functions that mimic the atomic orbitals, thereby introducing the atoms a priori. In QCT, the atoms can be obtained solely from the electron density, although one must then accept finding atoms without nucleus. These are the so-called nonnuclear attractors or pseudoatoms that appear occasionally. The appearance of such nonnuclear attractors can be an artifact from the use of small basis sets, although some can be genuine in that they survive at high level of theory [56]. These nonnuclear attractors do not fit well within our usual view of the molecular graph or molecular structure, but accepting QCT, one needs to accept the existence of pseudoatoms as well, and thus the concept of an atom without a nucleus because the atom has been defined as the union of the nucleus with its basin.

Finally, it is worth turning back to the question whether atoms in molecules can be defined uniquely. According to Parr et al., the AIM cannot be defined uniquely [6]. According to these authors, an AIM can neither be observed directly from experiment, nor could one gather enough properties of the AIM to define it unambiguously. They cannot conceive any experimental measurement that would show that one definition is uniquely correct, and so they accept the existence of several different approaches, all possibly useful in specific contexts. The AIM is therefore considered a noumenon: an object or pure thought not connected with sense perception. This point of view was scrutinized by Bader and Matta, claiming QCT to be the only correct and coherent way to define the AIM [40]. According to the latter authors, the QCT AIM is also confirmed by experiment.

Since Dalton, chemists have thought of molecules in terms of a collection of atoms. The entire discussion between the two different points of view, namely the noumenon view and Bader's view, is not about the usefulness of the AIM concept. There seems to be agreement on the usefulness of the concept. Rather, it is about pinpointing the AIM through a definition that excludes every other possibility. In any event, given the philosophy of Dirac's statement at the beginning of the chapter, QCT does have the advantage of allowing AIM to be derived from quantum mechanics. An atom was neither directly nor indirectly introduced at any stage. Yet, domains that do conform to the concept of an atom in a molecule resulted.

REFERENCES

1. Dirac, P.A.M. *Proc. Roy. Soc. (London)*, 1929, *123*, 714–733.
2. Tapia, O. *J. Math. Chem.*, 2006, *39*, 637–669.
3. Frankland, E.W. *J. Chem. Soc.*, 1866, *19*, 372–395.
4. Heitler, W. and London, F. *Z. Phys.*, 1927, *44*, 455–472.
5. Popelier, P.L.A. *Faraday Discuss*, 2007, *135*, 3–5.
6. Parr, R.G., Ayers, P.W., and Nalewajski, R.F. *J. Phys. Chem. A*, 2005, *109*, 3957–3959.
7. Francl, M.M., Carey, C., Chilian, L.E., and Gange, D.M. *J. Comput. Chem.*, 1996, *17*, 367–383.
8. Breneman, C.M. and Wiberg, K.B. *J. Comput. Chem.*, 1990, *11*, 361–373.
9. Francl, M.M. and Chirlian, L.E. *Rev. Comput. Chem.*, 2000, *14*, 1–31.
10. Singh, U.C. and Kollman, P.A. *J. Comput. Chem.*, 1984, *5*, 129–145.
11. Francl, M.M., Carey, C., Chirlian, L.E., and Gange, D.M. *J. Comput. Chem.*, 1996, *17*, 367–383.
12. Stone, A.J. and Price, S.L. *J. Phys. Chem.*, 1988, *92*, 3325–3335.
13. Popelier, P.L.A., Rafat, M., Devereux, M., Liem, S.Y., and Leslie, M. *Lecture Series on Computer and Computational Sciences*, 2005, *4*, 1251–1255.
14. Popelier, P.L.A. *Structure and Bonding. Intermolecular Forces and Clusters,* D.J.Wales (Ed.), vol. 115, pp. 1–56, Springer, Heidelberg, 2005.
15. Szabo, A. and Ostlund, N.S. *Modern Quantum Chemistry—Introduction to Advanced Electronic Structure Theory*, Dover Publications, New York, 1996.
16. Carbó-Dorca, R. and Bultinck, P. *J. Math. Chem.*, 2004, *36*, 201–210.
17. Mulliken, R.S. *J. Chem. Phys.*, 1955, *23*, 1833–1840.
18. Mulliken, R.S. *J. Chem. Phys.*, 1955, *23*, 1841–1846.
19. Mulliken, R.S. *J. Chem. Phys.*, 1955, *23*, 2338–2342.
20. Mulliken, R.S. *J. Chem. Phys.*, 1955, *23*, 2343–2346.
21. Jensen, F. *Introduction to Computational Chemistry*, John Wiley & Sons, Chichester, 1999.
22. Hirshfeld, F.L. *Theor. Chim. Acta*, 1977, *44*, 129–138.
23. Daudel, R. *Compt. Rend. Acad. Sci.*, 1952, *235*, 886–888.
24. Roux, M. and Daudel, R. *Compt. Rend. Acad. Sci.*, 1955, *240*, 90–92.
25. Roux, M., Besnainou, S., and Daudel, R. *J. Chim. Phys.*, 1956, *54*, 218–221.
26. Parr, R.G. and Yang, W. *Density Functional Theory of Atoms and Molecules*, Oxford University Press, New York, 1989.
27. Geerlings, P., De Proft, F., and Langenaeker, W. *Chem. Rev.*, 2003, *103*, 1793–1873.
28. Ayers, P.W. *J. Chem. Phys.*, 2000, *113*, 10886–10898.
29. Nalewajski, R.F., Parr, R.G. *Proc. Natl. Acad. Sci. U.S.A.* 2000, *97*, 8879–8882.
30. Nalewajski, R.F., Switka, E., and Michalak, A. *Int. J. Quantum Chem.* 2002, *87*, 198–213.
31. Nalewajski, R.F. *Chem. Phys. Lett.*, 2003, *372*, 28–34.
32. Nalewajski, R.F. and Broniatowska, E. *Int. J. Quantum Chem.*, 2005, *101*, 349–362.
33. Ayers, P.W. *Theor. Chem. Acc.*, 2006, *115*, 370–378.
34. Mayer, I. and Salvador, P. *Chem. Phys. Lett.*, 2004, *383*, 368–375.
35. Becke, A.D. *J. Chem. Phys.*, 1988, *88*, 2547–2553.
36. Fonseca Guerra, C., Handgraaf, J.-W., Baerends, E.J., and Bickelhaupt, F.M. *J. Comput. Chem.* 2003, *25*, 189–210.
37. Bultinck, P., Ayers, P.W., Fias, S., Tiels, K., and Van Alsenoy, C. *Chem. Phys. Lett.*, 2007, *444*, 205–208.
38. Davidson, E.R. and Chakravorty, S. *Theor. Chim. Acta*, 1992, *83*, 319–330.
39. Bader, R.F.W. and Matta, C.F. *J. Phys. Chem. A*, 2004, *108*, 8385–8394.
40. Matta, C.F. and Bader, R.F.W. *J. Phys. Chem. A*, 2006, *110*, 6365–6371.

41. Bultinck, P., Van Alsenoy, C., Ayers, P.W., and Carbó-Dorca, R. *J. Chem. Phys.*, 2007, *126*, 144111.
42. Bultinck, P. *Faraday Discussions*, 2007, *135*, 244–246.
43. Pierce, J.R. *An Introduction to Information Theory, Symbols, Signals and Noise,* Dover Publications, New York, 1980.
44. Rousseau, B., Peeters, A., and Van Alsenoy, C. *Chem. Phys. Lett.* 2000, *324*, 189–194.
45. Bader, R.F.W. *Atoms in Molecules: A Quantum Theory*, Clarendon, Oxford 1990.
46. Bader, R.F.W. *Chem. Rev.* 1991, *91*, 893–928.
47. Popelier, P. *Atoms in Molecules: An Introduction*, Prentice-Hall, Harlow, 2000.
48. Malcolm, N.O.J. and Popelier, P.L.A. *J. Comput. Chem.*, 2003, *24*, 437–442.
49. Rafat, M. and Popelier, P.L.A. *J. Comput. Chem.*, 2007, *28*, 2602–2617.
50. Popelier, P.L.A. *Comp. Phys. Commun.*, 1998, *108*, 180–190.
51. Popelier, P.L.A. *Theor. Chem. Acc.*, 2001, *105*, 393–399.
52. Popelier, P.L.A. *Molec. Phys.*, 1996, *87*, 1169–1187.
53. Bader, R.F.W. *Phys. Rev. B* 1994, *49*, 13348–13356.
54. Bader, R.F.W. *J. Phys. Chem., A* 2007, *111*, 7966–7972.
55. Pendás, A.M., Francisco, E., Blanco, M.A., and Gatti, C. *Chem. Eur. J.*, 2007, *13*, 9362–9371.
56. Gatti, C., Fantucci, P., and Pacchioni, G. *Theoret. Chim. Acta*, 1987, *72*, 433–458.

16 Molecular Quantum Similarity

*Patrick Bultinck, Sofie Van Damme,
and Ramon Carbó-Dorca*

CONTENTS

16.1 INTRODUCTION

Similarity between quantum systems, such as atoms and molecules, plays a very important role throughout chemistry. Probably the best example is the ubiquitously known periodic system of the elements. In this system, elements are arranged both horizontally and vertically in such a way that in both directions, elements have a high similarity to their neighbors. Another closely related idea is that of transferability. In chemistry, one speaks of transferability of an entity when its properties remain similar between different situations. An example is the transferability of the properties of a functional group between one molecule and another. The main motto of using similarity in chemistry is the assumption that similar molecules have similar properties.

Although chemical reasoning very often implies using the concept of similarity, the very definition of similarity depends often on the people involved. A physical chemist might base the judgment of a degree of similarity between two molecules A and B on a completely different basis from an organic chemist, and this may again be different from what a medicinal chemist might think of the degree of similarity. One could say that: "Similarity, like beauty, lies in the eyes of the beholder" [1]. A clear example is found in the field of human psychology, where it was shown that the degree of similarity perceived by an observer between different facial expressions depends on the observer's mental state [1–3]. This dependence of its perception upon the observer makes it clear that similarity is often considered as a qualitative concept.

Such dependence is naturally not acceptable if one wants to put similarity between quantum systems in a theoretical framework. As will be shown below, the so-called theory of molecular quantum similarity (MQS) does offer a solid basis. The aim of the present chapter is to introduce the basic aspects of the theory and to allow the reader to follow the literature. For applications and a more in-depth presentation of the mathematical aspects, the reader is referred to the review by Bultinck et al. [4].

16.2 MOLECULAR DESCRIPTORS

Prior to developing the MQS theory, it is worth introducing the concept of a molecular descriptor. A molecular descriptor effectively describes a certain feature of a molecule. Examples are the molecular weight, the number of atoms of a certain element, and its dipole moment. Totally, there are at least several hundreds of such descriptors in use, some derived purely from computation and some from experiment. These features are used to express the degree of similarity between molecules. The dependence of the degree of similarity between two molecules on the perception of the observer can now easily be translated in to the choice of molecular descriptor that chemists of different fields may apply to assess the similarity. A medicinal chemist may use a completely different set of descriptors, namely those of most importance to his field, whereas an inorganic chemist may use a different set of descriptors. Moreover, the same chemist may turn to different descriptors depending on the project he is working on. Molecular descriptors also play a crucial role in quantitative structure property or activity relationships (QSP(A)R). In such relationships one links an observable property or activity of a molecule to the set of descriptors in a quantitative way to be able to find more information on some mechanism underlying the property or activity and to be able to predict this property or activity for some molecule whose property or activity is not yet known. Reviews of the many types of molecular descriptors may be found in the works by Downs [5], Brown [6], Mason and Pickett [7], and Bajorath [8].

A specific class of molecular descriptors is the one based on quantum chemical calculations. These descriptors may or may not be observables themselves. They may correspond to a computed value for some experimentally verifiable quantity, or they may be purely conceptual descriptors. A review of quantum chemical molecular descriptors has been given by Karelson et al. [9,10].

Although one could consider the electron density as just one of the many quantum chemical descriptors available, it deserves special attention. In QSM, it is the only descriptor used for a number of reasons. The idea of using the electron density as the ultimate molecular descriptor is founded on the basic elements of quantum mechanics. First of all, it is the all-determining quantity in density functional theory (DFT) [11] and also holds a very close relation to the wave function. Convincing arguments were given by Handy and are attributed to Wilson [12], although initial ideas can also be traced back to Born [13] and von

Neumann [14]. The electron density $\rho(\mathbf{r})$ has several important features. First, it gives the number of electrons N,

$$\int \rho(\mathbf{r})d\mathbf{r} = N \qquad (16.1)$$

Secondly, information is obtained on the nature of the nuclei in the molecule from the cusp condition [11]. Thirdly, the Hohenberg–Kohn theorem points out that, besides determining the number of electrons, the density also determines the external potential that is present in the molecular Hamiltonian [15]. Once the number of electrons is known from Equation 16.1 and the external potential is determined by the electron density, the Hamiltonian is completely determined. Once the electronic Hamiltonian is determined, one can solve Schrödinger's equation for the wave function, subsequently determining all observable properties of the system. In fact, one can replace the whole set of molecular descriptors by the electron density, because, according to quantum mechanics, all information offered by these descriptors is also available from the electron density.

All this form a quite convincing argument to use the electron density as the most fundamental and in fact the only descriptor. According to Dean, similarity in the activity or properties of a molecule will occur whenever molecules have similar electron densities [16].

16.3 MOLECULAR QUANTUM SIMILARITY

In 1980, Carbó et al. were the first to express molecular similarity using the electron density [17]. They introduced a distance measure between two molecules A and B in the sense of a Euclidean distance in the following way:

$$d_{AB}^2 = \int d\mathbf{r}[\rho_A(\mathbf{r}) - \rho_B(\mathbf{r})]^2 \qquad (16.2)$$

where
  $\rho_A(\mathbf{r})$ is the electron density of molecule A
  $\rho_B(\mathbf{r})$ is the electron density of molecule B

Working out Equation 16.2 yields

$$d_{AB}^2 = \int d\mathbf{r}[\rho_A(\mathbf{r})]^2 + \int d\mathbf{r}[\rho_B(\mathbf{r})]^2 - 2\int d\mathbf{r}[\rho_A(\mathbf{r})\rho_B(\mathbf{r})] \qquad (16.3)$$

Introducing the notion of a molecular quantum similarity measure (MQSM) Z_{AB} as

$$Z_{AB} = \int d\mathbf{r}[\rho_A(\mathbf{r})\rho_B(\mathbf{r})] \qquad (16.4)$$

Equation 16.3 can be rewritten in terms of the so-called self-similarity and Z_{AB}

$$d_{AB} = Z_{AA} + Z_{BB} - 2Z_{AB} \qquad (16.5)$$

In case of perfect similarity between two molecules A and B, one would find $d_{AB} = 0$, and the more the two density functions differ, the larger will be the value of d_{AB} and the smaller will be Z_{AB}. Besides the distance measure of similarity between two electron densities, another index was developed which is known as the Carbó similarity index [17]. For two molecules A and B, this index is given as

$$C_{AB} = \frac{\int dr[\rho_A(r)\rho_B(r)]}{\sqrt{\int dr[\rho_A(r)]^2 \int dr[\rho_B(r)]^2}} = \frac{Z_{AB}}{\sqrt{Z_{AA}Z_{BB}}} \tag{16.6}$$

According to this index, a perfect similarity is indicated by $C_{AB} = 1$, and lower values indicate decreasing similarity.

16.4 EXTENSION TO OTHER OPERATORS

Till now, all MQSM rely on overlap integrals between electron densities. However, this is not the only possibility. Based on the development of the theory of vector semispaces, Carbó et al. have shown that one can extend the QSM theory to include different kinds of operators [18,19]. In general, a MQSM can be obtained via

$$Z_{AB} = \int dr_1 dr_2[\rho_A(r_1)\Omega(r_1, r_2)\rho_B(r_2)] \tag{16.7}$$

where till now, the operator $\Omega(r_1, r_2)$ corresponded to the Dirac delta function, i.e.,

$$\Omega(r_1, r_2) = \delta(r_1 - r_2) \tag{16.8}$$

This is certainly not the only possibility, and MQSM can be defined using many choices of $\Omega(r_1, r_2)$ provided it remains positive definite. Examples include the Coulomb MQSM and kinetic QSM. In case of the Coulomb operator, one does not perform the point-by-point similarity calculation as in Equation 16.4 but introduces weighting of the surrounding points using as an operator:

$$\Omega(r_1, r_2) = |r_1 - r_2|^{-1} \tag{16.9}$$

The case of a so-called kinetic MQSM corresponds to [20]

$$Z_{AB} = \int dr_1[\nabla\rho_A(r_1)\nabla\rho_B(r_1)] \tag{16.10}$$

In fact, one can even use a positive definite operator, the electron density of yet a third molecule X acting as template, to obtain [21,22]

$$Z_{AB} = \int dr_1[\rho_A(r_1)\rho_X(r_1)\rho_B(r_1)] \tag{16.11}$$

16.5 SIMILARITY MATRICES

Having established different possible operators to compute MQSM, one can pick one of these operators and compute the MQSM between each two molecules in a set. Putting all these MQSM in a $(N \times N)$ matrix, one obtains the so-called MQS matrix, denoted by $\mathbf{Z}$:

$$\mathbf{Z} = \begin{bmatrix} Z_{11} & \cdots & Z_{1N} \\ \vdots & \ddots & \vdots \\ Z_{N1} & \cdots & Z_{NN} \end{bmatrix} \tag{16.12}$$

Each of these columns of this symmetrical matrix may be seen as representing a molecule in the subspace formed by the density functions of the N molecules that constitute the set. Such a vector may also be seen as a molecular descriptor, where the infinite dimensionality of the electron density has been reduced to just N scalars that are real and positive definite. Furthermore, once chosen a certain operator in the MQSM, the descriptor is unbiased. A different way of looking at $\mathbf{Z}$ is to consider it as an N-dimensional representation of the operator within a set of density functions. Every molecule then corresponds to a point in this N-dimensional space. For the collection of all points, one can construct the so-called point clouds, which allow one to graphically represent the similarity between molecules and to investigate possible relations between molecules and their properties [23–28].

Similarity matrices have often been used to cluster molecules and to derive some molecular set taxonomy. Besides the point clouds described above, one can construct dendrograms in different ways. Starting from the MQS matrix $\mathbf{Z}^0$, Bultinck and Carbó-Dorca [29] first obtain a similarity matrix $\mathbf{C}$ holding as elements, the C_{AB} obtained using some similarity index transformation (for example the Carbó index [6]). Having done this, one can construct a sequential agglomerative hierarchical nonoverlapping dendrogram by searching the sequence of the largest values of C_{AB}. In the first step, the largest element of the matrix $\mathbf{C}$ is sought and the two molecules involved are clustered. This cluster of two molecules is represented by a new, virtual quantum object. Its descriptor in the quantum similarity matrix is obtained by averaging the vectors of both molecules in $\mathbf{Z}^0$ (and not $\mathbf{C}$). So the descriptor for the artificial molecule X is obtained as

$$\forall K = [1 \ldots N]: Z_{KX} = \frac{Z_{KA}^0 + Z_{KB}^0}{2} \tag{16.13}$$

where the values of Z_{KX} constitute a descriptor vector in a new MQS matrix $\mathbf{Z}$. From this $\mathbf{Z}$, a new matrix $\mathbf{C}$ is constructed and the entire process is repeated until all molecules are clustered. At every stage when two objects (molecules or clusters) A and B are clustered into an object X, a new column for X is constructed in a new matrix $\mathbf{Z}$ according to

$$\forall K = [1 \ldots N]: Z_{KX} = \frac{1}{N_A + N_B} \sum_{I \in \{A,B\}}^{N_A + N_B} Z_{KI}^0 \tag{16.14}$$

Where the sum runs over all the objects already contained in A and B.

16.6 COMPUTING MOLECULAR QUANTUM SIMILARITY MEASURES

Having established the most important concepts for MQS, the next step is to actually compute the numerical values associated with the quantum similarity measures. Electron densities can naturally be obtained from many quantum chemical methods such as DFT, Hartree–Fock, configuration interaction, and many more, even from experiment.

The electron density can quite generally be written as

$$\rho(\mathbf{r}) = \sum_{\nu} \sum_{\mu} D_{\mu\nu} \phi_{\nu}^{*}(\mathbf{r}) \phi_{\mu}(\mathbf{r}) \tag{16.15}$$

where
$D_{\mu\nu}$ is the charge and bond order matrix
$\phi_{\nu}(\mathbf{r})$ and $\phi_{\mu}(\mathbf{r})$ are the basis functions

It then follows that the MQSM are given as

$$Z_{AB} = \sum_{\nu \in A} \sum_{\mu \in A} \sum_{\sigma \in B} \sum_{\lambda \in B} \int d\mathbf{r}_1 d\mathbf{r}_2 D_{\mu\nu} D_{\lambda\sigma} [\phi_{\nu}^{*}(\mathbf{r}_1) \phi_{\mu}(\mathbf{r}_1) \Omega(\mathbf{r}_1, \mathbf{r}_2) \phi_{\sigma}^{*}(\mathbf{r}_2) \phi_{\lambda}(\mathbf{r}_2)] \tag{16.16}$$

This means that for the evaluation of the MQSM, one needs to solve many integrals involving four basis functions. Although there are many good algorithms for computing these integrals, their number can become limitative for routine application, especially when higher order angular momentum basis functions are involved or when the MQSM have to be evaluated for large numbers of large molecules. Another reason to find an alternative approach to the evaluation of the MQSM is that often the MQSM have to be calculated repeatedly, as in maximization of the MQSM, as it will be described later.

It is well-known that a superposition of isolated atomic densities looks remarkably much like the total electron density. Such a superposition of atomic densities is best known as a promolecular density, like it has been used by Hirshfeld [30] (see also the chapter on atoms in molecules and population analysis). Carbó-Dorca and coworkers derived a special scheme to obtain approximate electron densities via the so-called atomic shell approximation (ASA) [31–35]. Generally, for a molecule A with atoms N, a promolecular density is defined as

$$\rho_{A}^{\text{pro}}(\mathbf{r}) = \sum_{\alpha=1}^{N} \rho_{\alpha}^{0}(\mathbf{r}) \tag{16.17}$$

where $\rho_{\alpha}^{0}(\mathbf{r})$ is the density of the isolated atom α. The key feature of the ASA method is that the atomic electron densities are computed directly in terms of only s-type Gaussians, instead of the usual expansion of the molecular orbitals in terms of Gaussians. The key equation for the promolecular ASA density then becomes

$$\rho_A^{\text{pro}}(\mathbf{r}) = \sum_{\alpha=1}^{N} \rho_\alpha^0(\mathbf{r}) = \sum_{\alpha=1}^{N} P_\alpha \rho_\alpha^{\text{ASA}}(\mathbf{r}) \qquad (16.18)$$

with P_α the number of electrons attached to the atom α. In most applications, however, the value of P_α is chosen as the atomic number of the element involved. In some cases, a degree of internal charge transfer is introduced by putting P_α equal to the population of the atom according to an ab initio calculation and some specific choice of population analysis.

The ASA density for this atom is expressed in terms of the s-type Gaussians as

$$\rho_\alpha^{\text{ASA}}(\mathbf{r}) = \sum_{i=1}^{m} w_i |s_i(\mathbf{r})|^2 \qquad (16.19)$$

with w_i the weighting coefficients of the s-type Gaussians $s_i(\mathbf{r})$. Note that the ASA atomic densities possess a norm 1, although it is trivial to renormalize them to N. A special scheme has been developed to derive these ASA atomic densities by fitting an expansion of s-type Gaussians against true ab initio computed atomic densities. Much attention is paid to make sure that all w_i remain strictly positive definite, resulting in nonlinear optimizations. The reason for this stringent condition is naturally that no negative electron densities could arise within a finite volume. For more information on this optimization, the reader is referred to the works of Carbó-Dorca et al. [31–35].

Replacing ab initio densities with promolecular densities using the ASA expansion may seem a quite drastic approximation, but experience has shown that this is not the case [36–40]. The reason is that the ASA method very well captures those areas where the density is the highest, namely near the cores of the atoms. On the other hand, the valence region is characterized by a much smaller density and thus has no big influence on the MQSM so that the ASA approach is certainly viable from a computational point of view.

The main advantage of using the ASA method to obtain approximate electron densities is the very important gain in computational efficiency to compute MQSM and thus perform similarity analysis among the molecules of some molecular set. This is important for those cases where either the molecules are very numerous, or very big or a combination of both.

As stated above, the valence region contributes relatively little to the MQSM, hence the good performance of the ASA method. One might be concerned whether this is not disadvantageous for the use of molecular similarity in, for example, comparing molecular reactivity. It is well known that chemistry and chemical reactions involve changes mainly in the valence regions and therefore one would expect that these should have a bigger impact on the MQSM. Different ways to deal with this problem have been used. Among the more often used approaches, one can refer to the momentum space density similarities of Cooper and Allan allowing to reduce the dominating impact of the core regions on the MQSM [41–43]. Concerning the link to chemical reactivity, the work of Boon et al. [44–47] and Bultinck and Carbó-Dorca [48] is worth mentioning where MQSM computed over density functions were not used, but using different conceptual DFT quantities such as the Fukui function.

16.7   ROLE OF MOLECULAR ALIGNMENT

The MQSM values are quite strongly dependent on the alignment between the molecules whose similarity is being computed. It is clear that when two molecules are positioned very remotely from each other the MQSM will be very low. As an example, two hydrogen molecules put at a very large distance will yield a very low MQSM, whereas when aligning them perfectly the MQSM will be maximal. Likewise, the Carbó index will change from nearly zero to one. This has a profound effect on the use of the QSM theory for concluding degrees of similarity.

It is therefore highly desirable to establish an algorithm that allows to consistently perform molecular alignment before concluding any degree of similarity between molecules. Generally speaking, two approaches are used routinely.

The first approach involves structural alignment. The most often used algorithm for QSM is the topogeometrical superposition approach (TGSA), where the topologically most similar regions on the molecule are aligned so that the MQSM can be computed [49,50]. It is based on recognizing the largest common substructure. For TGSA only the molecular coordinates are required after which the chemical bonds are assigned. The chemical bonds between the heavy atoms constitute a set of dyads, and every dyad in the first molecule is compared to every dyad in the second molecule. The two dyads are considered sufficiently similar if their difference in bond length is below some threshold. From the set of dyads that meet this requirement, one can proceed to define sets of three atoms by adding a third atom bonded to one of the dyad atoms to this dyad. Then these three atomic sets are compared between the two molecules and if a successful match can be found, the molecules are aligned based on this match. If different alignments are found, for each alignment the MQSM is computed in the ASA approach and the one with the highest MQSM is retained for further use. The advantage of such topogeometrical alignment schemes is that chemical bonding, mainly a valence region effect, is used and as such the alignment produced corresponds usually to a chemist's alignment of the molecules.

The problem with such structural alignments is that different algorithms may yield different alignments and thus MQSM values. Also, these algorithms are largely limited to cases where some degree of structural similarity is encountered. An alternative procedure relies on maximizing the MQSM in terms of the alignment parameters. Two specific algorithms have been developed specifically for this goal, namely the MaxiSim [51] and QSSA [52] algorithms. The QSSA algorithm is the most general as it does not at any stage require any element of structural alignment. It simply computes the MQSM for some chosen initial alignment and then using the simplex method, locally optimizes the MQSM via changing the alignment parameters. Experience has shown that this maximization regularly gets stuck in local maxima so that QSSA also includes a genetic algorithm to enhance the chances of locating the global maximum. Both MaxiSim and QSSA make use of the ASA approximate densities as usually numerous calculations of the MQSM have to be performed.

The maximum similarity algorithms always give a similarity between two molecules that is equal to or larger than the value computed after structural alignment. However, structural alignment very often succeeds at giving chemically more relevant alignments. The maximum similarity approaches very often reach their high values because of the alignment of the heaviest atoms in the molecules, ignoring the kind and number of chemically similar bonds or functional groups.

Finally, it is worth mentioning that alignment-free methods for evaluating molecular similarity have been tested [45]. Recently, the question about the necessary positive definite nature the MQSM matrices must possess has been put forward and a building up algorithm for a set of molecular structures has been described [53].

16.8 SIMILARITY INDICES

Many applications of QSM use not the MQSM values as such but similarity indices. There exist many such indices and for a review, the reader is referred to Willett et al. [54]. One example that was touched upon above is the Carbó index. The Carbó index is one of the so-called C-class descriptors giving a value between 0 and 1 where a higher value indicates higher similarity. Although originally defined for overlap MQSM, it can be generalized to different operators Ω so that the Carbó index becomes

$$
\begin{aligned}
C_{AB} &= \frac{\int dr_1 dr_2 [\rho_A(r_1)\Omega(r_1,r_2)\rho_B(r_2)]}{\sqrt{\int dr_1 dr_2 [\rho_A(r_1)\Omega(r_1,r_2)\rho_A(r_2)] \int dr_1 dr_2 [\rho_B(r_1)\Omega(r_1,r_2)\rho_B(r_2)]}} \\
&= \frac{Z_{AB}(\Omega)}{\sqrt{Z_{AA}(\Omega)Z_{BB}(\Omega)}}
\end{aligned}
\tag{16.20}
$$

Besides the Carbó index another index that has been used in the field of QSM is the Hodgkin–Richards index, which uses an arithmetic mean instead of the geometric mean [55,56]:

$$
C_{AB} = \frac{2Z_{AB}(\Omega)}{Z_{AA}(\Omega) + Z_{BB}(\Omega)}
\tag{16.21}
$$

Other examples of similarity indices include the Petke [57] and Tanimoto [58] indices, both of which have been investigated in the QSM context. Several studies have shown that mutual relationships between these different indices exist [59–62].

A different class of indices is the D indices, where similarity is expressed as a distance. With these indices, perfect similarity is characterized by a zero distance. The best known is the Euclidean distance, introduced in Equation 16.5. Again, different connections have been found to exist between C- and D-class indices [59–62].

According to some authors, the similarity between shape functions might give more interesting information than using the electron density [63]. In this context, it should be mentioned that the Carbó index retains the same value. Moreover, it has been found that using the shape function instead of the electron density usually does not result in substantially different degrees of similarity, even when using other similarity indices. Only in case one goes to very diverse molecular sets, some differences may arise [48].

16.9 FRAGMENT SIMILARITY

It is well known, as it has been previously commented, that chemists rely quite often on the concept of transferability of the properties of atoms or functional groups between different molecules. In this context, a molecular fragment is more transferable if it is more similar to the same fragment in a different molecule.

From the perspective of MQS, this means that the similarity needs to be computed between fragments of a molecule. This requires methods to obtain a fragment density from a molecular electron density. Generally speaking, use is made of some operator w_f acting on the molecular density $\rho^{Mol}(\mathbf{r})$ to yield the fragment density as $\rho_f(\mathbf{r})$:

$$\rho_f(\mathbf{r}) = w_f \rho^{Mol}(\mathbf{r}) \tag{16.22}$$

Several forms of w_f have already been used within the field of MQS. These methods include the Hirshfeld partitioning [30], Bader's partitioning based on the virial theorem within atomic domains in a molecule [64], and the Mulliken approach [65]. For more information on all the three methods, refer to Chapter 15.

Quite generally the fragment similarity measure for a fragment f between two molecules A and B is then computed as

$$Z_{AB}^f(\Omega) = \int d\mathbf{r}_1 d\mathbf{r}_2 \left[w_f^A \rho_A(\mathbf{r}_1) \Omega(\mathbf{r}_1, \mathbf{r}_2) w_f^B \rho_B(\mathbf{r}_2) \right] \tag{16.23}$$

where w_f^A is the operator used to extract the fragment density from molecule A.

Different studies have appeared on fragment similarity using all the three partitioning schemes mentioned above, although in the case of the Hirshfeld method, the stockholder partitioning has also been applied directly to the similarity index [46,66,67].

Cioslowski et al. used fragment similarity indices to compute the degree of similarity between atoms of the same element in different molecules, where the atoms were those derived from Bader's atoms in molecules theory [68,69]. They introduced a novel atomic similarity index for atom A in molecule X and atom B in molecule Y defined as [70]

$$C_{A(X),B(Y)} = \begin{bmatrix} \dfrac{\int\limits_{\omega_{AB}} \rho_X(\mathbf{r})d\mathbf{r}}{\int\limits_{\omega_A} \rho_X(\mathbf{r})d\mathbf{r}} \end{bmatrix} \begin{bmatrix} \dfrac{\int\limits_{\omega_{BA}} \rho_Y(\mathbf{r})d\mathbf{r}}{\int\limits_{\omega_B} \rho_Y(\mathbf{r})d\mathbf{r}} \end{bmatrix} \tag{16.24}$$

where

  ω_A is the basin of atom A in molecule X

  $\omega_{AB} = \omega_{BA}$ is the intersection of this basin with the basin of atom B in molecule $Y(\omega_A)$

A case of specific interest in recent years addressed the validation of the holographic electron density theorem [71]. According to this theorem, all the information on any property of the molecule is not only contained in the global molecular electron density but also in every infinitesimal volume element of that density. This has been studied thoroughly in recent years in the specific case of quantifying a degree of chirality in a molecule. Considering a molecule with a chiral center, for example a carbon atom, Mezey et al. and Boon et al. studied the similarity between that atom in both enantiomers of a molecule [46,66,67,72]. It was found that not only near the center of chirality, the similarity index was significantly different from 1, but also for atoms further away from the center of chirality. This effect has been observed also in other cases. Moreover, the effect of chirality can even be transferred from a chiral center to a loosely bound achiral solvent molecule. Such an effect has recently been described in literature and confirmed experimentally using vibrational circular dichroism spectroscopy [73,74]. Another application of fragment similarity has been the study of the similarity between atoms to establish if the same trends as in the periodic system could be found [75–77].

The generality of similarity as a basic concept throughout chemistry makes it the theory of MQS very useful nearly in any field of chemistry including for instance biological activity. Not only is similarity by itself an interesting subject, it also opens the path to many related issues such as complexity and more generally many concepts in information entropy measures [78].

16.10   CONCLUSION

This chapter has dealt with introducing the main concepts within a theory called MQS. It has discussed the different steps to be taken to evaluate and quantify a degree of similarity between molecules in some molecular set but also fragments in molecules. QSM provides a scheme that relieves the arbitrariness of molecular similarity by using the electron density function as the sole descriptor, in agreement with the Hohenberg–Kohn theorems. It also addressed the different pitfalls that are present, for example the dependence on proper molecular alignment.

By its size, this chapter fails to address the entire background of MQS and for more information, the reader is referred to several reviews that have been published on the topic. Also it could not address many related approaches, such as the density matrix similarity ideas of Ciosloswki and Fleischmann [79,80], the work of Leherte et al. [81–83] describing simplified alignment algorithms based on quantum similarity or the empirical procedure of Popelier et al. on using only a reduced number of points of the density function to express similarity [84–88]. It is worth noting that MQS is not restricted to the most commonly used electron density in position space. Many concepts and theoretical developments in the theory can be extended to momentum space where one deals with the three components of linear momentum rather than with Cartesian coordinates. This has the advantage of reducing the bias of molecular alignment toward aligning the core electron distributions [41–43], it also proves valuable in providing information on theoretical studies of the chemical bond [78].

REFERENCES

1. Herndon, W.C. and Bertz, S.H. *J. Comput. Chem.* 1987, *8*, 367–374.
2. Tversky, A. *Psychol. Rev.* 1977, *84*, 327–352.
3. Rouvray, D.H. In *Concepts and Applications of Molecular Similarity*, Johnson, M.A., Maggiora, G.M. (Eds.), Wiley-Interscience, New York, 1990, pp. 15–43.
4. Bultinck, P., Girones, X., and Carbó-Dorca, R. In *Reviews in Computational Chemistry*, Lipkowitz, K.B., Larter, R., Cundari, T. (Eds.), John Wiley & Sons, Hoboken, NJ, 2005, Vol. 21, pp. 127–207.
5. Downs, G.M. In *Computational Medicinal Chemistry for Drug Discovery*, Bultinck, P., De Winter, H., Langenaeker, W., Tollenaere, J.P. (Eds.), Dekker, New York, 2003, pp. 364–386.
6. Brown, R.D. *Perspect. Drug Discov. Design* 1997, *7/8*, 31–49.
7. Mason, J.S. and Pickett, S.D. *Perspect. Drug Discov. Design* 1997, *7/8*, 85–114.
8. Bajorath, J. *J. Chem. Inf. Comput. Sci.* 2001, *41*, 233–245.
9. Karelson, M., Lobanov, V.S., and Katritzky, A.R. *Chem. Rev.* 1996, *96*, 1027–1043.
10. Karelson, M. In *Computational Medicinal Chemistry for Drug Discovery*, Bultinck, P., De Winter, H., Langenaeker, W., Tollenaere, J.P., (Eds.), Dekker, New York, 2003, pp. 641–667.
11. Parr, R.G. and Yang, W. *Density Functional Theory of Atoms and Molecules*, Oxford University Press, Oxford, 1989.
12. Handy, N.C. In *Lecture Notes in Quantum Chemistry II*, Roos, B.O. (Ed.), Springer, Heidelberg, 1994.
13. Born, M. *Atomic Physics*, Blackie & Son, London, 1935.
14. von Neumann J. *Mathematical Foundations of Quantum Mechanics*, Princeton University Press, Princeton, NJ, 1955.
15. Hohenberg, P. and Kohn, W. *Phys. Rev.* 1964, *136*, B864-B871.
16. Dean, P.M. In *Concepts and Applications of Molecular Similarity*, Johnson, M.A., Maggiora, G.M. (Eds.), Wiley-Interscience, New York, 1990, pp. 211–238.
17. Carbó, R., Arnau, J., and Leyda, L. *Int. J. Quantum Chem.* 1980, *17*, 1185–1189.
18. Carbó-Dorca, R. In *Advances in Molecular Similarity*, Vol. 2, Carbó-Dorca, R., Mezey, P.G. (Eds.), JAI Press, London, 1998, pp. 43–72.
19. Carbó-Dorca, R., Amat, Ll., Besalú, E., Gironés, X., Robert, D. In *Fundamentals of Molecular Similarity*, Carbó-Dorca, R., Gironés, X., Mezey, P.G. (Eds.), Kluwer Academic/Plenum Press, New York, 2001, pp. 187–320.
20. Carbó-Dorca, R. *J. Math. Chem.* 2002, *32*, 201–223.
21. Gironés, X., Gallegos, A., and Carbó-Dorca, R. *J. Chem. Inf. Comput. Sci.* 2000, *40*, 1400–1407.
22. Robert, D. and Carbó-Dorca, R. *J. Chem. Inf. Comput. Sci.* 1998, *38*, 620–623.
23. Robert, D., Gironés, X. and Carbó-Dorca, R. *Polycyc. Arom. Comp.* 2000, *19*, 51–71.
24. Carbó, R. and Calabuig, B. *Int. J. Quant. Chem.* 1992, *42*, 1695–1709.
25. Carbó, R. and Calabuig, B. *J. Chem. Inf. Comput. Sci.* 1992, *32*, 600–606.
26. Carbó, R. and Calabuig, B. *J. Mol. Struct. (THEOCHEM)* 1992, *254*, 517–531.
27. Carbó, R., Calabuig, B., Vera, L. and Besalú, E. *Adv. Quant. Chem.* 1994, *25*, 253–313.

28. Carbó, R. and Besalú, E. In *Molecular Similarity and Reactivity: From Quantum Chemical to Phenomenological Approaches*, Carbó, R. (Ed.), Kluwer Academic, Dordrecht, 1995, pp. 3–30.

29. Bultinck, P. and Carbó-Dorca, R. *J. Chem. Inf. Comput. Sci.* 2003, *43*, 170–177.

30. Hirshfeld, F.L. *Theoret. Chim. Acta* 1977, *44*, 129–138.

31. Constans, P. and Carbó, R. *J. Chem. Inf. Comput. Sci.* 1995, *35*, 1046–1053.

32. Constans, P., Amat, Ll., Fradera, X., and Carbó-Dorca, R. In *Advances in Molecular Similarity*, Vol. 1, Carbó-Dorca, R., Mezey, P.G. (Eds.), JAI Press, London, 1996, pp. 187–211.

33. Amat, Ll. and Carbó-Dorca, R. *J. Comput. Chem.* 1997, *18*, 2023–2039.

34. Amat, Ll. and Carbó-Dorca, R. *J. Comput. Chem.* 1999, *20*, 911–920.

35. Amat, Ll. and Carbó-Dorca, R. *J. Chem. Inf. Comput. Sci.* 2000, *40*, 1188–1198.

36. Gironés, X., Amat, Ll., and Carbó-Dorca, R. *J. Chem. Inf. Comput. Sci.* 2002, *42*, 847–852.

37. Amat, Ll. and Carbó-Dorca, R. *Int. J. Quant. Chem.* 2002, *87*, 59–67.

38. Gironés, X., Amat, Ll., and Carbó-Dorca, R. *J. Mol. Graph. Model.* 1998, *16*, 190–196.

39. Gironés, X., Carbó-Dorca, R., and Mezey, P.G. *J. Mol. Graph. Model.* 2001, *19*, 343–348.

40. Bultinck, P., Carbó-Dorca, R., and Van Alsenoy, C. *J. Chem. Inf. Comput. Sci.* 2003, *43*, 1208–1217.

41. Cooper, D.L. and Allan, N.L. In *Molecular Similarity and Reactivity: From Quantum Chemical to Phenomenological Approaches*, Carbó, R. (Ed.), Kluwer Academic Publishers, Dordrecht, 1995, pp. 31–55.

42. Al-Fahemi, J.H.A., Cooper, D.L., and Allan, N.L. *J. Mol. Graphics Mod.* 2007, *26*, 607–612.

43. Al-Fahemi, J.H.A., Cooper, D.L., and Allan, N.L. *Chem. Phys. Lett.* 2005, *416*, 376–380.

44. Boon, G., De Proft, F., Langenaeker, W., and Geerlings, P. *Chem. Phys. Lett.* 1998, *295*, 122–128.

45. Boon, G., Langenaeker, W., De Proft, F., De Winter, H., Tollenaere, J.P., and Geerlings, P. *J. Phys. Chem. A* 2001, *105*, 8805–8814.

46. Boon, G., Van Alsenoy, C., De Proft, F., Bultinck, P., and Geerlings, P. *J. Phys. Chem. A* 2003, *107*, 11120–11127.

47. Geerlings, P., Boon, G., Van Alsenoy, C., and De Proft, F. *Int. J. Quant. Chem.* 2005, *101*, 722–732.

48. Bultinck, P. and Carbó-Dorca, R. *J. Chem. Sci.* 2005, *117*, 425–435.

49. Gironés, X., Robert, D., and Carbó-Dorca, R. *J. Comput. Chem.* 2001, *22*, 255–263.

50. Gironés, X. and Carbó-Dorca, R. *J. Comput. Chem.* 2004, *25*, 153–159.

51. Constans, P., Amat, Ll., and Carbó-Dorca, R. *J. Comput. Chem.* 1997, *18*, 826–846.

52. Bultinck, P., Kuppens, T., Gironés, X., and Carbó-Dorca, R. *J. Chem. Inf. Comput. Sci.* 2003, *43*, 1143–1150.

53. Carbó-Dorca, R. *J. Math Chem.* DOI 10.1007/s10910–007–9305–z.

54. Willett, P., Barnard, J.M., and Downs, G.M. *J. Chem. Inf. Comput. Sci.* 1998, *38*, 983–996.

55. Hodgkin, E.E. and Richards, W.G. *Int. J. Quant. Chem. Quantum. Biol. Symp.* 1987, *14*, 105–110.

56. Hodgkin, E.E. and Richards, W.G. *Chem. Ber.* 1988, *24*, 1141.

57. Petke, J.D. *J. Comput. Chem.* 1993, *14*, 928–933.

58. Tou, J.T. and González, R.C. *Pattern Recognition Principles*, Addison-Wesley, Reading, MA, 1974.

59. Maggiora, G.M., Petke, J.D., and Mestres, J. *J. Math. Chem.* 2002, *31*, 251–270.

60. Carbó, R., Besalú, E., Amat, Ll., and Fradera, X. *J. Math. Chem.* 1996, *19*, 47–56.

61. Carbó-Dorca, R., Besalú, E., Amat, Ll., and Fradera, X. In *Advances in Molecular Similarity*, Carbó-Dorca, R., Mezey, P.G., (Eds.), JAI Press, London, 1996, Vol. 1, pp. 1–42.
62. Robert, D. and Carbó-Dorca, R. *J. Chem. Inf. Comput. Sci.* 1998, *38*, 469–475.
63. De Proft, F., Ayers, P.W., Sen, K.D., and Geerlings, P. *J. Chem. Phys.* 2004, *120*, 9969–9973.
64. Bader, R.F.W. *Atoms in Molecules: A Quantum Theory*, Clarendon Press, Oxford, 1990.
65. Mulliken, R.S. *J. Chem. Phys.* 1955, *23*, 1833–1840.
66. Boon, G., Van Alsenoy, C., De Proft, F., Bultinck, P., and Geerlings, P. *J. Mol. Struct. (THEOCHEM)* 2005, *727*, 49–56.
67. Boon, G., Van Alsenoy, C., De Proft, F., Bultinck, P., and Geerlings, P. *J. Phys. Chem. A* 2006, *110*, 5114–5120.
68. Stefanov, B.B. and Cioslowski, J. In *Advances in Molecular Similarity*, Vol. 1, Carbó-Dorca, R., Mezey, P.G., (Eds.), JAI Press, London, 1996, pp. 43–59.
69. Cioslowski, J., Stefanov, B., and Constans, P. *J. Comput. Chem.* 1996, *17*, 1352–1358.
70. Cioslowski, J. and Nanayakkara, A. *J. Am. Chem. Soc.* 1993, *115*, 11213–11215.
71. Mezey, P.G. *Mol. Phys.* 1999, *96*, 169–178.
72. Mezey, P.G., Ponec, R., Amat, Ll., and Carbó-Dorca, R. *Enantiomer* 1999, *4*, 371–378.
73. Debie, E, Jaspers, L., Bultinck, P., Herrebout, W., and Van Der Veken, B. *Chem. Phys. Lett.* 2008, *450*, 426–430.
74. Debie, E., Bultinck, P., Herrebout, W., and Van der Veken, B. *Phys. Chem. Chem. Phys.*, 2008, *10*, 3498–3508.
75. Robert, D. and Carbó-Dorca, R. *J. Math. Chem.* 1998, *23*, 327–351.
76. Robert, D. and Carbó-Dorca, R. *Int. J. Comput. Chem.* 2000, *77*, 685–692.
77. Borgoo, A., Godefroid, M., Sen, K.D., De Proft F., and Geerlings, P. *Chem. Phys. Lett.* 2004, *399*, 363–367.
78. Montgomery, H.E. and Sen, K.D. *Phys. Lett.*, 2008, 2271–2273.
79. Cioslowski, J. and Fleischmann, E.D. *J. Am. Chem. Soc.* 1991, *113*, 64–67.
80. Cioslowski, J. and Fleischmann, E.D. *Croat. Chem. Acta* 1993, *66*, 113–121.
81. Leherte, L. *J. Math. Chem* 2001, *29*, 47–83.
82. Leherte, L., Meurice, N., and Vercauteren, D.P. *J. Comput.-Aided Mol. Des.* 2005, *19*, 525–549.
83. Leherte, L. *J. Comp. Chem.* 2006, *27*, 1800–1816.
84. Popelier, P.L.A. *J. Phys. Chem. A* 1999, *103*, 2883–2890.
85. O'Brien, S.E. and Popelier, P.L.A. *Can. J. Chem.* 1999, *77*, 28–36.
86. O'Brien, S.E. and Popelier, P.L.A. *J. Chem. Inf. Comput. Sc.* 2001, *41*, 764–775.
87. O'Brien, S.E. and Popelier, P.L.A. *J. Chem. Soc. Perk. Trans 2* 2002, *3*, 478–483.
88. Popelier, P.L.A., Chaudry, U.A., and Smith P.J. *J. Chem. Soc. Perk. Trans 2* 2002, *7*, 1231–1237.

17 The Electrostatic Potential as a Guide to Molecular Interactive Behavior

Peter Politzer and Jane S. Murray

CONTENTS

17.1   COULOMB'S LAW AND THE ELECTROSTATIC POTENTIAL

Consider two stationary point charges, Q_1 and Q_2, separated by a distance R. Let $\mathbf{R}$ be the distance vector from Q_1 to Q_2. Then Coulomb's law states that the electrostatic force $\mathbf{F}$ exerted upon Q_2 by Q_1 is

$$\mathbf{F} = \frac{1}{4\pi\varepsilon_0} \frac{Q_1 Q_2}{R^2} \mathbf{i} \qquad (17.1)$$

where
     ε_0 is the permittivity of free space, which is a constant
     $\mathbf{i}$ is a unit vector in the direction of $\mathbf{R}$

If Q_1 and Q_2 have the same sign, then $\mathbf{F}$ is positive and therefore in the direction of $\mathbf{i}$ (and $\mathbf{R}$), meaning that Q_2 is being repelled away from Q_1. When Q_1 and Q_2 have different signs, $\mathbf{F}$ is negative and hence opposite to $\mathbf{i}$; Q_2 is now being attracted to Q_1. (An analogous analysis applies to the force exerted by Q_2 upon Q_1)

Suppose next that Q_2 is brought in infinitesimal increments $d\mathbf{R}$ from infinity to the separation R. The work involved is

$$W = \int_{\infty}^{R} \mathbf{F} \cdot d\mathbf{R} = \int_{\infty}^{R} |\mathbf{F}| \cos \theta |d\mathbf{R}| \tag{17.2}$$

where θ is the angle between $\mathbf{F}$ and $d\mathbf{R}$; the latter is opposite in direction to $\mathbf{R}$. If $\mathbf{F}$ is positive (repulsive), then $\theta = 180°$, $\cos \theta = -1$; for $\mathbf{F}$ negative (attractive), $\theta = 0°$, $\cos \theta = 1$. Substituting Equation 17.1 into Equation 17.2 and integrating yields $W = (1/4\pi\varepsilon_0)(|Q_1 Q_2|/R)$ when Q_1 and Q_2 have the same sign, and $W = -(1/4\pi\varepsilon_0)(|Q_1 Q_2|/R)$ when they have different signs. These two expressions can be combined into one by removing the absolute value signs:

$$W = \frac{1}{4\pi\varepsilon_0} \frac{Q_1 Q_2}{R} = \Delta E \tag{17.3}$$

If the process of bringing Q_1 and Q_2 together is adiabatic (no heat transfer), then $W = \Delta E$, the change in energy. Thus energy must be provided ($\Delta E > 0$) to bring Q_1 and Q_2 together when they repel, and it is released ($\Delta E < 0$) when they attract. ΔE, as given by Equation 17.3, is the interaction energy of Q_1 and Q_2.

Now divide ΔE in Equation 17.3 by Q_2. This gives $V(R)$, Equation 17.4, the electrostatic potential produced at the distance R by the charge Q_1:

$$V(R) = \frac{1}{4\pi\varepsilon_0} \frac{Q_1}{R} \tag{17.4}$$

The term "electrostatic" signifies that Q_1 is stationary. $V(R)$ is a scalar quantity, just like ΔE, and depends only upon Q_1. $V(R)$ can be viewed as indicating what is the "potential" for interaction at the distance R with any other charge Q_i; comparing the signs of $V(R)$ and Q_i will show whether the interaction would be attractive or repulsive, and the product $Q_i V(R)$ will give its strength, i.e., the interaction energy.

Equation 17.5 can easily be extended to a group of stationary point charges Q_i. If it is desired to know the net electrostatic potential that they produce at some point in space $\mathbf{r}$, this is given simply by

$$V(\mathbf{r}) = \frac{1}{4\pi\varepsilon_0} \sum_i \frac{Q_i}{R_i} \tag{17.5}$$

in which R_i is the distance from Q_i to the point $\mathbf{r}$.

Our interest in this chapter is in the electrostatic potentials that are produced by the nuclei and the electrons of atoms and molecules. The effect of the nuclei can be obtained by treating them as stationary point charges (invoking the Born–Oppenheimer approximation) and using Equation 17.5. The electrons are a greater problem, because they are in continual motion. However we do, in principle, have means of determining—either computationally or experimentally—the electronic density function $\rho(\mathbf{r})$, which gives the average number of electrons in the volume element $d\mathbf{r}$ around each point in space. If each $\rho(\mathbf{r})d\mathbf{r}$ is multiplied by the charge of an electron, $-e$, then we have the average electronic charge in that $d\mathbf{r}$ and can again use Equation 17.5, summing over each $-e\rho(\mathbf{r})d\mathbf{r}$. However the summation over this infinite number of charges is done by integration; thus,

$$V(\mathbf{r}) = \frac{1}{4\pi\varepsilon_0}\left[\sum_A \frac{Z_A e}{|\mathbf{R}_A - \mathbf{r}|} - e\int \frac{\rho(\mathbf{r}')d\mathbf{r}'}{|\mathbf{r}' - \mathbf{r}|}\right] \tag{17.6}$$

where
   Z_A is the atomic number of nucleus A
   $Z_A e$ is its charge
   $\mathbf{R}_A$ is the position of nucleus A
   $|\mathbf{R}_A - \mathbf{r}|$ is then its distance from the point $\mathbf{r}$

The electronic charge in each volume element $d\mathbf{r}'$ is $-e\,\rho(\mathbf{r}')d\mathbf{r}'$ and $|\mathbf{r}' - \mathbf{r}|$ is its distance to $\mathbf{r}$.

$V(\mathbf{r})$ is the electrostatic potential created at the point $\mathbf{r}$ by the nuclei and the electrons of an atom or a molecule. (For atoms, the summation in Equation 17.6 contains only one term.) It may seem inconsistent that Equation 17.6 came from Coulomb's law, which is for stationary charges, and electrons are not stationary. (Nor are nuclei, but we are working within the framework of the Born–Oppenheimer approximation.) This apparent inconsistency is resolved by noting that we treat the electronic density $\rho(\mathbf{r})$ as static; the average number of electrons in each volume element is constant, even though the electrons do not remain as the same ones. $V(\mathbf{r})$ can be either positive or negative in a given region, depending upon whether the effect of the nuclei or that of the electrons is dominant there. It is important to note that $V(\mathbf{r})$ is a measurable physical property, which can be determined experimentally, via diffraction techniques [1–3], as well as computationally.

For convenience, Equation 17.6 is often written in terms of atomic units, au, whereby it takes the form

$$V(\mathbf{r}) = \sum_A \frac{Z_A}{|\mathbf{R}_A - \mathbf{r}|} - \int \frac{\rho(\mathbf{r}')d\mathbf{r}'}{|\mathbf{r}' - \mathbf{r}|} \tag{17.7}$$

Even though $V(\mathbf{r})$ is energy or charge, as was seen above, it is customary to express it in energy units, e.g., hartrees (which are the atomic units of energy), kcal/mol, or kJ/mol. What is then being given is actually the energy of the interaction of the atom or molecule's charge distribution with a positive unit point charge (i.e., a proton) placed at the point $\mathbf{r}$. This is used as the measure of $V(\mathbf{r})$. Relevant conversion factors are hartrees $\times\, 627.5 = $ kcal/mol and kcal/mol $\times\, 4.184 = $ kJ/mol.

17.2   FEATURES OF ATOMIC AND MOLECULAR ELECTROSTATIC POTENTIALS

The electrostatic potential $V(\mathbf{r})$ is a local property; it has a separate value at each point $\mathbf{r}$ in the space of an atom or a molecule. However, it may only be of interest and necessary to evaluate it at certain points, depending upon the situation and purpose.

For spherically averaged atoms and monoatomic ions, the electrostatic potential depends only on the radial distance r from the nucleus. Hence, $V(\mathbf{r}) = V(r)$, and it could be determined only along one radian, since it will be the same along all others. For neutral ground-state atoms and monatomic positive ions, $V(r)$ is positive everywhere, the effect of the highly concentrated nucleus dominating that of the dispersed electrons; for these species, $V(r)$ decreases monotonically with increasing r, reaching zero at infinity [4,5].

For negative monatomic ions, on the other hand, $V(r)$ is positive near the nucleus but then decreases to a negative minimum at a radial distance r_{min} that is characteristic of the particular ion, after which it increases to zero at infinity. It has been shown that the electronic charge encompassed within the spherical volume between $r = 0$ and $r = r_{min}$ exactly equals the ion's nuclear charge [5], which is therefore, by Gauss' law, totally shielded. $V(r_{min})$ accordingly reflects the excess negative charge of the ion, and so is a determinant of the strength of its interactions. We have demonstrated that r_{min} values are good estimates of crystallographic ionic radii [5,6] and that the magnitudes of $V(r_{min})$ correlate well with lattice energies, for a given cation. The concept of the minimum of $V(r)$ defining a characteristic boundary surface for monatomic anions has been extended by Gadre and Shrivastava to polyatomic ones, establishing the boundary by applying the criterion $\nabla V(\mathbf{r}) \cdot \mathbf{s}(\mathbf{r}) = 0$, where $\mathbf{s}(\mathbf{r})$ is a unit vector perpendicular to the anion boundary surface at $\mathbf{r}$ [7].

While $V(r)$ is positive everywhere for neutral ground-state atoms, their interaction to form molecules normally produces some regions of negative electrostatic potential. These are most often found near (1) lone pairs of the more electronegative atoms (N, O, F, Cl, etc.), (2) π electrons of unsaturated molecules, and (3) strained C–C bonds. Every such negative region necessarily has one or more local minima, V_{min}, at which $V(\mathbf{r})$ reaches the most negative values in that region. It has been proven, however, that $V(\mathbf{r})$ has no local maxima other than at the positions of the nuclei [8]. Accordingly $V(\mathbf{r})$ decreases monotonically from each nucleus to the negative local minima, between which are saddle points. For a detailed topological analysis of $V(\mathbf{r})$ for a group of 18 molecules, see Leboeuf et al. [9].

As $V(\mathbf{r})$ is a local maximum at each nucleus in a molecule, if it is plotted along the internuclear axis z between two bonded atoms, there must be an *axial* minimum of $V(z)$ (usually positive) at some point z_m along that axis. This axial minimum has the interesting property that a point charge Q_i placed at z_m would feel no electrostatic force in either direction along the internuclear axis [10,11].

Proof: The electrostatic interaction energy of Q_i with the molecule would be $Q_i V(z_m)$. By Equations 17.2 and 17.3, the force along the axis in either direction depends upon $(\partial Q_i V(z)/\partial z)_{z_m}$, which is zero, since $V(z)$ is a minimum at z_m.

Thus, the point z_m appears to define a natural axial boundary between the two bonded nuclei. We have shown that the z_m, when determined for a large number of bonds, can indeed be used to establish an effective set of covalent radii [10,11].

17.3 MOLECULAR ELECTROSTATIC POTENTIALS AND REACTIVITY

17.3.1 BACKGROUND

In seeking to analyze and predict the reactivities of molecules in terms of their electrostatic potentials, it is necessary to take into account that $V(\mathbf{r})$ is based upon static average electronic charge distributions. Thus if it is desired to use the electrostatic potential of a molecule X to gain insight into how it may interact with some approaching Y, it is only relevant to consider $V(\mathbf{r})$ of X in its outer regions. There the perturbing effect of Y is quite small and the $V(\mathbf{r})$ of isolated X is still a meaningful indicator of how Y "sees" and "feels" X. As Y comes closer, its presence polarizes in some manner the $\rho(\mathbf{r})$ and hence the $V(\mathbf{r})$ of X, so that these are no longer what they were originally. (This is shown graphically by Francl [12] and Alkorta et al. [13].) There have been some attempts to correct for this via perturbation theory; this was reviewed some time ago by Politzer and Daiker [14], and more recently by Orozco and Luque [15].

What has been done more typically, however, is to evaluate $V(\mathbf{r})$ only at or beyond some appropriate distances from the nuclei. For instance, what was formerly often done was to compute two-dimensional contour plots of $V(\mathbf{r})$ in planes passing through a molecule (but omitting the regions near the nuclei) or in planes well removed from the nuclei, e.g., 1.75 Å [3,14,16,17]. This approach is of course more straightforward for planar molecules than for nonplanar ones. Another possibility that is sometimes utilized is to show just one, presumably important, three-dimensional outer contour of $V(\mathbf{r})$ [18]. This introduces the problem of choosing which one to show.

In the past 20 years, it has become common to determine $V(\mathbf{r})$ on a three-dimensional outer surface of the molecule. Such a surface can be established in various ways, such as overlapping spheres centered on the nuclei and having the van der Waals (or other) radii of the respective atoms [19,20]. Our preference is the widely followed suggestion by Bader et al. [21] that the surface be taken to be an outer contour of the molecule's electronic density $\rho(\mathbf{r})$, e.g., $\rho(\mathbf{r}) = 0.001$ au (electrons/bohr^3), which encompasses more than 95% of the molecule's electronic charge. Such a surface has the advantage that it reflects the actual features of the particular molecule such as lone pairs, π electronic charge, strained bonds, etc. One could of course choose some other outer contour of $\rho(\mathbf{r})$, such as the 0.002 or 0.0015 au. While this would change the magnitudes of the surface potential, our experience has been that the important trends and conclusions would be unaffected. We use the notation $V_S(\mathbf{r})$ to designate the electrostatic potential on a molecular surface.

To illustrate, Figure 17.1 shows $V_S(\mathbf{r})$ on the 0.001 au surface of 1-butanol, $H_3C-CH_2-CH_2-CH_2OH$, computed at the density functional B3PW91/6-31G(d,p) level. Looking first at the alkyl portion of the molecule, its $V_S(\mathbf{r})$ is fairly bland; the hydrogens are weakly positive and the carbons weakly negative. More interesting is the region of the hydroxyl group. The electrostatic potential on the surface of the oxygen is quite negative, reflecting its electronegativity and lone pairs; $V_S(\mathbf{r})$ reaches a negative extreme of $V_{S,min} = -36$ kcal/mol. In contrast, the hydrogen has lost electronic charge to the oxygen, and has a strongly positive surface potential, with a

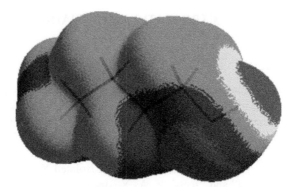

FIGURE 17.1 (See color insert following page 302.)   Electrostatic potential on the $\rho(\mathbf{r}) =$ 0.001 au molecular surface of 1-butanol, computed at the B3PW91/6-31G(d,p) level. Color ranges, in kcal/mol, are: red, more positive than 30; yellow, between 15 and 30; green, between 0 and 15; blue, between -20 and 0; purple, more negative than -20. The hydroxyl hydrogen is at the far right (red and yellow), and the oxygen is below it (purple and blue).

maximum, $V_{S,max}$, of $+43$ kcal/mol. (It was stated earlier that the electrostatic potential $V(\mathbf{r})$ throughout the total space of a molecule has maxima only at the nuclei. However the potential $V_S(\mathbf{r})$ on a surface of a molecule can have several local maxima $V_{S,max}$ as well as local minima $V_{S,min}$, the most negative values.)

It might seem that the $V_{S,min}$ and $V_{S,max}$ on a suitable molecular surface should indicate sites susceptible to electrophilic and nucleophilic reactants, respectively. Such reasoning has had some success in the past [3,14,16,17], but it is not reliable. For example, shown in Figure 17.2 is the electrostatic potential on a surface of anisole (methoxybenzene), **1**, which is well known to undergo electrophilic attack at the *ortho* and *para* positions.

However, the most negative surface potentials are at the methoxy oxygen, $V_{S,min} = -24$ kcal/mol and above and below the central portion of the ring, both having $V_{S,min} = -20$ kcal/mol; the latter are due to the aromatic π electrons. Thus $V_S(\mathbf{r})$ does not predict *ortho* and *para* reactivity. The problem is that the most negative regions are not necessarily the sites of the most reactive, least-strongly held electrons; to find these, some other indicator is needed, such as the local ionization energy [22,23], the Fukui [24], or related functions [25]. With respect to nucleophilic attack, the difficulty is that the most positive $V_S(\mathbf{r})$ are often associated with hydrogens. In general, therefore, the electrostatic potential on molecular surfaces is not a dependable guide to chemical reactivity.

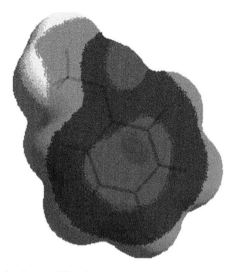

FIGURE 17.2 (See color insert following page 302.)   Electrostatic potential on the $\rho(\mathbf{r}) =$ 0.001 au molecular surface of anisole, **1**, computed at the B3PW91/6-31G(d,p) level. Color ranges, in kcal/mole, are: yellow, between 10 and 20; green, between 0 and 10; blue, between -15 and 0; purple, more negative than -15. The methoxy group is at the upper left. The most negative regions (purple) are by the oxygen and above and below the ring.

17.3.2 NONCOVALENT INTERACTIONS

Noncovalent interactions include, for example, those between the molecules in condensed stages (liquids, solids, solutions), hydrogen bonding, physical adsorption, the early stages of biological recognition processes, etc. Such interactions are primarily electrostatic in nature [18,26], and the participants are sufficiently far apart that polarization and charge transfer are minimal. The electrostatic potential $V_S(\mathbf{r})$ evaluated in an outer region or on an outer surface of a molecule should therefore be well suited for analyzing and predicting its noncovalent interactive behavior.

One approach, which has been used extensively in pharmacological areas [27], is to look for characteristic patterns of positive and negative regions that may enhance or inhibit a certain type of activity. The early stages of drug–receptor and enzyme–substrate interactions, in which the participants "recognize" each other through their outer electrostatic potentials, can be analyzed in this manner [28]. We were able, on this basis, to find qualitative trends in the toxicities of chlorinated dibenzo-*p*-dioxins and related systems [29].

More quantitative is to look specifically at the most positive and most negative values of the surface potential, the $V_{S,max}$ and $V_{S,min}$. For instance, one would anticipate that the $V_{S,min}$ might reveal hydrogen bond acceptor sites such as the oxygen in 1-butanol, Figure 17.1, while the high $V_{S,max}$ associated with the hydroxyl hydrogen might indicate potency as a donor. (Compare this hydrogen to the aromatic and alkyl ones in Figures 17.1 and 17.2.) These roles for $V_{S,min}$ and $V_{S,max}$ have been confirmed; good correlations were found between $V_{S,min}$ and $V_{S,max}$ and empirical measures of hydrogen-bond-accepting and -donating tendencies, respectively [30].

Another type of noncovalent interaction, in relation to which the molecular surface electrostatic potential has been very important, is σ-hole bonding. Some covalently bonded Group V, VI, and VII atoms are able to interact attractively in a highly directional manner with negative regions on other molecules, e.g., the lone pairs of Lewis bases [31–34]. When a Group VII atom is involved, it is often called "halogen bonding." Since many Group V–VII covalently bonded atoms are considered to be negative in character, it maybe surprising that they might interact attractively with negative sites. It has been found, however, originally by Brinck et al., for halogens [35], that such atoms can, in appropriate molecular environments, have regions of positive electrostatic potential (σ-holes) on the outer portions of their surfaces, on the extensions of their covalent bonds [31–36]. These positive σ-holes correspond to the electron-deficient outer lobes of half-filled p bonding orbitals.

This can be seen in Figure 17.3 for SBr$_2$. Each bromine has a positive region on its surface, $V_{S,max} = +22$ kcal/mol, on the extension of its S–Br bond, while the sulfur has two positive regions, each with $V_{S,max} = +20$ kcal/mol, on the extensions of the Br–S bonds. The remainders of the sulfur and bromine surfaces are negative; thus there can be several different attractive interactions between positive and negative portions of their surfaces: S---Br, S---S, Br---Br. Note that the last two are "like–like" interactions, between the same atom in each of two identical molecules [37]; this could not be explained in terms of atomic charges assigned to an atom as a whole.

FIGURE 17.3 (See color insert following page 302.)   Electrostatic potential on the $\rho(\mathbf{r}) = 0.001$ au molecular surface of SBr$_2$, computed at the B3PW91/6-31G(d,p) level. Color ranges, in kcal/mol, are: red, more positive than 15; yellow, between 10 and 15; green, between 0 and 10; blue, between −9 and 0. One of the two bromines is at the bottom, while the sulfur is at the upper right. The red regions are the positive σ-holes, on the extensions of the S–Br and Br–S bonds.

σ-Hole bonding, which is competitive with hydrogen bonding [38–40], has been observed both experimentally and computationally [31–34,36–41]. Its significance, particularly in molecular biology [36,41] and in crystal engineering [32,39,41], is recognized increasingly being.

Hydrogen bonding and σ-hole bonding can be treated in terms of the $V_{S,max}$ and $V_{S,min}$ on a molecular surface. To deal with noncovalent interactions in general, however, it is necessary to go beyond the qualitative pattern of $V_S(\mathbf{r})$ and its $V_{S,max}$ and $V_{S,min}$, and to fully access the whole range of information that $V_S(\mathbf{r})$ contains. This is done by analyzing it in terms of several statistically defined features, adopted over a period of years as we saw specific needs to be addressed. These features are

(1) The averages of the positive and negative potentials on the surface, $\bar{V}_S^+$ and $\bar{V}_S^-$,

$$\bar{V}_S^+ = \frac{1}{m} \sum_{i=1}^{m} V_S^+(\mathbf{r}_i) \quad \text{and} \quad \bar{V}_S^- = \frac{1}{n} \sum_{j=1}^{n} V_S^-(\mathbf{r}_j) \tag{17.8}$$

where the m points $\mathbf{r}_i$ are those at which $V_S(\mathbf{r})$ is positive, $V_S^+(\mathbf{r}_i)$, and the n $\mathbf{r}_j$ are those where it is negative, $V_S^-(\mathbf{r}_j)$.

(2) The average deviation, Π,

$$\Pi = \frac{1}{m+n} \sum_{k=1}^{m+n} |V_S(\mathbf{r}_k) - \bar{V}_S| \tag{17.9}$$

in which $\bar{V}_S$ is the overall average of $V_S(\mathbf{r})$, $\bar{V}_S = (m\bar{V}_S^+ + n\bar{V}_S^-)/(m+n)$. Π provides a measure of the internal charge separation in a molecule, more meaningful for this purpose than the dipole moment, which is so dependent on symmetry; for instance, the dipole moment of p-dinitrobenzene is zero, even though it is composed of highly polar components. Π has been found to correlate with several empirical indices of polarity [42].

(3) The positive, negative, and total variances, σ_+^2, σ_-^2, and σ_{tot}^2,

$$\sigma_{tot}^2 = \sigma_+^2 + \sigma_-^2 = \frac{1}{m} \sum_{i=1}^{m} [V_S^+(\mathbf{r}_i) - \bar{V}_S^+]^2 + \frac{1}{n} \sum_{j=1}^{n} [V_S^-(\mathbf{r}_j) - \bar{V}_S^-]^2 \tag{17.10}$$

These quantities indicate the variability of the positive, negative, and total surface potentials, i.e., $V_S^+(\mathbf{r}_i)$, $V_S^-(\mathbf{r}_j)$, and $V_S(\mathbf{r})$. Due to the terms in Equation 17.10 being squared, the variances are particularly sensitive to the extrema of $V_S(\mathbf{r})$, $V_{S,max}$, and $V_{S,min}$. It may seem that there is some similarity between σ_{tot}^2 and Π, but in fact the former is normally much larger in magnitude and they may not even show the same trend in a series of molecules [43].

(4) A balance parameter, ν,

$$\nu = \frac{\sigma_+^2 \sigma_-^2}{\left(\sigma_{tot}^2\right)^2} \tag{17.11}$$

which is intended to indicate the degree of balance between the positive and negative surface potentials. From Equation 17.11, ν reaches a maximum of 0.250 when $\sigma_+^2 = \sigma_-^2$, whether they may be large or small. Accordingly, the closer the ν is to 0.250, the better the ability of the molecule to interact through *both* its positive and negative regions, either strongly or weakly.

The quantities defined by Equations 17.8 through 17.11, plus the positive and negative surface areas A_S^+ and A_S^-, provide a detailed characterization of the electrostatic potential on a molecule's surface, and thereby a basis for analyzing and describing its noncovalent interactions. It has indeed been possible to develop good correlations between various combinations of these quantities and a number of condensed phase physical properties that depend upon noncovalent interactions, including heats of phase transitions, boiling points and critical constants, solubilities and solvation energies, partition coefficients, surface tensions, viscosities, diffusion constants, etc. The procedure is to establish an experimental database for the property of interest, to compute the quantities defined by Equations 17.8 through 17.11 for the respective molecules, and then to use multivariable statistical analyses to determine the best subset of the computed quantities for fitting the database. The resulting analytical relationship then allows the property to be predicted for compounds for which it is not known. This work has been reviewed on several occasions [43–45].

The quantities that best represent a particular property can often be rationalized on the basis of physical intuition. For example, those that reflect interactions between like molecules, such as heats of sublimation and vaporization, can be expressed well in terms of molecular surface area and the product $\nu\sigma_{tot}^2$. A large value for this product means that each molecule has both significantly positive and significantly negative surface potentials, which is needed to ensure strongly attractive intermolecular interactions, with consequently higher energy requirements for the solid → gas and liquid → gas transitions.

The analytical expression for a given property can be reparametrized, if desired, to apply to a particular class of compounds. Our tendency is usually to have, as general, a database as possible. But for example, Byrd and Rice desired to optimize the heat of sublimation and heat of vaporization equations specifically for nitro derivatives [46]. They retained the dependence on surface area and $\nu\sigma_{tot}^2$, but used a nitro compound database to obtain new coefficients for these quantities.

17.4   FUNDAMENTAL NATURE OF ELECTROSTATIC POTENTIAL

Our focus in this chapter has been upon the relationship of the electrostatic potential to molecular interactive behavior. However the significance of $V(\mathbf{r})$ goes far beyond, as we shall briefly point out. A more detailed overview, with relevant references, has been given by Politzer and Murray [47].

$V(\mathbf{r})$ is linked to the electronic density $\rho(\mathbf{r})$ by both Equation 17.7 and Poisson's equation,

$$\nabla^2 V(\mathbf{r}) = 4\pi\rho(\mathbf{r}) - 4\pi \sum_A Z_A \delta(\mathbf{r} - \mathbf{R}_A) \qquad (17.12)$$

According to the Hohenberg–Kohn theorem [48], the properties of a system of electrons and nuclei in its ground state are determined entirely by $\rho(\mathbf{r})$. Thus the total energy, for example, is a functional of $\rho(\mathbf{r})$, $E = \Im[\rho(\mathbf{r})]$.

According to the rigorous relationship between $\rho(\mathbf{r})$ and $V(\mathbf{r})$ mentioned above, it can be argued that $V(\mathbf{r})$ is also fundamental in nature. In addition, it has the advantage of lending itself better to further analytical development. For instance, it was shown long ago that $V(\mathbf{r})$ must decrease monotonically with radial distance from the nucleus for a ground-state atom [4]. It is known empirically that $\rho(\mathbf{r})$ does the same [4], but the proof of this remains elusive.

As another example, the functional expression for the energy in terms of $\rho(\mathbf{r})$ is known only approximately. However exact formulas have been developed that relate the energies of atoms and molecules to the electrostatic potentials at their nuclei [49–52]. This has been done as well for the chemical potentials (electronegativities) of atoms [53]. Thus, both the intrinsic significance and the practical applications of the electrostatic potential continue to be active areas of investigation.

REFERENCES

1. Stewart, R. F. *J. Chem. Phys.* 1972, *57*, 1664.
2. Stewart, R. F. *Chem. Phys. Lett.* 1979, *65*, 335.
3. Politzer, P. and Truhlar, D. G., Eds. *Chemical Applications of Atomic and Molecular Electrostatic Potentials*, Plenum Press, New York, 1981.
4. Weinstein, H., Politzer, P., and Srebrenik, S. *Theor. Chim. Acta* 1975, *38*, 159.
5. Sen, K. D. and Politzer, P. *J. Chem. Phys.* 1989, *90*, 4370.
6. Sen, K. D. and Politzer, P. *J. Chem. Phys.* 1989, *91*, 5123.
7. Gadre, S. R. and Shrivastava, I. H. *J. Chem. Phys.* 1991, *94*, 4384.
8. Pathak, R. K. and Gadre, S. R. *J. Chem. Phys.* 1990, *93*, 1770.
9. Leboeuf, M., Köster, A. M., Jug, K., and Salahub, D. R. *J. Chem. Phys.* 1999, *111*, 4893.
10. Wiener, J. M. M., Grice, M. E., Murray, J. S., and Politzer, P. *J. Chem. Phys.* 1996, *104*, 5109.
11. Politzer, P., Murray, J. S., and Lane, P. *J. Comput. Chem.* 2003, *24*, 505.
12. Francl, M. M. *J. Phys. Chem.* 1985, *89*, 428.
13. Alkorta, I., Perez, J. J., and Villar, H. O. *J. Mol. Graph.* 1994, *12*, 3.
14. Politzer, P. and Daiker, K. C. In: *The Force Concept in Chemistry*, Deb, B. M., Ed., Van Nostrand Reinhold, New York, 1981, Chapter 6.
15. Orozco, M. and Luque, F. J. In: *Molecular Electrostatic Potentials: Concepts and Applications*, Murray, J. S. and Sen, K., Eds., Elsevier, Amsterdam, 1996, Chapter 4.
16. Scrocco, E. and Tomasi, *J. Top. Curr. Chem.* 1973, *42*, 95.
17. Scrocco, E. and Tomasi, *J. Adv. Quantum Chem.* 1978, *11*, 115.
18. Naray-Szabo, G. and Ferenczy, G. G. *Chem. Rev.* 1995, *95*, 829.
19. Du, Q. and Arteca, G. A. *J. Comput. Chem.* 1996, *17*, 1258.
20. Connolly, M. L. In: *Encyclopedia of Computational Chemistry*, Vol. 3, Schleyer, P. v. R., Ed., Wiley, New York, 1998, p. 1698.
21. Bader, R. F. W., Carroll, M. T., Cheeseman, J. R., and Chang, C., *J. Am. Chem. Soc.* 1987, 109, 7968.
22. Sjoberg, P., Murray, J. S., Brinck, T., and Politzer, P. *Can. J. Chem.* 1990, *68*, 1440.
23. Politzer, P., Murray, J. S., and Concha, M. C. *Int. J. Quantum Chem.* 2002, *88*, 19.
24. Ayers, P. W. and Levy, M. *Theor. Chem. Acc.* 2000, *103*, 353.

25. Morell, C., Grand, A., and Toro-Labbé, A. *J. Phys. Chem. A* 2005, *109*, 205.
26. Hirschfelder, J. O., Curtiss, C. F., and Bird, R. B. *Molecular Theory of Gases and Liquids*, Wiley, New York, 1954.
27. Politzer, P., Laurence, P. R., and Jayasuriya, K. *Environ. Health Perspect.* 1985, *61*, 191.
28. Murray, J. S. and Sen, K., Eds. *Molecular Electrostatic Potentials: Concepts and Applications*, Elsevier, Amsterdam, 1996.
29. Sjoberg, P., Murray, J. S., Brinck, T., Evans, P., and Politzer, P. *J. Mol. Graphics* 1990, *8*, 81.
30. Hagelin, H., Murray, J. S., Brinck, T., Berthelot, M., and Politzer, P. *Can. J. Chem.* 1995, *73*, 483.
31. Politzer, P., Lane, P., Concha, M. C., Ma, Y., and Murray, J. S. *J. Mol. Model.* 2007, *13*, 305.
32. Politzer, P., Murray, J. S., and Concha, M. C. *J. Mol. Model.* 2007, *13*, 643.
33. Murray, J. S., Lane, P., Clark, T., and Politzer, P. *J. Mol. Model.* 2007, *13*, 1033.
34. Murray, J. S., Lane, P., and Politzer, P. *Int. J. Quantum Chem.* 2007, *107*, 2286.
35. Brinck, T., Murray, J. S., and Politzer, P. *Int. J. Quantum Chem., Quantum Biol. Symp.* 1992, *19*, 57.
36. Auffinger, P., Hays, F. A., Westhof, E., and Shing Ho, P. *Proc. Nat. Acad. Sci.* 2004, *101*, 16789.
37. Politzer, P., Murray, J. S., and Concha, M. C. *J. Mol. Model.*, 2008, 14, 689.
38. Di Paolo, T. and Sandorfy, C. *Can. J. Chem.* 1974, *52*, 3612.
39. Corradi, E., Meille, S. V., Messina, M. T., Metrangolo, P., and Resnati, G. *Angew. Chem. Int. Ed.* 2000, *39*, 1782.
40. Politzer, P., Murray, J. S., and Lane, P. *Int. J. Quantum Chem.* 2007, *107*, 3046.
41. Metrangolo, P., Neukirsch, H., Pilati, T., and Resnati, G. *Acc. Chem. Res.* 2005, *38*, 386.
42. Brinck, T., Murray, J. S., and Politzer, P. *Mol. Phys.* 1992, *76*, 609.
43. Murray, J. S. and Politzer, P. *J. Mol. Struct. (Theochem)* 1998, *425*, 107.
44. Politzer, P. and Murray, J. S. *Trends Chem. Phys.* 1999, *7*, 157.
45. Politzer, P. and Murray, J. S. *Fluid Phase Equil.* 2001, *185*, 129.
46. Byrd, E. F. C. and Rice, B. M. *J. Phys. Chem A* 2006, *110*, 1005.
47. Politzer, P. and Murray, J. S. *Theor. Chem. Acc.* 2002, *108*, 134.
48. Hohenberg, P. and Kohn, W. *Phys. Rev. B* 1964, *136*, 864.
49. Foldy, L. L. *Phys. Rev.* 1951, *83*, 397.
50. Wilson, E. B. Jr. *J. Chem. Phys.* 1962, *36*, 2232.
51. Politzer, P. and Parr, R. G. *J. Chem. Phys.* 1974, *61*, 4258.
52. Politzer, P. *Theor. Chem. Acc.* 2004, *111*, 395.
53. Parr, R. G., Donnelly, R. A., Levy, M., and Palke, W. E. *J. Chem. Phys.* 1978, *68*, 3801.

18 Fukui Function

Paul W. Ayers, Weitao Yang, and Libero J. Bartolotti

CONTENTS

18.1 WHAT IS THE FUKUI FUNCTION?

The Fukui function, denoted by $f(\mathbf{r})$, is defined as the differential change in electron density due to an infinitesimal change in the number of electrons [1]. That is,

$$f(\mathbf{r}) = \left(\frac{\partial \rho(\mathbf{r})}{\partial N}\right)_{v(\mathbf{r})}, \tag{18.1}$$

where $\rho(\mathbf{r})$ is the electron density and

$$N = \int \rho(\mathbf{r}) d\mathbf{r} \tag{18.2}$$

is the total number of electrons in the system. For isolated molecules at zero temperature, the Fukui function is ill-defined because of the derivative discontinuity [2–4]. To resolve this difficulty, Fukui functions from above and below are defined using the one-sided derivatives,

$$f^+(\mathbf{r}) = \left(\frac{\partial \rho(\mathbf{r})}{\partial N}\right)_{v(\mathbf{r})}^+ = \lim_{\varepsilon \to 0^+} \frac{\rho_{N+\varepsilon}(\mathbf{r}) - \rho_N(\mathbf{r})}{\varepsilon} \tag{18.3}$$

$$f^-(\mathbf{r}) = \left(\frac{\partial \rho(\mathbf{r})}{\partial N}\right)_{v(\mathbf{r})}^- = \lim_{\varepsilon \to 0^+} \frac{\rho_N(\mathbf{r}) - \rho_{N-\varepsilon}(\mathbf{r})}{\varepsilon}. \tag{18.4}$$

When a molecule accepts electrons, the electrons tend to go to places where $f^+(\mathbf{r})$ is large because it is at these locations that the molecule is most able to stabilize additional electrons. Therefore a molecule is susceptible to nucleophilic attack at sites where $f^+(\mathbf{r})$ is large. Similarly, a molecule is susceptible to electrophilic attack at sites where $f^-(\mathbf{r})$ is large, because these are the regions where electron removal destabilizes the molecule the least. In chemical density functional theory (DFT), the Fukui functions are the key regioselectivity indicators for electron-transfer controlled reactions.

This chapter will overview the historical development of the Fukui function concept from its origins in the work of Parr and Yang [1] to the present day. The recent review by Levy and one of the present authors provides a more mathematical perspective on the Fukui function [5].

18.2 CONTEXT

The genesis of chemical DFT can be traced back to the 1978 paper published by of Parr et al. [6]. That paper identified the electronic chemical potential as the derivative of the electronic energy with respect to the number of electrons at fixed molecular geometry:

$$\mu = \left(\frac{\partial E}{\partial N}\right)_{v(\mathbf{r})}. \tag{18.5}$$

The electronic chemical potential is constant for a system in its electronic ground state, which led Parr et al. to associate the chemical potential with minus one times the electronegativity, since the electronegativity is also equalized in the ground state [7]. This equalization of the chemical potential also suggests that electronic structure theory can be expressed in a way that resembles classical thermodynamics. Ergo, Parr et al. wrote the total differential of the energy as

$$dE_{v,N} = \left(\frac{\partial E_{v,N}}{\partial N}\right)_{v(\mathbf{r})} dN + \int \left(\frac{\delta E_{v,N}}{\delta v(\mathbf{r})}\right)_N \delta v(\mathbf{r}) d\mathbf{r}$$

$$= \mu_{v,N} dN + \int \rho_{v,N}(\mathbf{r}) \delta v(\mathbf{r}) d\mathbf{r}. \tag{18.6}$$

In 1982, Nalewajski and Parr took the thermodynamic analogy to its logical conclusion by extending the Legendre-transform structure of classical thermodynamics to DFT [8]. One of their results was the Maxwell relation for Equation 18.6,

$$\left(\frac{\delta \mu_{v,N}}{\delta v(\mathbf{r})}\right)_N = \left(\frac{\partial \rho_{v,N}(\mathbf{r})}{\partial N}\right)_{v(\mathbf{r})}. \tag{18.7}$$

By differentiating with respect to the number of electrons, the preceding equations presuppose that one can compute the energy (Equation 18.5) and the density (Equation 18.7) for systems with noninteger numbers of electrons. But how? Every real and finite system has an integer number of electrons.

This issue was resolved in 1982 by Perdew et al. [2,9], who showed how to formulate DFT for systems with fractional electron number by adapting the zero-temperature grand-canonical ensemble construction of Gyftopoulos and Hatsopoulos [10]. Their analysis reveals that the energy and electron density should be linearly interpolated between integer values:

$$E_{v,N} = (1 + \lfloor N \rfloor - N)E_{v,\lfloor N \rfloor} + (N - \lfloor N \rfloor)E_{v,\lceil N \rceil} \tag{18.8}$$

$$\rho_{v,N}(\mathbf{r}) = (1 + \lfloor N \rfloor - N)\rho_{v,\lfloor N \rfloor}(\mathbf{r}) + (N - \lfloor N \rfloor)\rho_{v,\lceil N \rceil}(\mathbf{r}) \tag{18.9}$$

Here, $\rho_{v,N}(\mathbf{r})$ and $E_{v,N}$ denote the ground-state density and electronic energy for N electrons bound by the external potential $v(\mathbf{r})$. $\lfloor N \rfloor$ and $\lceil N \rceil$ denote the nearest integer below and above N, respectively (e.g., $\lfloor 9.1 \rfloor = 9$; $\lceil 9.1 \rceil = 10$). Although the key equations were originally derived by thermodynamic arguments, they are much more fundamental. Indeed, Equations 18.8 and 18.9 can be derived directly from the properties of the Hohenberg–Kohn $F[\rho]$ functional [3,4].

18.3 ORIGINS

By 1984, it was clear that the derivatives that arise in this "DFT thermodynamics" contain chemically useful information. μ is -1 times the electronegativity. $\rho(\mathbf{r})$ is the electron density—fundamental in its own right, but also closely related to the electrostatic potential. If, in analogy to Equation 18.6, one writes the total differential for the chemical potential,

$$d\mu = \left(\frac{\partial \mu}{\partial N}\right)_{v(\mathbf{r})} dN + \int \left(\frac{\delta \mu}{\delta v(\mathbf{r})}\right)_N \delta v(\mathbf{r}) d\mathbf{r}$$
$$= \eta dN + \int f(\mathbf{r}) \delta v(\mathbf{r}) d\mathbf{r} \tag{18.10}$$

then the chemical hardness, η, appears. Already in 1983, Parr and Pearson had established the relevance of η for the hard or soft acid or base principle [11]. So it seems certain that the function, $f(\mathbf{r})$, that enters into Equation 18.10 is also relevant to chemistry. What does $f(\mathbf{r})$ mean? This is the question that Prof. Robert Parr posed to one of the present authors (W.Y.), early in 1984. And then Parr left for a conference.

Upon Parr's return, W.Y. conveyed the following: (a) a large change in the chemical potential should be favorable ("$d\mu$ big is good"), and so molecules should be most reactive where $f(\mathbf{r})$ is large. (b) Referring to the Maxwell relation, $f(\mathbf{r})$ is related to the change in density in response to changes in the number of electrons, N. (c) From Equation 18.9, it is apparent that one cannot differentiate the electron density with respect to N when N is an integer. Instead one has the one-sided derivatives,

$$f_{v,N}^+(\mathbf{r}) = \rho_{v,N+1}(\mathbf{r}) - \rho_{v,N}(\mathbf{r}) \tag{18.11}$$

$$f_{v,N}^-(\mathbf{r}) = \rho_{v,N}(\mathbf{r}) - \rho_{v,N-1} \tag{18.12}$$

Parr immediately pointed out that, in the frozen orbital approximation, these derivatives can be approximated with the squares of the lowest unoccupied (LUMO) and highest occupied molecular orbitals (HOMO):

$$f_{v,N}^+(\mathbf{r}) = \left|\phi_{v,N}^{\text{LUMO}}(\mathbf{r})\right|^2 = \rho_{v,N}^{\text{LUMO}}(\mathbf{r}) \tag{18.13}$$

$$f_{v,N}^-(\mathbf{r}) = \left|\phi_{v,N}^{\text{HOMO}}(\mathbf{r})\right|^2 = \rho_{v,N}^{\text{HOMO}}(\mathbf{r}) \tag{18.14}$$

Based on the orbital approximations, it is clear that $f(\mathbf{r})$ is the DFT analog of the frontier orbital regioselectivity for nucleophilic $(f_{v,N}^+(\mathbf{r}))$ and electrophilic $(f_{v,N}^-(\mathbf{r}))$ attack. It is then reasonable to define a reactivity indicator for radical attack by analogy to the corresponding orbital indicator,

$$f_{v,N}^0(\mathbf{r}) = \frac{1}{2}\left(f_{v,N}^+(\mathbf{r}) + f_{v,N}^-(\mathbf{r})\right) = \frac{1}{2}\left(\rho_{v,N+1}(\mathbf{r}) - \rho_{v,N-1}(\mathbf{r})\right)$$
$$\approx \frac{1}{2}\left(\rho_{v,N}^{\text{HOMO}}(\mathbf{r}) + \rho_{v,N}^{\text{LUMO}}(\mathbf{r})\right) \tag{18.15}$$

All of these regioselectivity indicators are called Fukui functions, in honor of Kenichi Fukui, who pioneered the analogous frontier orbital reactivity descriptors in the early 1950s [12–14]. The Fukui function and its twin, the local softness [15].

$$s(\mathbf{r}) = \left(\frac{\partial\rho(\mathbf{r})}{\partial\mu}\right)_{v(\mathbf{r})} = \left(\frac{\partial N}{\partial\mu}\right)_{v(\mathbf{r})} f(\mathbf{r}) = \frac{f(\mathbf{r})}{\eta},$$
$$\propto f(\mathbf{r}) \tag{18.16}$$

provide a DFT-based approach to many of the chemical phenomenon that are normally explained using frontier orbitals.

18.4   A BIT MORE THAN "JUST FMO THEORY"

Based on the foregoing discussion, one might suppose that the Fukui function is nothing more than a DFT-inspired restatement of frontier molecular orbital (FMO) theory. This is not quite true. Because DFT is, in principle, exact, the Fukui function includes effects—notably electron correlation and orbital relaxation—that are a priori neglected in an FMO approach. This is most clear when the electron density is expressed in terms of the occupied Kohn–Sham spin-orbitals [16],

$$\rho_{v,N}(\mathbf{r}) = \sum_{i=1}^{N}\left|\phi_{v,N}^{(i)}(\mathbf{r})\right|^2. \tag{18.17}$$

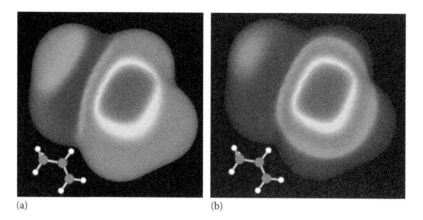

(a)   (b)

FIGURE 18.1 (See color insert following page 302.)   Propylene is susceptible to electrophilic attack on the double bond. This can be deduced by plotting (a) the Fukui function from below, $f^-(\mathbf{r})$, or (b) the HOMO density, $\rho^{\text{HOMO}}(\mathbf{r})$, on the van der Waals' surface of the molecule.

Differentiating this expression with respect to the number of electrons gives a frontier orbital approximation in Equation 18.13 plus a correct due to orbital relaxation [17,18],

$$f_{v,N}^+(\mathbf{r}) = \left|\phi_{v,N}^{(N+1)}(\mathbf{r})\right|^2 + \sum_{i=1}^{N}\left(\frac{\partial\left|\phi_{v,N}^{(i)}(\mathbf{r})\right|^2}{\partial N}\right)_{v(\mathbf{r})}$$

$$f_{v,N}^-(\mathbf{r}) = \left|\phi_{v,N}^{(N)}(\mathbf{r})\right|^2 + \sum_{i=1}^{N}\left(\frac{\partial\left|\phi_{v,N}^{(i)}(\mathbf{r})\right|^2}{\partial N}\right)_{v(\mathbf{r})}$$

(18.18)

In most cases, the orbital relaxation contribution is negligible and the Fukui function and the FMO reactivity indicators give the same results. For example, the Fukui functions and the FMO densities both predict that electrophilic attack on propylene occurs on the double bond (Figure 18.1) and that nucleophilic attack on BF_3 occurs at the Boron center (Figure 18.2). The rare cases where orbital relaxation effects are nonnegligible are precisely the cases where the Fukui functions should be preferred over the FMO reactivity indicators [19–22]. In short, while FMO theory is based on orbitals from an independent electron approximation like Hartree–Fock or Kohn–Sham, the Fukui function is based on the true many-electron density.

18.5   CONDENSED FUKUI FUNCTIONS

In chemistry, one is rarely interested in which "point" in a molecule is most reactive; rather one wishes to identify the atom in a molecule is most likely to react with an attacking electrophile or nucleophiles. This suggests that a coarse-grained atom-by-atom representation of the Fukui function would suffice for chemical purposes. Such a representation is called a condensed reactivity indicator [23].

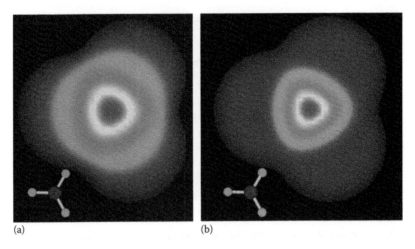

(a)   (b)

FIGURE 18.2 (See color insert following page 302.) BF_3 is susceptible to nucleophilic attack at the boron site. This can be deduced by plotting (a) the Fukui function from above, $f^+(\mathbf{r})$, or (b) the LUMO density, $\rho^{LUMO}(\mathbf{r})$, on the van der Waals' surface of the molecule.

To coarse-grain the Fukui function, consider an atomic partition of unity,

$$1 = \sum_\alpha w^{(\alpha)}(\mathbf{r}); \quad 0 \le w^{(\alpha)}(\mathbf{r}). \tag{18.19}$$

$w^{(\alpha)}(\mathbf{r})$ represents the fraction of the electron density (or any other molecular property density) at $\mathbf{r}$ that can be associated with atom α. If the atomic partition of unity is inserted inside the derivative that defines the Fukui function, one obtains

$$f_{v,N}^\pm(\mathbf{r}) = \left(\frac{\partial \sum_\alpha w^{(\alpha)}(\mathbf{r})\rho(\mathbf{r})}{\partial N}\right)_{v(\mathbf{r})}^\pm = \sum_\alpha \left(\frac{\partial w^{(\alpha)}(\mathbf{r})\rho(\mathbf{r})}{\partial N}\right)_{v(\mathbf{r})}^\pm$$

$$= \sum_\alpha \left(\frac{\partial \rho^{(\alpha)}(\mathbf{r})}{\partial N}\right)_{v(\mathbf{r})}^\pm \tag{18.20}$$

where $\rho^{(\alpha)}(\mathbf{r}) = w^{(\alpha)}(\mathbf{r})\rho(\mathbf{r})$ is the density of the atom in the molecule (AIM). Equation 18.20 suffices to decompose the Fukui function into a sum of atomic contributions; integrating over each of these contributions gives an expression for the condensed Fukui functions in terms of the atomic populations [23],

$$f_{v,N}^{+,\alpha} = \int \left(\frac{\partial \rho^{(\alpha)}(\mathbf{r})}{\partial N}\right)_{v(\mathbf{r})}^+ d\mathbf{r} = \left(\frac{\partial p^{(\alpha)}}{\partial N}\right)_{v(\mathbf{r})}^+ = p_{v,N+1}^{(\alpha)} - p_{v,N}^{(\alpha)} \tag{18.21}$$

$$f_{v,N}^{-,\alpha} = \int \left(\frac{\partial \rho^{(\alpha)}(\mathbf{r})}{\partial N}\right)_{v(\mathbf{r})}^- d\mathbf{r} = \left(\frac{\partial p^{(\alpha)}}{\partial N}\right)_{v(\mathbf{r})}^- = p_{v,N}^{(\alpha)} - p_{v,N-1}^{(\alpha)} \tag{18.22}$$

or, more conveniently, in terms of the atomic charges,

$$f_{v,N}^{+,\alpha} = q_{v,N}^{(\alpha)} - q_{v,N+1}^{(\alpha)} \tag{18.23}$$

$$f_{v,N}^{-,\alpha} = q_{v,N-1}^{(\alpha)} - q_{v,N}^{(\alpha)}. \tag{18.24}$$

One advantage of this "response of molecular fragment" approach [24] to condensed Fukui functions is that Equations 18.21 through 18.24 are easily evaluated from the population analysis data that accompanies the output of most quantum chemistry calculations.

An alternative construction, termed "fragment of molecular response" [24], inserts the atomic partition of unity outside the derivative,

$$f_{v,N}^{\pm}(\mathbf{r}) = \sum_{\alpha} w^{\alpha}(\mathbf{r}) \left(\frac{\partial \rho(\mathbf{r})}{\partial N} \right)_{v(\mathbf{r})}^{\pm} = \sum_{\alpha} w^{(\alpha)}(\mathbf{r}) f_{v,N}^{\pm}(\mathbf{r}), \tag{18.25}$$

and gives rise to an alterative, and generally inequivalent, expression for the condensed Fukui functions [25],

$$f_{v,N}^{\pm,\alpha} = \int w^{(\alpha)}(\mathbf{r}) f_{v,N}^{\pm}(\mathbf{r}) d\mathbf{r}. \tag{18.26}$$

There seems to be no mathematical reason to favor one formulation of the condensed Fukui function over the other.

The condensed Fukui functions for propylene and BF_3 are given below.

Atom	C1	C2	C3	H	H	H	H	H	H
$f_{propylene}^{-,\alpha}$	0.2796	0.2159	0.0718	0.0883	0.0732	0.0387	0.0732	0.0829	0.0763
Atom	B	F	F	F					
$f_{BF_3}^{+,\alpha}$	0.6049	0.1316	0.1316	0.1319					

For propylene, the condensed Fukui function not only predicts that the electrophilic attack occurs on one of the doubly bonded carbon atoms, but it also predicts that there is a preference for the terminal carbon, in accordance with Markonikov's rule. Nucleophilic attack is predicted at the boron atom in BF_3.

In these examples, the condensed Fukui functions were computed using Hirshfeld population analysis [26], which is unique among the commonly employed population analysis methods, because the same results are obtained from the "response of molecular fragment" and the "fragment of molecular response" approaches [24]. There are other arguments in favor of the Hirshfeld scheme too [27,28], many of them based on the tendency for the atom-condensed Hirshfeld Fukui functions to be nonnegative [25,29,30]. Nonetheless, condensed Fukui functions maybe computed using any population analysis method: common methods include the Mulliken [31–34], Bader [35,36], Voronoi, Hirshfeld-I [37,38], and electrostatic fitting approaches [39–41].

18.6 COMPUTATIONAL CONSIDERATIONS

There are four strategies for computing the Fukui function. The first is to use the result of an orbital-based method (Hartree–Fock or Kohn–Sham DFT) on the N-electron system to evaluate the FMO approximation in Equations 18.13 and 18.14. This approach cannot be recommended (it neglects the effects of orbital relaxation), but neither should it be denigrated (it is one of the easiest ways to compute the Fukui function, and it is usually effective). The second approach uses single-point calculations to calculate the density of the $N - 1$ and $N + 1$ electron systems and then computes the Fukui function using the equations for the Fukui function from exact theory (Equations 18.11 and 18.12). This approach gives accurate results, but it is logically inconsistent because it presupposes Equation 18.9, which is usually not true in approximate calculations (e.g., Kohn–Sham DFT with an approximate exchange-correlation functional, Hartree–Fock, or MP2) [42–45]. The third approach features mathematical constructions that allow one to exactly [46] or approximately [47] compute the derivatives in Equations 18.3 and 18.4. This approach is conceptually satisfying and generally reliable, but it is more difficult (it requires computer programming to implement) and computationally expensive. A fourth approach, which has been proposed but not yet implemented, would compute Fukui functions at the ab initio level using electron propagator theory [48,49] or the closely related extended Koopmans' theorem [50–54]. These approaches are computational costly and difficult to implement, but potentially valuable for benchmarking.

For small molecules, we recommend using the second approach. The computational cost of both the first and the second approaches is dominated by the geometry optimization on the N-electron species, so the computational cost of additional single-point calculations on the $N - 1$ and $N + 1$ electron species is negligible. In addition, even though most of the exchange-correlation functionals used in Kohn–Sham DFT calculations overstabilize systems with fractional numbers of electrons and thus do not satisfy Equations 18.8 and 18.9, these functionals are comparatively accurate for calculations with integer numbers of electrons. For this reason, it is better to use the $\Delta N = \pm 1$ formulas from the exact theory (Equations 18.11 and 18.12) than it is to use the infinitesimal change dN formulas.

For larger molecules, one is caught between the localization error of Hartree–Fock (which tends overstabilize subsystems with integer numbers of electrons, leading to Fukui functions that are localized on one part of the system) and the delocalization error of most approximate DFT functionals (which tend to overstabilize subsystems with fractional numbers of electrons, and thus favor delocalizing the Fukui function over the entire molecule) [55,56]. Since accurate ab initio calculations are computationally impracticable for large systems, we advise using a Kohn–Sham DFT calculation with an exchange-correlation functional (e.g., MCY3 [57]), that was designed to behave correctly for systems with fractional numbers of electrons. When such a functional is used, the second ($\Delta N = \pm 1$) and third ($\Delta N = \pm dN$) computational approaches will give equally reliable results.

For neutral molecules, the $\Delta N = +1$ electron calculation requires computing the energy of an anion, and it is possible that the additional electron is unbound. When that occurs, the molecule has zero electron affinity, and it is such a poor nucleophiles that the Fukui function from above, $f_{v,N}^+(\mathbf{r})$, does not exist. Some molecules with unbound anions have negative electron affinities associated with metastable resonances embedded in the continuum. In such cases, the response of the (negative) electron affinity to changes in external potential provides an ansatz for computing $f_{v,N}^+(\mathbf{r})$, though it is not clear exactly how the resulting Fukui function should be interpreted [58].

18.7 WHY DOES THE FUKUI FUNCTION WORK?

Why—and when—does the Fukui function work? The first restriction—already noted in the original 1984 paper—is that the Fukui function predicts favorable interactions between molecules that are far apart. This can be understood because when one uses the perturbation expansion about the separated reagent limit to approximate the interaction energy between reagents, one of the terms that arises is the Coulomb interaction between the Fukui functions of the electron–donor and the electron–acceptor [59,60],

$$J_f[AB] = -(\Delta N)^2 \iint \frac{f_A^+(\mathbf{r})f_B^-(\mathbf{r}')}{|\mathbf{r} - \mathbf{r}'|} d\mathbf{r}\,d\mathbf{r}' \tag{18.27}$$

This term can only control regioselectivity if the transition state occurs relatively early along the reaction path (so that the asymptotic expansion about the separated reagent limit is still relevant) and if the extent of electron-transfer is large compared to the electrostatic interactions between the reagents. The importance of Equation 18.27 for explaining the utility and scope of the Fukui function was first noted by Berkowitz in 1987 [59].

The utility of the Fukui function for predicting chemical reactivity can also be described using the variational principle for the Fukui function [61,62]. The Fukui function from the above discussion, $f_{v,N}^+(\mathbf{r})$, represents the "best" way to add an infinitesimal fraction of an electron to a system in the sense that the electron density $\rho_{v,N}(\mathbf{r}) + \varepsilon f_{v,N}^+(\mathbf{r})$ has lower energy than any other $N + \varepsilon$-electron density for this system. A Lewis base (also known as reducing agent, nucleophile, etc.) will attack the system in the place where it is most able to accept additional electrons. Thus, the Lewis base will attack the system where $f_{v,N}^+(\mathbf{r})$ is the largest. Similarly, the Fukui function below, $f_{v,N}^-(\mathbf{r})$ is the "least bad" way to remove an infinitesimal fraction of an electron from a system. A Lewis acid (also called oxidizing agent, electrophiles, etc.) will attack the system in the place where it is most willing to donate electrons. Thus the Lewis acid will attack the system where $f_{v,N}^-(\mathbf{r})$ is the largest. Although the mathematical details for this explanation of the Fukui function's predictive power are relatively modern (from 2000) [62], the basic ideas described in this paragraph date back to the first papers on the Fukui function [1,17].

To this point, we have always discussed the Fukui function as a reactivity indicator for reactions where electron-transfer effects make the dominant energetic contribution and thus control the reactivity. Electron transfer tends to be the dominant effect when a soft acid reacts with a soft base; such reactions are classified by Klopman as being "orbital-controlled" [63]. When a hard acid reacts with a hard base, the reaction is usually "charge controlled." In such reactions, electrostatic effects dominate. Nonetheless, there has been recent controversy about whether the Fukui function can be used as a reactivity indicator for reactions between hard acids and bases. Li and Evans argue that the Fukui function can be used, and that hard acids and hard bases tend to react at the site where the Fukui function is the *smallest* [64]. Melin et al. criticize this conclusion by pointing out hard–hard interactions tend to be electrostatically controlled, so that the best regioselectivity indicator for such reactions is not the Fukui function, but the electrostatic potential [65]. Finally, a detailed mathematical analysis was performed by Anderson et al. [60]. That analysis finds the middle ground between these two interpretations. The electrostatic potential is the dominant reactivity indicator for hard–hard interactions, but when there are two molecular sites that are electrostatically similar, the Fukui function enters as a "tie breaker" indicator. In most cases, the site with the largest Fukui function will be the most reactive. However, for reactions between extremely hard reagents (i.e., in the limit of strong electrostatic control), the site with the minimum Fukui function is preferred.

18.8 PROSPECTS

The Fukui function is certainly among the most fundamental and useful reactivity indicators in chemical DFT, and it will certainly continue to be widely used in routine applications. It would be interesting to see what might be learned from applying the Fukui function to problems beyond conventional small-molecule chemistry. For example, will the Fukui function reveal something about enzyme–ligand docking and enzymatic catalysis that current, electrostatically focused, methods miss? There are some favorable preliminary results along those lines [66–68].

On the theoretical side, however, there is room for skepticism: the rate of progress since the Berkowitz's 1987 paper has been astonishingly slow. One issue with the Fukui function—and with chemical DFT in general—is that there is rarely any reason to favor the Fukui function over the corresponding FMO descriptors. Given that most chemists find FMO theory more familiar and easier to use than the Fukui function, one may reasonably question the practical utility of the Fukui function. The Fukui function sometimes works even when FMO theory fails. Indeed, there are some chemical phenomena that FMO is intrinsically incapable of explaining. For example, there has been significant recent interest in redox-induced electron rearrangements (RIER), where *oxidization* of a metal complex induces the *reduction* of one or more metal centers (or vice versa) [69]. This phenomenon cannot be explained by FMO theory because when one removes an electron from the HOMO, the electron density decreases everywhere including the metal centers that are known, from experiment, to be reduced. By contrast, this phenomenon is readily explained by the Fukui function: the key insight is that orbital relaxation effects cause the Fukui function on the metal centers to be negative [22]. In fact, RIER can be considered to be a "success story" for the Fukui function: a theoretical study of the Fukui function predicted the existence of RIER even before the phenomenon was experimentally observed [21], and the current "best explanation" of RIER is based on studies of negative Fukui functions [21,22,25]. Nonetheless, if one wishes to make a compelling argument favoring Fukui functions over FMOs, one needs to identify more chemical phenomena where the Fukui function works but FMO theory fails.

A theoretical obstacle that prevents broader application of the Fukui function was alluded to in Section 18.6: How does one know whether a given reaction is electron-transfer controlled (so that the Fukui function is the relevant reactivity indicator), electrostatically controlled (so that the electrostatic potential is the relevant reactivity indicator), or somewhere in between? Most practitioners in chemical DFT have developed an intuition for "what indicator works when," but there do not seem to be any objective criteria for deciding which reactivity indicator to use. (Indeed, selecting the right indicator sometimes requires trial and error.) Transforming chemical DFT from a descriptive theory to a predictive theory requires developing tools for discerning when the Fukui function is the relevant reactivity indicator, and when something else is. This problem may not be insolvable because, unlike FMO theory, chemical DFT has an elegant and simple mathematical structure that is conducive to the systematic derivation of "reactivity rules" for chemical reactions.

ACKNOWLEDGMENTS

P.W.A. acknowledges the financial support from NSERC, the Canada Research Chairs, and Sharcnet; W.Y. acknowledges financial support from NSF; L.J.B. acknowledges financial support from RENCI@ECU.

REFERENCES

1. Parr, R. G. and Yang, W. T. *J. Am. Chem. Soc.* 1984, *106*, 4049.
2. Perdew, J. P., Parr, R. G., Levy, M., and Balduz, J. L., Jr. *Phys. Rev. Lett.* 1982, *49*, 1691.
3. Yang, W. T., Zhang, Y. K., and Ayers, P.W. *Phys. Rev. Lett.* 2000, *84*, 5172.
4. Ayers, P. W. *J. Math. Chem.* 2008, *43*, 285.
5. Ayers, P. W. and Levy, M. *Theor. Chem. Acc.* 2000, *103*, 353.
6. Parr, R. G., Donnelly, R. A., Levy, M., and Palke, W. E. *J. Chem. Phys.* 1978, *68*, 3801.
7. Sanderson, R. T. *Science* 1951, *114*, 670.
8. Nalewajski, R. F. and Parr, R. G. *J. Chem. Phys.* 1982, *77*, 399.
9. Zhang, Y. K. and Yang, W. T. *Theor. Chem. Acc.* 2000, *103*, 346.
10. Gyftopoulos, E. P. and Hatsopoulos, G. N. *Proc. Natl. Acad. Sci. U.S.A.* 1965, *60*, 786.
11. Parr, R. G. and Pearson, R. G. *J. Am. Chem. Soc.* 1983, *105*, 7512.
12. Fukui, K., Yonezawa, T., and Shingu, H. *J. Chem. Phys.* 1952, *20*, 722.
13. Fukui, K., Yonezawa, T., and Nagata, C. *J. Chem. Phys.* 1953, *21*, 174.
14. Fukui, K., Yonezawa, T., and Nagata, C. *Bull. Chem. Soc. Japan* 1954, *27*, 423.
15. Yang, W. T. and Parr, R. G. *Proc. Natl. Acad. Sci. U.S.A.* 1985, *82*, 6723.
16. Kohn, W. and Sham, L. J. *Phys. Rev.* 1965, *140*, A1133.
17. Yang, W. T., Parr, R. G., and Pucci, R. *J. Chem. Phys.* 1984, *81*, 2862.

18. Cohen, M. H. and Ganduglia-Pirovano, M. V. *J. Chem. Phys.* 1994, *101*, 8988.
19. Langenaeker, W., Demel, K., and Geerlings, P. *Theochem—J, Molec. Struct.* 1991, *80*, 329.
20. Bartolotti, L. J. and Ayers, P. W. *J. Phys. Chem. A* 2005, *109*, 1146.
21. Ayers, P. W. *PCCP* 2006, *8*, 3387.
22. Melin, J., Ayers, P. W., and Ortiz, J. V. *J. Phys. Chem. A* 2007, *111*, 10017.
23. Yang, W. T. and Mortier, W. J. *J. Am. Chem. Soc.* 1986, *108*, 5708.
24. Bultinck, P. F. S.; Van Alsenoy, C., Ayers, P. W., and Carbo-Dorca, R. *J. Chem. Phys.* 2007, *127*, 034102.
25. Ayers, P. W., Morrison, R. C., and Roy, R. K. *J. Chem. Phys.* 2002, *116*, 8731.
26. Hirshfeld, F. L. *Theor. Chim. Act.* 1977, *44*, 129.
27. Nalewajski, R. F. and Parr, R. G. *Proc. Natl. Acad. Sci.* U.S.A. 2000, *97*, 8879.
28. Ayers, P. W. *J. Chem. Phys.* 2000, *113*, 10886.
29. Roy, R. K., Pal, S., and Hirao, K. *J. Chem. Phys.* 1999, *110*, 8236.
30. Roy, R. K., Hirao, K., and Pal, S. *J. Chem. Phys.* 2000, *113*, 1372.
31. Mulliken, R. S. *J. Chem. Phys.* 1955, *23*, 1833.
32. Mulliken, R. S. *J. Chem. Phys.* 1955, *23*, 1841.
33. Mulliken, R. S. *J. Chem. Phys.* 1955, *23*, 2343.
34. Mulliken, R. S. *J. Chem. Phys.* 1955, *23*, 2338.
35. Bader, R. F. W., Nguyendang, T. T., and Tal, Y. *J. Chem. Phys.* 1979, *70*, 4316.
36. Bader, R. F. W. *Atoms in Molecules: A Quantum Theory*, Clarendon, Oxford, 1990.
37. Bultinck, P., Ayers, P. W., Fias, S., Tiels, K., and Van Alsenoy, C. *Chem. Phys. Lett.* 2007, *444*, 205.
38. Bultinck, P., Van Alsenoy, C., Ayers, P. W., and Carbo-Dorca, R. *J. Chem. Phys.* 2007, *126*, 144111.
39. Singh, U. C. and Kollman, P. A. *J. Comput. Chem.* 1984, *5*, 129.
40. Besler, B. H., Merz, K. M., and Kollman, P. A. *J. Comput. Chem.* 1990, *11*, 431.
41. Breneman, C. M. and Wiberg, K. B. *J. Comput. Chem.* 1990, *11*, 361.
42. Mori-Sanchez, P., Cohen, A. J., and Yang, W. T. *J. Chem. Phys.* 2006, *125*, 201102.
43. Ruzsinszky, A., Perdew, J. P., Csonka, G. I., Vydrov, O. A., and Scuseria, G. E. *J. Chem. Phys.* 2007, *126*, 104102.
44. Mandado, M., Van Alsenoy, C., Geerlings, P., De Proft, F., and Mosquera, R. A. *Chemphyschem* 2006, *7*, 1294.
45. Cohen, M. H. and Wasserman, A. *Isr. J. Chem.* 2003, *43*, 219.
46. Ayers, P. W., De Proft, F., Borgoo, A., and Geerlings, P. *J. Chem. Phys.* 2007, *126*, 224107.
47. Michalak, A., De Proft, F., Geerlings, P., and Nalewajski, R. F. *J. Phys. Chem. A* 1999, *103*, 762.
48. Ohrn, Y. and Born, G. *Adv. Quantum Chem.* 1981, *13*, 1.
49. Melin, J., Ayers, P. W., and Ortiz, J. V. *J. Chem. Sci.* 2005, *117*, 387.
50. Day, O. W., Smith, D. W., and Morrison, R. C. *J. Chem. Phys.* 1975, *62*, 115.
51. Smith, D. W. and Day, O. W. *J. Chem. Phys.* 1975, *62*, 113.
52. Ellenbogen, J. C., Day, O. W., Smith, D.W., and Morrison, R. C. *J. Chem. Phys.* 1977, *66*, 4795.
53. Morrell, M. M., Parr, R. G., and Levy, M. *J. Chem. Phys.* 1975, *62*, 549.
54. Ayers, P. W. and Melin, J. *Theor. Chem. Acc.* 2007, *117*, 371.
55. Mori-Sanchez, P., Cohen, A. J., and Yang, W. T. *Phys. Rev. Lett.* 2008, *100*, 146401.
56. Cohen, A. J., Mori-Sanchez, P., and Yang, W. T. *Phys. Rev. B* 2008, *77*, 115123.
57. Cohen, A. J., Mori-Sanchez, P., and Yang, W. T. *J. Chem. Phys.* 2007, *126*, 191109.
58. Tozer, D. J. and De Proft, F. *J. Chem. Phys.* 2007, *127*, 034108.
59. Berkowitz, M. *J. Am. Chem. Soc.* 1987, *109*, 4823.
60. Anderson, J. S. M., Melin, J., and Ayers, P. W. *J. Chem. Theor. Comput.* 2007, *3*, 358.

61. Chattaraj, P. K., Cedillo, A., and Parr, R. G. *J. Chem. Phys.* 1995, *103*, 7645.
62. Ayers, P. W. and Parr, R. G. *J. Am. Chem. Soc.* 2000, *122*, 2010.
63. Klopman, G. *J. Am. Chem. Soc.* 1968, *90*, 223.
64. Li, Y. and Evans, J. N. S. *J. Am. Chem. Soc.* 1995, *117*, 7756.
65. Melin, J., Aparicio, F., Subramanian, V., Galvan, M., and Chattaraj, P. K. *J. Phys. Chem. A* 2004, *108*, 2487.
66. Roos, G., Loverix, S., De Proft, F., Wyns, L., and Geerlings, P. *J. Phys. Chem. A* 2003, *107*, 6828.
67. Roos, G., Loverix, S., Brosens, E., Van Belle, K., Wyns, L., Geerlings, P., and Messens, *J. Chembiochem* 2006, 7, 981.
68. Beck, M. E. *J. Chem Inf. Model.* 2005, *45*, 273.
69. Min, K. S., DiPasquale, A. G., Golen, J. A., Rheingold, A. L., and Miller, J. S. *J. Am. Chem. Soc.* 2007, *129*, 2360.

19 Shape Function

Paul W. Ayers and Andrés Cedillo

CONTENTS

19.1   WHAT IS THE SHAPE FUNCTION?

The shape function, denoted as $\sigma(\mathbf{r})$, is defined as the electron density per particle,

$$\sigma(\mathbf{r}) = \frac{\rho(\mathbf{r})}{N}, \tag{19.1}$$

where $\rho(\mathbf{r})$ is the electron density and

$$N = \int \rho(\mathbf{r})d\mathbf{r} \tag{19.2}$$

is the total number of electrons in the system. The shape function is the one-electron probability distribution—the probability of observing a specific electron (say, the third) somewhere in the system. The shape function captures information about the relative abundance of the electrons from place to place; on its surface, it does not contain any information about the total number of electrons. The aptly named shape function allows one to consider the shape ($\sigma(\mathbf{r})$) and size (N) of the electron density separately.

     This chapter reviews the historical development of the shape function concept from its origins in the work of Parr and Bartolotti [1] to the present day. The recent review by Geerlings, De Proft, and one of the present authors provides an alternative perspective [2].

19.2 ORIGINS

The shape function had a role in theoretical chemistry and physics long before it was named by Parr and Bartolotti. For example, in x-ray measurements of the electron density, what one actually measures is the shape function—the relative abundance of electrons at different locations in the molecule. Determining the actual electron density requires calibration to a standard with known electron density. On the theoretical side, the shape function appears early in the history of Thomas–Fermi theory. For example, the Majorana–Fermi–Amaldi approximation to the exchange potential is just [3,4]

$$v_x^{MFA}(\mathbf{r}) = -\int \frac{\sigma(\mathbf{r}')}{|\mathbf{r} - \mathbf{r}'|} d\mathbf{r}'. \tag{19.3}$$

In the context of density functional theory (DFT), the shape function can be considered to be the fundamental variable in the Levy-constrained search [5],

$$F[\rho] = F_N[\sigma] = \min_{\sigma(\mathbf{r}_1)=\iint \cdots \int |\Psi(\mathbf{r}_1,\mathbf{r}_2,\dots\mathbf{r}_N)|^2 d\mathbf{r}_2 d\mathbf{r}_3\dots d\mathbf{r}_N} \langle \Psi | \hat{T} + V_{ee} | \Psi \rangle. \tag{19.4}$$

The contribution of Parr and Bartolotti is not diminished by these precedents; they were the first to recognize that $\sigma(\mathbf{r})$ is a quantity of interest in its own right, separate from the electron density [1]. They also deserve credit for coining the name, "shape function," which captures the essence of the quantity and provides an essential verbal handle that facilitated future work.

The paper of Parr and Bartolotti is prescient in many ways [1]. It defines the shape function and describes its meaning. It notes the previously stated link to Levy's constrained search. It establishes the importance of the shape function in resolving "ambiguous" functional derivatives in the DFT approach to chemical reactivity—the subdiscipline of DFT that Parr has recently begun to call "chemical DFT" [6–9]. Indeed, until the recent resurgence of interest in the shape function, the Parr–Bartolotti paper was usually cited because of its elegant and incisive analysis of the electronic chemical potential [10],

$$\mu = \left(\frac{\delta \mathcal{E}_v[\rho[v; N]]}{\delta \rho(\mathbf{r})} \right) = \left(\frac{\partial E[v; N]}{\partial N} \right)_{v(\mathbf{r})}, \tag{19.5}$$

where

$$\mathcal{E}_v[\rho] = F[\rho] + \int \rho(\mathbf{r})v(\mathbf{r})d\mathbf{r} \tag{19.6}$$

is the variational energy functional of Hohenberg and Kohn [11], and $\rho[v; N, \mathbf{r}]$ and

$$E[v; N] = \min_{\substack{\rho(\mathbf{r})\geq 0 \\ \langle\rho(\mathbf{r})\rangle=N}} \mathcal{E}_v[\rho] \tag{19.7}$$

are the ground-state electron density and the ground-state electronic energy, respectively, for N-electrons bound by the external potential $v(\mathbf{r})$. In the DFT variational principle, Equation 19.7, the only variations of the electron density that are allowed are those that preserve the normalization. This raises the issue of how one should deal with the constrained variational derivative,

$$\left(\frac{\delta\mathcal{E}_v[\rho[v; N]]}{\delta\rho(\mathbf{r})}\right)_N = \left(\frac{\delta\mathcal{E}_v[\rho[v; N]]}{\delta(N\sigma(\mathbf{r}))}\right)_N = \frac{1}{N}\left(\frac{\delta\mathcal{E}_v[N\cdot\sigma[v; N]]}{\delta\sigma(\mathbf{r})}\right)_N. \tag{19.8}$$

In particular, is it possible to determine the chemical potential (which obviously depends on how the energy responds to variations in the number of electrons) from the variation of the electron density at fixed electron number? Parr and Bartolotti show that this is not possible: the derivatives in Equation 19.8 are equal to an arbitrary constant and thus ill defined. One has to remove the restriction on the functional derivative to determine the chemical potential. Therefore, the fluctuations of the electron density that are used in the variational method are insufficient to determine the chemical potential.

Notice that how the shape function naturally enters this discussion. Because the number of electrons is fixed, the variational procedure for the electron density is actually a variational procedure for the shape function. So it is simpler to restate the equations associated with the variational principle in terms of the shape function. Parr and Bartolotti have done this, and note that because the normalization of the shape function is fixed,

$$1 = \int \sigma(\mathbf{r})d\mathbf{r}, \tag{19.9}$$

functional derivatives with respect to the shape function are determined only up to an additive constant,

$$\left(\frac{\delta X[\sigma; N]}{\delta\sigma(\mathbf{r})}\right)_N = N\left(\frac{\delta X[\rho]}{\delta\rho(\mathbf{r})}\right) + \text{constant}. \tag{19.10}$$

This constant is often chosen to be zero.

One oft-overlooked facet of the Parr–Bartolotti paper is its mathematical treatment of constrained functional derivatives. The problem of constrained functional derivatives [2,12–16] arises repeatedly in DFT—often in the exactly the same "number conserving" context considered by Parr and Bartolotti—but their work is rarely cited in that context. Much of the recent work on the shape function is related to its importance for evaluating the constrained functional derivatives associated with the DFT variational principle [13–15].

19.3 RESURGENCE: APPLICATIONS TO CHEMICAL DFT

After its introduction by Parr and Bartolotti, the shape function concept lay dormant for about a decade. It made occasional appearances; for example, the shape function makes a cameo in the definition of the local hardness in terms of the hardness kernel [17–19],

$$\eta(\mathbf{r}) = \int \sigma(\mathbf{r}')\eta(\mathbf{r}, \mathbf{r}')d\mathbf{r}'. \tag{19.11}$$

But there was no significant further development of the ideas set forth by Parr and Bartolotti until 1994, when one of the present authors (A.C.) realized that the shape function provided the key to resolve a mathematical difficulty that is inherent in the different Legendre-transform representations of chemical DFT [20].

The use of the electron density and the number of electrons as a set of independent variables, in contrast to the canonical set, namely, the external potential and the number of electrons, is based on a series of papers by Nalewajski [21,22]. A.C. realized that this choice is problematic because one cannot change the number of electrons while the electron density remains constant. After several attempts, he found that the energy per particle possesses the convexity properties that are required by the Legendre transformations. When the Legendre transform was performed on the energy per particle, the shape function immediately appeared as the conjugate variable to the external potential, so that the electron density was split into two pieces that can be varied independent: the number of electrons and their distribution in space.

The Legendre transform is a mathematical "trick" that provides four distinct but equivalent ways to understand changes in electronic structure [21,22]. The first distinction is based on whether one examines the system from an electron-following perspective (where the external potential changes, and the electron density adapts to those changes) or an electron-preceding perspective (where the electron density changes, and the external potential responds to those changes) [23–26]. The electron-following perspective is most conventional, and it is called the canonical representation. The electron-preceding perspective is called the isomorphic representation. The second distinction is based on whether one can determine the number of electrons in the system (e.g., an isolated molecule) or whether the number of electrons in the system fluctuates due to the presence of a reservoir of electrons with a specified chemical potential (e.g., molecules in solution, macroscopic samples of condensed matter). When the number of electrons is fixed, the system is said to be closed; when the number of electrons fluctuates, it is said to be open. The latter representations are associated with the grand canonical ensemble (open system; electron-following perspective) or the grand isomorphic ensemble (open system; electron-preceding perspective), in analogous to the corresponding ensembles in classical statistical mechanics.

Just as in classical statistical mechanics, the different pictures of electronic changes are related by Legendre transforms. The state function for closed systems in the electron-following picture is just the electronic ground-state energy, $E[v;N]$. The total differential for the energy provides reactivity indicators for describing how various perturbations stabilize or destabilize the system,

$$dE = \left(\frac{\partial E}{\partial N}\right)_{v(\mathbf{r})} dN + \int \left(\frac{\delta E}{\delta v(\mathbf{r})}\right)_N \delta v(\mathbf{r})d\mathbf{r}$$
$$= \mu dN + \int \rho(\mathbf{r})\delta v(\mathbf{r}). \tag{19.12}$$

Moving from the closed-electron-following picture (canonical ensemble) to the open-electron-following picture (grand canonical ensemble) is done by Legendre transform, generating the new state function:

$$\Omega[v; \mu] = E[v; N] - \left(\frac{\partial E}{\partial N}\right)_{v(\mathbf{r})} N = E - \mu N. \tag{19.13}$$

The total differential of the grand potential is

$$d\Omega = -N d\mu + \int \rho(\mathbf{r}) \delta v(\mathbf{r}) d\mathbf{r}. \tag{19.14}$$

The closed-electron-preceding picture (isomorphic ensemble) requires a Legendre transform to eliminate the external potential as a variable,

$$
\begin{aligned}
F[\rho; N] &= E[v; N] - \int \left(\frac{\delta E}{\delta v(\mathbf{r})}\right)_N v(\mathbf{r}) d\mathbf{r} \\
&= E[v; N] - \int \rho(\mathbf{r}) v(\mathbf{r}) d\mathbf{r}.
\end{aligned} \tag{19.15}
$$

As previously mentioned, this expression is conceptually clumsy because the number of electrons is a function of the electron density through Equation 19.2. As one cannot vary the number of electrons if the electron density is fixed, the naïve expression for the total differential

$$dF = \left(\frac{\partial F}{\partial N}\right)_{\rho(\mathbf{r})} dN + \int \left(\frac{\delta F}{\delta \rho(\mathbf{r})}\right)_N \delta\rho(\mathbf{r}) d\mathbf{r} \tag{19.16}$$

is nonsensical. Notice, however, that the second term in Equation 19.16 is precisely the sort of constrained functional derivative that Parr and Bartolotti expressed in terms of the shape function (cf. Equation 19.8). This suggests rewriting the total differential in terms of the shape function [27]:

$$dF = \left(\frac{\partial F}{\partial N}\right)_{\sigma(\mathbf{r})} dN + \int \left(\frac{\delta F}{\delta \sigma(\mathbf{r})}\right)_N \delta\sigma(\mathbf{r}) d\mathbf{r}. \tag{19.17}$$

The preceding expression is mathematically correct and conceptually useful: it separates energetic effects by increasing the number of electrons (dN) and of polarizing the electron density ($\delta\sigma(\mathbf{r})$).

Equation 19.17 is not the original way the $\{N, \rho(\mathbf{r})\}$ ambiguity was resolved [20]. As mentioned previously, the original paper on the shape function in the isomorphic representation performed the Legendre transform on the energy per particle. This gives an intensive, per electron, state function [20]:

$$\phi[\sigma; N] = \frac{F[N\sigma(\mathbf{r})]}{N} = \frac{E[v; N]}{N} - \int \sigma(\mathbf{r}) v(\mathbf{r}) d\mathbf{r} \tag{19.18}$$

for the isomorphic ensemble. Most recent work, however, is based on Equation 19.17.

The open-electron-preceding picture (grand isomorphic ensemble) is similarly obtained. One has the state function:

$$R[\rho; \mu] = \Omega[v; \mu] - \int \rho(\mathbf{r})v(\mathbf{r})d\mathbf{r}$$

$$= F[\rho; N] - \mu N \qquad (19.19)$$

with the total differential [27]:

$$dR = \left(\frac{\partial R}{\partial \mu}\right)_{\sigma(\mathbf{r})} d\mu + \int \left(\frac{\delta R}{\delta \sigma(\mathbf{r})}\right)_{\mu} \delta \sigma(\mathbf{r})d\mathbf{r}. \qquad (19.20)$$

Equations 19.17 and 19.20 provided the foundation for further progress on the shape function-based perspective on chemical DFT. The first extension, by Baekelandt et al. [27], laid out the mathematical structure associated with these new pictures and introduced new reactivity indicators. This paper reveals that the isomorphic ensemble provides a particularly useful approach to the "hardness" picture of chemical reactivity, and allows one to define a local hardness indicator,

$$h(\mathbf{r}) = \frac{1}{N} \left(\frac{\delta \mu}{\delta \sigma(\mathbf{r})}\right)_N, \qquad (19.21)$$

that is less ambiguous than the conventional Ghosh–Berkowitz–Parr form [17,18]. Referring to Equation 19.10, it is clear that the definition in Equation 19.21 is unique except for an arbitrary constant shift. The equivalence class of acceptable forms for the conventional definition of the local hardness, $\eta(\mathbf{r}) = (\delta \mu/\delta \rho \ (\mathbf{r}))_{v(\mathbf{r})}$, is far larger [12]. After this came a flurry of papers, providing further development of the hardness concept in the isomorphic ensemble [28] and further interpretation for the first-order [29] and higher-order response functions [30].

19.4   REINTERPRETATION: USING THE SHAPE FUNCTION AS THE FUNDAMENTAL DESCRIPTOR

This flurry of interest in the shape function was just beginning to dissipate in when one of the authors (P.W.A.) began graduate school. That June, Weitao Yang organized a satellite symposium on DFT for the 9th International Congress on Quantum Chemistry. Understanding the talks at this symposium required a competence far above his own, and during one of the morning sessions (most likely on June 4, 1997), he had a hypnagogic revelation: maybe you do not need the electron density; maybe the shape function has enough information in it all by itself. After the session, P.W.A. excitedly (and rudely) interrupted Prof. Robert Parr (who was talking to Prof. John Pople, though P.W.A. did not know this) during the stroll to lunch and expounded on this idea. Prof. Parr's response was to politely introduce P.W.A. to Prof. Pople and query "Can you prove it?" Which he could not.

But the proof is deceptively simple [31]. Because the shape function is proportional to the electron density, it inherits the characteristic electron-nuclear coalescence cusps at the positions of the atomic nuclei [32,33]. The location of those cusps determines the positions of the nuclei, $\mathbf{R}_\alpha$; the "steepness" of the cusps determines the atomic charges, Z_α. So the shape function determines the external potential for any molecular system [31].

It is more surprising that the shape function also determines the number of electrons. However, the number of electrons can be deduced from the asymptotic decay of the shape function [31,34–40],

$$\sigma(\mathbf{r}) \sim r^{2\beta} e^{-2\alpha r},$$

$$\alpha = \sqrt{2IP},$$

$$\beta = \frac{\sum_\alpha Z_\alpha - N + 1}{\alpha} - 1.$$

(19.22)

For a molecule, or more generally, any electronic system with a Coulombic external potential:

$$v(\mathbf{r}) = \sum_\alpha \frac{-q_\alpha}{|\mathbf{r} - \mathbf{R}_\alpha|},$$

(19.23)

where $\sigma(\mathbf{r})$ determines both $v(\mathbf{r})$ and N, and through them all properties of the system [31].

The preceding theorem falls well short of the Hohenberg–Kohn theorem because it is restricted to Coulombic external potentials. The theorem is not true for all external potentials. In fact, for any Coulombic system, there always exists a one-electron system, with external potential,

$$v(\mathbf{r}) = \frac{\nabla^2 \sqrt{\sigma(\mathbf{r})}}{2\sqrt{\sigma(\mathbf{r})}},$$

(19.24)

that has the same ground-state $\sigma(\mathbf{r})$. The theorem can be extended to excited states using the excited-state cusp conditions that are directly analogous to the ones used by Nagy in her DFT theory for single excited states [31,41–43]. Like the density, the shape function is also analytic except at the location of the atomic nuclei. This means that if one does not need to know the entire shape function: if one knows the shape function in any connected region with finite volume, this is sufficient to determine the shape function for the entire molecule [2,44]. This holographic shape function theorem is derived in precisely the same way as its analogue for the electron density [45–47].

Like the first Hohenberg–Kohn theorem, the preceding theorems are "existence" theorems; they say that "the shape function is enough" but they do not provide any guidance for evaluating properties based on the shape function alone. Once one knows that shape functionals exist, however, there are systematic ways to construct them using, for example, the moment expansion technique [48–51]. For atomic systems, moment-based shape functions have been developed for the number of electrons [31], the Kohn–Sham kinetic energy (T_s) [52], the exchange energy (E_x) [52], and the ionization potential [53]. The approximate functional for the number of electrons not only provides numerical evidence for the sufficiency of the shape function in property calculations, but also demonstrates that approximate shape functions (which generally do not have the correct asymptotic decay) are still adequate for approximating molecular properties. The shape functionals for the kinetic and exchange energies are key components of a variational shape function principle for determining the ground-state energy [54]. In these cases, however, the shape functionals are less accurate than the corresponding density functionals. Since the computational cost of evaluating shape functionals and density functionals is similar (the limiting step in both cases is the evaluation of numerical integrals), there would seem to be a little reason to prefer "shape-functional theory" to DFT. However, the shape functional for the ionization potential is more accurate than the analogous density functional [53]. Based on this result, it seems that the shape function is a preferable descriptor to the electron density when one is attempting to model periodic properties, or other atomic or molecular properties that do not depend strongly on the number of electrons. On the other hand, when computing properties that grow in tandem to the number of electrons (N, T_s, E_x, etc.), the electron density is preferable. A more detailed analysis about when shape functionals are superior or inferior to their density-functional analogues may be found in Ref. [53]. As a rule of thumb, shape functionals are better for "chemical" properties (i.e., properties that depend strongly on the column in the periodic table) while density functionals are better for "physical" properties (properties that depend strongly on the number of electrons).

19.5   SHAPE FUNCTION AS A DESCRIPTOR OF ATOMIC AND MOLECULAR SIMILARITY

Because of its utility for describing the chemical properties of systems, the shape function has proved to be very useful for studies of atomic [55–58] and molecular similarity [54,59–61]. For example, the Carbó indicator of molecular similarity is in fact a shape functional [59]:

$$Z_{AB}^{\sigma} = \frac{\int \sigma_A(\mathbf{r})\sigma_B(\mathbf{r})d\mathbf{r}}{\sqrt{\int \sigma_A^2(\mathbf{r})d\mathbf{r}}\sqrt{\int \sigma_B^2(\mathbf{r})d\mathbf{r}}}. \tag{19.25}$$

This similarity indicator, in fact, precedes Parr and Bartolotti's introduction of the "shape function" terminology [59]. In general, it seems that the shape function is preferred to the electron density as a descriptor of molecular similarity whenever one is interested in chemical similarity. Similarity measures that use the electron density will typically predict that "fluorine resembles chlorine less than it resembles sodium, oxygen, or neon" using the shape function helps one to avoid conflating similarity of electron number with chemical similarity [53,57].

The shape function is also used to measure the similarity between an atom in a molecule (AIM) and an isolated atom or ion [62,63],

$$I\left[\rho_A | \rho_A^0\right] = \int \rho_A(\mathbf{r}) \ln\left(\frac{\rho_A(\mathbf{r})}{\rho_A^0(\mathbf{r})}\right) d\mathbf{r}$$
$$= N_A \left(\ln\left(\frac{N_A}{N_A^0}\right) + \int \sigma_A(\mathbf{r}) \ln\left(\frac{\sigma_A(\mathbf{r})}{\sigma_A^0(\mathbf{r})}\right) d\mathbf{r}\right). \qquad (19.26)$$

The first term in this expression is an "entropy of mixing term" related to electron transfer; the second term is the information loss due to polarization of the AIM. Minimizing the information loss per atom results in the Hirshfeld population analysis [64,65] and many other results in the broad field of chemical information theory [26,66–75]. Zeroing the entropy of mixing term by choosing a reference ion that has the same number of electrons as the AIM, one obtains the Hirshfeld-I population analysis [76,77].

19.6 PROSPECTS

Whither the shape function? The shape function concept is unquestionably durable, and so the main question is whether the shape function will remain an incidental component of other theories or whether, eventually, the shape-functional theory will emerge as an important field of inquiry in its own right. The essential theoretical foundations for a chemical shape-functional theory are already well established at this stage: one has total differentials that show how key state functions respond to changes in the number of electrons and polarization of the electron density (cf. Section 19.3), and one has the fundamental existence theorem which states that all properties of molecule can be deduced from knowledge of its state function (cf. Section 19.4). But there is little current progress in these directions. Further progress will almost certainly require computational rather than theoretical developments, because the theory cannot progress much further without the concrete guidance of computation provides. One nominee for such computational scrutiny would be the shape-Fukui function [27,44]:

$$f_\sigma^\pm(\mathbf{r}) = \left(\frac{\partial \sigma(\mathbf{r})}{\partial N}\right)_{v(\mathbf{r})} = \frac{f^\pm(\mathbf{r}) - \sigma(\mathbf{r})}{N}. \qquad (19.27)$$

This quantity is trivially computed from the Fukui function, $f^\pm(\mathbf{r})$ [78–80], and the shape function, and it has a simple interpretation: the shape Fukui function measures where the relative abundance of electrons increases or decreases when electrons are added to (or removed from) a system. In our experience, plotting $f_\sigma^\pm(\mathbf{r})$ often provides a simpler and easier way to interpret picture of chemical reactivity than the Fukui function itself. Perhaps this is because $\sigma(\mathbf{r})$ is the local density approximation (LDA) to the Fukui function [81]. Since the numerator in Equation 19.27, $f^\pm(\mathbf{r}) - \sigma(\mathbf{r})$, is the "post–LDA correction" to the Fukui function [81], the shape

Fukui function represents the deviation of the system's response to adding or subtracting electrons from electron gas behavior.

Currently, the most exciting work on the shape function is being done in the fields of atomic and molecular similarity. One certainly expects the shape function to continually appear as those fields continue to progress. However, the role of the shape function in those fields is presently incidental: the shape function appears, but underlying theory plays no essential role. It would be interesting to develop molecular similarity measures that exploit the insights from the shape-functional pictures of electronic changes.

ACKNOWLEDGMENTS

P.W.A. acknowledges NSERC, Sharcnet, and the Canada Research Chairs. A.C. acknowledges CONACYT grant 49057-F.

REFERENCES

1. Parr, R. G. and Bartolotti, L. J. *J. Phys. Chem.* 1983, *87*, 2810.
2. Geerlings, P., De Proft, F., and Ayers, P. W. Theoretical aspects of Chemical Reactivity, Toro-Labbé, A. (Ed.), Elsevier, Amsterdam, 2007.
3. Majorana, E. *Il Nuovo Cimento* 1929, *6*, xiv.
4. Fermi, E. and Amaldi, E. *Accad. Ital. Rome* 1934, *6*, 117.
5. Levy, M. *Proc. Natl. Acad. Sci. U.S.A.* 1979, *76*, 6062.
6. Parr, R. G. and Yang, W. *Density-Functional Theory of Atoms and Molecules*, Oxford University Press, New York, 1989.
7. Geerlings, P., De Proft, F., and Langenaeker, W. *Chem. Rev.* 2003, *103*, 1793.
8. Ayers, P. W., Anderson, J. S. M., and Bartolotti, L. J. *Int. J. Quantum Chem.* 2005, *101*, 520.
9. Chermette, H. *J. Comput. Chem.* 1999, *20*, 129.
10. Parr, R. G., Donnelly, R. A., Levy, M., and Palke, W. E. *J. Chem. Phys.* 1978, *68*, 3801.
11. Hohenberg, P. and Kohn, W. *Phys. Rev.* 1964, *136*, B864.
12. Harbola, M. K., Chattaraj, P. K., and Parr, R. G. *Isr. J. Chem.* 1991, *31*, 395.
13. Gal, T. *Phys. Rev. A* 2001, *6302*.
14. Gal, T. *J. Math. Chem.* 2007, *42*, 661.
15. Gal, T. *J. Phys. A—Math. Gen.* 2002, *35*, 5899.
16. Luo, J. *J. Phys. A—Math. Gen.* 2006, *39*, 9767.
17. Berkowitz, M., Ghosh, S. K., and Parr, R. G. *J. Am. Chem. Soc.* 1985, *107*, 6811.
18. Ghosh, S. K. and Berkowitz, M. *J. Chem. Phys.* 1985, *83*, 2976.
19. Berkowitz, M. and Parr, R. G. *J. Chem. Phys.* 1988, *88*, 2554.
20. Cedillo, A. *Int. J. Quantum Chem.* 1994, 231.
21. Nalewajski, R. F. and Parr, R. G. *J. Chem. Phys.* 1982, *77*, 399.
22. Nalewajski, R. F. *J. Chem. Phys.* 1983, *78*, 6112.
23. Nakatsuji, H. *J. Am. Chem. Soc.* 1974, *96*, 24.
24. Nakatsuji, H. *J. Am. Chem. Soc.* 1974, *96*, 30.
25. Nakatsuji, H. *J. Am. Chem. Soc.* 1973, *95*, 345.
26. Nalewajski, R. F. *Adv. Quantum Chem.* 2006, *51*, 235.
27. Baekelandt, B. G., Cedillo, A., and Parr, R. G. *J. Chem. Phys.* 1995, *103*, 8548.
28. De Proft, F., Liu, S. B., and Parr, R. G. *J. Chem. Phys.* 1997, *107*, 3000.
29. De Proft, F. and Geerlings, P. *J. Phys. Chem. A* 1997, *101*, 5344.
30. Liu, S. B. and Parr, R. G. *J. Chem. Phys.* 1997, *106*, 5578.

31. Ayers, P. W. *Proc. Natl. Acad. Sci.* U.S.A. 2000, *97*, 1959.
32. Kato, T. *Commun. Pure Appl. Math.* 1957, *10*, 151.
33. Steiner, E. *J. Chem. Phys.* 1963, *39*, 2365.
34. Morrell, M. M., Parr, R. G., and Levy, M. *J. Chem. Phys.* 1975, *62*, 549.
35. Katriel, J. and Davidson, E. R. *Proc. Natl. Acad. Sci.* U.S.A. 1980, *77*, 4403.
36. Levy, M., Perdew, J. P., and Sahni, V. *Phys. Rev. A* 1984, *30*, 2745.
37. Almbladh, C. O. and Von Barth, U. *Phys. Rev. B* 1985, *31*, 3231.
38. Hoffmann-Ostenhof, M. and Hoffmann-Ostenhof, T. *Phys. Rev. A* 1977, *16* 1782.
39. Ahlrichs, R., Hoffmann-Ostenhof, M., Hoffmann-Ostenhof, T., and Morgan, J. D., III *Phys. Rev. A* 1981, *23*, 2106.
40. Patil, S. H. *J. Phys. B: At., Mol. Opt. Phys.* 1989, *22*, 2051.
41. Nagy, A. *Int. J. Quantum Chem.* 1998, *70*, 681.
42. Nagy, A. and Sen, K. D. *Chem. Phys. Lett.* 2000, *332*, 154.
43. Nagy, A. and Sen, K. D. *J. Phys. B* 2000, *33*, 1745.
44. De Proft, F., Ayers, P. W., Sen, K. D., and Geerlings, P. *J. Chem. Phys.* 2004, *120*, 9969.
45. Riess, J. and Munch, W. *Theor. Chim. Act.* 1981, *58*, 295.
46. Mezey, P. G. *Mol. Phys.* 1999, *96*, 169.
47. Bader, R. F. W. and Becker, P. *Chem. Phys. Lett.* 1988, *148*, 452.
48. Liu, S., Nagy, A., and Parr, R. G. *Phys. Rev. A* 1999, *59*, 1131.
49. Nagy, A., Liu, S. B., and Parr, R. G. *Phys. Rev. A* 1999, *59*, 3349.
50. Liu, S. B., De Proft, F., Nagy, A., and Parr, R. G. *Adv. Quantum Chem.* 2000, *36*, 77.
51. Ayers, P. W., Lucks, J. B., and Parr, R. G. *Acta Chim. Phys. Debricina* 2002, *34–35*, 223.
52. Ayers, P. W. *Phys. Rev. A* 2005, *71*, 062506.
53. Ayers, P. W., De Proft, F., and Geerlings, P. *Phys. Rev. A* 2007, *75*, 012508.
54. Bultinck, P. and Carbo-Dorca, R. *J. Math. Chem.* 2004, *36*, 191.
55. Borgoo, A., De Proft, F., Geerlings, P., and Sen, K. D. *Chem. Phys. Lett.* 2007, *444*, 186.
56. Borgoo, A., Godefroid, M., Indelicato, P., De Proft, F., and Geerlings, P. *J. Chem. Phys.* 2007, *126*, 044102.
57. Borgoo, A., Godefroid, M., Sen, K. D., De Proft, F., and Geerlings, P. *Chem. Phys. Lett.* 2004, *399*, 363.
58. Sen, K. D., De Proft, F., Borgoo, A., and Geerlings, P. *Chem. Phys. Lett.* 2005, *410*, 70.
59. Carbo, R., Leyda, L., and Arnau, M. *Int. J. Quantum Chem.* 1980, *17*, 1185.
60. Bultinck, P. and Carbo-Dorca, R. *J. Chem. Sci.* 2005, *117*, 425.
61. Bultinck, P., Girones, X., and Carbo-Dorca, R. Reviews in computational chemistry, Volume 21, Lipkowitz, K. B., Larter, R. and Cundari, T. R. (Eds.), Wiley, New York, 2005.
62. Ayers, P. W. *Theor. Chem. Acc.* 2006, *115*, 370.
63. Parr, R. G., Ayers, P. W., and Nalewajski, R. F. *J. Phys. Chem. A* 2005, *109*, 3957.
64. Nalewajski, R. F. and Parr, R. G. *Proc. Natl. Acad. Sci.* U.S.A. 2000, *97*, 8879.
65. Nalewajski, R. F. and Parr, R. G. *J. Phys. Chem. A* 2001, *105*, 7391.
66. Nalewajski, R. F. and Switka, E. *PCCP* 2002, *4*, 4952.
67. Nalewajski, R. F. *PCCP* 2002, *4*, 1710.
68. Nalewajski, R. F., Switka, E., and Michalak, A. *Int. J. Quantum Chem.* 2002, *87*, 198.
69. Nalewajski, R. F. *J. Phys. Chem. A* 2003, *107*, 3792.
70. Nalewajski, R. F. *Chem. Phys. Lett.* 2003, *372*, 28.
71. Nalewajski, R. F. and Broniatowska, E. *J. Phys. Chem. A* 2003, *107*, 6270.
72. Nalewajski, R. F. and Broniatowska, E. *Chem. Phys. Lett.* 2003, *376*, 33.
73. Nalewajski, R. F. *Adv. Quantum Chem.* 2003, *43*, 119.
74. Nalewajski, R. F. *Struct. Chem.* 2004, *15*, 391.
75. Nalewajski, R. F. *Mol. Phys.* 2005, *103*, 451.
76. Bultinck, P., Van Alsenoy, C., Ayers, P. W., and Carbo-Dorca, R. *J. Chem. Phys.* 2007, *126*, 144111.

77. Bultinck, P., Ayers, P. W., Fias, S., Tiels, K., and Van Alsenoy, C. *Chem. Phys. Lett.* 2007, *444*, 205.
78. Ayers, P. W. and Levy, M. *Theor. Chem. Acc.* 2000, *103*, 353.
79. Parr, R. G. and Yang, W. *J. Am. Chem. Soc.* 1984, *106*, 4049.
80. Yang, W., Parr, R. G., and Pucci, R. *J. Chem. Phys.* 1984, *81*, 2862.
81. Chattaraj, P. K., Cedillo, A., and Parr, R. G. *J. Chem. Phys.* 1995, *103*, 10621.

20 An Introduction to the Electron Localization Function

P. Fuentealba, D. Guerra, and A. Savin

CONTENTS

"Of course, perhaps Lewis put it the right way..."

Agatha Christie
The Hollow

The role of quantum theory in chemistry has a history of almost 100 years, and the advances have been important. Nowadays, it is possible to do quantitative predictions with chemical accuracy for middle-size molecules, and some type of calculations, especially density functional-based methodologies, are routinely done in many chemical laboratories. One very important aspect on the influence of quantum theory in chemistry is the one of understanding. There are many chemical concepts which can be understood only through the laws of quantum mechanics. This chapter is about conceptual understanding and is not about the other very important issue of computing with chemical accuracy.

One of the most important models to understand chemistry is the electron pair concept of Lewis [1]. He put forward the model where in an atom or molecule "...each pair of electrons has a tendency to be drawn together." This very important model, which is at the very beginning of any general chemistry textbook, has however an important problem. It goes against the Coulomb's law. Lewis noticed it and he went on further to even affirm that perhaps "...Coulomb's law of inverse squares must fail at small distances." This remarkable fact is to our knowledge never discussed in the textbooks. At that time Lewis had no knowledge of the development of quantum mechanics, and already in the 1930s, he retracted this statement [2]. Now, the explanation of why the electrons have a tendency to be drawn together, even against the Coulomb's law, is found in the Pauli exclusion principle and the influence of the kinetic energy. The Pauli exclusion principle is not only the reason for this tendency but also for the existence of the periodic table of elements. Hence, it is the Pauli exclusion principle that makes chemistry as we know it.

Let us start with the Schrödinger equation

$$\mathcal{H}\Psi = E\Psi \tag{20.1}$$

Of course, the Coulomb interaction appears in the Hamiltonian operator, $\mathcal{H}$, and is often invoked for interpreting the chemical bond. However, the wave function, Ψ, must be antisymmetric, i.e., must satisfy the Pauli exclusion principle, and it is the only fact which explains the Lewis model of an electron pair. It is known that all the information is contained in the square of the wave function, $|\Psi|^2$, but it is in general much complicated to be analyzed as such because it depends on too many variables. However, there have been some attempts [3]. Lennard-Jones [4] proposed to look at a quantity which should keep the chemical significance and nevertheless reduce the dimensionality. This simpler quantity is the reduced second-order density matrix

$$P_2(x_1, x_2) = \frac{N(N-1)}{2} \int dx_3 \cdots dx_N |\Psi(x_1, x_2, x_3, \ldots, x_N)|^2 \tag{20.2}$$

which depends only on three spatial coordinates, $\vec{r}$, plus spin, σ, for each of the electrons of the pair (x stands for the couple $\vec{r}\sigma$). Hence, $P_2(\vec{r}_1\sigma_1, \vec{r}_2\sigma_2)$ times an infinitesimal volume element squared is interpreted as the probability to find one electron with spin σ_1 in a volume element around $\vec{r}_1$, and another electron with spin σ_2 in a volume element around $\vec{r}_2$. The prefactor comes from the fact that electrons are indistinguishable. Notice the analogy with the one-particle density $\rho(x)$,

$$\rho(x) = N \int dx_2 \cdots dx_N |\Psi(x_1, x_2, x_3, \ldots, x_N)|^2 \tag{20.3}$$

which is related to the probability to find a particle with spin σ around $\vec{r}$.

It is evident that Lennard-Jones was following the track opened by Lewis, by concentrating on the pair of electrons. To get some insight into $P_2(x_1, x_2)$, it is natural to start with the simplest antisymmetric wave function, a Slater determinant constructed by real orbitals. In this case, one obtains

$$P_{2,\text{det}}(x_1, x_2) = \frac{1}{2}\left[\rho_{\sigma_1}(\vec{r}_1)\rho_{\sigma_2}(\vec{r}_2) - \delta_{\sigma_1,\sigma_2}\gamma_{\sigma_1}(\vec{r}_1, \vec{r}_2)^2\right] \tag{20.4}$$

where ρ_σ is the σ-spin component of the electron density:

$$\rho_\sigma(\vec{r}) = \gamma_\sigma(\vec{r}, \vec{r}) \tag{20.5}$$

and

$$\gamma_\sigma(\vec{r}_1, \vec{r}_2) = \sum \phi_i(\vec{r}, \sigma)\phi_i(\vec{r}', \sigma) \tag{20.6}$$

where the ϕ_i are the spin orbitals making the Slater determinant. To understand the features of $P_{2,\text{det}}$, it is useful to consider that the ϕ_i are localized. There is no loss of generality as $P_{2,\text{det}}$ is invariant with respect to rotations among the orbitals. A further simplification makes it particularly easy to see how $P_{2,\text{det}}$ behaves. Imagine that the space can be divided into regions Ω_i, such that the localized orbitals ϕ_i satisfy the following relationship:

$$\phi_i(\vec{r}, \sigma) = \begin{cases} \sqrt{\rho_\sigma(\vec{r})} & \text{for } \vec{r} \in \Omega_i \\ 0 & \text{for } \vec{r} \notin \Omega_i \end{cases} \tag{20.7}$$

In this case

$$\gamma_\sigma(\vec{r}, \vec{r}') = \begin{cases} \sqrt{\rho_\sigma(\vec{r})\rho_\sigma(\vec{r}')} & \text{for } \vec{r}, \vec{r}' \in \Omega_i \\ 0 & \text{otherwise} \end{cases} \tag{20.8}$$

Now, using Equation 20.4, one can construct the reduced second-order density matrix. For $\sigma_1 \neq \sigma_2$, $P_{2,\text{det}}$ is quite boring:

$$P_{2,\text{det}}(\vec{r}_1\sigma_1, \vec{r}_2\sigma_2) = \frac{1}{2}\rho_{\sigma_1}(\vec{r}_1)\rho_{\sigma_2}(\vec{r}_2) \tag{20.9}$$

Hence, the probability of finding one electron with spin α around $\vec{r}_1$ and another with spin β around $\vec{r}_2$ is just the product of the probabilities of finding one particle in the given positions. The probabilities are independent, which means that the behavior of electrons is not correlated. However, for the case of two electrons with the same spin:

$$P_{2,\text{det}}(\vec{r}_1\sigma, \vec{r}_2\sigma) = \begin{cases} 0 & \text{for } \vec{r}, \vec{r}' \in \Omega_i \\ \frac{1}{2}\rho_\sigma(\vec{r})\rho_\sigma(\vec{r}_2) & \text{otherwise} \end{cases} \tag{20.10}$$

the probability changes dramatically when one of $\vec{r}_1$, $\vec{r}_2$ share the same Ω or not. Imagine exploring space with a probe electron in $\vec{r}_1$. P_2 is zero for a whole region $\vec{r}_2 \in \Omega_i$, for all $\vec{r}_1 \in \Omega_i$. The moment $\vec{r}_1$ leaves this region, all probabilities change because a new region is defined. Thus, the electrons with spin α or β partition the space into regions Ω_i. For a closed shell system, the picture is even more simple, as localized orbitals are the same for both spins, the regions Ω_i will be the same for both spins. In such a case, each region is occupied by a pair of electrons, one with α-spin and the other with β-spin. Moreover, each electron with spin α or β "excludes" another electron with the same spin from that region. This is the ultimate explanation for the Lewis electron pair model. There is not a new attractive force between a pair of electrons. It is just a repulsion between electrons of the same spin due to the Pauli exclusion principle, which explains the electron pair model of Lewis.

Let us now consider a simple example: four noninteracting fermions in a one-dimensional box, $x \in [0, \pi]$. The wave function is a Slater determinant with two doubly occupied orbitals:

$$\phi_k(x) = \sqrt{\frac{2}{\pi}}\sin(kx), \quad k = 1, 2 \tag{20.11}$$

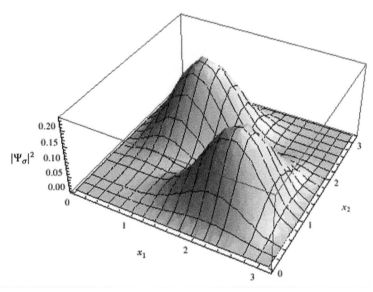

FIGURE 20.1 (See color insert following page 302.)   Plot of $|\Psi_\sigma|^2$, $\sigma = \alpha$, β for four independent fermions in a box.

To analyze $|\Psi|^2$, it is sufficient to consider $|\Psi_\alpha(x_1, x_2)|^2$ and $|\Psi_\beta(x_1, x_2)|^2$, where the index indicates the spin. As there are only two particles of a given spin, moving in one dimension, it is possible to plot $|\Psi_\sigma|^2$, as shown in Figure 20.1 [5].

One finds two maxima independent of the spin: one electron around $x \approx 0.955$ and the other one around $x \approx 2.186$. It is easier to understand the origin of the maxima by considering localized orbitals, $\Psi = \frac{1}{\sqrt{2}}(\phi_1 \pm \phi_2)$. Both couples of orbitals are shown in Figure 20.2, and in Figure 20.3 the one-particle density is shown.

Remember that the square of the wave function, or any of the reduced density matrices, are independent of a unitary transformation of the orbitals. Hence, any pair of orbitals is as good as the other. However, the chemical picture of molecular orbitals is easily understood for most of the chemists. In this case, it is easier looking

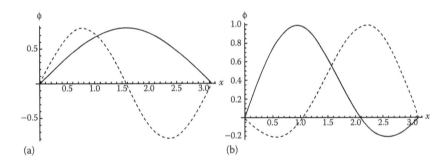

FIGURE 20.2   (a) Two lowest energy orbitals, ϕ_1, ϕ_2, for particles in a box and (b) two localized orbitals, for four independent fermions in a box.

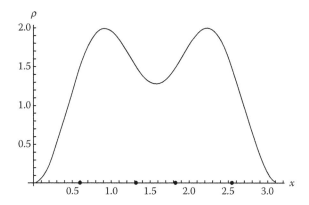

FIGURE 20.3 One-particle density for four noninteracting fermions in a one-dimensional box. The dots on the abscissa show different positions in which the density has the same value.

at the localized orbitals. The plot of $|\Psi_\sigma|^2$ shows that the maxima are located close to the maximum of each orbital, and taking into account that each orbital is occupied by two electrons, one of each spin, the two maxima show us the regions where is most probable to find an electron pair in the sense of Lewis.

Let us now look at the one particle density and compare it with the pair density $P_2(x_1\sigma, x_2\sigma')$. We have to examine now the two possibilities, both electrons with the same spin or with different spin. In Figure 20.4, we have the pair density for the

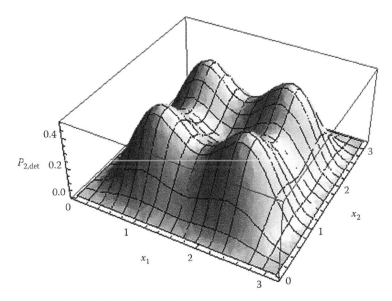

FIGURE 20.4 (See color insert following page 302.) $P_{2,\text{det}}(x_1, x_2)$ for fermions in a box, for $\sigma_1 = \alpha$ and $\sigma_2 = \beta$.

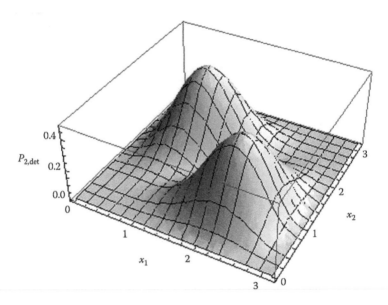

FIGURE 20.5 (See color insert following page 302.) $P_{2,\text{det}}(x_1, x_2)$ for fermions in a box, for $\sigma_1 = \sigma_2 = \alpha$.

case of different spin. It is clear that its structure corresponds to one of one-particle density because the probabilities are not correlated. However, for the case of the two electrons with the same spin, the picture looks different (Figure 20.5).

Now, the probability of finding both electrons with the same spin around the same point, $x_1 = x_2$, has vanished, as the consequence of the Pauli principle. The box has been partitioned into its left and its right parts. As long as one particle with one spin is in one part, it will impose the other electron of the same spin to be in the other half of the box. For example, we can consider the points chosen among those marked by dots on the abscissa on the Figure 20.3 showing the one-particle density. While for different spin, P_2 will be the same for any pair of positions selected; this is not the case when both electrons have the same spin. When the spin is the same, and the electrons are on the points marked on the same half of the box, P_2 will be very small. However, if one point belongs to those on the left, the other to those of the right, P_2 is as large as for electrons with different spin.

Although the reduction of dimensionality by reducing $|\Psi|^2$ to P_2 is, in general, enormous, having to work in six dimensions is still difficult for the human mind, and probably not needed for the analysis of the chemical bond: we see molecules in three dimensions. A way to further reduce dimensions has been noticed by Luken and Culberson [6] and by Becke and Edgecombe [7]. The idea is simple: as $P_2(\vec{r}_1\sigma, \vec{r}_2\sigma)$ is relatively insensitive as $\vec{r}_1$ moves within a given region Ω_i, but changes suddenly as it moves to another region Ω_j, one could concentrate on the change of P_2. A simple way to look at it is to consider a small sphere, which moves together with the reference electron in r_1 [8]. If the radius of the sphere is R, the probability to find two electrons of spin σ in this small sphere is

$$\int_{\text{small sphere}} P_2(\vec{r}_1\sigma, \vec{r}_2\sigma)\mathrm{d}^3 r_2 = C(\vec{r}_1)\int r_{12}^2\mathrm{d}^3 r_{12} + \cdots \tag{20.12}$$

where we have used the expansion:

$$P_2(\vec{r}_1\sigma, \vec{r}_2\sigma) = C(\vec{r}_1)r_{12}^2 + \cdots \tag{20.13}$$

We see immediately that as long as the sphere stays in one Ω_i, the probability of having two electrons with the same spin in it is very small. When the sphere overlaps with two different Ω_i, the probability increases. How large should the sphere be? It turns out that it is useful to choose the sphere in such a way that the probability of having two electrons with opposite spin is the same, independently of $\vec{r}_1$ [9]. In other words, using the expression of $P_{2,\text{det}}(\vec{r}_1\sigma_1, \vec{r}_2\sigma_2)$ for $\vec{r}_1 = \vec{r}_2$, $\sigma_1 \neq \sigma_2$,

$$\left(\frac{4\pi R^3}{3}\right) \propto 1/\rho(\vec{r}_1) \tag{20.14}$$

where we restricted the formula to the closed shell case. Thus, we get for the quantity of interest,

$$C(\vec{r})\rho(\vec{r})^{-5/3} \tag{20.15}$$

Kohout [10] used this function as an electron localization indicator (ELI). In the electron localization function (ELF), this function is scaled:

$$\eta(\vec{r}) = \left(1 + \kappa(C(\vec{r})\rho(\vec{r})^{-5/3})^2\right)^{-1} \tag{20.16}$$

where the constant κ comes from the proportionality relationship of Equation 20.14, but has now a well-defined value [7]. In this way, the values of ELF range between 0 and 1. A large value of ELF, close to 1, occurs when ELI is small, and it means a region where probability exists for finding an electron pair. A small value of ELF corresponds to a large value of ELI. Referring to our example of four independent particles in a box, one can see both functions in Figure 20.6.

It is clear that both functions are in some way the inverse of each other, but the interpretation is the same, the electron tends to localize at the borders of the box.

The final form of the ELF is

$$\eta(\vec{r}) = \left(1 + \left(\frac{D_P}{D_F}\right)\right)^{-1} \tag{20.17}$$

where

$$D_F = c_F\rho(\vec{r})^{5/3} \tag{20.18}$$

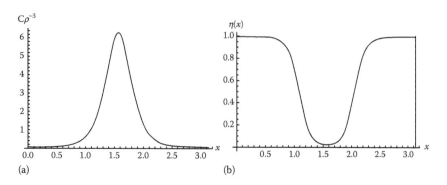

FIGURE 20.6 (a) $C\rho^{-3}$ for four particles in a one-dimensional box and (b) $\eta(x)$ for four particles in a one-dimensional box.

and

$$D_P = \frac{1}{2} \sum_i |\nabla_{\phi_i}|^2 - \frac{1}{8} \frac{|\nabla_\rho|^2}{\rho} \qquad (20.19)$$

the first term on the right-hand side represents the kinetic energy density of the noninteracting system and the second one is the von Weizsäcker kinetic energy density. This is the most important ingredient of the ELF, and the one which allows us a qualitative understanding of the relationship between the ELF and the exclusion principle of Pauli. The von Weizsäcker kinetic energy is the exact kinetic energy functional for a system composed of noninteracting bosons, particles which do not follow the Pauli exclusion principle. It is also exact for the hydrogen atom or any other one-particle system, the most localized system in a rigorous quantum mechanics sense. It is exact for the helium atom or any other two electron system in the Hartree–Fock approximation, the best examples of an electron pair. Hence for all those examples the term D_P will be exactly 0 and the value of the ELF will be 1. After helium atom, the best examples of localized electron pair are surely the most inner electrons of any other molecular or atomic system, i.e., electrons with a configuration very close to $1s^2$, the kinetic energy density of these pair of electron is surely very well approximated by the von Weizsäcker term. Hence in the regions very near to nuclei, the ELF will also have a value close to 1. One can then hypothesize that the von Weizsäcker term will be a very good approximation in all regions where there is a localized electron pair. This is what we found in our example in Figure 20.6b. Hence, the ELF appears to be a measure of the excess of kinetic energy density due to the exclusion principle [11]. The other terms of the function can be thought as a way to map a function, which goes for $-\infty$ to $+\infty$ to a better behaved function, which goes from 0 to 1.

The ELF was proposed by Becke and Edgecombe [7] in 1990 and very soon extensively applied to a variety of systems ranging from atoms to inorganic and organic molecules to solids [9]. In 1994, a topological analysis of the ELF was developed [12], which permits it to perform a more quantitative analysis of the three-dimensional function. From then on, the ELF has been extensively applied to a great amount of systems and has also been used to quantify other chemical concepts like strength of the hydrogen bond [13] and aromaticity [14]. Beside the first review article [9,15] there are other more recent review articles [16] where the reader can find a variety of different applications. In the final part of this article, we concentrate on one simple application of the ELF. Mainly we will analyze the ELF of the series of diatomic molecules of the type E_2 with $E = C$, Si, Ge, Sn, and Pb on their triplet and singlet lowest states.

The atoms of the group 14 present a diverse chemistry. Whereas the first member of the series, carbon atom, is unique in the variety of bonding it forms, the other members change the bonding nature also because of the predominance of relativistic effects, mainly the spin–orbit coupling, in the last member of the series. For instance, the existence of a double bond in the family of molecules of the type $R_2C = CR_2$ is something obvious for any chemistry student. However, for the rest of the series $R_2E = ER_2$ with $E = Si$, Ge, Sn, and Pb is by no way obvious. In fact, the most simple member of the series with $R = H$ does not exist. A nice application of the ELF to understand this type of bonds can be found in Ref. [17]. Within this context, it appears interesting to look at the ELF of the homonuclear dimers E_2 in both the singlet and the triplet states. However, it is necessary to be forewarned that theoretical calculation of the electronic structure of the dimers of the group 14 is a very difficult task. It is one of this situations where the smallest molecules are the most difficult ones. Starting from C_2, which presents an important multideterminantal character of the wave function, making it hard to obtain quantitative results with any Kohn–Sham methodology, and finishing in Pb_2 where the spin–orbit effects are very important, making it hard to obtain quantitative results using any one-component scheme. Therefore, the present analysis is only qualitative in an attempt to interpret the results obtained using a density constructed with Kohn–Sham orbitals calculated using the Stuttgart pseudopotential [18] for all atoms. For the heavier atoms of Ge, Sn, and Pb, the small core pseudopotentials were used. The studied configurations are $\sigma_g^2 \sigma_u^2 \pi_u^1 \pi_u^1 \sigma_g^2$, $^3\Sigma_g$, for the triplets and $\sigma_g^2 \sigma_u^2 \pi_u^2 \pi_u^2$, $^1\Sigma_g$, for the singlet in all molecules. Note that the configuration of the singlet state is unusual. The two σ orbitals do not contribute to the bonding because they are a bonding–antibonding pair adding nothing to the bond order. Therefore, the state is stable only due to the existence of two bonding π-orbitals, and the molecules could be classified as "only π-bonding." This point has interesting consequence in the form of the ELF as will be discussed below.

In Table 20.1, one can see the calculated values of dissociation energy, highest occupied molecular orbital (HOMO)–lowest unoccupied molecular orbital (LUMO) gap, triplet–singlet gap, and bond length for the triplet and singlet states. In all molecules, the triplet state is the lowest in energy in agreement with the experimental evidence. However, as stated above, the values are only a rough estimate of the experimental dissociation energies, which are 6.2, 3.21, 2.65, 2.04, and 0.86 eV, for the dimers of C_2, Si_2, Ge_2, Sn_2, and Pb_2, respectively. The HOMO–LUMO gap is not so small in any case, but the presence of low-lying states is known. It is also interesting to observe that the bond lengths of the singlet states are shorter than

TABLE 20.1

Dissociation Energy, HOMO–LUMO Gap, Triplet–Single Gap, and Bond Length for the Triplet and Singlet States of the Studied Molecules

Dimer	Multiplicity	D_e	Gap$_{H-L}$	Gap$_{T-S}$	Bond Length
C_2	Triplet	5.33	2.55	0.62	1.372
	Singlet	4.71	1.87		1.258
Si_2	Triplet	2.96	1.91	1.10	2.304
	Singlet	1.86	0.77		2.068
Ge_2	Triplet	2.87	1.92	1.22	2.406
	Singlet	1.65	0.60		2.176
Sn_2	Triplet	2.40	1.73	1.22	2.781
	Singlet	1.18	0.42		2.543
Pb_2	Triplet	2.21	1.72	1.29	2.912
	Singlet	0.93	0.33		2.678

Note: All energy values in eV and bond lengths in Å.

the ones of the triplet state, which is the strongest bond. Hence those molecules do not obey the simple rule: the stronger the bond, the shorter the bond length. They do not follow the simple bond order of molecular orbital theory, because all of them present a bond order of 2 in the singlet as well as in the triplet state.

Let us look at the ELF of those molecules in Figure 20.7. Looking at a three-dimensional function presents some ambiguities in the way the function is shown. We have chosen isosurfaces instead of maps of contours, and the value of the isosurface is arbitrary. However, for molecules of the second and third periods, it is generally accepted that any value between 0.7 and 0.9 is good enough to represent the regions where it is most probable to find localized electrons. For the heavier atoms where the d-electrons or f-electrons play an important role, the ELF values are lower and the maxima are not greater than 0.7–0.8. Hence the isosurfaces to show the interesting regions are approximately 0.3–0.5. The small values because of the presence of d-electrons, were first noticed by Kohout and Savin [19] and later on discussed by Burdett and McCornick [20] and Kohout et al. [21]. The colors are also arbitrary. We have chosen red for the isosurfaces representing the core electrons, yellow for the bonding electron pair, blue for the lone pairs, and green for the "not obvious" bonding electrons in the singlet states. For C_2 and Si_2 there are no core basins because of the use of pseudopotential. Let's start analyzing the triplet state. There is no evidence of a triple bond because of the existence of a lone pair behind the atoms (blue regions). The form of the isosurface for the bonding electrons is similar in all studied molecules, but it is getting smaller when one goes down the periodic table. For Pb_2, the isosurface of the basin representing the bond region is so small that it does not have the character-istics of the basins of the other molecules, and the form of the isosurface is more similar to the ones of the singlet than to the other molecules in a triplet state. The comparison from Ge_2 to Pb_2 shows clearly a change in the topology of the basin. It is interesting to note that for the lighter atoms, the HOMO orbital is clearly the double occupied σ_g, whereas for Pb_2, it is the degenerate pair π_u, which are the HOMO in all the singlet states. This could explain the similarity in the form of the localization

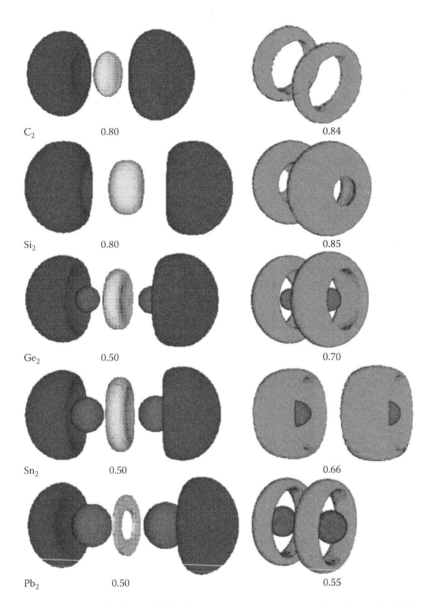

C$_2$ 0.80   0.84

Si$_2$ 0.80   0.85

Ge$_2$ 0.50   0.70

Sn$_2$ 0.50   0.66

Pb$_2$ 0.50   0.55

FIGURE 20.7 (See color insert following page 302.) ELF isosurfaces for triplet (left) and singlet (right) of the different molecules. The values of the isosurfaces are indicated below each picture.

domains. Remember that in the singlet state there is no σ bond. It is only a π bond with a nodal surface through the line connecting both atoms. The basins are cylindrical in shape and of other nature. They represent a pure π bond. One can then speculate that in the singlet state the bond length is shorter but weaker than in the triplet state because it does not have any σ bond.

ACKNOWLEDGMENT

Part of this work has been supported by Fondecyt (Chile), Grant No. 1080184.

REFERENCES

1. G. N. Lewis, *J. Am. Chem. Soc.* 38, 1916, 762.
2. G. N. Lewis, *J. Chem. Phys.* 1, 1933, 17.
3. E. Chamorro, P. Fuentealba, and A. Savin, *J. Comp. Chem.* 24, 2003, 496.
4. J. E. Lennard-Jones, *J. Chem. Phys.* 20, 1952, 1024.
5. S. Wolfram, *The Mathematica Book*, Wolfram Media, Cambridge, 1996.
6. W. L. Luken and J. C. Culberson, *Int. J. Quantum Chem.* 16, 1982, 265.
7. A. D. Becke and K. E. Edgecombe, *J. Chem. Phys.* 92, 1990, 5397.
8. J. F. Dobson, *J. Chem. Phys.* 94, 1991, 4328.
9. A. Savin, R. Nesper, S. Wengert, and Th. F. Faessler, *Angew. Chem.*, 109, 1997, 1892; *Angew. Chem. Int. Ed. Engl.* 36, 1997, 1808.
10. M. Kohout, *Int. J. Quantum Chem.* 97, 2004, 651.
11. A. Savin, 0. Jepsen, J. Flad, O. K. Andersen, H. Preuss, and H. G. von Schnering, *Angew. Chem.* 104, 1992, 187; *Angew Chem. Int. Ed. Engl.* 31, 1992, 187.
12. B. Silvi and A. Savin, *Nature* 371, 1994, 683.
13. F. Fuster and B. Silvi, *Theor. Chem. Acc.* 104, 2000, 13.
14. J. C. Santos, W. Tiznado, R. Contreras, and P. Fuentealba, *J. Chem. Phys.* 120, 2004, 1670; J. C. Santos, J. Andres, A. Aizman, and P. Fuentealba, *J. Chem. Theory Comput.* 1, 2005, 83.
15. B. Silvi, *J. Phys. Chem.* A107, 2003, 3081.
16. P. Fuentealba, E. Chamorro, and J. C. Santos, in *Theoretical Aspects of Chemical Reactivity*, Ed. A. Toro, Elsevier, Amsterdam, 2007; J. Poater, M. Duran, M. Sola, and B. Silvi, *Chem. Rev.* 105, 2005, 3911.
17. H. Grutzmacher and T. F. Fassler, *Chem. Eur. J.* 6, 2000, 2317.
18. B. Metz, H. Stoll, and M. Dolg, *J. Chem. Phys.* 113, 2000, 2563.
19. M. Kohout and A. Savin, *J. Comp. Chem.* 18, 1997, 1431.
20. J. Burdett and T. McCornick, *J. Phys. Chem.* A102, 1998, 6366.
21. M. Kohout, F. Wagner, and Y. Grin, *Theor. Chem. Acc.* 108, 2002, 150.

21 Reaction Force: A Rigorously Defined Approach to Analyzing Chemical and Physical Processes

Alejandro Toro-Labbé, Soledad Gutiérrez-Oliva, Peter Politzer, and Jane S. Murray

CONTENTS

21.1 POTENTIAL ENERGY CURVES AND REACTION COORDINATES

A convenient and effective technique for following the course of a chemical or physical process is by means of a two-dimensional diagram depicting the change in the energy of the system as it proceeds from its initial to final state. This is commonly called a potential energy plot, even though it is the total energy that is being shown, because the latter is typically varying in accordance with the relative positions of the atoms that comprise the system. (In general, potential energy is associated with position, kinetic energy with motion.)

The other axis in such a plot is the "reaction coordinate," which is simply some variable in terms of which the progress of the process can be measured. For a bond dissociation, $XY \rightarrow X + Y$, an obvious choice for the reaction coordinate is the X–Y separation. But in many processes, there maybe several possibilities, some better than others. For example, in the addition of Cl_2 to ethylene, one option

293

might be the C–C distance, as the bond changes from double to single. Another might be one of the H–C–C–H dihedral angles, reflecting the transition of the carbons from planar to tetrahedral configurations.

In a one-step chemical reaction, the reactants and products are typically separated by an energy barrier, the maximum of which corresponds to the transition state; this is in fact a unique structure along the energy profile. A universal reaction coordinate for any such process can be established by using the classical equations of motion to obtain the paths of lowest potential energy leading from the transition structure to the reactants and to the products. When these paths are described in terms of mass-weighted Cartesian coordinates, they represent what is designated as the "intrinsic reaction coordinate" [1,2]. For a multistep reaction, perhaps involving several intermediates and transition states, the same procedure can be followed with respect to each of the latter.

Figure 21.1a is a representative plot of the variation of the potential energy $V(\mathbf{R}_c)$ along the intrinsic reaction coordinate $\mathbf{R}_c$ for a one-step process $A \rightarrow B$. Note that $\mathbf{R}_c$ is treated as a vector, always in the direction from reactants to products. For the reverse process, therefore, $\mathbf{R}_c$ would increase in the opposite direction. $V(\mathbf{R}_c)$ goes through a maximum at the point $|\mathbf{R}_c| = R_c = \beta$, which is the position of the

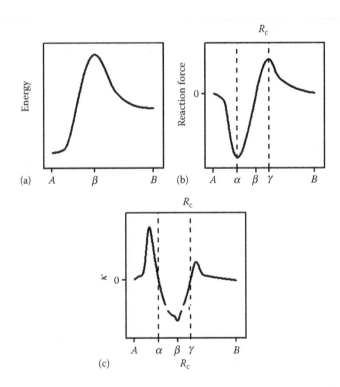

FIGURE 21.1 Profiles of (a) energy, (b) reaction force, and (c) reaction force constant for a generic endoenergetic elementary step. Vertical dashed lines indicate the limits of the reaction regions defined in the text.

transition structure. The energy barrier (activation energy) for the forward reaction, $A \rightarrow B$, is therefore $\Delta E_{act,f} = V(\beta) - V(A)$; for the reverse, $B \rightarrow A$, it would be $\Delta E_{act,r} = V(\beta) - V(B)$. In the example shown, $V(B) > V(A)$, meaning that the forward reaction $A \rightarrow B$ must absorb energy, $\Delta E_f = V(B) - V(A) > 0$. The reverse, $B \rightarrow A$, accordingly releases energy, $\Delta E_r = V(A) - V(B) < 0$. The ensuing discussion will be independent of whether a reaction has $\Delta E > 0$ or $\Delta E < 0$.

21.2   REACTION FORCE

From classical physics, the negative gradient of a potential energy is a force: $-\Delta V(\mathbf{r}) = \mathbf{F}(\mathbf{r})$. It follows that a force can also be associated with a process described by a potential energy $V(\mathbf{R}_c)$:

$$\mathbf{F}(\mathbf{R}_c) = -\frac{\partial V(\mathbf{R}_c)}{\partial \mathbf{R}_c} \qquad (21.1)$$

$\mathbf{F}(\mathbf{R}_c)$ is labeled the "reaction force." For a $V(\mathbf{R}_c)$ profile such as that in Figure 21.1a, $\mathbf{F}(\mathbf{R}_c)$ has the form shown in Figure 21.1b. From A to β, $V(\mathbf{R}_c)$ is increasing and so $\mathbf{F}(\mathbf{R}_c)$ is negative, by Equation 21.1. At $R_c = \alpha$, $V(\mathbf{R}_c)$ has an inflection point, and $\mathbf{F}(\mathbf{R}_c)$ passes through a minimum:

$$\frac{\partial^2 V(\mathbf{R}_c)}{\partial \mathbf{R}_c^2} = 0 = -\frac{\partial \mathbf{F}(\mathbf{R}_c)}{\partial \mathbf{R}_c} \qquad (21.2)$$

At β, $V(\mathbf{R}_c)$ has a maximum, and $\mathbf{F}(\mathbf{R}_c)$ is therefore zero:

$$\frac{\partial V(\mathbf{R}_c)}{\partial \mathbf{R}_c} = 0 = -\mathbf{F}(\mathbf{R}_c) \qquad (21.3)$$

After β, $V(\mathbf{R}_c)$ is decreasing and $\mathbf{F}(\mathbf{R}_c)$ is positive, with a maximum at γ, where $V(\mathbf{R}_c)$ has another inflection point. Since the reactants and products correspond to minima of $V(\mathbf{R}_c)$, $\mathbf{F}(A) = \mathbf{F}(B) = 0$. For the reverse process, $B \rightarrow A$, in which $\mathbf{R}_c$ increases from B to A, $\mathbf{F}(\mathbf{R}_c)$ would be the negative (i.e., mirror image) of that in Figure 21.1b; it would have a minimum at γ and a maximum at α.

The minimum and maximum of the reaction force, in a natural and universal manner, divide any process having $V(\mathbf{R}_c)$ such as that in Figure 21.1a into three reaction regions along the intrinsic reaction coordinate: $A \rightarrow \alpha$, $\alpha \rightarrow \gamma$, and $\gamma \rightarrow B$, as shown in Figure 21.1b. What is the significance of these regions? Our answers to this come out of our experience with a number of chemical reactions and conformational transformations [3–13].

In the first, "preparative" or "reactant" region, from the reactants at A to the force minimum at α, what occurs are primarily structural distortions in the reactants, such as bond stretching, angle bending, rotations, etc. These are a preparation for what is to follow. The resistance of the system (i.e., the reactants) to these changes manifests itself in a negative, retarding reaction force. (Keep in mind that the positive direction is from A to B.) Overcoming this retarding force requires an energy $\Delta E(A \rightarrow \alpha)$, where

$$\Delta E(A \to \alpha) = V(\alpha) - V(A) = -\int_A^\alpha \mathbf{F}(\mathbf{R}_c) \cdot d\mathbf{R}_c \tag{21.4}$$

$\Delta E(A \to \alpha)$ is just the negative of the area under the $\mathbf{F}(\mathbf{R}_c)$ curve between $R_c = A$ and $R_c = \alpha$. At α, the system can be said to consist of distorted states of the reactants.

It is in the second region, from the force minimum at α to its maximum at γ, that the main part of the transition to products takes place, the changes in electronic density distributions and related properties. For example, new bonds may begin to form. These factors produce an increasingly positive driving force that, starting at $R_c = \alpha$, gradually overcomes the retarding one (between α and β); thus the resultant $\mathbf{F}(\mathbf{R}_c)$ becomes less negative. At the transition structure, at $R_c = \beta$, the two opposing forces exactly balance. $V(\mathbf{R}_c)$ has been increasing all the while, and here, it reaches its maximum. The driving force continues to increase between β and γ and is now dominant; $\mathbf{F}(\mathbf{R}_c) > 0$. It reaches its maximum at $R_c = \gamma$. Thus, only the transition to products region is characterized (throughout) by an increasing driving force. The energy associated with this region can be expressed as the sum of the contributions from its two zones, $\alpha \to \beta$ and $\beta \to \gamma$:

$$\Delta E(\alpha \to \gamma) = \Delta E(\alpha \to \beta) + \Delta E(\beta \to \gamma) = -\int_\alpha^\beta \mathbf{F}(\mathbf{R}_c) \cdot d\mathbf{R}_c - \int_\beta^\gamma \mathbf{F}(\mathbf{R}_c) \cdot d\mathbf{R}_c$$

$$\tag{21.5}$$

$\Delta E(\alpha \to \beta)$ is positive, since $\mathbf{F}(\mathbf{R}_c)$ is opposite in direction to $\mathbf{R}_c$ between α and β, while $\Delta E(\beta \to \gamma)$ is negative, because $\mathbf{F}(\mathbf{R}_c)$ is now in the same direction as $\mathbf{R}_c$.

Note that the activation energy is composed of two components, consisting of the energies associated with the reactant region and the first zone of the transition region:

$$\Delta E_{act,f} = \Delta E(A \to \beta) = \Delta E(A \to \alpha) + \Delta E(\alpha \to \beta) \tag{21.6}$$

The significance of this will be discussed later in this chapter.

At $R_c = \gamma$, the system can be described as being in a distorted state of the products. At last, "product" region, $\gamma \to B$, reaction force diminishes as the system relaxes structurally to its final state. An energy $\Delta E(\gamma \to B)$ is released:

$$\Delta E(\gamma \to B) = -\int_\gamma^B \mathbf{F}(\mathbf{R}_c) \cdot d\mathbf{R}_c \tag{21.7}$$

For the reverse reaction, $B \to A$, for which $\mathbf{F}(\mathbf{R}_c)$ would have the opposite signs, the ΔE corresponding to each region would be obtained simply by reversing the limits in the integrals in Equations 21.4, 21.5, and 21.7.

To summarize, structural changes occur throughout a process, but in the first and third regions, they are the dominant factor. The electronic properties that are affected by the process, which can include electrostatic potentials, ionization energies, electronic populations, etc., are, in general, more restricted in their variation, which tends to be slow and gradual in the reactant and product regions but strikingly rapid and extensive in the two zones ($\alpha \rightarrow \beta$ and $\beta \rightarrow \gamma$) of the transition region.

As an example, consider the S_N2 substitution [9,12],

$$H_3C-Cl + H-OH \rightarrow H_3C-OH + H-Cl \tag{21.8}$$

In the reactant region, before the force minimum at $R_c = \alpha$, the major event is stretching of the C—Cl bond, which is accompanied by the CH_3 group becoming more planar. In the first zone of the transition region, between α and β, the C–O bond begins to form, while the chlorine, still moving away from the carbon, becomes increasingly negative. The second transition zone, $\beta \rightarrow \gamma$, sees the stretching of an H–OH bond and the beginning of the H–Cl covalent bond formation. In the product region, $\gamma \rightarrow B$, occur the final separation of H from OH and the relaxation of the C–O and H–Cl bonds to their equilibrium lengths in the products.

The calculated (B3LYP/6-31G*) electrostatic potential associated with oxygen illustrates the often dramatic variation in electronic properties that occurs primarily in the transition region [9]. The value is $-36\,\text{kcal/mol}$ in H–OH and increases only slightly in the reactant region, to $-34\,\text{kcal/mol}$ at $R_c = \alpha$. In the transition region, however, between α and γ, it goes all the way to $-7\,\text{kcal/mol}$, before leveling off to a final $-20\,\text{kcal/mol}$ in H_3C–OH.

Consider the proton transfer [13],

$$\underset{HC-C-SH}{\overset{O\ \ O}{\overset{\|\ \ \|}{}}} \quad \rightleftharpoons \quad \underset{HC-C=S}{\overset{O\ \ OH}{\overset{\|\ \ |}{}}} \tag{21.9}$$

for which the profiles of global electronic properties (dipole moment and chemical hardness) together with local bond electronic populations have been calculated at the B3LYP/6-311G** level. These properties are displayed in Figure 21.2. It can be observed in Figure 21.2a that the dipole moment, a measure of the polarity of the system, remains practically constant in the reactant region. Then, it starts to increase rapidly in the transition region, where it changes dramatically by about 2D. Hardness, a global electronic property that is related to the reactivity of molecular systems [14], also remains constant in the reactant region, but then, upon entering the transition region, it decreases sharply until reaching the product region where it converges smoothly to the product value. Also interesting is the behavior of the CO and CS bond electronic populations. Within the reactant region, the former keeps quite consistently maintains its double bond character and the latter conserves its single bond character. In the transition region, the CO and CS bond populations exhibit strong changes and cross each other at exactly the transition structure, where a maximum delocalization within the reactive OCS backbone should be expected. In the product region, the CO and CS bond populations converge smoothly to their final values, with CS now having double bond character and CO being a single bond.

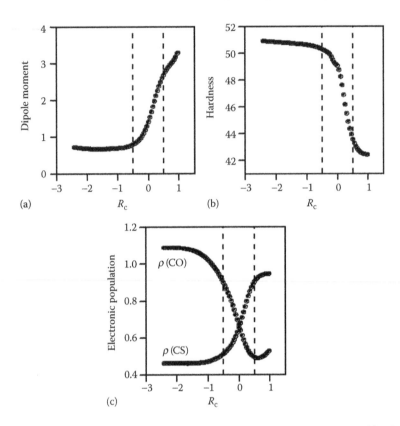

FIGURE 21.2 Profiles of (a) dipole moment (in D), (b) chemical hardness (in kcal/mol), and (c) CO and CS bond electronic populations for the reaction shown in Equation 21.9. Vertical dashed lines indicate the limits of the reaction regions defined in the text.

21.3 TWO COMPONENTS OF THE ACTIVATION ENERGY

As mentioned earlier, the minimum reaction force at $R_c = \alpha$ divides the activation energy into two components, Equation 21.6. $\Delta E(A \rightarrow \alpha)$ and $\Delta E(\alpha \rightarrow \beta)$ are, respectively, the energies needed for the preparative portion of the reaction in the reactant region $A \rightarrow \alpha$, and for the initial phase of the transition to products in the zone between $R_c = \alpha$ and $R_c = \beta$. For the reverse process, $B \rightarrow A$, the minimum $F(R_c)$ would be at $R_c = \gamma$, and the activation energy would be

$$\Delta E_{act,r} = \Delta E(B \rightarrow \gamma) + \Delta E(\gamma \rightarrow \beta) \tag{21.10}$$

We have usually found the energy required in the reactant region ($A \rightarrow \alpha$ in $A \rightarrow B$ and $B \rightarrow \gamma$ in $B \rightarrow A$) to be larger than that needed for the first phase of the transition to products ($\alpha \rightarrow \beta$ in $A \rightarrow B$ and $\gamma \rightarrow \beta$ in $B \rightarrow A$) [3,4,6–13], although on occasion, they may be similar [5].

The natural division of an activation energy into two components by the force minimum can provide useful insight into the activation process. The activation energy for the forward proton transfer reaction (Reaction 21.9), 35.7 kcal/mol, is higher than for the reverse, 31.2 kcal/mol, which seems surprising since it would be expected that the hydrogen would prefer the more electronegative oxygen. Analysis of the two components of $\Delta E_{act,f}$ and $\Delta E_{act,r}$ clarifies this apparent anomaly. The preparative stage of the forward process in the reactant region requires considerably more energy than that of the reverse, 25.8 vs. 16.2 kcal/mol. The second component of the activation energy, corresponding to the first zone of the transition region, does indeed demand more energy for the reverse process, 15.0 kcal/mol, than for the forward reaction, 9.9 kcal/mol.

Looking at the two components of the activation energy can be particularly helpful in elucidating the role of an external agent such as a solvent or a catalyst. For example, we have looked at the S_N2 substitution given in Equation 21.8 in both, the gas phase and in aqueous solution [12]. Table 21.1 shows that the presence of the solvent lowers the activation energy in both, the forward and reverse directions, by 7.8 kcal/mol for the former and 8.3 kcal/mol for the latter. What is noteworthy, however, is that this lowering occurs primarily in the preparative stage of each process, the reactant region before the force minimum. The energies required in the initial zones of the transition regions are decreased by only 1.3 and 2.5 kcal/mol, respectively. This indicates that the effect of the solvent is not to stabilize the transition state, as might be inferred from considering only the overall changes in the activation barriers, but rather to facilitate the structural effects in the preparative stages of the processes.

Table 21.1 contains computed activation data for the keto $\rightleftharpoons$ enol tautomerization of thymine [10], Equation 21.11, both, in the presence and absence of

$$(21.11)$$

Mg(II) ion. The Mg(II) promotes keto $\rightarrow$ enol, decreasing $\Delta E_{act,f}$ by 5.4 kcal/mol, but inhibits enol $\rightarrow$ keto, increasing $\Delta E_{act,r}$ by 13.3 kcal/mol. In both instances, however, it is again the preparative components of the activation energies that are, by far, the most affected.

In both examples, the external agent—whether solvent or catalyst—has affected primarily, the preparative phase of the reaction. More studies are needed in this area, and are, in fact, in progress [15].

21.4   REACTION FORCE CONSTANT

The second derivative of a potential energy along any given path is the corresponding force constant. Thus, we can introduce the reaction force constant, $\kappa(\mathbf{R}_c)$, just as we did the reaction force $\mathbf{F}(\mathbf{R}_c)$ earlier [17]:

TABLE 21.1

Computed Activation Energies and Their Components, in kcal/mol, for Reactions 21.9,[a] 21.8,[b] and 21.11[c]

Reaction

(21.9)

$$H_3C-Cl + H-OH \rightleftharpoons H_3C-OH + H-Cl \qquad (21.8)$$

(21.11)

	Forward			**Reverse**		
Reaction	$\Delta E_{act,f}$	$\Delta E(A \rightarrow \alpha)$	$\Delta E(\alpha \rightarrow \beta)$	$\Delta E_{act,r}$	$\Delta E(B \rightarrow \gamma)$	$\Delta E(\gamma \rightarrow \beta)$
21.9	35.7	25.8	9.9	31.2	16.2	15.0
21.8, gas phase	66.6	39.1	27.5	61.6	34.9	26.7
21.8, aqueous	58.8	32.6	26.2	53.3	29.1	24.2
21.11, without Mg(II)	49.6	32.6	17.0	30.5	23.5	7.0
21.11, with Mg(II)	44.2	27.1	17.1	43.8	34.4	9.4

[a] B3LYP/6-311G**, Ref. [13].
[b] CCSD(T)/aug-cc-pVTZ, Ref. [12].
[c] B3LYP/6-311++G**, Ref. [10].

$$\kappa(\mathbf{R}_c) = \frac{\partial^2 V(\mathbf{R}_c)}{\partial \mathbf{R}_c^2} = -\frac{\partial F(\mathbf{R}_c)}{\partial \mathbf{R}_c} \qquad (21.12)$$

The variation of $\kappa(\mathbf{R}_c)$ along $\mathbf{R}_c$ can be seen in Figure 21.1c. $\kappa(\mathbf{R}_c)$ is positive in the reactant region $A \rightarrow \alpha$, where $F(\mathbf{R}_c)$ is decreasing, and it has a maximum at the first inflection point of $F(\mathbf{R}_c)$. $\kappa(\mathbf{R}_c)$ then passes through zero at the minimum $F(\mathbf{R}_c)$ and is negative throughout the transition to products region, in which $F(\mathbf{R}_c)$ is increasing. $\kappa(\mathbf{R}_c)$ has a negative minimum at the second inflection point of $F(\mathbf{R}_c)$, and then goes to zero at the maximum of the latter. In the product region, $\gamma \rightarrow B$, $\kappa(\mathbf{R}_c)$ is again positive, due to $F(\mathbf{R}_c)$ decreasing, and has another maximum at the third inflection point of $F(\mathbf{R}_c)$.

What is particularly significant is that $\kappa(\mathbf{R}_c)$ is negative throughout the transition region, between α and γ. A nonlinear system has $3N - 6$ internal degrees of freedom (N being the number of atoms), which correspond to normal modes of vibration. For the reactants and products, the energy is minimum with respect to all these degrees of freedom, while at the transition structure ($R_c = \beta$), it is maximum. For the latter, the force constant for one of these degrees of freedom is negative, the corresponding vibration frequency being imaginary; this is indeed how transition states are commonly identified computationally. Miller et al. have discussed the procedure for the vibrational analysis of the nonstationary states along $\mathbf{R}_c$ [16]. For these, one of the $3N - 6$ degrees of freedom should be along $\mathbf{R}_c$, with the others corresponding to normal modes orthogonal to $\mathbf{R}_c$. Movement along $\mathbf{R}_c$ and infinitesimal rotations and translations must be projected out of the force constant matrix.

We have carried out this type of analysis for the proton transfer [17],

$$HO - N = S \rightarrow O = N - SH \qquad (21.13)$$

We found that the force constant for movement along $\mathbf{R}_c$, $k_R(\mathbf{R}_c)$, has a profile strikingly similar to that of $\kappa(\mathbf{R}_c)$ (Figure 21.1c). $k_R(\mathbf{R}_c)$ is positive and has maxima in the reactant and product regions, goes to zero very near $R_c = \alpha$ and $R_c = \gamma$, and is negative with a minimum in the transition region. It was further noted that the average, $\langle k \rangle$, of all $3N - 6$ force constants had sharp and deep minima at α and γ. Thus, significant features of the results of the vibrational analysis coincide with the key points defined by the reaction force, its minimum and maximum. What is also very significant is that both, $\kappa(\mathbf{R}_c)$ and $k_R(\mathbf{R}_c)$ are negative not only at $R_c = \beta$, the transition structure, but throughout the $\alpha \rightarrow \gamma$ region, thus providing support for linking the entire region to the transition to products, rather than focusing only upon the transition state itself.

21.5   DISCUSSION AND SUMMARY

The reaction force naturally and universally divides a process into well-defined regions along the reaction coordinate. When the latter is chosen to be the intrinsic reaction coordinate, each of these reaction regions emphasizes a particular aspect of the process. While our focus in this chapter has been upon one-step chemical reactions having an energy barrier in both, the forward and reverse directions, more complex processes—perhaps having one or more intermediates and additional transition states—can be treated in the same manner. $F(\mathbf{R}_c)$ will then have several minima and maxima; see, for example, Herrera and Toro-Labbé [11]. An interesting special case is bond dissociation or formation, for which we have found a rather remarkable uniformity in the position of the $F(\mathbf{R}_c)$ extremum (which is a minimum for dissociation, maximum for formation) [18]. We have found, in general, that structural factors tend to be dominant in those regions in which $F(\mathbf{R}_c)$ is decreasing, whereas rapid and extensive changes in electronic properties are likely to be associated with regions in which $F(\mathbf{R}_c)$ is increasing.

An important consequence of reaction force analysis is that it shows that an activation energy is composed of two components, representing the energies required in the reactant (preparative) region of the process and the first zone of the transition to products. This can provide considerable insight into what is happening during the activation portion of the process, especially into the roles of any external agents such as solvents, catalysts, inhibitors, etc.

The full potential of the reaction force as a rigorously defined approach to analyzing chemical and physical processes has yet to be realized. It continues to be explored.

ACKNOWLEDGMENTS

The authors wish to thank financial support from FONDECYT grants N°1060590 and N°11070197; FONDAP grant N° 11980002 (CIMAT). The authors wish to thank interesting and stimulating discussions on the reaction force concept with Professors P.W. Ayers (Canada), P.K. Chattaraj (India), P. Geerlings (Belgium), A. Grand (France), P. Jaque (Belgium), Ch. Morell (France), A. Vela (Mexico) and M. Yañez (Spain).

REFERENCES

1. Fukui, K. *Acc. Chem. Res.* 1981, *14*, 363.
2. Gonzalez, C. and Schlegel, H. B. *J. Phys. Chem.* 1990, *94*, 5523.
3. Toro-Labbé, A. *J. Phys. Chem. A* 1999, *103*, 4398.
4. Jaque, P. and Toro-Labbé, A. *J. Phys. Chem. A* 2000, *104*, 995.
5. Martínez, J. and Toro-Labbé, A. *Chem. Phys. Lett.* 2004, *392*, 132.
6. Toro-Labbé, A., Gutiérrez-Oliva, S., Concha, M. C., Murray, J. S., and Politzer, P. *J. Chem. Phys.* 2004, *121*, 4570.
7. Politzer, P., Toro-Labbé, A., Gutiérrez-Oliva, S., Herrera, B., Jaque, P., Concha, M. C., and Murray, J. S. *J. Chem. Sci.* 2005, *117*, 467.
8. Gutiérrez-Oliva, S., Herrera, B., Toro-Labbé, A., and Chermette H. *J. Phys. Chem. A* 2005, *109*, 1748.
9. Politzer, P., Burda, J. V., Concha, M. C., Lane, P., and Murray, J. S. *J. Phys. Chem. A* 2006, *110*, 756.
10. Rincón, E., Jaque, P., and Toro-Labbé, A. *J. Phys. Chem. A* 2006, *110*, 9478.
11. Herrera, B. and Toro-Labbé, A. *J. Phys. Chem. A* 2007, *111*, 5921.
12. Burda, J. V., Toro-Labbé, A., Gutiérrez-Oliva, S., Murray, J. S., and Politzer, P. *J. Phys. Chem. A* 2007, *111*, 2455.
13. Toro-Labbé, A., Gutiérrez-Oliva, S., Murray, J. S., and Politzer, P. *Mol. Phys.* 2007, *105*, 2619.
14. Pearson, R. G. *Chemical Hardness, Applications from Molecules to Solids,* 1997, Wiley-VCH, Weinheim.
15. Burda, J. V., Toro-Labbé, A., Gutiérrez-Oliva, S., Murray, J. S., and Politzer, P. *J. Phys. Chem. A*, Submitted.
16. Miller, W. H., Handy, N. C., and Adams, J. E. *J. Chem. Phys.* 1980, *72*, 99.
17. Jaque, P., Toro-Labbé, A., Politzer, P., and Geerlings, P. *Chem. Phys. Lett.*, 2008, in press.
18. Politzer, P., Murray, J. S., Lane, P., and Toro-Labbé, A. *Int. J. Quantum Chem.* 2007, *107*, 2153.

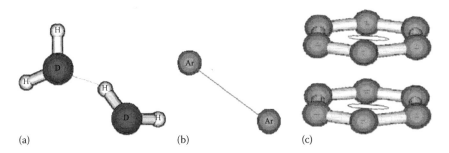

(a)   (b)   (c)

FIGURE 3.4 The prototypical noncovalent interactions between (a) water dimer, (b) Ar dimer, and (c) benzene dimer.

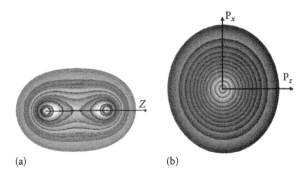

(a)   (b)

FIGURE 5.4 Coordinate and momentum space charge densities of H_2 molecule illustrating the bond directionality principle. Isosurfaces from 0.04 to 0.01 a.u. are plotted for the coordinate space charge density (a). Isosurfaces from 1.0 to 0.01 a.u. are plotted for the momentum space charge density (b).

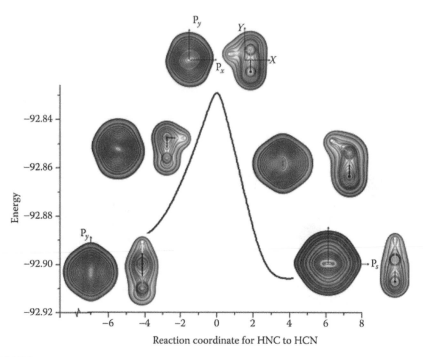

FIGURE 5.6 Momentum and coordinate space charge density profiles for the reaction path from HNC to HCN.

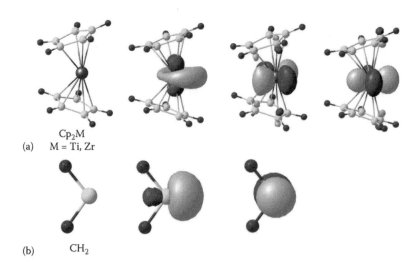

(a) Cp$_2$M
    M = Ti, Zr

(b) CH$_2$

FIGURE 14.1 Frontier molecular orbitals of (a) bent metallocenes (Cp$_2$M; M = Ti, Zr) and (b) carbene.

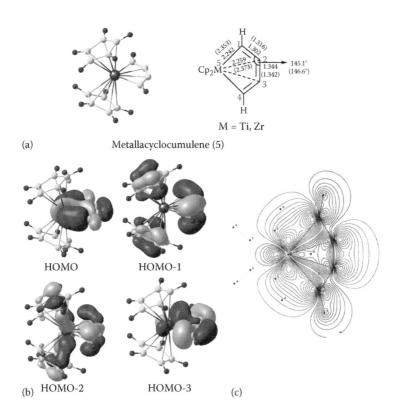

(a) Metallacyclocumulene (5)

HOMO   HOMO-1

(b) HOMO-2   HOMO-3   (c)

FIGURE 14.2 (a) Structure and important geometrical parameters of metallacyclocumulene (**5**). (b) molecular orbitals of metallacyclocumulene (**5**). The values in normal font correspond to M = Ti and the values in parenthesis correspond to M = Zr. (c) A contour diagram of the HOMO of metallacyclocumulene (**5**) showing the in-plane interaction of metal and the ligand. (Reproduced from Bach, M.A., Parameswaran, P., Jemmis, E.D., Rosenthal, U., *Organametallics*, 26, 2149, 2007; Jemmis, E.D., Phukan, A.K., Jiao, H., and Rosenthal, U., *Organometallics*, 22, 4958, 2003. With permission.)

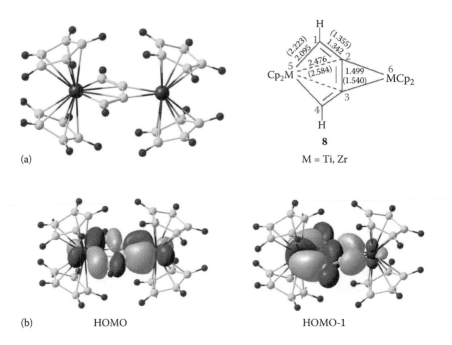

(a)

M = Ti, Zr

(b)　　　　　HOMO　　　　　　　　　　HOMO-1

FIGURE 14.5 (a) The structure and important geometrical parameters of the bimetallic complex **8**. The values in the normal font correspond to M = Ti and the values in parenthesis correspond to M = Zr. (b) Molecular orbitals of the complex **8** showing the in-plane delocalization.

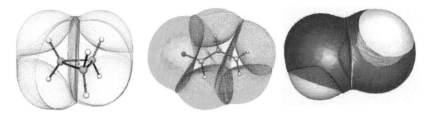

FIGURE 15.1 Examples of the QCT partitioning of the electron density. (left) All atoms in cyclopropane (except for the front methylene group); (middle) acrolein; and (right) a water dimer (global minimum).

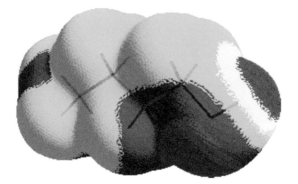

FIGURE 17.1 Electrostatic potential on the $\rho(\mathbf{r}) = 0.001$ au molecular surface of 1-butanol, computed at the B3PW91/6-31G(d,p) level. Color ranges, in kcal/mole, are: red, more positive than 30; yellow, between 15 and 30; green, between 0 and 15; blue, between −20 and 0; purple, more negative than −20. The hydroxyl hydrogen is at the far right (red and yellow), and the oxygen is below it (purple and blue).

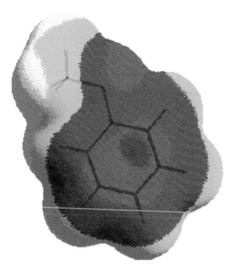

FIGURE 17.2 Electrostatic potential on the $\rho(\mathbf{r}) = 0.001$ au molecular surface of anisole, **1**, computed at the B3PW91/6-31G(d,p) level. Color ranges, in kcal/mole, are: yellow, between 10 and 20; green, between 0 and 10; blue, between −15 and 0; purple, more negative than −15. The methoxy group is at the upper left. The most negative regions (purple) are by the oxygen and above and below the ring.

FIGURE 17.3 Electrostatic potential on the $\rho(\mathbf{r}) = 0.001$ au molecular surface of SBr_2, computed at the B3PW91/6-31G(d,p) level. Color ranges, in kcal/mole, are: red, more positive than 15; yellow, between 10 and 15; green, between 0 and 10; blue, between -9 and 0. One of the two bromines is at the bottom, while the sulfur is at the upper right. The red regions are the positive σ-holes, on the extensions of the S–Br and Br–S bonds.

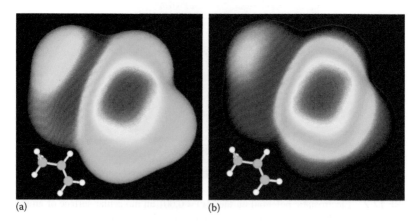

(a)   (b)

FIGURE 18.1 Propylene is susceptible to electrophilic attack on the double bond. This can be deduced by plotting (a) the Fukui function from below, $f^-(\mathbf{r})$, or (b) the HOMO density, $\rho^{HOMO}(\mathbf{r})$, on the van der Waals' surface of the molecule.

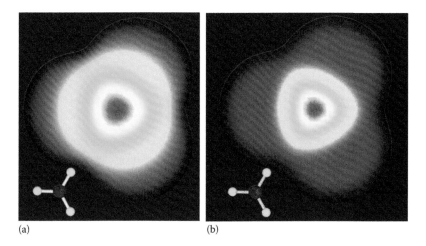

(a)   (b)

FIGURE 18.2   BF$_3$ is susceptible to nucleophilic attack at the boron site. This can be deduced by plotting (a) the Fukui function from above, $f^+(\mathbf{r})$, or (b) the LUMO density, $\rho^{LUMO}(\mathbf{r})$, on the van der Waals' surface of the molecule.

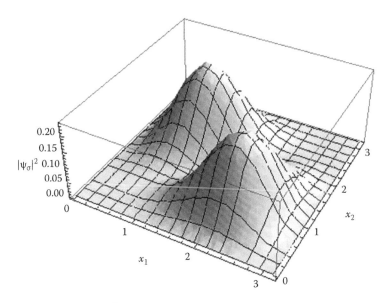

FIGURE 20.1   Plot of $|\Psi_\sigma|^2$, $\sigma = \alpha$, β for four independent fermions in a box.

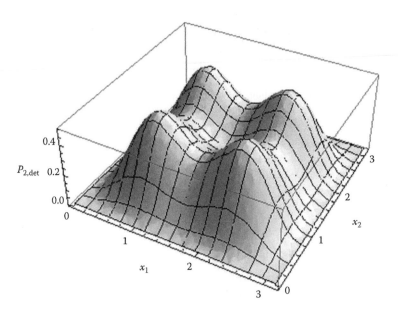

FIGURE 20.4 $P_{2,\text{det}}(x_1, x_2)$ for fermions in a box, for $\sigma_1 = \alpha$ and $\sigma_2 = \beta$.

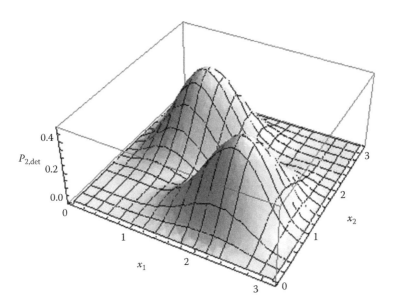

FIGURE 20.5 $P_{2,\text{det}}(x_1, x_2)$ for fermions in a box, for $\sigma_1 = \sigma_2 = \alpha$.

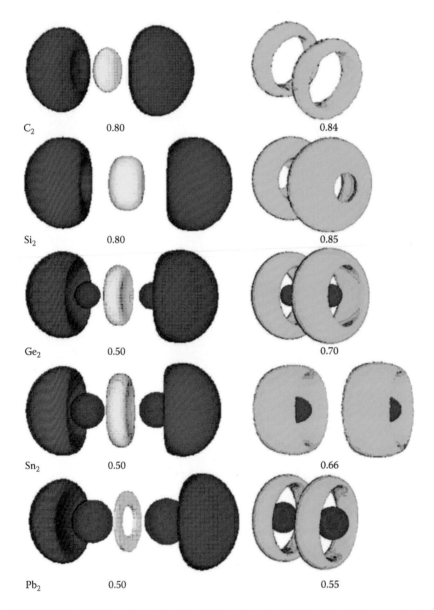

C$_2$	0.80	0.84
Si$_2$	0.80	0.85
Ge$_2$	0.50	0.70
Sn$_2$	0.50	0.66
Pb$_2$	0.50	0.55

FIGURE 20.7 ELF isosurfaces for triplet (left) and singlet (right) of the different molecules. The values of the isosurfaces are indicated below each picture.

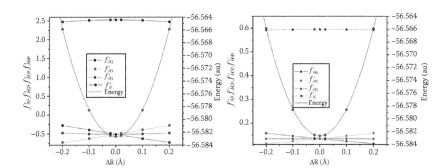

FIGURE 23.2 Profiles of (a) energy and FF values for nucleophilic attack and (b) for electrophilic attack for distortion in the N–H bond distance of ammonia.

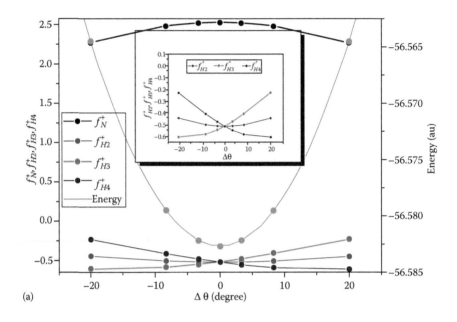

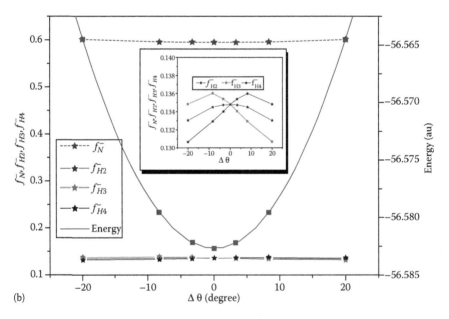

FIGURE 23.3 Profiles of (a) energy and FF for nucleophilic attack and (b) electrophilic attack for distortion in the H–N–H bond angle of ammonia.

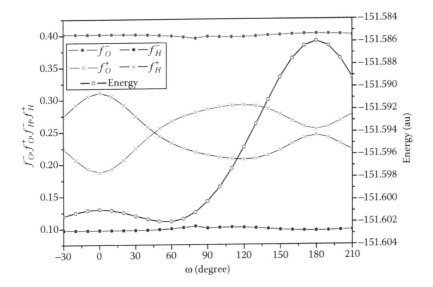

FIGURE 23.4 Profiles of energy and FF values due to the internal rotation in H_2O_2 molecule.

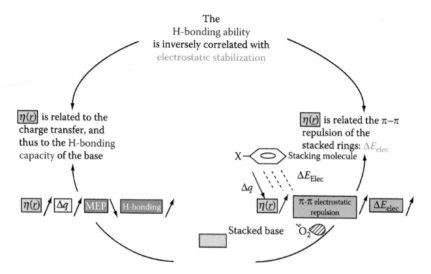

FIGURE 27.6 $\eta(\underline{r})$ can be used for the estimation of the electrostatic interaction and the hydrogen bonding ability. (Reprinted from Mignon, P., Loverix, S., Steyaert, J., and Geerlings, P., *Nucl. Acids Res.*, 33, 1779, 2005. With permission.)

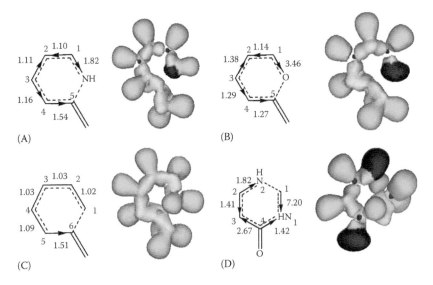

FIGURE 28.4 Scheme of TS structures together with relative electron fluctuation magnitudes and ELF = 0.60 pictures for reactions A–D.

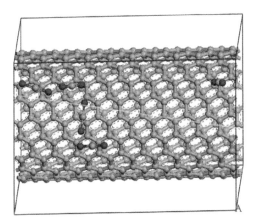

FIGURE 32.4 The localized minima as obtained after the GCMS simulation with carbon dioxide adsorption over single-wall CNT with a fixed fugacity of 100 kPa.

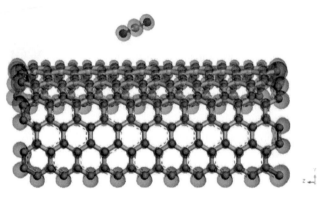

FIGURE 32.6 The electrophilic Fukui function of carbon dioxide adsorbed over single-wall CNT is plotted as an isosurface with a grid of 0.2 Å.

22 Characterization of Changes in Chemical Reactions by Bond Order and Valence Indices

György Lendvay

CONTENTS

22.1 INTRODUCTION

One of the most important questions of chemical reactivity is what happens during chemical reactions. This question can be understood in reaction dynamical experiments and calculations simulating them. In the theoretical description, the motion of atoms is followed in time. The motion of atoms depends on what kind of forces act between the atoms at various relative arrangements, which is determined by the potential energy surface (PES) of the reaction. This is a multidimensional surface (hypersurface), giving the energy as a function of the $3N$ coordinates of the N atoms participating in the reaction. If the conditions ensuring the applicability of the Born–Oppenheimer approximation are valid, then the forces acting between atoms do not depend on the velocities of the atomic nuclei, and the potential energy is given by the solution of the electronic Schrödinger equation at each molecular arrangement.

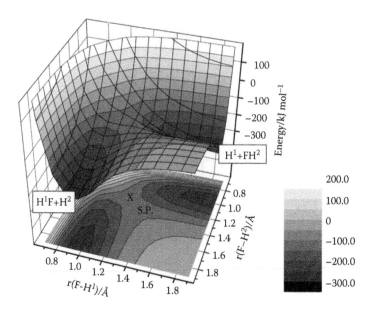

FIGURE 22.1 A perspective view of the potential surface of the $H^1F + H^2 \rightarrow H^1 + FH^2$ reaction in the collinear H^1–F–H^2 arrangement.

An example of a potential surface of a reacting system in Figure 22.1. This kind of plot shows the potential energy for a special (collinear or bent) arrangement of atoms during a bimolecular reaction in a triatomic system:

$$A + BC \rightarrow AB + C \qquad (22.1)$$

The full multidimensional surface cannot be visualized in its entirety. Even in the simplest case, a triatomic system, there are three independent coordinates specifying the arrangement of atoms. However, in three-dimensional space available for us, one axis is needed to show the energy, and only two dimensions are left for the independent variables. In fortunate cases, this is satisfactory to understand the topology of the PES, but in general, we need to constrain our attention on the qualitatively important features. The main features of a potential surface can be recognized in Figure 22.1, where the potential energy is plotted against the A–B and B–C distances. The reactant limit corresponds to large reactant separation (r_{AB}) and small B–C distance; here, the relevant section of the reactants' potential surface around their equilibrium geometry is visible, in this simple case, that of the BC vibration. The same characterizes the product limit (large r_{BC}, small r_{AB}). If r_{AB} decreases from the reactant limit or r_{BC} from the product limit, the potential does not change until the reactants start to "notice" each other. This way the reactant valley and the product valley are formed, which are connected. At smaller separation, the valleys bend and their bottom increases. (Note that in unimolecular reactions the partners never go far away and the valleys are reduced to basins closed on all sides.) When the reactants (or products from the other side) are close together, the potential energy at the bottom increases, and, along the line corresponding to the bottom of the valley passes through a maximum. The line traced by the coordinates corresponding to the bottom of the valley is called minimum energy path (MEP). In topological terms, the maximum along the MEP corresponds to a saddle point (marked as SP in Figure 22.1), which is a maximum along the direction of the MEP and a minimum along that perpendicular to it (in multi-dimensional systems along all other coordinates perpendicular to the MEP) akin to a mountain pass separating two valleys. The structure corresponding to the saddle point is called transition structure (TS). Reaction occurs if the atoms in real three-dimensional space move along a path that corresponds to motion along the MEP. Obviously, the shape of the potential surface (height, curvature) at the saddle point and its neighborhood is the bottleneck for the reaction, as this forms a barrier for the system that needs to be passed when going from the reactant valley to the product valley. The rate of the reaction can be approximately calculated in the following way. Let us assume that there is a (multi)dimensional dividing surface that separates the reactants from products, in the sense that if the system passes this surface from one side to the other, reaction occurs. A "good" dividing surface is such that if the system passes it, it will never come back (point of no return). Because of this property, the dividing surface is called transition state (for which also the TS abbreviation is used), which corresponds to a whole set of geometrical arrangements that are in-between the reactant and product limits. The rate of reaction is given by the flux through this surface which can be calculated using the principles of statistical mechanics. This method is called transition-state theory (TST). In variational TST, the location of the dividing surface is varied, treating the ensemble of reacting molecules either as a microcanonical or a canonical one. As a first approximation, a plausible location for the dividing surface is a plane perpendicular to the MEP at the saddle point, but this is only approximate because, in general, it is not a point of no return (this version of the theory is called conventional TST). In this case, the transition state is located at the TS (note that although both are abbreviated as TS, the two are conceptually different, the former being a dynamical and the latter a topological concept). Chemists tend to think in terms of conventional TST and created the smart combination transition-state structure (see e.g., Ref. [1]).

To understand the factors determining the reactivity in a system, one needs to know what determines the shape of the potential surface. Note that in the discussion of reactions in terms of the shape of the potential surface, we often say what happens "first" and what happens later when we walk along the MEP, visualizing that, as the reaction proceeds, the system moves from the reactant side to the product side strictly along the MEP. It should be noted that this has nothing to do with the actual temporal progress of a reaction as the point representing the system never follows strictly the MEP, it can make turns and back-and-forth oscillations. When time is used in these discussions, one should keep in mind that it only means the position along the MEP.

22.2   CALCULATION OF BOND ORDER AND VALENCE INDICES

In a chemical reaction, some bonds are broken and some new ones are made. Bond breaking requires energy, whereas during bond formation, energy is released. This is what gives rise to the barrier along the MEP. In elementary chemical reactions, the breakage of the "old" bond and formation of the "new" one take place simultaneously. As a consequence, the energy to be invested is not as high as that needed to completely break a bond. Chemists would like to understand how the properties of atoms influence the energetic consequences of the changes of chemical bonds. A very useful tool for this purpose is to follow the degree of bonding between the atoms of the system changes during the reaction. The degree of bonding can be measured by the multiplicity of a bond, called bond number if integer, or bond order if fractional [2,3]. The bond order cannot be guessed by chemical intuition for a geometrical arrangement intermediate between reactants and products. Fortunately, it can be calculated from ab initio wave functions.

In this chapter, we use the definitions of bond order and valence indices provided by Mayer [4–6] (for a historical account, see Ref. [6a] and for other types of bond indices, see Ref. [6b]). In terms of electronic structure theory, they represent an extension to Mulliken's population analysis. The bond order is defined as

$$B_{AB} = \sum_{\mu \in A} \sum_{\nu \in B} \left[(PS)_{\mu\nu}(PS)_{\nu\mu} + (P^sS)_{\mu\nu}(P^sS)_{\nu\mu} \right] \tag{22.2}$$

and the valence of an atom is

$$V_A = \sum_{B \neq A} B_{AB} \tag{22.3}$$

In open-shell systems, the sum of the bond orders of all bonds in which atom A participates differs from the valence defined according to Equation 22.2; the difference accounts for the "unsatisfied" bonding capability of the atom, the free valence, expressed as

$$F_A = \sum_{\mu,\nu \in A} (P^sS)_{\mu\nu}(P^sS)_{\nu\mu}, \tag{22.4}$$

then, the total valence of atom A is

$$V_A = \sum_{B \neq A} B_{AB} + F_A \tag{22.5}$$

The calculation of the indices requires the overlap matrix $\mathbf{S}$ of atomic orbitals and the first-order density (or population) matrix $\mathbf{P}$ (in open-shell systems in addition the spin density matrix $\mathbf{P}^s$). The summations refer to all atomic orbitals μ centered on atom A, etc. These matrices are all computed during the Hartree–Fock iteration that determines the molecular orbitals. As a result, the three indices can be obtained essentially at no cost after the self-consistent field (SCF) calculation because the few matrix multiplications and summations involved are done extremely fast.

Note that the bond order index defined by Mayer accounts for the covalent contribution to the bond (this is why of late it is often mentioned as "shared electron pair density index," SEDI). As such, the index cannot be expected to produce the integer values corresponding to the Lewis picture if a bond has a significant ionic contribution. The bond order index defined in this way measures the degree of correlation of the fluctuation of electron densities on the two atoms in question [7].

One has to note that the formulas are based on the atomic orbital basis set used in the SCF calculation. The atom is, in a sense, identified by its nucleus and the ensemble of atomic orbitals centered on it. (This is often expressed as the "Hilbert space description" of an atom [8], and the analysis of properties is referred to as "Hilbert space analysis.") These basis sets are centered on the atoms and are developed so that they could describe, in the most efficient way, the molecular orbitals in hopefully all molecules of the selected atom. As a consequence, they differ from the traditional s, p, d orbitals, which are so fruitfully used in the interpretation of bonding. One complication that arises from this difference is easy to visualize as follows. The atomic orbitals in a good basis set have a significant contribution from diffuse constituents, which are necessary to get good molecular orbitals, but it is hard to assign chemical content to them. The diffuse orbitals have significant amplitude at large distances from the nucleus of the atom they are centered on, even in the neighborhood of the other nuclei in the molecule. From Equation 22.1, one can see that the bond order index contains contributions from different atomic orbitals. The diffuse orbitals do contribute to the electron density in the close neighborhood of another nucleus they penetrate, which clearly belongs to that atom, but this contribution is assigned to their "mother" atom (as if they "stole" electron density from another atom). If the mutual contribution of atoms to each other's internal "affairs" is not balanced, the bond order index will also be distorted. Experience shows that the most reasonable bond order values can be obtained if one uses a basis set that is close to chemists' way of thinking. Such basis sets are minimal basis sets in which the atom is assumed to have as many s, p, d, . . . type orbitals as are needed to accommodate all electrons and at the same time assure that the basis set is spherically symmetric. (This means that if a subshell is occupied by any number of electrons, all orbitals belonging to the subshell has to be included — this guarantees spherical symmetry.) The prototype and most generally used version of such basis sets is the STO-3G set designed by Pople and coworkers [9,10], which was proposed to be used in the calculation of bond order and valence indices [11].

Tables 22.1 and 22.2 show how the general principles sketched above are manifested in real systems. The C–C and C–H bond order indices and the C and H valence indices were calculated for ethane, ethene, ethyne, and benzene at the $HF/6-31G^*$ geometry with various basis sets. The bond order of the C–H bond is close to unity in all cases. The carbon–carbon bonds have bond orders close to one, two and three in ethane, ethene, and ethyne, respectively. In benzene, all C–C bonds have the same bond order, which is close to 1.5. Note that definition (Equation 22.1) yields nonzero bond orders between nonbonded atoms also, and in certain cases,

TABLE 22.1

Bond Order Indices Calculated for Prototype Hydrocarbon Molecules Calculated from Hartree–Fock Wave Functions at the 6-31G* Equilibrium Geometry

| Molecule | C_2H_6 | | C_2H_4 | | C_2H_2 | | C_6H_6 | |
Bond	C–C	C–H	C–C	C–H	C–C	C–H	C–C	C–H
STO-3G	1.01	0.98	2.02	0.98	3.00	0.98	1.45	0.94
6-31G	0.92	0.96	1.95	0.96	3.43	0.82	1.45	0.94
6-31G*	0.97	0.96	1.98	0.96	3.19	0.87	1.45	0.96
6-31G**	0.97	0.98	1.97	0.98	3.19	0.89	1.82	0.77
6-31+G*	0.83	0.96	2.18	0.93	3.67	0.76	1.87	0.74
6-31++G**	1.01	0.97	2.26	0.93	3.62	0.78	2.08	0.65

TABLE 22.2

Valence Indices Calculated for Prototype Hydrocarbon Molecules Calculated from Hartree–Fock Wave Functions at the 6-31G* Equilibrium Geometry

| Molecule | C_2H_6 | | C_2H_4 | | C_2H_2 | | C_6H_6 | |
Atom	C	H	C	H	C	H	C	H
STO-3G	3.97	1.00	3.98	0.98	3.99	0.98	3.86	1.00
6-31G	3.76	0.93	3.84	0.96	4.29	0.82	3.91	0.93
6-31G*	3.82	0.93	3.86	0.96	4.08	0.87	3.92	0.93
6-31G**	3.86	0.95	3.89	0.98	4.12	0.89	4.42	0.95
6-31+G*	3.64	0.92	4.05	0.93	4.55	0.76	4.59	0.92
6-31++G**	3.90	0.95	4.19	0.93	4.53	0.78	4.87	0.95

small negative values are also obtained, but no chemical meaning is to be assigned to such values. Also, note that it does not make sense to consider and present bond orders with many-digit accuracy. Two digits beyond the decimal point are accurate enough in the study of reactions and in the analysis of special structures. Three digits can be used if a tendency is to be presented in a series of similar systems or reactions. More "accurate" values do not make sense for two reasons. Firstly, the value can change more than 0.0001 by a geometry change of 0.001 Å. Secondly, and more importantly, we are calculating a qualitative index, which is not the expectation value of an operator, but the product of a reasonable but ad hoc definition. This property of bond order and valence indices explains why they have to be treated in a way different from physical quantities like molecular geometry, relative energies, vibrational frequencies, etc. One can compare this situation to that of Hückel theory vs. ab initio calculations: with the former, nice clear pictures of bonding, reactivity tendencies, etc. can be generated; in the latter, the simplicity is lost in the wealth of data. The physical properties are the better the closer we approach the accurate solution of the molecule's Schrödinger equation. In the regular linear combination of atomic orbital–molecular orbital (LCAO–MO) approach, this is done in terms of basis sets that are designed to represent the most "useful" dimensions of an abstract infinite-dimensional space (the "unit vectors" of the Hilbert space), and not to reflect qualitative, chemical properties. A consequence is that it does not make too much sense to search for "converged" bond order and valence indices as can clearly be seen in Table 22.1: the "convergence" of the indices with the increase of the quality of the basis set is erratic and changes from molecule to molecule. One can see that the minimal STO-3G basis set provides the most reasonable values, which is also observed in other systems.

One way of getting rid of distortions and basis set dependence could be that one switches to the formalism developed by Bader [12] according to which the three-dimensional physical space can be partitioned into domains belonging to individual atoms (called atomic basins). In the definition of bond order and valence indices according to this scheme, the summation over atomic orbitals will be replaced by integration over atomic domains [13]. This topological scheme can be called "physical space analysis." Table 22.3 shows some examples of bond order indices obtained with this method. Experience shows that the bond order indices obtained via Hilbert space and physical space analysis are reasonably close, and also that the basis set dependence is not removed by the physical space analysis.

The disadvantage of the physical space analysis is that the calculation of atomic basins and the subsequent integration is not always straightforward, and definitely requires much more time than the Hilbert space analysis (recall the latter is instantaneous). Our experience shows that the latter analysis does provide satisfactory information so that it is not necessary to perform the physical space analysis.

The calculation of Mayer bond orders and valence indices is straightforward using the codes that Mayer has made available on the Internet [14]. The formulas have been built into several electronic structure codes, like HONDO [15], Gaussian 03 [16] (note that in Gaussian only the closed-shell formulas have been coded), ADF [17], and some versions of GAMESS [18].

Formulas 21.1 through 21.3 are designed for Hartree–Fock wave functions. There are some attempts to define similar indices using wave functions obtained via methods including electron correlation [19]. Similarly, to the situation with respect to basis set improvement, the results based on correlated wave functions do not necessarily make the qualitative picture of bonding easier to understand. An exception is when there is a significant nondynamical correlation in the system,

TABLE 22.3
Bond Order Indices Calculated Using the Topological Definition

Molecule	C_2H_6		C_2H_4		C_2H_2		C_6H_6	
Bond	C–C	C–H	C–C	C–H	C–C	C–H	C–C	C–H
6-31G*	1.013	0.966	1.984	0.972	2.885	0.953	1.399	0.963
6-31 ++G**	0.987	0.966	1.881	0.984	2.860	0.977	1.386	0.977

Source: Ángyán, J.G., Loos, M., and Mayer, I. *J. Phys. Chem.* 98, 5244, 1994.

i.e., a multideterminant wave function is required to characterize the bonding in the molecule qualitatively correctly.

22.3 ANALYSIS OF SIMPLE REACTIONS

Equipped with these tools one can investigate how bond orders change in a simple chemical reaction like dissociation of the H_2 molecule. Note that this type of calculation requires spin polarization, i.e., unrestricted Hartree–Fock or multiconfiguration treatment. In Figure 22.2 the bond order and the free valence of the atoms as well as the energy in the H_2 molecule are plotted as a function of the bond length, obtained with the Hartree–Fock method and the STO-3G basis set. The restricted Hartree–Fock wave function is the only solution at small distances; the unrestricted solution splits off at a bond length of about 1.2 Å. Accordingly, the bond order is unity in the neighborhood of the equilibrium bond length, 0.74 Å, and decreases as the bond is stretched, as one expects. The decrease first is observed at a separation of 1.2 Å, where the UHF solution first appears, and the bond order approaches zero slower and slower, just as the energy converges to its asymptotic value. Note that the bond order does not increase above 1 if the bond is compressed. This is not surprising as the two electrons are all involved in the bond, and a single bond is made by one pair of electrons. (Actually, the bond order that can be calculated using Mayer's definition for diatomic molecules sooner or later decreases when the bond length is reduced because larger and larger parts of the atomic orbitals go not only

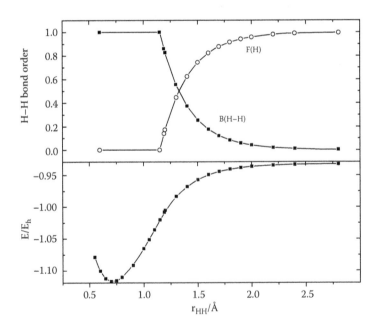

FIGURE 22.2 The changes of bond order and free valence indices as well as the energy during the dissociation of the H_2 molecule obtained in an UHF/STO-3G calculation.

near but even beyond the other nucleus, and contribute to the electron density there.) The development of the free valence is complementary to that of the bond order index, as during the whole process, the total valence of the two atoms, in this case, the sum of the bond order and the free valence index, is constant, unity at the (U)HF/STO-3G level. At the equilibrium bond length, the two atoms have zero free valence, indicating that the electrons are all "occupied" in bonding. With increasing bond length, nonzero free valence appears on both atoms and at the limit of two dissociated atoms, both have a free valence of unity. The spin density localized on each atom changes analogously.

Dissociation of polyatomic molecules shows a slightly different picture, as can be seen for the C–C rupture in ethane in Figure 22.3. The decrease of the bond length is much faster and follows more closely the change of energy. In the H_2 molecule, the bond order is still around 0.5 at a nuclear separation of 1.5 Å when in terms of bond energy, the molecule has dissociated to as much as 75%. The behavior of other diatomic molecules is similar. The reason for the slow decrease of the bond order in diatomic molecules is that the bonding electrons necessarily remain between the two atoms, as there is no other bond they could move to. In dissociating polyatomic molecules and especially during chemical reactions, the electrons originally participating in a chemical bond can be shifted into other bonds and the bond order index can follow more closely the actual status of bonding. In turn, for the same reason, when the bond is compressed in polyatomic molecules, the bond order index can increase above unity even for formally single bonds.

In a bimolecular reaction, at least two bonds change simultaneously. Figure 22.4 shows how bond orders and valences change in the reaction of methane with a H atom (shown in the deuterium-labeled form that can be studied experimentally)

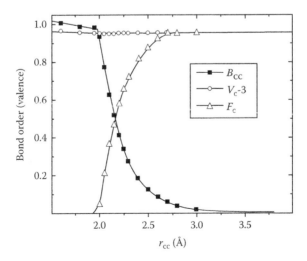

FIGURE 22.3 The bond order of the C–C bond, the total and the free valence of the C atoms as a function of the C–C distance (obtained in a relaxed potential surface scan using the UHF method).

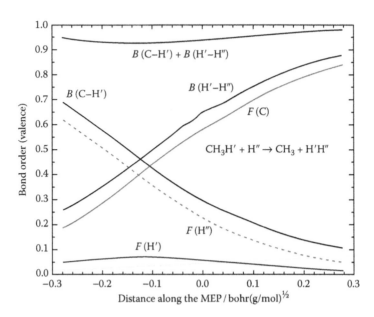

FIGURE 22.4 Bond order and valence indices calculated along the MEP of an atom transfer reaction.

$$CD_3H' + H \rightarrow CD_3 + H'H \qquad (22.6)$$

The figure shows the indices along the MEP of the reaction. The abscissa shows the distance along the MEP measured in both directions from the saddle point. One can see that the bond order of the breaking C–H' bond decreases, that of the forming H'–H bond increases at the same pace as one moves along the MEP from the reactant to product direction. Note that at the saddle point the bond order of the C–H' and H–H' bonds is 0.3 and 0.65, respectively. This means that this reaction is not "halfway" between reactants and products at the saddle point, the TS is product-like: both the breakage of the old bond and the formation of the new bond is roughly 70% complete. The free valence of H decreases as the atom gets more and more involved in bonding, while V_C increases. All these processes take place at a harmonious parallel pace. A consequence of this concerted bond development along the MEP is that the H' atom does not exhibit any significant free valence during the reaction, its bonding capability remains saturated along the MEP. The figure shows that the sum of the bond orders of the bonds which atom H' forms, i.e., of C–H' and H–H', remains constant along the MEP, and is essentially unity. This is called the principle of bond order conservation. Bond order is found to be conserved in many H atom transfer reactions, not only in bimolecular, but also in intramolecular reactions like 1.5 H atom transfer isomerization of alkyl radicals, as well as in numerous other reactions. The principle was heuristically formulated in the 1960s before bond orders were possible to get from molecular wave functions, and some semiempirical methods of generating potential surfaces: the bond energy–bond order (BEBO) method [20–22] and the related bond strength–bond length (BSBL) method [23,24] were formulated based on it.

22.4 MODELING THE BEHAVIOR OF A SERIES OF REACTIONS

What we noticed above, namely, that the TS, while being intermediate between the reactants and products may resemble one of them, makes bond orders a useful tool to judge whether a TS is reactant- or product-like. The question whether a TS resembles more a reactant or the product is important when studying a series of reactions. Understanding tendencies manifested in a series of related systems is one of the most important tasks in chemistry. A simple model illustrates how one can get information on the tendencies expected for a series of reactions if the parameter characterizing the "location" of a member reaction in the series is the reaction enthalpy. Utilizing the principle of bond order conservation and the correlation of bond energy and bond order, one can approximately determine the potential profile along the MEP and locate the barrier, the maximum on this profile. The parameters are changing in the series of reactions, and, as a consequence, the location and height of the barrier is shifted. The model [25] starts from the principle of bond order conservation for an $A + BC \rightarrow AB + C$ type reaction:

$$B_{AB} + B_{BC} = 1 \tag{22.7}$$

At the beginning, $B_{AB} = 1$ and $B_{BC} = 0$. As the reaction progresses (i.e., we walk along the MEP from the reactant limit to the product limit), B_{AB} decreases and B_{BC} increases. The smaller the bond order is the less strong is the bond. Various assumptions can be made about the BEBO correlation, based on the comparison of experimental binding energies and bond orders [20,21] or by investigating the change of bond order and the energy with the variation of bond length of a diatomic molecule in ab initio calculations [25], the most generally valid being

$$V_{XY} = -D_{XY}B_{XY}^{p} \tag{22.8}$$

where
V_{XY} is the energy of the X—Y bond
D_{XY} is the binding energy of the diatomic molecule at its equilibrium geometry
B_{XY} the bond order of the stretched bond

The exponent p varies [25] between 1 and 2.

If one assumes that the energy change due to the A—C interaction is much smaller than that due to the breaking and forming of a bond, the energy along the MEP will be the sum of the contributions from the A—B and B—C bonds, the zero of energy being set to the reactant level:

$$V_{MEP} = D_{BC} - D_{AB}B_{AB}^{p} - D_{BC}B_{BC}^{p} \tag{22.9}$$

In the experimental studies of series of reactions, one reactant is generally held constant and the other is varied. The model is applicable if the dominant change

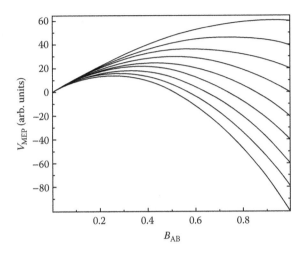

FIGURE 22.5 The potential profiles along the MEP of a series of reactions in which the member processes are distinguished by the reaction enthalpy, calculated from Equation 22.9 with $p = 1.8$.

due to changing a reactant is the dissociation energy of the breaking or forming bond. The change of the dissociation energy implies that the reaction heat changes. Figure 22.5 shows how the profile looks, for several values of the energy difference $D_{BC} - D_{AB}$. The shape of these curves is concave everywhere, in contrast to the commonly seen bell-shaped picture where the potential energy along the MEP first increases slowly, then speeds up where the interaction becomes strong, and slows down again when passing the barrier. The reason is that our variable, the bond order, differs from the bond distance type geometrical variables. The bond order, as we shall see later, is exponentially related to the length of the bond, i.e., when the bond order changes by 0.01 in the left side of Figure 22.5, the bond length changes by a large amount, while in the middle part, the same change of the bond order is connected to a much smaller bond length change (for a more detailed discussion of the BO—BL correlation, see the section near the end of this chapter). If the plot is converted to bond length coordinates, we get back the usual bell shape. The maximum of each curve in Figure 22.5 corresponds to the saddle point on the PES, and is shifted from left to right if the reaction energy is increased from very negative (very exothermic limit) to large positive values (very endothermic limit). For very exothermic reactions B_{BC} at the maximum is small, and, accordingly, $B_{AB} = 1 - B_{BC}$ is close to one, which means that the formation of the B–C bond and the rupture of the A—B bond is in an early stage: the TS is reactant-like. At the same time, the barrier is low. The opposite is true for the very endothermic limit: the TS is product-like, and the barrier of the A + BC reaction is high. These two statements conform very well with Hammond's principle [1,26], which states that the TS resembles more the structure that is energetically closer to it. It is interesting to see how the height of the barrier changes as a function of the reaction energy. This is shown in Figure 22.6.

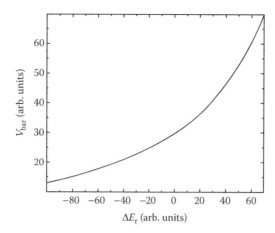

FIGURE 22.6   The height of the barrier along the MEP in a series of reactions as a function of the reaction enthalpy.

The curve is close to linear in both, the very exothermic and very endothermic limits, which can be connected to two general principles of physical organic chemistry. One of them is the linear free energy relationship, exemplified by the Evans–Polanyi rule [27], which states that there exists a linear correlation between the activation energy (which is closely related, but not identical, to the barrier height) and the reaction enthalpy. As noted above, the barrier height vs. reaction energy curve is linear at the two extremes, so that one can expect that a linear free energy relationship can be observed for those a series of reactions, which are all either very exothermic or very endothermic, but not for an entire series covering the whole reaction energy range. A similar conclusion is obtained from the application of the Marcus theory to this kind of reactions [28]. The other principle is the reactivity–selectivity principle [29] according to which large reactivity is characterized by small selectivity and vice versa. In terms of reaction rates, it can be worded as: in a series of reactions, there is a large difference between the rates of neighboring members of the series if the reactions are slow, while if the reactions are fast small rate difference can be observed between two members of the series separated by the same reaction heat difference. It is exactly what one can see in Figure 22.6: the slope of the curve (which determines the difference between two neighbors) is small on the exothermic side and large on the endothermic side. This means that if two very exothermic reactions are separated by a reaction heat difference of 1 kcal/mol, the barrier height differs much less than when both reactions are very endothermic. If the barrier heights are small, the activation energies are also small, so that the reaction rates are high. In other words, the reactivity is large, the rate difference is small. On the other end of the series, the slope of the barrier height–reaction heat curve is large, the barrier height itself is also large; the rate is small and the rate differences are large. The simple model based on bond orders reproduces a number of experimental observations.

22.5 MULTIBOND REACTIONS

The most exciting application of bond order indices concerns the description of chemical reactions involving the simultaneous change of several bonds. An example is the unimolecular decomposition of ethanol, which can happen at high temperature or IR multiphoton excitation of the molecule. Out of the possible dissociation channels, the lowest barrier characterizes the concerted water loss of the molecule, yielding ethene and H_2O [30].

Scheme 22.1 shows the bond orders in the reactant, the TS and the products, calculated at the B3LYP/6-31G* geometry. Note that this geometry is essentially the same as that obtained with complete active space multiconfiguration SCF calculations. In ethanol, all bonds are single, and the calculated bond orders are very close to unity. Concerning the products, we have already seen that the bond order of the carbon–carbon double bond is very close to 2. Water has two O–H bonds both with a bond order close to 1. In the TS, the O atom is separated from the carbon atom it was connected to in ethanol by about 1.88 Å, the H atom is at 1.45 Å from the other carbon, the length of the newly formed O–H bond is 1.25 Å. The bond orders of the bonds involved in the reaction are: C–O 0.57, C–C 1.27, C–H 0.43, and O–H 0.47. At first glance, it is not easy to see whether this TS is reactant- or product-like. One can, however, bring the progress of the formation/transformation of the bonds on equal footing by calculating the progress variable [11] which shows the degree of completion of the formation, rupture or bond transformation at the TS:

$$X_{AB}^{TS} = \frac{B_{AB}^{TS} - B_{AB}^{R}}{B_{AB}^{P} - B_{AB}^{R}} \tag{22.10}$$

where B_{AB}^{R}, B_{AB}^{TS}, and B_{AB}^{P} denote the bond order of the A–B bond in the reactant, TS, and the product, respectively. These indices are: C–O 0.58, C–C 0.75, C–H 0.44, and O–H 0.50. This means that the transformation of the C–C bond from single to double is about 75% complete and the rupture of the C–O bond, the formation of the O–H bond and the rupture of the C–H bond is somewhat in a less advanced stage at the TS, around 50%. It implies that in the TS the degree of completion of some bonds differs from that of some others, and the TS can be considered to be neither reactant- nor product-like. This is not uncommon in multibond reactions.

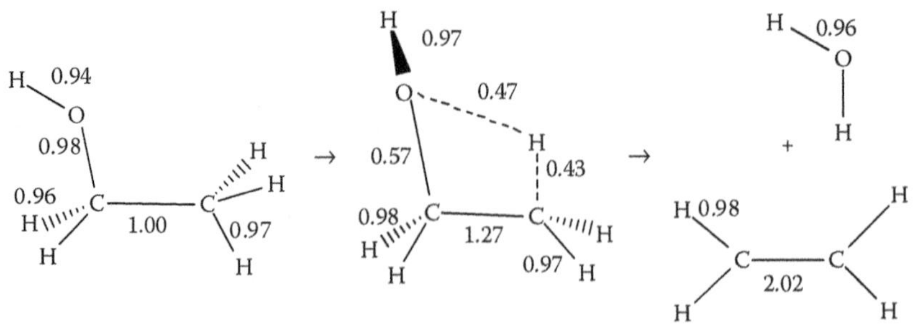

SCHEME 22.1

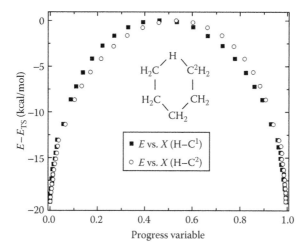

FIGURE 22.7 The potential profile as a function of the progress of the breakage of the H–C^1 and the making of the H–C^2 bond.

The definition of the progress variable can be extended to any point along the MEP, and it can be used to decide more precisely the progress of the development of bonds, both, with respect to the energetics and even with respect to each other. As an example, Figure 22.7 shows the potential energy in the 1,5-H atom transfer reaction in the 1-pentyl radical, which takes place via a six-member cyclic TS. The two curves correspond to the energy plotted against the progress variables calculated using the definition (Equation 22.10) for the breaking C^1–H and for the forming C^2–H bond. In principle, the two curves should coincide, but the bond orders do not develop perfectly in-phase, which causes the slight asymmetry. From this plot, one can see the "natural" accuracy of tendencies that can be achieved via the calculation of bond orders. In the light of the discussion on the semiquantitative nature of bond orders, the slight mismatch between the forward and backward progress of an identity reaction is not surprising and has to be accepted as an intrinsic property.

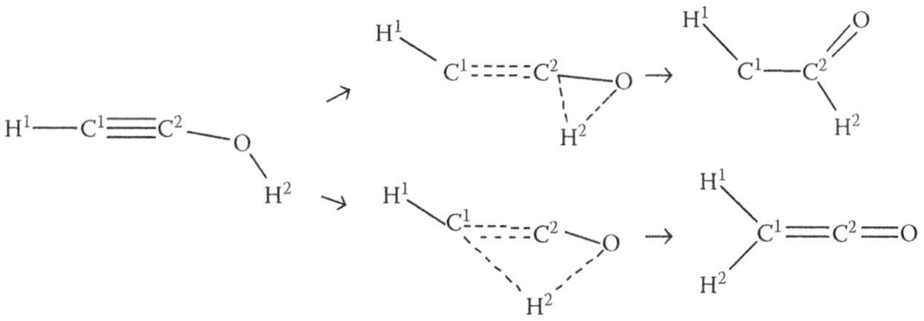

SCHEME 22.2

As another example, a detailed analysis of the isomerization reactions in the C_2H_2O system [31] (ketene–ethynol–oxirene–formyl-methylene) can be found in Ref. [11], in terms of bond orders calculated along the MEP of the possible reactions. Here, we present two cases, representing two typical scenarios, a 1,2- and a 1,3-H atom transfer reaction of ethynol. As can be seen in Scheme 22.2, in the first reaction, the alcoholic H atom moves from the O to the neighboring carbon atom, in the second to the terminal carbon. Both reactions are concerted: there is a single barrier separating the reactants from products on the MEP of both reactions. From the analysis we shall see, however, that the two differ in the microscopic mechanism: the 1,2-H atom transfer is synchronous, the 1,3-H-transfer is asynchronous. The meaning of this term will become clear soon. Shown in Figure 22.8 are the bond orders of the C–C, C–H, and O–H bonds as a function of the angle that characterizes best the change of the molecular geometry during the two reactions, namely, the H–O–C angle. (From the point of view of the nature of the TS and the progress of reaction the actual choice of the geometrical parameter is not important as long as all bond orders are plotted against the same coordinate.) During the 1,2-H atom transfer from the O atom to carbon C^1 the O–H bond breaks, a new C^1–H bond is formed and the triple carbon–carbon bond of ethynol turns into a single bond. The terminal carbon C^2 is tetravalent at the beginning and bivalent at the end: in singlet carbenes, in general, carbon can form two σ bonds, the other two valence electrons form a singlet pair and do not participate in bonding. As the left panel of Figure 22.8a shows, these processes take place simultaneously, each of them being in the same stage along the MEP. In the 1,3-H atom transfer, in addition to the breakage of the O–H bond and the formation of a new C^2–H bond, the triple C^1–C^2 bond is degraded from triple to double, and the C–O bond is converted from single to double. In the right panel of Figure 22.8a, one can see that during the reaction, the O–H^2 rupture takes place earlier than the build-up of the C^2–H^2 bond. The triple-to-double C–C conversion happens in phase with the O–H rupture, and, with a small delay with respect to these, the C–O single to double conversion takes place in phase with the formation of the C^2–H bond. Although the old bond that connected the H atom which is transferred breaks earlier than the new C–H bond is formed, the bonding capacity of H remains saturated during the reaction. The price for this is that it forms a temporary bond with the central C^2 atom it passes near by. This is not surprising: the C^2–H^2 distance is almost constant along the central part of the PES, and is shorter than either the breaking or the forming bond. The fact that the development of bonds in this reaction is not in the same phase as the reaction progresses is referred to as nonsynchronicity, a term proposed by Dewar [32]. The progress variable defined earlier makes the visualization of nonsynchronicity easy. In Figure 22.8b, the progress variables of the bonds are plotted against the same reaction coordinate as in Figure 22.8a. In the left panel the curves showing the progress of bond transformation run close together, while in the ethynol–ketene reaction (right panel) two distinct groups are obtained, representing two groups of bonds developing together. In another representation the potential energy can be plotted as a function of the progress variable of various bonds (as shown in Ref. [11]). The barrier will be at an early position on this plot for the group of bonds that change in the first stage of the reaction and at a late position for those that are completed in a later phase of the reaction.

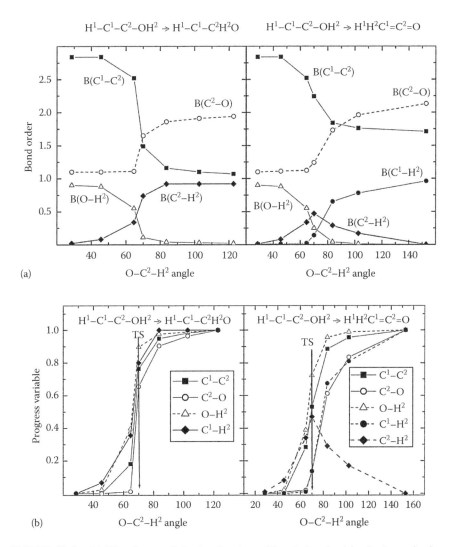

FIGURE 22.8 (a) The change of the bond orders of bonds involved in the isomerization reaction of ethynol to formyl-methylene (left panel) and ketene (right panel). (b) The progress of the transformation of the bonds involved in the isomerization reaction of ethynol to formyl-methylene (left panel) and ketene (right panel). The arrows indicate the location of the TSs along the respective MEP.

22.6 CORRELATION BETWEEN BOND ORDER AND BOND LENGTH

Bond order, as the index showing the degree of bonding between two atoms, is a measure of the strength of the interaction between them. The larger the bond order is, the stronger the bond is. On the other hand, the strength of the bond is reflected in the length of the bond: the stronger the bond, the shorter it is expected to be, and, in fact,

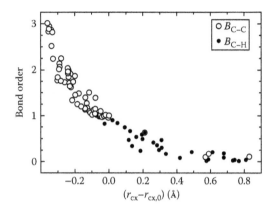

FIGURE 22.9 The bond order of the C–C and C–H bonds in the TSs of many chemical reactions as a function of the extension of the respective bond lengths with respect to the equilibrium single bond length.

can be observed be so. Obviously, there must be a correlation between the length and the bond order of a bond. The existence of such a correlation was so clear already in the early days of quantum chemistry that Pauling empirically derived a correlation, and defined the quantity called chemist's bond order via

$$n_{XY} = \exp\left[-\beta\left(R_{XY} - R_{XY}^0\right)\right] \tag{22.11}$$

where R_{XY} and R_{XY}^0 are the actual and the equilibrium single bond distances for the X–Y atom pair. Definition (22.2) enables us to calculate bond orders for a variety of molecules in a variety of reacting systems. Figure 22.9 shows the correlation for C–C and C–H bonds taken from the reactants, products, TSs and points along the MEP of many reactions. The good correlation between the ab initio bond orders and Pauling's formula indicate that an exponential correlation is rather reasonable. This helps to understand why chemist's bond orders can be used so successfully in many applications, like in the BEBO method mentioned earlier and in other approximate methods for constructing potential surfaces [33], in generating interatomic force fields in modeling surface processes, as well as in the deciding which atoms are connected when creating a visual image of a molecule.

22.7 SUMMARY

Bond order and valence indices are easy to calculate from wave functions generated in computational chemistry. The examples presented in this chapter indicate that the little computational cost can bring valuable information about changes in bonding during chemical reactions. Bond order indices help us to characterize the reactant- or product-like nature of the TS of reactions, follow the progress of the conversion of bonds, and find out whether the reaction is synchronous or not.

ACKNOWLEDGMENT

Financial support from the Hungarian Scientific Research Fund (T 43558 and T 49257) is acknowledged.

REFERENCES

1. Jencks, W. P. *Chem. Rev.* 1985, *85*, 511.
2. Pauling, L. *The Nature of the Chemical Bond*, 3rd. ed., Cornell University Press, Ithaca, NY, 1960.
3. Huheey, J. E. *Inorganic Chemistry*, 3rd. ed., Harper & Row, New York, 1983.
4. Mayer, I. *Chem. Phys. Lett.* 1983, *97*, 270, 1985, *117*, 396 (addendum).
5. Mayer, I. *Int. J. Quantum Chem.* 1986, *29*, 73, *Int. J. Quantum Chem.* 1986, *29*, 477.
6. (a) Mayer, I. *J. Comp. Chem.* 2007, *28*, 204; (b) Roy, D. R., Bultinck, P., Subramanian, V., and Chattaraj, P. K. *J. Mol. Struc. (THEOCHEM)* 2008, *854*, 35 and references therein.
7. de Giambiagi, M. S., Giambiagi, M., and Jorge, F. E. *Theor. Chim. Acta* 1985, *68*, 337.
8. Hall, G. G. Chairman's remarks, Fifth International Congress on Quantum Chemistry, Montreal, 1985; quoted in Ref. [3].
9. Hehre, W. J., Stewart, R. F., and Pople, J. A. *J. Chem. Phys.* 1969, *51*, 2657.
10. Collins, J. B., Schleyer, P. V. R., Binkley, J. S., and Pople, J. A. *J. Chem. Phys.* 1976, *64*, 5142.
11. Lendvay, G. *J. Phys. Chem.* 1994, *98*, 6098.
12. Bader, R. F. W. *Atoms in Molecules—A Quantum Theory*, Oxford University Press, Oxford, 1990.
13. Ángyán, J. G., Loos, M., and Mayer, I. *J Phys Chem.* 1994, *98*, 5244.
14. Mayer, I. Program Bondord, available at http://occam.chemres.hu
15. Dupuis, M., Marquez, A., and Davidson, E. R. HONDO 2000.
16. Gaussian 03, Revision B.03; Gaussian, Inc.: Pittsburgh, PA. Gaussian 98, Revision A.7; Gaussian, Inc.: Pittsburgh, PA.
17. Amsterdam Density Functional (ADF) program, www.scm.com
18. GAMESS, www.msg.chem.iastate.edu/gamess/
19. Ángyán, J. G., Rosta, E., and Surján, P. *Chem. Phys. Lett.* 1999, *299*, 1.
20. Johnston, H. S. *Adv. Chem. Phys.* 1960, *3*, 131.
21. Parr, C. and Johnston, H. S. *J. Am. Chem. Soc.* 1963, *85*, 2544.
22. Johnston, H. S. *Gas Phase Reaction Rate Theory*, Ronald Press, New York, 1966.
23. Bérces, T. *Int. J. Chem. Kinet.* 1980, *12*, 123.
24. Bérces, T. *Int. J. Chem. Kinet.* 1980, *12*, 183.
25. Lendvay, G. *J. Phys. Chem.* 1989, *93*, 4422.
26. Hammond, G. S. *J. Am. Chem. Soc.* 1955, *77*, 334.
27. Evans, M. G. and Polanyi, M. *Trans. Faraday Soc.* 1938, *34*, 11.
28. Marcus, R. A. *J. Phys. Chem.* 1968, *72*, 791.
29. Johnson, C. D. *Chem. Rev.* 1975, *75*, 755.
30. Yamabe, T., Koizumi, M., Yamashita, K., and Tachibana, A. *J. Am. Chem. Soc.* 1984, *106*, 2255.
31. Tanaka, K. and Yoshimine, M. *J. Am. Chem. Soc.* 1980, *102*, 7655.
32. Dewar, M. J. S. *J. Am. Chem. Soc.* 1984, *106*, 209.
33. Laganà, A. *J. Chem. Phys.* 1991, *95*, 2216.

23 Variation of Local Reactivity during Molecular Vibrations, Internal Rotations, and Chemical Reactions

S. Giri, D. R. Roy, and P. K. Chattaraj

CONTENTS

23.1 INTRODUCTION

Conceptual density functional theory (DFT) [1–7] has been quite successful in explaining chemical bonding and reactivity through various global and local reactivity descriptors as described in the previous chapters. The Fukui function (FF) [4,5] is an important local reactivity descriptor that is used to describe the relative reactivity of the atomic sites in a molecule. The FF [4,5] is defined as

$$f(\vec{r}) = [\delta\mu/\delta v(\vec{r})]_N = [\partial\rho(\vec{r})/\partial N]_{v(\vec{r})} \qquad (23.1)$$

where $\rho(\vec{r})$ is the density of an N-electron system and the chemical potential and external potential are denoted μ and $v(\vec{r})$, respectively. As proposed by Yang and Mortier [6], the condensed FFs, $\{f_k^\alpha\}$, may be expressed using a finite difference method, giving:

$$f_k^+ = q_k(N+1) - q_k(N) \quad \text{for nucleophilic attack} \qquad (23.2a)$$

$$f_k^- = q_k(N) - q_k(N-1) \quad \text{for electrophilic attack} \qquad (23.2b)$$

$$f_k^o = [q_k(N+1) - q_k(N-1)]/2 \quad \text{for radical attack} \quad (23.2c)$$

where q_k is the electronic population of atom k. The electronic populations of the $N+1$-electron and $N-1$-electron species are computed at the optimized geometry of the N-electron molecule, so that the condensed FF measures the change in atomic population at fixed molecular geometry.

23.2   LOCAL REACTIVITY PROFILES

In this chapter, we study the variation in the FF during asymmetric stretching and bending in ammonia, internal rotation in H_2O_2, and along the intrinsic reaction coordinate (IRC) of three prototypical examples of chemical reactions, viz., (1) a thermoneutral reaction, such as a symmetrical gas-phase S_N2 type nucleophilic substitution:

$$F_a^- + CH_3 - F_b \rightarrow F_a - CH_3 + F_b^- \quad (23.3)$$

(2) an endothermic reaction

$$HCN \rightarrow CNH \quad (23.4)$$

and (3) an exothermic reaction

$$CH_3F + H' \rightarrow CH_3 + H'F. \quad (23.5)$$

In Chapter 22, the variation in bond order is described. In order to understand the change in local reactivity during vibration, rotation, and reaction, the energy profiles and transition states (TSs) of the aforementioned reactions are determined by DFT calculations (using B3LYP/6–311 + G(d,p) and the Gaussian 03 and 98 [8] programs). After the TS structures for Reactions 23.3 through 23.5 were determined, the minimum energy path (MEP) was constructed using Fukui's steepest-descent method from the TS [9]. All along the MEP, the condensed FFs [6], f_k^α, were computed using Equation 23.2. The atomic populations are calculated using the Mulliken population analysis (MPA) [10] scheme. For reactions 23.3 through 23.5, respectively, the bond making and bond breaking processes are breaking of C–F$_b$, H–C, and C–F bonds and making of F$_a$–C, N–H, and F–H' bonds.

Figure 23.1 depicts the optimized structures of NH_3 and H_2O_2. The asymmetric stretching and bending modes of NH_3 are studied. For this purpose, the distortions in bond length (ΔR) and bond angle ($\Delta\theta$) are performed along the symmetry modes as described by Pearson and Palke [11]. The profiles of the energy and the FFs for the stretching and bending modes are presented in Figures 23.2 and 23.3, respectively. Due to symmetry, all H-atoms are equally reactive at the equilibrium ($\Delta R = 0$, $\Delta\theta = 0$) configuration. Reactivity of H_3 and H_4 (alternately N–H_3 and N–H_4 bond lengths are increasing and decreasing) varies in the opposite sense during the asymmetric stretching whereas that of N and H_2 remains more or less constant. For the bending mode, f_k^+ and f_k^- show opposing trends for all three

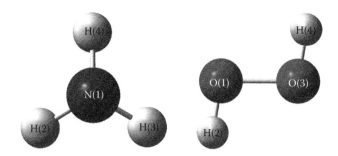

FIGURE 23.1 Optimized geometries of the NH_3 and H_2O_2 molecules.

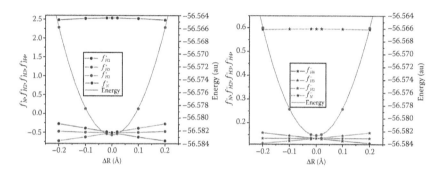

FIGURE 23.2 (See color insert following page 302.) Profiles of (a) energy and FF values for nucleophilic attack and (b) for electrophilic attack for distortion in the N–H bond distance of ammonia.

H-centers. Moreover, both, f^+ and f^- of H_3 and H_4 exhibit changes in the opposite directions.

As shown in Figure 23.4, the energy profile for internal rotation in H_2O_2 [12] has a local maximum at $\omega = 0°$, a local minimum at $\omega = 60°$, and a global maximum at $\omega = 180°$. The susceptibility to electrophilic attack at the O and H centers does not change appreciably. A global maximum (minimum) is observed for f_H^+ (f_O^+) at $\omega = 0°$ and a local maximum (minimum) is observed for f_H^+ (f_O^+) at $\omega = 180°$. While $f_H^+ > f_O^+$ at $\omega = 0°$, $f_O^+ > f_H^+$ at $\omega = 180°$. This crossover, however, does not take place exactly at $\omega = 60°$.

Figure 23.5 shows the profiles of $f_{F_a}^-$ and $f_{F_b}^-$ for the thermoneutral reaction (Reaction 23.3) as a function of the IRC, the distance along the MEP. In this reaction, the nucleophile F_a^- first forms a strongly electrostatically bound complex in which the bonds in CH_3F_b are intact, and a similar complex is formed between the products. These complexes are taken to be reactants and products in the figure. The profiles of bond order [13] and the energy are also provided. The maximum energy structure (IRC = 0) corresponds to the TS. The effectiveness of electrophilic attack (i.e. nucleophilicity) at the F_a^- (F_b^-) center gradually decreases (increases) and passes through an inflection point around the TS and then levels off. The reactivity of F_a^- is

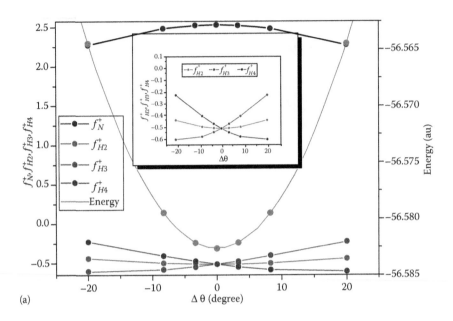

(a)

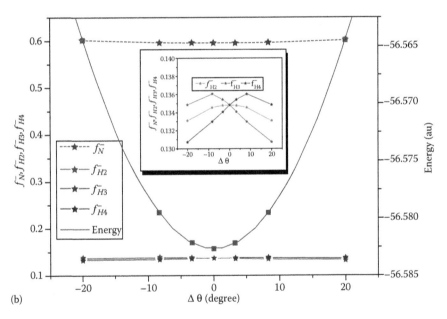

(b)

FIGURE 23.3 (See color insert following page 302.) Profiles of (a) energy and FF for nucleophilic attack and (b) electrophilic attack for distortion in the H–N–H bond angle of ammonia.

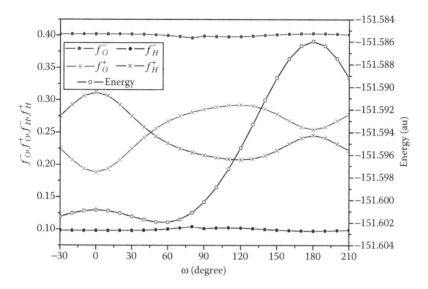

FIGURE 23.4 (See color insert following page 302.) Profiles of energy and FF values due to the internal rotation in H_2O_2 molecule.

the maximum before the reaction and decreases in the presence of CH_3–F_b due to the incipient formation of a bond with the carbon atom. In the course of the reaction, the reactivity of F_b increases because it is in the process of breaking a bond

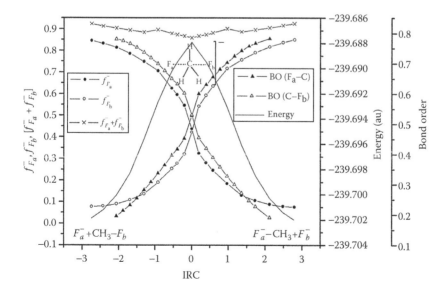

FIGURE 23.5 Profiles of different local reactivity descriptors (electrophilic attack) along the path of the gas phase S_N2 substitution: $F_a^- + CH_3$—$F_b \rightarrow F_a$—$CH_3 + F_b^-$. Profiles of energy and bond order are also shown. (Reprinted from Chattaraj, P.K. and Roy, D.R., *J. Phys. Chem. A*, 110, 11401, 2006. With permission.)

and becoming an anion. In the TS, the reactivities of both, the F_a and F_b are equal. Therefore, the reactivity can complement the pictures of bonding and interaction. The intersection point between $f_{F_a}^-$ and $f_{F_b}^-$ coincides with the saddle point of this S_N2 reaction and in addition to the energy and bond order profiles, it may be used in locating the TS (IRC $= 0$). The coincidence is, however, due to symmetry and neither bond order nor Fukui indices can be used to strictly locate the TS of an arbitrary reaction. Like the bond order conservation [14,15], a principle of reactivity conservation appears to exist: [7] $[f_{F_a}^- + f_{F_b}^-]$ remains approximately constant along the IRC.

The profiles of f_N^- and f_C^- for the endothermic reaction (Reaction 23.4) are shown in Figure 23.6. Also superimposed are the profiles of bond orders of the C–H and H–N bonds and the energy. It may be noted that the point of intersection between f_N^- and f_C^- lies slightly left to the TS (toward the reactant). At the TS, the bond orders of C–H and H–N bonds are closer to their values at the product limit than to the reactant limit, i.e., the TS is product-like, in agreement with the Hammond postulate [16]. Based on this, one can expect that as both bond breaking and bond forming are more than 50% complete, the intersection point between lines showing the progress of these two processes lies slightly left to the TS (toward the reactant) and at the TS, it already crossed the intersection point conforming to the product reactivity pattern.

The sum of the reactivities of the C and N atoms, $[f_N^- + f_C^-]$, remains more or less conserved when compared to the variations in f_N^-, f_C^-, and E. The principle of reactivity conservation, however, seems to be obeyed less well than in the thermoneutral reaction (Reaction 23.3).

Figure 23.7 presents the profiles of $f_C^-, f_{H'}^-, [f_C^- + f_{H'}^-]$, E, and the F–H' and C–F bond orders along the MEP of the exothermic reaction. The point of intersection between f_C^- and $f_{H'}^-$ lies slightly right to the TS (toward the product). The net reactivity given by $[f_C^- + f_{H'}^-]$ remains more or less conserved. For an exothermic

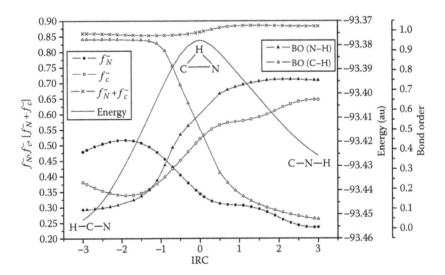

FIGURE 23.6   Same as Figure 23.5 but for the gas phase endothermic reaction: HCN → CNH.

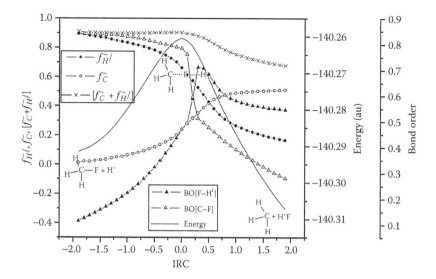

FIGURE 23.7 Same as Figure 23.5 but for the gas phase exothermic reaction: $CH_3F + H' \rightarrow CH_3 + H'F$.

reaction, the TS lies closer to the reactant in energy and geometry and the corresponding intersection point lies slightly right to the TS, in accordance with the Hammond postulate [16].

Therefore, for a thermoneutral reaction, the intersection point between the bond order profiles for the bond making and the bond breaking processes coincides with the TS; the reactivity of the two reacting atoms also equalizes at the TS, as can be seen from the intersection of their FF profiles. These intersection points of the associated bond orders and condensed FFs lie toward the left (right) of the TS for an endothermic (exothermic) reaction, in agreement with the Hammond postulate.

ACKNOWLEDGMENTS

We thank Council of Scientific and Industrial Research, New Delhi, for financial assistance and Professors G. Lendvay and P. W. Ayers for helpful discussions.

REFERENCES

1. Parr, R. G. and Yang, W. *Density functional Theory of Atoms and Molecules*, Oxford University Press, New York, 1989.
2. Geerlings, P., De Proft, F., and Langenaeker, W. *Chem. Rev.* 2003, *103*, 1793.
3. Chattaraj, P. K., Nath, S., and Maiti, B. Reactivity descriptors, in *Computational Medicinal Chemistry for Drug Discovery*, Tollenaere, J., Bultinck, P., Winter, H. D., and Langenaeker, W. (Eds.), Marcel Dekker, New York, 2004, Chapter 11, pp. 295–322.
4. Parr, R. G. and Yang, W. *J. Am. Chem. Soc.* 1984, *106*, 4049.
5. Ayers, P. W. and Levy, M. *Theor. Chem. Acc.* 2000, *103*, 353.

6. Yang, W. and Mortier, W. J. *J. Am. Chem. Soc.* 1986, *108*, 5708.

7. Chattaraj, P. K. and Roy, D. R. *J. Phys. Chem. A* 2006, *110*, 11401.

8. Gaussian 03, Revision B.03; Gaussian, Inc.: Pittsburgh, PA. Gaussian 98, Revision A.7; Gaussian, Inc.: Pittsburgh, PA.

9. Fukui, K. *Acc. Chem. Res.* 1981, *14*, 363.

10. Mulliken, R. S. *J. Chem. Phys.* 1955, *23*, 1833.

11. Pearson, R. G. and Palke, W. E. *J. Phys. Chem.* 1992, *96*, 3283.

12. Gutiérrez-Oliva, S., Letelier, J. R., and Toro-Labbé, A. *Mol. Phys.* 1999, *96*, 61 and references therein.

13. Ángyán, J. G., Loos, M., and Mayer, I. *J. Phys. Chem.* 1994, *98*, 5244.

14. Lendvay, G. *J. Phys. Chem.* 1989, *93*, 4422; Lendvay, G. *THEOCHEM* 1988, *167*, 331.

15. Maity, D. K. and Bhattacharyya, S. P. *J. Am. Chem. Soc.* 1990, *112*, 3223.

16. Hammond, G. S. *J. Am. Chem. Soc.* 1955, *77*, 334.

24 Reactivity and Polarizability Responses

Patrick Senet

CONTENTS

24.1 INTRODUCTION

Density functional theory (DFT) [1,2] is widely applied to evaluate atomic and molecular polarizabilities. The interested reader may consult, for instance, excellent books [3,4] covering the main formal and technical aspects of this topic and its main applications to chemistry. Our purpose in this chapter is different: we emphasize the relations between the responses of a molecule to a potential and its chemical reactivity. After an elementary introduction to polarizabilities, we review the relations between atomic and molecular polarizabilities and DFT reactivity descriptors [5–8], namely the chemical hardnesses and softnesses [9–13], and the Fukui functions [14,15]. We also present new definitions and interpretations of these concepts starting from the first principles.

DFT reactivity theory is a perturbative approach in which the chemical descriptors are defined for isolated chemical moieties. For this reason, they provide only an indirect and approximative description of the actual reaction between two reagents, which approach each other at a chemical bond distance in well-defined physicochemical conditions (temperature, pressure, salt, etc.). It is legitimate to have doubts about the relevance of the DFT descriptors for chemistry. Are the DFT descriptors really useful? The answer to this provocative question is yes for the following reasons.

From a theoretical point of view, a knowledge of the many-body free energy surface is required for a complete microscopic description of a chemical reaction between reagents. By including potential of mean forces (describing the effective interactions between the atoms at finite temperature) and a heat bath, one may, in principle, simulate the formation and the break of chemical bonds by using a formalism "à la Langevin." Such a complete statistical approach is rarely feasible in practice. Indeed, an essential ingredient of the free energy surface is the underlying potential energy surface. Calculations of accurate potential energy surface in the Born–Oppenheimer approximation is not an easy task. A potential between atoms is a small number compared to the value of the total energy of each isolated atom. An accurate evaluation of differences between large total energies is a difficult numerical task and computationally expensive. An alternative road to a direct calculation of potential energies is the application of perturbation theory [8]. Only the properties of isolated systems are computed. The propensity of an atom or molecular fragment to "react" is then evaluated by a response function, i.e., a derivative of the total energy. These derivatives are called "descriptors of the chemical reactivity" [5,6] and are discussed in many chapters of this book. The simplest descriptor is maybe the electronic density itself [16,17]. Another example is the molecular electrostatic potential used in structural biology to compare ligand–substrate interactions [16,17]. DFT descriptors describe, to a certain extent, how the energy varies when a local interaction is switched on and they help us to understand the local reactivity of large molecules.

On the other hand, there is considerable interest to quantify the similarities between different molecules, in particular, in pharmacology [7]. For instance, the search for a new drug may include a comparative analysis of an active molecule with a large molecular library by using combinatorial chemistry. A computational comparison based on the similarity of empirical data (structural parameters, molecular surfaces, thermodynamical data, etc.) is often used as a prescreening. Because the DFT reactivity descriptors measure intrinsic properties of a molecular moiety, they are in fact chemical fingerprints of molecules. These descriptors establish a useful scale of similarity between the members of a large molecular family (see in particular Chapter 15) [18–21].

The (nonlocal) polarizabilities are important DFT reactivity descriptors. But, how are polarizabilities related to chemistry? As stated above, an essential ingredient of the free energy surface is the potential energy surface and, in particular, its gradients. In a classical description of the nuclei, they determine the many possible atomic trajectories. Thanks to Feynman, one knows a very elegant and exact formulation of the force between the atoms namely [22,23]

$$\mathbf{F}_I = Z_I e^2 \left[\sum_{J \neq I} Z_J \frac{(\mathbf{R}_I - \mathbf{R}_J)}{|\mathbf{R}_I - \mathbf{R}_J|^3} + \int d\mathbf{r}' \, \rho(\mathbf{r}') \frac{(\mathbf{r}' - \mathbf{R}_I)}{|\mathbf{r}' - \mathbf{R}_I|^3} \right], \tag{24.1}$$

where

Z_I is the nuclear charge
e is the elementary charge
$\rho(\mathbf{r}')$ is the ground-state electronic density of the molecule

This famous electrostatic theorem states that the internuclear forces, which drive the formation and breakage of bonds, are equal to the forces computed from electrostatics. Of course, the simplicity of this formula hides a formidable difficulty: $\rho(\mathbf{r}')$ depends itself on the positions of all atoms, it is an unknown functional of the electron–nuclei potential $v_{\text{ext}}(\mathbf{r})$:

$$v_{\text{ext}}(\mathbf{r}) = - \sum_J \frac{e^2 Z_J}{|\mathbf{r} - \mathbf{R}_I|}. \tag{24.2}$$

$\rho(\mathbf{r}')$ must be evaluated for each nuclear configuration along the (classical) trajectory. An alternative is the calculation of its variation $\delta\rho(\mathbf{r}')$ for (small) atomic displacements by using perturbation theory, i.e., by using the polarizability responses.

They are, in fact, two possible perturbative approaches. In the first approach, one considers two (or any number) reagents A and B as a single system where A and B are two fragments in interactions. The total density is $\rho(\mathbf{r}') = \rho_A(\mathbf{r}') + \rho_B(\mathbf{r}')$ [24]. Displacement of A relative to B is an internal mechanical deformation of the complete system: it is related to response function of $A + B$. For instance, a small displacement $\mathbf{U}_K$ of an atom K of B will modify the force on the other atoms by

$$\Delta\mathbf{F}_I = Z_I Z_K e^2 \left[\frac{(\mathbf{R}_I - \mathbf{R}_K - \mathbf{U}_K)}{|\mathbf{R}_I - \mathbf{R}_K - \mathbf{U}_K|^3} - \frac{(\mathbf{R}_I - \mathbf{R}_K)}{|\mathbf{R}_I - \mathbf{R}_K|^3} \right]$$
$$+ Z_I e^2 \int d\mathbf{r}' \, \delta\rho(\mathbf{r}') \frac{(\mathbf{r}' - \mathbf{R}_I)}{|\mathbf{r}' - \mathbf{R}_I|^3} \tag{24.3}$$

where $I \neq K$ and in which $\delta\rho(\mathbf{r}')$ is the variation of the density induced by the motion of K. $\delta\rho$ can be computed by perturbation [25–27]. To the first perturbation order, one has

$$\delta^{(1)}\rho(\mathbf{r}) = \int d\mathbf{r}' \left[\frac{\delta\rho(\mathbf{r})}{\delta v_{\text{ext}}(\mathbf{r}')} \right]_N \delta v_{\text{ext}}(\mathbf{r}') \equiv \int d\mathbf{r}' \chi_1(\mathbf{r}, \mathbf{r}') \delta v_{\text{ext}}(\mathbf{r}'), \tag{24.4}$$

in which $\chi_1(\mathbf{r}, \mathbf{r}')$ is the linear polarizability kernel [25] of the A–B system. The functional derivative in Equation 24.4 is carried out at constant electron number N. $\delta^{(1)}\rho$ is the part of the density deformation, which is proportional to the perturbation. In this approximation, one has

$$\delta^{(1)}\rho(\mathbf{r}) = +e^2 Z_k \int d\mathbf{r}' \chi_1(\mathbf{r}, \mathbf{r}') \left[\frac{1}{|\mathbf{r}' - \mathbf{R}_K|} - \frac{1}{|\mathbf{r}' - \mathbf{R}_K - \mathbf{U}_K|} \right]$$

$$\simeq -e^2 Z_K \int d\mathbf{r}' \chi_1(\mathbf{r}, \mathbf{r}') \left[\frac{(\mathbf{r}' - \mathbf{R}_K)}{|\mathbf{r}' - \mathbf{R}_K|^3} \right] \cdot \mathbf{U}_K \qquad (24.5)$$

For larger displacement $\mathbf{U}_K$, the variation of $\chi_1(\mathbf{r}, \mathbf{r}')$ relative to $\mathbf{U}_K$ can be computed by using the nonlinear polarizability kernels χ_n defined below [26] (see Section 24.4). Forces and nonlocal polarizabilities are thus intimately related.

In a second approach of the reactivity, one fragment A is represented by its electronic density and the other, B, by some "reactivity probe" of A. In the usual approach, which permits to define chemical hardness, softness, Fukui functions, etc., the probe is simply a change in the total number of electrons of A. [5,6,8] More realistic probes are an electrostatic potential ϕ, a pseudopotential (as in Equation 24.102), or an electric field $\mathbf{E}$. For instance, let us consider a homogeneous electric field $\mathbf{E}$ applied to a fragment A. How does this field modify the intermolecular forces in A? Again, the Hellman–Feynman theorem [22,23] tells us that for an instantaneous nuclear configuration, the force on each atom changes by

$$\Delta \mathbf{F}_I = e^2 \sum_J Z_J \int d\mathbf{r}' \, \delta\rho(\mathbf{r}') \frac{(\mathbf{r}' - \mathbf{R}_J)}{|\mathbf{r}' - \mathbf{R}_J|^3}, \qquad (24.6)$$

where $\delta\rho(\mathbf{r}')$ is the density induced by the field. For small electric fields, one has (see Section 24.2)

$$\delta^{(1)}\rho(\mathbf{r}) = e \int d\mathbf{r}' \chi_1(\mathbf{r}, \mathbf{r}') \mathbf{r}' \cdot \mathbf{E}, \qquad (24.7)$$

where $\chi_1(\mathbf{r}, \mathbf{r}')$ is now the linear polarizability kernel of the isolated A fragment. One may conclude that linear and nonlinear polarizabilities kernels are directly related to the forces between atoms and are thus probes of "chemical reactivity." Nonlinear responses also describe the effects of electric fields due to the environment on the linear chemical descriptors (see Chapter 25). Finally, the (nonlocal) polarizabilities are also descriptors of the similarity between molecules and fragments [28].

Polarizabilities are responses to a potential (the gradient of which is a field). On the contrary, Fukui functions, chemical hardness and softness are responses to a transfer or removal of an integer number of electrons. Both responses are DFT descriptors but the responses which involve a change in the number of electrons are called "chemical electronic responses" [26]. Formal relations between polarizabilities and chemical electronic responses have been derived in the literature [5,8,29–33]. An intriguing question arises from these works: the polarizabilities represent a polarization where no electron number is changed. How are they related to chemical electronic responses that involve an actual charge transfer? A change in the number of electrons implies a change of Hamiltonian but the polarizabilities are computed from only one well-defined Hamiltonian. The paradox is solved below by introducing a new concept: a polarization Fukui function (Equation 24.44). This response describes an internal charge transfer from a part of a molecule to another. Because an internal charge transfer does not change the total number of electrons, this Fukui function and its corresponding polarization hardness are continuous derivatives of the total energy relative to an "electron number" (partial charge) (see Section 24.3).

This chapter aims to present the fundamental formal and exact relations between polarizabilities and other DFT descriptors and is organized as follows. For pedagogical reasons, we present first the polarizability responses for simple models in Section 24.2. In particular, we introduce a new concept: the dipole atomic hardnesses (Equation 24.20). The relationship between polarizability and chemical reactivity is described in Section 24.3. In this section, we clarify the relationship between the different Fukui functions and the polarizabilities, we introduce new concepts as, for instance, the polarization Fukui function, and the interacting Fukui function and their corresponding hardnesses. The formulation of the local softness for a fragment in a molecule and its relation to polarization is also reviewed in detail. Generalization of the polarizability and chemical responses to an arbitrary perturbation order is summarized in Section 24.4.

24.2 PERTURBATION THEORY: ELEMENTARY MODELS

24.2.1 Atom in an Electric Field

General properties and definitions of polarizabilities can be introduced without invoking the complete DFT formalism by considering first an elementary model: the dipole of an isolated, spherical atom induced by a uniform electric field. The variation of the electronic density is represented by a simple scalar: the induced atomic dipole moment. This coarse-grained (CG) model of the electronic density permits to derive a useful explicit energy functional where the functional derivatives are formulated in terms of polarizabilities and dipole hardnesses.

24.2.1.1 Polarizabilities

The ground-state electronic density $\rho(\mathbf{r})$ is uniquely related to the external potential $v_{\text{ext}}(\mathbf{r})$ as stated by the fundamental theorems of DFT [1,2,8]. At zero field, the external potential of an atom is due to its nuclei and $v_{\text{ext}}(r) = -Ze^2/r$ where Z is the nuclear charge. It is shifted by the quantity V

$$V(x; E) = exE, \tag{24.8}$$

when a field E is applied along the x Cartesian direction ($e = 1.602 \times 10^{-19}$ C is the elementary charge). This variation of the potential induces a deformation of the electronic density, which is a functional of V and a function of E. The variation, $\delta\rho(\mathbf{r}; E) = \rho(\mathbf{r}; E) - \rho(\mathbf{r}; 0)$ can be quantified by its first moment, i.e., by the atomic dipole moment p induced by the field:

$$p(E) \equiv -e \int dx \, x \, \delta\rho(\mathbf{r}; E). \tag{24.9}$$

The polarizabilities α_n are defined by the derivatives of the density (represented here by the dipole p) relative to the potential applied (represented here by the field E):

$$\alpha_n \equiv \left[\frac{\partial^n p}{\partial E^n} \right]_0, \tag{24.10}$$

where the derivative is evaluated at zero field. Because we consider the response to a field applied along one direction, the polarizabilities are simple scalars instead of tensors or kernels (see below). (For a complete description of the tensorial character of the polarizabilities, the reader may consult, for instance, Ref. [4].) The dipole-induced $p(E)$ is computed by using a Taylor expansion around the reference state ($E = 0$):

$$p(E) = \sum_{n=1}^{\infty} \frac{\alpha_n}{n!} E^n \equiv \sum_{n=1}^{\infty} \delta p_n. \tag{24.11}$$

α_n is called the response of nth order because the variation δp_n is proportional to the nth power of the perturbation. The leading term in Equation 24.11 is in general the linear response ($n = 1$), the nonlinear terms ($n \neq 1$) are called "hyperpolarizabilities" and are important in nonlinear optics [34].

The polarizabilities are useful only if the series Equation 24.11 converges, i.e., in general for values of E not too large. In this case, the series can be truncated to the lowest orders. For the purpose of this chapter, we consider nonlinear responses up to the order 5:

$$p(E) = \alpha_1 E + \frac{\alpha_3}{3!} E^3 + \frac{\alpha_5}{5!} E^5 + O(E^7). \tag{24.12}$$

The expansion (Equation 24.12) does not contain even powers of the field because of the spherical symmetry of an isolated atom. Indeed for an atom, the even derivatives in Equation 24.10 are zero as well as for any molecule having an inversion center. Note that α_3 and α_5 are, in fact, the components of tensors, respectively of the so-called second and fourth hyperpolarizabilities [4].

A variation of the density (p) implies a variation of the energy (ε). The variation of the energy of the atom is the work done by the field

$$\Delta\varepsilon(E) \equiv - \int_0^E dE' \, p(E'). \tag{24.13}$$

Using Equation 24.12 in Equation 24.13, one finds

$$\Delta\varepsilon(E) = -\frac{1}{2}\alpha_1 E^2 - \frac{\alpha_3}{4!}E^4 - \frac{\alpha_5}{6!}E^6 + O(E^8). \qquad (24.14)$$

Equation 24.14 provides an alternative definition of the electronic responses: they are derivatives of the energy ε relative to the field E. Note that the response of order n, the nth derivative of the response to the perturbation, is the $n+1$th derivative of the energy relative to the same perturbation. Hence, the linear response α_1 is a second derivative of the energy. Because the potential (E) and the density (p) are uniquely related to each other, the field can be formulated as a function of the dipole moment p. The expansion of the field in function of p can be obtained from Equation 24.12 which can be easily inverted to give

$$E \equiv \eta_1 p + \frac{\eta_3}{3!}p^3 + \frac{\eta_5}{5!}p^5 + O(p^7), \qquad (24.15)$$

in which we have defined

$$\eta_1 = \frac{1}{\alpha_1},$$

$$\eta_3 = -\frac{\alpha_3}{(\alpha_1)^4}, \qquad (24.16)$$

$$\eta_5 = \frac{10(\alpha_3)^2 - \alpha_1\alpha_5}{(\alpha_1)^7}.$$

As it can be easily checked, Equations 24.12 through 24.16 are consistent. $\Delta\varepsilon$ in Equation 24.14 represents the energy gained by an atom in a field E. It can be reformulated in terms of the deformation of its electronic density (p) by using Equations 24.15 and 24.16 in Equation 24.14:

$$\Delta\varepsilon(E) = \Delta\varepsilon(p) = -\frac{h_1}{2}p^2 - \frac{h_3}{4!}p^4 - \frac{h_5}{6!}p^6 + \cdots, \qquad (24.17)$$

where we have introduced new quantities

$$h_1 = \eta_1,$$

$$h_3 = 3\eta_3, \qquad (24.18)$$

$$h_5 = 5\eta_5.$$

24.2.1.2 Dipole Hardnesses

Equation 24.17 shows that the energy gained by the system when a field E is applied is a function of the electronic density represented by p. According to the variational principle of DFT, the energy in the ground state (in the absence of a field) is minimum [1,2].

Therefore, at zero field, any variation of the p relative to its ground-state value cost an energy we note as $\tilde{\Delta}\varepsilon$. The energy cost is a function of p and is exactly

$$\tilde{\Delta}\varepsilon(p;0) = -\Delta\varepsilon(E) = \frac{h_1}{2}p^2 + \frac{h_3}{4!}p^4 + \frac{h_5}{6!}p^6 + O(p^8). \qquad (24.19)$$

Equation 24.19 is valid for any dipole p which is "E-representable": For an isolated atom, all dipoles are E-representable because any dipole p can be viewed as induced by some uniform electric field E. We call the derivatives of the energy relative to the dipole p

$$h_n \equiv \frac{\partial^n \tilde{\varepsilon}}{\partial p^n}, \qquad (24.20)$$

the dipole hardnesses. They are analogous to the so-called linear ($n = 1$) and nonlinear ($n \neq 1$) hardness kernels [5,8,26,30] introduced in the DFT theory of reactivity (see Section 24.3).

The hardness h_n are intimately related to the linear and nonlinear electronic responses as shown explicitly in Equation 24.18. In particular, h_1 is simply the inverse of the linear polarizability: it is well known in chemistry that a "hard atom" has a low polarizability. The nonlinear terms $h_{n \neq 1}$, could allow to better quantify the hardness/softness and polarizability relations (see Section 24.2.2). Note that for an atom in a molecule, the contribution of α_2 has to be considered as well in Equation 24.12 through Equation 24.18. On the other hand, Equation 24.18 shows that all the polarizabilities can be formulated in terms of the linear one, if the derivatives h_n, which are function of p, are known:

$$\alpha_3 = -h_3(\alpha_1)^4/3,$$
$$\alpha_5 = \frac{10(\alpha_3)^2 - h_5/5(\alpha_1)^7}{\alpha_1}. \qquad (24.21)$$

Equations 24.21 are very particular cases of a general theorem for the responses demonstrated previously [26]. It is interesting to note that the evaluation of nonlinear hyperpolarizabilities is a stringent test of the validity and robustness of exchange-correlation functionals [35]. Equation 24.21 permits to explain qualitatively why: the electrostatic part of energy does not contribute at all to h_3, which depends only on the exchange-correlation functional. On the contrary, h_1 is dominated by the Coulomb propagator.

24.2.2 Linear and Nonlinear Dipole Hardnesses: Numerical Results

As the formation of a covalent bond between two atoms implies a (dipolar) deformation of the density, polarizability and reactivity must be related. Indeed, Nagle demonstrated an empirical relation between the atomic polarizabilities (response to a field) and the scales of electronegativities (reactivity) [36]. More precisely, various scales of atomic electronegativities are reproduced by the following linear relation [36]:

$$\chi = A \left[\frac{n}{\alpha_1} \right]^{1/3} + B, \qquad (24.22)$$

where the exact values of the constants A and B depend on the electronegativity scale chosen and where n is the number of valence electrons. The first term in Equation 24.22 can be interpreted as the inverse of an atomic radius R (as $\alpha_1 \simeq R^3$).

The concept of dipole hardness permit to explore the relation between polarizability and reactivity from first principles. The physical idea is that an atom is more reactive if it is less stable relative to a perturbation (here the external electric field). The atomic stability is measured by the amount of energy we need to induce a dipole. For very small dipoles, this energy is quadratic (first term in Equation 24.19). There is no linear term in Equation 24.19 because the energy is minimum relative to the dipole in the ground state (variational principle). The curvature h_1 of $E(p)$ is a first measure of the stability and is equal exactly to the inverse of the polarizability. Within the quadratic approximation of $E(p)$, one deduces that a low polarizable atom is expected to be more "stable" or "less reactive" as it does in practice. But if the dipole is larger, it might be useful to consider the next perturbation order:

$$\tilde{\Delta}\varepsilon(p;0) \simeq \frac{h_1}{2}p^2 + \frac{h_3}{4!}p^4 = \left[\frac{1}{2\alpha_1} \right]p^2 - \left[\frac{\alpha_3}{8(\alpha_1)^4} \right]p^4. \qquad (24.23)$$

We have computed Equation 24.23 for various elements of the Mendeliev table by using the recommended theoretical values of the atomic polarizabilities α_1 and α_3 published recently (see Tables 3 and 10 in Chapter IV of Ref. [4]). The results are presented in Figures 24.1 and 24.2. In the later figure, we have assumed $\alpha_3 = 0$. By comparing Figures 24.1 and 24.2, we observe that the nonlinearity (h_3) has a major effect only for hydrogen (curve with diamonds) and helium (curve with squares), at least in the range of dipole moment considered (<0.5 D). For these two light elements, there is an "activation barrier": the energy cost to create a dipole has a maximum at about $p = 0.36$ D (0.141 a.u.). For He, this value corresponds to a static field of about 0.014 a.u. (Equation 24.15) enough to induce (weak) nonlinear effects [37]. Such a maximum exists probably also for the other elements but at higher dipole moments. However, for large dipole moments (not represented), we should probably include the higher order terms (h_5, h_7, ...), which are unknown. In Figure 24.1, the value of the energy at $p = 0.5$ D, for instance, decreases in the following order: Ne, F, O, N, Ar, C, Cl, S, B, P, Si–Be, Al, Mg, Na–Li. The pairs of atoms Si–Be and Li–Na follow nearly the same energy curves. It is interesting to note that Ne is the most stable atom and Na–Li is less stable. The scale of stability is not modified if we neglect h_3 (Figure 24.2), the energy is the highest for Ne (most stable) and the lowest for Li–Na (less stable). The stability, expected for H and He, is therefore determined by α_1 only.

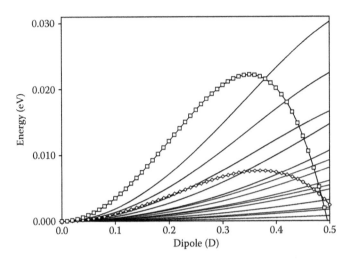

FIGURE 24.1 $E(p)$ function as computed from Equation 24.23 in the text. The curve with squares (diamonds) represent the result for He (H). The other curves are for the elements Ne, F, O, N, Ar, C, Cl, S, B, P, Si–Be, Al, Mg, Na–Li in the order of decreasing energy. The curves are for Li and Na (Si and Be) cannot be distinguished at the scale drawn.

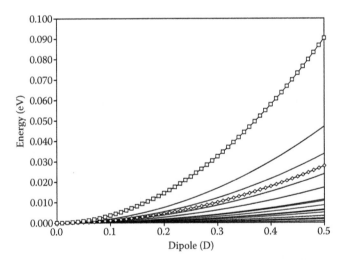

FIGURE 24.2 $E(p)$ function as computed from Equation 24.23 in the text with $h_3 = 0$. The curve with squares (diamonds) represent the result for He (H). The other curves are for the elements Ne, F, O, N, Ar, C, Cl, S, B, P, Si–Be, Al, Mg, Na–Li in the order of decreasing energy. The curves for Li and Na (Si and Be) cannot be distinguished at the scale drawn.

The peculiar behavior of H might be relevant to understand the hydrogen bond, which deforms the electronic cloud of the proton. On the other hand, it is surprising to discover an "anomalous" behavior for a closed-shell atom like He. However, it has been demonstrated in helium-atom-scattering that interactions between He atoms and metal surfaces are not simple: He is not an "inert" probe [27,38]. There is hybridization of the valence electronic states of He with the electronic wave function at the Fermi level of metals at several angstroms from the surface and He behaves very differently from Ne [38].

Figure 24.1 is the first application of a concept of "dipole hardness." There is now an entire field to explore: what will this concept tell us about reactivity of molecules? How are dipole hardnesses related to vibrational properties the energy of which is the order of magnitude considered in Figure 24.1?

24.2.3 A MOLECULE IN AN ELECTRIC FIELD: THE BARE AND SCREENED RESPONSE FUNCTIONS

24.2.3.1 Theory

One uses a simple CG model of the linear responses ($n = 1$) of a molecule in a uniform electric field E in order to illustrate the physical meaning of the screened electric field and of the bare and screened polarizabilities. The screened nonlocal CG polarizability is analogous to the exact screened Kohn–Sham response function χ_s (Equation 24.74). Similarly, the bare CG polarizability can be deduced from the nonlocal polarizability kernel χ_1 (Equation 24.4). In DFT, χ_1 and χ_s are related to each other through another potential response function (PRF) (Equation 24.36). The latter is represented by a dielectric matrix in the CG model.

The CG model represents the deformation of the electronic density by a collection of atomic dipoles. Although the definition of the dipole moment of an atom $\mathbf{p}$ in a molecule is not unique, the partitioning of the molecular dipole moment in atomic contributions can be built rigorously by partitioning the electronic density [39,40] in fragments (see Chapter 14). This permits to compute the polarizability of an atom in a molecule (see, for instance, Refs. [41–43]). In fact, all atomic multipoles (in particular, atomic charges [44]) can be properly defined using a density partitioning [45].

In the present model, each atom numbered by $J = 1, 2, \ldots$ has a screened nonlocal polarizability tensor, $\overset{\leftrightarrow}{\tilde{\alpha}}(K, J)$ defined by

$$\tilde{\alpha}_{uv}(J, K) = \left[\frac{\partial p_u(J)}{\partial \tilde{E}_u(K)} \right]_0, \tag{24.24}$$

in which $p_u(J)$ is the Cartesian component u of the dipole moment of atom J induced by the field $\mathbf{E}$ applied. The derivative in Equation 24.24 is evaluated at zero applied field and $\tilde{E}_v(K)$ is the local field in the direction Cartesian v evaluated at the position of the atom K. The local screened polarizability is the diagonal element of the nonlocal polarizability matrix: $\tilde{\alpha}_{uv}(K, K) \equiv \tilde{\alpha}_{uv}(K)$. Obviously, for an atom in a molecule, $\overset{\leftrightarrow}{\tilde{\alpha}}(K)$ is different from the polarizability of an isolated atom $\overset{\leftrightarrow}{\alpha}_1$ (Equation 24.10) because the atom is bonded. In addition, the response of an atom to a field is nonlocal (Equation 24.24) in a molecule.

The screened field, $\tilde{E}_v(K)$, corresponds to the total electric field evaluated at the atom K, which is not, in general, simply the field applied because of the influence of the other atoms. Indeed, each atom bears a dipole moment induced by the applied field. Each dipole generates itself a field (proportional to the magnitude of the dipole). All these fields add to define the local field felt by an atom within the molecule. The mutual polarization of the atoms is described by the following general relation:

$$\tilde{E}_u(K) = E_u(K) + \sum_{L \neq K} \sum_u U_{vu}(K, L) E_u(L), \tag{24.25}$$

or in matrix notation

$$\tilde{\mathbf{E}} = \mathbf{E} + \overset{\leftrightarrow}{\mathbf{U}} \mathbf{E}, \tag{24.26}$$

where E_v, (K) is the value of the field applied at site K and $\overset{\leftrightarrow}{\mathbf{U}}$ is a propagator represented by a tensor:

$$U_{vu}(K, L) \equiv \left[\frac{\partial \tilde{E}_u(K)}{\partial E_u(L)}\right]_0 \quad \forall K \neq L, \tag{24.27}$$

which measures the influence of the applied field at site L on the local field at site K. In other words, $\overset{\leftrightarrow}{\mathbf{U}}$ "propagates" the perturbation of the atom L to the atom K.

Equation 24.25 can be written as

$$\tilde{\mathbf{E}} = \overset{\leftrightarrow}{\mathbf{M}}^{-1} \mathbf{E}, \tag{24.28}$$

where $\overset{\leftrightarrow}{\mathbf{M}}$ is a nonlocal field response, the inverse of which is given by

$$M_{vu}^{-1} = \left[\frac{\partial \tilde{E}_u(K)}{\partial E_u(L)}\right]_0 = \delta_{vu}\delta_{JK} + U_{vu}(J, K). \tag{24.29}$$

The physical meaning of Equation 24.28 is clear: the local field felt by an atom is not the bare applied field but a field screened by $\overset{\leftrightarrow}{\mathbf{M}}^{-1}$, which is the inverse of a molecular nonlocal field response. The bare nonlocal polarizability $\overset{\leftrightarrow}{\alpha}(L, J)$ describing the response of the atom (dipole induced at) L to the bare field $\mathbf{E}$ applied at the atom J is also related to $\overset{\leftrightarrow}{\mathbf{M}}^{-1}$ by

$$\alpha_{vu}(K, J) = \frac{\partial p_u(K)}{\partial E_u(J)} = \sum_L \sum_t \frac{\partial p_u(K)}{\partial \tilde{E}_t(L)} \frac{\partial \tilde{E}_t(L)}{\partial E_u(J)}, \tag{24.30}$$

$$\overset{\leftrightarrow}{\alpha} = \overset{\leftrightarrow}{\tilde{\alpha}} \overset{\leftrightarrow}{\mathbf{M}}^{-1}.$$

24.2.3.2 Numerical Application

A large number of dipole models have been developed in the past to predict the molecular polarizabilities [46,47]. In these models, two atoms interact via Coulomb forces between dipoles centered on their nuclei. The atomic dipoles are point or extended dipoles [47]. From these models, one deduces:

$$M_{vu}(K,L) = \delta_{vu}\delta_{KL} - (1 - \delta_{KL})$$

$$\times \left[\lambda_5 \frac{3(R_{Ku} - R_{Lu})(R_{Kv} - R_{Lv})}{R_{KL}^5} - \lambda_3 \frac{\delta_{uv}}{R_{KL}^3}\right] \tilde{\alpha}_{uv}(K,L), \quad (24.31)$$

where λ_5 and λ_3 are damping constants equal to 1 for point dipoles. Equation 24.31 can be applied to any molecule. In simple models, $\tilde{\alpha}_{uv}(K,L) = \delta_{uv}\delta_{KL}\tilde{\alpha}$ [46]. Let us apply Equation 24.31 to a homopolar diatomic molecule with a bond along the Cartesian axis X (bond length $= L$). Each atom has an isotropic (screened) atomic polarizability $\tilde{\alpha}_{uv}(K,L) = \delta_{uv}\delta_{KL}\tilde{\alpha}$. For this simple model, one finds a block diagonal matrix $\overleftrightarrow{\mathbf{M}}$

$$\overleftrightarrow{\mathbf{M}} = \begin{bmatrix} 1 & 0 & 0 & \chi & 0 & 0 \\ 0 & 1 & 0 & 0 & -\frac{\chi}{2} & 0 \\ 0 & 0 & 1 & 0 & 0 & -\frac{\chi}{2} \\ \chi & 0 & 0 & 1 & 0 & 0 \\ 0 & -\frac{\chi}{2} & 0 & 0 & 1 & 0 \\ 0 & 0 & -\frac{\chi}{2} & 0 & 0 & 1 \end{bmatrix}, \quad (24.32)$$

where $\chi \equiv -2\tilde{\alpha}/L^3$. The inverse is

$$\overleftrightarrow{\mathbf{M}}^{-1} = \begin{bmatrix} \frac{1}{1-\chi^2} & 0 & 0 & -\frac{\chi}{1-\chi^2} & 0 & 0 \\ 0 & \frac{1}{1-\chi^2/4} & 0 & 0 & \frac{\chi}{2(1-\chi^2/4)} & 0 \\ 0 & 0 & \frac{1}{1-\chi^2/4} & 0 & 0 & \frac{\chi}{2(1-\chi^2/4)} \\ -\frac{\chi}{1-\chi^2} & 0 & 0 & \frac{1}{1-\chi^2} & 0 & 0 \\ 0 & \frac{\chi}{2(1-\chi^2/4)} & 0 & 0 & \frac{1}{1-\chi^2/4} & 0 \\ 0 & 0 & \frac{\chi}{2(1-\chi^2/4)} & 0 & 0 & \frac{1}{1-\chi^2/4} \end{bmatrix}. \quad (24.33)$$

One deduces from Equation 24.30:

$$\alpha_{xx}(K) = \alpha_{xx}(L) = \frac{\tilde{\alpha}}{1 - \frac{2\tilde{\alpha}}{L^3}},$$

$$\alpha_{yy}(K) = \alpha_{yy}(L) = \frac{\tilde{\alpha}}{1 + \frac{\tilde{\alpha}}{L^3}}, \quad (24.34)$$

$$\alpha_{zz}(K) = \alpha_{yy}(K) = \alpha_{yy}(L)$$

For a Cl atom $\tilde{\alpha} = 1.934$ Å^3 in a Cl$_2$ molecule ($L = 1.988$ Å) [46]. According to Equation 24.34, the atomic polarizability (response to the bare field) along the bond is $\alpha_{xx} = 3.81$ Å^3 and the transverse polarizability (response to the bare field) is $\alpha_{yy} = 1.05$ Å^3. The measured atomic polarizabilities in Cl$_2$ (equal to half of the components of the molecular polarizability) are close: $\alpha_{xx} = 3.30$ Å^3 and $\alpha_{yy} = 1.81$ Å^3 [46].

The atomic polarizability along the bond is increased relative to $\tilde{\alpha}$: the field felt by the atom is enhanced by the field of the other atom in this direction [46].

24.3 NONLOCAL POLARIZABILITY AND CHEMICAL REACTIVITY

24.3.1 POTENTIAL RESPONSE FUNCTION AND FUKUI FUNCTIONS

In DFT, the PRF [31–33] is a nonlocal response similar to the matrix $\overleftrightarrow{\mathbf{M}}$ introduced in the CG model described above (Equation 24.28). Indeed, in Kohn–Sham theory [1,2,8], (the gradient of) the external potential $v_{ext}(\mathbf{r})$ is analogous to a bare field $\mathbf{E}$ and (the gradient of) the Kohn–Sham potential $v_{KS}(\mathbf{r})$ is analogous to a screened field $\tilde{E}$. A perturbation of the external potential $\delta v_{ext}(\mathbf{r})$ (as induced by a field applied or due to a change of the molecular geometry) implies a variation of the Kohn–Sham potential $v_{KS}(\mathbf{r})$,

$$\delta v_{KS}(\mathbf{r}) \equiv \int d\mathbf{r}'\, \kappa^{-1}(\mathbf{r}, \mathbf{r}') \delta v_{ext}(\mathbf{r}'), \tag{24.35}$$

where the inverse PRF response is simply

$$\kappa^{-1}(\mathbf{r}, \mathbf{r}') \equiv \left[\frac{\delta v_{KS}(\mathbf{r})}{\delta v_{ext}(\mathbf{r}')}\right]_N, \tag{24.36}$$

in which the functional derivative is carried out at constant electron number N.

The PRF and its inverse play a fundamental role in DFT of reactivity where it is related to the Fukui functions [5,32]. The Fukui functions $F^{\pm}(\mathbf{r})$ (see Chapter 17) are reactivity indices, which measure the propensity of a region in a molecule to accept $(+)$ or donate $(-)$ electrons in a chemical reaction [8,15]:

$$F^{\pm}(\mathbf{r}) = \left[\frac{\partial \rho(\mathbf{r})}{\partial N}\right]^{\pm}. \tag{24.37}$$

The Fukui functions generalize the concept of frontier orbitals by including the relaxation of the orbital upon the net addition or removal of one electron. Because the number of electrons of an isolated system can only change by discrete integer number, the derivative in Equation 24.37 is not properly defined. Only the finite difference approximation of Equation 24.37 allows to define these Fukui functions (noted here by capital letters) $F^{\pm}(\mathbf{r})$

$$F^{\pm}(\mathbf{r}) \equiv \pm \rho(\mathbf{r}, N \pm 1) \mp \rho(\mathbf{r}, N). \tag{24.38}$$

Equation 24.38 implies that $F^{\pm}(\mathbf{r})$ are properties of both, the ground-state of the molecule and its single charged ions.

An alternative formulation of the Fukui functions based on the mutual polarization of the reagents can be constructed as follows [24,32]. In a chemical reaction, the description of a local electron transfer does not need to involve any actual change in the total electron numbers N of the reagents. It is enough to consider the local modification of the electronic density in some local volume around a reference site (for instance, the nuclei of an atom) as the result of a local polarization. The modification of the external potential and of the Kohn–Sham self-consistent potential are chemical perturbations, which do not change the total number of electrons N. In this polarization formulation, the Fukui functions are equal to Kohn–Sham frontier orbitals screened by the inverse potential response [24,32]. For instance, the Fukui function $f^-(\mathbf{r})$ (response of the molecule related to the loss of an electron) is related to the highest occupied molecular orbital (HOMO) by the following relation:

$$f^-(\mathbf{r}) = \int d\mathbf{r}' \rho_{\text{HOMO}}(\mathbf{r}') \kappa^{-1}(\mathbf{r}', \mathbf{r}). \tag{24.39}$$

We call the Fukui function $f^-(\mathbf{r})$ the HOMO response. Equation 24.39 is demonstrated as follows. The $\rho_{\text{HOMO}}(\mathbf{r})$ is the so-called Kohn–Sham Fukui function denoted as $f_s^-(\mathbf{r})$ [32]. According to the first-order perturbation theory, one has

$$f_s^-(\mathbf{r}) = \left[\frac{\delta \varepsilon_{\text{HOMO}}}{\delta v_{\text{KS}}(\mathbf{r})}\right]_N = \rho_{\text{HOMO}}(\mathbf{r}). \tag{24.40}$$

We define similarly the HOMO response $f^-(\mathbf{r})$ by

$$f^-(\mathbf{r}) \equiv \left[\frac{\delta \varepsilon_{\text{HOMO}}}{\delta v_{\text{ext}}(\mathbf{r})}\right]_N. \tag{24.41}$$

By using the chain rule for functional derivatives, we prove Equation 24.39

$$f^-(\mathbf{r}) = \int d\mathbf{r}' \left[\frac{\delta \varepsilon_{\text{HOMO}}}{\delta v_{\text{KS}}(\mathbf{r}')}\right]_N \left[\frac{\delta v_{\text{KS}}(\mathbf{r}')}{\delta v_{\text{ext}}(\mathbf{r})}\right]_N. \tag{24.42}$$

The same reasoning can be applied to the lowest unoccupied molecular orbital (LUMO) response $f^+(\mathbf{r})$ (response of the molecule related to the gain of an electron) using

$$f^+(\mathbf{r}) \equiv \left[\frac{\delta \varepsilon_{\text{LUMO}}}{\delta v_{\text{ext}}(\mathbf{r})}\right]_N. \tag{24.43}$$

The definitions in Equations 24.40 and 24.43 do not involve any derivative relative to the number of electrons N and can thus always be properly defined for any molecular system. One should emphasize that the $\varepsilon_{\text{LUMO}}$ is often badly described in any one-electron orbital theory. One cannot therefore expect $f^+(\mathbf{r})$ to be a meaningful quantity. On the contrary, $f^-(\mathbf{r})$ is involved in all charge reorganizations of a molecular system, including a charge transfer between two of its fragments. But, it should be clear that $f^-(\mathbf{r})$ can only approximately represent any property of the molecular cation being a property of the ground-state system.

24.3.2 Polarization and Chemical Hardness Fukui Function

As mentioned in [Section 24.1], and as already demonstrated in Equation 24.39, the Fukui functions as well as the chemical hardness of an isolated system can be properly defined without invoking any change in its electron number. We define a new Fukui function called polarization Fukui function, which very much resembles the original formulation of the Fukui function but with a different physical interpretation. Because of space limitation, only a brief presentation is given here. More details will appear in a forthcoming work [33]. One assumes a potential variation $\delta u_{ext}(\mathbf{r})$, which induces a deformation of the density $\delta\rho(\mathbf{r})$. A normalized polarization Fukui function is defined by

$$\delta\rho^{(1)}(\mathbf{r}) = \lambda f_P(\mathbf{r}) + \Delta\rho(\mathbf{r}), \tag{24.44}$$

where $\delta\rho^{(1)}(\mathbf{r})$ is the first-order density response. In Equation 24.44, λ is an internal (non-integer) charge which can be computed ab initio using Hartree–Fock or Kohn–Sham orbital theories as demonstrated in our recent work [33]. More precisely, one has

$$\lambda = -\int d\mathbf{r}' \Delta\rho(\mathbf{r}'). \tag{24.45}$$

Both λ and $\Delta\rho(\mathbf{r})$ depend on the perturbation $\delta u_{ext}(\mathbf{r})$. The linear response χ_1 (Equation 24.4) is obtained by functional derivative of Equation 24.44:

$$\chi_1(\mathbf{r}, \mathbf{r}') = \left[\frac{\delta\rho^{(1)}(\mathbf{r})}{\delta v_{ext}(\mathbf{r}')}\right]_N,$$

$$= \left[\frac{\delta\lambda}{\delta v_{ext}(\mathbf{r}')}\right]_N f_P(\mathbf{r}) + \chi_1^{\lambda=0}(\mathbf{r}, \mathbf{r}'), \tag{24.46}$$

where $\chi_1^{\lambda=0}(\mathbf{r}, \mathbf{r}')$ is the response at zero internal charge transfer describing the $\Delta\rho(\mathbf{r})$ deformation. Because χ_1 is a symmetric kernel, one deduces finally

$$f_P(\mathbf{r}) = C\left[\frac{\delta\lambda}{\delta v_{ext}(\mathbf{r})}\right]_N, \tag{24.47}$$

in which C is a constant. The value of C is obtained by using the conservation of the number of electrons (a potential cannot create a net charge but only an internal charge transfer λ). By integrating Equation 24.46 and using Equations 24.45 and 24.47, we obtain

$$C = \frac{-1}{\int d\mathbf{r} \int d\mathbf{r}' \chi_1^{\lambda=0}(\mathbf{r}, \mathbf{r}')} \equiv \frac{1}{\eta_P}, \tag{24.48}$$

where we define a polarization hardness η_P. From Equations 24.46 and 24.47, one also finds

$$f_P(\mathbf{r}) = \frac{\int d\mathbf{r}' \chi_1^{\lambda=0}(\mathbf{r}', \mathbf{r})}{\int d\mathbf{r}'' \int d\mathbf{r}''' \chi_1^{\lambda=0}(\mathbf{r}'', \mathbf{r}''')}. \tag{24.49}$$

The physical meaning and value of λ depend on the choice of $f_P(\mathbf{r})$. We may choose $f_P(\mathbf{r})$ as being the solution of the following equation:

$$\int d\mathbf{r}' h(\mathbf{r}, \mathbf{r}') f_P(\mathbf{r}') = \eta_P. \tag{24.50}$$

Equation 24.50 is an important property of the usual Fukui function in DFT [48,49]. We show here that this Fukui function can be formulated as a polarization Fukui function and does not involve any change in electron number. Equation 24.50 implies

$$\chi_1^{\lambda=0}(\mathbf{r}, \mathbf{r}') = -h^{-1}(\mathbf{r}, \mathbf{r}'). \tag{24.51}$$

From Equations 24.46 through 24.51, we deduce

$$\chi_1(\mathbf{r}, \mathbf{r}') = -h^{-1}(\mathbf{r}, \mathbf{r}') + \frac{f_P(\mathbf{r}) f_P(\mathbf{r}')}{\eta_P}, \tag{24.52}$$

which is analogous to the so-called Berkowitz–Parr relation [30] but with the Fukui function and hardness interpreted as responses to a potential (polarization) and not to a change in electron number N. In addition, the charge transfer involved λ, being an internal charge transfer, is a continuous variable. In particular, we have exactly [28]

$$\eta_P = \left[\frac{\partial^2 \varepsilon}{\partial \lambda^2} \right]_0. \tag{24.53}$$

The internal charge transfer λ depends also on the potential. It should be emphasized that λ_P and η_P can be defined in principle for any arbitrary choice of $f_P(\mathbf{r})$ in Equation 24.44. Derivation of $f_P(\mathbf{r})$ does not involve any chemical potential. It is an unique property of the Hamiltonian with N electrons [28,33].

24.3.3 EXACT RELATIONS BETWEEN THE DIFFERENT FUKUI FUNCTIONS OF AN ISOLATED SYSTEM

24.3.3.1 Introduction

In addition to popular finite difference approximations of Fukui functions of an isolated system (Equation 24.38), at least six other different Fukui functions can be defined as responses to a potential. These later concepts do not depend on a net change in the number of electrons and are related by exact relations to the different polarizability kernels. In this section, we derive the exact relations between them.

We introduce the following notations for the energy functional $\varepsilon[\rho; v_{ext}]$ and its components [1,2,8]:

$$\varepsilon[\rho; v_{ext}] = T_s[\rho] + E_I[\rho] + \int d\mathbf{r} \; \rho(\mathbf{r}) v_{ext}(\mathbf{r}), \tag{24.54}$$

where $T_s[\rho]$ is the Kohn–Sham kinetic energy functional and $E_I[\rho]$ the electron–electron energy component. We have

$$T_s[\rho] = -\frac{\hbar^2}{2m} \sum_i n_i \int d\mathbf{r}' \; \phi_i^*(\mathbf{r}') \nabla^2 \phi_i(\mathbf{r}'), \tag{24.55}$$

in which the sum is over N (the number of electrons) occupied microstates of the molecule. The occupation number $n_i = 0$ or 1 and $\phi_i(\mathbf{r}')$ is the spatial part of the spin–orbital i. Each microstate is doubly degenerate when the molecular density is not polarized in spin. The difference between the interaction energy $\varepsilon_I[\rho]$ and the classical Coulomb electron–electron repulsion is the exchange-correlation energy $\varepsilon_{xc}[\rho]$

$$\varepsilon_{xc}[\rho] = \varepsilon_I[\rho] - \frac{1}{2} \int d\mathbf{r} \; \rho(\mathbf{r}) \int d\mathbf{r}' \; \rho(\mathbf{r}') \left(\frac{e^2}{[\mathbf{r} - \mathbf{r}']} \right). \tag{24.56}$$

The Kohn–Sham potential is [2]

$$v_{KS}(\mathbf{r}) = v_{ext}(\mathbf{r}) + \frac{\delta \varepsilon_I}{\delta \rho(\mathbf{r})}. \tag{24.57}$$

24.3.3.2 Six Fukui Functions and Three Hardnesses of an Isolated System

We demonstrate now that six Fukui functions at constant electron number can be defined for an isolated molecule. The two Kohn–Sham Fukui functions are [32]

$$f_s^\pm(\mathbf{r}) \equiv \left[\frac{\delta \varepsilon_\pm}{\delta v_{KS}(\mathbf{r})} \right]_0, \tag{24.58}$$

where $\varepsilon_+(\varepsilon_-)$ is $\varepsilon_{HOMO}(\varepsilon_{LUMO})$. According to the first-order perturbation theory f_s^+ (f_s^-) is equal to $\rho_{HOMO}(\mathbf{r})$ $[\rho_{LUMO}(\mathbf{r})]$. We define two other HOMO and LUMO responses by [32]

$$f^\pm(\mathbf{r}) \equiv \left[\frac{\delta \varepsilon_\pm}{\delta u_{ext}(\mathbf{r})} \right]_0. \tag{24.59}$$

On the other hand, in addition to the polarization Fukui response $f_P(\mathbf{r}')$ obeying Equation 24.50, a sixth electron–electron Fukui response $f_I(\mathbf{r}')$ is now introduced as the solution of

$$\int d\mathbf{r}' h_I(\mathbf{r}, \mathbf{r}') f_I(\mathbf{r}') = \eta_I, \qquad (24.60)$$

where h_I is an electron–electron hardness kernel or propagator [32]

$$h_I(\mathbf{r}, \mathbf{r}') = \left[\frac{\delta^2 \varepsilon_I}{\delta\rho(\mathbf{r})\delta\rho(\mathbf{r}')} \right]_0, \qquad (24.61)$$

and η_I is a new interacting hardness

$$\eta_I = \frac{1}{\int d\mathbf{r}'' \int d\mathbf{r}' h_I^{-1}(\mathbf{r}', \mathbf{r}'')}. \qquad (24.62)$$

To complete this presentation, we define a Kohn–Sham hardness (gap) by

$$\eta_s = \varepsilon_{LUMO} - \varepsilon_{HOMO}, \qquad (24.63)$$

and a Kohn–Sham "chemical potential" μ_s as

$$\mu_s = \frac{\varepsilon_{LUMO} + \varepsilon_{HOMO}}{2}. \qquad (24.64)$$

The derivatives of these Kohn–Sham quantities relative to the potentials are other reactivity descriptors:

$$\begin{aligned}
\left[\frac{\delta\eta_s}{\delta v_{KS}(\mathbf{r})} \right]_N &= \rho_{LUMO}(\mathbf{r}) - \rho_{HOMO}(\mathbf{r}), \\
\left[\frac{\delta\eta_s}{\delta v_{ext}(\mathbf{r})} \right]_N &= f^+(\mathbf{r}) - f^-(\mathbf{r}), \\
\left[\frac{\delta\mu_s}{\delta v_{KS}(\mathbf{r})} \right]_N &= \frac{\rho_{LUMO}(\mathbf{r}) + \varepsilon_{HOMO}(\mathbf{r})}{2}, \\
\left[\frac{\delta\mu_s}{\delta v_{ext}(\mathbf{r})} \right]_N &= \frac{f^+(\mathbf{r}) + f^-(\mathbf{r})}{2}.
\end{aligned} \qquad (24.65)$$

All these functional derivatives are well defined and do not involve any actual derivative relative to the electron number. It is remarkable that the derivatives of the Kohn–Sham chemical potential μ_s gives the so-called "radical" Fukui function [8] either in a frozen orbital approximation or by including the relaxation of the KS band structure. On the other hand, the derivative of the Kohn–Sham HOMO–LUMO gap (defined here as a positive quantity) is the so-called nonlinear Fukui function $f'(\mathbf{r})$ [26,32,50] also called Fukui difference [51].

24.3.3.3   Finite Difference Approximations of Fukui Functions as Potential Derivatives

Although we will not discuss the finite difference approximations $F^{\pm}(\mathbf{r})$ further in this section, it is useful to note that these functions can be defined also as potential derivatives of a "gap" and a chemical potential. The gap is the usual definition of the chemical hardness [8,9],

$$\eta = I - A, \tag{24.66}$$

and the chemical potential is the usual DFT electronegativity (change of sign) [8,52]

$$\mu = -\left(\frac{I+A}{2}\right). \tag{24.67}$$

The ionization potential and electron affinity of the molecule are I and A, respectively. By construction, these definitions involve three Hamiltonians ($N-1$, N, $N+1$). However, one may define Fukui functions without invoking any actual derivative relative to the number of electrons by using the derivative of the chemical potential relative to the potential [8]

$$
\begin{aligned}
\mathcal{F}^{-}(\mathbf{r}) &\equiv -\left[\frac{\delta I}{\delta v_{\text{ext}}(\mathbf{r})}\right]_{N} \\
&= \left[\frac{\delta E(N)}{\delta v_{\text{ext}}(\mathbf{r})}\right]_{N} - \left[\frac{\delta E(N-1)}{\delta v_{\text{ext}}(\mathbf{r})}\right]_{N-1}, \\
&= \rho(\mathbf{r}; N) - \rho(\mathbf{r}; N-1) = F^{-}(\mathbf{r}).
\end{aligned}
\tag{24.68}
$$

and similarly,

$$\mathcal{F}^{+}(\mathbf{r}) \equiv -\left[\frac{\delta A}{\delta v_{\text{ext}}(\mathbf{r})}\right]_{N} = \rho(\mathbf{r}; N+1) - \rho(\mathbf{r}; N) = F^{+}(\mathbf{r}). \tag{24.69}$$

The potential derivatives of the gap and of the chemical potential are therefore

$$\left[\frac{\delta \eta}{\delta v_{\text{ext}}(\mathbf{r})}\right]_{N} = \mathcal{F}^{+}(\mathbf{r}) - \mathcal{F}^{-}(\mathbf{r}), \tag{24.70}$$

$$\left[\frac{\delta \mu}{\delta v_{\text{ext}}(\mathbf{r})}\right]_{N} = \frac{\mathcal{F}^{-}(\mathbf{r}) + \mathcal{F}^{+}(\mathbf{r})}{2}. \tag{24.71}$$

Equations 24.68 and 24.69 represent the response to electrophilic and nucleophilic reagents, respectively [8]. On the other hand, Equation 24.70 is a nonlinear Fukui function [51] and Equation 24.71 represents a radical Fukui function [8]. All these functions are computed by using different molecular Hamiltonians.

24.3.3.4 Theorems for the Fukui Functions of an Isolated System

The frontier orbitals responses (or bare Fukui functions) $f^{\pm}(\mathbf{r})$ and the Kohn–Sham Fukui functions (or screened Fukui functions) $f_s^{\pm}(\mathbf{r})$ are related by Dyson equations obtained by using the PRF and its inverse [32]. Indeed, by using Equation 24.57 and the chain rule for functional derivatives in Equation 24.36, one obtains

$$\kappa^{-1}(\mathbf{r},\mathbf{r}') = \delta(\mathbf{r}-\mathbf{r}') + \frac{\delta}{\delta v_{\text{ext}}(\mathbf{r}')}\left[\frac{\delta \varepsilon_I}{\delta \rho(\mathbf{r})}\right],$$
$$= \delta(\mathbf{r}-\mathbf{r}') + \int d\mathbf{r}'' h_1(\mathbf{r},\mathbf{r}'')\chi_1(\mathbf{r}'',\mathbf{r}'), \qquad (24.72)$$

and

$$\kappa(\mathbf{r},\mathbf{r}') = \delta(\mathbf{r}-\mathbf{r}') - \frac{\delta}{\delta v_{\text{KS}}(\mathbf{r}')}\left[\frac{\delta \varepsilon_I}{\delta \rho(\mathbf{r})}\right],$$
$$= \delta(\mathbf{r}-\mathbf{r}') - \int d\mathbf{r}'' h_1(\mathbf{r},\mathbf{r}'')\chi_s(\mathbf{r}'',\mathbf{r}'), \qquad (24.73)$$

where χ_s is the noninteracting response function (analogous to the screened polarizability response of the CG model above, Equation 24.24)

$$\chi_s(\mathbf{r},\mathbf{r}') = \left[\frac{\delta \rho(\mathbf{r})}{\delta v_{\text{KS}}(\mathbf{r}')}\right]_N. \qquad (24.74)$$

Using Equations 24.72 and 24.39, one obtains

$$f^{\pm}(\mathbf{r}') = f_s^{\pm}(\mathbf{r}') + \int d\mathbf{r}'' \int d\mathbf{r} f_s^{\pm}(\mathbf{r})h_1(\mathbf{r},\mathbf{r}'')\chi_1(\mathbf{r}'',\mathbf{r}'). \qquad (24.75)$$

On the other hand, using Equation 24.73, one finds

$$f_s^{\pm}(\mathbf{r}') = f^{\pm}(\mathbf{r}') - \int d\mathbf{r}'' \int d\mathbf{r} f^{\pm}(\mathbf{r})h_1(\mathbf{r},\mathbf{r}'')\chi_s(\mathbf{r}'',\mathbf{r}'). \qquad (24.76)$$

Equations 24.75 and 24.76 generalize the Dyson equations derived previously [32]. The differences between $f^{\pm}$ and the frontier orbitals $f_s^{\pm}$ have nice and simple physical interpretations [24,32] as either the variation of the electronic density induced by effective external potentials $\delta v_{f_s^{\pm}}$

$$\delta\rho_{\text{ee}}^{\pm}(\mathbf{r}) \equiv f^{\pm}(\mathbf{r}) - f_s^{\pm}(\mathbf{r}) = \int d\mathbf{r}' \chi_1(\mathbf{r},\mathbf{r}')\delta v_{f_s^{\pm}}(\mathbf{r}'), \qquad (24.77)$$

in which

$$\delta v_{f_s^{\pm}}(\mathbf{r}) = \int d\mathbf{r}' h_1(\mathbf{r},\mathbf{r}')f_s^{\pm}(\mathbf{r}'), \qquad (24.78)$$

or as the variation of the electronic density induced by an effective Kohn–Sham potential δv_f

$$\delta\rho_{ee}^{\pm}(\mathbf{r}) = \int d\mathbf{r}' \chi_s(\mathbf{r},\mathbf{r}')\delta v_f(\mathbf{r}'), \qquad (24.79)$$

in obvious notations. Without the exchange-correlation energy, these effective potentials are reduced to the electrostatic potentials generated by the Fukui functions $f^{\pm}$ and $f_s^{\pm}$, respectively. They are related to the covalent atomic radii [24,53].

It is worth noting that screened response $\chi_s(\mathbf{r}, \mathbf{r}')$ can be computed from the Kohn–Sham orbital wave functions and energies using standard first-order perturbation theory [3]

$$\chi_s(\mathbf{r},\mathbf{r}') = \sum_{i=1} n_i \sum_{k\neq i} \frac{\phi_i^*(\mathbf{r})\phi_k(\mathbf{r})\phi_k^*(\mathbf{r}')\phi_i(\mathbf{r}')}{\varepsilon_i - \varepsilon_k}, \qquad (24.80)$$

where the sum is over all Kohn–Sham microstates. $n_i = 1$ for an occupied state and 0 for an empty state. The calculation of the bare response $\chi_1(\mathbf{r}, \mathbf{r}')$ is more difficult as we have to solve the following equation (analogous to Equation 24.30 of the CG dielectric model used above):

$$\chi_1(\mathbf{r},\mathbf{r}') = \int d\mathbf{r}'' \chi_s(\mathbf{r},\mathbf{r}'')\kappa^{-1}(\mathbf{r}'',\mathbf{r}'), \qquad (24.81)$$

which reduces to the famous Bethe–Salpter equation [54]

$$\chi_1(\mathbf{r},\mathbf{r}') = \chi_s(\mathbf{r},\mathbf{r}') + \int d\mathbf{r}'' \int d\mathbf{r}''' \chi_s(\mathbf{r},\mathbf{r}'')h_1(\mathbf{r}'',\mathbf{r}''')\chi_1(\mathbf{r}''',\mathbf{r}'). \qquad (24.82)$$

A possible computational strategy is to calculate $\chi_s(\mathbf{r}, \mathbf{r}')$ first using the standard sum-over states formula (Equation 24.80). Equation 24.75 can be used next to generate successive "Born approximations" of the functions $f^{\pm}(\mathbf{r})$. For instance, the first Born approximation would be

$$f^{\pm}(\mathbf{r}') = f_s^{\pm}(\mathbf{r}') + \int d\mathbf{r}'' \int d\mathbf{r} f_s^{\pm}(\mathbf{r})h_1(\mathbf{r},\mathbf{r}'')\chi_s(\mathbf{r}'',\mathbf{r}'). \qquad (24.83)$$

It is worth noting that an alternative computational scheme to solve the Dyson equations Equations 24.75 and 24.76 has appeared recently [55,56]. Simple models of the response χ_s have also been developed and applied to reactivity [57].

We end this section by demonstrating that the interacting and polarization Fukui functions are eigenvectors of bare response functions. Indeed, one proves that $f_1(\mathbf{r})$, (Equation 24.60), is an (left) eigenvector of the PRF by multiplying Equations 24.72 and 24.73 by $f_1(\mathbf{r})$ and by integrating

$$\int d\mathbf{r}\, f_1(\mathbf{r})\kappa^{-1}(\mathbf{r},\mathbf{r}') = \int d\mathbf{r}\, f_1(\mathbf{r})\kappa(\mathbf{r},\mathbf{r}') = f_1(\mathbf{r}'), \qquad (24.84)$$

where the eigenvalue is equal to 1. For $f_P(\mathbf{r})$, defined by Equation 24.50, we find

$$\int d\mathbf{r}\, f_P(\mathbf{r})\alpha^{-1}(\mathbf{r},\mathbf{r}') = \int d\mathbf{r}\, f_P(\mathbf{r})\alpha(\mathbf{r},\mathbf{r}') = f_P(\mathbf{r}'), \qquad (24.85)$$

where the new function $\alpha^{-1}(\mathbf{r},\mathbf{r}')$ is defined as follows:

$$\alpha^{-1}(\mathbf{r},\mathbf{r}') \equiv \frac{\delta}{\delta v_{ext}(\mathbf{r}')}\left[\frac{\delta\varepsilon}{\delta\rho(\mathbf{r})}\right]_N,$$

$$= \delta(\mathbf{r}-\mathbf{r}') + \int d\mathbf{r}''h(\mathbf{r},\mathbf{r}'')\chi_1(\mathbf{r}'',\mathbf{r}'). \qquad (24.86)$$

Using $h = h_I + h_s$, one obtains the relation between the inverse PRF and α^{-1},

$$\alpha^{-1}(\mathbf{r},\mathbf{r}') = \kappa^{-1}(\mathbf{r},\mathbf{r}') + \int d\mathbf{r}''h_s(\mathbf{r},\mathbf{r}'')\chi_1(\mathbf{r}'',\mathbf{r}'). \qquad (24.87)$$

24.3.4 SOFTNESSES, COULOMB HOLE, AND MOLECULAR FRAGMENTS

The different Fukui functions are defined on the entire space and are normalized. Therefore, their average values decrease with the system size. In order to compare the properties of a molecular group in molecules of different sizes, one must compare instead, the corresponding local softnesses. For each Fukui function, there exists a local softness $s(\mathbf{r})$ defined simply by [13]

$$s(\mathbf{r}) = \frac{f(\mathbf{r})}{\eta}, \qquad (24.88)$$

where η is the corresponding hardness [12]. The integral of the local softness is the global softness S [13]

$$S = \frac{1}{\eta}. \qquad (24.89)$$

One of the applications of the theory of reactivity is to compare the similarities and differences of a set of molecules using DFT descriptors as Fukui functions and local softnesses [6,18,58]. But the comparison of continuous functions $s(\mathbf{r})$ in 3D is not an easy task. Chemistry proceeds in general by partitioning a large molecule into fragments; each fragment having its own intrinsic reactivity modulated by the groups to which it is bonded. The softness $s(\mathbf{r})$ can be partitioned as well into molecular fragments. These fragments descriptors are the so-called "condensed reactivity indices" [59–64]. The lost of information due to the condensation is not critical in general, and simplifies the treatment of a large set of molecules. There is, therefore, a considerable scope for applications in chemistry and pharmacology and a growing interest to develop the DFT theory for fragments descriptors and its numerical implementation.

In recent papers, the local softness (or Fukui function) of a molecular fragment reformulated in terms of polarization was demonstrated [18,29]. Application of these fragments descriptors to biomolecules is very promising. The descriptors are introduced by dividing a molecule into two regions 1 and 2 (the generalization to any number of fragments is straightforward). One starts with the property [48]

$$\int d\mathbf{r}' \, \mathcal{H}(\mathbf{r}, \mathbf{r}') S(\mathbf{r}') = 1, \tag{24.90}$$

where $\mathcal{H}$ and S are generic quantities, they are either equal to (see Equation 24.50)

$$\mathcal{H}(\mathbf{r}, \mathbf{r}') = h(\mathbf{r}, \mathbf{r}')$$

$$S(\mathbf{r}') = s_P(\mathbf{r}') = \frac{f_P(\mathbf{r}')}{\eta_P}, \tag{24.91}$$

$$S = \int d\mathbf{r}' S(\mathbf{r}') = S_P,$$

or (see Equation 24.60)

$$\mathcal{H}(\mathbf{r}, \mathbf{r}') = h_1(\mathbf{r}, \mathbf{r}')$$

$$S(\mathbf{r}') = s_1(\mathbf{r}') = \frac{f_1(\mathbf{r}')}{\eta_1}, \tag{24.92}$$

$$S = \int d\mathbf{r}' S(\mathbf{r}') = S_1.$$

The integral in Equation 24.90 is the sum of an integration on each molecular fragment which occupies a volume Ω_1 and Ω_2, respectively

$$\int_{\Omega_1} d\mathbf{r}' \mathcal{H}(\mathbf{r}, \mathbf{r}') S(\mathbf{r}') + \int_{\Omega_2} d\mathbf{r}' \, \mathcal{H}(\mathbf{r}, \mathbf{r}') S(\mathbf{r}') = 1. \tag{24.93}$$

We define the hardness kernels $\mathcal{H}_1$ and $\mathcal{H}_2$ of the two fragments by

$$\mathcal{H}(\mathbf{r}, \mathbf{r}') = \mathcal{H}_1(\mathbf{r}, \mathbf{r}') \quad \text{if } \mathbf{r} \ni \Omega_1 \text{ and } \mathbf{r}' \ni \Omega_1, \tag{24.94}$$

$$\mathcal{H}(\mathbf{r}, \mathbf{r}') = \mathcal{H}_2(\mathbf{r}, \mathbf{r}') \quad \text{if } \mathbf{r} \ni \Omega_2 \text{ and } \mathbf{r}' \ni \Omega_2, \tag{24.95}$$

Using the definitions, we deduce

$$\int_{\Omega_1} d\mathbf{r}' \, \mathcal{H}_1(\mathbf{r}, \mathbf{r}') S(\mathbf{r}') + \int_{\Omega_2} d\mathbf{r}' \, \mathcal{H}(\mathbf{r}, \mathbf{r}') S(\mathbf{r}') = 1, \tag{24.96}$$

$$S(\mathbf{r}''') + \int_{\Omega_1} d\mathbf{r} \, \mathcal{H}_1^{-1}(\mathbf{r}''', \mathbf{r}) \int_{\Omega_2} d\mathbf{r}'' \, \mathcal{H}(\mathbf{r}, \mathbf{r}'') S(\mathbf{r}'') = s_1(\mathbf{r}'''), \tag{24.97}$$

where we define a regional softness and its integral as

$$s_1(\mathbf{r}''') \equiv \int_{\Omega_1} d\mathbf{r} \; \mathcal{H}_1^{-1}(\mathbf{r}''',\mathbf{r}), \tag{24.98}$$

$$\mathcal{S}_1^R = \int_{\Omega_1} d\mathbf{r}''' \; s_1(\mathbf{r}'''). \tag{24.99}$$

The integral of the regional softness is a property built from the hardness kernel (more precisely its inverse) defined entirely in the region Ω_1. Equation 24.97 can be reformulated as

$$S(\mathbf{r}''') - S_2 \int_{\Omega_1} d\mathbf{r} \; \mathcal{H}_1^{-1}(\mathbf{r}''',\mathbf{r})v_2(\mathbf{r}) = s_1(\mathbf{r}'''), \tag{24.100}$$

where we have defined the softness of a fragment

$$S_2 \equiv \int_{\Omega_2} d\mathbf{r}'''' S(\mathbf{r}''''), \tag{24.101}$$

and a pseudopotential

$$v_2(\mathbf{r}) \equiv \frac{\int_{\Omega_2} d\mathbf{r}'' \; \mathcal{H}(\mathbf{r},\mathbf{r}'')S(\mathbf{r}'')}{S_2}. \tag{24.102}$$

By integrating Equation 24.100 and by using the property

$$S_2 + S_1 = S, \tag{24.103}$$

one finally obtains a "softness–softness" relation [18,28]

$$S = S_1^R + S_2(1 - Q^0(1)), \tag{24.104}$$

where $Q^0(1)$ is an internal charge transfer (analogous to λ in Equation 24.45)

$$Q^0(1) = -\int_{\Omega_1} d\mathbf{r}'''' \int_{\Omega_1} d\mathbf{r} \; \mathcal{H}_1^{-1}(\mathbf{r}''',\mathbf{r})v_2(\mathbf{r}). \tag{24.105}$$

Note that Equation 24.104 can be formulated in terms of Fukui functions instead of softnesses.

The physical meaning of Equation 24.104 is: The global molecular softness S and the softnesses of its parts S_1^R (regional) and S_2 are related through a linear relation where $Q^0(1)$ represents the net charge induced on fragment 1 due to a pseudopotential $v_2(\mathbf{r})$ generated by a normalized charge distribution on fragment 2 equal to

TABLE 24.1

Global Softness and Local Softness (Side Chains) of Amino Acids Computed at MP2/6–311G(d,p) in Ref. [18]

Name	S_2 (a.u.)	S (a.u.)
Aliphatic and hydroxyl residues	0.3583	1.8393
Acidic residues	0.6675	2.0454
Amide residues	0.7784	1.8562
Basic residues	1.1929	2.2212
Sulfur-containing residues	1.5875	2.0505
Histidine	1.7234	2.1717
Aromatic residues	2.2704	2.3621

$S(\mathbf{r}'')/S_2$. Indeed, this charge distribution $S(\mathbf{r}'')/S_2$ induces a deformation of the density $\Delta\rho_1$ on the fragment 1 via the potential v_2 (Equation 24.102)

$$\Delta\rho_1(\mathbf{r}) = -\int_{\Omega_1} d\mathbf{r}'\,\mathcal{H}_1^{-1}(\mathbf{r},\mathbf{r}')v_2(\mathbf{r}'),$$

$$= \int_{\Omega_1} d\mathbf{r}'\,\chi_1^{\lambda=0}(\mathbf{r},\mathbf{r}')v_2(\mathbf{r}'), \qquad (24.106)$$

where $\chi_1^{\lambda=0}$ is a response function at zero intramolecular charge transfer (Equation 24.51). In the usual interpretation of the softnesses, $\Delta\rho_1(\mathbf{r})$ would correspond to the density deformation on fragment 1 when the fragment 2 is "charged" with one electron distributed according to $S(\mathbf{r})/S_2$. $\Delta\rho_1(\mathbf{r})$ is what we called the "Coulomb hole" in Refs. [18,29] In a CG model, one demonstrates that this quantity can be understood as the polarization charge accompanying one test electron in a molecular system [29]. When $S(\mathbf{r}') = s_l(\mathbf{r}')$ [Equation 24.92], it corresponds to the density generated on the fragment 1 by one electron localized on the fragment 2 due to electron–electron interactions.

Equation 24.104 can be used to compute ab initio the similarity and differences between the members of a molecular family. In the case of amino acids, for instance, [18], S_1^R is chosen as a constant fragment (the amino-carboxyl part) and S_2 corresponds to the local softness of the side chain. In Table 24.1, we report the average value of local softness (S_2) computed for each group of amino acids: local softness only permits to clearly distinguish between the different types of amino acids which have all nearly the same global softness. Calculations of S and S_2 for each of the twenty amino acids permits to prove the linear relationship (Equation 24.104) and deduces the values of S_1^R and $Q^0(1)$ [18].

24.4 LINEAR AND NONLINEAR RESPONSES

24.4.1 NONLOCAL POLARIZABILITIES

Because of limitations of space, we will give only a brief introduction to the nonlinear responses here and show how these derivations are related to the new polarization and interacting Fukui functions f_P and f_I. The interested reader may consult Refs. [26,32] where the extension of the DFT theory of reactivity to non-linear responses is explained in detail. The first derivation of nonlinear descriptors can be found in Ref. [50].

We consider a variation of the external potential $\delta v_{ext}(\mathbf{r})$ at constant electron number N. The formal expression of the energy variation due to this perturbation can be found by a direct application of the Hellman–Feynman theorem [22,23,26].

$$\Delta\varepsilon[\delta\rho;\delta v_{ext}] = \int d\mathbf{r}'\rho_0(\mathbf{r}')\delta v_{ext}(\mathbf{r}')$$
$$+ \sum_{n=1}^{\infty} \frac{1}{n+1} \int d\mathbf{r}'\delta\rho_n(\mathbf{r}')\delta v_{ext}(\mathbf{r}'), \tag{24.107}$$

where the nth-order variation of the electronic density, $\delta\rho_n$, is proportional to the nth power of the perturbation δv_{ext} and is given by

$$\delta\rho_n(\mathbf{r}) = \frac{1}{n!} \int d\mathbf{r}_1 \int d\mathbf{r}_2 \cdots \int d\mathbf{r}_n \chi_n(\mathbf{r},\mathbf{r}_1,\mathbf{r}_2,\ldots,\mathbf{r}_n)$$
$$+ \delta v_{ext}(\mathbf{r}_1)\delta v_{ext}(\mathbf{r}_2) \cdots \delta v_{ext}(\mathbf{r}_n). \tag{24.108}$$

The linear ($n = 1$) and nonlinear ($n \neq 1$) nonlocal polarizability responses χ_n are

$$\chi_n(\mathbf{r},\mathbf{r}_1,\mathbf{r}_2,\ldots,\mathbf{r}_n) \equiv \left[\frac{\delta^n\rho(\mathbf{r})}{\delta v_{ext}(\mathbf{r}_1) \cdots \delta v_{ext}(\mathbf{r}_n)} \right]_N. \tag{24.109}$$

Equations 24.108 and 24.109 are generalization of Equations 24.11 and 24.10 introduced to represent the polarizabilities of an atom in a uniform electric field above.

The linear response χ_1 plays a fundamental role. It can be evaluated using the Bethe–Salpter equation (Equation 24.82) where the screened response χ^s is evaluated from Kohn–Sham equations (Equation 24.80). It is remarkable that any non-linear response can be computed using the linear one and the hardness kernels [26,32]. For instance, $\chi_3(\mathbf{r},\mathbf{r}_1,\mathbf{r}_2,\mathbf{r}_3)$ (see diagram 52a in Ref. [26]) is

$$\chi_3(\mathbf{r},\mathbf{r}',\mathbf{r}'',\mathbf{r}''') = 3 \cdot \chi_1(\mathbf{r},\mathbf{r}_1) \cdot \chi_1(\mathbf{r}',\mathbf{r}_2) \cdot \chi_1(\mathbf{r}_4,\mathbf{r}_3)h(\mathbf{r}_1,\mathbf{r}_2,\mathbf{r}_3)$$
$$\times \cdots \chi_1(\mathbf{r}'',\mathbf{r}_5)\chi_1(\mathbf{r}''',\mathbf{r}_6)h(\mathbf{r}_4,\mathbf{r}_5,\mathbf{r}_6)$$
$$+ \cdot \chi_1(\mathbf{r},\mathbf{r}_1) \cdot \chi_1(\mathbf{r}',\mathbf{r}_2) \cdot \chi_1(\mathbf{r}'',\mathbf{r}_3) \cdot \chi_1(\mathbf{r}''',\mathbf{r}_4)h(\mathbf{r}_1,\mathbf{r}_2,\mathbf{r}_3,\mathbf{r}_4). \tag{24.110}$$

In Equation 24.110, each filled dot corresponds to an integral over one $\mathbf{r}_i$ coordinate.

Equation 24.110 permits to formulate the nonlinear dipole hardness h_3 of an atom in an electric field (Equation 24.18) in terms of the responses χ_n. We assume a field applied along the x Cartesian direction. The nonlinear polarizability α_3 (Equation 24.10) is

$$\alpha_3 = \cdots \chi_3(\mathbf{r},\mathbf{r}',\mathbf{r}'',\mathbf{r}''')x\,x'\,x''\,x''', \tag{24.111}$$

and the linear one is

$$\alpha_1 = \cdots \chi_1(\mathbf{r}, \mathbf{r}')x\, x'. \tag{24.112}$$

Therefore, we conclude that

$$
\begin{aligned}
h_3 = &- 9\, p_1 p_1 p_1 p_1 h(\mathbf{r}_1, \mathbf{r}_2, \mathbf{r}_3)\chi_1 h(\mathbf{r}_1, \mathbf{r}_5, \mathbf{r}_6) \\
&- 3 p_1 p_1 p_1 p_1 h(\mathbf{r}_1, \mathbf{r}_2, \mathbf{r}_3, \mathbf{r}_4).
\end{aligned}
\tag{24.113}
$$

In Equation 24.113, all variables are integrated and p_1 is a local dipole:

$$p_1(\mathbf{r}) \equiv \frac{\int d\mathbf{r}_1 \chi_1(\mathbf{r}, \mathbf{r}_1)x_1}{\int d\mathbf{r}_1 \int d\mathbf{r}_2 x_1 \chi_1(\mathbf{r}_1, \mathbf{r}_2)x_2}. \tag{24.114}$$

The hardness kernels in Equation 24.110 depend on the kinetic energy functional as well as on the electron–electron interactions. Thomas–Fermi models can be used to evaluate the kinetic part of these hardness kernels and can be combined with a band structure calculation of the linear response χ_1.

On the other hand, functional derivatives of the Bethe–Salpeter equation allows to evaluate the nonlinear responses using the interaction kernels h_I only (which depend on the Hartree and exchange-correlation energies). The relations between the screened nonlinear responses and the bare ones are derived by using nonlinear PRF [32].

$$h_n(\mathbf{r}, \mathbf{r}_1, \mathbf{r}_2, \dots, \mathbf{r}_n) = \left[\frac{\delta^n v_{\text{ext}}(\mathbf{r})}{\delta v_{\text{KS}}(\mathbf{r}_1) \dots \delta v_{\text{KS}}(\mathbf{r}_n)} \right]_N. \tag{24.115}$$

It is emphasized that the PRF and their inverse involve only quantities which can be computed in the KS formulation of DFT, i.e., the responses χ_n and the kernels h_I [32].

24.4.2   NONLINEAR CHEMICAL ELECTRONIC RESPONSES

The generalization of the Fukui functions to nonlinear and nonlocal chemical responses is done in Refs. [26,32] by using N derivatives and the KS perturbation equations. In this section, we propose a brief survey of a complementary derivation based on the concept of the internal charge transfer λ introduced above. A more detailed discussion, including computational schemes, will be presented elsewhere.

The present formulation does not involve any global change in the number of electrons of a molecule and can be properly defined for an isolated system. We consider a variation $\delta\rho(\mathbf{r})$ induced by a potential (which does not need to be small) $\delta v_{\text{ext}}(\mathbf{r})$ and generalize the formula Equation 24.44 to an arbitrary perturbation order

$$\delta\rho(\mathbf{r}) = \sum_{n=1}^{\infty} \delta\rho_n(\mathbf{r}) = \sum_{n=1}^{\infty} \frac{\lambda^n}{n!} f_P^{(n)}(\mathbf{r}) + \Delta^{(n)}\rho(\mathbf{r}), \tag{24.116}$$

with $f_P^{(1)}(\mathbf{r}) = f_P(\mathbf{r})$ and the nonlinear polarization Fukui functions equal to $f_P^{(n \neq 1)} = (\mathbf{r})$. Because a potential does not induce a change in electron number, one must have

$$\int d\mathbf{r}\, f_P^{(n)}(\mathbf{r}) = \int d\mathbf{r}\, \frac{\partial^n \delta\rho(\mathbf{r})}{\partial \lambda^n} = 0 \quad \forall n \neq 1. \tag{24.117}$$

Linear and nonlinear polarization hardnesses are defined by

$$\eta_P^{(n)} = \frac{\partial^n \varepsilon}{\partial \lambda^n}. \tag{24.118}$$

The relations between the polarization chemical electronic responses $\left(f_P^{(n)}(\mathbf{r}),\ \eta_P^{(n)},\ldots\right)$ and the polarizability responses χ_n are similar to the exact equations we derived earlier when $f_p(\mathbf{r})$ is defined by Equation 24.50 [26]. For instance, the expression of the first nonlinear hardness $\eta_P^{(2)}$ is obtained by deriving the linear equation (Equation 24.50) relative to λ, and by using again the chain rule for functional derivatives:

$$\eta_P^{(2)} = \int d\mathbf{r}'\, h(\mathbf{r},\mathbf{r}') f_P^{(2)}(\mathbf{r}') + \int d\mathbf{r}' \int d\mathbf{r}''\, h(\mathbf{r},\mathbf{r}',\mathbf{r}'') f_P(\mathbf{r}'') f_P(\mathbf{r}'),$$

$$= \int d\mathbf{r} \int d\mathbf{r}' \int d\mathbf{r}''\, f_P(\mathbf{r}) h(\mathbf{r},\mathbf{r}',\mathbf{r}'') f_P(\mathbf{r}'') f_P(\mathbf{r}'). \tag{24.119}$$

The nonlinear Fukui function is also easily obtained by deriving Equation 24.50 relative to the external potential

$$f_P^{(2)}(\mathbf{r}') = \left[\frac{\delta \eta_P}{\delta v_{\text{ext}}(\mathbf{r})}\right]_N = \int d\mathbf{r}'\, h(\mathbf{r},\mathbf{r}') \xi_P(\mathbf{r}',\mathbf{r})$$

$$+ \int d\mathbf{r}' \int d\mathbf{r}''\, h(\mathbf{r},\mathbf{r}',\mathbf{r}'') \chi_1(\mathbf{r}'',\mathbf{r}) f_P(\mathbf{r}'),$$

$$= \int d\mathbf{r} \int d\mathbf{r}' \int d\mathbf{r}''\, f_P(\mathbf{r}) h(\mathbf{r},\mathbf{r}',\mathbf{r}'') \chi_1(\mathbf{r}'',\mathbf{r}) f_P(\mathbf{r}'). \tag{24.120}$$

Finally, one may define a nonlocal Fukui (nonlinear) function $\xi_P(\mathbf{r}',\mathbf{r})$ as

$$\xi_P(\mathbf{r}',\mathbf{r}) \equiv \left[\frac{\delta f_P(\mathbf{r})}{\delta v_{\text{ext}}(\mathbf{r})}\right]_N. \tag{24.121}$$

Similar relations can be obtained for the nonlinear f_1 functions. Kohn–Sham orbital formulations of these nonlinear responses can be constructed along the lines described previously [32] and will be presented elsewhere.

REFERENCES

1. P. Hohenberg and W. Kohn. *Phys. Rev.* **136**: B864, 1964.
2. W. Kohn and L. J. Sham. *Phys. Rev.* **140**: A1133, 1965.
3. G. D. Mahan and K. R. Subbaswamy. *Local Density Theory of Polarisability.* London: Plenum, 1990.
4. G. Maroulis (Ed.), *Atoms, Molecules and Cluster in Electric Fields.* London: Imperial College Press, 2006.
5. H. Chermette. *J. Comput. Chem.* **20**: 129, 1999.
6. P. Geerlings, F. De Proft, and W. Langenaeker. *Chem. Rev.* **103**: 1793, 2003.
7. P. K. Chattaraj, U. Sakar, and D. R. Roy, *Chem. Rev.* **106**: 2085, 2006.
8. R. G. Parr and W. Yang. *Density-Functional Theory of Atoms and Molecules.* New York: Oxford University Press, 1989.
9. R. G. Pearson. *Science* **151**: 172, 1963.
10. R. G. Pearson. *Hard and Soft Acids and Bases.* Stroundenboury, PA: Dowden, Hutchinson & Ross, 1973.
11. R. G. Pearson. *J. Chem. Educ.* **45**: 981, 1986.
12. R. G. Parr and R. G. Pearson. *J. Am. Chem. Soc.* **105**: 7512, 1983.
13. R. G. Parr and W. Yang. *Proc. Natl. Acad. Sci. USA* **82**: 6723, 1985.
14. K. Fukui. *Science* **218**: 747, 1982.
15. R. G. Parr and W. Yang. *J. Am. Chem. Soc.* **106**: 4049, 1984.
16. R. Carbo, L. Leyda, and M. Arnau. *Int. J. Quantum Chem.* **17**: 1185–1189, 1980.
17. R. L. Carbo-Dorca. *Molecular Similarity and Reactivity: From Quantum Chemical to Phenomenological Approaches.* Dordrecht: Kluwer Academic, 1995.
18. P. Senet and F. Aparicio. *J. Chem. Phys.* **126**: 145105, 2007.
19. P. Geerlings, G. Boon, C. Van Alsenoy, and F. De Proft. *Int. J. Quant. Chem.* **101**: 722, 2005.
20. A. Borgoo, M. Torrent-Sucarrat, F. De Proft, and P. Geerlings, *J. Chem. Phys.* **126**: 234104, 2007.
21. R. F. Nalewajski and R. G. Parr, *Proc. Natl. Acad. Sci.* **97**: 8879, 2000.
22. R. P. Feynman. *Phys. Rev.* **56**: 340, 1939.
23. H. Hellmann. *Einfuhrung in die Quantumchemie.* Liepzig: Deuticke, 1937.
24. P. Senet. *Chem. Phys. Lett.* **275**: 527, 1997.
25. D. Pines and P. Nozieres. *The Theory of Quantum Liquids.* Vol. I. New-York: Benjamin, 1966.
26. P. Senet. *J. Chem. Phys.* **105**: 6471, 1996.
27. P. Senet, G. Benedek, and J. P. Toennies. *Europhys. Lett.* **57**: 430, 2002.
28. A. Kristhal, P. Senet, and C. Van Alsenoy, submitted.
29. P. Senet and M. Yang. *J. Chem. Sci.* **117**: 411, 2005.
30. M. Berkowitz and R. G. Parr. *J. Chem. Phys.* **88**: 2554, 1988.
31. M. H. Cohen, M. V. Ganduglia-Pirovano, and J. Kudrnovsky. *J. Chem. Phys.* **101**: 8988, 1994; M. H. Cohen, M. V. Ganduglia-Pirovano, and J. Kudrnovsky. *J. Chem. Phys.* **103**: 3543, 1995.
32. P. Senet. *J. Chem. Phys.* **107**: 2516, 1997.
33. P. Senet et al. In preparation.
34. See for instance, J. Guthmuller and B. Champagne. *J. Chem. Phys.* **127**: 164507, 2007.
35. See for instance, F. A. Bulat, A. Toro-Labbé, B. Champagne, B. Kirtman, and W. Yang. *J. Chem. Phys.* **123**: 014319, 2005.
36. J. K. Nagle. *J. Am. Chem. Soc.* **112**: 4741, 1990.
37. P. K. Chattaraj and B. Maiti. *J. Phys. Chem. A* **105**: 169, 2001.
38. M. Petersen, S. Wilke, P. Ruggerone, B. Kohler, and M. Scheffler. *Phys. Rev. Lett.* **76**: 995, 1996.

39. R. F. W. Bader. *Chem. Rev.* **91**: 893, 1991.
40. F. L. Hirschfeld. *Theoret. Chem. Acta (Berl)* **44**: 129, 1977.
41. M. Yang, P. Senet, and C. Van Alsenoy. *Int. J. Quantum Chem.* **101**: 535, 2005.
42. A. Krishtal, P. Senet, M. Yang, and C. Van Alsenoy. *J. Chem. Phys.* **125**: 034312, 2006.
43. A. Krishtal, P. Senet, and C. Van Alsenoy. *J. Chem. Theory and Comp.* in press 2008.
44. See for instance, R. F. Nalewajski, E. Switka, and A. Michalak. *Int. J. Quantum Chem.* **87**:198, 2002.
45. B. Rousseau, A. Peeters, and C. Van Alsenoy. *Chem. Phys. Lett.* **324**: 189, 2000.
46. J. Appelquist, J. K. Carl, and K. K. Fung. *J. Am. Chem. Soc.* **94**: 2952, 1972.
47. B. T. Thole. *Chem. Phys.* **59**: 341, 1981.
48. S. K. Ghosh. *Chem. Phys. Lett.* **172**: 77, 1990.
49. P. K. Chattaraj, A. Cedillo, and R. G. Parr. *J. Chem. Phys.* **103**:7645, 1995.
50. P. Fuentealba and R. G. Parr. *J. Chem. Phys.* **94**: 5559, 1991.
51. C. Morell, A. Grand, and A. Toro-Labbé. *Chem. Phys. Lett.* **425**: 342, 2006.
52. R. G. Parr, R. Donelly, M. Levy, and W. E. Palke. *J. Chem. Phys.* **68**: 3801, 1978.
53. P. K. Chattaraj, A. Cedillo, and R. G. Parr. *J. Chem. Phys.* **103**: 10621 1995.
54. J. Callaway. *Quantum Theory of the Solid State.* San Diego: Academic Press, 1974.
55. P. W. Ayers, F. De Proft, A. Borgoo, and P. Geerlings. *J. Chem. Phys.* **126**: 224107, 2007.
56. N. Sablon, F. De Proft, P. W. Ayers, and P. Geerlings. *J. Chem. Phys.* **126**: 224108, 2007.
57. P. Fuentealba and A. Cedillo, *J. Chem. Phys.* 110: 9807, 1999.
58. A. Baeten, F. De Proft, and P. Geerlings. *Int. J. Quantum Chem.* **60**, 931, 1996.
59. W. Yang and W. J. Mortier. *J. Am. Chem. Soc.* **108**, 5708, 1986.
60. P. W. Ayers, R. C. Morrison, and R. K. Roy. *J. Chem. Phys.* **116**, 8731, 2002.
61. F. De Proft, R. Vivas-Reyes, A. Peeters, C. Van Alsenoy, and P. Geerlings. *J. Comput. Chem.* **24**, 463, 2003.
62. See for instance, S. Damoun, G. Van de Woude, F. Méndez, and P. Geerlings. *J. Phys. Chem. A* **101**, 886, 1997.
63. P. Bultinck, S. Fias, C. Van Alsenoy, P. W. Ayers, and R. Carbo-Dorca. *J. Chem. Phys.* **127**: 034102, 2007.
64. F. De Proft, W. Langenaeker, and P. Geerlings. *J. Phys. Chem. A* **97**, 1826, 1993.

25 External Field Effects and Chemical Reactivity

Rahul Kar and Sourav Pal

CONTENTS

25.1 INTRODUCTION

In the past decade, conceptual density functional theory (DFT) has been of considerable interest to the community of computational chemists [1–3]. Recently, the area has been extensively reviewed by Geerlings et al. [4]. The energy and electron density of an atom or a molecule contain the information on stability and reactivity, respectively. Hence, the behavior of electron density and its variation with respect to the perturbations can reveal many interesting features concerning the reactivity pattern of the atomic and molecular systems [5–8]. Due to small perturbations, the electron density of the interacting systems will be redistributed and hence, would signify the reactivity pattern. Therefore, it is important to study such external field perturbations on the energy and electron density as they would enrich us with the knowledge of the interaction of a molecule with the external field. The derivatives of the energy and electron density with respect to the number of electrons collectively form the set of reactivity descriptors. These offer satisfactory information about the stability and reactivity of the molecular systems under external perturbation.

This chapter is intended to provide basic understanding and application of the effect of electric field on the reactivity descriptors. Section 25.2 will focus on the definitions of reactivity descriptors used to understand the chemical reactivity, along with the local hard–soft acid–base (HSAB) semiquantitative model for calculating interaction energy. In Section 25.3, we will discuss specifically the theory behind the effects of external electric field on reactivity descriptors. Some numerical results will be presented in Section 25.4. Along with that in Section 25.5, we would like to discuss the work describing the effect of other perturbation parameters. In Section 25.6, we would present our conclusions and prospects.

25.2 THEORY

The time-independent nonrelativistic electronic Hamiltonian, under the Born–Oppenheimer approximation can be written as [1]

$$\hat{H} = -\frac{1}{2}\sum_{i=1}^{N}\nabla_i^2 - \sum_{i,A=1}^{N}\frac{Z_A}{r_{iA}} + \sum_{i<j}^{N}\frac{1}{r_{ij}} \tag{25.1}$$

where the notations have their usual meaning. The external field, electric or magnetic field, can be introduced as a perturbation parameter. The dipolar interaction of the Hamiltonian with the electric field, F can be written as,

$$\hat{H} = \hat{H}_0 - \hat{d}\cdot\vec{F} \tag{25.2}$$

where
    $\hat{H}_0$ is the unperturbed Hamiltonian
    d is the dipole moment operator

The total energy of the system can be written as

$$E = E_0 - \sum_i d_i F_i - \frac{1}{2}\sum_{i,j}\alpha_{ij}F_iF_j - \frac{1}{6}\sum_{i,j,k}\beta_{ijk}F_iF_jF_k - \frac{1}{3}\sum_{i,j}\Theta_{ij}F_{ij} - \cdots \tag{25.3}$$

where
    d_i is the permanent dipole moment
    Θ_{ij} is the quadrupole moment
    α_{ij}, β_{ijk} are the dipole polarizability and hyperpolarizability, respectively
    F_{ij} is the component of the field gradient

However, in general, the energy of a perturbed system can be expanded in a Taylor series as

$$E(F) = E_0 + \sum_i \left(\frac{\partial E}{\partial F_i}\right)_0 F_i + \frac{1}{2}\sum_{i,j}\left(\frac{\partial^2 E}{\partial F_i \partial F_j}\right)_0 F_i F_j + \frac{1}{6}\sum_{i,j,k}$$

$$\times \left(\frac{\partial^3 E}{\partial F_i \partial F_j \partial F_k}\right) F_i F_j F_k + \cdots \tag{25.4}$$

where the subscript 0 means that the derivatives are evaluated at zero field. Thus, we have

$$d_i = -\left(\frac{\partial E}{\partial F_i}\right)_0; \quad \alpha_{ij} = -\left(\frac{\partial^2 E}{\partial F_i \partial F_j}\right)_0; \quad \beta_{ijk} = -\left(\frac{\partial^3 E}{\partial F_i \partial F_j \partial F_k}\right)_0 \tag{25.5}$$

Similarly, the interaction with a magnetic field can be written in terms of a magnetic dipole, quadrupole, and other moments. However, the first derivative in the Taylor series expansion, in the presence of magnetic field, is the permanent magnetic dipole moment; the second derivative is the magnetizability. For a further study on the electrical and magnetic properties, one can refer to Dykstra [9].

25.2.1 REACTIVITY DESCRIPTORS

The basic theorem in DFT is the Hohenberg–Kohn theorem, where the ground-state energy is defined as a functional of electron density and is given by [10]

$$E[\rho] = F_{HK}[\rho] + \int v(r)\rho(r)dr \tag{25.6}$$

The global reactivity descriptors, such as chemical potential and chemical hardness, are the derivative of energy with respect to the number of electrons. The formal expressions for chemical potential (μ) and chemical hardness (η) are [1,11]

$$\mu = \left(\frac{\partial E[\rho]}{\partial N}\right)_{v(r)} = \frac{-I - A}{2} \tag{25.7}$$

$$\eta = \frac{1}{2}\left(\frac{\partial \mu}{\partial N}\right)_{v(r)} = \frac{I - A}{2} \tag{25.8}$$

The working equations, due to the finite difference approximation, are given alongside where ionization energy and electron affinity are represented as I and A, respectively. Electronegativity, chemical potential, and hardness have been computed using this finite difference approximation. Global softness, S, is the inverse of hardness. These descriptors are used to describe the system globally. However, the local reactivity descriptors (LRD) such as Fukui function (FF), local softness [12b], etc., are derivatives of electron density and are used to understand the reactivity of an atom in a molecule. Fukui function ($f(r)$) is defined as, [12a]

$$f(r) = \left(\frac{\partial \rho(r)}{\partial N}\right)_{v(r)} = \left(\frac{\delta \mu}{\delta v(r)}\right)_N \tag{25.9}$$

It is to be noted that $f(r)$ is normalized to unity. Due to discontinuity problem in the number of electrons [13] in atoms and molecules, the right- and left-hand side derivatives at a fixed number of electrons introduces the concepts of FF for nucleophilic and electrophilic attack, respectively. Introducing the finite difference approximation and the concept of atom condensed Fukui function (CFF) [14], the working equations are

$$f_k^+ \approx q_k^{N_0+1} - q_k^{N_0} \tag{25.10}$$

$$f_k^- \approx q_k^{N_0} - q_k^{N_0-1} \tag{25.11}$$

Along the same line, the CFF for radical attack is defined as the average of f_k^+ and f_k^-. Descriptors such as electrophilicity index and its local counterpart are also a useful quantity [15]. Several other reactivity descriptors have been proposed to explain the reactivity of chemical species [16,17].

25.2.2 INTERACTION ENERGY USING LOCAL HSAB PRINCIPLE

Let us consider two systems A and B, which interact through their atoms. If the interaction between the systems occur through the atom x of A with the atom k of the molecular system B, one can express the total interaction energy from the local point of view as [18,19]

$$(\Delta E_{int})_{Ax-Bk} = \frac{-(\mu_A - \mu_B)^2}{2} \left(\frac{S_A f_{Ax} S_B f_{Bk}}{S_A f_{Ax} + S_B f_{Bk}} \right)_v - \frac{1}{4} \left(\frac{\lambda}{S_A f_{Ax} + S_B f_{Bk}} \right)_\mu \tag{25.12}$$

where the notations have their usual significance. For a detailed mathematical derivation, one can refer to Ref. [18]. The first term in the interaction energy arises due to the equalization of chemical potential of A and B, where the normal global softness has been replaced by the local softnesses of the reacting site. The second term is due to the maximization of hardness among equichemical potential ensembles. This term involves the knowledge of the complex and the expression for the interaction energy has been modeled in terms of charge transfer term λ, which has been proposed as the difference in the sum of the atomic charges of the system A or B before and after the interaction [20]. This development was extended later to interactions occurring through more than a pair of atoms or sites in A and B [21].

25.3 EFFECT OF ELECTRIC FIELD: THEORY

The effect of external field on reactivity descriptors has been of recent interest. Since the basic reactivity descriptors are derivatives of energy and electron density with respect to the number of electrons, the effect of external field on these descriptors can be understood by the perturbative analysis of energy and electron density with respect to number of electrons and external field. Such an analysis has been done by Senet [22] and Fuentealba [23]. Senet discussed perturbation of these quantities with respect to general local external potential. It can be shown that since $\rho(r) = \delta E/\delta v_{ext}$, Fukui function can be seen either as a derivative of chemical potential with respect to the external potential or of electron density with respect to the number of electrons. Further, response of $f(r)$ with N provides the same results as response of η with respect to v_{ext}

$$\frac{\delta f(r)}{\delta N} = \frac{\delta^3 E}{\delta v_{ext}\delta N\delta N} = \frac{\delta\eta}{\delta v_{ext}} \tag{25.13}$$

Higher-order derivatives with respect to external potential define $\chi_1(r, r')$, $\chi_2(r, r', r'')$, etc., and their response with N define $\xi_1(r, r')$, $\xi_2(r, r', r'')$, etc. This chain of derivatives is diagrammatically depicted in Figure 25.1 [22]. Thus, an exact one-electron formulation of all chemical responses (linear and nonlinear; hardness, FF) in terms of Kohn–Sham orbital of the unperturbed system was derived [22b].

With specific reference to a homogeneous electric field as external potential, one can derive similar response. It can be shown that

$$\frac{\delta\mu}{\delta\vec{F}} = \frac{\delta^2 E}{\delta\vec{F}\delta N} = \frac{\delta\vec{D}}{\delta N} \tag{25.14}$$

where $\vec{F}$ is the homogeneous external field and $\vec{D}(\equiv\delta E/\delta\vec{F})$ is dipole operator. Following Senet, similar nonlinear response with respect to $\vec{F}$ can be derived.

It may be recalled that specifically, response of chemical hardness with respect to external field was discussed by Pal and Chandra [24] quite long back. Using a finite field approximation of η as $E_{N+1} + E_{N-1} - 2E_N$, it can be shown that

$$\frac{\delta\eta}{\delta\vec{F}} = \vec{D}_{N+1} + \vec{D}_{N-1} - 2\vec{D}_N \tag{25.15}$$

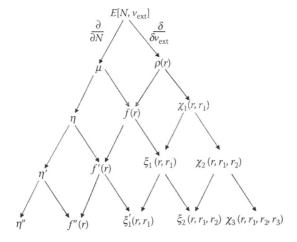

FIGURE 25.1 Energy derivatives are defined by increasing the order of perturbation from 0 to 3. Left arrow represents derivative with respect to number of electrons while right arrow designates derivatives with respect to external potential. (Reprinted from Senet, P., *J. Chem. Phys.*, 107, 2516, 1997. With permission.)

By a numerical observation, it was found that

$$\frac{\delta \eta}{\delta \vec{F}}\Big|_{\vec{F}=0}$$

is a linearly decreasing function of cube root of polarizability, a linear relation between cube root of polarizability of N-electron system with dipole moments of $N+1$, $N-1$, and N electron system was derived.

Few years later, Fuentealba and Cedillo [23] has shown that the variation of the Kohn–Sham FF with respect to the external perturbation depends on the knowledge on the highest occupied molecular orbital (HOMO) density and a mean energy difference of all of the occupied and unoccupied orbital. The quantity, mean energy difference, has been approximately interpreted as hardness. Under this approximation, it has been stated that the greater the hardness, the smaller the variation of the FF, under the external perturbation. This statement then signifies that the system will become less reactive as the hardness of the system increases due to the external perturbation.

These external fields can affect the physical properties and reactivity of the molecular systems. Such effects are especially important in ordered crystalline environments such as solid oxides and biological macromolecules [25–28]. These local electrostatic fields play an important role in catalytic functions and in governing the stabilization of many biomolecular systems [29–32]. These environmental effects cause dramatic changes in the reactivity, which can be different from the gas phase.

However, another study concluded that the changes of the hydrogen-bond stability may be important in biological processes. For these, the influence of local electric fields created by Li^+, Na^+, and Mg^{2+} ions on the properties and reactivity of hydrogen bonds in HF and HCl dimer has been carried out by means of ab initio self-consistent field (SCF) method [33]. A few years later, the effect of intensity and vector direction of the external electric field on activation barriers of unimolecular reactions were studied using the semiempirical MINDO/3 method [34]. However, both semiempirical and ab initio calculations were performed to study the multiplicity change for carbene-like systems in external electric fields of different configurations (carbene and silylene) and the factor that determines the multiplicity and hence the reactivity of carbene-like structures is the nonuniformity of the field [35].

Among the recently published works, the one which showed that the cyclic structures of water clusters open up to form a linear structure above a certain threshold electric field value a was a systematic ab initio study on the effect of electric field on structure, energetics, and transition states of trimer, tetramer, and pentamer water clusters (both cyclic and acyclic) [36]. Considering cis-butadiene as a model system, the strength and the direction of a static electric field has been used to examine the delocalization energy, the probabilities of some local electronic structures, the behavior of electron pairs, and the electronic fluctuations [37]. Another recent work performed by Rai et al. focused on the studies using the DFT and its time-dependent counterpart of effects of uniform static electric field on aromatic and aliphatic hydrocarbons [38].

25.4 EFFECT OF ELECTRIC FIELD: NUMERICAL RESULTS

The effect of external electric field on reactivity descriptors was recently presented by some numerical analysis [39–41]. All these studies concentrated on the effect of uniform electric field on the reactivity descriptors and were performed with Becke's three-parameter hybrid functional [42] (B3) combined with the electron correlation functional of Lee, Yang, and Parr (LYP) [43]. One of those considered formaldehyde and acetaldehyde molecules with $6-31+G^*$ basis set using G98W [44] and DMOL3 packages [45]. It was observed that the variation of the reactivity descriptors with electric field strength was nonlinear and nonuniform [39]. This may be due to the field range chosen, i.e., the electric field varies from 0.0 to 1.0 a.u. [1 a.u. electric field strength $\cong 51.42$ V/Å]. They observed that the changes in the pattern of reactivity indices are strongly influenced by the electric field, thus leading to nonlinear dependence of the descriptors at larger values of the field. However, the field was only applied parallel to the C–H bond in formaldehyde and C–C bond in acetaldehyde, i.e., only along the bond axis. The effect of electric field on the perpendicular directions would have been an important issue, as there is induced polarization perpendicular to the plane of these nonlinear molecules. This fact has been dealt in our recent work, where we apply electric field in all the three perpendicular directions as the direction of the field is an important factor in deciding the stability and reactivity [41]. Moreover, a clear understanding on the variation of the local reactivity indices with the field was essential.

Now, let us try to analyze systematically the behavior of the reactivity descriptors in the presence of electric field. It is always a good idea to start with some simple linear molecules [40] such as HF, HCN, CO, and C_2H_2 to understand the response of the field on these descriptors. All these calculations were performed by GAMESS software [46] with 6-31G(d,p) basis set and a comparatively weak field range was chosen to vary from 0.000 to 0.012 a.u. In the following, we summarize the results obtained in Refs. [40,41].

25.4.1 RESPONSE OF REACTIVITY DESCRIPTORS FOR LINEAR MOLECULES

When the electric field is directed toward the more electronegative atom, along the bond, the chemical potential of all the above molecules, except CO, decreases with the increase in the field strength (Figure 25.2a). This behavior of CO can be attributed to its small dipole moment [40]. However, hardness values decrease with the increase in the field values for all the molecules (Figure 25.2a). Due to the symmetrical nature of acetylene, the hardness is invariant to the change in the field strength (Figure 25.2a).

However, the behavior of the local descriptors of reactivity is somewhat different. The local quantities are calculated using the Löwdin-based method of population analysis [47]. It was observed that when the electric field is applied toward the more electronegative atom of the systems HF and HCN, the value of its nucleophilicity (CFF for electrophilic attack) decreases with increasing field strength (Figure 25.2c). The variation of FF can be well supported with the variation in the induced dipole moment, given in Table 25.1, of the species in

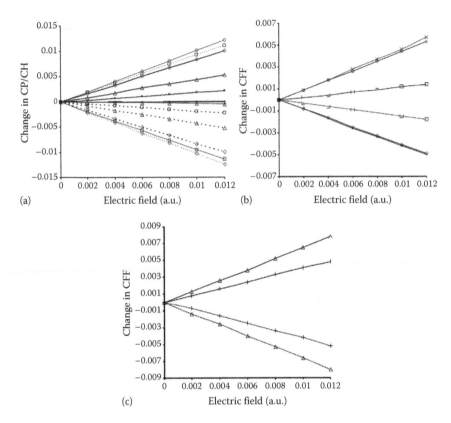

FIGURE 25.2 (a) Variation of changes in chemical potential and chemical hardness (both in a.u.) with respect to zero field versus electric field (in a.u.) for linear molecules HF, HCN, CO, and C_2H_2. $\rightarrow$ chemical potential/chemical hardness for HF; $\triangle$ chemical potential/chemical hardness for HCN; $\boxdot$ chemical potential/chemical hardness for CO; and $+$ chemical potential/chemical hardness for HCCH. Bold lines represent change in hardness while thinner ones represent values of change in chemical potential. (b) Variation of changes in CFF for nucleophilic attack with respect to zero fields versus electric field (in a.u.) for linear molecules HF, HCN and C_2H_2. $\boxdot$ f_H^+ of HF; $\times$ f_H^+ of HCN; and $\rightarrow$ f_H^+ of HCCH. (c) Variation of changes in CFF for electrophilic attack with respect to zero fields versus electric field (in a.u.) for linear molecules HF and HCN. $\times$ f_F^- of HF; $\rightarrow$ f_N^- of HCN. In all three figures, dashed line represents the direction of the electric field when it is toward the more electronegative atom while the solid line represents the electric field when it is in opposite direction. (Reprinted from Kar, R., Chandrakumar, K.R.S., and Pal, S., *J. Phys. Chem. A*, 111, 375, 2007. With permission.)

that field direction. As an example, when the field is applied toward N and F atom in HCN and HF, respectively, their nucleophilicity and induced dipole moment decrease (Figure 25.2c; Table 25.1). The exception to this is provided by CO. The induced dipole moment of C^-O^+ decreases as the field is applied toward C atom, however, the nucleophilicity of C atom increases marginally. The anomaly

TABLE 25.1

Variation of Dipole Moment (in D) of HF, HCN, and CO with Electric Field

Electric Field (in a.u.)	Dipole Moment (H)[a] of HF	Dipole Moment (F) of HF	Dipole Moment (C) of CO	Dipole Moment (O) of CO	Dipole Moment (H) of HCN	Dipole Moment (N) of HCN
0	1.7972	1.7972	0.1412	0.1412	2.8794	2.8794
0.002	1.8194	1.7748	0.0793	0.2033	2.9773	2.7812
0.004	1.8414	1.7523	0.0176	0.2655	3.075	2.6827
0.006	1.8634	1.7297	−0.044	0.3279	3.1724	2.584
0.008	1.8851	1.7068	−0.1054	0.3905	3.2696	2.4851
0.01	1.9068	1.6839	−0.1667	0.4532	3.3666	2.3858
0.012	1.9282	1.6607	−0.2278	0.5161	3.4633	2.2863

Source: Reprinted from Kar, R., Chandrakumar, K.R.S., and Pal, S., *J. Phys. Chem. A*, 111, 375, 2007. With permission.

[a] Atom in parentheses means that the electric field is toward that atom.

can be explained by the small dipole moment of CO, which is very sensitive to the calculation [40]. On the other hand, the electrophilicity (CFF for nucleophilic attack) of the electropositive atom in HF increases when the field is applied toward that atom, (Figure 25.2b). However, on the same line, the electrophilicity of H and O atom in HCN and CO, respectively, decreases. This behavior of CO is anomalous with the behavior of induced dipole moment and is expected to be due to its small dipole moment. With the acetylene molecule, the electrophilicity of the H atom, toward which the field is applied, decreases (Figure 25.2b).

At this point, it is significant to mention that the direction of the electric field plays an important role in deciding the chemical reactivity of a species (Figure 25.2a through c), i.e., when the field direction is reversed, the behavior of these descriptors reverses. Moreover, it can be observed that the variation of reactivity descriptors is linear and uniform in this field range.

It should be noticed that for the same field direction, the hardness as well as the CFF decreases (discussed above). A qualitative explanation for the above results is that the variation of the hardness parameter in the presence of external perturbation is actually dependent on the net effect exhibited by all the atoms present in the molecule [40].

25.4.2 RESPONSE OF STABILIZATION ENERGY CALCULATED THROUGH LOCAL HSAB MODEL

The variation of the interaction energy (both local HSAB and quantum chemical) of NCH–CO, NCH–OC, HCCH–CO, HCCH–OC, FH–CO, FH–OC, HCCH–NCH complexes (reactive atoms are bold) with electric field, as obtained in Ref. [40] is

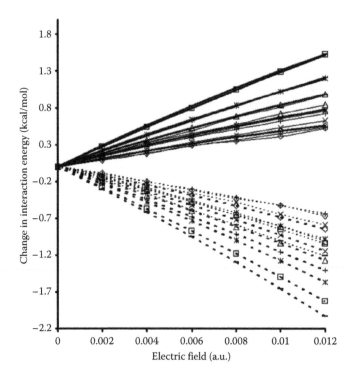

FIGURE 25.3 Variation of the change in interaction energy (kcal/mol) for **FH–CO, FH–OC, NCH–CO, NCH–OC** and **HCCH–CO, HCCH–OC,** and **HCCH–NCH** complexes with respect to zero field versus electric field (in a.u.). ⊟ FH–CO; + FH–OC; ⬠ NCH–OC; ✳ NCH–CO; ◇ HCCH–OC; ✻ HCCH–CO; and — HCCH–NCH. Dashed lines signify that the field is applied along CO while solid lines represent the interaction energy values when the field direction is reversed. For interactions occurring through N atom, dashed lines signify that the field is applied along NCH while solid lines represent the interaction energy values when the field direction is reversed. Bold lines represent the values of quantum chemical interaction energy calculated as $\Delta E_{QC} = E_{AB} - (E_A + E_B)$. (Reprinted from Kar, R., Chandrakumar, K.R.S., and Pal, S., *J. Phys. Chem. A*, 111, 375, 2007. With permission.)

depicted in Figure 25.3. Before discussing the variation of interaction energy with the electric field, we would like to throw light on the charge transfer term mentioned in Equation 25.12. This charge transfer term is calculated using the Löwdin-based method of population analysis. It is observed that when the field is directed toward CO (considering the interaction through either C or O atom), the intermolecular charge transfer term λ always increases. With HCN, when the field is directed toward it and if the interaction occurs through N atom with HCCH or HF, there is an increase in the λ value with the increase in field strength. It also shows that the direction of the external field is important for the intermolecular charge transfer, trends reverse in opposite direction, which will eventually affect the mutual interaction between the monomer systems and hence, the stability of the complexes [40].

It is observed that the stability of the above complexes, formed with CO (either through C or O atom), increases when the field is applied along the CO direction (Figure 25.3). The nucleophilicity of C is stronger than O in CO, hence, the HF gets more stabilized with the former than the latter. This is well supported by the quantum chemical calculations (Figure 25.3). However, the local HSAB interaction energy of C_2H_2 and HCN with CO does not follow with the quantum chemical energy. This anomaly can be attributed to the combined effect of both the values of charge transfer term and FF of the reactive atoms. But, as far as the complex HCCH–NCH interacting through the N atom of HCN is concerned, they get stabilized when the field is applied toward HCN, whereas the interaction decreases when the field direction is reversed. The actual quantum chemical calculation for all these complexes also shows this similar trend [40]. Now, if we compare the local HSAB interaction energy of the above complexes, we can conclude that the complexes **FH–CO**, **NCH–OC**, and **HCCH–NCH** are more stabilized than the others and their stability increases with the increase in field strength at a particular direction (Figure 25.3).

25.4.3 RESPONSE OF REACTIVITY DESCRIPTORS FOR SOME C_{2v} MOLECULES

Similarly, the response of the reactivity descriptors on some C_{2v} molecules was studied very recently [41]. Again, some simple molecules belonging to C_{2v} point group symmetry such as H_2O, CH_2S, HCHO, SO_2, and O_3 were considered. The principal axis of all these molecules is the C_2 axis of symmetry which passes through the central atoms. All the calculations were performed with 6-31++G (d, p) basis set. However, now the field is varied from 0.000 to 0.006 a.u. because it is found that for higher field values sometimes is unable to converge to the correct state for $(N \pm 1)$ systems. In order to verify whether the applied field is strong enough to really mimic the molecular interaction and reaction, we calculated the difference in the energy of the molecule with 0.000 and 0.006 a.u. field. The above molecular systems in this field range, applied along the principal axis and in perpendicular direction, clearly suggests that there is a weak interaction with the external electric field [41].

It is observed that when the field is applied toward the central atom (say from the line joining the two H atoms to the O atom along the principal axis, in case of H_2O) of all the species, and if the central atom is more electronegative, the chemical potential and hardness decrease with increasing field values [41]. For example, when the electric field is applied toward the O atom in H_2O, the chemical potential and hardness decrease. On the other hand, when the field is applied along the principal axis toward the central atom, the nucleophilicity decreases for that atom, provided the central atom is more electronegative. The nucleophilicity of O in H_2O decreases when the electric field is applied toward that atom. Moreover, the electrophilicity of H atoms in H_2O decreases. This trend is the same as observed in case of linear molecules. Here, the results of only one molecule (H_2O) are presented. For details, the reader may refer to our recent paper [41].

Besides, we calculated the field at which the linearity in the variation of reactivity descriptors breaks down and tried to put forward some perturbative analysis. In general, the Fukui function in the presence of external electric field (F) can be written as

$$f(r) = f_0(r) + f'(r)F + \frac{1}{2!}f''(r)F \cdot F + \cdots \qquad (25.16)$$

The first term $f_0(r)$ is the FF at zero field. The second term $f'(r) = \delta^2\rho(r)/\delta N \delta F$ and the third term $f''(r)$ is equal to $\delta^3\rho(r)/\delta N \delta F \delta F$ can be related, respectively, to the linear Fukui response and nonlinear Fukui response (Figure 25.1). At rather stronger fields, the linearity in the variation of the reactivity descriptors breaks down [41]. Electrophilicity index and its local counterpart have been used along with the other indices to study the response of electric field [39].

25.5 EFFECT OF OTHER PERTURBATION PARAMETERS

Apart from the external electric field, one can think of magnetic field, geometry, etc., as a perturbation parameter. Tanwar et al. proposed a principle named minimum magnetizability principle (MMP) [48], in line with the minimum polarizability principle (MPP) [49] to extend the domain of application of conceptual DFT to magnetochemistry. It was shown that there is a minimum value of magnetizability at equilibrium geometry. Otherwise, HSAB principle [50] and maximum hardness principle (MHP) [51] are also extensively used. Apart from this, the effects of an external magnetic field, the electron–nucleus (hyperfine) interaction, and the lifetime of the radicals on the yield of the cage recombination products, the chemically induced polarization of the nuclei in weak magnetic fields, and the magnetic isotope effect were examined [52].

However, Nalewajski used DFT-based charge sensitivity approach for an analysis of the ground-state couplings between the geometrical and population degrees of freedom of open molecular or reactive systems, which includes how the hardness changes with the system geometry, coupling between the electronic and geometrical structure parameters etc. [53]. Another interesting work by Chattaraj et al. focused on the study of the dynamical behavior of the global and local reactivity indices within the framework of quantum fluid DFT [54].

In addition to these external electric or magnetic field as a perturbation parameter, solvents can be another option. Solvents having different dielectric constants would mimic different field strengths. In the recent past, several solvent models have been used to understand the reactivity of chemical species [55,56]. The well-acclaimed review article on solvent effects can be exploited in this regard [57]. Different solvent models such as conductor-like screening model (COSMO), polarizable continuum model (PCM), effective fragment potential (EFP) model with mostly water as a solvent have been used in the above studies.

25.6 CONCLUSION AND PROSPECTS

In this chapter, we have tried to give an introduction to the response of reactivity descriptors to the external perturbation. We have systematically studied the response of external field, mainly the electric field, to linear molecules by means of the conceptual DFT. It is demonstrated that the global and local indices have different behavior and both have to be considered for a complete understanding of the response of external field. However, complexes formed by these linear molecules are more stabilized at the higher field applied in a specified direction than at the zero fields. Moreover, it is interesting to monitor that although the value of hardness increases or decreases for all the systems in the presence of the electric field, there is a further enhancement or decrease of the bond strength of these complexes at the higher field values. This interesting feature can be ascribed to the increase or decrease in the value of the FF indices and the parameter λ due to the applied electric field in a particular direction.

Presently, we have only discussed the effect of the uniform field on the reactivity descriptors. However, in case of the nonuniform field, the terms arising due to higher moments such as quadrupole, octupole, etc., moments would become important. It would also be of great interest to discuss and understand the influence of external electric field on the reactivity of important biological molecules such as deoxyribonucleic acid bases and the Watson–Crick-type base pair interaction energies. Moreover, the behavior of the reactivity descriptors can be analyzed in the presence of magnetic field. Combining the effects of electric and magnetic field, one may be able to relate intuitively the spectroscopic properties to the reactivity descriptors.

ACKNOWLEDGMENTS

Rahul Kar acknowledges the financial support of the Council of Scientific and Industrial Research (CSIR), New Delhi.

REFERENCES

1. Parr, R. G. and Yang, W. *Density Functional Theory of Atoms and Molecules*, Oxford University Press, New York, 1989.
2. Nalewajski R. F., Ed. *Topics in Current Chemistry: Density Functional Theory IV: Theory of Chemical Reactivity*, Springer, Berlin, 1996.
3. Gazquez, J. L. *Chemical Hardness; Structure and Bonding 80*, Sen, K. D., Ed., Springer-Verlag; Berlin, 1993.
4. Geerlings, P., De Proft, F., and Langenaeker, W. *Chem. Rev.* 2003, *103*, 1793.
5. (a) Perez, P., Contreras, R., and Aizman, A. *Chem. Phys. Lett.* 1996, *260*, 236; (b) ibid. *J. Mol. Struct. THEOCHEM* 1997, *290*, 169.
6. Rico, J. F., Lopez, R., Ema, I., and Ramirez, G. *J. Chem. Phys.* 2002, *116*, 1788.
7. Dykstra, C. E. *J. Mol. Struct. THEOCHEM* 2001, *573*, 63.
8. Buckingham, A. D., Fowler, P. W., and Hutson, J. M., *Chem. Rev.*1988, *88*, 963.
9. Dykstra, C. E. *Ab-Initio Calculations of the Structures and Properties of Molecules*, Elsevier Science, Amsterdam, 1988.
10. Hohenberg, K. and Kohn, W. *Phys, Rev. B* 1964, *136*, 864.
11. (a) Parr, R. G. and Pearson, R. G. *J. Am. Chem. Soc.* 1983, *105*, 7512. (b) Pearson, R. G. *J. Am. Chem. Soc.* 1985, *107*, 6801 (c). Parr, R. G., Donnelly, R. A., Levy, M., and Palke, W. E. *J. Chem. Phys.* 1978, *68*, 3801.
12. (a) Parr, R. G. and Yang, W. *J. Am. Chem. Soc.* 1984, *106*, 4049; (b) Yang, W. and Parr, R. G. *Proc. Natl. Acad. Sci.* 1985, *82*, 6723.

13. (a) Perdew, J. P., Parr, R. G., Levy, M., and Balduz, J. L., Jr. *Phys. Rev. Lett.* 1982, *49*, 1691. (b). Zhang, Y. and Yang, W. *Theor. Chem. Acc.* 2000, *103*, 346.

14. Yang, W. and Mortier, W. J. *J. Am. Chem. Soc.* 1986, *108*, 5708.

15. (a) Parr, R. G., Szentpaly, L. V., and Liu, S. *J. Am. Chem. Soc.* 1999, *121*, 1922; (b) Chattaraj, P. K., Maiti, B., and Sarkar, U. *J. Phys. Chem. A* 2003, *107*, 4973; (c) Chattaraj, P. K., Sarkar, U., and Roy, D. R. *Chem. Rev.* 2006, *106*, 2065.

16. (a) Shetty, S., Kar, R., Kanhere, D. G., and Pal, S. *J. Phys. Chem. A* 2006, *110*, 252 (b) Roy, R. K., Krishnamurti, S., Geerlings, P., and Pal, S. *J. Phys. Chem. A* 1998, *102*, 3746.

17. (a) Morell, C., Grand, A., and Toro-Labbe, A. *J. Phys. Chem. A* 2005, *109*, 205; (b) Tanwar, A., Bagchi, B., and Pal, S. *J. Chem. Phys.* 2006, *125*, 214304.

18. (a) Mendez, F. and Gazquez, J. L. *J. Am. Chem. Soc.* 1994, *116*, 9298; (b) Gazquez, J. L. and Mendez, F. *J. Phys. Chem.* 1994, *98*, 4591.

19. Gazquez, J. L. and Mendez, F. *Proc. Indian Acad. Sci.* 1994, *106*, 183.

20. Pal, S. and Chandrakumar, K. R. S. *J. Am. Chem. Soc.* 2000, *122*, 4145.

21. (a) Chandrakumar, K. R. S. and Pal, S. *J. Phys. Chem. B* 2001, *105*, 4541; (b) Chandrakumar, K. R. S. and Pal, S. *J. Phys. Chem. A* 2002, *106*, 5737.

22. (a) Senet, P. *J. Chem. Phys.* 1996, 105, 6471; (b) Senet, P. *J. Chem. Phys.* 1997, 107, 2516.

23. Fuentealba, P. and Cedillo, A. *J. Chem. Phys.* 1999, *110*, 9867.

24. Pal, S. and Chandra, A. K. *J. Phys. Chem.* 1995, *99*, 13865.

25. (a) Kreuzer, H. J. and Wang, L. C. *J. Chem. Phys.* 1990, *93*, 6065; (b) Ernst, N., Drachset, W., Li, Y., and Block, J. H. *Phys. Rev. Lett.* 1986, *57*, 2686 (c) Bragiel, P. *Surf. Sci.* 1992, *266*, 35.

26. (a) Cerveau, G., Corriu, R. J. P., Framery, E., Ghosh, S., and Nobili, M. *Angew. Chem. Int. Ed.* 2002, *41*, 594 (b) Pethica, B. A., *Langmuir* 1998, *14*, 3115. (c) Hochstrasser, R. M. *Acc. Chem. Res.* 1973, *6*, 263.

27. (a) Eckert, M. and Zundel, G. *J. Phys. Chem.* 1987, *91*, 5170; (b) Eckert, M. and Zundel, G. *J. Phys. Chem.* 1988, *92*, 7016; (c) Hill, T. L. *J. Am. Chem. Soc.* 1958, *80*, 2142. (d) Hobza, P., Hofmann, H., and Zahradhik, R. *J. Phys. Chem.* 1983, 87, 573; (e) Xu, D., Phillips, J. C., and Schulten, K., *J. Phys. Chem.* 1996, *100*, 12108.

28. Lippard, S. J. and Berg, J. M. *Principles of Bioinorganic Chemistry*, University Science Books, Mill Valley, CA, 1994.

29. (a) Chandrakumar, K. R. S., Pal, S., Goursot, A., and Vetrivel, R. In: Murugesan, V., Arabindoo, B., Palanichamy, M., Eds., *Recent Trends in Catalysis*, Narosa Publishing House, New Delhi, 1999, p. 197.

30. (a) Cohen de Lara, E., Kahn, R., and Seloudoux, R. *J. Chem. Phys.* 1985, *83*, 2646; (b) Cohen de Lara, E. and Kahn. R. *J. Phys. (Paris)* 1981, *42*, 1029. (c) Cohen de Lara, E. and Kahn, R. *J. Phys. Lett.* 1984, *45*, 255.

31. (a) Li, P., Xiang, Y., Grassian,V. H., and Larsen, S. C., *J. Phys. Chem. B* 1999, *103*, 5058; (b) Bordiga, S., Garrone, E., Lamberti, C., Zecchina, A., Arean, C., Kazansky, V., and Kustov, L. *J. Phys. Chem.* 1994, *90*, 3367. (c) Ferrari, A. M., Ugliengo, P., and Garrone, E. *J. Chem. Phys.* 1996, *105*, 4129. (d) Gruver, V. and Fripiat, J. J. *J. Phys. Chem.* 1994, *98*, 8549.

32. Olivera, P. P. and Patrito, E. M. *Electrochimica. Acta* 1998, *44*, 12477.

33. Hobza, P., Hofmann, H. J., and Zahradni, K. R. *J. Phys. Chem.* 1983, *87*, 573.

34. Lobanov, V. V. and Bogillo, V. I. *Langmuir* 1996, *12*, 5171.

35. Vorontsova, I. K., Mikheikin, I. D., Rakitina, V. A., and Abronin, I. A. *Int. J. Quan. Chem.* 2004, *100*, 573.

36. Choi, Y. C., Pak, C., and Kim, K. S. *J. Chem. Phys.* 2006, *124*, 94308.

37. Karafiloglou, P. *J. Comp. Chem.* 2006, *27*, 1883.

38. Rai, D., Joshi, H., Kulkarni, A. D., Gejji, S. P., and Pathak, R. K. *J. Phys. Chem. A* 2007, *111*, 9111.
39. Parthasarathi, R., Subramanian, V., and Chattaraj, P. K., *Chem. Phys. Lett.* 2003, *382*, 48.
40. Kar, R., Chandrakumar, K. R. S., and Pal, S. *J. Phys. Chem. A* 2007, *111*, 375.
41. Kar, R. and Pal, S. *Theo. Chem. Acc.* 2008, 120, 375.
42. Becke, A. D. *J. Chem. Phys.* 1993, *98*, 5648.
43. Lee, C., Yang, W., and Parr, R. G. *Phys. Rev. B* 1988, *37*, 785.
44. Frisch, M. J. et al. *GAUSSIAN 98, revision A.7*, Gaussian Inc, Pittsburgh, PA, 1998.
45. DMOL3, Accelrys Inc., San Diego, CA.
46. Schmidt, M. W. et al. *J. Comput. Chem.* 1993, *14*, 1347.
47. (a) Lowdin, P. O. *J. Chem. Phys.* 1953, *21*, 374; (b) Lowdin, P. O. *J. Chem. Phys.* 1950, *18*, 365.
48. Tanwar, A., Pal, S., Roy, D. R., and Chattaraj, P. K. *J. Chem. Phys.* 2006, *125*, 56101.
49. (a) Ghanty, T. K. and Ghosh, S. K. *J. Phys. Chem.* 1996, *100*, 12295; (b) Chattaraj, P. K. and Sengupta, S. *J. Phys. Chem.* 1996, *100*, 16126.
50. (a) Pearson, R. G. *J. Am. Chem. Soc.* 1963, *85*, 3533; (b) Chattaraj, P. K., Lee, H., and Parr, R. G. *J. Am. Chem. Soc.* 1991, *113*, 1856.
51. (a) Pearson, R. G. *J. Chem. Educ.*, 1987, *64*, 561; (b) Pearson, R. G. *J. Chem. Educ.*, 1999, *76*, 267; (c) Parr, R. G. and Chattaraj, P. K. *J. Am. Chem. Soc.* 1991, *113*, 1854.
52. Berdinskii, V. L., Yasina, L. L., and Buchachenko, A. L. *Russ. J. Phys. Chem.* 2003, *77*, 1522.
53. Nalewajski, R. F. *Phys. Chem. Chem. Phys.* 1999, *1*, 1037.
54. Chattaraj, P. K. and Maiti, B. *J. Phys. Chem. A* 2001, *105*, 169.
55. (a) Padmanabhan, J., Parthasarathi, R., Subramanian, V., and Chattaraj, P. K. *J. Phys. Chem. A* 2006, *110*, 2739; (b) San, N., Kilic, M., and C lnar, Z. *J. Adv. Oxd. Tech.* 2007, *10*, 51.
56. (a) Ciofini, I., Hazebroucq, S., Joubert, L., and Adamo, C. *Theo.Chem. Acc.* 2004, *111*, 188; (b) Balawender, R., Safi, B., and Geerlings, P. *J. Phys. Chem. A* 2001, *105*, 6703.
57. Tomasi, J. and Perisco, M. *Chem. Rev.* 1994, *94*, 2027.

26 Solvent Effects and Chemical Reactivity*

V. Subramanian

CONTENTS

26.1 INTRODUCTION

It is well known that the surrounding solvent environment plays a crucial role in a chemical reaction. For example, the formation of tetraethylammonium iodide has been studied in many nonpolar and polar solvents. It is found that the rate of the reaction is quite sensitive to the solvent. From the least polar (hexane) to the most polar (nitrobenzene) solvent, the rate constant increases by 2700 times [1]. The polar transition state of this reaction is stabilized in a high dielectric constant medium. Since the

---

* Dedicated to my research supervisor and mentor, Dr. T. Ramasami, Secretary, Department of Science and Technology, Government of India, on his 60th birthday.

solvent effect is environmental, its dielectric constant, dipole moment, viscosity, size, etc. may affect the rate of a reaction.

Understanding the effect of solvation on chemical reactions was one of the main objectives of research in the last century [2–11] as evident from the wealth of literature available on this subject; numerous experimental and theoretical studies have been made in the past to assess the importance of solvent environment on the structure, stability, spectra, and reactivity. It is a well-known fact that accurate treatments of solvent effect on various chemical systems are particularly challenging for theoretical chemistry [3–5]. The reaction is primarily determined by changes in the electronic structure of various species involved. In addition, if reactions are occurring in solution, the change in the electronic energy is similar to that of the solvation energy. Hence, the solvent effect is an important factor of the chemical reaction. It is also essential to realize that the solvent significantly influences the electronic structure of various species involved in the chemical reaction. As a consequence, it is necessary to treat the electronic structure and the solvation together.

26.2   TREATMENT OF SOLVENT MOLECULES AND THERMODYNAMICS BACKGROUND

There are several ways by which the solvent molecules can influence the chemical reactivity. In some reactions, solvent molecules are directly involved in chemical reactions and are tightly bound to the solute molecules. In these cases, solvent molecules effectively form an integral part of the solute and hence these solvent molecules should be treated explicitly. In other systems, the solvent molecules do not interact directly with the solute and provide an environment that strongly influences the behavior of the solute molecules. The typical characteristic of this category of systems is high anisotropic environment around liquid crystal or lipid bilayer. In these cases, the environment strongly influences the conformations of the dissolved solutes. Here, it is not necessary to treat the solvent molecules explicitly but they are modeled using mean field theories proposed by Marcelja [12]. These methods provide a useful compromise between realism and simplicity. Using this approach, meaningful simulations on the required timescale can be carried out. Only a single molecule or a chain is treated explicitly in terms of a realistic all-atom description, and the interactions with the other part of the system (e.g., the surrounding lipid and solvent) are parametrized using a potential of mean force and appropriate random forces. This makes it possible to sample the large number of configurations that are needed to accurately represent the equilibrium state of the bilayer systems [5].

In the third model, solvent molecules act as a bulk medium and significantly modify the solute properties. In this type, solute–solvent interaction is modeled using the continuum approach [8–11]. A variety of models have been proposed in the literature to treat solvent molecules in different situations.

The solvation free energy is defined as the free energy change to transfer a solute molecule from vacuum to solvent [13–15]. It can be considered to have three components:

$$\Delta G_{\text{solvation}} = \Delta G_{\text{electrostatic}} + \Delta G_{\text{vdw}} + \Delta G_{\text{cavity}} \qquad (26.1)$$

where the first term is the electrostatic contribution to the solvation free energy. The electrostatic free energy contributes significantly to polar solvents and charged solutes. The second term is the van der Waals interaction between the solute and the solvent. This can be further divided into $\Delta G_{repulsive}$ and $\Delta G_{dispersive}$. The third term is the free energy required to form solute cavity within the solvent. It is made up of the entropic penalty associated with the solvent reorganization around the solute and the energy required to create cavity. In addition to these terms, $\Delta G_{hydrogen\text{-}bonding}$ can be included to treat the hydrogen bonding interaction between the solute and the solvent. In modeling the effect of solvation, different methods based on classical and quantum mechanics (QM) are used. Methods of treating solvation range from a detailed description at the atomic level to reaction field-based continuum methods. Some of the salient features of the various models are described in Table 26.1.

It can be seen from Table 26.1 that various methods used to model the effect of a solvent can be broadly classified into three types: (1) those which treat the solvent as continuous medium, (2) those which describe the individual solvent molecules (discrete/explicit solvation), and (3) combinations of (1) and (2) treatments. The following section provides a brief introduction to continuum models.

TABLE 26.1
Different Kinds of Solvation Models

Features	Models		
	Explicit Model	**Continuum Model**	**Combined Model**
Representation	Solvent molecules are explicitly treated	Represented as a solvent continuous medium	Combined representation of solute–solvent
Merits	It is generally more accurate and detailed atomic interaction of solute–solvent	Simple, inexpensive to calculate	Atomic interaction of solute–solvent and bulk effects are included. Generally it gives better results than pure continuum models
Disadvantages	Computationally expensive	Ignore specific hydrogen bonding interactions	Computationally expensive when compared to continuum models and less demanding when compared to full explicit treatment

26.3   CONTINUUM MODEL

The treatment of solvation effects based on continuum models has dominated this field since the early nineteenth century. The seminal works of Born [16] on the electrostatic free energy of insertion of a monatomic ion in a continuum dielectric [16] and Onsager [17] as well as Kirkwood [18] on the electrostatic free energy of insertion of dipolar solutes have formed the basis for the development of various continuum solvation models. Continuum models are included in molecular mechanics (MM) and classical molecular dynamics approaches (force field-based simulations) to understand the solvent effect on large systems, specifically biomolecules [5]. These methods are successful to understand the effect of solvation on structure, stability, spectra, and reactivity [8–11]. Both quantum and classical formulations have different options to describe electrostatic contributions to solvation energy [8–11]. The approaches are (1) multipole expansion, (2) apparent surface charge, (3) image charge, (4) finite difference, and (5) finite elements. The quantum formulations of the first three approaches have been implemented in the electronic structure calculations [19] and all the five approaches are used in classical models.

In continuum model, the solvent is described as a uniform polarizable dielectric medium (ε) and the solute of suitable shaped cavity is placed in the dielectric medium [8–11]. By definition, the continuum can be considered as a configuration-averaged or time-averaged solvent environment, where the averaging is Boltzmann-weighted at the temperature of interest. The creation of the solute cavity in the solvent continuum is a destabilization process. On the other hand, the dispersive interaction (van der Waals interaction) between the solute and the solvent is attractive in nature, which stabilizes the system. The electrostatic interaction of a solute with the solvent depends upon the charge distribution and polarizability of the solute. The presence of solute molecules in the dielectric continuum further polarizes the medium, which in turn repolarizes the solute molecules, and therefore electrostatic interaction between the solute and the solvent stabilizes the entire system. This interaction is called as reaction field. A general strategy employed in the continuum model is explained below.

The solute molecule with electronic distribution ρ_M can be written as the sum of contribution from nuclear ρ_{nuc} and that from electron ρ_{el}:

$$\rho_M(\mathbf{r} \colon \{\mathbf{R}\}) = \rho_{nuc}(\mathbf{r} \colon \{\mathbf{R}\}) + \rho_{el}(\mathbf{r} \colon \{\mathbf{R}\}) \tag{26.2}$$

$$\rho_{nuc}(\mathbf{r} \colon \{\mathbf{R}\}) = \sum Z_\alpha \delta(r - R_\alpha) \tag{26.3}$$

where $\mathbf{R}$ represents the position of nuclei that govern electron distribution parametrically. The contribution from electron can be obtained from wave function. Using the basic definition of the dielectric continuum model, the solute is located in the cavity and immersed in the dielectric medium. The basic equation for dielectric continuum model is Poisson–Laplace equation. The electrostatic fields in the cavity and outside the cavity are ϕ_{in} and ϕ_{out}. These quantities can be obtained by solving the following equations with appropriate boundary conditions.

$$\nabla^2 \varphi_{in} = -4\pi\rho_M(r) \tag{26.4}$$

$$\nabla^2 \phi_{out} = 0 \tag{26.5}$$

The electrostatic potential in the cavity (ϕ_{in}) can be separated into the direct contribution from solute (ϕ_{solute}) and continuum medium (ϕ_{medium}). The electrostatic energy between the solute and the continuum is calculated as

$$W_{MS} = \int d\tau \rho_M \phi_{medium} \tag{26.6}$$

The solvation free energy is the work done (W_{MS}) by moving infinitely small charges:

$$\Delta G_{sol} = \frac{1}{2} W_{MS} \tag{26.7}$$

In the classical approach, it is relatively simple to calculate the solvation energies. However, in the quantum mechanical formulations, the electronic structure of the solute molecule depends on the reaction field and the reaction field in turn depends on the structure of the solute. It is a typical nonlinear problem and has to be solved iteratively. Several approaches have been proposed for solving this problem [8–11]. All of them are based on the modification of the Hamiltonian in the following equation:

$$[H_0 + H_{RF}]\Psi = E\Psi \tag{26.8}$$

where H_0 is the usual Hamiltonian for an isolated molecule and H_{RF} is the Hamiltonian representing the interaction between the solute and the continuum medium:

$$\langle\Psi|H_{RF}|\Psi\rangle = W_{MS} \tag{26.9}$$

In the case of the Hartree–Fock method for a closed shell systems, the new Fock matrix is written as

$$F_{\nu\mu}^1 = F_{\nu\mu}^{(0)} + \langle\mu|H_{RF}|\nu\rangle \tag{26.10}$$

where $F^{(0)}$ is the Fock matrix of isolated molecule. The basic differences in the various continuum models are presented in the following section.

Various components of the interactions are calculated using different formalisms. In fact, the shape and size of the cavity are defined differently in various versions of the continuum models. It is generally accepted that the cavity shape should reproduce that of the molecule. The simplest cavity is spherical or ellipsoidal. Computations are simpler and faster when simple molecular shapes are used. In Born model, with simplest spherical reaction field, the free energy of difference between vacuum and a medium with a dielectric constant is given as [16]

$$\Delta G_{\text{elec}} = -\frac{q^2}{2a}\left(1 - \frac{1}{\varepsilon}\right) \tag{26.11}$$

where ΔG_{elec} is the work done to transfer the ion from vacuum to the solvent phase, q is the molecular charge, and a is the radius of the cavity. In this model, the charge and net dipole moment of the solute molecule are taken into account, and cavity and dispersive interaction are neglected. The model based on spherical cavity with dipole is known as Onsager model [17]. The free energy of solvation is written as

$$\Delta G_{\text{elec}}^{(\mu)} = -\frac{\varepsilon - 1}{2\varepsilon + 1}\frac{\mu^2}{a^3} \tag{26.12}$$

where μ is the dipole moment. In the Kirkwood model, the multipole expansion in a cavity is used [16]. The same model with ellipsoidal cavity is referred as Kirkwood–Westheimer model [20]. This model does not include the back polarization effect. By taking into account of the back polarization effect of the medium on the solute molecules, the free energy of solvation is given by

$$\Delta G_{\text{elec}}^{(\mu)} = -\frac{\varepsilon - 1}{2\varepsilon + 1}\frac{\mu^2}{a^3}\left[1 - \frac{\varepsilon - 1}{2\varepsilon + 1}\frac{2\alpha}{a^3}\right]^{-1} \tag{26.13}$$

where α is the polarizability of the molecule.

26.4 SELF-CONSISTENT REACTION FIELD METHOD

Self-consistent reaction field (SCRF) models are the most efficient way to include condensed-phase effects into quantum mechanical calculations [8–11]. This is accomplished by using SCRF approach for the electrostatic component. By design, it considers only one physical effect accompanying the insertion of a solute in a solvent, namely, the bulk polarization of the solvent by the mean field of the solute. This approach efficiently takes into account the long range solute–solvent electro-static interaction and effect of solvent polarization. However, by design, this model cannot describe local solute–solvent interactions.

The salient features of quantum formulation of Onsager reaction field model (dipole model) is described here. In this method, the reaction field is treated as perturbation to the Hamiltonian of the isolated molecule. If H_0 is the Hamiltonian of the isolated molecule and H_{RF} is the reaction field [21], the Hamiltonian of the whole system (H_{tot}) is represented as

$$H_{\text{tot}} = H_0 + H_{\text{RF}} \tag{26.14}$$

$$H_{\text{RF}} = -\mu^{\text{T}}\frac{2(\varepsilon - 1)}{(2\varepsilon + 1)a^3}\langle\psi|\bar{\mu}|\psi\rangle \tag{26.15}$$

By solving Equation 26.14, the electrostatic contribution to the solvation free energy is obtained as

$$\Delta G_{\text{elec}} = \langle \psi | H_{\text{tot}} | \psi \rangle - \langle \psi_0 | H_0 | \psi_0 \rangle + \frac{1}{2} \frac{2(\varepsilon + 1)}{(2\varepsilon + 1)} \frac{\mu^2}{a^3} \quad (26.16)$$

where μ is the dipole moment operator. One of the limitations of this method is the use of a spherical cavity: molecules are rarely in exact spherical shape. However, a spherical representation is the first approximation to the shape of many molecules. It is also possible to use an ellipsoidal cavity, which may be the more appropriate shape for some molecules. In the case of a spherical cavity, the radius can be calculated from the molecular volume:

$$a^3 = \frac{3V_{\text{m}}}{4\pi N_{\text{A}}} \quad (26.17)$$

where V_{m} is the molecular volume and N_{A} is the Avogadro's number. The molecular volume can be calculated from (1) ratio of molecular weight/density, (2) largest distance within the molecule, and (3) electron density contour.

Pisa group of Tomasi and coworkers have made significant contributions to the development and implementation of these solvation models [8–11]. They have developed a variety of quantum formulations of continuum models. In polarizable continuum method (PCM), more realistic cavity shape based on the van der Waals radii of the atoms in solute is defined [22]. In this method, cavity surface is divided into a large number of small surface elements with point charges. This system of point charges represents the polarization of solvent, and the magnitude of each surface charge is proportional to the electric gradient at that point. The total electrostatic potential at each surface element equals to the sum of potential due to solute and that of other surface charges. Various versions of PCM approach have been developed [8–11]. Dielectric PCM (DPCM) [23], isodensity PCM (IPCM) [24], and self-consistent isodensity PCM (SCIPCM) [25] are some of the versions of PCM method. In addition, integral equation formalism (IEF) has also been developed to model solvation process [26–28].

In addition to the Pisa group, extensive and systematic work on solvation has been carried out by the Barcelona group [10]. Luque, Orozco, and coworkers have reformulated the DPCM model. They have referred this method as Miertus, Scrocco, and Tomasi (MST) model [10,22]. This method has been applied to almost all aspects of solvation problems with special reference to organic and biological systems. In this method, various contributions to the free energy of solvation such as (1) cavitation, (2) van der Waals, and (3) electrostatic have been calculated. In the quantum mechanical framework, the electrostatic contribution is determined by adding the perturbation operator to the solute Hamiltonian and self-consistently solving the corresponding nonlinear Schrodinger equation.

Conductor-like screening model (COSMO) is one of variants of PCM method [29]. In this method, the cavity is considered to be embedded in a conductor with an infinite dielectric constant [29]. An extension to this method, called COSMO-RS (conductor-like screening model for real solvents), has been proposed by Klamt [30]. In COSMO-RS method, the interactions in a fluid is described as local contact interactions of molecular surfaces and the screening charge densities from molecular contact is used to quantify the interaction energies. COSMO-RS has become a predictive method for the thermodynamic properties of pure and mixed fluids.

26.5 SOLVATION MODELS

In the development of solvation models, Cramer and Truhalar have made several noteworthy contributions [8–11]. Most of the implicit solvation models do not include the effect of first solvation shell on the solute properties. This can be satisfactorily treated by finding the "best" effective radii within implicit models. In addition to the first-solvent-shell effects, dispersion interactions and hydrogen bonding are also important in obtaining realistic information on the solvent effect of chemical systems.

The magnitude of the free energy effect associated with any first-solvation-shell phenomenon is approximately proportional to the number of solvent molecules in the first solvation shell. It can be calculated from the solvent-accessible surface area (SASA). This concept of SASA is introduced by Lee and Richards [31] and independently by Hermann [32]. The SASA is calculated as the area traced out by the center of a ball rolling over the surface of a solute, where the radius of the ball is the effective half width of the first solvent shell (1–2 Å for water). In models (SM_x) developed by the Cramer and Truhalar, the concept of SASA has been used with empirical surface tensions [9]. This term is included along with the electrostatic term in the continuum model. This model has the advantage of bulk-dielectric model along with the first solvation effects. Cramer and Truhalar have refined the SM_x by parameterization of atomic surface tensions not only in terms of properties of the atoms of the solute but also in terms of solvent properties [9]. By using widely available solvent descriptors, this has allowed the development of several SM_x solvation models that are applicable not only to water but to any organic solvent also. These models are called as universal solvation models [9].

26.6 NONELECTROSTATIC CONTRIBUTION
TO THE SOLVATION FREE ENERGY

In the previous section, various methods employed to calculate the electrostatic contributions to the free energy of solvation have been presented. However, it is important to provide some ideas about the calculation of nonelectrostatic contributions. These factors are essential for solutes, which are neither charged nor polar. The cavity and van der Waals terms can be combined and represented as [5]

$$\Delta G_{cav} + \Delta G_{vdw} = \gamma A + b \qquad (26.18)$$

where A is the SASA and γ and b are constants. The cavity terms arise due to the solvent pressure and reorganization of solvent molecules around the solute. The solvent molecules, which are present in the first solvation shell, are the most affected by reorganization. As mentioned in the previous section, the number of solvent molecules present in the first shell is proportional to the SASA of the solute. Similarly, the solute–solvent van der Waals interactions depend on the number of solvent molecules present in the first solvation shell. Hence, both cavity and van der Waals contributions are modeled in terms of SASA.

26.7 QM/MM MODEL

The combined QM/MM methods are used to study the solvation of the systems that (1) are too large for QM treatments, (2) cannot be appropriately modeled by classical methods because they involve large electron density redistributions, and (3) involve breaking or formation of chemical bonds [10]. A variety of schemes have been proposed in the literature to combine QM/MM regions. The coupling scheme proposed by Morokuma and coworkers is generally known by the acronym ONIOM (our own n-layered integrated molecular orbital + molecular mechanics) [33]. The continuum solvation models are implemented in the ONIOM approach. These models are applied to study the properties of solvated molecular systems. It is evident from the results that these methods appear to be a promising tool for accurate calculations on large molecules in solution [34,35].

26.8 SOLVENT EFFECT AND ELECTRON CORRELATION

With sufficiently large basis set, the Hartree–Fock (HF) method is able to account for $\sim$99% of the total energy of the chemical systems. However, the remaining $\sim$1% is often very important for describing chemical reaction. The electron correlation energy is responsible for the same. It is defined as the difference between the exact nonrelativistic energy of the system (ε_0) and Hartree–Fock energy (E_0) obtained in the limit that the basis set approaches completeness [36]:

$$E_{\text{corr}} = \varepsilon_0 - E_0 \tag{26.19}$$

The dispersive force arises due to the intermolecular electron correlation between the solute and the solvent. Further, it is also important to include the changes in intramolecular and intermolecular solvent electron correlation upon insertion of the solute in the solvent continuum. Further, electron correlation affects the structure of the solute and its charge distribution. Hence, the wave function obtained from the calculation with electron correlation provides a more accurate description of reaction field.

26.9 APPLICATIONS OF SOLVATION MODELS

The solvation models are used to predict the properties of small molecules and large biomolecules employing different levels of theory. In the prediction of solvent effect using electronic structure calculation, semiempirical, HF, post-HF, and DFT-based hybrid methods have been widely used [2–11]. Since a wealth of literature is available on the applications of solvation models, it is a mammoth task to present the complete details. Some of the important applications are presented in the next section.

26.10 SOLVENT EFFECT ON EQUILIBRIA

Many research groups have applied the continuum models to understand the effect of solvation on conformational, tautomeric, acid/base, or other equilibria. The utility of continuum solvation models can be understood with the following example. The keto–enol tautomerism 2-pyridone to 2-hydroxypyridine have been estimated using HF/6–31G* geometries at QCISD/6–31 + G** levels of calculation employing Onsager dipole model [37]. The calculated change in free energy of reaction (ΔG_r^0) in gas phase, cyclohexane, and acetonitrile are −0.6, 0.4, and 2.3 kcal/mol, respectively. These are in reasonable agreement with the corresponding experimental values of −0.8, 0.3, and 3.0 kcal/mol. It can be observed from the results that (1) the level of theory, (2) size of the basis sets, and (3) rigorousness in solvent representation are the important factors that determine the quality of results.

26.11 REACTION MECHANISMS

The effect of solvent environment on the chemical reactivity is well known. However, it is a challenging problem for theoretical chemists to predict the effect of the solvent on the chemical reactivity. With the confidence gained in understanding the chemical reaction mechanism in vacuum using various electronic structure calculation methods, several attempts have been made to probe the reactivity in solvent medium. The success of solvation models in predicting the S_N2 reactions in solvent environments is illustrated [8–11,38].

26.12 SOLVENT EFFECT ON SPECTRA

Along with a variety of electronic structure methods, the continuum solvation models have been used to predict the electronic, vibrational, and nuclear magnetic resonance (NMR) spectra [8–11]. It is well known that carbonyl stretching frequency undergoes shift in the solvent environment. Several calculations have been carried out on prediction of stretching frequency in different solvent environments using various continuum models. It is possible to establish a linear relationship between the solvent-induced shifts and the other physicochemical properties in solution. Recently, the solvent-induced shifts of the carbonyl (C = O) stretching frequency of acetone in 21 organic solvents have been studied [39]. Results of the multiple regression analysis have shown that four descriptors, namely (1) the solvation free energy of solute in continuous dielectric medium, (2) the global interaction energy of the solute–solvent system, (3) the maximum electrostatic potential on the hydrogen atom of the solvent molecule, and (4) the maximum condensed nucleophilic Fukui functions are important descriptors in predicting the shifts in various solvents. Tomasi and coworkers have demonstrated the effects of solvent environment on NMR shielding parameters [11]. Using high-level quantum chemical calculations with coupled description of solvent environment with continuum and discrete solvation model provided insight into the solvent-induced shifts. Recently, Klein et al. have carried out ab initio calculations on the ^{17}O NMR chemical shifts for water using PCM model [40]. They have investigated the limitations of PCM approach in predicting the NMR shifts of water clusters. It is clearly evident from the literature that it is possible to gain insight into the effect of solvent on the spectra of molecules using continuum and combined models.

26.13   SOLVENT EFFECT ON THE CONCEPTUAL DFT-BASED REACTIVITY DESCRIPTORS

Many important concepts of chemical reactivity have been defined in density functional theory (DFT) via electron density of the chemical system [41]. Both global and local reactivity descriptors have been used to understand the global chemical reactivity and site selectivity. Electronegativity (χ), chemical potential (μ), chemical hardness (η), and electrophilicity (ω) are the global descriptors used to understand the various qualitative concepts in chemical reactivity [42–45]. The local reactivity and site selectivity can be quantified with the help of various local descriptors such as Fukui functions, local hardness, local softness, and local philicity. Chemical potential measures the escaping tendency of an electron cloud. It is negative of electronegativity. Small or zero hardness means that the substance is highly reactive. Softness is reciprocal of hardness. Fukui function is derivative of electronic chemical potential and it is space dependent (local) function. It describes the sensitivity of system's chemical potential to external perturbation at a particular point. With the help of various local descriptors, it is possible to gain insight into the site selectivity for nucleophilic, electrophilic, and radical attack. Since chemical reactivity is influenced by the solvent environment, it is of interest to study the effect of the same on various descriptors. In the following section, the effect of solvent on various descriptors is described.

26.14   SOLVENT EFFECT ON ELECTRONEGATIVITY, HARDNESS, AND SOFTNESS

Lipinski and Komorowski have predicted the effect of solvent on the electronegativity and hardness in a homogeneous polar medium using a virtual charge model [46]. It is found that the hardness of ions decreases with increasing solvent polarizability, whereas the electronegativity index decreases for cations and increases for anions. The calculated χ and η for molecules showed minor dependencies with solvent polarity. Safi et al. [47] have employed the continuum approach for the first time to study the influence of solvent on group electronegativity and hardness values of CH_2F, CH_2Cl, CH_3, $CH_3–CH_2$, and $C(CH_3)_3$. It is observed and concluded that the groups become less electronegative and less hard with increasing dielectric constant.

In addition to the continuum models, the explicit solvation has also been used to quantify the reactivity [48]. In this study, the effect of solvent on the electronegativity, hardness, condensed Fukui function, and atomic softness for a set of molecules and ions have been studied. All anions show a significant change in the chemical potential. Both HOMO and LUMO energy levels decrease in the solvent phase when compared to the gas phase. For acids, the increase in the LUMO orbital energy is larger than that of HOMO energy. For the group of salts, it is interesting to note that LUMO energies increase with a decrease in the HOMO energy levels. This leads to a small change in the chemical potential.

Menses et al. have studied the solvent effects on two main global descriptors of reactivity, namely the electronic chemical potential (the negative of electronegativity) and the chemical hardness by taking neutral molecules and ions [49]. It is evident from this study that the electronic potential of cations increases by solvation, and therefore their solution phase electronegativity decreases as a consequence of charge transfer from the solvent to the solute. For anions, the electronic chemical potential decreases by solvation and therefore their electronegativity increases as a consequence of charge transfer from the solute to the solvent. It is also important to mention that the chemical hardness always decreases upon solvation because the electrostatic potential decreases as the effective radius of the solute increases. However, it is very difficult to give an order of decreasing pattern in hardness for cations versus anions because this quantity strongly depends on the actual structure of the charged solute and its solvation layer.

26.15   SOLVENT EFFECT ON ELECTROPHILICITY

In a recent review, Chattaraj et al. have provided a detailed account of aspects of electrophilicity index [45]. Perez et al. have systematically investigated the continuum solvent effect on the electrophilicity index using the self-consistent isodensity-polarized continuum model (SCI-PCM) [50]. It is interesting to observe a linear relationship between the change in electrophilicity index and the solvation energy. Further, it is found that solvation enhances the electrophilicity power of neutral electrophilic ligands and decreases the power in charged and ionic electrophiles. Depending on the nature of molecule, the elctrophilicity varies with different solvent environments. Theoretical study of the $trans$-$N_2H_2 \rightarrow cis$-N_2H_2 and $F_2S_2 \rightarrow FSSF$ intramolecular rearrangement reactions in gas and solution phases has been carried out by Chattaraj et al. [51]. In both reactions, the electrophilicity decreases in the presence of the solvent, and all the species associated with these reactions become less electrophilic in the solution phase. The partitioning of the changes in electrophilicity has been made. It is evident from the results that changes in electrophilicity in solvent medium is primarily determined by the changes in the electronic chemical potential and the amount of charge transfer.

A detailed analysis of the global and local reactivity patterns of neutral and charged peroxides have been studied using continuum IPCM and SCIPCM models [52]. It can be seen from the results that the energy barriers for the (1,2) hydrogen shift are modified by the presence of the solvent environment. In fact, it is interesting to note that both implicit and explicit solvation models provide similar trend about the reactivity. Nucleophilic cyclopropane ring opening in duocarmycin derivatives by methanol under acid conditions has been studied in both gas and solvent phase

TABLE 26.2

Calculated Fukui Function for Formaldehyde in Gas and Solvent (Water) Phases at HF/6–31G** Level Using Mulliken Population Analysis

Atom	Gas Phase			Solvent Phase		
	f_k^+	f_k^-	f_k^0	f_k^+	f_k^-	f_k^0
C	0.305	0.018	0.161	0.324	0.020	0.162
O	0.271	0.508	0.389	0.242	0.509	0.376
H	0.212	0.237	0.225	0.227	0.236	0.231
H	0.212	0.237	0.225	0.227	0.236	0.231

[53]. It is evident from the calculated values that solvent effect decreases the electrophilicity. However, the presence of same environment reduces the hardness value considerably, thereby making the reaction feasible.

26.16   SOLVENT EFFECT ON LOCAL DESCRIPTORS

Continuum model has been applied for the first time to predict the Fukui functions of formaldehyde, methanol, acetone, and formamide in water medium [54]. The results reveal that the potential for electrophilic and nucleophilic attack increases when passing from the gas phase to an aqueous medium. The calculated Fukui functions for formaldehyde at Hartree–Fock (HF) level of theory are presented in Table 26.2.

It can be noted that solvent has only marginal effect on the Fukui functions and hence local site selectivity. A systematic investigation has been made to study the effect of solvation on the local philicity indices of carbonyl compounds using B3LYP scheme employing direct calculation method [55]. It is possible to observe from the results that solvation marginally influences the local reactivity profiles.

Recently, a comprehensive analysis has been made to probe the effect of solvation on the reactivity and toxicity of the complete series of chlorobenzenes through the conceptual DFT-based global and local descriptors [56]. Using the reactivity values in gas and solvent phases, quantitative structure toxicity relationship (QSTR) has been developed for selected chlorobenzene against *Rana japonica* tadpoles. It is interesting to observe that reactivity descriptors obtained from solvation calculation provides good correlation between the experimental and the predicted toxicity values. Overall, the solvent effect significantly influences the global electrophilicity than its local counterpart.

26.17   SUMMARY

It is established that a detailed understanding of chemical or biochemical systems is impossible without an accurate description of their solvent effects. Hence, tremendous effort has been made in the past to develop solvation models. In this chapter, a brief introduction to continuum solvation models and their applications are presented. Continuum models reasonably predict solvent effect on chemical systems in which solvent molecules act as a bulk medium. The results have clearly shown that the level of quantum chemistry calculation, quality of force field in the classical molecular mechanics molecular dynamics as well as rigorousness in the description of the solvent effect are very important factors in the prediction of results. Solvation models cannot describe the specific hydrogen bonding interaction between the solute and the solvent. In order to account for specific solute–solvent interactions, the combined model is an attractive alternative approach. The combined solvation model realistically describes the effect of first solvation shell and bulk dielectric environment.

ACKNOWLEDGMENTS

The work presented in this chapter was in part supported by the grant received from Council of Scientific and Industrial Research (CSIR), New Delhi, and Central Leather Research Institute (CLRI), Chennai, India. The author is thankful to Prof. P. K. Chattaraj for his collaboration and kind invitation to write this chapter. The author is grateful Dr. R. Parthasarathi, Dr. J. Padmanabhan, Mr. M. Elango, Dr. U. Sarkar, Mr. D. Roy, Mr. S. Sundar Raman, Mr. R. Vijayaraj, Mr. R. Gopalakrishnan, and others for their scientific contributions and help.

REFERENCES

1. Espenson, J. H. *Chemical Kinetics and Reaction Mechanisms*, 2nd ed., McGraw-Hill, New York, 1995, p. 29.
2. Tapia O. and Bertran, J., (eds.), *Solvent Effects and Chemical Reactivity*, Kluwer Academic, Dordrecht, 1996.
3. Jensen, F. *Introduction to Computational Chemistry*, Wiley, Chichester, 1999.
4. Levine, I. *Quantum Chemistry*, 5th ed., Prentice Hall, Upper Saddle River, NJ, 1999.
5. Leach, A. R. *Molecular Modelling: Principles and Applications*, 2nd ed., Prentice Hall, Upper Saddle River, NJ, 2001.
6. Hirata, F. *Molecular Theory of Solvation*, Kluwer Academic Press, New York, 2003.
7. Cramer, C. J. *Essentials of Computational Chemistry: Theories and Models*, 2nd ed., Wiley, Chichester, 2004.
8. Tomasi, J. and Persico, M. *Chem. Rev.* 1994, 94, 2027.
9. Cramer, C. J. and Truhlar, D. G. *Chem. Rev.* 1999, 99, 2161.
10. Orozco, M. and Luque, F. J. *Chem. Rev.* 2000, 100, 4187.
11. Tomasi, J., Mennucci, B., and Cammi, R. *Chem. Rev.* 2005, 105, 2999.
12. Marcelja, S. *Nature* 1973, 241, 451.
13. Ben-Naim, A. *J. Phys. Chem.* 1978, 82, 792.
14. Ben-Naim, A. and Marcus, Y. *J. Chem. Phys.* 1984, 81, 2016.
15. Ben-Naim, A. *Solvation Thermodynamics*, Plenum, New York, 1987.
16. Born, M. *Z. Physik.* 1920, 1, 45.
17. Onsager, L. *J. Am. Chem. Soc.* 1936, 58, 1486.
18. Kirkwood, J. G. *J. Chem. Phys.* 1939, 7, 911.
19. Foresman, J. B. and Frisch, A. E., *Exploring Chemistry with Electronic Structure Methods*, 2nd ed., Gaussian, Inc., Pittsburgh, PA, 1996.
20. Kirkwood, J. G. and Westheimer, F. H. *J. Chem. Phys.* 1938, 6, 506.
21. Tapia, O. and Goscinski, O. *Mol. Phys.* 1975, 29, 1653.
22. Miertus, S., Scrocco, E., and Tomasi, J. *Chem. Phys.* 1981, 55, 117.

23. Cammi, R. and Tomasi, J. *J. Comput. Chem.* 1995, 16, 1449.
24. Foresman, J. B., Keith, T. A., Wiberg, K. B., Snoonian, J., and Frisch, M. J. *J. Phys. Chem.* 1996, 100, 16098.
25. Wiberg, K. B., Keith, T. A., Frisch, M. J., and Murcko, M. *J. Phys. Chem.* 1995, 99, 9072.
26. Cance's, E., Mennucci, B., and Tomasi, J. *J. Chem. Phys.* 1997, 107, 3032.
27. Mennucci, B., Cance's, E., and Tomasi, J. *J. Phys. Chem. B* 1997, 101, 10506.
28. Cance's, E. and Mennucci, B. *J. Math. Chem.* 1998, 23, 309.
29. Klamt, A.and Schüürmann, G. *J. Chem. Soc., Perkin Trans.* 1993, 2, 799.
30. Klamt, A. *J. Phys. Chem.* 1995, 99, 2224.
31. Lee, B. and Richards, F. M. *J. Mol. Biol.* 1971, 55, 379.
32. Hermann, R. B. *J. Phys. Chem.* 1972, 76, 2754.
33. Maseras, F. and Morokuma, K. *J. Comp. Chem.* 16, 1170 (1995).
34. Vreven, T., Mennucci, B., Silva, C. O. d., and Morokuma, K. *J. Chem. Phys.* 2001,115, 62.
35. Mennucci, B., Martinez, J. M., and Tomasi, J. *J. Phys. Chem. A* 2001, *105*, 7287.
36. Szabo, A. and Ostlund, N. S. Modern Quantum Chemistry: Introduction to Advanced Electronic Structure Theory, Dover Publications, New York, 1996.
37. Wong, M. W., Wiberg, K. B., and Frisch, M. J. *J. Am. Chem. Soc.* 1992, 114, 1645–1652.
38. Pomelli, C. S. and Tomasi, J. *J. Phys. Chem. A* 1997,101,3561.
39. Chang, C. M. *J. Phys. Chem. A* 2008, 112, 2482–2488.
40. Klein, R. A., Mennucci, B., and Tomasi, J. *J. Phys. Chem. A* 2004, 108, 5851–5863.
41. Parr, R. G. and Yang, W. *Density Functional Theory of Atoms and Molecules*, Oxford University Press, Oxford, 1989.
42. Geerlings, P., De Proft, F., and Langenaeker, W. *Chem. Rev.* 2003, 103, 1793.
43. Chermette, H. *J. Comput. Chem.* 1999, 20, 129.
44. Chattaraj, P. K. and Parr, R. G. Density functional theory of chemical hardness. Sen, K.D., Mingos, D. M. P. (eds.), in *Chemical Hardness, Structure and Bonding*, Springer-Verlag, Berlin, 1993, Vol. 80, pp. 11–25.
45. Chattaraj, P. K., Sarkar, U., and Roy, D. R. *Chem. Rev.* 2006, 106, 2065–2091.
46. Lipinski, J. and Komorowski, L. *Chem. Phys. Lett.* 1996, 262, 449.
47. Safi, B., Choho, K., De Proft, F., and Geerlings, P. *J. Phys. Chem. A* 1998, 102, 5253.
48. Safi, B., Balawender, R., and Geerlings, P. *J. Phys. Chem. A* 2001, 105, 11102–11109.
49. Meneses, L., Fuentealba, P., and Contreras, R. *Chem. Phys. Lett.* 2006, 433, 54–57.
50. Perez, P., Toro-Labbe, A., and Contreras, R. *J. Am. Chem. Soc.* 2001, 123, 5527–5531.
51. Chattaraj, P. K., Pe'rez, P., Zevallos, J., and Toro-Labbe', A. *THEOCHEM* 2002, 580, 171.
52. Aparicio, F., Contreras, R., Galva'n, M., and Cedillo, A. *J. Phys. Chem. A* 2003, 107, 10098.
53. Cimino, P., Improta, R., Bifulco, G., Riccio, R., Gomez-Paloma, L., and Barone, V. *J. Org. Chem.* 2004, 69, 2816.
54. Sivanesan, D., Amutha, R., Subramanian, V., Nair, B. U., and Ramasami, T. *Chem. Phys. Lett.* 1999, 308, 223.
55. Parthasarathi, R., Padmanabhan, J., Elango, M., Subramanian, V., and Chattaraj, P. K. *Chem. Phys. Lett.* 2004, 394, 225.
56. Padmanabhan, J., Parthasarathi, R., Subramanian, V., and Chattaraj, P. K. *J. Phys. Chem. A* 2006, 110, 2739–2745.

27 Conceptual Density Functional Theory: Toward an Alternative Understanding of Noncovalent Interactions

Paul Geerlings

CONTENTS

27.1 INTRODUCTION

The advent of density functional theory (DFT) [1,2] has had a profound impact on quantum and computational chemistry. The ingenious proof, given in 1964 by Hohenberg and Kohn [1], that the wave function of a many-electron system (atom, molecule, etc.), a function of the three spatial and a single spin coordinate of all N electrons of that system, could be replaced as basic carrier of information by the electron density function $\rho(\underline{r})$, a function of only three spatial coordinates, offered a perspective for dramatic computational simplification of electronic structure calculations. The price to be paid was (and still is) the unknown exchange correlation potential $v_{XC}(\underline{r})$ also appearing in the working equations of DFT, the celebrated Kohn Sham equations [3], being the counterpart of the Hartree–Fock equations in wave function theory. Due to the efforts of many leading quantum chemists, exchange correlation potentials of ever increasing performance were presented in the past two decades (although sometimes suffering from heavy parameterization) [4] so that at this moment DFT is undoubtedly the main workhorse for computational studies on geometrical and electronic characteristics of molecular ground states and their evolution upon a chemical reaction, for molecules involving not too heavy main or transition group elements, representable by a single configuration [5]. On the other hand, since the pioneering work by Parr in the late 1960s [6], DFT turned out to be a highly valuable instrument for describing and interpreting chemical reactivity starting from sharper definitions of various traditional chemical concepts such as electronegativity, hardness, and softness. This branch of DFT, termed conceptual density functional theory [7], plays a fundamental role in understanding reactions on the basis of the properties of the individual reactions following Parr's dictum "to calculate a molecule is not to understand it" [7] (for reviews see Refs. [8–12]). The basic ingredient is the perturbation expansion [13] of the energy of a system in terms of the two variables characterizing the Hamiltonian: the number of electrons N and the external potential $v(\underline{r})$, i.e., the potential felt by the electrons due to the nuclei (the influence of external electromagnetic fields will be left out of consideration) [13–15]. The final aim is to describe the interaction between two systems A and B in terms of the coefficients $\left(\partial^{n}E/\partial N^{m}\delta v(r)^{m'}\right)$ $(n=m+m')$ of the isolated reactants A and B when expanding the $E=E[N, v(\underline{r})]$ functional. The latter quantities can easily be looked upon as response functions and can be global in nature (i.e., $\underline{r}$-independent, e.g., the electronegativity $\chi = -(\partial E/\partial N)_{v}$), local (i.e., $\underline{r}$-dependent, e.g., the density itself $\rho(\mathbf{r}) = (\delta E/\delta v(\underline{r})_{N})$, or nonlocal (i.e., dependent on two or more positions in space, e.g., the linear response function $\chi(\underline{r},\underline{r}') = \left(\delta^{2}E/\delta v(\underline{r})\delta v(\underline{r}')\right)_{N}$ (vide infra)).

    These descriptors have been widely used for the past 25 years to study "chemical reactivity," i.e., the propensity of atoms, molecules, surfaces to interact with one or more reaction partners with formation or rupture of one or more covalent bonds. Kinetic and/or thermodynamic aspects, depending on the (not always obvious and even not univoque) choice of the descriptors were hereby considered. In these studies, the reactivity descriptors were used "as such" or within the context of some principles of which Sanderson's electronegativity equalization principle [16], Pearson's hard and soft acids and bases (HSAB) principle [17], and the maximum hardness principle [17,18] are the three best known and popular examples.

    In this context, an avalanche of studies were devoted to acid–base reactions in their broadest sense (i.e., the Lewis picture), also involving complexation reactions, to the typical organic reactions of addition, substitution, and elimination types, involving nucleophilic and electrophilic reagents including the case of radicalar reactions and excited states (for a review see Ref. [11]) in which our group has been active for nearly 15 years [11,19]. In recent contributions, our particular attention has been focused on the formulation of selection rules for pericyclic reactions, the famous Woodward Hoffmann rules [20], in terms of density-based reactivity descriptors [21] and to the description of redox reactions, which maybe unexpected due to the dominant role of the change in number of electrons, hardly received attention in this context [22].

A domain of fundamental importance in chemistry and biochemistry hardly treated in the conceptual DFT context are intermolecular interactions not leading to a new constellation of covalent bonds, the so called noncovalent interactions [23]. This type of interactions (e.g., H-bonding, dispersion interaction, etc.) has only recently been touched upon in the literature mainly by our group. The aim of the present chapter is to present an eye opener when and how these interactions can be described in terms of the DFT-based reactivity descriptors. After a brief summary of the basic theory and calculation of the descriptors in Section 27.2, we will subsequently describe their application in Section 27.3, as interpretational tools, in the study of H-bonding (Section 27.3.1) and π–π stacking interactions, and their repercussion on H-bonding (Section 27.3.2). In Section 27.3.3, we show how DFT-based descriptors may be of interest when evaluating dispersion interactions hereby turning conceptual DFT from an interpretative tool to a computational tool.

27.2 CONCEPTUAL DFT AS A SOURCE OF INTERACTION DESCRIPTORS

As stated in the introduction, conceptual DFT is based on a series of reactivity descriptors mostly originating from a functional Taylor expansion of the $E = E[N, v(\underline{r})]$ functional. These $\left(\partial^n E / \partial N^m \delta v(\underline{r})^{m'}\right)$ quantities can be considered as response functions quantifying the response of a system for a given perturbation in N and/or $v(\underline{r})$. In the case of molecular interactions (leading to a new constellation of covalent bonds or not), the perturbation is caused by the reaction partner. In Scheme 27.1 an overview of the interaction descriptors up to $n = 2$ (for a more complex tabulation and discussion of descriptors up to $n = 3$, see Refs. [11,12]) is given.

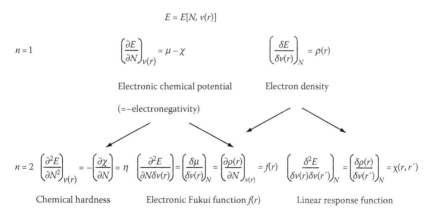

SCHEME 27.1 Response functions $\partial^n E / \partial N^m \delta v(\underline{r})^{m'}$ up to $n = 2$.

The preponderant role of the electronegativity and density itself as first-order response functions is clear: the identification of the electronegativity as the negative or the electronic chemical potential [24], appearing in the DFT analogue of the time-independent Schrödinger equation [2], links the basics of DFT and conceptual DFT. The simplest of the second-order response functions is the global descriptor $(\partial^2 E / \partial N^2)_v$, identified by Parr and Pearson as the chemical hardness η [25]. It turns out to be natural to define its counterpart as chemical softness: $S = 1/\eta$. These concepts were put forward by Pearson in the early 1960s in his HSAB theory on an empirical basis with criteria derived from experimental data on the strength of interaction between generalized acids and bases [25].

The second member of the $n = 2$ series is a mixed derivative $(\partial^2 E / \partial N \delta v(\underline{r}))$, which can easily be seen to be equal to $(\partial \rho(\underline{r}) / \partial N)_v$. This quantity indicates how the electron density at a given point $\underline{r}$ changes when the total number of electrons of a given system changes. In view of its reduction to the highest occupied molecular orbital (HOMO) or lowest unoccupied molecular orbital (LUMO) density (when the left or right derivative is considered), this quantity has been termed Fukui function $f(\underline{r})$ [26]. Multiplied by the total softness it turns the global quantity S into its local counterpart $s(\underline{r})$, the local softness. The search for a local counterpart of the hardness $\eta(\underline{r})$ has been much less trivial [27] and is still the subject of debate in today's literature [28]. In the present contribution, we will adopt an intuitive approach and use $\eta(\underline{r})$ as a measure of charge concentration in a given point $\underline{r}$ in analogy with the global hardness, which is well known to be high in the case of highly charged, small systems (i.e., small, highly charged cations or small anions [in the gas phase being singly charged at most]). A popular working equation involves the negative of the electronic part [29] of the molecular electrostatic potential [30].

$$\eta(\underline{r}) = \frac{1}{2N} V_{el}(\underline{r}) = -\frac{1}{2N} \int \frac{\rho(\underline{r}')}{|r - r'|} d\underline{r}' \tag{27.1}$$

Note that recently Ayers and coworkers have shown how the Molecular Electrostatic Potential (MEP)

$$\text{MEP}(\underline{r}) = \sum_A \frac{Z_A \underline{R}_A}{|\underline{R}_A - \underline{r}|} - \int \frac{\rho(\underline{r}')}{|r' - r|} d\underline{r}' \tag{27.2}$$

where the summation over A runs over all nuclei with charge Z_A and position $\underline{R}_A$ could be considered as an integral part of the hard–hard interactions [31].

In the examples to be discussed in Section 27.3, the toolbox of descriptors will be relatively compact, containing η, S, $f(\underline{r})$, $s(\underline{r})$, and $\eta(\underline{r})$. The evaluation of these quantities will, in most cases, be done by the finite difference approach (for a review see Ref. [10]) leading to the working equations

$$\eta = I - A \quad S = \frac{1}{I - A} \tag{27.3}$$

where I and A are the vertical ionization energy and electron affinity, respectively. In the same vein $f(\underline{r})$ will be evaluated as

$$f^+(\underline{r}) = \rho_{N+1}(\underline{r}) - \rho_N(\underline{r}) \tag{27.4a}$$

or

$$f^-(\underline{r}) = \rho_N(\underline{r}) - \rho_{N-1}(\underline{r}) \tag{27.4b}$$

when considering nucleophilic or electrophilic interaction types, and combined with S to yield the corresponding $s^-(\underline{r})$ or $s^+(\underline{r})$ functions

$$s^+(\underline{r}) = Sf^+(r) \quad s^-(\underline{r}) = Sf^-(r) \tag{27.5}$$

Details on the numerical evaluation of the descriptors will be given in the individual cases but in most cases a computational DFT approach is used, with a hybrid functional of the B3LYP type [32]. Condensation of $f(r)$ or $s(r)$ is done with conventional population analysis techniques (Mulliken [33], Natural Population Analysis (NPA) [34]) or with the Hirshfeld technique [35], often used by our group [36].

27.3 CASE STUDIES OF CONCEPTUAL DFT IN NONCOVALENT INTERACTIONS

27.3.1 HYDROGEN BONDING: THE ROLE OF (LOCAL) SOFTNESS IN (VERY) STRONG HYDROGEN BONDING

Hydrogen bonding [37] is a unique type of inter- and intramolecular interaction not only for its fundamental role in the vital biological and chemical processes, but also for the amount of ambiguity in its operative range. In reality, the spectrum of hydrogen bond strengths extend from 1 to 4 kcal/mol for weak bonds to 4–15 kcal/mol for moderate bonds, and 15–40 kcal/mol for strong bonds [37]. Various models have been developed in order to reveal the mysterious nature of this wide range of interactions. Hydrogen bonding, schematically represented as A—$H \cdots B$, involves electronegative proton donor (A) and acceptor atoms (B). Therefore, the first models were developed on a purely electrostatic basis. Later on, Gilli et al. would qualify this model as the simple electrostatic model (SEM) [38]. By Coulson's [39] introduction of valence-bond (VB) theory into hydrogen bonding, the electrostatic picture was further modulated by delocalizational, repulsive, and dispersive contributions. This idea of partitioning the interaction energy into its components was revisited by Morokuma and others using molecular orbital Theory [40a,b] and used to demonstrate the importance of covalency in hydrogen bonding by Fonseca Guerra et al. and Poater et al. [40c–f]. The inadequacy of the SEM in describing the resonance-assisted hydrogen bonding (RAHB), among others observed in O—$H \cdots O$ type of bonds, has led Gilli et al. to focus more on the covalent nature of the hydrogen bonding, which was already suggested occasionally on the basis of both x-ray and neutron diffraction experiments [41], and ab initio and semiempirical calculations [42–44]. The conclusion stating that "forces determining the hydrogen bond strength are a mixture of both electrostatic and covalent contributions" forms the basis for the electrostatic–covalent hydrogen bond model (ECHBM) [38] According to this model, weak hydrogen bonds are electrostatic in nature. As the strength of the interaction increases, the covalent character of the bond also increases and very strong hydrogen bonds are actually three-center-four-electron covalent bonds. In a vital synthesis, Gilli et al. finally classified the hydrogen bonds as strong (with subclasses negative charge-assisted ((−)CAHB), positive charge-assisted, ((+) CHAB), and resonance-assisted (RAHB)), moderate (with one subclass of polarization-assisted hydrogen bonds (PAHB)), and weak [38,45].

Concentrating on strong and moderate H-bonds and taking, e.g., a homonuclear H-bond of the type O—H $\cdots$ O, the different cases can be represented as $[-O \cdots H \cdots O-]^-$ ((−)CAHB), $[=O \cdots H \cdots O=]^+$ ((+)CAHB), $-O-H \cdots O=$ with the two oxygens connected by a π conjugated system of variable length, and

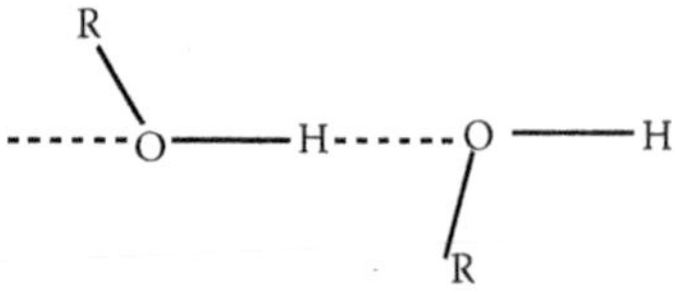

(PAHB).

Gilli et al. pinpointed to the lack of general chemical rules or a unified hydrogen-bond theory as the H-bond puzzle [38]. An alternative view on solving this puzzle using the tools of conceptual DFT such as local hardness, $\eta(\underline{r})$ and local softness, $s(\underline{r})$ is shown below. A series of both homonuclear and heteronuclear resonance-assisted hydrogen bonds of the O—H$\cdots$N, N—H$\cdots$O, N—H$\cdots$N, and O—H$\cdots$O type with strength varying from weak to very strong have been studied (Figure 27.1) for this purpose, all of them being the intramolecular hydrogen bond type [46].

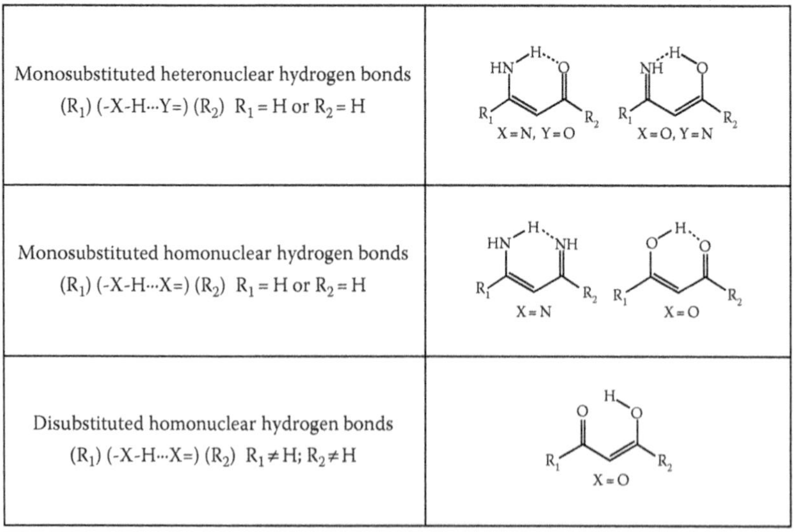

Monosubstituted heteronuclear hydrogen bonds (R_1) (-X-H$\cdots$Y=) (R_2) R_1 = H or R_2 = H	
Monosubstituted homonuclear hydrogen bonds (R_1) (-X-H$\cdots$X=) (R_2) R_1 = H or R_2 = H	
Disubstituted homonuclear hydrogen bonds (R_1) (-X-H$\cdots$X=) (R_2) $R_1 \neq$ H; $R_2 \neq$ H	

FIGURE 27.1 Heteronuclear and homonuclear mono- and disubstituted hydrogen bonded structures studied. (Reprinted from Ozen, A.S., Aviyente, V., De Proft, F., and Geerlings, P., *J. Phys. Chem.*, A110, 5860, 2006. With permission.)

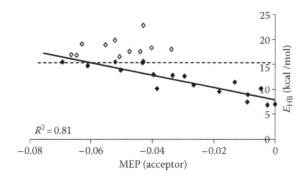

FIGURE 27.2 Hydrogen bond energy (E_{HB}) vs. MEP of acceptor atom in the moderate and moderate-to-strong hydrogen bond region for O—H$\cdots$O type of hydrogen bonds (Unfilled bullets above the plateau are not incorporated in the regression). (Reprinted from Ozen, A.S., Aviyente, V., De Proft, F., and Geerlings, P., *J. Phys. Chem.*, A110, 5860, 2006. With permission.)

The electrostatic component in this picture can be translated into a conceptual DFT framework by introducing the negative charge concentration on the acceptor atom as revealed, e.g., in the NPA charge [47,48], the MEP [49], or local hardness [27b,27c,27d,29]. On the other hand, the highly charged H-atom may stabilize the system by polarizing A and B, and local softness values may be employed to probe the propensity of polarization of A and B, and this might provide a measure of covalency. As an indication that electrostatics is not telling the whole story in the case of strong hydrogen bonding we plot in Figure 27.2 the hydrogen bond energy vs. the MEP at the acceptor atom.

For weak and moderate H-bonding the usual pattern is observed, i.e., the more negative the MEP in the acceptor region, the stronger the H-bond (higher E_{HB}) (filled bullets in Figure 27.2). A reasonable linear correlation can be drawn ($R^2 = 0.81$) until a kind of plateau of 15 kcal/mol is reached. It is, however, seen that in the strong hydrogen bonding regime above 15 kcal/mol the data points (unfilled bullets) do not show any correlation at all. This suggests that above the plateau the hydrogen bonding loses its dominant electrostatic nature and acquires a partial covalent nature. As a result of the relationship between covalent bonding and soft interactions [12,50] the softness values s^0 (average of s^+ and s^-) were evaluated both for these acceptor and donor atoms.

In Figure 27.3, the relationship between the hydrogen bond energy and the product of the local softness values of donor and acceptor atom (A and B) is given for a series of closely related O—H$\cdots$O types of bonds (NR$_2$, NO$_2$ family, etc.).

It is clear that above 15 kcal/mol, a region of saturation of the first-order electrostatic effect is entered, and a second-order effect can be discerned becoming the discriminating factor of hydrogen bond strength.

The appearance and dominance of this second-order effect can be interpreted through the strong polarization of the acceptor and donor atoms. It should also be noted that a special approach to softness matching [11,19,51] has been adapted in this study. According to this method, which was originally derived from Pearson's HSAB principle, [17] the most favorable interaction between the sites A and B

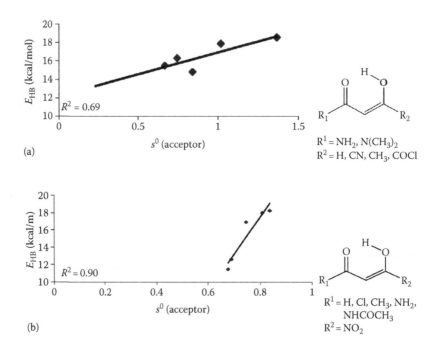

FIGURE 27.3   (a) E_{HB} (kcal/mol) vs. local softness of acceptor for NR_2 family. (With permission.) (b) E_{HB} (kcal/mol) vs. local softness of donor for NO_2 family. (Reprinted from Ozen, A.S., Aviyente, V., De Proft, F., and Geerlings, P., *J. Phys. Chem.*, A110, 5860, 2006. With permission.)

occurs when $\Delta s = s_A - s_B = 0$ or in other words, when $s_A = s_B$. However, in the present strong hydrogen bonded systems, the spirit of softness matching has changed from looking for a minimal Δs value to a maximal value which is the product $s_A \times s_B$. This procedure can be justified in the following way: the local version of the HSAB principle states that soft–soft interactions occur preferentially between sites of the same softness. However, if both values are small at the local level, one can hardly expect a matching to represent an ideal situation for soft–soft interactions. On the other hand, the product of local softness values combines the idea that the difference should be small, but at the same time the individual values should be large and therefore, is a better approach to the soft–soft interactions in the present systems.

Since there is a one-to-one correspondence between covalent and soft inter-actions, the local-softness-related trends obtained in the strong hydrogen-bonding region might be promising as a sign of the covalent character involved, supporting the ECHBM. In conclusion and as a first application of conceptual DFT in studying noncovalent interactions, these results illustrate the electrostatic versus covalent aspect of hydrogen bonding, not in a wave function context (VB theory) but in a conceptual DFT context.

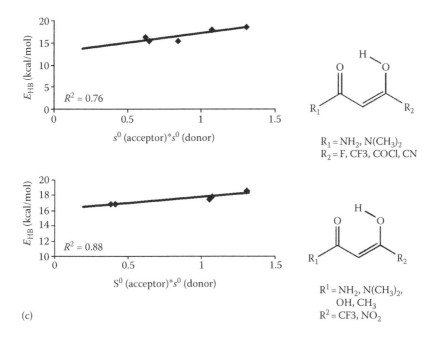

FIGURE 27.3 (continued) (c) Combined effect of the local softness values of the acceptor and donor atoms on hydrogen bonding energy. (Reprinted from Ozen, A.S., Aviyente, V., De Proft, F., and Geerlings, P., *J. Phys. Chem.*, A110, 5860, 2006. With permission.)

27.3.2 π–π Stacking: The Role of Local Hardness and Electrostatics and Its Interplay with Hydrogen Bonding

Intermolecular interactions involving aromatic rings are common in various areas of chemistry, biochemistry, and biology. In proteins, e.g., π stacking has been the subject of systematic search in crystal structures, showing that the side chains preferentially interact in a parallel-displaced orientation [52]. Although the dispersion energy is known to be the principal source of stabilization of stacked complexes, the electrostatic component of the interaction is not negligible, though its role is still controversial [53].

A series of experimental studies revealed that the interaction between phenyl rings increases monotonically when passing from an electron-donating to an electron-withdrawing substituent [54]. In line with these results, Hunter and coworkers [55] proposed a set of rules stating that the aromatic ring can be described as "a positively charged σ-framework between two regions of negatively charged π-electron density." According to this electrostatics-based model, an electron-donating substituent on one of the interacting molecules should increase the negative charge of the π-cloud and thus the repulsion between the two stacked aromatic cycles, whereas electron withdrawing substituents should show the inverse behavior. However, in contrast to Hunter–Sanders rules and experimental results, several recent high-level computational studies of substituent effects on π–π interactions showed that in the parallel-stacked benzene dimers, substituted benzenes with electron-withdrawing or -donating substituents bind stronger to benzene than unsubstituted benzene [56,57]. It was stated that electrostatics, dispersion, induction, and exchange-repulsion are significant to the overall binding energies. In our latest studies, substituted benzenes were found to bind stronger to aromatic nitrogen bases than unsubstituted benzene [58,59].

In the case study discussed below [60], part of our ongoing interest in the application of DFT reactivity descriptors to biosystems [61–66], the interaction between cytosine and substituted benzenes is studied. Cytosine, a nucleobase derived from pyridine, was chosen because of its small size and because it possesses both a nitrogen and an oxygen atom as H-bond acceptors sites.

MP2 calculations [67] were performed on cytosine stacked with a series of seven monosubstituted benzenes (Ph–X; $X = NO_2$, CHO, F, H, CH_3, OH, NH_2). Although it is known that MP2 overestimates the π–π interactions as compared to higher level CCSD(T) calculations [68], trends can be expected to be correctly reproduced. After correction for the basis set superposition error (BSSE) [69], the total interaction energy ΔE_{MP2} can be expressed as the sum of the Hartree–Fock interaction energy ΔE_{HF} and the correlation contribution to the interaction energy corresponding to the dispersion energy. The electrostatic component of the interaction energy was extracted via Stone's distributed multipole analysis [70]. The geometry of the complexes was optimized in a parallel-displaced arrangement, with the substituent X located as far as possible from the H-bond acceptor atoms of cytosine, avoiding direct interactions between the substituent and these atoms (Figure 27.4).

In Table 27.1, the properties of the optimized complexes of cytosine and the substituted benzenes are given. It is seen that the correlation part of the interaction energy, i.e., the dispersion energy, constitutes the major source of stabilization of the complexes though the electrostatic term exhibits negative values, which are of the order of 50% on the average indicating that electrostatics is playing a

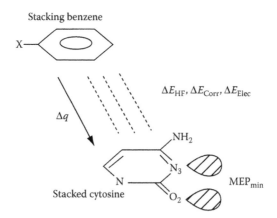

FIGURE 27.4 Geometry of the stacking interaction between substituted benzenes and cytosine. (Reprinted from Mignon, P., Loverix, S., Steyaert, J., and Geerlings, P., *Nucl. Acids Res.*, 33, 1779, 2005. With permission.)

TABLE 27.1

Properties for the Optimized Complexes of Cytosine and Substituted Benzenes (Ph–X): Interaction Energy Components: ΔE_i (kcal/mol), Charge Transfer to the Pyrimidine, Δq (a.u.), Molecular Electrostatic Potential Minimum around the Nitrogen and Oxygen Atoms (see Figure 27.4) (a.u.), Dihedral Angle between the Ring Planes, Dihedral Angle between the Stacked Rings, Φ (degree), Distance between the Rings, r (Å). Properties of the Isolated Substituted Benzenes: Substituent α_x and total Substituted Benzene Polarizabilities α (α_{Ph} Is Calculated according to Equation 27.8); Global Hardness η and Local Hardness $\eta(r)$ (1.7 Å above the Center of the Ring)

	Complex Properties									Properties of the Isolated Substituted Benzenes				
Substituent	ΔE_{MP2}	ΔE_{HF}	ΔE_{Corr}	ΔE_{Elec}	Δq	MEP N$_3$	MEP O$_2$	Φ	r	α	α_x	α_{Ph}	η	$\eta(r)$
NO$_2$	−6.51	2.40	−8.91	−5.22	0.0087	−0.0828	−0.1311	18.9	3.56	76.84	18.08	58.76	0.392	0.0887
CHO	−6.37	3.31	−9.68	−4.84	−0.0088	−0.0881	−0.1355	10.2	3.47	64.91	9.64	55.26	0.359	0.0970
F	−5.33	1.81	−7.15	−4.35	−0.0057	−0.0883	−0.1361	25.7	3.70	59.45	2.41	57.04	0.346	0.1033
H	−5.32	4.30	−9.62	−3.18	−0.0232	−0.0941	−0.1421	13.5	3.46	58.38	0.26	58.13	0.343	0.1062
CH$_3$	−5.05	3.51	−8.56	−3.72	−0.0180	−0.0911	−0.1421	17.7	3.59	70.71	12.12	58.59	0.331	0.1057
OH	−5.41	1.92	−7.33	−3.84	−0.0143	−0.0919	−0.1398	21.6	3.66	63.94	5.58	58.36	0.326	0.1050
NH$_2$	−5.33	3.92	−9.25	−2.71	−0.0350	−0.0986	−0.1461	12.7	3.49	66.88	9.11	57.78	0.339	0.1130

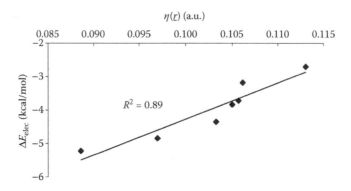

FIGURE 27.5 Electrostatic interaction energy (ΔE_{elec}) between cytosine and the substituted benzenes Ph–X (kcal/mol) vs. the local hardness $\eta(\underline{r})$. (Reprinted from Mignon, P., Loverix, S., Steyaert, J., and Geerlings, P., *Nucl. Acids Res.*, 33, 1779, 2005. With permission.)

nonnegligible role in the stabilization process. It is seen that the charge transfer Δq between the stacking benzene and the stacked cytosine increases with decreasing hardness of the substituted benzene in agreement with the simple electronegativity equalization models showing that charge transfer in an acid–base interaction is inversely proportional to the sum of the hardnesses of acid and base [11,17b]. Increasing charge transfer invariably leads to deeper (i.e., more negative) MEP values both at N_3 and O_2, i.e., stronger H-bonding.

These results can be related to the evolution of the local hardness $\eta(\underline{r})$ evaluated at a distance of 1.7 Å above the isolated benzene rings (about half the distance between the rings in the optimized complexes).

Table 27.1 indicates that the overall local hardness increases with decreasing electron-withdrawing character of the substituents. Consequently, a larger repulsion between the π-electron clouds of the two stacked rings is expected, yielding a smaller contribution to the electrostatic component of the stabilization energy ΔE_{elec} as revealed also in the table.

Figure 27.5 nicely illustrates this trend. Once the system is given the freedom to interact with its partner and to get rid of this accumulation of negative charges, a (larger) charge transfer occurs with increasing H-bond capacity at the donor atom.

In summary, $\eta(\underline{r})$ turns out to be a key index in connecting the electrostatic component of the stacking interaction energy with the hydrogen bonding capacity of cytosine as schematically represented is Figure 27.6.

27.3.3 ROLE OF DFT DESCRIPTORS IN THE (EVALUATION OF) DISPERSION INTERACTION: FROM LOCAL POLARIZABILITY TO LOCAL SOFTNESS

27.3.3.1 Introduction

As stated in Section 27.3.2, dispersion interactions are fundamental building blocks in intermolecular interactions. Their accurate calculation in the case of larger systems (say π stacking of aromatic rings) remains, however even today, despite the

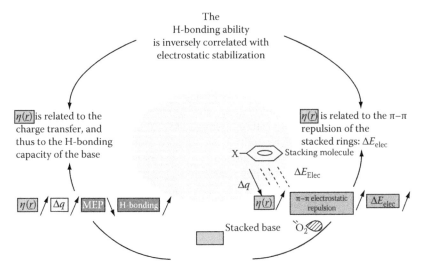

$\eta(\underline{r})$ is a key index: Measures the negative charge
accumulation above the aromatic cycle

FIGURE 27.6 (See color insert following page 302.) $\eta(\underline{r})$ can be used for the estimation of the electrostatic interaction and the hydrogen bonding ability. (Reprinted from Mignon, P., Loverix, S., Steyaert, J., and Geerlings, P., *Nucl. Acids Res.*, 33, 1779, 2005. With permission.)

spectacular evolution in quantum chemical methodologies and computing power, an almost unsurmountable task. Hartree–Fock and DFT dramatically fail in reproducing these effects [71].

In wave function theory, highly correlated calculation levels should be used, at least second-order Møller–Plesset perturbation theory, or preferentially higher level methods such as coupled cluster (e.g., CCSD(T)) [72]. These methods only in recent years became applicable to system of the size relevant for the study of aromatic ring stacking [73]. The exponential increase in computing power will bring systems of ever-increasing size into reach but the unfavorable scaling of even MP2 [72] makes the progress go slow. This drawback at the wave function approach and the spectacular success of DFT incited considerable effect in recent years to construct DFT methods with a correct description of dispersion interactions using conventional functionals [74–76] or with dedicated correlation functionals [77–80] taking long-range dispersion interactions into account. An alternative is the inclusion of empirical dispersion energy corrections, analogous to the ansatz in molecular mechanics force fields. The latter approach received considerable attention in recent years because of its modest computational cost [81–85].

Empirically corrected DFT theories almost invariably go back to second-order perturbation theory with expansion of the interaction Hamiltonian in inverse powers of the intermolecular distance, leading to R^{-6}, R^{-8}, and R^{-10} corrections to the energy in an isotropic treatment (odd powers appear if anisotropy is taken into account [86]).

Using an average energy or Unsold approximation [87] an expression of the type

$$E \sim \frac{U_1 U_2}{U_1 + U_2} \frac{1}{R^6} \alpha_1 \alpha_2 \tag{27.6}$$

is obtained with α_1 and α_2 being the average polarizabilities of the molecules and U_1 and U_2 being the fixed energies associated to the molecules in their considered electronic state (e.g., their ionization energy) [88]. When considering large molecules and realizing that polarizability is an additive property [89,90], the idea that some parts of the molecules contribute to a larger extent to the dispersion energy (6) than others, shows up. The local polarizability $\alpha(\underline{r})$ concept [91,92] with

$$\int \alpha(\underline{r})d\underline{r} = \alpha \tag{27.7}$$

may be taken as a basis to split up the molecular polarizability in group contributions [93]. In the case of stacked aromatic rings, this splitting allows us to estimate the contribution to the dispersion energy (and its evolution, e.g., upon substitution) of the stacked ring, eliminating the direct influence of the constituents on the dispersion energy.

27.3.3.2 Toward Aromatic Ring Polarizabilities

To estimate the dispersion interaction contribution to the overall interaction between stacked rings we studied, in parallel with the role of electrostatics in Section 27.3.2, the interaction between monosubstituted benzenes Ph–X (X = H, F, NH_2, Cl, CH_3, OH, CN, COOH, CHO, NO_2) and pyridine, a simpler case with only one H-bond accepting site [58], and cytosine (with two H-bond accepting sites). An offset parallel geometry was chosen, keeping the X group and the N atom of pyridine at maximal distance, the level used being MP2/6–31G*(0.25)//MP2/6–31G$^-$ with BSSE correction [68].

In order to eliminate the direct influence of the substituents the polarizability of the substituted benzenes Ph–X was split up, as mentioned above, as

$$\alpha(\text{Ph–X}) = \alpha_{\text{Ph}} + \alpha_X \tag{27.8}$$

where α_X was evaluated as the polarizability of the isolated radical corresponding to the substituent (e.g., $^\bullet CH_3$ for $-CH_3$) in line with our previous work on the evaluation of group properties in a conceptual DFT context [93]. Equation 27.6, when referring to the benzene ring contribution to the dispersion energy, can then be simplified to

$$\Delta E_{\text{disp}}\left(R^{-6}\right) \sim -\frac{\alpha_{\text{Ph}}}{R^6} \tag{27.9}$$

where α_{Ph} is the polarizability of the benzene ring in the substituted benzene (data given in Tables 27.1 and 27.2). In Figure 27.7 we plot the contribution of the

TABLE 27.2

Interaction Energies at MP2 Level, ΔE_{MP2}, Their HF Counterparts ΔE_{HF}, and Their Dispersion (ΔE_{corr}) and Electrostatic (ΔE_{elec}) Contributions for Pyridine and Substituted Benzenes (Ph–X) (kcal/mol)

X	ΔE_{MP2}	ΔE_{HF}	ΔE_{Corr}	ΔE_{Elec}
NO_2	−3.79	6.89	−10.67	−0.57
CN	−4.13	4.99	−9.12	−0.55
COOH	−3.49	7.87	−11.35	−0.45
CHO	−3.90	5.18	−9.08	−0.41
Cl	−3.35	8.12	−11.47	−0.63
F	−2.89	7.92	−10.81	−0.34
H	−2.78	7.89	−10.67	−0.15
CH_3	−3.34	5.68	−9.02	−0.02
OH	−2.69	8.39	−11.08	−0.21
NH_2	−3.20	7.66	−10.86	−0.51

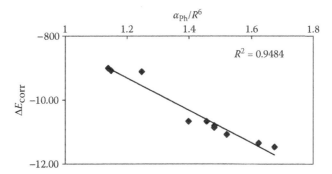

FIGURE 27.7 Correlation part of the interaction energy ΔE_{corr} between pyridine and the substituted benzenes (Ph–X) (kcal/mol) vs. the benzene ring polarizability α_{Ph} divided by R^6 (see Equation 27.8) (a.u.). (Reprinted from Mignon, P., Loverix, S., De Proft, F., and Geerlings, P., *J. Phys. Chem. A*, 108, 6038, 2004. With permission.)

dispersion energy to the interaction energy, obtained as the difference between the MP2 an HF energies vs. $\alpha_{Ph}R^{-6}$. A striking linear relationship is obtained, confirmed in the case of cytosine (Table 27.1) (Figure 27.8), the correlation coefficient being in both cases 0.94.

Both correlations indicate that polarizability at local level can be used to quantify the dispersion energy. In the next section, we take a further step within direction, moving to an atoms-in-molecules level, thereby linking local polarizability to local softness.

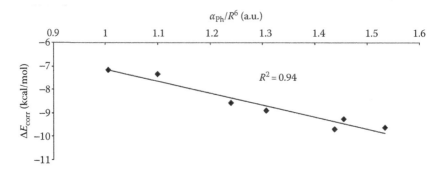

FIGURE 27.8 Correlation part of the interaction energy (ΔE_{Corr}) between cytosine and the substituted benzenes Ph–X (kcal/mol) vs. the benzene ring polarizability divided by R^6 (see Equation 27.8) (a.u.). (Reprinted from Mignon, P., Loverix, S., Steyaert, J., and Geerlings, P., *Nucl. Acids Res.*, 33, 1779, 2005. With permission.)

27.3.4 ROLE OF DFT DESCRIPTORS IN THE (EVALUATION OF) DISPERSION INTERACTIONS: FROM ATOMS-IN-MOLECULES POLARIZABILITY TO LOCAL SOFTNESS

Recently Becke and Johnson [94] presented an elegant model for the evaluation of the C_6, C_8, and C_{10} coefficients characterizing the dispersion energy between two nonoverlapping systems A and B at distance R:

$$E_{disp} = -\left(\frac{C_6}{R^6} + \frac{C_8}{R^8} + \frac{C_{10}}{R^{10}}\right) \qquad (27.10)$$

starting from the instantaneous dipole moment of an electron and its exchange role. At reference point $\underline{r}_1$ and for an electron with spin σ, this quantity can be written as

$$d_{x\sigma}(\underline{r}_1) = \left[\frac{1}{\rho_\sigma(\underline{r}_1)}\sum_{ij}\varphi_{i\sigma}(\underline{r}_1)\varphi_{j\sigma}(\underline{r}_1) \times \int \underline{r}_2\varphi_{i\sigma}(\underline{r}_2)\varphi_{j\sigma}(\underline{r}_2)d\underline{r}_2\right] - \underline{r}_1 \qquad (27.11)$$

where $\varphi_{i\sigma}$ and $\varphi_{j\sigma}$ are the occupied orbitals and ρ_σ denotes the electron density associated with spin σ. It was shown that C_6, C_8, C_{10} could be expressed in terms of the polarizability and the ground state expectation values of the square of the multipole moments associated with $d_{x\sigma}$.

For example,

$$C_6 = \frac{\alpha_A\alpha_B\langle M_1^2\rangle_A\langle M_1^2\rangle_B}{\alpha_A\langle M_1^2\rangle_B + \alpha_B\langle M_1^2\rangle_A} \qquad (27.12)$$

where the expectation values of M_l^2 are approximated as

$$\langle M_l^2 \rangle = \sum_\sigma \int \rho_\sigma(\underline{r}) \left[\underline{r}^l - (\underline{r} - \underline{d}_{\chi\sigma})^l \right]^2 d\underline{r} \qquad (27.13)$$

The dispersion coefficients can, to a very good approximation, be expressed in terms of atom–atom interactions between the atoms constituting A and B as (taking again C_6 as example)

$$C_{6,AB} = \sum_a^A \sum_b^B C_{6,ab} \qquad (27.14)$$

with

$$C_{6,ab} = \frac{\alpha_a \alpha_b \langle M_l^2 \rangle_a \langle M_l^2 \rangle_b}{\alpha_a \langle M_l^2 \rangle_b + \alpha_b \langle M_l^2 \rangle_a} \qquad (27.15)$$

indicating the need for "optimal" atomic polarizability values, preferentially dependent on the molecular environment for a given atom. Such atoms-in-molecules [95] polarizabilities were recently presented by Krishtal et al. [96] based on a Hirshfeld-type [35] partitioning scheme, as a sequel to previous work, in collaboration with our group, on electron density partitioning [97]. In the Hirshfeld scheme the total density $\rho(\underline{r})$ at a given point, $\underline{r}$ is written as a sum of atom contributions $\rho_a(\underline{r})$ where each atom gets a weight factor equal to its contribution to the promolecular density, i.e., the density written as a sum of isolated atomic densities with the positions of the nuclei being chosen as those in the real molecule.

$$\rho(\underline{r}) = \sum_a w_a(\underline{r})\rho_a(\underline{r}) \quad w_a(\underline{r}) = \rho_a^0(\underline{r}) \bigg/ \sum_b \rho_b^0(\underline{r}) \qquad (27.16)$$

It was proven [96] that the molecular polarizability can be written as a sum of intrinsic atomic polarizabilities of the atoms in the molecule and a charge delocalization term. Thus, the xy element of the molecular polarizability tensor of molecule A can be decomposed as

$$\alpha_{xy}^A = \sum_{a(A)} \left(\alpha_{xy}^a + x_a q_a^{(y)} \right) \qquad (27.17)$$

with x_a being the Cartesian coordinate of atom a in the x-direction. The atomic intrinsic polarizability is obtained using the following numerical integration:

$$\alpha_{xy}^a = \int w_a(\underline{r}) x_a \rho_a^{(y)}(\underline{r}) d\underline{r} \qquad (27.18)$$

where $w_a(\underline{r})$ is the Hirshfeld weight defined in Equation 27.16 and $\rho^{(y)}(\underline{r})$ is the first-order perturbed molecular density in the y-direction [98], which is calculated by methods such as coupled perturbed HF or KS. $q_a^{(y)}$ in Equation 27.17 is the first-order perturbed atomic charge on atom a, given by

$$q_a^{(y)} = \int w_a(\underline{r})\rho^{(y)}(\underline{r})\mathrm{d}\underline{r} \tag{27.19}$$

The intrinsic atom-in-molecule polarizabilities were tested by us [99] for their performance in the calculation of the dispersion energy for a set of Van der Waals complexes, at their equilibrium geometry using a DFT- B3LYP computational ansatz combined with an aug-cc-pVTZ basis set for the calculation of the $\langle M_l^2 \rangle$ values using Van Alsenoy's STOCK program, also used to partition the polarizabilities [100].

In Table 27.3 and Figure 27.9 the results are summarized and compared with the "best" values extracted from the literature [94d].

It is seen that, except for the p-benzene complex, which is clearly an outlier, and which is also problematic in Becke's approach (for a detailed discussion see Ref. [96]), the dispersion energy evaluated with our method and including up to the C_{10} term shows a fair correlation with the high-level results. Figure 27.9 shows a linear correlation of 0.97, with a slope (0.97) close to 1, indicating that with a two-parameter equation, the dispersion energy can be obtained with 3% and a systematical error of 0.12 kcal/mol. Note that if the regression line is forced to pass through the origin, the correlation only slightly decreases to 0.95 whereas the slope changes to 1.08, allowing with a one parameter equation the estimation of the

---

TABLE 27.3
Dispersion Energy Contributions from the C_6, C_8, and C_{10} Terms Compared with Values from High Level Calculations from the Literature

Complex[a]	$E_{disp}(C_6)$	$E_{disp}(C_8)$	$E_{disp}(C_{10})$	$E_{disp}(C_6 + C_8 + C_{10})$	HLev
ac–ac	0.26 (63)	0.11 (27)	0.04 (10)	0.41	0.35
met–met	0.56 (71)	0.17 (22)	0.06 (7)	0.79	0.56
met–et	0.32 (68)	0.11 (23)	0.04 (9)	0.47	0.42
met–Ne	0.16 (64)	0.07 (28)	0.02 (8)	0.25	0.23
benz–Ne	0.34 (62)	0.15 (27)	0.06 (11)	0.55	0.44
met-benz	0.76 (63)	0.31 (26)	0.13 (11)	1.20	1.00
CO_2–CO_2	0.41 (60)	0.19 (28)	0.08 (12)	0.68	0.53
t-benz	1.12 (60)	0.51 (27)	0.25 (13)	1.88	1.90
p-benz	0.98 (69)	0.33 (24)	0.10 (7)	1.41	2.93

Note: Percentage contribution of the individual terms as compared to their sum, $C_6 + C_8 + C_{10}$, is quoted in parenthesis. All Data in kcal/mol.

[a] ac, acetylene; met, methane; et, ethylene; benz, benzene; T, T-shaped benzene dimer; p, parallel face-to-face benzene dimer; HLev, High level.

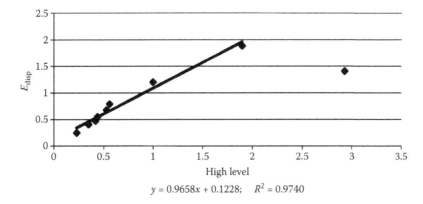

$$y = 0.9658x + 0.1228; \quad R^2 = 0.9740$$

FIGURE 27.9 Dispersion energy calculated using Equation 27.10 vs. high-level values from the literature (all values in kcal/mol). (Reprinted from Olasz, A., *J. Chem. Phys.*, 127, 224105, 2007. With permission.)

dispersion energy within 8%. Finally the importance of the C_8 and C_{10} coefficients should be noticed as can be judged from their percentage contributions: C_6 roughly accounts for 65%, C_8 for 25%, and C_{10} for 10% of the dispersion energy. The simplicity of our ansatz easily permits to take these higher order terms into account.

As a further step currently under investigation, the relationship between local polarizability and local softness is studied with the aim to substitute atom-in-molecule polarizabilities by atom-condensed softness values. In this way, conceptual DFT could be exploited in a computational strategy, an ansatz rarely used until now, the best known example being the electronegativity equalization method [101].

27.4 CONCLUSIONS

The case studies presented in this chapter highlight the possibilities of transferring the use of conceptual DFT descriptors from reactivity studies, leading to alterations in the covalent bonding pattern, toward the much less exploited, but extremely important field of noncovalent intermolecular interactions. Properties like local softness and local hardness turned out to provide useful information on and insight in noncovalent intermolecular interactions of the H-bond, electrostatic, and dispersion types. The covalent character of strong hydrogen bonds is remarkably well described through the local softness and the electrostatic compound of π stacking by local hardness. The dispersion component of this interaction can be quantified starting from the idea of the additivity of the polarizability and the use of aromatic ring-in-molecules polarizability values. The idea ultimately leads to an atom-in-molecule approach for dispersion interactions in Van der Waals complexes, using intrinsic atoms-in-molecules values. On the basis of the well known polarizability–global softness relationship, the success of the method is an incentive to further work on the possibility of the use of condensed atomic softness values in the evaluation of dispersion interactions.

ACKNOWLEDGMENTS

This chapter results from the research in a particular subdomain of conceptual DFT carried out by my group in recent years. Numerous present and past collaborators of the ALGC group, at pre- and postdoctoral level, coauthors of the case studies discussed in the text, deserve my most sincere thanks: Prof. Frank De Proft, Dr. Pierre Mignon, Dr. Stefan Loverix, Dr. Kenno Vanommeslaeghe, not to forget Prof. Jan Steyaert from the Biotechnology Department. A preponderant role has also been played by a number of colleagues of the Antwerp Quantum Chemistry group (Prof. Christian Van Alsenoy, Dr. Elisa Kirshtal) and by Dr. Andras Olasz and Prof. Tamas Veszprémi (Budapest) and Dr. Alimet Ozen and Prof. V. Aviyente (Istanbul). It was a particular pleasure to host Andras and Alimet during several stays in our group.Finally, I would like to thank Prof. Pratim Chattaraj, with whom I share now already for many years a deep interest in conceptual DFT for his kind invitation to contribute to this volume. Diane Sorgeloos is gratefully acknowledged for her meticulous work in typing and styling this manuscript.

REFERENCES

1. Hohenberg, P. and Kohn, W. *Phys. Rev. B*, 1964, *136*, 864.
2. Parr, R.G. and Yang, W. *Density Functional Theory of Atoms and Molecules*, Oxford University Press, New York, 1989.
3. Kohn, W. and Sham, L.J. *Phys. Rev. A*, 1965, *140*, 1133.
4. For a recent and critical account on the relative merits of the various v_{xc} presented, see Sousa, S.F., Fernandez, P.A., and Ramos, M.J. *J. Phys. Chem. A*, 2007, *111*, 10439.
5. Koch, W. and Holthausen, M.C. *A Chemist Guide to Density Functional Theory*, 2nd ed., Wiley-VCH, Weinheim, 2001.
6. Parr, R.G. *Annu. Rev. Phys. Chem.*, 1983, *34*, 631.
7. Parr, R.G. in *Density Functional Methods in Physics*, Dreizler, R.M. and J. da Providencia, eds., Plenum, New York, 1985, p. 141.
8. Parr, R.G. and Yang, W. *Annu. Rev. Phys. Chem.*, 1995, *46*, 79.
9. Geerlings, P., De Proft, F., and Langenaeker, W. *Adv. Quant. Chem.*, 1999, *33*, 303.
10. Chermette, H. *J. Comput. Chem.*, 1999, *20*, 129.
11. Geerlings, P., De Proft, F., and Langenaeker, W. *Chem. Rev.*, 2003, *103*, 1793.
12. Geerlings, P. and De Proft, F. *Phys. Chem. Chem. Phys.*, 2008, *10*, 3028.
13. Ayers, P.W. and Parr, R.G. *J. Am. Chem. Soc.*, 2001, *123*, 2007.
14. Liu, S. *Reviews of Modern Quantum Chemistry, A Celebration to the Contributions of R.G.Parr*, K.D. Sen, ed., World Scientific, Hackensack, NJ, 2003, 493.
15. Ayers, P.W., Anderson, J.S.M., and Bartolotti, L.J. *Int. J. Quant. Chem.*, 2005, *101*, 520.
16. (a) Sanderson, R.T. *Science*, 1952, *116*, 41. (b) Sanderson, R.T. *Polar Covalence*, Academic Press, New York, 1983.
17. (a) Pearson, R.G. *J. Am. Chem. Soc.*, 1963, *85*, 3533. (b) Pearson, R.G. *Chemical Hardness*, Wiley-VCH, Weinheim, 1997.
18. Pearson, R.G. *J. Chem. Educ.*, 1968, *45*, 981.
19. Geerlings, P. and De Proft, F. *Int. J. Quant. Chem.*, 2000, *80*, 227.
20. Woodward, R.B. and Hoffmann, R. *The Conservation of Orbital Symmetry*, Academic Press, New York, 1970.
21. (a) Ayers, P.W., Morell, C., De Proft, F., and Geerlings, P. *Chem. Eur. J.*, 2007, *13*, 8240. (b) De Proft, F., Ayers, P.W., Fias, S., and Geerlings, P. *J. Chem. Phys.*, 2006,

125, 214101. (c) De Proft, F., Chattaraj, P.K., Ayers, P.W., Torrent-Sucarrat, M., Elango, M., Subramanian, V., Giri, S., and Geerlings, P. *J. Chem. Theory. Comp.*, 2008, *4*, 1065.

22. (a) Moens, J., Geerlings, P., and Roos, G. *Chem. Eur. J.*, 2007, *13*, 8174. (b) Moens, J., Roos, G., Jaque, P., De Proft, F., and Geerlings, P. *Chem. Eur. J.*, 2007, *13*, 9331. (c) Moens, J., Jaque, P., De Proft, F., and Geerlings, P. *J. Phys. Chem. A*, 2008, *112*, 6023.

23. Pullman, B. (ed.), *Intermolecular Interaction: From Diatomics to Biopolymers*, Pullman, B. (ed.), John Wiley & Sons, New York, 1978.

24. Parr, R.G., Donnelly, R.A., Levy, M., and Palke, W.E. *J. Chem. Phys.*, 1978, *68*, 3801.

25. Parr, R.G. and Pearson, R.G. *J. Am. Chem. Soc.*, 1983, *105*, 7512.

26. Parr, R.G. and Yang, W. *J. Am. Chem. Soc.*, 1984, *106*, 4049.

27. (a) Ghosh, S.K. and Berkowitz, M. *J. Chem. Phys.*, 1985, *83*, 864. (b) Harbola, M.K., Chattaraj, P.K., and Parr, R.G. *Isr. J. Chem.*, 1991, *31*, 395. (c) Ghosh, S.K. *Chem. Phys. Lett.*, 1990, *172*, 77. (d) Langenaeker, W., De Proft, F., and Geerlings, P. *J. Phys. Chem.*, 1995, *99*, 6424. (e) De Proft, F., Geerlings, P., Liu, S., and Parr, R.G. *Polish J. Chem.*, 1998, *72*, 1737. (f) Torrent-Sucarrat, M. and Geerlings, P. *J. Chem. Phys.*, 2006, *125*, 244101. (g) Torrent-Sucarrat, M., Salvador, P., Geerlings, P., and Solà, M. *J. Comput. Chem.*, 2007, *28*, 574. (h) Torrent-Sucarrat, M., Salvador, P., Geerlings, P., and Solà, M. *J. Comput. Chem.*, 2008, *29*, 1064.

28. Chattaraj, P.K., Roy, D.R., Torrent-Sucarrat, M., and Geerlings, P. *Theor. Chem. Acc.*, 2007, *118*, 923.

29. Berkowitz, M., Ghosh, S.K., and Parr, R.G. *J. Am. Chem. Soc.*, 1985, *107*, 6811.

30. Bonaccorsi, R., Scrocco, E., and Tomasi, J. *J. Chem. Phys.*, 1970, *52*, 5270.

31. Anderson, J.S.M., Melin, J., and Ayers, P.W. *J. Chem. Theor. Comput.*, 2007, *3*, 358.

32. (a) Becke, A.D. *J. Chem. Phys.*, 1993, *98*, 5648. (b) Lee, C., Yang, W., and Parr, R.G. *Phys. Rev. B*, 1998, *37*, 785.

33. Mulliken, R.S. *J. Chem. Phys.*, 1955, *23*, 1833, 1841, 2338, 2343.

34. Reed, A.E., Curtiss, L.A., and Weinhold, F. *Chem. Rev.*, 1988, *88*, 894.

35. Hirshfeld, F.L. *Theor. Chim. Acta*, 1977, *44*, 129.

36. De Proft, F., Van Alsenoy, C., Peeters, A., Langenaeker, W., and Geerlings, P. *J. Comput. Chem.*, 2002, *23*, 1198.

37. Jeffrey, G.A. *An Introduction to Hydrogen Bonding*, Oxford University Press, New York, 1997.

38. Gilli, G. and Gilli, P. *J. Mol. Struct.*, 2000, *552*, 1.

39. Coulson, C.A. in *Hydrogen Bonding*, D. Hadzi, ed., Pergamon Press, New York, 1959.

40. (a) Morokuma, K. *J. Chem. Phys.*, 1971, *55*, 1236. (b) Kitaura, K. and Morokuma, K. *Int. J. Quant. Chem.*, 1976, *10*, 325. (c) Fonseca Guerra, C. and Bickelhaupt, M. *Angew. Chem. Int. Ed.*, 1999, *38*, 2942. (d) Fonseca Guerra, C. and Bickelhaupt, M. *Angew. Chem. Int. Ed.*, 2002, *41*, 2092. (e) Fonseca Guerra, C., Bickelhaupt, M., Snijders, J.G., and Baerends, E.J. *Chem. Eur. J.*, 1999, *5*, 3581. (f) Poater, J., Fradera, X., Solà, M., Duran, M., and Simon, S. *Chem. Phys. Lett.*, 2003, *369*, 248.

41. Stevens, E.D., Lehmann, M., and Coppens, P. *J. Am. Chem. Soc.*, 1977, *99*, 2829.

42. Pimentel, G.C. *J. Chem. Phys.*, 1951, *19*, 446.

43. Reid, C. *J. Chem. Phys.*, 1959, *30*, 182.

44. Kollman, P.A. and Allen, L.C. *J. Am. Chem. Soc.*, 1970, *92*, 619.

45. (a) Gilli, G., Bellucci, F., Ferretti, V., and Bertolasi, V. *J. Am. Chem. Soc.*, 1989, *111*, 1023. (b) Gilli, P., Bertolasi, V., Ferretti, V., and Gilli, G. *J. Am. Chem. Soc.*, 1994, *116*, 909. (c) Gilli, P., Bertolasi, V., Ferretti, V., and Gilli, G. *J. Am. Chem. Soc.*, 2000, *122*, 10405. (d) Gilli, P., Bertolasi, V., Pretto, L., Ferretti, V., and Gilli, G. *J. Am. Chem. Soc.*, 2004, *126*, 3845.

46. Ozen, A.S., Aviyente, V., De Proft, F., and Geerlings, P. *J. Phys. Chem.*, 2006, *A110*, 5860.

47. Hocquet, A., Toro-Labbé, A., and Chermette, H. *J. Mol. Struct. (THEOCHEM)*, 2004, *686*, 213.

48. Geerlings, P., Vos, A., and Schoonheydt, R. *J. Mol. Struct. (THEOCHEM)*, 2006, *762*, 69.

49. Kollmann, P., Kelvey, J.M.C., Johansson, A., and Rothenberg, S. *J. Am. Chem. Soc.*, 1975, *97*, 955.

50. Klopman, G. *J. Am. Chem. Soc.*, 1968, *90*, 223.

51. (a) Gazquez, J.L. and Mendez, F. *J. Phys. Chem.*, 1994, *98*, 4591. (b) Damoun, S., Van De Woude, G., Mendez, F., and Geerlings, P. *J. Phys. Chem. A*, 1997, *101*, 886.

52. Mc.Gaughey, G.B., Gagne, M., and Rappe, A.K. *J. Biol. Chem.*, 1988, *273*, 15458.

53. Sinnokrot, M.O. and Sherill, C.D. *J. Am. Chem. Soc.*, 2004, *126*, 7690.

54. (a) Cozzi, F., Cinquini, M., Annunziata, R., Dwyer, T., and Siegel, J.S. *J. Am. Chem. Soc.*, 1992, *114*, 5729. (b) Rashkin, M.J. and Waters, M.L. *J. Am. Chem. Soc.*, 2002, *124*, 1860.

55. (a) Hunter, C.A. and Sanders, J.K.M. *J. Am. Chem. Soc.*, 1990, *112*, 5525. (b) Hunter, C.A., Lawson, K.R., Perkins, J., and Urch, C.J. *J. Chem. Soc. Perkin Trans.*, 2, 2001, 651.

56. Sinnokrot, M.O. and Sherill, C.D. *J. Phys. Chem. A*, 2003, *107*, 8377.

57. Sinnokrot, M.O. and Sherill, C.D. *J. Am. Chem. Soc.*, 2004, *126*, 7690.

58. Mignon, P., Loverix, S., De Proft, F., and Geerlings, P. *J. Phys. Chem. A*, 2004, *108*, 6038.

59. Mignon, P., Loverix, S., and Geerlings, P. *Chem. Phys. Lett.*, 2005, *401*, 40.

60. Mignon, P., Loverix, S., Steyaert, J., and Geerlings, P. *Nucl. Acids Res.*, 2005, *33*, 1779.

61. Baeten, A., Maes, D., and Geerlings, P. *J. Theor. Biol.*, 1998, *195*, 27.

62. Mignon, P., Steyaert, J., Loris, R., Geerlings, P., and Loverix, S. *J. Biol. Chem.*, 2002, *277*, 36770.

63. Roos, G., Loverix, S., De Proft, F., Wyns, L., and Geerlings, P. *J. Phys. Chem. A*, 2003, *107*, 6828.

64. Roos, G., Messens, J., Loverix, S., Wyns, L., and Geerlings, P. *J. Phys. Chem. B*, 2004, *108*, 17216.

65. Roos, G., Buts, L., Van Belle, K., Brosens, E., Geerlings, P., Loris, R., Wyns, L., and Messens, J. *J. Mol. Biol.*, 2006, *360*, 826.

66. Roos, G., Loverix, S., Brosens, E., Van Belle, K., Wyns, L., Geerlings, P., and Messens, J. *Chem. BioChem.*, 2006, *7*, 981.

67. Møller, C. and Plesset, M.S. *Phys. Rev.*, 1934, *46*, 618.

68. (a) Tsuzuki, S., Honda, K., Uchimaru, T., Mikami, M., and Tanabe, K. *J. Am. Chem. Soc.*, 2002, *124*, 104. (b) Sinnokrot, M.O., Valeev, E.F., and Sherrill, C.D., *J. Am. Chem. Soc.*, 2002, *124*, 10887. (c) Tsuzuki, S., Honda, K., and Azumi, R. *J. Am. Chem. Soc.*, 2002, *124*, 12200.

69. Boys, S.F. and Bernardi, F. *Mol. Phys.*, 1970, *10*, 553.

70. (a) Stone, A.J. *Chem. Phys. Lett.*, 1981, *83*, 233. (b) Stone, A.J. and Alderton, M. *Mol. Phys.*, 1985, *56*, 1047.

71. Hobza, P. and Sponer, J. *Chem. Rev.*, 1999, *99*, 3247.

72. For an account of these methods, see Jensen, F., *Introduction of Computational Chemistry*, Wiley, New York, 1999.

73. Sinnokrot, M.O., Valeev, E.F., and Sherill, C.D. *J. Am. Chem. Soc.*, 2002, *124*, 10887.

74. Xu, X. and Goddard, W.A. *Proc. Natl. Acad. Sci.*, 2004, *101*, 267.

75. Zhao, Y. and Truhlar, D.G. *J. Chem. Theory Comput.*, 2005, *1*, 415.

76. Zhao, Y. and Truhlar, D.G. *J.Chem.Theory Comput.*, 2007, *3*, 289.

77. Dion, M., Rydberg, H., Schröder, E., Langreth, D.C., and Lundqvist B.I. *Phys. Rev. Lett.*, 2004, *92*, 246401.

78. Ikura, H., Tsuneda, T., Yanai, T., and Hirao, K. *J. Chem. Phys.*, 2001, *115*, 3540.

79. Sato, T., Tsuneda, T., and Hirao, K. *J. Chem. Phys.*, 2007, *126*, 234114.

80. Kamiya, M., Tsuneda, T., and Hirao, K. *J. Chem. Phys.*, 2002, *117*, 6010.
81. Grimme, S. *J. Comput. Chem.*, 2004, *25*, 1463.
82. Grimme, S. *J. Comput. Chem.*, 2006, *27*, 1787.
83. Anthony, J. and Grimme, S. *Phys. Chem. Chem. Phys.*, 2006, *8*, 5287.
84. Grimme, S., Anthony, J., Schwabe, T., and Muck-Lichtenfeld, C. *Org. Biomol. Chem.*, 2007, *5*, 741.
85. Jurečka, P., Černy, J., Hobza, P., and Salahub, D.R. *J. Comput. Chem.*, 2007, *28*, 555.
86. Buckingham, A.D. *Adv. Chem. Phys.*, 1967, *13*, 107.
87. Unsöld, A. *Z. Physik*, 1927, *43*, 563.
88. London, F. *Z. Physik*, 1930, *63*, 245.
89. Denbigh, K.G. *Trans. Faraday Soc.*, 1940, *36*, 936.
90. Vickery, B.C. and Denbigh, K.G. *Trans. Faraday Soc.*, 1944, *45*, 61.
91. Stott, M.J. and Zaremba, E. *Phys. Rev. A*, 1980, *21*, 12.
92. Vela, A. and Gazquez, J.L. *J. Am. Chem. Soc.*, 1990, *112*, 1490.
93. De Proft, F., Langenaeker, W., and Geerlings, P. *J. Phys. Chem.*, 1993, *97*, 1826.
94. (a) Becke, A.D. and Johnson, E.R. *J. Chem. Phys.*, 2005, *122*, 154104. (b) Johnson, E.R. and Becke, A.D. *J. Chem. Phys.*, 2005, *123*, 024101. (c) Becke, A.D. and Johnson, E.R. *J. Chem. Phys.*, 2005, *123*, 154101. (d) Johnson, E.R. and Becke, A.D. *J. Chem. Phys.*, 2006, *124*, 174104. (e) Becke, A.D. and Johnson, E.R. *J. Chem. Phys.*, 2006, *124*, 014104.
95. Bader, R.F.W. *Atoms in Molecules: A Quantum Theory*, Clarendon Press, Oxford, 1990.
96. Krishtal, A., Senet, P., Yang, M., and Van Alsenoy, C. *J. Chem. Phys.*, 2006, *125*, 034312.
97. De Proft, F., Vivas-Reyes, R., Peeters, A., Van Alsenoy, C., and Geerlings, P. *J. Comput. Chem.*, 2003, *24*, 463.
98. Weeny, R.Wc. *Rev. Mod. Phys.*, 1960, *32*, 335.
99. Olasz, A., Vanommeslaeghe, K., Krishtal, A., Veszprémi, T., Van Alsenoy, C., and Geerlings, P. *J. Chem. Phys.*, 2007, *127*, 224105.
100. Rousseau, B., Peeters, A., and Van Alsenoy, C. *Chem. Phys. Lett.*, 2000, *324*, 189.
101. Mortier, W.J. in *Electronegativity*, Sen, K.D. and Jorgenson, C.K. eds., *Structure and Bonding. 66*, Springer-Verlag, Berlin, 1987, p. 125.

28 Aromaticity and Chemical Reactivity

Eduard Matito, Jordi Poater, Miquel Solà,
and Paul von Ragué Schleyer

CONTENTS

28.1 INTRODUCTION

Michael Faraday's seminal 1825 paper, reporting the isolation of benzene (dicarburet of hydrogen) by distillation, noted that it was much less reactive than "monocarburet of hydrogen" (*trans*-2-butene) [1]. Such decreased reactivity has been taken as an experimental characteristic of aromaticity ever since. Kekulé applied the term "aromatic," originally denoting a characteristic odor or fragrance, to classify derivatives of benzene generally.

Despite nearly two centuries of intense scrutiny, aromaticity remains a unique research stimulus in chemistry. The concept of aromaticity is elusive; it is not directly observable. Numerous indirect measures have been devised, based on the manifestations and ramifications of aromaticity. One of the most recent and widely accepted definitions [2] described aromaticity as "a manifestation of electron delocalization in

419

closed circuits, either in two or three dimensions. This results in energy lowering, often quite substantial, and a variety of unusual chemical and physical properties. These include a tendency toward bond equalization, unusual reactivity, and characteristic spectroscopic features. Since aromaticity is related to induced ring currents, magnetic properties are particularly important for its detection and evaluation."

Given that aromaticity is not a directly measurable property, it cannot be defined in an unambiguously quantitative manner. The evaluation of the aromaticity of a whole molecule or its parts (e.g., individual rings in a polycyclic arene) is usually done indirectly by measuring some physicochemical property that reflects some manifestation of its aromatic character [3–5]. This leads to the myriad of classical structural [6], magnetic [2,7], energetic [8], and electronic-based [9] measures of aromaticity. All currently available descriptors of aromaticity represent approximations (sometimes arbitrary) to the problem of measuring this phenomenon and no single property that could be taken as a direct measure of aromaticity exists. Consequently, it is widely accepted that the concept of aromaticity should be analyzed by employing a multiplicity of measures [2,10].

Surprisingly, given the fuzzy character of this concept [11], there exist very simple chemical models that can account for many aspects related to the aromaticity of organic molecules. One of these extremely straightforward and powerful models is the Clar's π-sextet rule [12,13]. Since we will refer to this model several times throughout this chapter, it is convenient to describe it now briefly. According to Clar, the Kekulé resonance structure with the largest number of disjoint aromatic π-sextets, i.e., benzene-like moieties, is the most important for the characterization of properties of polycyclic aromatic hydrocarbons (PAHs). Aromatic π-sextets are defined as six π-electrons localized in a single benzene-like ring separated from adjacent rings by formal C–C single bonds. For instance, application of this rule to phenanthrene indicates that the resonance structure **2** is more important than resonance structure **1** (Scheme 28.1). Therefore, outer rings in phenanthrene are expected to have a larger local aromaticity than the central ring. This result has been confirmed by several measures of local aromaticity [14–17]. It is also generally recognized [18], with some exceptions [19,20], that the known 5 kcal · mol^{-1} greater thermochemical stability of phenanthrene over anthracene is related to differences in their aromaticity. Phenanthrene has two Clar π-sextets (Scheme 28.1) but anthracene has only one (Scheme 28.2).

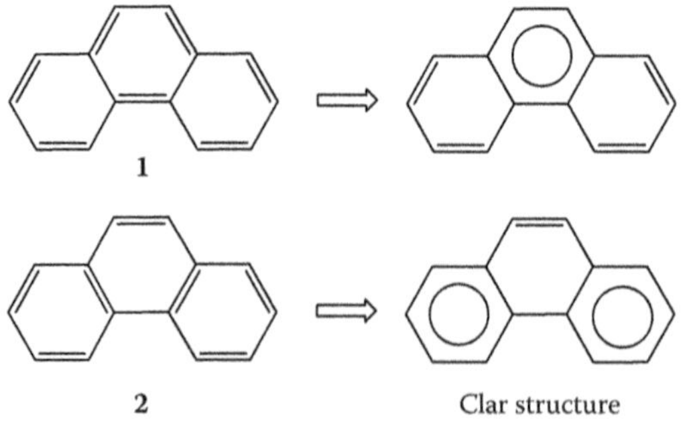

SCHEME 28.1   Phenanthrene.

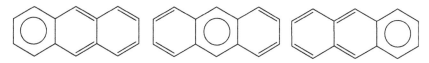

SCHEME 28.2

Like phenanthrene, some PAHs have a unique Clar structure, whereas several alternative Clar structures are possible for other PAHs [21]. Thus, Clar's rule does not designate the resonance structure mainly responsible for the aromaticity of anthracene. Clar's model cannot differentiate between its outer and inner ring (Scheme 28.2).

Benzene is the archetype of a two-dimensional aromatic molecule that exhibits all typical structural and chemical manifestations of aromaticity as, for instance, substantial energy stabilization, bond length equalization, and characteristic spectroscopic features as well as distinctive magnetic properties related to strong induced ring currents. Benzene also presents the traditional (but unusual for unsaturated organic compounds) reactivity of aromatic compounds. Thus, benzene reacts through electrophilic aromatic substitutions rather than additions [3]. However, this kind of reactivity behavior traditionally related to aromatic compounds has many important exceptions. For instance, phenanthrene and anthracene add bromine like olefins [3,22], the reaction rate being faster for anthracene than phenanthrene [23]. And 3-D aromatic compounds such as fullerenes (e.g., C_{60} or C_{70}) have a rich and extensive addition chemistry but no substitution reactions at all [24,25]. Thus, it is clear that the reactivity of aromatic species is relatively complex and, for this reason, to our knowledge, no index of aromaticity has been defined based on chemical reactivity properties.

Aromaticity and reactivity are two deeply connected concepts, although their relationship is complex. Obviously, not all reactions are affected by aromaticity (e.g., S_N2, E2...), but taking into account that among the approximately 20 million compounds known, two-thirds are fully or partially aromatic [26], it is clear that many chemical reactions will be influenced by the increase or decrease in aromaticity all over the reaction. The most important group of this kind of reactions are the pericyclic ones, which were categorized by Woodward and Hoffmann into five groups: sigmatropic shifts, cycloaddition, electrocyclic, cheletropic, and group transfer reactions [27,28]. Chemical structures (reactants, intermediates, and products) and transition states (TSs) are often influenced by aromatic stabilization or antiaromatic destabilization. For an elementary reaction, it may happen that all chemical structures and TSs or just one or two of them are aromatic or antiaromatic. In pericyclic reactions, it is found that most thermally allowed reactions take place through aromatic TSs, while TSs of thermally forbidden reactions are usually less aromatic or antiaromatic (vide infra). On the other hand, reactions can also be driven by deantiaromatization of the reactants. The dimerization of cyclobutadiene is a good example; it takes place spontaneously even at very low temperature to give the *syn*-dimer, in which only two new C–C bonds form in a [4 + 2] cycloaddition reaction [29]. The reaction starts from two antiaromatic molecules (cyclobutadiene), proceeds through a quite stable aromatic TS, and gives a nonaromatic product.

In addition to pericyclic processes, many other reactions are influenced by aromaticity. In electrophilic substitution reactions [30], for instance, formation of the Wheland intermediate reduces but may not annihilate the aromaticity of the initial reactants completely [31]. In fact, protonated benzene has significant aromaticity due to hyperconjugation [32]. Aromaticity loss also occurs in the dearomatization of *ortho-* and *para*-substituted phenols to form cyclohexadienones under oxidizing conditions [33]. Last but not least, the presence of (anti)aromatic rings and metallacycles in many organometallic and biochemical processes is also ubiquitous (consider, for example, the large number of existing transition-metal-mediated cycloaddition reactions such as the catalyzed trimerization of alkynes [34,35], reactions concerning porphyrins [36], the Dötz benzannulation reaction and the haptotropic rearrangements [37,38], or the reactivity of metallocarbohedrenes [39] among many others).

The aim of this chapter is to illustrate with some representative examples the relationship between the two deeply connected phenomena of aromaticity and chemical reactivity. The coverage of the considerable theoretical work done on the relationship between aromaticity and chemical reactivity is extensive, but is not exhaustive. Rather, we confine our discussion to the most relevant chemical reactions. Section 28.2 is devoted to the current and most widely used descriptors of aromaticity. The complex relation between aromaticity and stability is analyzed in Section 28.3. Section 28.4 discusses the most relevant chemical reactions for which one of the main driving forces is aromaticity. And finally, in Section 28.5, the most important findings are briefly summarized.

Let us conclude this section with two comments. Firstly, many of the studies (but not all) that are discussed along this chapter have been carried out in the framework of the density functional theory (DFT) using different functionals. Although the DFT has become increasingly popular in the last decade, especially for calculating large- and medium-size organometallic and bioinorganic transition metal (TM) compounds [38], there are several recent reports that warn about the use of this methodology in organic reactions, because it fails to describe accurately the energies of saturated and unsaturated hydrocarbons [40,41]. The authors of these works discourage the uncritical use of older density functionals for computing the energies of organic molecules. Secondly, it has long been recognized that main group inorganic compounds, like the carboranes, are aromatic. They undergo electrophilic substitution, just like benzene and are thermally much more stable. In 2001, Boldyrev et al. [42] noted aromaticity in an all-metal compound, Al_4^{2-}, for the first time. Many newly observed inorganic clusters have aromatic characteristics [43]. It is now recognized that the aromaticity concept can be applied to the entire periodic table. However, the reactivities are too diverse to relate to aromaticity meaningfully. Hence, this chapter focuses mostly on "classical" organic aromatic compounds.

28.2 MEASURES OF AROMATICITY

The numerous aromaticity measures proposed in recent decades have widened the number of descriptors in the literature considerably [2,8,9,44,45]. Aromaticity descriptors can be classified as magnetic, energetic, electronic, and structural [46].

Some indices aim to measure the "local" aromaticity of an individual ring in a polycyclic system, while others measure the "global" aromaticity of the whole molecule.

Nucleus-independent chemical shift (NICS) [47] is the most popular magnetic index of aromaticity [2]. NICS is a local aromaticity measure defined originally as the negative value of the absolute shielding computed at the ring center [NICS(0)] or at some other point of interest in the system [47]. For instance, it is desirable to calculate the index at a point 1 Å above a ring center [NICS(1)] in order to minimize the σ-electron effects [48]. Judging from comparisons with energetic, geometric, and other magnetic aromaticity measures, rings with appreciably negative NICS values are aromatic; the more negative the NICS values, the more aromatic the rings. But such NICS values are isotropic, whereas the aromatic π-electron ring currents induced by the applied external magnetic field influence the zz NICS tensor component perpendicular to the ring selectively. Hence, NICS(1)$_{zz}$ and especially NICS $(0)_{\pi,zz}$ (based on only the zz contributions of the π molecular orbitals [MOs]) are recommended as the best NICS-based indicators of aromaticity [2,49,50].

Accounting for aromaticity of a given molecule from its geometry is computationally a nondemanding task, usually carried out by means of harmonic oscillator model of aromaticity (HOMA) [6], which is the most widely used geometrical index of aromaticity. HOMA is based both on the degree of bond length alternation in a ring as well as the divergence of the average bond length from a value characterizing aromatic molecules. Because HOMA has the shortcoming of relying on arbitrary references, it has been shown recently, not to be suitable for the study of chemical reactions (involving unusual geometries) [51]. More details are given in the next section.

Lately the electronic indices of aromaticity have gained a prominent role to account for the aromaticity of organic molecules. In 2000, Giambiagi and coworkers [52], based on the MO multicenter index [53] that gives the electronic population shared by a given set of atoms, proposed I_{ring} as an electronic aromaticity measure. Three years later, Poater et al. [54] introduced the *para*-delocalization index (PDI), obtained from the delocalization index as defined in the framework of the atoms in molecules (AIM) theory, which is calculated by double integration of the exchange-correlation density over the basins of atoms A and B, thus giving a measure of the number of electrons delocalized or shared between these two atoms. PDI has proven to be a measure of aromaticity for six-membered rings, based on the independent findings of Fulton [55] and Bader [56] that benzene possesses larger electron sharing for *para*- than for *meta*-disposed atoms despite the larger distance. In 2005, Matito et al. [57] introduced the FLU index, which is constructed considering the amount of electron sharing between contiguous atoms, which should be substantial in aromatic molecules, and also taking into account the similarity of electron sharing between adjacent atoms. Thus, FLU was inspired in the HOMA structural index, in order to provide an electronic index that could measure aromaticity in rings of arbitrary size. Since FLU depends on reference values, it suffers from the same drawbacks as HOMA, and it is not suitable for the study of chemical reactivity. Bultinck and coworkers [58] proposed the multicenter index (MCI), which is calculated by the summation of I_{ring} contributions from all permutations of the atoms in the ring. Recently some of works [59] showed that neither I_{ring} nor MCI are size-consistent. Consequently normalized versions of MCI and I_{ring} were proposed, namely, I_{NB} and I_{NG}. In addition, I_{NG} was shown to closely match the topological resonance energy per π-electron (TREPE) [60] at the Hückel molecular orbital (HMO) level of theory (topological resonance energy [TRE] values only depend on topology of a conjugated system, and do not contain additional parameters outside the MO method used). Similar arguments were used to propose a normalized version of FLU index [61]. Note that all these electronic descriptors are local aromaticity measures.

In contrast to the previous criteria, energy-based indices are usually measures of global aromaticity. The energetic criteria may well be the most important because of their close relationship to reactivity trends and the chemical behavior of molecules. Historically, the concept of resonance energy (RE) was conceived to estimate the stability of aromatic species with respect to its acyclic, cyclic olefinic, or conjugated unsaturated analogues. Improving the concept of RE, TREPE was proposed by Gutman and coworkers [60], and it was further improved by the definition of the aromatic stabilization energy (ASE). ASE was defined as the reaction energy of homodesmotic and isodesmic reactions involving the species in question [8,62]. In particular, isodesmic reaction schemes demand only equal numbers of formal single and double bonds in products and reactants, while homodesmotic schemes require that there must the same number of bonds between given atoms in each state of hybridization both in products and reactants. It is important to note here that the widely diverging literature evaluations of the RE and ASE of benzene (ASE values reported for benzene range by over 50 $kcal \cdot mol^{-1}$) [8] can be adjusted to ca. 65 and 26 $kcal \cdot mol^{-1}$, respectively, after considering the necessary corrections for conjugation, hyperconjugation, and protobranching [63]. In a very recent work [64], enthalpies of hydrogenation reactions have been used to define a new global aromaticity scale.

28.3 AROMATICITY AND STABILITY

One of the important features of cyclic aromatic compounds is their enhanced energetic (thermodynamic) and chemical (kinetic) stability as compared to their linear counterparts. In connection with the thermodynamic stability (kinetic stability will be discussed in next sections), Chattaraj et al. [65,66] have shown that aromatic compounds exhibit negative changes in energy and polarizability but positive changes in hardness in agreement with the principles of minimum energy, minimum polarizability, and maximum hardness [67]. In many cases, among different isomers, the most aromatic compound is the most stable both thermodynamically and kinetically. Anthracene and phenantrene (and, in general, acenes and phenacenes) represent a paradigmatic example of this property. Although some authors have attributed the higher thermodynamic stability of phenantrene to H$\cdots$H attraction between the bay H atoms in phenantrene [68], a more likely reason for the lower energy of phenanthrene is its higher aromaticity [18,69,70]. Not only phenanthrene but also, in general, all kinked benzenoids are more stable and more aromatic than their linear polyacene isomers [18].

It is wrong, however, to conclude that among a series of isomers the most stable is always the most aromatic. Aromaticity is one of the many factors that affect the relative energies of isomers. Other aspects such as strain energy, hyperconjugation, topological charge localization, the presence of hydrogen bonds, long-range interactions, etc. may have, in some case, greater influence than aromaticity in determining the final relative energies of isomers. Many literature examples confirm this point. For instance, in a series of cyclopentafused pyrene congeners, Havenith et al. [71,72] found that the least stable isomer had the highest aromatic character and that the relative order of stability does not follow the trend given by the resonance energies. Similarly, Subramanian and coworkers [73] showed that there is no direct relationship between the thermodynamic stability of heterobicyclic isomers and their aromaticity. From aromaticity analyses based on experimental endohedral shieldings, Bühl and Hirsch [74,75] also concluded that there is no relation between relative stabilities of fullerenes and their aromatic character. Another interesting example concerns the *ortho-*, *meta-*, and *para*-benzyne species, the three possible biradicals generated by removing two H atoms from benzene. For these systems, the order of stability is *o*-benzyne > *m*-benzyne > *p*-benzyne while the aromaticity order of these biradicals is exactly opposite: *p*-benzyne > *m*-benzyne > *o*-benzyne [76,77]. A recent paper shows that aromaticity and stability of aminomethylbenzoic acids [78] can have opposite trends. The authors of this study found that the more extensive conjugation between substituents in the isomers increases the stability but decreases the ring aromaticity.

28.4 CHEMICAL REACTIONS INFLUENCED BY CHANGES IN AROMATICITY

28.4.1 CONJUGATED HYDROCARBONS

The present section is organized as follows. Firstly, the reactivity and aromaticity of the different rings that compose an acene system as a reactant is analyzed, and secondly, the aromaticity of the TS structures of pericyclic and pseudopericyclic reactions is discussed.

28.4.1.1 Acenes

Acenes are PAHs consisting of linearly fused benzene rings. Electrophilic substitution on these molecules is the most common reaction, whereas nucleophilic substitution onto acenes is much more rare and occurs only in extreme cases of strongly activated acenes [79]. As the number of rings increases, the members of the acene family become increasingly reactive, so that the higher members cannot be characterized experimentally [80]. Therefore, the acene series constitutes a good example that aromatic compounds are not necessarily kinetically stable. Indeed, the higher members of the acene series are extremely reactive despite being aromatic [81]. Although benzene and naphthalene are quite unreactive toward addition reactions [82], the central ring of anthracene is protonated, brominated and put through Diels–Alder reactions readily. Tetracene and pentacene participate in even more remarkable 1,4-cycloadditions [80,83]. Despite the large strides in the sophistication of the theoretical studies, the HOMO–LUMO gap, electronic structure, stability or reactivity, and aromaticity of acenes are still controversial [79,84]. The successive reduction in the band gap and reduction of the ionization potentials, as well as the increasing proton electron affinities, are also examples of the monotonic behavior in the acene series. Such progressions in acene properties appear to coincide with the sequential loss of benzenoid character (aromaticity) predicted by several MO treatments and Clar's qualitative sextet concept [12,85].

Schleyer et al. [80] studied the aromaticity and the Diels–Alder reactions of acenes with acetylene and found that the aromaticity correlates with the overall reaction energy. There is no significant decrease in relative aromatic stabilization along the acene series based on NICS and RE calculations (see Figure 28.1). Moreover, the more reactive inner rings are actually more aromatic than the less reactive outer rings. At the same time, the HOMO coefficients are consistent with the regioselectivity of Diels–Alder reactions that prefer the middle rings, despite the greater aromaticity [82,86,87]. A more recent study by Portella et al. [17] also confirms that the more reactive inner rings are more aromatic than the outer rings by means of PDI, HOMA, and NICS measures of aromaticity. This trend is opposite to that found for the equivalent phenacenes, known experimentally to be more stable energetically than the isomeric acenes and also the helicenes, in which the outer rings have the largest aromaticities [17].

Addition reactions to acenes provide a clear example that the local aromaticity of various rings in a PAH does not control the regioselectivity of the reactions: more aromatic rings can be more reactive [80]. The ground state energy of the PAH is just the same, whatever site reacts. The differences in TS energies (which parallel the product energies closely) decide the preferential position of attack. In anthracene, for instance, the central ring is the most reactive, even though many methods show this ring to be the most aromatic. The reason for the higher reactivity of the inner ring is readily explained by using the Clar's π-sextet rule. As shown in Scheme 28.3, the product of the addition to the central is more aromatic (two Clar π-sextet rings) than the product resulting from the addition to the outer ring (only one Clar ring) [88,89].

Recent quantum chemical studies of addition reactions to acenes and other PAHs involve the Diels–Alder reactions of ethene [86] and of $^1\Delta_g$ oxygen [87] as well as

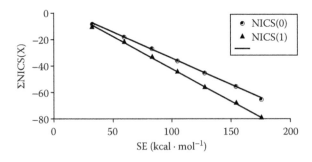

FIGURE 28.1 Correlations ($R^2 = 0.998$) of stabilization energies (SE) of acenes versus the total NICS(0) and NICS(1) sums.

SCHEME 28.3

the uncatalyzed 1,4-hydrogenation reaction [90]. The regiochemistry for all these reactions may be broadly predicted based on the extent of disruption to aromaticity resulting from addition via various possible pathways. The meso-carbon atoms are the most reactive position of linear acenes, since addition at a multiple ring junction disrupts the aromaticity of several rings and results in a greater reduction in aromatic stabilization and higher reaction barriers.

28.4.1.2 Pericyclic Reactions

Already in 1938, Evans and Warhurst [91] noted the analogy between the π-electrons of benzene and the six delocalized electrons in the cyclic TS of the Diels–Alder reaction of butadiene and ethylene. Taking into account the relationship between stability and reactivity, they concluded that conjugated molecules are thermochemically more stable and, in some cases, they behave more reactive than nonconjugated molecules [92]. In addition, the higher the mobility of the π-electrons in the TS, the greater the lowering of the activation energy. Generalized through the Woodward–Hoffmann rules [27] and the Hückel–Möbius concept by Zimmerman [93], thermally allowed pericyclic reactions are considered to take place preferentially through concerted aromatic TSs [94]. The aromaticity of pericyclic TSs was first analyzed systematically by Schleyer et al. [95–100] on the basis of geometric-, energetic-, and magnetic-based criteria, in all cases proving the electronic delocalization of these structures. Ponec [101,102] explored pericyclic reactions with molecular similarity measures and, more recently, Mandado and coworkers [103] analyzed pericyclic reactions by means of multicenter electron indices, namely a quantity closely resembling I_{ring}. Research characterizing the aromaticity of several pericyclic TSs is discussed below.

28.4.1.2.1 Diels–Alder Reactions
The well-known Diels–Alder reaction [95,104–106] is a standard method for forming substituted cyclohexenes through the thermally allowed 4s + 2s cycloaddition of alkenes and dienes. In particular, the reaction between ethene and 1,3-butadiene to yield cyclohexene is the prototype of a Diels–Alder reaction (Scheme 28.4). It is now well recognized that this reaction takes place via a synchronous and concerted mechanism through an aromatic boatlike TS [105].

SCHEME 28.4 Diels–Alder Reaction

The aromatic nature of this TS has been confirmed theoretically using magnetic-based indices such as NICS and the magnetic susceptibility exaltations [100], as well as geometry-based indices like HOMA, or electronic-based indices like PDI [51,107]. The performance of a series of aromatic measures has been analyzed by applying them to the simplest Diels–Alder between ethene and 1,3-butadiene. The evolution of electronic indices (PDI, FLU, MCI, I_{ring}, I_{NG}, and I_{NB}) along the reaction path is depicted in Figure 28.2, whereas Figure 28.3 shows the evolution of the different NICS measures and the geometric HOMA.

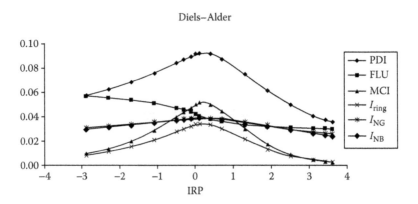

FIGURE 28.2 Plot of PDI (electrons), FLU$^{1/2}$ (values divided by 10), MCI (electrons), I_{ring} (electrons), I_{NB}, and I_{NG} versus the reaction coordinate (IRP in amu$^{1/2}$ Bohr). Negative values of the IRP correspond to the reactants side of the reaction path and positive values to the product side of the butadiene–ethylene DA cycloaddition.

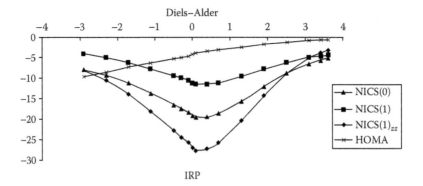

FIGURE 28.3 Plot of NICS(0) (ppm), NICS(1) (ppm), NICS(1)$_{zz}$ (ppm), and HOMA (values divided by 15) versus the the reaction coordinate (IRP in amu$^{1/2}$ Bohr). Negative values of the IRP correspond to the reactants side of the reaction path and positive values to the product side of the DA cycloaddition. NICS(0) and NICS(1) values have been computed at the RCP and at 1 Å above the RCP, respectively.

It is seen how all indicators of aromaticity correctly predict that a structure close to the TS is the most aromatic species along the reaction path, except HOMA and FLU indices, that are unsuccessful to account for the aromaticity of the TS. The reason for the failure of HOMA and FLU is that both values measure variances of the structural and electronic patterns, respectively, with respect to a reference value. Therefore, HOMA and FLU might fail if they are not applied to stable species because, while reactions are occurring, structural and electronic parameters suffer major changes. It is worth noting that the isotropic NICS(0) and NICS(1) values of the TSs are larger than those of benzene, but this is not the case with the more sophisticated NICS indices.

The synthetic utility of the Diels–Alder reaction has been significantly expanded by the development and use of a wide variety of dienes and dienophiles that contain masked functionalities. For instance, o-quinodimethanes are exceedingly reactive because cycloaddition establishes a benzenoid ring and results in aromatic stabilization [108]. Therefore, aromatization determines the energetically most favorable reaction path. In this case, DFT ^{13}C, ^{1}H NMR, NICS, and MO-NICS (single MO contributions to NICS) calculations indicate that the increase of aromatic character of the developing benzenoid ring along the reaction path is especially pronounced after the TS is reached.

28.4.1.2.2 1,3-Dipolar Cycloadditions

The 1,3-dipolar cycloaddition was defined as a general type of reaction by Huisgen [109,110], who developed an impressive research program to explore the preparative possibilities of this reaction, as well as its mechanistic aspects. 1,3-Dipolar cycloaddition reactions are one of the best and more general methods for the construction of five-membered rings in a convergent and stereo-controlled manner. Among the 1,3-dipoles involving second-period elements, nitrile oxides and nitrones have proved to be among the most useful and versatile reagents [111]. The wide range of dipolarophiles has allowed the chemical synthesis of a considerable number of nitrogen and oxygen-containing heterocycles in both inter- and intramolecular processes [111].

Different experiments allowed the elucidation of the mechanism of 1,3-dipolar cycloaddition reactions, which go through a six-electron supra–supra concerted mechanism in accordance with the Woodward–Hoffmann rules. As expected for such a thermally allowed pericyclic reaction, this [3 + 2] cycloaddition takes place through an aromatic TS. The NICS values confirm the high aromatic character of the corresponding TSs, thus validating that the 1,3-dipolar reaction between carbon–carbon multiple bonds and nitrile oxides or nitrones takes place via an in-plane aromatic TS. The large values of the NICS computed at the (3, +1) critical points of electron density (extremum in the electron density, a point where the gradient of the electron density is zero) are compatible with a ring current circulating along the molecular plane. In addition, the regiochemistry of the reaction between nitrile oxides and substituted alkenes is not determined by the aromaticity of the possible TSs. Instead, favorable electrostatic interactions between atoms or groups can stabilize the more asynchronous and less aromatic TSs [111].

28.4.1.2.3 [2 + 2 + 2] Cycloadditions

Although the energetic advantages of aromatic delocalization in cyclic TSs are well known, even thermally allowed and strongly exothermic reactions may have substantial activation barriers. One example is the Woodward–Hoffmann thermally allowed [2 + 2 + 2] trimerization of acetylene to yield benzene (see Scheme 28.5). In the course of this reaction, three acetylenic π-bonds are converted into C–C σ-bonds to form benzene. This reaction is highly exothermic (exp. $\Delta H_r^0 = -142.8$ kcal $\cdot$ mol^{-1}) [112], but has an unexpectedly large enthalpy barrier of about 50 kcal $\cdot$ mol^{-1} for a thermally allowed process [113]. Although there is a lack of consensus of how aromaticity changes during the transformation from reactants to product [100,112,114,115], it is widely accepted that this thermally allowed reaction has an aromatic TS. Most authors found that the system evolves from localized σ- and π-electrons in the reactants to the well known π-delocalization in benzene through a TS, which has mainly in-plane σ-electron delocalization with only minor π-electron delocalization. So, it seems reasonable to think that the aromaticity of the six-membered ring being formed increases from reactants to the TS and it decreases slightly before reaching the aromatic benzene product. This was found in a recent study [107] by means of a series of electronic-based aromaticity criteria. NICS [100] gives a maximum of aromaticity at the TS, although, more importantly, when NICS is separated between the NICS(π) and NICS(σ) components, it is observed how the π component increases from the reactant to the product, whereas the σ component is highly diamagnetic at the TS but strongly paramagnetic for benzene. This confirms that at the TS the in-plane contribution is larger, although the π delocalization is also important [100].

28.4.1.2.4 Other Examples

In addition to the numerous pericyclic aromatic TSs, other reactions deserve attention. These include the Cope and Claisen rearrangements, the pericyclic reactions with Möbius TSs, the Bergman cyclizations [77,116], and the TSs for 1,5-H shifts [100,117].

28.4.1.3 Pseudopericyclic Reactions

Woodward and Hoffmann provided an understanding of pericyclic reaction mechanisms based on conservation of orbital symmetry. A few years later, Ross et al. [118] coined the term pseudopericyclic for a set of reactions they discovered, which were not explained by the Woodward–Hoffmann rules (like the oxidation of tricyclic

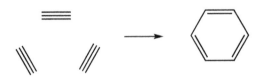

SCHEME 28.5

thiirane to a rearranged sulfoxide). Birney's group has carried out the most extensive work on pseudopericyclic reactions [119]. Birney [120] describes a pseudopericyclic reaction as a reaction with a low or a nonexistent barrier, planar TS and cyclic overlap disconnections. Unlike pericyclic reactions, pseudopericyclic reactions do not occur through aromatic TS. These criteria provide guidance to discern between pericyclic and pseudopericyclic reactions, but unfortunately they are not clear cut. Ambiguities are apparent. Aromatic compounds are usually planar but aromatic pericyclic reaction TSs usually are not. In contrast, pseudopericyclic reactions occur through planar, but nonaromatic TSs.

In a pericyclic reaction, the electron density is spread among the bonds involved in the rearrangement (the reason for aromatic TSs). On the other hand, pseudo-pericyclic reactions are characterized by electron accumulations and depletions on different atoms. Hence, the electron distributions in the TSs are not uniform for the bonds involved in the rearrangement. Recently some of us [121,122] showed that since the electron localization function (ELF), which measures the excess of kinetic energy density due to the Pauli repulsion, accounts for the electron distribution, we could expect connected (delocalized) pictures of bonds in pericyclic reactions, while pseudopericyclic reactions would give rise to disconnected (localized) pictures. Thus, ELF proves to be a valuable tool to differentiate between both reaction mechanisms.

A set of electrocyclic ring closures is the subject of recent controversy because their mechanism lies in the borderline between pericyclic and pseudo-pericyclic reactions [123–127]. The mechanisms were clarified by means of ELF analyses [121,122]. As shown in Figure 28.4, connected patterns (C) are

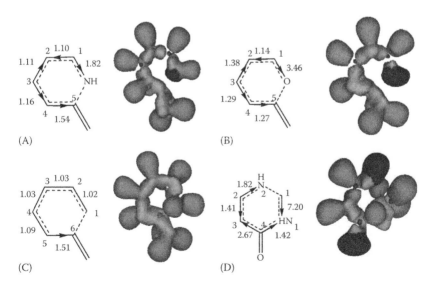

FIGURE 28.4 (See color insert following page 302.) Scheme of TS structures together with relative electron fluctuation magnitudes and ELF = 0.60 pictures for reactions A–D.

characteristic of aromatic TSs, and thus pericyclic reactions, while disconnected patterns (A, B, and D) are typical of the nonaromatic TSs involved in pseudopericyclic reactions.

28.4.2 Inorganic and Organometallic Compounds

28.4.2.1 Reactivity of All-Metal Aromatic Species

All-metal aromatic clusters are among the most appealing species discovered [43] since the turn of the century. Li and coworkers [42,128] were the first to observe and analyze the Al_4Li^-, Al_4Na^-, and Al_4Cu^- series in the gas phase. Their main structural unit, Al_4^{2-}, was shown to exhibit both π and σ aromaticity. Since then, many other aromatic species like XAl_3^- ($X = Si$, Ge, Sn, Pb) [129] or Al_3^- [130], among others, have received extensive attention.

All-metal clusters present several common features with π-conjugated analogues. They exhibit exalted linear and nonlinear optical properties such as higher polarizability or higher second hyperpolarizability [131]. They are also stabilized by complexation with transition metals [132] such as (η^4-Al_4M_4)-$Fe(CO)_3$ systems ($M = Li$, Na, K). Such complexes exhibit very small bond length alternations supporting the interpretation that the Al_4M_4 units are actually six π-electrons $Al_4M_4^{2-}$ species. Some of these metal clusters also show stabilization upon sandwich complexation by metals [133–137], for example, in the $(Al_4M_4)_2Ni$ ($M = Li$, Na, and K) series. All-metal clusters can also substitute their complexed organic counterparts, as, for instance, in (Al_4M_4)-$Fe(CO)_3$, which is obtained upon substitution on (C_4H_4)-$Fe(CO)_3$ [132]. Moreover, all-metal clusters are stabilized by forming superclusters, such as Al_6^{2-} or $(Al_4^{2-}Ca^{2+})$ [138].

Like D_{3h} $C_6H_3^+$, the first species exhibiting double aromaticity (π and in plane σ aromaticity) [139], but unlike classical conjugated organic compounds, which present only π-(anti)aromaticity, all-metal clusters and, in general, inorganic compounds might present not only the π-(anti)aromaticity, but also σ- or δ-(anti)aromaticity. Therefore, quantification of aromaticity in all-metal clusters is much more complex than in organic compounds. Recently some of us [140] have demonstrated that NICS profiles calculated in the perpendicular direction of the ring served the purpose of classifying a series of monocyclic inorganic compounds as aromatic, nonaromatic, or antiaromatic. The profiles give much more information than the single-point NICS calculation, enabling a clear classification of inorganic compounds.

Such NICS profiles were used to show that while the cyclo-$[M_3]^{2-}$ ($M = Be$, Mg, Ca) possesses σ-aromaticity the complex formed by the stabilization of an alkalimetal cation enhances its π-aromaticity, leading to an unprecedented change from σ-aromaticity to π-aromaticity. Furthermore, some clusters of these series M_3X_2 ($M = Be$, Mg, Ca; $X = Li$, Na, K) undergo dramatic alterations by changing the distance from X to the center of M_3 ring (Na_2Mg_3, Li_2Mg_3, and X_2Ca_3) while others keep its aromaticity along this process (X_2Be_3 and K_2Mg_3). This finding may open the way to reactivity control processes based on this aromaticity tuning [141].

28.4.2.2   Reactions Involving Metallabenzenes

Transition metal (TM) compounds exhibiting aromaticity were first considered in Hoffmann's pioneering work [142]. Metallabenzenes or metallacyclohexatrienes are derived from benzene, where one C–H moiety has been replaced by a fragment containing a TM. The first example, osmabenzene (osmium as the TM), was reported by Elliot and coworkers in 1982 [143]. Since then a plethora of metalla-benzenes have been synthesized, including metallabenzenes with two substitutions of a C–H fragment by a TM moiety [144]. Metallacyclic molecules differ from regular aromatic compounds in that the π-bonding requires the involvement of the metal $d(\pi)$ orbitals instead of $p(\pi)$ orbitals of main group elements. So far there have been three interpretations given for bonding in metallabenzenes. Thorn and Hoffmann [142] suggested that metallabenzenes possess 6 π-electrons, four of which come from the occupied orbitals in $C_5H_5^-$ and two from the occupied d_{xz} orbital of the metal. On the other hand, one of us (PvR.S.) [145] suggested that the occupied d_{yz} metal orbital also significantly contributes to the π-orbital inter-actions in metallabenzenes, thus there are actually 8π-electrons in metallabenzenes, which become stabilized by a Möbius type of aromaticity. Finally, Frenking's group [146] analyzed the chemical bonding in these species from an energy decomposition analysis perspective. As indicated by their findings, the 16- and 18-electron metallabenzene complexes are 10π-electron systems. According to their π-bonding strength, metallabenzenes should be considered as being aromatic, but less than benzene. However, some of these compounds exhibit ASE values close to benzene.

28.4.2.3   Haptotropic Changes

A characteristic of π-coordinated metal complexes is their ability to undergo hapto-tropic rearrangements, which involve the movement of a metal ligand, such as chromium tricarbonyl, between two different π-coordinating sites. These rearrange-ments are appealing for their potential applications as molecular switches, especially when the rearrangements can be thermally or photochemically reversed. One of us (M.S.) [147] has recently reviewed the η^6,η^6-interring haptotropic rearrangement of tricarbonylchronium $(Cr(CO)_3)$ in several PAHs, where π-coordinated $Cr(CO)_3$ migrates between different six-membered rings. Upon coordination of $Cr(CO)_3$, the PAHs experience a loss of aromaticity in both the coordinated ring and its neighbors. NICS(0), which indicates enhanced aromaticity of the coordinated ring upon com-plexation, gives misleading result, while $NICS(1)_{zz}$ yields the correct trend. The failure of NICS(0) to show a decreased aromaticity of the aromatic ring after coordination of the $Cr(CO)_3$ moiety is due to the extra ring current generated by the electron pairs responsible for coordination of $Cr(CO)_3$ in the organic substrate [148]. On the other hand, the haptotropic migration occurs in a single step or stepwise depending on the orbital interaction between $Cr(CO)_3$ and the PAH fragment. Migration of $Cr(CO)_3$ is favored by PAHs size whereas it is slowed down by the effect of its curvature.

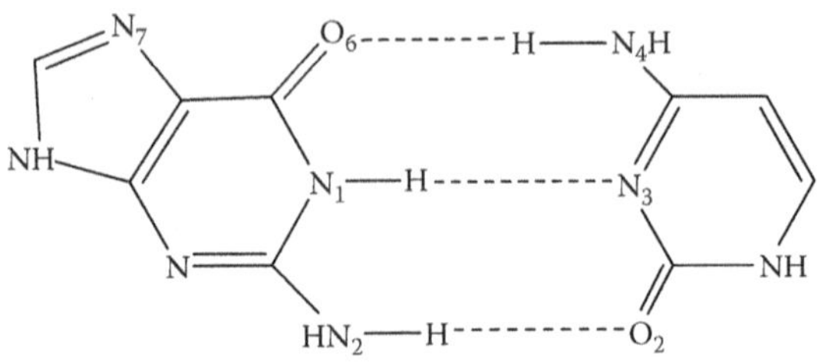

SCHEME 28.6

28.4.2.4 Guanine–Cytosine Base-Pair Interacting with Metal Cations

Physicochemical properties of DNA are affected by the addition of metal cations. The double helix is stabilized through the interaction with metal cations, which neutralize the negatively charged backbone phosphate groups. The metal cations also interact specifically with the nitrogenous bases. This modifies the hydrogen bonds and the aromatic stacking interactions. It is now well established that the N_7 position of the guanine is the preferred binding site (see Scheme 28.6).

Recently some of us [149] analyzed the influence of some metal cations ($M = Cu^+$, Cu^{2+}, and Ca^{2+}) coordinated to the N_7 of guanine on hydrogen bonding and aromaticity of the guanine–cytosine base pair. The analysis showed that the strengthening of the $N_1 \cdots N_3$ and $N_2 \cdots O_2$ hydrogen bonds and the weakening of the $O_6 \cdots N_4$ hydrogen bond is essentially caused by the modification of donor–acceptor interactions rather than to electrostatic interactions. The interaction of Cu^+ and Ca^{2+} results into strengthening of hydrogen bonding in the guanine–cytosine pair, which increases the aromaticity of the pyrimidinic ring in cytosine and the purinic ring in guanine. On the other hand, interaction with Cu^{2+} or ionization removes a π-electron, breaking the π-electron distribution and thus triggering a reduction of aromaticity in both the rings of guanine.

28.5 CONCLUSIONS

Aromaticity remains a concept of central importance in chemistry. It is very useful to rationalize important aspects of many chemical compounds such as the structure, stability, spectroscopy, magnetic properties, and last but not the least, their chemical reactivity. In this chapter, we have discussed just a few examples in which the presence of chemical structures (reactants, intermediates, and products) and TSs with aromatic or antiaromatic properties along the reaction coordinate have a profound effect on the reaction. It is clear that many more exciting insights in this area, especially from the newly developed aromatic inorganic clusters, can be expected in the near future from both experimental and theoretical investigations.

REFERENCES

1. Faraday, M. *Philos. Trans. R. Soc. Lond.* 1825, *115*, 440–446.
2. Chen, Z., Wannere, C. S., Corminboeuf, C., Puchta, R., and Schleyer, P. v. R. *Chem. Rev.* 2005, *105*, 3842–3888.
3. Wiberg, K. B. In *Theoretical and Computational Chemistry*, Maksic, Z. B., Orville-Thomas, W. J., Eds., Elsevier, Amsterdam, 1999, Vol. 6, pp. 519–536.
4. Krygowski, T. M., Cyrański, M. K., Czarnocki, Z., Häfelinger, G., and Katritzky, A. R. *Tetrahedron* 2000, *56*, 1783–1796.
5. Katritzky, A. R., Jug, K., and Oniciu, D. C. *Chem. Rev.* 2001, *101*, 1421–1449.
6. Krygowski, T. M. and Cyrański, M. K. *Chem. Rev.* 2001, *101*, 1385–1419.
7. Mitchell, R. H. *Chem. Rev.* 2001, *101*, 1301–1315.
8. Cyrański, M. K. *Chem. Rev.* 2005, *105*, 3773–3811.
9. Poater, J., Duran, M., Solà, M., and Silvi, B. *Chem. Rev.* 2005, *105*, 3911–3947.
10. Poater, J., García-Cruz, I., Illas, F., and Solà, M. *Phys. Chem. Chem. Phys.* 2004, *6*, 314–318.
11. Baldridge, K. K. and Siegel, J. S. *J. Phys. Org. Chem.* 2004, *17*, 740–742.
12. Clar, E. *Polycyclic Hydrocarbons*, Academic Press, London, Vols. 1, 2, 1964.
13. Clar, E. *The Aromatic Sextet*, Wiley, New York, 1972.
14. Cyrański, M. K., Stepién, B. T., and Krygowski, T. M. *Tetrahedron* 2000, *56*, 9663–9967.
15. Schulman, J. M. and Disch, R. L. *J. Phys. Chem. A* 1999, *103*, 6669–6672.
16. Portella, G., Poater, J., and Solà, M. *J. Phys. Org. Chem.* 2005, *18*, 785–791.
17. Portella, G., Poater, J., Bofill, J. M., Alemany, P., and Solà, M. *J. Org. Chem.* 2005, *70*, 2509–2521.
18. Poater, J., Visser, R., Solà, M., and Bickelhaupt, F. M. *J. Org. Chem.* 2007, *72*, 1134–1142.
19. Santos, J. C., Tiznado, W., Contreras, R., and Fuentealba, P. *J. Chem. Phys.* 2004, *120*, 1670–1673.
20. Santos, J. C., Andres, J., Aizman, A., and Fuentealba, P. *J, Chem. Theor. Comput.* 2005, *1*, 83–86.
21. Randić, M. *Chem. Rev.* 2003, *103*, 3449–3605.
22. Wiberg, K. B. *J. Org. Chem.* 1997, *62*, 5720–5727.
23. Altschuler, L. and Berliner, E. *J. Am. Chem. Soc.* 1966, *88*, 5837–5845.
24. Haddon, R. C. *Science* 1993, *261*, 1545–1550.
25. Taylor, R. and Walton, D. R. M. *Nature* 1993, *363*, 685–693.
26. Balaban, A. T., Oniciu, D. C., and Katritzky, A. R. *Chem. Rev.* 2004, *104*, 2777–2812.
27. Woodward, R. B. and Hoffmann, R. *Angew. Chem. Int. Ed. Engl.* 1969, *8*, 781–782.
28. Woodward, R. B. and Hoffmann, R. *The Conservation of Orbital Symmetry*, Verlag Chemie, Weinheim, 1970.
29. Halevi, E. A. *Orbital Symmetry and Reaction Mechanism. The OCAMS View*, Springer-Verlag, Berlin, 1992.
30. Fuster, F., Sevin, A., and Silvi, B. *J. Phys. Chem. A* 2000, *104*, 852–858.
31. Arrieta, A. and Cossío, F. P. *J. Mol. Struct. (Theochem)* 2007, *811*, 19–26.
32. Sieber, S. Schleyer, P. v. R., and Gauss, J. *J. Am. Chem. Soc.* 1993, *115*, 6987–6988.
33. Vo, N. T., Pace, R. D. M., O'Hara, F., and Gaunt, M. J. *J. Am. Chem. Soc.* 2008, *130*, 404–405.
34. Lautens, M., Klute, W., and Tam, W. *Chem. Rev.* 1996, *96*, 49–92.
35. Frühauf, H.-W. *Chem. Rev.* 1997, *97*, 523–596.
36. Loew, G. H. and Harris, D. L. *Chem. Rev.* 2000, *100*, 407–419.
37. Dötz, K. H., Wenzel, B., and Jahr, H. C. *Top. Curr. Chem.* 2004, *248*, 63–103.
38. Torrent, M., Solà, M., and Frenking, G. *Chem. Rev.* 2000, *100*, 439–493.
39. Rohmer, M.-M., Bénard, M., and Poblet, J. M. *Chem. Rev.* 2000, *100*, 495–542.

40. Izgorodina, E. I., Brittain, D. R. B., Hodgson, J. L., Krenske, E. H., Lin, C. Y., Namazian, M., and Coote, M. L. *J. Phys. Chem. A* 2007, *111*, 10754–10768.

41. Wodrich, M. D., Corminboeuf, C., Schreiner, P. R., Fokin, A. A., and Schleyer, P. v. R. *Org. Lett.* 2007, *9*, 1851–1854.

42. Li, X., Kuznetsov, A. E., Zhang, H.-F., Boldyrev, A. I., and Wang, L.-S. *Science* 2001, *291*, 859–861.

43. Boldyrev, A. I. and Wang, L.-S. *Chem. Rev.* 2005, *105*, 3716–3757.

44. Schleyer, P. v. R. *Chem. Rev.* 2001, *101*, 1115–1118.

45. Schleyer, P. v. R. *Chem. Rev.* 2005, *105*, 3433–3435.

46. Schleyer, P. v. R. and Jiao, H. *Pure Appl. Chem.* 1996, *68*, 209–218.

47. Schleyer, P. v. R., Maerker, C., Dransfeld, A., Jiao, H., and van Eikema Hommes, N. J. R. *J. Am. Chem. Soc.* 1996, *118*, 6317–6318.

48. Schleyer, P. v. R. Jiao, H. J., Hommes, N., Malkin, V. G., and Malkina, O. L. *J. Am. Chem. Soc.* 1997, *119*, 12669–12670.

49. Schleyer, P. v. R., Manoharan, M., Wang, Z. X., Kiran, B., Jiao, H. J., and Puchta, R., van Eikema Hommes, N. J. R. *Org. Lett.* 2001, *3*, 2465–2468.

50. Fallah-Baher-Shaidaei, H., Wannere, C. S., Corminboeuf, C., Puchta, R., and Schleyer, P. v. R. *Org. Lett.* 2006, *8*, 863–866.

51. Matito, E., Poater, J., Duran, M., and Solà, M. *J. Mol. Struct. (Theochem)* 2005, *727*, 165–171.

52. Giambiagi, M., de Giambiagi, M. S., dos Santos, C. D., and de Figueiredo, A. P. *Phys. Chem. Chem. Phys.* 2000, *2*, 3381–3392.

53. Giambiagi, M., de Giambiagi, M. S., and Mundim, K. C. *Struct. Chem.* 1990, *1*, 423–427.

54. Poater, J., Fradera, X., Duran, M., and Solà, M. *Chem. Eur. J.* 2003, *9*, 400–406.

55. Fulton, R. L. *J. Phys. Chem.* 1993, *97*, 7516–7529.

56. Bader, R. F. W., Streitwieser, A., Neuhaus, A., Laidig, K. E., and Speers, P. *J. Am. Chem. Soc.* 1996, *118*, 4959–4965.

57. Matito, E., Duran, M., and Solà, M. *J. Chem. Phys.* 2005, *122*, 014109.

58. Bultinck, P., Ponec, R., and Van Damme, S. *J. Phys. Org. Chem.* 2005, *18*, 706–718.

59. Cioslowski, J., Matito, E., and Solà, M. *J. Phys. Chem. A* 2007, *111*, 6521–6525.

60. Gutman, I., Milun, M., and Trinajstic, N. *J. Am. Chem. Soc.* 1977, *99*, 1692–1704.

61. Matito, E., Feixas, F., and Solà, M. *J. Mol. Struct. (Theochem)* 2007, *811*, 3–11.

62. Cyrański, M. K., Schleyer, P. v. R., Jiao, H., and Hohlneicher, G. *Tetrahedron* 2003, *59*, 1657.

63. Wodrich, M. D., Wannere, C. S., Mo, Y., Jarowski, P. D., Houk, K. N., and Schleyer, P. v. R. *Chem. Eur. J.* 2007, *13*, 7731–7744.

64. Mucsi, Z., Viskolcz, B., and Csizmadia, I. G. *J. Phys. Chem. A* 2007, *111*, 1123–1132.

65. Chattaraj, P. K., Sarkar, U., and Roy, D. R. *J. Chem. Educ.* 2007, *84*, 354–357.

66. Chattaraj, P. K., Roy, D. R., Elango, M., and Subramanian, V. *J. Phys. Chem. A* 2005, *109*, 9590–9597.

67. Pearson, R. G. *Chemical Hardness: Applications from Molecules to Solids*, Wiley-VCH, Oxford, 1997.

68. Matta, C. F., Hernández-Trujillo, J., Tang, T.-H., and Bader, R. F. W. *Chem. Eur. J.* 2003, *9*, 1940–1951.

69. Poater, J., Solà, M., and Bickelhaupt, F. M. *Chem. Eur. J.* 2006, *12*, 2889–2895.

70. Poater, J., Solà, M., and Bickelhaupt, F. M. *Chem. Eur. J.* 2006, *12*, 2902–2905.

71. Havenith, R. W. A., van Lenthe, J. H., Dijkstra, F., and Jenneskens, L. W. *J. Phys. Chem. A* 2001, *105*, 3838–3845.

72. Havenith, R. W. A., Jiao, H., Jenneskens, L. W., van Lenthe, J. H., Sarobe, M., Schleyer, P. v. R., Kataoka, M., Necula, A., and Scott, L. T. *J. Am. Chem. Soc.* 2002, *124*, 2363–2370.

73. Subramanian, G., Schleyer, P. v. R., and Jiao, H. *Angew. Chem. Int. Ed. Engl.* 1996, *35*, 2638–2641.
74. Bühl, M. *Chem. Eur. J.* 1998, *4*, 734–739.
75. Bühl, M. and Hirsch, A. *Chem. Rev.* 2001, *101*, 1153–1183.
76. De Proft, F., Schleyer, P. v. R., van Lenthe, J. H., Stahl, F., and Geerlings, P. *Chem. Eur. J.* 2002, *8*, 3402–3410.
77. Poater, J., Bickelhaupt, F. M., and Solà, M. *J. Phys. Chem. A* 2007, *111*, 5063–5070.
78. Lima, C. F. R. A. C., Gomes, L. R., and Santos, L. M. N. B. F. *J. Phys. Chem. A* 2007, *111*, 10598–10603.
79. Reddy, A. R., Fridman-Marueli, G., and Bendikov, M. *J. Org. Chem.* 2007, *72*, 51–61.
80. Schleyer, P. v. R., Manoharan, M., Jiao, H. J., and Stahl, F. *Org. Lett.* 2001, *3*, 3643–3646.
81. Aihara, J. *Phys. Chem. Chem. Phys.* 1999, *1*, 3193–3197.
82. Biermann, D. and Schmidt, W. *J. Am. Chem. Soc.* 1980, *102*, 3163–3173.
83. Sarova, G. H. and Berberan-Santos, M. N. *Chem. Phys. Lett.* 2004, *397*, 402–407.
84. Poater, J., Bofill, J. M., Alemany, P., and Solà, M. *J. Phys. Chem. A* 2005, *109*, 10629–10632.
85. Suresh, C. H. and Gadre, S. R. *J. Org. Chem.* 1999, *64*, 2505–2512.
86. Cheng, M. F. and Li, W. K. *Chem. Phys. Lett.* 2003, *368*, 630–638.
87. Chien, S. H., Cheng, M. F., Lau, K. C., and Li, W. K. *J. Phys. Chem. A* 2005, *109*, 7509–7518.
88. Dabestani, R. and Ivanov, I. N. *Photochem. Photobiol.* 1999, *70*, 10–34.
89. Glidewell, C. and Lloyd, D. *Tetrahedron* 1984, *40*, 4455–4472.
90. Zhong, G., Chan, B., and Radom, L. *J. Mol. Struct. (Theochem)* 2007, *811*, 13–17.
91. Evans, M. G. and Warhurst, E. *Trans. Faraday Soc.* 1938, *34*, 0614–0624.
92. Evans, M. G. *Trans. Faraday Soc.* 1939, *35*, 0824–0834.
93. Zimmerman, E. H. *Acc. Chem. Res.* 1971, *4*, 272–280.
94. Dewar, M. J. S. *Angew. Chem. Int. Ed. Engl.* 1971, *10*, 761–776.
95. Herges, R., Jiao, H. J., and Schleyer, P. v. R. *Angew. Chem. Int. Ed.* 1994, *33*, 1376–1378.
96. Jiao, H. J. and Schleyer, P. v. R. *Angew. Chem. Int. Ed. Engl.* 1993, *32*, 1763–1765.
97. Jiao, H. J. and Schleyer, P. v. R. *J. Chem. Soc. Perkin Trans. 2* 1994, 407–410.
98. Jiao, H. J. and Schleyer, P. v. R. *J. Am. Chem. Soc.* 1995, *117*, 11529–11535.
99. Jiao, H. J. and Schleyer, P. v. R. *Angew. Chem. Int. Ed. Engl.* 1995, *34*, 334–337.
100. Jiao, H. J. and Schleyer, P. v. R. *J. Phys. Org. Chem.* 1998, *11*, 655–662.
101. Ponec, R. *Top. Curr. Chem.* 1995, 1–26.
102. Ponec, R. and Strnad, M. *J. Math. Chem.* 1991, *8*, 103.
103. Mandado, M., González-Moa, M. J., and Mosquera, R. A. *Chem. Phys. Chem.* 2007, *8*, 696–702.
104. Bernardi, F., Celani, P., Olivucci, M., Robb, M. A., and Suzzivalli, G. *J. Am. Chem. Soc.* 1995, *117*, 10531–10536.
105. Houk, K. N., Gonzalez, J., and Li, Y. *Acc. Chem. Res.* 1995, *28*, 81–90.
106. Houk, K. N., Li, Y., Storer, J., Raimondi, L., and Beno, B. *J. Chem. Soc., Faraday Trans.* 1994, *90*, 1599–1604.
107. Feixas, F., Matito, E., Poater, J., and Solà, M. *J. Comput. Chem.* 2008, 29, 1543–1554.
108. Corminboeuf, C., Heine, T., and Weber, J. *Org. Lett.* 2003, *5*, 1127–1130.
109. Huisgen, R. *Angew. Chem. Int. Ed. Engl.* 1963, *2*, 566–633.
110. Huisgen, R. *Angew. Chem. Int. Ed. Engl.* 1963, *2*, 633–645.
111. Cossío, F. P., Morao, I., Jiao, H. J., and Schleyer, P. v. R. *J. Am. Chem. Soc.* 1999, *121*, 6737–6746.
112. Morao, I. and Cossío, F. P. *J. Org. Chem.* 1999, *64*, 1868–1874.
113. Cioslowski, J., Liu, G. H., and Moncrieff, D. *Chem. Phys. Lett.* 2000, *316*, 536–540.
114. Havenith, R. W. A., Fowler, P. W., Jenneskens, L. W., and Steiner, E. *J. Phys. Chem. A* 2003, *107*, 1867–1871.

115. Santos, J. C., Polo, V., and Andrés, *J. Chem. Phys. Lett.* 2005, *406*, 393–397.

116. Jones, R. R. and Bergman, R. G. *J. Am. Chem. Soc.* 1972, *94*, 660–661.

117. Sakai, S. *J. Phys. Chem. A* 2006, *110*, 6339–6344.

118. Ross, J. A., Seiders, R. P., and Lemal, D. M. *J. Am. Chem. Soc.* 1976, *98*, 4325–4327.

119. Birney, D. M. *J. Am. Chem. Soc.* 2000, *122*, 10917–10925.

120. Zhou, C. and Birney, D. M. *J. Am. Chem. Soc.* 2002, *124*, 5231–5241

121. Matito, E., Poater, J., Duran, M., and Solà, M. *Chem. Phys. Chem.* 2006, *7*, 111–113.

122. Matito, E., Solà, M., Duran, M., and Poater, J. *J. Phys. Chem. B* 2005, *109*, 7591–7593.

123. Chamorro, E. E. and Notario, R. *J. Phys. Chem. A* 2004, *108*, 4099–4104.

124. de Lera, A. R., Alvarez, R., Lecea, B., Torrado, A., and Cossio, F. P. *Angew. Chem. Int. Ed.* 2001, *40*, 557–561.

125. de Lera, A. R. and Cossio, F. P. *Angew. Chem. Int. Ed.* 2002, *41*, 1150–1151.

126. Rodriguez-Otero, J. and Cabaleiro-Lago, E. M. *Angew. Chem. Int. Ed.* 2002, *41*, 1147–1150.

127. Rodriguez-Otero, J. and Cabaleiro-Lago, E. M. *Chem. Eur. J.* 2003, *9*, 1837–1843.

128. Li, X., Kuznetsov, A. E., Zhang, H.-F., Boldyrev, A. I., and Wang, L.-S. *Science* 2001, *291*, 841–842.

129. Li, X., Zhang, H. F., Wang, L.-S., Kuznetsov, A. E., Cannon, N. A., and Boldyrev, A. I. *Angew. Chem. Int. Ed. Engl.* 2001, *40*, 1867–1870.

130. Kuznetsov, A. E., Boldyrev, A. I., Zhai, H.-J., Li, X., and Wang, L.-S. *J. Am. Chem. Soc.* 2002, *124*, 11791–11801.

131. Datta, A. and Pati, S. K. *J. Phys. Chem. A* 2004, *108*, 9527–9530.

132. Datta, A., Mallajosyula, S. S., and Pati, S. K. *Acc. Chem. Res.* 2007, *40*, 213–221.

133. Li, Z., Zhao, C., and Chen, L. *J. Mol. Struct. (Theochem)* 2007, *810*, 1–6.

134. Mercero, J. M., Formoso, E., Matxain, J. M., Eriksson, L. A., and Ugalde, J. M. *Chem. Eur. J.* 2006, *12*, 4495–4502.

135. Mercero, J. M. and Ugalde, J. M. *J. Am. Chem. Soc.* 2004, *126*, 3380–3381.

136. Yang, L. M., Ding, Y. H., and Sun, C. C. *Chem. Phys. Chem.* 2006, *7*, 2478–2482.

137. Yang, L. M., Ding, Y. H., and Sun, C. C. *Chem. Eur. J.* 2007, *13*, 2546–2555.

138. Kuznetsov, A. E., Boldyrev, A. I., Zhai, H. J., Li, X., and Wang, L. S. *J. Am. Chem. Soc.* 2002, *124*, 104–112.

139. Chandrasekhar, J., Jemmis, E. D., and Schleyer, P. v. R. *Tetrahedron Lett.* 1979, *20*, 3707.

140. Jiménez-Halla, J. O.C., Matito, E., Robles, J., and Solà, M. *J. Organomet. Chem.* 2006, *691*, 4359–4366.

141. Jiménez-Halla, J. O. C., Matito, E., Robles, J., Blancafort, L., and Solà, M. Submitted 2008.

142. Thorn, D. L. and Hoffmann, R. *Nouv. J. Chem.* 1979, *3*, 39.

143. Elliot, G. P., Roper, W. R., and Waters, J. M. *J. Chem. Soc., Chem. Commun.* 1982, 811.

144. Profilet, R. D., Fanwick, P. E., and Rothwell, I. P. *Angew. Chem. Int. Ed. Engl.* 1992, *31*, 1261.

145. Schleyer, P. v. R. and Wang, Z.-X. Personal Communication cited in J. R. Bleeke, *Chem. Rev.* 2001, 101, 1205.

146. Fernández, I. and Frenking, G. *Chem. Eur. J.* 2007, *13*, 5873–5884.

147. Jiménez-Halla, J. O. C., Robles, J., and Solà, M. *J. Phys. Chem. A* 2008, *112*, 1202–1213.

148. Feixas, F., Jiménez-Halla, J. O. C., Matito, E., Poater, J., and Solà, M. *Polish J. Chem.* 2007, *81*, 783–797.

149. Poater, J., Sodupe, M., Bertran, J., and Solà, M. *Mol. Phys.* 2005, *103*, 163–173.

29 Multifold Aromaticity, Multifold Antiaromaticity, and Conflicting Aromaticity: Implications for Stability and Reactivity of Clusters

Dmitry Yu. Zubarev, Alina P. Sergeeva, and Alexander I. Boldyrev

CONTENTS

29.1 INTRODUCTION

Despite its unsaturated nature, benzene with its sweet aroma, isolated by Michael Faraday in 1825 [1], demonstrates low chemical reactivity. This feature gave rise to the entire class of unsaturated organic substances called aromatic compounds. Thus, the aromaticity and low reactivity were connected from the very beginning. The aromaticity and reactivity in organic chemistry is thoroughly reviewed in the book by Matito et al. [2]. The concepts of aromaticity and antiaromaticity have been recently extended into main group and transition metal clusters [3–10]. The current chapter will discuss relationship among aromaticity, stability, and reactivity in clusters.

Aromaticity/antiaromaticity in cluster systems has certain peculiarities when compared with organic compounds. The striking feature of chemical bonding in cluster systems is the multifold nature of aromaticity, antiaromaticity, and conflicting aromaticity [3–10]. Double aromaticity (the simultaneous presence of σ- and π-aromaticity) was introduced in chemistry by Schleyer and coworkers [11] in the late 1970s to explain properties of 3,5-dehydrophenyl cation. Double aromaticity and antiaromaticity was first used by Martin-Santamaria and Rzepa [12] to explain chemical bonding in small carbon rings. Berndt and coworkers [13] have shown that small carborane molecules containing three- and four-membered rings also exhibit both σ- and π-aromaticity. The amount of possible combinations of aromaticity and antiaromaticity is even greater in clusters [9]. When only s-atomic orbitals (AOs) are involved in chemical bonding, one may expect only σ-aromaticity or σ-antiaromaticity. If p-AOs are involved, σ-tangential (σ_t), σ-radial (σ_r), and π-aromaticity/antiaromaticity could coexist [5]. In this case, multiple (σ- and π-) aromaticity, multiple (σ- and π-) antiaromaticity, and conflicting aromaticity (simultaneous σ-aromaticity and π-antiaromaticity or σ-antiaromaticity and π-aromaticity) can be encountered. If d-AOs are involved in chemical bonding, σ_t-, σ_r-, π-tangential (π_t-), π-radial (π_r-), and δ-aromaticity/antiaromaticity can occur. In this case, there can be multiple (σ-, π-, and δ-) aromaticity, multiple (σ-, π-, and δ-) antiaromaticity, and conflicting aromaticity (simultaneous aromaticity and antiaromaticity among the three types of σ, π, and δ bonds) [9,14,15]. The next challenge is to find φ-aromaticity, which may occur in multinuclear and cyclic f-metal systems. Involvement of f-AOs in chemical bonding might result in multiple (σ-, π-, δ-, and φ-) aromaticity, multiple (σ-, π-, δ-, and φ-) antiaromaticity, and conflicting aromaticity (simultaneous aromaticity and antiaromaticity among the four types of σ, π, δ, and φ bonds).

From the definitive point of view, both aromaticity and antiaromaticity are very vague concepts. The most recent discussion of aromaticity and antiaromaticity can be found in Refs. [16–18]. Various aromaticity indices are used to probe aromaticity such as *para*-delocalization index (PDI) [19], the aromatic fluctuation index (FLU) [20], MO multicenter bond index (MCI) [21,22], etc. Ponec and coworkers [23,24] proposed six-center bond index as a measure of aromaticity. Ponec et al. [25] further demonstrated that the MCI can be used for the quantitative characterization of homoaromaticity, nonhomoaromatic, and antihomoaromatic systems. Cioslowski et al. [26] developed normalized variants of MCI. Chattaraj and coworkers [27] studied efficiency of the multicenter indices in providing insights into the bonding, reactivity, and aromaticity in all-metal aromatic and antiaromatic compounds. Chattaraj and coworkers [28] used maximum hardness principle and the minimum polarizability principle to describe the stability and reactivity of aromatic and antiaromatic compounds. There are probes for aromaticity/antiaromaticity based on the response to the presence of external magnetic field such as nuclear-independent chemical shifts (NICS) pioneered by Schleyer and coworkers [29,30], the aromatic ring-current shieldings (ARCS) [31], and the gauge-including magnetically induced current (GIMIC) [32] proposed by Juselius and Sundholm, as well as maps of current density induced by a perpendicular magnetic field developed by Steiner and Fowler [33]. The comparison of a given system with prototypical one within MO theory is often enough to assign some certain type of the delocalized bonding, as it is done in the case of main-group elements or transition-metal clusters [5,9].

Boron clusters are the best understood clusters of the main group elements [7,8]. Today we are capable of explaining and predicting their geometric structures and other molecular and spectroscopic properties, because of the recent advances in developing chemical bonding model for these systems [34–58]. There is a considerable amount of experimental data on the stability and reactivity of boron cations reported in the pioneering works by Anderson and coworkers [59–66]. These two factors are the reason for us to focus on the relationship between aromaticity or antiaromaticity and reactivity and stability of the family of cationic boron clusters in the present chapter.

Anderson and coworkers [59–66] produced boron cluster cations B_2^+–B_{13}^+ in molecular beams using laser vaporization and studied their chemical reactivity and fragmentation properties. The structures of B_3^+–B_{13}^+ cations have been established computationally (see review [7] for details) represented in Figure 29.1. In this chapter, we are discussing stability and reactivity of B_3^+ – B_{13}^+ cations on the basis of their multifold aromaticity, multifold antiaromaticity, and conflicting aromaticity.

29.2 THEORETICAL METHODS

The structures of B_3^+–B_{13}^+ cations shown in Figure 29.1 were taken from the Ref. [7], in which they were considered as the most stable structures previously reported in the literature. They were reoptimized using hybrid density functional method known in the literature as B3LYP [67–69] with the 6-311 +G* basis set [70–72] as implemented in Gaussian 03 program [73]. There is no guarantee that all the considered structures here are indeed global minimum structures.

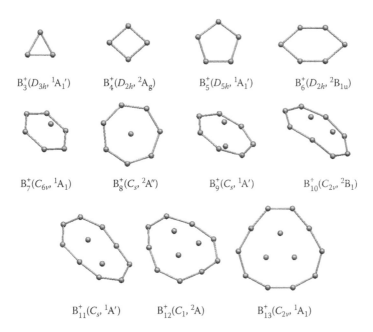

$B_3^+(D_{3h}, {}^1A_1')$ $B_4^+(D_{2h}, {}^2A_g)$ $B_5^+(D_{5h}, {}^1A_1')$ $B_6^+(D_{2h}, {}^2B_{1u})$

$B_7^+(C_{6v}, {}^1A_1)$ $B_8^+(C_s, {}^2A'')$ $B_9^+(C_s, {}^1A')$ $B_{10}^+(C_{2v}, {}^2B_1)$

$B_{11}^+(C_s, {}^1A')$ $B_{12}^+(C_1, {}^2A)$ $B_{13}^+(C_{2v}, {}^1A_1)$

FIGURE 29.1 Geometric structures of B_3^+–B_{13}^+ clusters optimized at B3LYP/6–311 +G*.

Chemical bonding in B_3^+–B_{13}^+ cations was elucidated using the recently developed adaptive natural density partitioning (AdNDP) method [74], which is an extension of the popular natural bond orbital (NBO) analysis. The AdNDP method allows one to partition the charge density into elements with the highest possible degree of localization of electron pairs. If some part of the density cannot be localized in this manner, it is left "delocalized" and is represented by the objects closely reminding canonical molecular orbitals. Thus, chemical bonding is described in terms of n-center two electron (nc-2e) bonds. These bonds are characterized by the occupation numbers (ON), which should be equal to 2.0 $|e|$ in the limiting case. Thus, AdNDP incorporates naturally the idea of delocalized (globally aromatic) bonds and achieves seamless description of chemical bonding in the most general sense. The nc-2e bonds obtained through AdNDP analysis were visualized using the MOLEKEL 4.3 program [75].

29.3   STABILITY AND REACTIVITY OF B_3^+–B_{13}^+ CATIONS

Anderson and coworkers [61] conducted measurements of absolute collision-induced dissociation (CID) cross sections, fragment appearance potentials (AP), and fragmentation branching ratios. They interpreted obtained data to evaluate stabilities and ionization potentials (IP). If the lowest AP is taken as close upper bound on the stability of the cluster ion (Figure 29.2a) then the following clusters B_5^+, B_7^+, and B_{13}^+ should be considered as particularly stable and B_6^+ and B_9^+ as particularly unstable. At the lower collision energy (10 eV), the only significant fragmentation channel for parent clusters from B_3^+ to B_5^+ is the loss of B^+ and production of B_{n-1} neutral species. For larger clusters, the charge usually remains on the B_{n-1}^+ cluster fragment. We calculated dissociation energy at the B3LYP/6-311 + G* level of theory along the $B_n^+ \rightarrow B_{n-1}^+ + B$ and $B_n^+ \rightarrow B_{n-1} + B^+$ channels (Figure 29.2b). According to our calculations, B_5^+, B_7^+, B_{10}^+, and B_{13}^+ clusters have high dissociation energies relative to their neighbors. B_6^+, B_9^+, and B_{12}^+ clusters have relatively low dissociation energies. We relate high dissociation energy to high stability and low dissociation energy to low stability of clusters. It should be kept in mind that this relationship should not be considered as precise and well defined. Our results on relative dissociation energy of cationic boron clusters generally agree with the experimental data on lowest fragmentation AP [61]. The question now is whether we can establish connection between the assignment of stability and chemical bonding in B_3^+–B_{13}^+. This question is positively resolved in Section 29.4.

29.4   CHEMICAL BONDING ANALYSIS OF B_3^+–B_{13}^+ CATIONS

In Table 29.1, we summarized our results of electronic structure calculations for B_3^+–B_{13}^+. We reported symmetry, spectroscopic state, valence electronic configuration, number of 2c-2e peripheral B–B σ-bonds, number of delocalized σ- and π-bonds, and assignment of global aromaticity/antiaromaticity. The description of chemical bonding in terms of 2c-2e peripheral B–B σ-bonds and nc-2e delocalized σ- and π-bonds was obtained via AdNDP method at

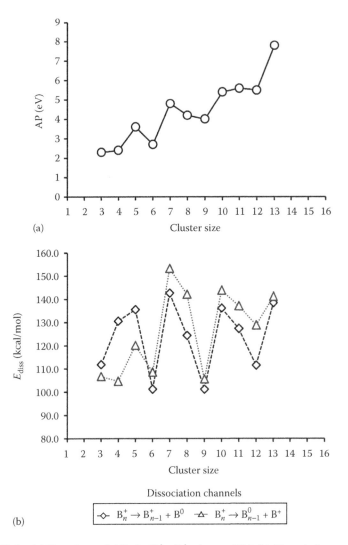

FIGURE 29.2 (a) Experimental APs for $B_3^+ - B_{13}^+$ clusters [61]. (b) Dissociation energies for $B_3^+ - B_{13}^+$ clusters calculated for the $B_n^+ \rightarrow B_{n-1}^+ + B$ (squares) and $B_n^+ \rightarrow B_{n-1} + B^+$ (triangles) fragmentation channels at B3LYP/6–311 + G*.

B3LYP/3-21G//B3LYP/6-311 + G*. From our experience, we know that the pattern of nc-2e bonding remains qualitatively the same, when we use STO-3G, 3-21G, or 6-31G* basis sets. The only difference is the variation of the values in the ON, which does not exceed 0.1 |e| [74]. For open shell systems, this analysis was performed for the closest closed shell species with doubly occupied HOMO at the geometry of the initial cation. The bonding patterns for $B_3^+ - B_{13}^+$ are presented in Figures 29.3 through 29.13. The 2c-2e peripheral B–B σ-bonds are all placed on the single cluster framework.

TABLE 29.1
Concise Chemical Bonding Analysis of B_3^+ – B_{13}^+ Cationic Clusters

Cluster	Valence Electronic Configuration	Number of 2c-2e B–B Peripheral Bonds	Number of Delocalized σ-bonds	Number of Delocalized π-bonds	Aromaticity/Antiaromaticity
B_3^+ $(D_{3h},\ {}^1A_1')$	$1a_1'^2 1e'^4 1a_2''^2$	3	0	1	π-Aromatic
B_4^+ $(D_{2h},\ {}^2A_g)$	$1a_g^2 1b_{1u}^2 1b_{2u}^2 1b_{3g}^2 1b_{3u}^2 2a_g^1$	4	1	1	π-Aromatic "1/2" σ-Aromatic
B_5^+ $(D_{5h},\ {}^1A_1')$	$1a_1'^2 1e_1'^4 1e_2'^4 1a_2''^2 2a_1'^2$	5	1	1	π-Aromatic σ-Aromatic
B_6^+ $(D_{2h},\ {}^2B_{1u})$	$1a_g^2 1b_{1u}^2 1b_{2u}^2 2a_g^2 1b_{3g}^2$ $2b_{2u}^2 1b_{3u}^2 3a_g^2 2b_{1u}^1$	6	2	1	π-Aromatic "1/2" σ-Antiaromatic
B_7^+ $(C_{6v},\ {}^1A_1)$	$1a_1^2 1e_1^4 1e_2^4 2a_1^2 1b_1^2 3a_1^2 2e_1^4$	6	3	1	π-Aromatic σ-Aromatic
B_8^+ $(C_s,\ {}^2A'')$	$1a'^2 2a'^2 3a'^2 4a'^2 5a'^2 6a'^2$ $7a'^2 1a''^2 8a'^2 9a'^2 10a'^2 2a''^1$	7	3	2	"1/2" π-Antiaromatic σ-Aromatic
B_9^+ $(C_s,\ {}^1A')$	$1a'^2 2a'^2 1a''^2 3a'^2 2a''^2 4a'^2$ $5a'^2 3a''^2 6a'^2 7a'^2 4a''^2 8a'^2 5a''^2$	7	4	2	π-Antiaromatic σ-Antiaromatic
B_{10}^+ $(C_{2v},\ {}^2B_1)$	$1a_1^2 1b_2^2 1b_1^2 2a_1^2 1a_2^2 3a_1^2 2b_2^2$ $2b_1^2 4a_1^2 2a_2^2 3b_1^2 3b_2^2 5a_1^2 4b_2^2 4b_1^1$	8	4	3	"1/2" π-Aromatic σ-Antiaromatic
B_{11}^+ $(C_s,\ {}^1A')$	$1a'^2 2a'^2 1a''^2 3a'^2 2a''^2 4a'^2 3a''^2 4a''^2 5a'^2 6a'^2$ $7a'^2 5a''^2 8a'^2 9a'^2 10a'^2 6a''^2$	9	4	3	π-Aromatic σ-Antiaromatic
B_{12}^+ $(C_s,\ {}^2A')$	$1a'^2 1a''^2 2a'^2 3a'^2 2a''^2 4a'^2 3a''^2 5a'^2 6a'^2 4a''^2 4a''^2$ $7a'^2 5a''^2 8a'^2 9a'^2 6a''^2 10a'^2 7a''^2 11a'^1$	9	5 + 1[a]	3	σ-Antiaromatic "1/2" σ-Aromatic
B_{13}^+ $(C_{2v},\ {}^1A_1)$	$1a_1^2 1b_2^2 2a_1^2 3a_1^2 2b_2^2 4a_1^2 3b_2^2 5a_1^2 4b_2^2 6a_1^2$ $7a_1^2 5b_2^2 1b_1^2 8a_1^2 9a_1^2 6b_2^2 1a_2^2 2b_1^2 10a_1^2$	10	5 + 1[a]	3	π-Aromatic σ-Aromatic

[a] Number of delocalized σ-MOs is counted separately for the internal cycle and that found between the external and internal cycles.

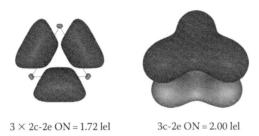

3 × 2c-2e ON = 1.72 lel   3c-2e ON = 2.00 lel

FIGURE 29.3 Localization pattern for B_3^+ obtained via AdNDP procedure. Here and hereafter 2c-2e peripheral σ-bonds are placed at the cluster framework.

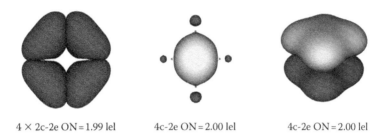

4 × 2c-2e ON = 1.99 lel   4c-2e ON = 2.00 lel   4c-2e ON = 2.00 lel

FIGURE 29.4 Localization pattern for B_4^+ obtained via AdNDP procedure.

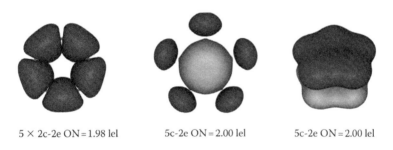

5 × 2c-2e ON = 1.98 lel   5c-2e ON = 2.00 lel   5c-2e ON = 2.00 lel

FIGURE 29.5 Localization pattern for B_5^+ obtained via AdNDP procedure.

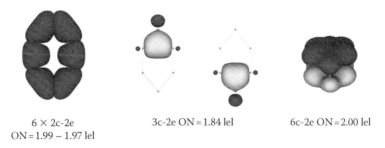

6 × 2c-2e   3c-2e ON = 1.84 lel   6c-2e ON = 2.00 lel
ON = 1.99 – 1.97 lel

FIGURE 29.6 Localization pattern for B_6^+ obtained via AdNDP procedure.

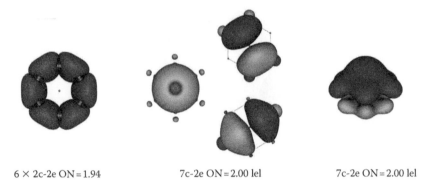

6 × 2c-2e ON = 1.94   7c-2e ON = 2.00 lel   7c-2e ON = 2.00 lel

FIGURE 29.7   Localization pattern for B_7^+ obtained via AdNDP procedure.

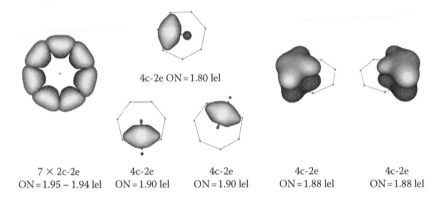

| 7 × 2c-2e | 4c-2e | 4c-2e | 4c-2e | 4c-2e |
| ON = 1.95 – 1.94 lel | ON = 1.90 lel | ON = 1.90 lel | ON = 1.88 lel | ON = 1.88 lel |

4c-2e ON = 1.80 lel

FIGURE 29.8   Localization pattern for B_8^+ obtained via AdNDP procedure.

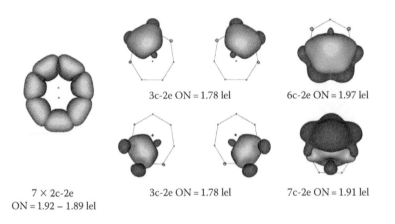

3c-2e ON = 1.78 lel   6c-2e ON = 1.97 lel

7 × 2c-2e
ON = 1.92 – 1.89 lel   3c-2e ON = 1.78 lel   7c-2e ON = 1.91 lel

FIGURE 29.9   Localization pattern for B_9^+ obtained via AdNDP procedure.

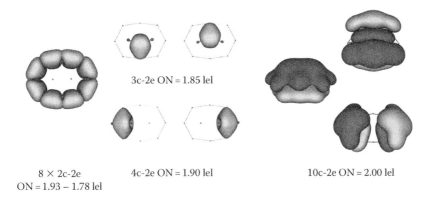

3c-2e ON = 1.85 lel

8 × 2c-2e
ON = 1.93 – 1.78 lel

4c-2e ON = 1.90 lel

10c-2e ON = 2.00 lel

FIGURE 29.10 Localization pattern for B_{10}^+ obtained via AdNDP procedure.

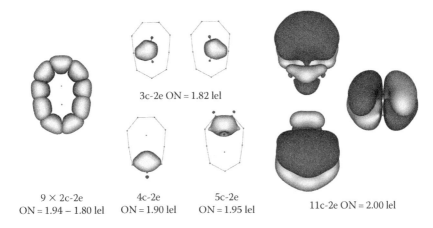

3c-2e ON = 1.82 lel

9 × 2c-2e
ON = 1.94 – 1.80 lel

4c-2e
ON = 1.90 lel

5c-2e
ON = 1.95 lel

11c-2e ON = 2.00 lel

FIGURE 29.11 Localization pattern for B_{11}^+ obtained via AdNDP procedure.

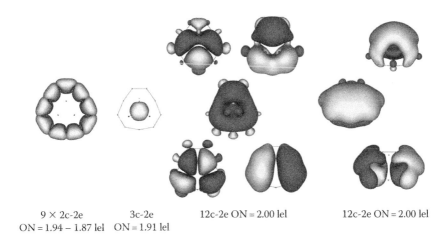

9 × 2c-2e
ON = 1.94 – 1.87 lel

3c-2e
ON = 1.91 lel

12c-2e ON = 2.00 lel

12c-2e ON = 2.00 lel

FIGURE 29.12 Localization pattern for B_{12}^+ obtained via AdNDP procedure.

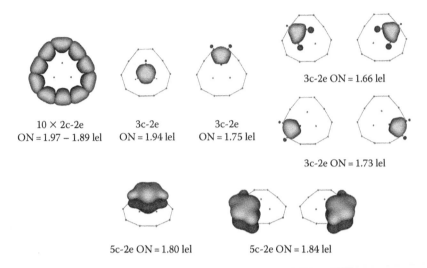

FIGURE 29.13   Localization pattern for B_{13}^+ obtained via AdNDP procedure.

According to our analysis, B_3^+ is a π-aromatic system with three 2c-2e peripheral B–B σ-bonds (Figure 29.3). B_4^+ is π-aromatic and "1/2" σ-aromatic ("1/2" is used as a label and means the σ-system with singly occupied delocalized $2a_g$-HOMO is half way down to being σ-aromatic in the neutral B_4 cluster) with four 2c-2e peripheral B–B σ-bonds (Figure 29.4). B_5^+ is doubly π- and σ-aromatic with five 2c-2e B–B σ-bonds (Figure 29.5). The global minimum structure of B_5^+ has actually C_{2v} symmetry, but after zero-point energy (ZPE) correction, it is effectively a perfect pentagon D_{5h} considered in the present study. B_6^+ is π-aromatic and "1/2" σ-antiaromatic ("1/2" is used as a label and means the σ-system with singly occupied delocalized $2b_{1u}$-HOMO is half way down to being σ-antiaromatic in the neutral B_6 cluster) with six 2c-2e peripheral B–B σ-bonds (Figure 29.6). Strictly speaking, the pyramidal C_{6v} ($^1A_1$) structure of B_7^+ should not be discussed in terms of σ-/π-aromaticity, but the deviation from planarity is rather small (Figure 29.1), and one can see in Figure 29.7 that delocalized bonds still can be approximately recognized as "σ-" and "π-" bonds. The same reasoning is used for the assignment of bonds in other (B_9^+, B_{10}^+, B_{11}^+, and B_{12}^+) nonplanar cations. So, B_7^+ is doubly π- and σ-aromatic with six 2c-2e B–B σ-bonds. B_8^+ is "1/2" π-antiaromatic ("1/2" is used as as a label and means that the π-system with singly occupied delocalized 2a-HOMO is half way down to being π-antiaromatic in the neutral B_8 cluster) and σ-aromatic with seven 2c-2e peripheral B–B σ-bonds (Figure 29.8). B_9^+ is doubly π- and σ-antiaromatic with seven 2c-2e B–B σ-bonds (Figure 29.9). B_{10}^+ is "1/2" π-aromatic ("1/2" is used as a label and means the π-system with singly occupied delocalized $4b_1$-HOMO is half way down to being π-aromatic in the neutral B_{10} cluster) and σ-antiaromatic with eight 2c-2e peripheral B–B σ-bonds (Figure 29.10). B_{11}^+ is π-aromatic and σ-antiaromatic with nine 2c-2e B–B σ-bonds (Figure 29.11). B_{12}^+ is π-aromatic and "1/2" σ-aromatic ("1/2" is used as a label and means the σ-system with singly occupied delocalized $11a'$-HOMO is halfway down to being σ-aromatic in the neutral B_{12} cluster) with nine 2c-2e B–B σ-bonds (Figure 29.12). There are totally six delocalized 3c-2e σ-bonds. One 3c-2e σ-bond is responsible for delocalized σ-bonding in the internal three-atomic cycle and five 3c-2e σ-bonds are responsible for delocalized σ-bonding between the external and internal cycles. Thus, we count the mentioned two sets of delocalized σ-bonds separately with the Huckel's rule holding for five 3c-2e σ-bonds ($4n + 2$ with $n = 2$) and for one 3c-2e σ-bond ($4n + 2$ with $n = 0$). B_{13}^+ is doubly π- and σ-aromatic with 10 2c-2e B–B σ-bonds (Figure 29.13). Like in the case of B_{12}^+, six 3c-2e σ-bonds of B_{13}^+ are split into two sets with five bonds responsible for bonding between external and internal cycles and one bond responsible for bonding within the internal cycle.

Summarizing the assignment of aromaticity/antiaromaticity in cationic boron clusters, we consider B_3^+ as π-aromatic, B_4^+ and B_{12}^+ as partially doubly ("1/2" σ- and π-) aromatic, B_5^+, B_7^+, and B_{13}^+ as doubly (σ- and π-) aromatic, B_9^+ as doubly (σ- and π-) antiaromatic, and B_6^+, B_8^+, and B_{10}^+ as partially conflicting aromatic, and B_{11}^+ as conflicting aromatic.

With patterns of localized bonding in hand, we now can establish the relationship between stability and aromaticity in a straightforward manner. Indeed, exceptional stability of B_5^+, B_7^+, and B_{13}^+ found in our calculations and experimental data is consistent with their double (σ- and π-) aromatic nature. The low reactivity of B_{13}^+ towards water, D_2, and oxygen found by Anderson and coworkers [60,62,63] is also consistent with its remarkable stability due to double aromaticity. They also stated that B_9^+ appears unusually reactive toward oxygen with respect to its nearest neighbors [60]. This observation is consistent with the double antiaromatic nature of B_9^+. Low stability of B_6^+ and B_9^+ found in our calculations and experimental data should be seen as a result of their double (σ- and π-) antiaromatic nature. Apparently, there are three clusters, B_{10}^+, B_{11}^+ and B_{12}^+, which do not follow these trends. The B_{10}^+ cation exhibits relatively high stability, both in theoretical calculations and experimental data, despite its partially conflicting aromatic nature. This can be attributed to the increase of the number of 2c-2e peripheral σ-bonds (from seven in B_9^+ to eight in B_{10}^+) and increase of π-aromatic character (from π-antiaromaticity in B_9^+ to "1/2" π-aromaticity in B_{10}^+). The pattern of chemical bonding in B_{11}^+ and B_{12}^+ (they both have nine peripheral 2c-2e σ-bonds, they are both π-aromatic, and their σ-aromatic character increases from σ-anti aromaticity in B_{11}^+ to "1/2" σ-aromaticity in B_{12}^+) is consistent (they both have nine peripheral 2c-2e σ-bonds, with the their stability evaluated on the basis of experimentally observed APs. The disagreement with computational results might be due to the deficiency of the estimated dissociation energies as a measure of stability.

The presented consideration of the family of cationic boron clusters exemplifies that the assessment of stability and reactivity of clusters can be performed at the qualitative level using multifold aromaticity, multifold antiaromaticity, and variety of conflicting aromaticities.

REFERENCES

1. Faraday, M. *Philos. Trans. R. Soc. Lond.* 1825, *115*, 440–446.
2. Matito, E., Poater, J., Solà, M., and Schleyer, P. v. R., Chapter 28, this volume.

3. Li, X., Kuznetsov, A. E., Zhang, H. F., Boldyrev, A. I., and Wang, L. S. *Science* 2001, *291*, 859–861.

4. Kuznetsov, A. E., Birch, K. A., Boldyrev, A. I., Li, X., Zhai, H.-J., and Wang, L. S. *Science* 2003, *300*, 622–625.

5. Boldyrev, A. I. and Wang, L. S. *Chem. Rev.* 2005, *105*, 3716–3757 and references therein.

6. Tsipis, C. A. *Coord. Chem. Rev.* 2005, *249*, 2740–2762 and references therein.

7. Alexandrova, A. N., Boldyrev, A. I., Zhai, H. J., and Wang, L. S. *Coord. Chem. Rev.* 2006, *250*, 2811–2866 and references therein.

8. Zubarev, D. Yu. and Boldyrev, A. I. *J. Comput. Chem.* 2007, *28*, 251–268 and references therein.

9. Zubarev, D. Yu., Averkiev, B. B., Zhai, H. -J., Boldyrev, A. I., and Wang, L. S. *Phys. Chem. Chem. Phys.* 2008, *10*, 257–267 and references therein.

10. Datta, A., Mallajosyula, S. S., and Pati, S. K. *Acc. Chem. Res.* 2007, *40*, 213–221.

11. Chandrasekhar, J., Jemmis, E. D., and Schleyer, P. v. R. *Tetrahedron Lett.* 1979, *39*, 3707–3710.

12. Martin-Santamaria, S. and Rzepa, H. S. *Chem. Commun.* 2000, 1503–1504.

13. Präsang, C., Mlodzianowska, A., Sahin, Y., Hofmann, M., Geiseler, G., Massa, W., and Berndt. A. *Angew. Chem. Int. Ed.* 2002, *41*, 3380–3382.

14. Zhai, H.-J., Averkiev, B. B., Zubarev, D. Yu., Wang, L. S, and Boldyrev, A. I. *Angew. Chem. Int. Ed.* 2007, *46*, 4277–4280.

15. Averkiev, B. B. and Boldyrev, A. I. *J. Phys. Chem. A* 2007, *111*, 12864–12866.

16. Special edition on aromaticity. Schleyer, P. v. R. ed. *Chem. Rev.* 2001, *101*, No. 5.

17. Special edition on heterocycles. Katrizky, A. ed. *Chem. Rev.* 2004, *104*, No. 5.

18. Special edition on delocalization pi and sigma. Schleyer, P. v. R. ed. *Chem. Rev.* 2005, *105*, No. 10.

19. Poater, J., Fradera, X., Duran, M., and Sola, M. *Chem. Eur. J.* 2003, *9*, 1113–1122.

20. Matito, E., Duran, M., and Sola, M. *J. Chem. Phys.* 2005, *122*, 014109.

21. Giambiagi, M. S., Giambiagi, M., and Fortes, M. S. *J. Mol. Struc. THEOCHEM,* 1997, *391*, 141–150.

22. Giambiagi, M., Giambiagi, M. S., Santos Silva, C. D., and Figueiredo, A. P. *Phys. Chem. Chem. Phys.* 2000, *2*, 3381–3392.

23. (a) Bultinck, P., Ponec R., and Dammme, S. V. *J. Phys. Org. Chem.* 2005, *18*, 706–718, (b) Bultinck, P., Fias, S., Ponec, R. *Chem. Eur. J.* 2006, *12*, 8813–8818.

24. (a) Bultinck, P., Ponec, R., and Carbo-Dorca, R. *J. Comput. Chem.* 2007, *28*, 152–160, (b) Bultinck, P., Rafat, M., Ponec, R., Gheluwe, B. V., Carbo-Dorca, R., Popelier, P. *J. Phys. Chem. A* 2006, *110*, 7642–7648.

25. Ponec, R., Bultinck, P., and Saliner, A. G. *J. Phys. Chem. A* 2005, *109*, 6606–6609.

26. Cioslowski, J., Matito, E., and Sola, M. *J. Phys. Chem.* 2007, *111*, 6521–6525.

27. (a) Chattaraj, P. K., Roy, D. R., Elango, M., and Subramanian, V., *J. Mol. Struct. THEOCHEM*, 2006, *759*, 109–110. (b) Roy, D. R., Bultinck, P., Subramanian, V., Chattaraj, P. K. *J. Mol. Struct. THEOCHEM,* 2008, *854*, 35–39.

28. (a) Chattaraj, P. K., Roy, D. R., Elango, M., and Subramanian, V. *J. Phys. Chem. A* 2005, *109*, 9590–9597, (b) Chattaraj, P. K., Giri, S. *J. Phys. Chem. A* 2007, *111*, 11116–11121, (c) Chattaraj, P. K., Sarkar, U., Roy, D. R. *J. Chem. Ed.* 2007, *84*, 354–357, (d) Roy, D. R., Chattaraj, P. K. *J. Phys. Chem. A,* 2008, *112*, 1612–1621.

29. Schleyer, P. v. R., Maerker, C., Dransfeld, A., Jiao, H., Hommes, N. J. R. van Eikema *J. Am. Chem. Soc.* 1996, *118*, 6317–6318.

30. For recent review of NICS application, see Chen, Z., Wannere, C. S., Corminboeuf, C., Puchta, R., Schleyer, P. v. R. *Chem. Rev.* 2005, *105*, 3842–3888.

31. Juselius, J. and Sundholm, D. *Phys. Chem. Chem. Phys.* 1999, *1*, 3429–3435.

32. Juselius, J. Sundholm, D., and Gauss, J. *J. Chem. Phys.* 2004, *121*, 3952–3963.

33. Steiner, E. and Fowler, P. W. *J. Phys. Chem. A* 2001, *105*, 9553–9562.
34. Fowler, J. E. and Ugalde, J. M. *J. Phys. Chem. A* 2000, *104*, 397–403.
35. Aihara, J. *J. Phys. Chem. A* 2001, *105*, 5486–5489.
36. Zhai, H. J., Wang, L. S., Alexandrova, A. N., and Boldyrev, A. I. *J. Chem. Phys.* 2002, *117*, 7917–7924.
37. Alexandrova, A. N., Boldyrev, A. I., Zhai, H. J., Wang, L. S., Steiner, E., and Fowler, P. W. *J. Phys. Chem. A* 2003, *107*, 1359–1369.
38. Zhai, H. J., Wang, L. S., Alexandrova, A. N., Boldyrev, A. I., and Zakrzewski, V. G. *J. Phys. Chem. A* 2003, *107*, 9319–9328.
39. Kuznetsov, A. E. and Boldyrev, A. I. *Struct. Chem.* 2002, *13*, 141–148.
40. Zhai, H. J., Alexandrova, A. N., Birch, K. A., Boldyrev, A. I., and Wang, L. S. *Angew. Chem. Int. Ed.* 2003, *42*, 6004–6008.
41. Alexandrova, A. N., Boldyrev, A. I., Zhai, H. J., and Wang, L. S. *J. Phys. Chem. A* 2004, *108*, 3509–3517.
42. Alexandrova, A. N., Zhai, H. J., Wang, L. S., and Boldyrev, A. I. *Inorg. Chem.* 2004, *43*, 3552–3554.
43. Alexandrova, A. N., Boldyrev, A. I., Zhai, H. J., and Wang, L. S. *J. Chem. Phys.* 2005, *122*, 054313.
44. Zhai, H. J., Kiran, B., Li, J., and Wang, L. S. *Nat. Mater.* 2003, *2*, 827–833.
45. Kiran, B., Bulusu, S., Zhai, H. J., Yoo, S., Zeng, X. C., and Wang, L. S. *Proc. Natl. Acad. Sci. USA* 2005, *102*, 961–964.
46. Aihara, J., Kanno, H., and Ishida, T. *J. Am. Chem. Soc.* 2005, *127*, 13324–13330.
47. Li, Q. S. and Jin, H. W. *J. Phys. Chem. A* 2002, *106*, 7042–7047.
48. Havenith, R. W. A., Fowler, P. W., and Steiner, E. *Chem. Eur. J.* 2002, *8*, 1068–1073.
49. Li, Q. S., Jin, Q., Luo, Q., Tang, A. C., Yu, J. K., and Zhang, H. X. *Int. J. Quant. Chem.* 2003, *94*, 269–278.
50. Jin, H. W. and Li, Q. S. *Phys. Chem. Chem. Phys.* 2003, *5*, 1110–1115.
51. Li, Q. S. and Jin, Q. *J. Phys. Chem. A* 2003, *107*, 7869–7873.
52. Ma, J., Li, Z., Fan, K., and Zhou, M. *Chem. Phys. Lett.* 2003, *372*, 708–716.
53. Li, Q. S. and Jin, Q. *J. Phys. Chem. A* 2004, *108*, 855–860.
54. Li, Q. S. and Gong, L.-F. *J. Phys. Chem. A* 2004, *108*, 4322–4325.
55. Li, Q. S., Gong, L.-F., and Gao, Z.-M. *Chem. Phys. Lett.* 2004, *390*, 220–227.
56. Li, Q., Zhao, Y., Xu, W., and Li, N. *Int. J. Quant. Chem.* 2005, *101*, 219–229.
57. Li, S.-D., Gou, J.-C., Miao, C.-Q., and Ren, G.-M. *Angew. Chem. Int. Ed.* 2005, *44*, 2158–2161.
58. Fowler, P. W. and Gray, B. R. *Inorg. Chem.* 2007, *46*, 2892–2897.
59. Hanley, L. and Anderson, S. L. *J. Phys. Chem.* 1987, *91*, 5161–5163.
60. Hanley, L. and Anderson, S. L. *J. Chem. Phys.* 1988, *89*, 2848–2860.
61. Hanley, L., Whitten, J. L., and Anderson, S. L. *J. Phys. Chem.* 1988, *92*, 5803–5812.
62. Hintz, P. A., Ruatta, S. A., and Anderson, S. L. *J. Chem. Phys.* 1990, *92*, 292–303.
63. Ruatta, S. A., Hintz, P. A., and Anderson, S. L. *J. Chem. Phys.* 1991, *94*, 2833–2847.
64. Hintz, P. A., Sowa, M. B., Ruatta, S. A., and Anderson, S. L. *J. Chem. Phys.* 1991, *94*, 6446–6458.
65. Sowa-Resat, M. B., Smolanoff, J., Lapiki, A., and Anderson, S. L. *J. Chem. Phys.* 1997, *106*, 9511–9522.
66. La Placa, S. J., Roland, P. A., and Wynne, J. J. *Chem. Phys. Lett.* 1992, *190*, 163–168.
67. Becke, A. D. *J. Chem. Phys.* 1993, *98*, 5648–5652.
68. Vosko, S. H., Wilk, L., and Nusair, M. *Can. J. Phys.* 1980, *58*, 1200–1211.
69. Lee, C., Yang, W., and Parr, R. G. *Phys. Rev. B* 1988, *37*, 785–789.
70. Binkley, J. S., Pople, J. A., and Hehre, W. J. *J. Am. Chem. Soc.* 1980, *102*, 939–947.
71. Gordon, M. S., Binkley, J. S., Pople, J. A., Pietro, W. J., and Hehre, W. J. *J. Am. Chem. Soc.* 1982, *104*, 2797–2803.

72. Pietro, W. J., Francl, M. M., Hehre, W. J., Defrees, D. J., Pople, J. A., and Binkley, J. S. *J. Am. Chem. Soc.* 1982, *104*, 5039–5048.

73. Gaussian 03, Revision C.02, Frisch, M. J., Trucks, G. W., Schlegel, H. B., Scuseria, G. E., Robb, M. A., Cheeseman, J. R., Montgomery, Jr., J. A., Vreven, T., Kudin, K. N., Burant, J. C., Millam, J. M., Iyengar, S. S., Tomasi, J., Barone, V., Mennucci, B., Cossi, M., Scalmani, G., Rega, N., Petersson, G. A., Nakatsuji, H., Hada, M., Ehara, M., Toyota, K., Fukuda, R., Hasegawa, J., Ishida, M., Nakajima, T., Honda, Y., Kitao, O., Nakai, H., Klene, M., Li, X., Knox, J. E., Hratchian, H. P., Cross, J. B., Bakken, V., Adamo, C., Jaramillo, J., Gomperts, R., Stratmann, R. E., Yazyev, O., Austin, A.J., Cammi, R., Pomelli, C., Ochterski, J. W., Ayala, P. Y., Morokuma, K., Voth, G. A., Salvador, P., Dannenberg, J. J., Zakrzewski, V. G., Dapprich, S., Daniels, A. D., Strain, M. C., Farkas, O., Malick, D. K., Rabuck, A. D., Raghavachari, K., Foresman, J. B., Ortiz, J. V., Cui, Q., Baboul, A.G., Clifford, S., Cioslowski, J., Stefanov, B. B., Liu, G., Liashenko, A., Piskorz, P., Komaromi, I., Martin, R. L., Fox, D. J., Keith, T., Al-Laham, M. A., Peng, C. Y., Nanayakkara, A., Challacombe, M., Gill, P. M. W., Johnson, B., Chen, W., Wong, M. W., Gonzalez, C., and Pople, J. A., Gaussian, Inc., Wallingford CT, 2004.

74. Zubarev, D. Yu. and Boldyrev, A. I. *Phys. Chem. Chem. Phys.* 2008, 10, 5207–5217.

75. Portmann, S. *MOLEKEL*, Version 4.3., CSCS/ETHZ, 2002.

30 Probing the Coupling between Electronic and Geometric Structures of Open and Closed Molecular Systems

Roman F. Nalewajski

CONTENTS

30.1 INTRODUCTION

Note: The atomic units are used throughout the paper; in the adopted notation **P** denotes the square or rectangular matrix, *P* stands for the row vector, and *P* represents the scalar quantity.

The internal degrees-of-freedom of molecular systems are of either electronic or nuclear (geometric) origins. In the Born–Oppenheimer (BO) approximation, the equilibrium (ground) state of the externally closed molecule is specified by the overall number of electrons N (integer) in the system and the external potential $v(r; Q)$ due to the nuclei located in the parametrically specified locations corresponding to the internal geometric coordinates Q. Alternatively, the state-parameters N and Q uniquely identify the system (Coulombic) Hamiltonian $\hat{H}(N, v) = \hat{H}(N, Q)$, its ground state $\Psi[N, v] = \Psi(N, Q)$, the electronic energy $E[N, v] = \langle \Psi[N, v] | \hat{H}(N, v) | \Psi[N, v] \rangle = E(N, Q)$, and the BO potential $W(N, Q) = E(N, Q) + V_{nn}(Q)$, where $V_{nn}(Q)$ stands for the nuclear repulsion energy. One similarly specifies the equilibrium state of an externally open-system characterized by the fractional average number of electrons, which is coupled to an external electron reservoir controlling the system chemical potential $\mu = \partial E(N, \boldsymbol{Q})/\partial N$. In the geometrically rigid molecule this equilibrium state is identified by μ and $\boldsymbol{Q}$. The corresponding equilibrium (relaxed) geometries are identified by the vanishing forces $\boldsymbol{F} = -[\partial W(N, \boldsymbol{Q})/\partial \boldsymbol{Q}]^{\mathrm{T}} = \boldsymbol{0}$, giving rise to the alternative sets of state-parameters, $(N, \boldsymbol{F})$ and $(\mu, \boldsymbol{F})$, in the electronically closed and open (relaxed) systems, respectively (e.g., Chattaraj and Parr, 1993; Cohen, 1996; Geerlings et al., 2003; Nalewajski, 1993, 1995, 1997, 2002a, 2003, 2006a; Nalewajski and Korchowiec, 1997; Nalewajski et al., 1996; Parr and Yang, 1989).

Each molecular process involves the mutually coupled displacements in the distribution of the system electrons and positions of its nuclei. In chemistry, the mutual interaction between the electronic and geometric structures of molecules or reactive systems plays a vital role in diagnosing their behavior in different environments. Therefore, designing the adequate descriptors of this coupling and establishing principles for a qualitative prediction of its structural and reactivity manifestations constitute a challenging problem in theoretical chemistry (e.g., Baekelandt et al., 1995; Cohen, 1996; Nalewajski, 1993, 1995, 1997, 1999, 2000, 2002a,b, 2003, 2006a,b; Nalewajski and Korchowiec, 1997; Nalewajski and Michalak, 1995, 1996, 1998; Nalewajski and Sikora, 2000; Nalewajski et al., 1996, 2008). Indeed, the rules governing this subtle interplay between the electronic and geometric degrees-of-freedom in molecular systems constitute an important part of the structural chemistry and reactivity theory (e.g., Ayers and Parr, 2000, 2001; Fujimoto and Fukui, 1974; Fukui, 1975, 1987; Gutmann, 1978; Klopman, 1968, 1974; Nalewajski, 1984). They reflect effects of the mutual interaction between an internal polarization (P) and/or external charge transfer (CT) on one side and the concomitant geometrical relaxation on the other side, e.g., in molecular subsystems of the donor–acceptor (DA) complexes. The Gutmann rules of structural chemistry (Gutmann, 1978) and their extension provided by the so-called mapping relations formulated within the charge sensitivity analysis (CSA) of molecular systems (Bakelandt et al., 1995; Nalewajski, 1995, 2006b; Nalewajski and Korchowiec, 1997; Nalewajski and Michalak, 1995, 1996, 1998; Nalewajski and Sikora, 2000; Nalewajski et al., 1996), allow for a qualitative and semiquantitative predictions, respectively, of such relaxational effects.

Another example is provided by the minimum energy coordinates (MECs) of the compliant approach in CSA (Nalewajski, 1995; Nalewajski and Korchowiec, 1997; Nalewajski and Michalak, 1995, 1996, 1998; Nalewajski et al., 1996), in the spirit of the related treatment of nuclear vibrations (Decius, 1963; Jones and Ryan, 1970; Swanson, 1976; Swanson and Satija, 1977). They all allow one to diagnose the molecular electronic and geometrical responses to hypothetical electronic or nuclear displacements (perturbations). The "thermodynamical" Legendre-transformed approach (Nalewajski, 1995, 1999, 2000, 2002b, 2006a,b; Nalewajski and Korchowiec, 1997; Nalewajski and Sikora, 2000; Nalewajski et al., 1996, 2008) provides a versatile theoretical framework for describing diverse equilibrium states of molecules in different chemical environments.

The essence of such a combined linear response treatment of the electronic and geometric state-variables is that all their mutual interactions are explicitly taken into account in the generalized electronic–nuclear Hessian. The relevant coupling terms are represented by the off-diagonal elements in such generalized "force constant" or "compliance" tensors, for the admissible selections of the system state-parameters, which specify the equilibria in the externally closed or open molecules, for their rigid or relaxed geometries. In such an approach, the overall number of electrons, N, and its energy-conjugate, the chemical potential μ, determine the electronic state-variables in the externally closed and open molecular systems, respectively. Accordingly, the internal coordinates Q or their energy-conjugates, the vanishing forces $F = 0$, describe the rigid or relaxed molecular geometries.

This theoretical framework thus unites the so called electron-following (EF) and electron-preceding (EP) perspectives (Nakatsuji, 1973, 1974a,b) on molecular changes, in the spirit of the BO approximation and the Hellmann–Feynman theorem, respectively. In the former the electron distribution responds to the geometrical (nuclear) perturbation, a displacement in nuclear positions, while the latter implies the system geometrical relaxation following a test displacement in the system electronic state-parameters. Such generalized "polarizabilities" of molecules are generated within CSA of molecular systems. They provide reliable criteria in the reactivity theories (e.g., Chattaraj and Parr, 1993; Cohen, 1996; Geerlings et al., 2003; Nalewajski, 1993, 1995, 1997, 2002a, 2003, 2006a; Nalewajski and Korchowiec, 1997; Nalewajski et al., 1996; Parr and Yang, 1989) based upon the modern density functional theory (DFT) (Dreizler and Gross, 1990; Hohenberg and Kohn, 1964; Kohn and Sham, 1965; Nalewajski, 1996; Parr and Yang, 1989). In the EF outlook the adjustment $\Delta\rho$ in the electron distribution represents the unconstrained (dependent) local state-variable of the molecular system in question: $\Delta\rho = \Delta\rho[N, \Delta v] = \Delta\rho(N, \Delta Q)$ or $\Delta\rho = \Delta\rho[\mu, \Delta v] = \Delta\rho(\mu, \Delta Q)$. In other words, the electron density responds ("follows") the displacements ΔQ of the system nuclei. This selection of the dependent (ρ) and independent (v) local state-variables generates the chemical softness kernel $\sigma(r, r')$ of the reactivity theory and has been classified as the chemical softness representation of molecular states (Nalewajski, 2006a,b).

These roles are reversed in the EP perspective of DFT, which can also be referred to as the chemical hardness representation (Nalewajski, 2006a,b), since it defines another key concept of the electronic structure and reactivity theories—the chemical hardness kernel $\eta(r, r')$, the inverse of $\sigma(r, r')$. In the EP approach, the displacement in the system electron density, $\Delta\rho$, effected either by the controlled change in the system number of electrons $\Delta N = \int \Delta\rho \, dr$ or a displacement in the chemical potential of the external reservoir, $\Delta\rho = \Delta\rho(\Delta N, F = 0)$ or $\Delta\rho = \Delta\rho(\Delta\mu, F = 0)$, respectively, is now regarded as the controlling, independent parameter of state, while the external potential responds to the specified redistribution of electrons, thus representing a dependent (unconstrained) state-variable $\Delta v = \Delta v[\Delta\rho]$. The shifts in the electron distribution thus "precede" the movements of nuclei, $\Delta Q = \Delta Q(\Delta N, F = 0)$ or $\Delta Q = \Delta Q(\Delta\mu, F = 0)$, in the spirit of the familiar Hellmann–Feynman (force) theorem. This way of approaching molecular displacements is quite common in the chemical reactivity theory. Indeed, chemists often envisage the key manipulation of the system electronic structure as the primary cause of the desired reconstruction of the molecular geometry, e.g., the breaking/forming of bonds in the molecule.

One requires both these perspectives to tackle all issues in the theory of electronic structure of molecules and their chemical reactivity. The wave function and density functional formulations of the quantum theory of electronic structure in molecular systems, thus emerge as the complementary descriptions, which "together" provide theoretical framework of the "complete" theory of chemical reactivity. The emergence of the modern DFT has provided the EP perspective and generated new approaches to many classical problems in chemistry. It offers an alternative point of view, from which one can approach the diverse physical/ chemical properties and processes involving atomic, molecular, and reactive systems. This novel perspective is in the spirit of the Sanderson's electronegativity equalization description of the equilibrium distribution of electrons in molecular systems (Sanderson, 1951, 1976). Examples of the reactivity indices quantifying the electronic–geometric coupling are provided by the electronic Fukui function (Korchowiec and Uchimaru, 1998; Michalak et al., 1999; Nalewajski et al., 1996; Parr and Yang, 1984; Yang and Parr, 1985; Yang et al., 1984) and its nuclear analog (Cohen, 1996; Cohen et al., 1994, 1995).

In this chapter, the diverse coupling constants and MEC components identified in the combined electronic–nuclear approach to equilibrium states in molecules and reactants are explored. The reactivity implications of these derivative descriptors of the interaction between the electronic and geometric aspects of the molecular structure will be commented upon within both the EP and EF perspectives. We begin this analysis with a brief survey of the basic concepts and relations of the generalized compliant description of molecular systems, which simultaneously involves the electronic and nuclear degrees-of-freedom. Illustrative numerical data of these derivative properties for selected polyatomic molecules, taken from the recent computational analysis (Nalewajski et al., 2008), will also be discussed from the point of view of their possible applications as reactivity criteria and interpreted as manifestations of the LeChâtelier–Braun principle of thermodynamics (Callen, 1962).

30.2 PERTURBATION–RESPONSE RELATIONS

We first examine the coupling relations within the canonical geometric representation, which corresponds to the BO potential $\widetilde{W}[N, v(\boldsymbol{Q})] = \widetilde{E}[N, v(\boldsymbol{Q})] + V_{nn}(\boldsymbol{Q})$, combining the molecular electronic energy $\widetilde{E}[N, v(\boldsymbol{Q})]$ and the nuclear repulsion term $V_{nn}(\boldsymbol{Q})$, in which the implicit dependence on the internal nuclear coordinates $\boldsymbol{Q} = \{Q_s\}$ consisting of bond lengths and angles through the external potential $v(r; \boldsymbol{Q})$ has been replaced by the explicit dependence on molecular geometry: $\widetilde{W}[N, v(\boldsymbol{Q})] \equiv W(N, \boldsymbol{Q}) = E(N, \boldsymbol{Q}) + V_{nn}(\boldsymbol{Q})$.

The purely geometrical derivatives of $W(N, \boldsymbol{Q})$ include the forces acting on nuclei along the internal geometric coordinates

$$\boldsymbol{F}(N, \boldsymbol{Q}) = -[\partial W(N, \boldsymbol{Q})/\partial \boldsymbol{Q}]^{\mathrm{T}} = \{F_s\}, \qquad (30.1)$$

and the geometric Hessian:

$$\mathbf{H} = \frac{\partial^2 W(N, \boldsymbol{Q})}{\partial \boldsymbol{Q} \partial \boldsymbol{Q}} = -\frac{\partial \boldsymbol{F}(N, \boldsymbol{Q})}{\partial \boldsymbol{Q}}. \qquad (30.2)$$

The purely electronic derivatives, calculated for the rigid molecular geometry, determine the system electronic chemical potential

$$\mu = \partial E(N, \boldsymbol{Q})/\partial N = \partial W(N, \boldsymbol{Q})/\partial N \tag{30.3}$$

and the electronic hardness

$$\eta = \frac{\partial^2 E(N, \boldsymbol{Q})}{\partial N^2} = \frac{\partial^2 W(N, \boldsymbol{Q})}{\partial N^2} = \frac{\partial \mu(N, \boldsymbol{Q})}{\partial N}. \tag{30.4a}$$

It should be recalled that η stands for the rigid geometry measure of the electronic global hardness, the inverse of the system global softness $S = (\partial N/\partial \mu)_{\boldsymbol{Q}}$:

$$\eta = \left(\frac{\partial \mu}{\partial N}\right)_{\boldsymbol{Q}} = S^{-1}. \tag{30.4b}$$

Finally, the mixed second derivatives define the nuclear Fukui function (NFF) indices:

$$\varphi(N, \boldsymbol{Q}) = -\left(\frac{\partial^2 W(N, \boldsymbol{Q})]}{\partial N\, \partial \boldsymbol{Q}}\right)^{\mathrm{T}} = \left(\frac{\partial F(N, \boldsymbol{Q})}{\partial N}\right)_{\boldsymbol{Q}} = -\left(\frac{\partial \mu(N, \boldsymbol{Q})}{\partial \boldsymbol{Q}}\right)^{\mathrm{T}}_{N}, \tag{30.5}$$

coupling the electronic state-parameter N with the geometric coordinates $\boldsymbol{Q}$.

The independent displacements (ΔN, $\Delta \boldsymbol{Q}$) (perturbations) in this canonical geometric representation give rise to the first differential of its "thermodynamic" potential $W(N, \boldsymbol{Q})$:

$$dW(N, \boldsymbol{Q}) = (\partial W/\partial N)_{\boldsymbol{Q}} dN + d\boldsymbol{Q}(\partial W/\partial \boldsymbol{Q})_N = \mu dN - \boldsymbol{F} d\boldsymbol{Q}^{\mathrm{T}}$$
$$\equiv \mu dN - \sum_s F_s dQ_s. \tag{30.6}$$

The generalized Hessian **H** transforming these perturbations into the linear responses of the corresponding energy conjugate quantities, representing the unconstrained electronic and geometric state-variables grouped in the generalized gradient $\boldsymbol{g} = (\mu, -\boldsymbol{F})$,

$$(\Delta\mu - \Delta F) \equiv \Delta\boldsymbol{g} = (\Delta N, \Delta \boldsymbol{Q})\mathbf{H}, \tag{30.7}$$

includes the following blocks defined by the generalized "force constants" of Equations 30.2, 30.4, and 30.5:

$$\mathbf{H} = \begin{bmatrix} \left(\frac{\partial \mu}{\partial N}\right)_{\boldsymbol{Q}} & -\left(\frac{\partial F}{\partial N}\right)_{\boldsymbol{Q}} \\ \left(\frac{\partial \mu}{\partial \boldsymbol{Q}}\right)_{N} & -\left(\frac{\partial F}{\partial \boldsymbol{Q}}\right)_{N} \end{bmatrix} \equiv \begin{bmatrix} H_{N,N} & H_{N,\boldsymbol{Q}} \\ H_{\boldsymbol{Q},N} & H_{\boldsymbol{Q},\boldsymbol{Q}} \end{bmatrix} = \begin{bmatrix} \eta & -\varphi \\ -\varphi^{\mathrm{T}} & \mathbf{H} \end{bmatrix}. \tag{30.8}$$

The overall transformation (30.7) thus combines the following partial electronic and geometric ground-state relations:

$$\Delta\mu = \Delta N\eta - \Delta Q\varphi^{\mathrm{T}} \quad \text{and} \quad -\Delta F = \Delta N\varphi + \Delta Q\mathbf{H}. \qquad (30.9)$$

The inverse of $\mathbf{H}$ determines the geometric compliance matrix (Nalewajski, 1993, 1995, 1997, 1999, 2000, 2002b, 2006a,b; Nalewajski and Korchowiec, 1997; Nalewajski et al., 1996, 2008) describing the open system in the (μ, F)-representation. The relevant thermodynamic potential is defined by the total Legendre transform of the system BO potential, which replaces the state-parameters (N, Q) with their energy conjugates (μ, F), respectively:

$$\Sigma(\mu, F) = W - N(\partial W/\partial N)_Q - Q(\partial W/\partial Q)_N = W - N\mu + QF^{\mathrm{T}}, \qquad (30.10)$$

$$d\Sigma = -Nd\mu + QdF^{\mathrm{T}} \quad \text{or} \quad -N = (\partial\Sigma/\partial\mu)_F \quad \text{and} \quad Q = (\partial\Sigma/\partial F)_\mu^{\mathrm{T}}. \qquad (30.11)$$

Setting $F = 0$ then identifies the properties for the equilibrium (relaxed) molecular geometry.

The generalized compliance matrix, combining the system electronic and geometric degrees-of-freedom,

$$\mathbf{S} = \mathbf{H}^{-1} = \begin{bmatrix} -\left(\frac{\partial N}{\partial\mu}\right)_F & \left(\frac{\partial Q}{\partial\mu}\right)_F \\ -\left(\frac{\partial N}{\partial F}\right)_\mu & \left(\frac{\partial Q}{\partial F}\right)_\mu \end{bmatrix} \equiv \begin{bmatrix} S_{\mu,\mu} & S_{\mu,F} \\ S_{F,\mu} & S_{F,F} \end{bmatrix}, \qquad (30.12)$$

relates the displacements (perturbations) of the representation independent variables $(\Delta\mu, \Delta F)$ with the conjugate responses in the unconstrained conjugate quantities $(-\Delta N, \Delta Q)$:

$$(-\Delta N, \Delta Q) = (\Delta\mu, \Delta F)\,\mathbf{S}. \qquad (30.13)$$

It summarizes the responses in the system average number of electrons and its geometry:

$$-\Delta N = \Delta\mu S_{\mu,\mu} + \Delta Q S_{F,\mu}, \quad \Delta Q = \Delta\mu S_{\mu,F} + \Delta F S_{F,F}. \qquad (30.14)$$

A reference to Equation 30.12 shows that the diagonal element $S_{\mu,\mu}$ represents the relaxed geometry analog of the negative global softness of Equation 30.4b, with the latter being defined for the rigid molecular geometry. It follows from the second of Equation 30.14 that a change in the chemical potential of an open system induces an extra relaxation of the molecular frame. This geometric "softness" effect is described by the derivatives of the row vector $S_{\mu,F} = \{S_{\mu,s}\} = S_{F,\mu}^{\mathrm{T}} \equiv \mathbf{S} = \{S_s\}$.

One can explicitly express the compliance matrix in terms of the elements of the principal charge sensitivities defining the generalized electronic–nuclear "hardness" matrix $\mathbf{H}$ of Equation 30.8, by eliminating ΔN and ΔQ from Equation 30.9:

$$-\Delta N = -\Delta\mu(\eta - B)^{-1} + \Delta F H^{-1}\varphi^{T}(\eta - B)^{-1}, \quad B = \varphi H^{-1}\varphi^{T};$$

$$\Delta Q = \Delta\mu\varphi H^{-1}C - \Delta F H^{-1}C\eta, \qquad\qquad C = \left(\eta I - \varphi^{T}\varphi H^{-1}\right)^{-1}. \quad (30.15)$$

or in the combined matrix form:

$$
S = \begin{bmatrix} -\left(\frac{\partial N}{\partial\mu}\right)_{F} = -(\eta - B)^{-1} \equiv -S^{rel} & \left(\frac{\partial Q}{\partial\mu}\right)_{F} = \varphi H^{-1}\, C \equiv S \\ -\left(\frac{\partial N}{\partial F}\right)_{\mu} = H^{-1}\varphi^{T}(\eta - B)^{-1} = S^{T} & \left(\frac{\partial Q}{\partial F}\right)_{\mu} = -H^{-1}C\eta \equiv G^{rel} \end{bmatrix}, \quad (30.16)
$$

where $S^{rel} = (\eta^{rel})^{-1}$ stands for the geometrically relaxed softness, inverse of the relaxed hardness, and G^{rel} denotes the electronically relaxed geometrical compliant matrix, which differs from its closed system analog $G = -H^{-1} = (\partial Q/\partial F)_{N}$.

Let us now turn to the mixed, partly inverted (N, F)-representation describing the geometrically relaxed, but externally closed molecular system. The relevant thermodynamic potential is now defined by the partial Legendre transformation of $W(N, Q)$ which replaces Q by F in the list of the system parameters of state:

$$\Theta(N, F) = W - Q(\partial W/\partial Q)_{N} = W + QF^{T}; \quad (30.17)$$

$$d\Theta = \mu dN + Q dF^{T} \quad \text{or} \quad \mu = (\partial\Theta/\partial N)_{F} \quad \text{and} \quad Q = (\partial\Theta/\partial F)_{N}^{T}. \quad (30.18)$$

Eliminating $\Delta\mu$ from the first Equation 30.14 and inserting it into the second Equation 30.14 then gives the following transformation of the representation independent displacements $(\Delta N, \Delta F)$ into the linear responses of their energy conjugates $(\Delta\mu, \Delta Q)$,

$$(\Delta\mu, \Delta Q) = (\Delta N, \Delta F)V, \quad (30.19)$$

where the relevant Hessian V expressed in terms of the principal compliance coefficients of Equation 30.12 reads

$$
V = \begin{bmatrix} \left(\frac{\partial\mu}{\partial N}\right)_{F} & \left(\frac{\partial Q}{\partial N}\right)_{F} \\ \left(\frac{\partial\mu}{\partial F}\right)_{N} & \left(\frac{\partial Q}{\partial F}\right)_{N} \end{bmatrix} \equiv \begin{bmatrix} V_{N,N} & V_{N,F} \\ V_{F,N} & V_{F,F} \end{bmatrix} = \begin{bmatrix} -S_{\mu,\mu}^{-1} & -S_{\mu,F}S_{\mu,\mu}^{-1} \\ -S_{F,\mu}S_{\mu,\mu}^{-1} & S_{F,F} - S_{F,\mu}S_{\mu,F}S_{\mu,\mu}^{-1} \end{bmatrix}.
$$

$$(30.20)$$

Again, the diagonal element $V_{N,N}$ represents the molecular hardness estimated for the relaxed geometry of the molecule, a companion parameter of the rigid geometry hardness of Equations 30.4a,b. The two partial relations for the electronic and geometric responses in Equation 30.19

$$\Delta\mu = \Delta N V_{N,N} + \Delta F V_{F,N} \quad \text{and} \quad \Delta Q = \Delta N V_{N,F} + \Delta F V_{F,F}, \quad (30.21)$$

again imply that there is an additional geometry relaxation due to a finite external CT between the externally open molecule and its electron reservoir, besides the usual term for constant N, due to the forces acting on the system nuclei. This extra relaxation of the molecular frame is described by the vectors $V_{N,F} = V_{F,N}^T$ of coupling derivatives.

The four blocks of V can be alternatively expressed in terms of the principal geometric derivatives defining the generalized Hessian of Equation 30.8. This can be accomplished first by expressing ΔQ as function of ΔN and ΔF, using the second Equation 30.9, and then by inserting the result into the first Equation 30.9:

$$\Delta\mu = \Delta N(\eta - B) + \Delta F H^{-1}\varphi^T \quad \text{and} \quad \Delta Q = \Delta N\varphi H^{-1} - \Delta F H^{-1}. \quad (30.22)$$

A comparison between Equations 30.21 and 30.22 then gives

$$V = \begin{bmatrix} \left(\frac{\partial\mu}{\partial N}\right)_F = (\eta - B) = \eta^{rel} & \left(\frac{\partial Q}{\partial N}\right)_F = \varphi H^{-1} \equiv f \\ \left(\frac{\partial\mu}{\partial F}\right)_N = H^{-1}\varphi^T \equiv f^T & \left(\frac{\partial Q}{\partial F}\right)_N = -H^{-1} \equiv G \end{bmatrix}, \quad (30.23)$$

where the row vector $f = \{f_s = (\partial Q_s/\partial N)_F = (\partial\mu/\partial F_s)_N\}$ groups the so-called geometric Fukui function (GFF) indices (Nalewajski, 2006b).

Finally, the remaining (μ, Q) representation describing the equilibrium state of an externally open molecular system with the "frozen" nuclear framework is examined. The relevant partial Legendre transform of the total electronic energy, which replaces N by μ in the list of independent state-parameters, defines the BO grand-potential:

$$\Xi(\mu, Q) = W - N(\partial W/\partial N)_Q = W - N\mu, \quad (30.24)$$

$$d\Xi = -Nd\mu - FdQ^T \quad \text{or} \quad -N = (\partial\Xi/\partial\mu)_Q \quad \text{and} \quad -F = (\partial\Xi/\partial Q)_\mu^T. \quad (30.25)$$

Eliminating ΔF from the second Equation 30.14 and inserting the result into the first of these two equations give the following transformation of the representation independent perturbations $(\Delta\mu, \Delta Q)$ into the linear responses of their conjugates $(-\Delta N, -\Delta F)$, expressed in terms of the matrix elements of the compliance matrix S of Equation 30.12:

$$-(\Delta N, \Delta F) = (\Delta\mu, \Delta Q)G, \quad (30.26)$$

$$G = \begin{bmatrix} -\left(\frac{\partial N}{\partial\mu}\right)_Q & -\left(\frac{\partial F}{\partial\mu}\right)_Q \\ -\left(\frac{\partial N}{\partial Q}\right)_\mu & -\left(\frac{\partial F}{\partial Q}\right)_\mu \end{bmatrix} \equiv \begin{bmatrix} G_{\mu,\mu} & G_{\mu,Q} \\ G_{Q,\mu} & G_{Q,Q} \end{bmatrix} = \begin{bmatrix} S_{\mu,\mu} - S_{\mu,F}S_{F,F}^{-1}S_{F,\mu} & S_{\mu,F}S_{F,F}^{-1} \\ S_{F,F}^{-1}S_{F,\mu} & -S_{F,F}^{-1} \end{bmatrix}.$$

$$(30.27)$$

The above matrix transformation combines the following electronic and geometric relations:

$$-\Delta N = \Delta\mu G_{\mu,\mu} + \Delta Q G_{Q,\mu} \quad \text{and} \quad -\Delta F = \Delta\mu G_{\mu,Q} + \Delta Q G_{Q,Q}. \quad (30.28)$$

The elements of $\mathbf{G}$ can be alternatively expressed in terms of the generalized hardness matrix of Equation 30.8, by eliminating ΔN from the first Equation 30.9 and by subsequent insertion of the result into the second of these equations:

$$
\begin{aligned}
-\Delta N &= -\Delta\mu\eta^{-1} - \Delta\boldsymbol{Q}\boldsymbol{\varphi}^{\mathrm{T}}\eta^{-1} \equiv -\Delta\mu S - \Delta\boldsymbol{Q}\mathbf{s}^{\mathrm{T}}, \\
-\Delta\boldsymbol{F} &= -\Delta\mu\mathbf{s} + \Delta\boldsymbol{Q}(\mathbf{H} - \boldsymbol{\varphi}^{\mathrm{T}}S\boldsymbol{\varphi}) = -\Delta\mu\mathbf{s} + \Delta\boldsymbol{Q}(\mathbf{H} - \boldsymbol{\varphi}^{\mathrm{T}}\mathbf{s}).
\end{aligned} \tag{30.29}
$$

where S is the electronic rigid geometry measure of the system global softness of Equation 30.4b, and the row vector $\mathbf{s}$ of the geometric softnesses (Nalewajski, 2006b) is defined as the product of the global softness and NFF vector:

$$
\mathbf{s} = S\boldsymbol{\varphi} = (\partial\boldsymbol{F}/\partial\mu)_{\boldsymbol{Q}} = (\partial N/\partial\boldsymbol{Q})_{\mu}^{\mathrm{T}} = (\partial\boldsymbol{F}/\partial N)_{\boldsymbol{Q}}/(\partial\mu/\partial N)_{\boldsymbol{Q}} = \left\{s_s \equiv (F_s)_{\mu}\right\}. \tag{30.30}
$$

A reference to the second Equation 30.29 shows that the effective geometrical Hessian of an open molecular system differs from that of the closed system (Equation 30.8) by the extra CT contribution involving the geometrical softnesses and NFF. One finally identifies the corresponding blocks of $\mathbf{G}$ by comparing the general relations of Equation 30.28 with the explicit transformations of Equation 30.29,

$$
\mathbf{G} = \begin{bmatrix}
-\left(\dfrac{\partial N}{\partial\mu}\right)_{\boldsymbol{Q}} = -S & -\left(\dfrac{\partial\boldsymbol{F}}{\partial\mu}\right)_{\boldsymbol{Q}} = -\mathbf{s} \\
-\left(\dfrac{\partial N}{\partial\boldsymbol{Q}}\right)_{\mu} = -\mathbf{s}^{\mathrm{T}} & -\left(\dfrac{\partial\boldsymbol{F}}{\partial\boldsymbol{Q}}\right)_{\mu} = \mathbf{H} - \boldsymbol{\varphi}^{\mathrm{T}}\mathbf{s} \equiv \mathbf{H}^{rel}
\end{bmatrix}, \tag{30.31}
$$

where $\mathbf{H}^{rel}$ denotes the electronically relaxed geometrical Hessian, which differs from its closed system analog $\mathbf{H} = -(\partial\boldsymbol{F}/\partial\boldsymbol{Q})_N$. The $\mathbf{G}$-matrix thus involves the negative, rigid geometry electronic softness as diagonal element associated with the electronic state-variable μ, the off-diagonal blocks representing the geometric softnesses, and the open-system (electronically relaxed) geometrical Hessian as the nuclear diagonal block; the latter differs from the closed-system (electronically rigid) Hessian $\mathbf{H}$ by the LeChâtelier–Braun (Callen, 1962) softening contribution:

$$
\begin{aligned}
\boldsymbol{\varphi}^{\mathrm{T}}\mathbf{s} &= \boldsymbol{\varphi}^{\mathrm{T}}S\boldsymbol{\varphi} = (\partial\mu/\partial\boldsymbol{Q})_N(\partial\boldsymbol{F}/\partial\mu)_{\boldsymbol{Q}} = \left[(\partial N/\partial\boldsymbol{Q})_{\mu}(\partial\boldsymbol{F}/\partial N)_{\boldsymbol{Q}}\right]^{\mathrm{T}} \\
&= (\partial\boldsymbol{F}/\partial\boldsymbol{Q})_N - (\partial\boldsymbol{F}/\partial\boldsymbol{Q})_{\mu}.
\end{aligned} \tag{30.32}
$$

Several geometrical quantities introduced in this section provide natural descriptors measuring a strength of the mutual coupling between the molecular electronic and geometric structures. In the canonical geometrical representation $W(N, \boldsymbol{Q})$ the diagonal blocks of the generalized Hessian $\mathbf{H}$, measuring the system rigid geometry hardness η and force constants $\mathbf{H}$, describe the decoupled aspects of the electronic and geometric structures, respectively. Thus, the NFF vector $\boldsymbol{\varphi}$, defining the off-diagonal blocks in $\mathbf{H}$, reflects the coupling between the electronic and nuclear aspects of the molecular structure. These derivatives describe the influence of geometrical displacements in the closed system on the molecular chemical potential, or alternatively, the effect of an external CT on geometrical forces. The two sets of geometric softnesses, $\mathbf{S} = \{S_s\}$ (Equation 30.16) and $\mathbf{s} = \{s_\alpha\}$ (Equation 30.30), reflect similar couplings in the externally open molecular systems. It should be stressed, however, that in the geometric compliance matrix $\mathbf{S}$ the interaction between these two facets of molecular structure enters the diagonal blocks as well, as explicitly indicated in Equations 30.15 and 30.16.

A similar effect of the system electronic or nuclear "softnening," due to its opening relative to a reservoir or a relaxation of its geometry, is seen in the diagonal blocks of the partial-compliant matrices $\mathbf{V}$ and $\mathbf{G}$. This spontaneous relaxation of the system electronic–nuclear structure reflects the LeChâtelier–Braun principle of "moderation" in the ordinary thermodynamics (Callen, 1962). Indeed, the extra electronic relaxation $\delta N(\Delta \mathbf{Q})$ induced by the the primary nuclear perturbation $\Delta \mathbf{Q}$ in the externally open system, in which a spontaneous CT between the molecule and its electron reservoir is allowed, effectively lowers the increase in the magnitude of forces on the system nuclei, compared to those in the externally closed system: $|\Delta \mathbf{F}(\Delta \mathbf{Q})|_N > |\Delta \mathbf{F}(\Delta \mathbf{Q})|_\mu$ (see Equation 30.32). The induced effect of the spontaneous geometry relaxation $\delta \mathbf{Q}(\Delta N)$ induced by the primary electronic perturbation ΔN similarly lowers the increase in the system chemical potential, compared to that in the rigid system: $\Delta \mu (\Delta N)_Q > \Delta \mu (\Delta N)_{F=0}$.

It should also be realized that the generalized softness matrix of Equations 30.12 and 30.16 represents the compliant description of the electronic "coordinate" N coupled to the system geometric relaxations (see Section 30.3). Indeed, the relaxed geometry global softness of the geometrical representation,

$$-S_{\mu,\mu} = (\partial N / \partial \mu)_F = (\eta - B)^{-1} = \left(\eta - \boldsymbol{\varphi}\mathbf{H}^{-1}\boldsymbol{\varphi}^{\mathrm{T}}\right)^{-1} \equiv S^{rel} \equiv \left(\eta^{rel}\right)^{-1} > \eta^{-1}$$
$$= S = (\partial N / \partial \mu)_Q > 0, \tag{30.33}$$

where the last inequality states the familiar LeChâtelier stability requirement (Callen, 1962; Nalewajski, 1993, 1995, 2006b; Nalewajski and Korchowiec, 1997; Nalewajski et al., 1996), differs from the conventional definition of the electronic global softness S, which invokes the rigid geometry constraint (Equation 30.4b). The geometric hardness contribution B in Equation 30.33 effectively softens the electronic distribution via the relaxation of nuclei, reflected by the negative purely geometric compliant $\mathbf{H}^{-1}$, and the "weighting" factors provided by the NFF $\boldsymbol{\varphi}$ reflecting the relative geometric softness of the molecule. The other diagonal block of the generalized geometrical compliants, which contains the electron–nuclear couplings,

$$\mathbf{S}_{F,F} = (\partial \mathbf{Q} / \partial \mathbf{F})_\mu = -\mathbf{H}^{-1}\mathbf{C}\eta = -\mathbf{H}^{-1}\left(\eta \mathbf{I} - \boldsymbol{\varphi}^{\mathrm{T}}\boldsymbol{\varphi}\mathbf{H}^{-1}\right)^{-1}\eta \neq -\mathbf{H}^{-1} = \mathbf{V}_{F,F} \tag{30.34}$$

is also seen to differ from the purely geometrical compliant $\mathbf{V}_{F,F}$ by the additional factor exhibiting both the electronic and nuclear origins. The mixture of the electronic and nuclear inputs is also seen to determine the off-diagonal blocks $\mathbf{S}_{\mu,F}$ and $\mathbf{S}_{F,\mu}$ of the geometric compliant matrix, respectively measuring the effect of the chemical potential on the relaxed nuclear positions ($\mathbf{S}_{\mu,F}$) or the influence of the forces on the effective charge of an open molecule ($\mathbf{S}_{F,\mu}$).

Let us now examine the compliance descriptors of the externally closed system in the $\Theta(N, \mathbf{F})$ representation, defined by the corresponding blocks of the geometric charge sensitivities $\mathbf{V}$ (Equation 30.20). Again, the first diagonal derivative in this matrix, $V_{N,N} = (\partial \mu / \partial N)_F = \eta - B$, allows the geometry of the system to relax, after an addition or removal of an electron, until the forces on nuclei exactly vanish. The electronic–geometric interaction is also detected in the coupling blocks $\mathbf{V}_{N,F} = (\partial \mathbf{Q} / \partial N)_F$ and $\mathbf{V}_{F,N} = (\partial \mu / \partial \mathbf{F})_N$. A reference to Equation 30.23 indicates that they are determined by the purely nuclear compliants $\mathbf{V}_{F,F} = -\mathbf{H}^{-1}$ and NFF.

The electronic–nuclear coupling in molecules is also detected in the other partial Legendre-transformed representation $\Xi(\mu, \mathbf{Q})$, which defines the combined Hessian $\mathbf{G}$ of Equation 30.27. Its first diagonal derivative,

$$G_{\mu,\mu} = -(\partial N / \partial \mu)_Q = -(\partial N / \partial \mu)_v = -S, \tag{30.35}$$

represents the purely electronic, global compliant reflecting the negative softness of the rigid system. The off-diagonal blocks $\mathbf{G}_{\mu,Q} = -(\partial \mathbf{F} / \partial \mu)_Q$ and $\mathbf{G}_{Q,\mu} = -(\partial N / \partial \mathbf{Q})_\mu$ represent the geometric softnesses of Equation 30.30. They thus measure the effect of the system chemical potential on forces on nuclei ($\mathbf{G}_{\mu,Q}$) or the influence of nuclear displacements on the effective molecular charge, for the rigid nuclear frame. Since in this representation the molecular system is externally open one detects in the geometrical Hessian of this representation the contribution due to external CT triggered by nuclear displacements:

$$\mathbf{G}_{Q,Q} = -(\partial \mathbf{F} / \partial \mathbf{Q})_\mu = \mathbf{H} - \boldsymbol{\varphi}^{\mathrm{T}} \mathbf{s} = \mathbf{H} - \boldsymbol{\varphi}^{\mathrm{T}} S \boldsymbol{\varphi} \neq \mathbf{H} = -(\partial \mathbf{F} / \partial \mathbf{Q})_N. \tag{30.36}$$

This block thus contains the electronically relaxed force constants along the system internal coordinates.

30.3 COMPLIANCE CONSTANTS AND MINIMUM ENERGY COORDINATES

This section begins with a brief summary of the compliance approach to nuclear motions (Decius, 1963; Jones and Ryan, 1970; Swanson, 1976; Swanson and Satija, 1977). The inverse of the nuclear force constant matrix $\mathbf{H}$ of Equation 30.2, defined in the purely geometric $\mathbf{Q}$-representation,

$$\mathbf{H} = \left\{ H_{s,s'} = \frac{\partial^2 W(N, \mathbf{Q})}{\partial Q_s\, \partial Q_{s'}} = -\left(\frac{\partial F_{s'}}{\partial Q_s}\right)_{Q_{t \neq s}} \right\}, \tag{30.37}$$

determines the geometric compliance matrix of the "reverse" $\mathbf{F}$-representation:

$$\mathbf{G} = \frac{\partial^2 \Theta(N, \mathbf{F})}{\partial \mathbf{F}\, \partial \mathbf{F}} = -\mathbf{H}^{-1} = \left\{ G_{s,s'} = \left(\frac{\partial^2 \Theta(N, \mathbf{F})}{\partial F_s\, \partial F_{s'}}\right) = \left(\frac{\partial Q_{s'}}{\partial F_s}\right)_{F_{t \neq s}} \right\}. \tag{30.38}$$

Here $\Theta(N, F)$ (see Equation 30.17) stands for the Legendre transform of the BO potential energy surface $W(N, Q)$, in which the nuclear-position coordinates Q are replaced by the corresponding forces F in the list of the parameters of state. Indeed, for the fixed number of electrons N,

$$[d\Theta(N,F)]_N = QdF^{\mathrm{T}} \quad \text{and} \quad [\partial^2\Theta(N,F)/\partial F_s F_{s'}]_N = (\partial Q_{s'}/\partial F_s)_{N,F'}. \quad (30.39)$$

The constraint of $F' = \{F_{t \neq s}\} = 0'$ in these derivatives implies that the remaining internal coordinates of the nuclear frame are free to relax the atomic positions until the forces associated with these geometrical degrees-of-freedom vanish, thus marking the minimum of the system energy with respect to $\{Q_{t \neq s}\}$.

The ratio of the matrix elements in sth row of $\mathbf{G}$, $\mathbf{G}_s = \{G_{s,s'}, s' = 1, 2, \ldots\}$, to the diagonal element $G_{s,s}$ determines kth vector of nuclear (geometric) interaction constants (Decius, 1963; Jones and Ryan, 1970):

$$(s')_s = G_{s,s'}/G_{s,s} = \left(\frac{\partial Q_{s'}}{\partial F_s}\right)_{F_{t \neq s}} \left(\frac{\partial Q_s}{\partial F_s}\right)_{F_{t \neq s}}^{-1} = \left(\frac{\partial Q_{s'}}{\partial Q_s}\right)_{F_{t \neq s}}, \quad s' = 1, 2, \ldots \quad (30.40)$$

These indices describe the minimum-energy responses, determined for $F'_s = \{F_{t \neq s} = 0\}$, of the remaining nuclear-position variables $\{Q_{s' \neq s}\}$ per unit displacement of sth nuclear coordinate. Thus they determine the sth geometric MEC (Decius, 1963; Jones and Ryan, 1970; Swanson, 1976; Swanson and Satija, 1977). This compliant concept can be used to predict the equilibrium responses of the system geometric structure to a given displacement (perturbation) ΔQ_s of the selected nuclear coordinate from the initial, equilibrium geometry of the molecule, which accounts for all couplings between geometric coordinates:

$$dQ(\Delta Q_s)|_{F'_s=0} = \{(s')_s \Delta Q_s\}. \quad (30.41)$$

In this section, several concepts of the compliant desciption of the combined electronic–nuclear treatment of molecular systems (Nalewajski, 1993, 1995, 2006b; Nalewajski and Korchowiec, 1997; Nalewajski et al., 1996, 2008) will be discussed. As remarked before, there are two types of geometrical constraints, which can be imposed on the molecule: (i) the rigid geometry Q and (ii) the condition of the vanishing forces $F = 0$ giving rise to the system equlibrium (relaxed) geometry. The latter description amounts to the compliant formalism of nuclear motions, in which one allows the system to relax all its remaining (electronic and/or nuclear) degrees-of-freedom in response to the probing displacements in the number of electrons in the system or positions of its constituent atoms. The $\Theta(N, F)$ and $\Sigma(\mu, F)$ representations correspond to such a nuclear-compliant treatment of the molecular geometrical structure, while the $W(N, Q)$ and $\Xi(\mu, Q)$ representations adopt the rigid geometry approach. The Legendre-transformed approach to geometric representations of molecular states provides the complete set of quantities, which can be used to monitor (or index) the electronic–geometric couplings in molecular systems, covering both the externally open and externally closed molecular systems.

The MEC can also be introduced in the combined electron–nuclear treatment of the geometric representations of the molecular structure (Nalewajski, 1993, 1995, 2006b; Nalewajski and Korchowiec, 1997; Nalewajski et al., 1996, 2008). Consider, for example, the generalized interaction constants defined by the electronic–nuclear softness matrix $\mathbf{S}$. The ratios of the matrix elements in $\mathbf{S}_{\mu,F} = \{S_{\mu,s'}\}$ to $S_{\mu,\mu}$ define the following interaction constants between the nuclear coordinates and the system average number of electrons:

$$(s')_N = S_{\mu,s'}/S_{\mu,\mu} = (\partial Q_{s'}/\partial\mu)_{F=0}/(\partial N/\partial\mu)_{F=0} = (\partial Q_{s'}/\partial N)_{F=0} \equiv V_{N,s'}. \tag{30.42}$$

Therefore, they reflect the minimum energy responses of the system geometrical coordinates per unit displacement in the system number of electrons. The GFF vector of Equation 30.23

$$\mathbf{f} = \{(s')_N\} = (\partial\mathbf{Q}/\partial N)_{F=0} = \mathbf{V}_{N,F} = (\partial\mu/\partial\mathbf{F})_N^{\mathrm{T}} = \mathbf{V}_{F,N}^{\mathrm{T}}$$
$$= (\partial\mathbf{Q}/\partial\mu)_{F=0}/(\partial N/\partial\mu)_{F=0} = \mathbf{S}/S^{rel} \tag{30.43}$$

can be interpreted as an alternative set the NFF indices, which diagnose the normalized effect of changing the oxidation state of the molecular system as a whole on its geometry. These indices define the following MEC grouping responses in nuclear coordinates due to a finite inflow or outflow of electrons $\Delta N \neq 0$:

$$dQ(\Delta N)|_{\mu,F=0} = \Delta N\mathbf{f}. \tag{30.44}$$

It should be realized at this point (see Equations 30.5 and 30.30) that NFF can also be interpreted as the MEC reflecting the rigid geometry response in forces per unit displacement in the system number of electrons:

$$\boldsymbol{\varphi} = \{(F_s)_N\} = (\partial\mathbf{F}/\partial N)_Q = \mathbf{H}_{N,Q} = (\partial\mu/\partial\mathbf{Q})_N^{\mathrm{T}} = \mathbf{H}_{Q,N}^{\mathrm{T}}$$
$$= (\partial\mathbf{F}/\partial\mu)_Q/(\partial N/\partial\mu)_Q = \mathbf{s}/S. \tag{30.45}$$

The geometric softnesses $\mathbf{s}$ also represent the rigid geometry interaction between forces $\mathbf{F}$ and the system chemical potential. The remaining interaction constants defined in this representation are given by the ratios

$$(N)_{s,\mu} = S_{\mu,s}/S_{s,s} = (\partial N/\partial Q_s)_{\mu,F'_s=0} \quad \text{and} \quad (s')_{s,\mu} = S_{s,s'}/S_{s,s} = (\partial Q_{s'}/\partial Q_s)_{\mu,F'_s=0}. \tag{30.46}$$

In the open molecule coupled to an external electron reservoir, which fixes the system chemical potential, they combine the minimum-energy responses in the system number of electrons and the remaining nuclear coordinates to a unit displacement of Q_s. The associated MECs,

$$dN(\Delta Q_s)|_{F'_s=0} = \{(N)_{s,\mu}\Delta Q_s\} \quad \text{and} \quad dQ(\Delta Q_s)|_{\mu,F'_s=0} = \{(s')_{s,\mu}\Delta Q_s\}, \tag{30.47}$$

add to a variety of descriptors of the electronic and geometric structures of molecular systems. The $(N)_{s,\mu}$ coupling constants can be used to probe trends in the chemical oxidation or reduction of the open molecule, which follows a given geometrical deformation of the molecule. These probing displacements allow one to identify nuclear changes, which are most effective in bringing about this electronic transformation of the molecule. The other set $\{(s')_{s,\mu}\}$ tests the geometrical consequences of a hypothetical nuclear position perturbation of the open molecule, thus facilitating a search for the most effective geometric manipulation of the molecular system in question, which is required to bring about the desired overall change in the system nuclear framework.

The partial-compliant matrix $\mathbf{V}$ of the $\Theta(N, \mathbf{F})$-representation defines analogous interaction constants for the N-controlled (externally closed) molecules:

$$(s')_\mu = V_{N,s'}/V_{N,N} = (\partial Q_{s'}/\partial N)_{F=0}/(\partial \mu/\partial N)_{F=0} = (\partial Q_{s'}/\partial \mu)_{F=0} = S_{\mu,s'} = S_{s'},$$

$$(30.48)$$

where $\{V_{N,s'}\} = \mathbf{V}_{N,F}$, and

$$(\mu)_{s,N} = V_{s,N}/V_{s,s} = (\partial \mu/\partial F_s)_N/(\partial Q_s/\partial F_s)_N = (\partial \mu/\partial Q_s)_{N,F_s=0},$$

$$(s')_{s,N} = V_{s',s}/V_{s,s} = (\partial Q_{s'}/\partial F_s)_{N,F_s'=0}/(\partial Q_s/\partial F_s)_{N,F_s'=0} = (\partial Q_{s'}/\partial Q_s)_{N,F_s'=0},$$

$$(30.49)$$

with $\{V_{s,N}\} \in \mathbf{V}_{F,N}$. The corresponding MECs

$$dQ(\Delta \mu)|_{N,F=0} = \{(s')_\mu \Delta \mu\},$$

$$d\mu(\Delta Q_s)|_{N,F_s'=0} = \{(\mu)_{s,N} \Delta Q_s\}, \, dQ(\Delta Q_s)|_{N,F_s'=0} = \{(s')_{s,N} \Delta Q_s\},$$

$$(30.50)$$

reflect the equilibrium responses in the system chemical potential and geometrical coordinates due to finite shifts in the system chemical potential or selected geometrical coordinates.

Finally, in the $\Xi(\mu, \mathbf{Q})$ representation, in which the generalized partial-compliant matrix $\mathbf{G}$ is defined, one obtains the following coupling constants:

$$(F_s)_N = G_{\mu,s'}/G_{\mu,\mu} = (\partial F_{s'}/\partial \mu)_Q/(\partial N/\partial \mu)_Q = (\partial F_s/\partial N)_Q = \varphi_s; \quad (30.51)$$

$$(N)_{F_s,\mu} = G_{s,\mu}/G_{s,s} = (\partial N/\partial Q_s)_\mu/(\partial F_s/\partial Q_s)_\mu = (\partial N/\partial F_s)_{\mu,Q_s'},$$

$$(F_{s'})_{F_s,\mu} = G_{s',s}/G_{s,s} = (\partial F_{s'}/\partial Q_s)_{\mu,Q_s'}/(\partial F_s/\partial Q_s)_{\mu,Q_s'} = (\partial F_{s'}/\partial F_s)_{\mu,Q_s'}. \quad (30.52)$$

These interaction constants determine the following MEC:

$$dF(\Delta N)|_{\mu,Q} = \{(F_s)_N \Delta N\}, \quad dF(\Delta F_s)|_{\mu,Q_s'} = \{(F_{s'})_{F_s,\mu} \Delta F_s\},$$

$$dN(\Delta F_s)|_{\mu,Q_s'} = \{(N)_{F_s,\mu} \Delta F_s\}.$$

$$(30.53)$$

They can be used to predict responses in forces due to an electron inflow or outflow to or from the open molecule or a displacement of selected force component, or the effect of such a force perturbation on the system average number of electrons.

30.4   ILLUSTRATIVE EXAMPLES

The recent extensive numerical analysis (Nalewajski et al., 2008) of the joint electronic–nuclear compliants in selected polyatomics including H_2O, NO_2, H_2O_2, ClF_3, and NH_2CHO, has generated representative coupling quantities and the MEC data. These molecules exhibit a variety of internal geometric degrees-of-freedom, bond lengths, and angles, which are specified in the Figure 30.1. In what follows we discuss some of these results, generated using the simplest Hartree–Fock (HF) theory [GAUSSIAN software (Frisch et al., 2004)] in the extended 6-31++G** basis set of the Gaussian orbitals, including the split-valence and polarization functions. In all derivative properties the angles are measured in radians; all these quantities correspond to the ground-state equilibrium geometries in the adopted basis set.

This compliant analysis has used the analytical forces and Hessians, and the finite difference estimates of the corresponding N-derivatives. The NFF have been calculated for both the electron-accepting $(\Delta N = +1)$ and electron-donating $(\Delta N = -1)$ processes, when the system acts as a Lewis acid and base, respectively, relative to the attacking nucleophilic (N) and electrophilic (E) agents. The Mulliken scheme for the neutral system approached by the radical agent (R), of the unbiased N-derivative given by the arithmetic average of these two estimates, has also been examined. These R-estimates are reported in Tables 30.1 through 30.4. The global hardness, which measures the curvature of the ground-state BO potential energy

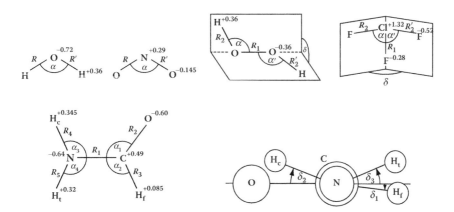

FIGURE 30.1   The internal coordinates in five representative molecular systems and the Mulliken net charges of bonded atoms (from HF calculations). The last diagram defines the dihedral angles in formamide, relative to the NCO plane, determining the out-of-plane displacements of the *cis* (H_c), *trans*-(H_t) and formyl (H_f) hydrogens, respectively.

TABLE 30.1

Comparison of the Molecular Hardness and Softness Quantities for the Rigid and Relaxed Geometries

Molecule	η	η^{rel}	S	S^{rel}
H_2O	0.448	0.446	2.231	2.240
NO_2	0.422	0.319	2.369	3.133
H_2O_2	0.468	0.394	2.139	2.541
ClF_3	0.471	0.395	2.125	2.533
NH_2CHO	0.360	0.355	2.779	2.815

Source: Adapted from Nalewajski, R. F., Błażewicz, D., and Mrozek, J., *J. Math. Chem.*, 2008, 44: 325–364.

Note: The reported relaxed ($^{rel.}$) quantities are averages of predictions using the $\Delta N = +1$ and $\Delta N = -1$ estimates of NFF; the same convention applies to the molecular compliant data reported in the remaining tables.

TABLE 30.2

Comparison of Selected Molecular Fukui Function and Softness Compliants

Compliant Constant	H_2O	NO_2	H_2O_2	ClF_3	NH_2CHO
$(F_R)_N = \varphi_R$ or $\varphi_{R_1} = (F_{R_1})_N$	−0.018	0.112	0.060	0.147	0.024
$\varphi_{R_2} = (F_{R_2})_N$			−0.015	0.063	−0.035
$(F_\alpha)_N = \varphi_\alpha$ or $\varphi_{\alpha_2} = (F_{\alpha_2})_N$	−0.012	−0.171	0.000	0.022	−0.019
$(F_R)_\mu = S_R$ or $S_{R_1} = (F_{R_1})_\mu$	−0.040	0.265	0.128	0.312	0.067
$S_{R_2} = (F_{R_2})_\mu$			−0.031	0.134	−0.099
$(F_\alpha)_\mu = S_\alpha$ or $S_{\alpha_1} = (F_{\alpha_1})_\mu$	−0.028	−0.405	0.000	0.046	−0.005
$(R)_\mu = S_R$ or $S_{R_1} = (R_1)_\mu$	−0.058	0.368	0.251	0.884	0.169
$S_{R_2} = (R_2)_\mu$			−0.039	0.489	−0.134
$(\alpha)_\mu = S_\alpha$ or $S_{\alpha_1} = (\alpha_1)_\mu$	−0.144	−1.407	0.153	0.019	0.044
$(R)_N = f_R$ or $f_{R_1} = (R_1)_N$	−0.026	0.117	0.099	0.349	0.060
$f_{R_2} = (R_2)_N$			−0.015	0.193	−0.048
$(\alpha)_N = f_\alpha$ or $f_{\alpha_2} = (\alpha_2)_N$	−0.064	−0.449	0.060	0.008	−0.076
$(\delta)_N = f_\delta$ or $f_{\delta_1} = (\delta_1)_N$			−2.672	0.717	0.000

Source: Adapted from Nalewajski, R. F., Błażewicz, D., and Mrozek, J., *J. Math. Chem.*, 2008, 44: 325–364.

surface along the N-coordinate, has been similarly estimated by interpolating the energies for the set of hypothetical electronic displacements $\Delta N = (-1, 0, +1)$.

In Table 30.1, we have compared the electronically relaxed hardness and softness descriptors for the geometrically rigid and relaxed molecules, respectively. As intuitively expected, relaxing the nuclear positions decreases the electronic hardness (increases softness) of the molecular system under consideration. This electronic

TABLE 30.3

Selected Interaction Constants for the Closed or Open H_2O and NO_2

	$(R')_R$ $(R')_{R,N}$	$(\alpha)_R$ $(\alpha)_{R,N}$	$(R)_\alpha$ $(R)_{\alpha,N}$	$(R')_{R,\mu}$	$(\alpha)_{R,\mu}$	$(R)_{\alpha,\mu}$	$(\mu)_R$	$(\mu)_{\alpha,N}$	$(N)_{R,\mu}$	$(N)_{\alpha,\mu}$
H_2O	0.017	−0.168	−0.044	0.018	−0.165	−0.044	0.016	0.011	−0.037	−0.024
NO_2	−0.243	−0.112	−0.051	−0.198	−0.248	−0.094	−0.104	0.182	0.313	−0.455

Source: Adapted from Nalewajski, R. F., Błażewicz, D., and Mrozek, J., *J. Math. Chem.*, 2008, 44: 325–364.

TABLE 30.4

Electronic–Nuclear Coupling Constants for H_2O_2 and ClF_3

	$(\mu)_{R_1,N}$	$(\mu)_{R_2,N}$	$(\mu)_{\alpha,N}$	$(\mu)_{\delta,N}$	$(N)_{R_1,\mu}$	$(N)_{R_2,\mu}$	$(N)_{\alpha,\mu}$	$(N)_{\delta,\mu}$
H_2O_2	−0.039	−0.009	−0.014	0.024	0.098	−0.024	0.037	−0.053
ClF_3	−0.133	−0.044	−0.004	−0.148	0.301	0.109	0.010	0.423

Source: Adapted from Nalewajski, R. F., Błażewicz, D., and Mrozek, J., *J. Math. Chem.*, 2008, 44: 325–364.

softening effect reflects the LeChâtelier–Braun moderation of the chemical potential response to the primary (electronic) perturbation defining the derivative, by the indirectly induced adjustments in the system geometry. A similar "softening" influence due to the system "opening" is observed in the electronically relaxed (diagonal) force constants (Nalewajski et al., 2008).

Let us now examine these electronic–nuclear coupling effects in more detail. The moderating exchange of electrons between the molecule and its hypothetical electron reservoir determines the effects of the electronic–nuclear coupling in the open molecular systems. Let us assume the initial electronic and geometric equilibria in such an initially open system: $\mu^0 = \mu_{res.}$ and $F^0 = 0$. The LeChâtelier stability criteria of these two (decoupled) facets of the molecular structure requires that the conjugate "forces" $\Delta\mu(\Delta N)$ or $\{\Delta F_s(\Delta Q_s)\}$ created by the primary electronic $(\Delta N > 0)$ or nuclear $\{\Delta Q_s > 0\}$ displacements,

$$\Delta\mu(\Delta N) = \eta\Delta N \quad \text{and} \quad \Delta F_s(\Delta Q_s) = H_{s,s}\Delta Q_s,$$

will subsequently trigger the directly coupled, spontaneous responses of the system, $\delta N(\Delta N)$ and $\delta Q_s(\Delta Q_s)$, which act in the direction to restore the initial equilibria. Therefore the latter must diminish the forces created by the primary displacement, when the hypothetical internal and external barriers effecting the displacements are lifted:

$$\delta\mu[\delta N(\Delta N)] = \eta \, \delta N(\Delta N) = -\Delta\mu(\Delta N) = -\eta\Delta N$$

and

$$\delta F_s[\delta Q_s(\Delta Q_s)] = H_{s,s}\delta Q_s(\Delta Q_s) = -\Delta F_s(\Delta Q_s) = -H_{s,s}\Delta Q_s,$$

or

$$\delta N(\Delta N) = -\Delta N \quad \text{and} \quad \delta Q_s(\Delta Q_s) = -\Delta Q_s.$$

This is assured by the positive character of the electronic hardness and the diagonal nuclear force constants, $\eta > 0$ and $H_{s,s} > 0$, since then $\Delta\mu(\Delta N) > 0$ implies $\delta N(\Delta N) < 0$, while $\Delta F_s(\Delta Q_s) > 0$ gives rise to $\delta Q_s(\Delta Q_s) < 0$.

However, due to the electron–nuclear coupling in molecules a given displacement in one aspect of the molecular structure creates forces in the complementary aspect:

$$\Delta\mu(\Delta Q_s) = \varphi_s\Delta Q_s \quad \text{and} \quad \Delta F_s(\Delta N) = \varphi_s\Delta N.$$

They trigger the indirectly coupled, spontaneous relaxations $\delta N(\Delta Q_s)$ and $\delta Q_s(\Delta N)$, which also act toward diminishing the directly coupled forces $\Delta\mu(\Delta N) > 0$ and $\Delta F_s(\Delta Q_s) > 0$, in accordance with the LeChâtelier–Braun principle (Callen, 1962):

$$\delta\mu[\delta N(\delta Q_s)] = \varphi_s\delta Q_s < 0 \quad \text{and} \quad \delta F_s[\delta Q_s(\delta N)] = \varphi_s\delta N < 0.$$

Hence, these indirectly induced electronic and/or nuclear relaxations must exhibit the opposite signs with respect to the corresponding NFF indices.

In Table 30.2, we have compared the geometric softnesses **s** [of $(\mu, \mathbf{Q})$-representation] and **S** [of $(\mu, \mathbf{F})$-representation], as well as the alternative Fukui function indices: NFF (φ) [of $(N, \mathbf{Q})$-representation] and GFF (f) [of $(N, \mathbf{F})$-representation]. They measure the electronic–nuclear interaction in the externally open or closed molecules. It follows from the table that the signs of the given NFF index and the corresponding softness component are the same. Indeed, the former represents the scaled version of the latter, with the relevant global hardness (positive) providing the scaling factor, so that these two sets of coupling quantities carry the same physical description of molecular responses. As explicitly indicated in Table 30.2, the reported quantities represent the relevant compliant constants. The **s** and φ vectors collect the force compliants in the open and closed molecular systems, respectively, while **S** and f data constitute the related coordinate compliants.

The selected MEC data for the two triatomic molecules are listed in Table 30.3. In the decoupled treatment, the interaction constant $(R')_R = (R')_{R,N}$ reflects the equilibrium linear response in R' per unit displacement in R, $(\alpha)_R = (\alpha)_{R,N}$ measures a similar response in the bond angle created by such "normalized" bond elongation, while $(R)_\alpha = (R)_{\alpha,N}$ stands for the linear bond length readjustment per unit (1 radian) change in the bond angle. It follows from these purely geometric entries that, e.g., in the ground state of the water molecule an increase in one bond length generates a small elongation of the other bond and a decrease in the bond angle. The latter coupling effect is also reflected by the negative character of the $(R)_\alpha$ index, which implies a bond shortening following the primary increase in the bond angle. The opposite sign of the coupling constants between two bond lengths is detected in NO_2.

Consider next the effects of the electronic opening on these equilibrium responses of geometric parameters to such geometric displacements. The corresponding electronically relaxed compliants, of the open molecule, are listed in the data columns 4 to 6 of Table 30.3. It follows from this comparison of the HF results for water molecule that $(s')_{s,\mu} > (s')_{s,N}$, for $Q_s \neq Q_{s'} = \{R, R', \alpha\}$. Therefore, in this approximation an elongation of one bond in response to lengthening of the other bond becomes more emphasized in the open molecule. Indeed, a reference to Table 30.2 indicates that $\varphi_R < 0$ and $\varphi_\alpha < 0$ imply an inflow of electrons $\delta N(\Delta Q_s > 0) > 0$ from the reservoir, for $Q_s = (R, \alpha)$, which starts populating the antibonding MO, thus giving an extra weakening of the other bond R' and hence its larger elongation.

The final four columns in Table 30.3 measure the effect of the specified geometrical displacement on the electronic state-parameters. The $\{(N)_{s,\mu}\}$ indices show that in H_2O a hypothetical bond elongation or increase in the bond angle both create an outflow of electrons from the system to the reservoir, in accordance with the signs of the previously reported NFF indices. The $(\mu)_{s,N}$ indices reflect a direct effect of a hypothetical shift in the coordinate Q_s of the closed system on its chemical potential, when the remaining geometrical degrees-of-freedom are fully relaxed. As shown in Table 30.3 both these indices, for $Q_s = (R, \alpha)$, are positive in water molecule. In other words, both longer bonds and larger angle in this system imply an increase in the system chemical potential. In NO_2 the opposite bond-elongation effect is predicted. One also detects changes in the signs of $(N)_{R,\mu}$ and $(R')_{R,\mu}$ indices of the open NO_2, compared to H_2O. It thus follows from these interaction constants that elongating one bond in NO_2 results in an inflow of electrons to this molecule and shortening the other bond.

Selected electronic compliants for H_2O_2 and ClF_3, due to geometric perturbations, are reported in Table 30.4. In the closed molecules, increases in bond lengths and angles are predicted to lower the system electronic chemical potential, while the opposite effects due to the dihedral angle is detected. In the externally open (electronically relaxed) ClF_3 the same perturbations generate an electron inflow to the molecular system from the reservoir, while in H_2O_2 increases in the $R_2 = R(O—H)$ and the dihedral angle trigger an electron outflow from the molecule.

30.5　USE OF COMPLIANT CONSTANTS AS REACTIVITY INDICES

The four geometric representations introduced in Section 30.2 describe alternative scenarios encountered in the theory of chemical reactivity. For example, the closed reactants in the opening stage of a reaction in the gas phase, geometry-rigid or geometry-relaxed, can be indexed by the derivative properties defined in the (N, Q) and (N, F) approaches, while the properties of the chemisorbed (externally open) reactants of the heterogeneous catalysis can be characterized using descriptors generated within the (μ, Q) and (μ, F) theoretical frameworks.

Both the EP and EF perspectives are respectively covered by the canonical (N, Q) representation and its inverse, the generalized softness representation (μ, F). Therefore, the speculative considerations about either the electronic and nuclear primary causes of chemical reactions can be enhanced and indeed quantified using the derivative quantities generated in the present development. This information can be applied indirectly, by using the respective sets of differential charge sensitivities, or directly, in terms of the relevant geometric MEC. For example, the fully relaxed MEC components defined in the totally inverted (μ, F) representation provide information about the equilibrium responses of the effective oxidation state and geometry of the chemisorbed reactants per unit displacement $\Delta\mu$ in the electronic chemical potential of the catalyst or the probing molecular deformations. This should facilitate an ultimate identification of the crucial electronic/geometrical requirements for the desired reaction pathway, thus aiding a search for the most effective catalyst of the surface reaction.

Clearly, the molecular compliants can be used directly as the one-reactant reactivity criteria, which allow one to diagnose the preferred sites of an attack by the approaching agents when this molecule becomes a part of the reactive system. However, combinations of these molecular descriptors can also be applied in the *decoupled* inter-reactant approach, e.g., in the DA complexes B---A, when the acidic or basic characters of the two subsystems are known beforehand. Indeed, the $\Delta N_A > 0$ and $\Delta N_B < 0$ displacements of the two reactants are predetermined by their electronegativity differences, and so are the associated responses in the chemical potentials to these primary perturbations: $\Delta\mu_A > 0$ and $\Delta\mu_B < 0$. These displacements can be subsequently applied to predict the geometrical changes of the two mutually open reactants, at the CT stage of the reaction, using the relevant $(s)_N$ or $(s)_\mu$ compliants, which fully account for the relaxation of the remaining, unconstrained molecular degrees-of-freedom.

The two-reactant *coupled* approach (Nalewajski and Korchowiec, 1997; Nalewajski et al., 1996, 2008; Nalewajski, 1993, 1995, 1997, 2002a, 2003, 2006a,b) can also be envisaged, but the relevant compliant and MEC data would require extra calculations on the reactive system A---B as a whole, with the internal coordinates Q now including those specifying the internal geometries of two subsystems and their mutual orientation in the reactive system. The two-reactant Hessian would then combine the respective blocks of the molecular tensors introduced in Section 30.2. The supersystem relations between perturbations and responses in the canonical geometric representation then read:

$$(\Delta\mu_A, \Delta\mu_B, -\Delta F) \equiv (\Delta\mu, -\Delta F) = (\Delta N_A, \Delta N_B, \Delta Q)\mathbf{H}(A\text{---}B) \equiv (\Delta N, \Delta Q)\mathbf{H}(A\text{---}B),$$

$$(30.54)$$

where the principal electronic–nuclear Hessian of the whole reactive system includes the hardnesses of the separate reactants as diagonal blocks and the geometric Hessian of the whole system. The off-diagonal hardnesses $\eta_{A,B} = \eta_{B,A}$ measure the (rigid geometry) response in the chemical potential of one reactant per unit shift in the number of electrons on the other reactant, while the rectangular NFF matrix

$$\varphi = \begin{bmatrix} \varphi_A \\ \varphi_B \end{bmatrix}$$

determines the coupling between the electronic and geometric degrees-of-freedom:

$$
\mathbf{H}(A\text{---}B) =
\begin{bmatrix}
\left(\frac{\partial \mu_A}{\partial N_A}\right)_Q & \left(\frac{\partial \mu_B}{\partial N_A}\right)_Q & -\left(\frac{\partial F}{\partial N_A}\right)_Q \\[4pt]
\left(\frac{\partial \mu_A}{\partial N_B}\right)_Q & \left(\frac{\partial \mu_B}{\partial N_B}\right)_Q & -\left(\frac{\partial F}{\partial N_B}\right)_Q \\[4pt]
\left(\frac{\partial \mu_A}{\partial Q}\right)_N & \left(\frac{\partial \mu_B}{\partial Q}\right)_N & -\left(\frac{\partial F}{\partial Q}\right)_N
\end{bmatrix}
=
\begin{bmatrix}
\eta_A & \eta_{A,B} & -\varphi_A \\
\eta_{B,A} & \eta_B & -\varphi_B \\
-\varphi_A^T & -\varphi_B^T & \mathbf{H}
\end{bmatrix}
$$

$$
=
\begin{bmatrix}
\eta & -\varphi \\
-\varphi^T & \mathbf{H}
\end{bmatrix}.
\tag{30.55}
$$

In a practical implementation of this combined treatment of the electronic and nuclear state-variables one could use as much of the intrareactant data generated in calculations on single reactants as possible.

The fully inverted compliance matrix, $\mathbf{S}(A\text{---}B) = \mathbf{H}(A\text{---}B)^{-1}$, which determines the inverse transformation

$$
(-\Delta N_A, -\Delta N_B, \Delta Q) \equiv (-\Delta N, \Delta Q) = (\Delta \mu_A, \Delta \mu_B, \Delta F)\mathbf{S}(A\text{---}B)
$$
$$
\equiv (\Delta \mu, \Delta F)\mathbf{S}(A\text{---}B),
\tag{30.56}
$$

exhibits the following block structure:

$$
\mathbf{S}(A\text{---}B) =
\begin{bmatrix}
-\left(\frac{\partial N_A}{\partial \mu_A}\right)_F & -\left(\frac{\partial N_B}{\partial \mu_A}\right)_F & \left(\frac{\partial Q}{\partial \mu_A}\right)_F \\[4pt]
-\left(\frac{\partial N_A}{\partial \mu_B}\right)_F & -\left(\frac{\partial N_B}{\partial \mu_B}\right)_F & \left(\frac{\partial Q}{\partial \mu_B}\right)_F \\[4pt]
-\left(\frac{\partial N_A}{\partial F}\right)_\mu & -\left(\frac{\partial N_B}{\partial F}\right)_\mu & \left(\frac{\partial Q}{\partial F}\right)_\mu
\end{bmatrix}
=
\begin{bmatrix}
-S_A^{rel} & -S_{A,B}^{rel} & S_A \\
-S_{B,A}^{rel} & -S_B^{rel} & S_B \\
S_A^T & S_B^T & G^{rel}
\end{bmatrix}
$$

$$
=
\begin{bmatrix}
-\mathbf{S}^{rel} & \mathbf{S} \\
\mathbf{S}^T & \mathbf{G}^{rel}
\end{bmatrix}.
\tag{30.57}
$$

Here, the relaxed softness matrix $\mathbf{S}^{rel}$ groups the equilibrium, fully relaxed responses in the subsystem numbers of electrons, following the displacements in the chemical potentials of their (separate) electron reservoirs, the relaxed geometric softness matrix

$$
\mathbf{S} =
\begin{bmatrix}
S_A \\
S_B
\end{bmatrix}
$$

group the related adjustments in the geometry of the reactive system, while the (relaxed) geometric compliant matrix $\mathbf{G}^{rel}$ collects the responses in the internal geometric coordinates to displacements in forces, of the externally open reactants coupled to their (separate) electron reservoirs.

Obviously, the partly inverted Legendre-transformed representations for reactive systems would similarly generate descriptors of the partially relaxed (electronically or geometrically) reactive systems.

30.6 CONCLUSION

All chemical or conformational changes involve both nuclear displacements and the concomitant electron redistributions. At a given stage of a general "displacement" of a molecular or reactive system, depending on what is considered as "perturbation" and what as the equilibrium "response" to it, the EF or the EP approaches can be adopted. In this chapter, we have presented a theoretical framework which covers both these perspectives of the reactivity theory. We have also reported illustrative numerical values of alternative derivative quantities describing molecular responses to both the electronic and nuclear perturbations, within the geometric Legendre-transformed representations defining the EP and EF perspectives on the molecular structure, in which the geometric coordinates Q replace the external potential $v(r; Q)$ in the list of the system state-parameters. A brief survey of the derivative descriptors of the externally closed and externally open molecular systems has been given and the basic relations between displacements of the representation state-parameters (perturbations) and responses in the conjugate (unconstrained) variables have been summarized for both the rigid and relaxed system geometries. Specific quantities reflecting the interaction between the geometrical and electronic structures of molecular systems and the MEC components have been identified and their physical content has been commented upon.

The relaxed (compliance) quantities of both the electronic and nuclear origin measure the generalized "softnesses" of molecules, which complement the corresponding hardness data. Indeed, the electronic softness (electronically relaxed quantity defined for the rigid geometry Q) and the purely nuclear compliants (geometrically relaxed, defined for the closed system at constant N) are examples of such complementary quantities to the more familiar electronic hardness and the nuclear force constant descriptors, respectively. This decoupled treatment neglects the mutual interaction between the electronic (N) and nuclear (Q) degrees-of-freedom or their energy conjugates, the electronic chemical potential μ, attributed to an external electron reservoir, and the forces F acting on the system nuclei, respectively.

The coupling between the electronic and geometrical structures of molecular systems is embodied in the potential energy surface of the adiabatic approximation. In the present development, both the molecular compliants, reflecting the electronic and/or nuclear adjustments, have been determined in the coupled treatment of the generalized linear responses of molecular systems, which admits the simultaneous electronic and nuclear relaxation of a molecule. In the principal (N, Q) representation this interaction is measured by the NFF. Together with the electronic hardness and geometric Hessian, it defines the generalized matrix of the system electronic–nuclear "force" constants. By its partial or total inversions, all relevant compliance data have been determined. Such a coupled description of the complementary aspects of the molecular structure provides the complete treatment of the (adiabatic) linear responses in molecules, which addresses alternative scenarios encountered in the theory of chemical reactivity. For example, the MEC reflecting the electronic–nuclear interaction provide a semiquantitative measure of responses in quantities describing one aspect of the molecular structure per unit displacement in quantities describing the other aspect.

The HF results generated for representative polyatomic molecules have used the N-derivatives estimated by finite differences, while the Q-derivatives have been calculated analytically, by standard methods of quantum chemistry. We have examined the effects of the electronic and nuclear relaxations on specific charge sensitivities used in the theory of chemical reactivity, e.g., the hardness, softness, and Fukui function descriptors. New concepts of the GFFs and related softnesses, which include the effects of molecular electronic and/or nuclear relaxations, have also been introduced.

The electronic–nuclear interaction has been examined by comparing the corresponding rigid and relaxed hardness or softness and FF data. These compliants reflect the influence of the nuclear relaxation on the system electronic hardnesses and softnesses, and the effect of the electronic relaxation on the nuclear force constants and vibration compliants. Of particular importance are the MEC components, which provide the ground-state "matching" relations between the hypothetical perturbations of molecular systems and their conjugated equilibrium responses. This should allow one to diagnose the effects of the electronic and nuclear perturbation, which is the most efficient in facilitating the chemical reaction or conformational change of interest. Such applications of this coupled electronic–nuclear treatment of reactants will be the subject of future investigations.

Finally, a possible use of these coupling constants as reactivity indices has been commented upon in both the one- and two-reactant approaches. In the interreactant decoupled applications the molecular compliants, obtained from calculations on separate reactants, can be used directly to qualitatively predict the intrareactant effects resulting from the interreactant CT. The building blocks of the combined electronic–nuclear Hessian for the two-reactant system have been discussed. The corresponding blocks of the generalized compliance matrix have also been identified. In such a complete, two-reactant treatment of reactants in the combined system, the additional calculations on the reactive system as a whole would be required.

REFERENCES

Ayers, P. W. and R. G. Parr. 2000. Variational principles for site selectivity in chemical reactivity: The Fukui function and chemical hardness revisited. *J. Am. Chem. Soc.* 122: 2010–2018.

Ayers, P. W. and R. G. Parr. 2001. Variational principles for describing chemical reactions: Reactivity indices based on the external potential. *J. Am. Chem. Soc.* 123: 2007–2017.

Baekelandt, B. G., Janssens, G. O. A., Toufar, H., Mortier W. J., Schoonheydt, R. A., and R. F. Nalewajski. 1995. Mapping between electron population and vibrational modes within the charge sensitivity analysis. *J. Phys. Chem.* 99: 9784–9794.

Callen, H. B. 1962. *Thermodynamics: An Introduction to the Physical Theories of Equilibrium Thermostatics and Irreversible Thermodynamics.* New York: Wiley.

Chattaraj, P. K. and R. G. Parr. 1993. Density functional theory of chemical hardness. *Struct. Bond.* 80: 11–25.

Cohen, M. H. 1996. Strengthening the foundations of chemical reactivity theory. *Top. Curr. Chem.* 183: 143–184.

Cohen, M. H., Ganguglia-Pirovano, M. V., and J. Kurdnovsky. 1994. Electronic and nuclear chemical reactivity. *J. Chem. Phys.* 101: 8988–8997.

Cohen, M. H., Ganguglia-Pirovano, M. V., and J. Kurdnovsky. 1995. Reactivity kernels, the normal modes of chemical reactivity, and the hardness and softness spectra. *J. Chem. Phys.* 103: 3543–3551.

Decius, J. C. 1963. Compliance matrix and molecular vibrations. *J. Chem. Phys.* 38: 241–248.

Dreizler R. M. and E. K. U. Gross. 1990. *Density Functional Theory: An Approach to the Quantum Many-Body Problem.* Berlin: Springer-Verlag.

Frisch, M. J., Trucks, G. W., Schlegel, H. B., Scuseria, G. E., Robb, M. A., Cheeseman, J. R., Montgomery, Jr., J. A., Vreven, T., Kudin, K. N., Burant, J. C., Millam, J. M., Iyengar, S. S., Tomasi, J., Barone, V., Mennucci, B., Cossi, M., Scalmani, G., Rega, N., Petersson, G. A., Nakatsuji, H., Hada, M., Ehara, M., Toyota, K., Fukuda, R., Hasegawa, J., Ishida, M., Nakajima, T., Honda, Y., Kitao, O., Nakai, H., Klene, M., Li, X., Knox, J. E., Hratchian, H. P., Cross, J. B., Bakken, V., Adamo, C., Jaramillo, J., Gomperts, R., Stratmann, R. E., Yazyev, O., Austin, A. J., Cammi, R., Pomelli, C., Ochterski, J. W., Ayala, P. Y., Morokuma, K., Voth, G. A., Salvador, P., Dannenberg, J. J., Zakrzewski, V. G., Dapprich, S., Daniels, A. D., Strain, M. C., Farkas, O., Malick, D. K., Rabuck, A. D., Raghavachari, K., Foresman, J. B., Ortiz, J. V., Cui, Q., Baboul, A. G., Clifford, S., Cioslowski, J., Stefanov, B. B., Liu, G., Liashenko, A., Piskorz, P., Komaromi, I., Martin, R. L., Fox, D. J., Keith, T., Al-Laham, M. A., Peng, C. Y., Nanayakkara, A., Challacombe, M., Gill, P. M. W., Johnson, B., Chen, W., Wong, M. W., Gonzalez, C., and J. A. Pople. 2004. *Gaussian 03, Revision D.01,* Wallingford CT: Gaussian, Inc.

Fujimoto, H. and K. Fukui. 1974. Intermolecular interactions and chemical reactivity. In *Chemical Reactivity and Reaction Paths,* (Ed.) G. Klopman, pp. 23–54. New York: Wiley-Interscience.

Fukui, K. 1975. *Theory of Orientation and Stereoselection.* Berlin: Springer-Verlag.

Fukui, K. 1987. Role of frontier orbitals in chemical reactions. *Science* 218: 747–754.

Geerlings, P., De Proft, F., and W. Langenaeker. 2003. Conceptual density functional theory. *Chem. Rev.* 103: 1793–1873.

Gutmann, V. 1978. *The Donor–Acceptor Approach to Molecular Interactions.* New York: Plenum.

Hohenberg, P. and W. Kohn. 1964. Inhomogeneous electron gas. *Phys. Rev.* 136B: 864–871.

Jones, L. H. and R. R. Ryan. 1970. Interaction coordinates and compliance constants. *J. Chem. Phys.* 52: 2003–2004.

Klopman, G. 1968. Chemical reactivity and the concept of charge and frontier-controlled reactions. *J. Am. Chem. Soc.* 90: 223–234

Klopman, G. 1974. The generalized perturbational theory of chemical reactivity and its applications. In *Chemical Reactivity and Reaction Paths,* (Ed.) G. Klopman, pp. 55–165. New York: Wiley-Interscience.

Kohn, W. and L. J. Sham. 1965. Self-consistent equations including exchange and correlation effects. *Phys. Rev.* 140A: 1133–1138.

Korchowiec, J. and T. Uchimaru. 1998. The charge transfer Fukui function: Extension of the finite-difference approach to reactive systems. *J. Phys. Chem. A* 102: 10167–10172.

Michalak, A., De Proft, F., Geerlings, P., and R. F. Nalewajski. 1999. Fukui functions from the relaxed Kohn–Sham orbitals. *J. Phys. Chem. A* 103: 762–771.

Nakatsuji, H. 1973. Electrostatic force theory for a molecule and interacting molecules. I. Concept and illustrative applications. *J. Am. Chem. Soc.* 95: 345–254.

Nakatsuji, H. 1974a. Common natures of the electron cloud of the systems undergoing change in nuclear configuration. *J. Am. Chem. Soc.* 96, 24–30.

Nakatsuji, H. 1974b. Electron-cloud following and preceding and the shapes of molecules. *J. Am. Chem. Soc.* 96: 30–37.

Nalewajski, R. F. 1984. Electrostatic effects in interactions between hard (soft) acids and bases. *J. Am. Chem. Soc.* 106: 944–945.

Nalewajski, R. F. 1993. The hardness based molecular charge sensitivities and their use in the theory of chemical reactivity. *Struct. Bond.* 80: 115–186.

Nalewajski, R. F. 1995. Charge sensitivity analysis as diagnostic tool for predicting trends in chemical reactivity. In *Proceedings of the NATO ASI on Density Functional Theory*, (Eds.) E. K. U. Gross, and R. M. Dreizler, pp. 339–389. New York: Plenum Press.

Nalewajski, R. F. (Ed.) 1996. *Density Functional Theory I–IV; Topics in Current Chemistry*, pp. 180–183. Berlin: Springer-Verlag.

Nalewajski, R. F.1997. Consistent two-reactant approach to chemisorption complexes in charge sensitivity analysis. In *Developments in the Theory of Chemical Reactivity and Heterogeneous Catalysis*. (Eds.) W. M. Mortier, and R. A. Schoonheydt, pp. 135–196. Trivandrum: Research Signpost.

Nalewajski, R. F. 1999. A coupling between the equilibrium state variables of open molecular and reactive systems. *Phys. Chem. Chem. Phys.* 1: 1037–1049.

Nalewajski, R. F. 2000. Coupling relations between molecular electronic and geometrical degrees of freedom in density functional theory and charge sensitivity analysis. *Computers Chem.* 24: 243–257.

Nalewajski, R. F. 2002a. Charge sensitivities of molecules and their fragments. In *Reviews of Modern Quantum Chemistry: A Celebration of the Contributions of Robert G. Parr*, Vol. II., (Ed.) K. D. Sen, 1071–1105. Singapore: World Scientific.

Nalewajski, R. F. 2002b. Studies of the nonadditive kinetic energy functional and the coupling between electronic and geometrical structures. In *Recent Advances in Density Functional Methods*, Part III., (Eds.) V. Barone, A. Bencini, and P. Fantucci, pp. 257–277. Singapore: World Scientific.

Nalewajski, R. F. 2003. Electronic structure and chemical reactivity: Density functional and information theoretic perspectives. *Adv. Quantum Chem.* 43: 119–184.

Nalewajski, R. F. 2006a. *Information Theory of Molecular Systems*. Amsterdam: Elsevier.

Nalewajski, R. F. 2006b. Probing the interplay between electronic and geometric degrees-of-freedom in molecules and reactive systems. *Adv. Quant. Chem.* 51: 235–305.

Nalewajski, R. F. and Korchowiec, J. 1997. *Charge Sensitivity Approach to Electronic Structure and Chemical Reactivity*. Singapore: World-Scientific.

Nalewajski, R. F. and A. Michalak. 1995. Use of charge sensitivity analysis in diagnosing chemisorption clusters: Minimum-energy coordinate and Fukui function study of model toluene-[V_2O_5] Systems. *Int. J. Quantum Chem.* 56: 603–613.

Nalewajski, R. F. and A. Michalak. 1996. Charge sensitivity and bond-order analysis of reactivity trends in allyl-[MoO_3] Systems: Two-reactant approach. *J. Phys. Chem.* 100: 20076–20088.

Nalewajski, R. F. and A. Michalak. 1998. Charge sensitivity/bond-order analysis of reactivity trends in allyl-[MoO_3] chemisorption systems: A comparison between (010)- and (100)-surfaces. *J. Phys. Chem. A* 102: 636–640.

Nalewajski, R. F. and O. Sikora. 2000. Electron-following mapping transformations from the electronegativity equalization principle. *J. Phys. Chem. A* 104: 5638–5646.

Nalewajski, R. F., Korchowiec J., and A. Michalak. 1996. Reactivity criteria in charge sensitivity analysis. *Top. Curr. Chem.* 183: 25–141.

Nalewajski, R. F., Błażewicz, D., and J. Mrozek. 2008. Compliance approach to coupling between electronic and geometric structures of open and closed molecular systems. *J. Math. Chem.*, 44: 325–364.

Parr, R. G. and W. Yang. 1984. Density functional approach to the frontier-electron theory of chemical reactivity. *J. Am. Chem. Soc.* 106: 4049–4050.

Parr, R. G. and W. Yang. 1989. *Density-Functional Theory of Atoms and Molecules*. New York: Oxford Univ. Press.

Sanderson, R. T. 1951. An interpretation of bond lengths and classification of bonds. *Science* 114: 670–672; 1976. *Chemical Bonds and Bond Energy*, 2nd edn. New York: Academic Press.

Swanson, B. I. 1976. Minimum energy coordinates. A relationship between molecular vibrations and reaction coordinates. *J. Am. Chem. Soc.* 98: 3067–3071.

Swanson, B. I. and S. K. Satija. 1977. Molecular vibrations and reaction pathways. Minimum energy coordinates and compliance constants for some tetrahedral and octahedral complexes. *J. Am. Chem. Soc.* 99: 987–991.

Yang, W. and R. G. Parr. 1985. Hardness, softness and the Fukui function in the electronic theory of metals and catalysis. *Proc. Natl. Acad. Sci. USA* 82: 6723–6726.

Yang, W., Parr, R. G., and R. Pucci. 1984. Electron density, Kohn–Sham frontier orbitals, and Fukui functions. *J. Chem. Phys.* 81: 2862–2863.

31 Predicting Chemical Reactivity and Bioactivity of Molecules from Structure

Subhash C. Basak, Denise Mills, Ramanathan Natarajan, and Brian D. Gute

CONTENTS

31.1 INTRODUCTION

Understanding the pattern of reactivity between chemicals and their biological targets is important not only from the viewpoint of fundamental chemistry and biochemistry, but also from the practical, day-to-day activities of regulatory agencies and for pharmaceutical drug design. Living processes, both in sickness and health, are to a large extent guided by the interaction of various small molecular ligands with the biocatalysts and receptors within the cell that determine the manifold aspects of physiological and pathological states. The chemicals not formed in our body, e.g., drugs and xenobiotics, affect their target organs by virtue of being recognized by critical macromolecules.

The basic paradigm underlying the field of research broadly referred to as quantitative structure–activity relationship (QSAR) modeling is that the structure of the chemical determines its activity:

$$P = f(S) \qquad\qquad (31.1)$$

where P represents a physicochemical, biochemical, or toxicological property and S symbolizes the salient aspects of molecular structure related to property/activity/toxicity.

There are many subfields of QSAR. One of the prominent areas of QSAR, often called Hansch analysis [1], is derived from physical organic chemistry following the leading work of Hammett [2]. Hammett derived a way of attaching "electronic" factors to various molecular substituents. Following a similar reasoning, Taft developed the "steric" index, which attempts to associate the extent to which a substituent around a reaction center hinders the reactivity of attaching groups [3]. Subsequently, in 1964, Hansch and Fujita came up with a multiparameter approach to QSAR in which they suggested the simultaneous use of electronic, steric, and hydrophobic parameters derived from physical organic model systems [4]. The basic assumption behind the so-called Hansch approach is as follows: The most important aspects of interactions between the ligand and the biotarget can be quantified in terms of steric, electronic, and hydrophobic factors associated with the ligand. The solvatochromic approach of Taft and coworkers [5] emphasized the use of such factors as molecular volume, dipolarity, hydrogen bond donor acidity, and hydrogen bond acceptor basicity as the fundamental molecular factors underlying bioactivity and toxicity of chemicals. It is worth mentioning that in many QSARs, Hansch et al. also used calculated electronic descriptors, e.g., energy of the highest occupied molecular orbital (E_{HOMO}) and energy of the lowest unoccupied molecular orbital (E_{LUMO}), instead of Hammett sigma-type descriptors [6].

QSARs based on physical organic models are actually property–property correlations, where independent variables (steric, electronic, hydrophobic, and hydrogen bonding descriptors) derived from one set of experimental data are used to predict another group of more complex properties such as pharmacological action or toxicity. Such correlations hardly give us any structural basis of the property under investigation. Another problem is that these so-called substituent constants are derived from physical organic model systems which are applicable to specific situations where the chemical reacts with the biotargets using the same or very similar mechanisms. It is noteworthy that, in the majority of practical cases, we are interested to develop predictive QSARs for large and structurally diverse sets of chemical substances. Some of these substances could be very unstable or even not yet synthesized. In such cases, experimentally based physical organic models will be of limited utility, if any. A QSAR approach based directly on molecular structure would be desirable in such cases.

31.2   THEORETICAL MOLECULAR DESCRIPTORS FOR QSAR

From the viewpoint of biology, the ligands can be broadly divided into two classes: (a) nonspecific and (b) specific. Chemicals which are called nonspecific, e.g., general anesthetics or narcotic environmental pollutants, interact with cellular protoplasmic structures reversibly [1,7]. By their mere presence in the cellular milieu, such substances cause a low-grade action for which a specific structural feature is difficult to find. Such action of chemicals has been found to be broadly correlated with their hydrophobicity, e.g., octanol:water partition coefficient. Specific chemical substances such as drugs or xenobiotics possess specific structural features, which perturb biological systems in specific targets and bring about strong action. The QSAR literature shows that the prediction of such specific effects requires not only hydrophobic, but also steric and electronic indices.

Three classes of calculated molecular descriptors, viz., topological and substructural descriptors, geometrical (3-D) indices, and quantum chemical (QC) indices, have been extensively used in QSAR studies pertaining to drug discovery and environmental toxicology [8–12].

31.2.1   Graph Theoretical or Topological Indices

Topological indices (TIs) are derived from the representation and characterization of molecules by properly weighted molecular graphs and the various invariants derived from them. The list of TIs used in QSARs reported in this chapter and their definitions are given in Table 31.1. They include topostructural (TS) and topochemical (TC) descriptors. Specifically, the descriptors used in our recent studies include connectivity [13,14], Triplet [15], neighborhood complexity [16], hydrogen bonding, and electrotopological indices [17], which are calculated with the software programs POLLY v. 2.3 [18], Triplet [15], and Molconn-Z v. 3.5 [19]. For more information on these descriptors, please refer to our previous publications [8,20,21].

31.2.2   Three-Dimensional and Quantum Chemical Descriptors

Three-dimensional (3-D) descriptors of molecules quantify their shape, size, and other structural characteristics which arise out of the 3-D disposition and orientation of atoms and functional groups of molecules in space. A special class of 3-D indices is quantitative descriptors of chirality. If a molecule has one or more chiral centers, the spatial disposition of atoms can produce enantiomers, many of which will have the same magnitude of calculated and experimental physicochemical properties having, at the same time, distinct bioactivity profiles. Basak and coworkers [22] have developed quantitative chirality indices to discriminate such isomers according to their structural invariants which are based on the Cahn–Ingold–Prelog (CIP) rules.

A review of literature would show that a suite of QC descriptors have also been used in QSARs for biological and toxicological correlations. Such indices have been derived both from semiempirical and ab initio (Hartree Fock and density functional theory) methods. In particular, in our QSAR studies, we have used the following levels of QC indices: local and global electrophilicity indices [11],

TABLE 31.1
Symbols, Definitions and Classification of Calculated Molecular Descriptors

TS

I_D^W	Information index for the magnitudes of distances between all possible pairs of vertices of a graph
$\bar{I}_D^W$	Mean information index for the magnitude of distance
W	Wiener index = half-sum of the off-diagonal elements of the distance matrix of a graph
I^D	Degree complexity
H^V	Graph vertex complexity
H^D	Graph distance complexity
$\overline{IC}$	Information content of the distance matrix partitioned by frequency of occurrences of distance h
M_1	A Zagreb group parameter = sum of square of degrees over all vertices
M_2	A Zagreb group parameter = sum of cross product of degrees over all neighboring (connected) vertices
$^h\chi$	Path connectivity index of order $h = 0$–10
$^h\chi_C$	Cluster connectivity index of order $h = 3$–6
$^h\chi_{PC}$	Path–cluster connectivity index of order $h = 4$–6
$^h\chi_{Ch}$	Chain connectivity index of order $h = 3$–10
P_h	Number of paths of length $h = 0$–10
J	Balaban's J index based on topological distance
nrings	Number of rings in a graph
ncirc	Number of circuits in a graph
DN^2S_y	Triplet index from distance matrix, square of graph order (# of nonhydrogen atoms), and distance sum; operation $y = 1$–5
DN^21_y	Triplet index from distance matrix, square of graph order, and number 1; operation $y = 1$–5
$AS1_y$	Triplet index from adjacency matrix, distance sum, and number 1; operation $y = 1$–5
$DS1_y$	Triplet index from distance matrix, distance sum, and number 1; operation $y = 1$–5
ASN_y	Triplet index from adjacency matrix, distance sum, and graph order; operation $y = 1$–5
DSN_y	Triplet index from distance matrix, distance sum, and graph order; operation $y = 1$–5
DN^2N_y	Triplet index from distance matrix, square of graph order, and graph order; operation $y = 1$–5
ANS_y	Triplet index from adjacency matrix, graph order, and distance sum; operation $y = 1$–5
$AN1_y$	Triplet index from adjacency matrix, graph order, and number 1; operation $y = 1$–5
ANN_y	Triplet index from adjacency matrix, graph order, and graph order again; operation $y = 1$–5
ASV_y	Triplet index from adjacency matrix, distance sum, and vertex degree; operation $y = 1$–5
DSV_y	Triplet index from distance matrix, distance sum, and vertex degree; operation $y = 1$–5
ANV_y	Triplet index from adjacency matrix, graph order, and vertex degree; operation $y = 1$–5

TC

O	Order of neighborhood when IC_r reaches its maximum value for the hydrogen-filled graph
O_{orb}	Order of neighborhood when IC_r reaches its maximum value for the hydrogen-suppressed graph
I_{orb}	Information content or complexity of the hydrogen-suppressed graph at its maximum neighborhood of vertices
IC_r	Mean information content or complexity of a graph based on the rth ($r = 0$–6) order neighborhood of vertices in a hydrogen-filled graph
SIC_r	Structural information content for rth ($r = 0$–6) order neighborhood of vertices in a hydrogen-filled graph

TABLE 31.1 (continued)
Symbols, Definitions and Classification of Calculated Molecular Descriptors

CIC_r	Complementary information content for rth ($r = 0$–6) order neighborhood of vertices in a hydrogen-filled graph
$^h\chi^b$	Bond path connectivity index of order $h = 0$–6
$^h\chi_C^b$	Bond cluster connectivity index of order $h = 3$–6
$^h\chi_{Ch}^b$	Bond chain connectivity index of order $h = 3$–6
$^h\chi_{PC}^b$	Bond path–cluster connectivity index of order $h = 4$–6
$^h\chi^v$	Valence path connectivity index of order $h = 0$–10
$^h\chi_C^v$	Valence cluster connectivity index of order $h = 3$–6
$^h\chi_{Ch}^v$	Valence chain connectivity index of order $h = 3$–10
$^h\chi_{PC}^v$	Valence path–cluster connectivity index of order $h = 4$–6
J^B	Balaban's J index based on bond types
J^X	Balaban's J index based on relative electronegativities
J^Y	Balaban's J index based on relative covalent radii
AZV_y	Triplet index from adjacency matrix, atomic number, and vertex degree; operation $y = 1$–5
AZS_y	Triplet index from adjacency matrix, atomic number, and distance sum; operation $y = 1$–5
ASZ_y	Triplet index from adjacency matrix, distance sum, and atomic number; operation $y = 1$–5
AZN_y	Triplet index from adjacency matrix, atomic number, and graph order; operation $y = 1$–5
ANZ_y	Triplet index from adjacency matrix, graph order, and atomic number; operation $y = 1$–5
DSZ_y	Triplet index from distance matrix, distance sum, and atomic number; operation $y = 1$–5
DN^2Z_y	Triplet index from distance matrix, square of graph order, and atomic number; operation $y = 1$–5
nvx	Number of nonhydrogen atoms in a molecule
nelem	Number of elements in a molecule
fw	Molecular weight
si	Shannon information index
totop	Total TI t
sumI	Sum of the intrinsic state values I
sumdelI	Sum of delta-I values
tets2	Total topological state index based on electrotopological state indices
phia	Flexibility index (kp1*kp2/nvx)
IdCbar	Bonchev–Trinajstić information index
IdC	Bonchev–Trinajstić information index
Wp	Wienerp
Pf	Plattf
Wt	Total Wiener number
knotp	Difference of chi-cluster-3 and path/cluster-4
knotpv	Valence difference of chi-cluster-3 and path/cluster-4
nclass	Number of classes of topologically (symmetry) equivalent graph vertices
numHBd	Number of hydrogen bond donors
numwHBd	Number of weak hydrogen bond donors
numHBa	Number of hydrogen bond acceptors
SHCsats	E-state of C sp^3 bonded to other saturated C atoms
SHCsatu	E-state of C sp^3 bonded to unsaturated C atoms
SHvin	E-state of C atoms in the vinyl group, $=$CH-
SHtvin	E-state of C atoms in the terminal vinyl group, $=$CH$_2$

(continued)

TABLE 31.1 (continued)
Symbols, Definitions and Classification of Calculated Molecular Descriptors

SHavin	E-state of C atoms in the vinyl group, $=$CH—, bonded to an aromatic C
SHarom	E-state of C sp^2 which are part of an aromatic system
SHHBd	Hydrogen bond donor index, sum of hydrogen E-state values for —OH, $=$NH, —NH2, —NH—, -SH, and #CH
SHwHBd	Weak hydrogen bond donor index, sum of C—H hydrogen E-state values for hydrogen atoms on a C to which a F and/or Cl are also bonded
SHHBa	Hydrogen bond acceptor index, sum of the E-state values for –OH, $=$NH, —NH2, —NH–, $>$N–, —O–, –S–, along with –F and –Cl
Qv	General polarity descriptor
NHBint$_y$	Count of potential internal hydrogen bonders ($y = 2$–10)
SHBint$_y$	E-state descriptors of potential internal hydrogen bond strength ($y = 2$–10)
	Electrotopological State index values for atoms types:
	SHsOH, SHdNH, SHsSH, SHsNH2, SHssNH, SHtCH, SHother, SHCHnX, Hmax Gmax, Hmin, Gmin, Hmaxpos, Hminneg, SsLi, SssBe, Sssss, Bem, SssBH, SsssB, SsssssBm, SsCH3, SdCH2, SssCH2, StCH, SdsCH, SaaCH, SsssCH, SddC, StsC, SdssC, SaasC, SaaaC, SssssC, SsNH3p, SsNH2, SssNH2p, SdNH, SssNH, SaaNH, StN, SsssNHp, SdsN, SaaN, SsssN, SddsN, SaasN, SssssNp, SsOH, SdO, SssO, SaaO, SsF, SsSiH3, SssSiH2, SsssSiH, SssssSi, SsPH2, SssPH, SsssP, SdsssP, SssssP, SsSH, SdS, SssS, SaaS, SdssS, SddssS, SssssssS, SsCl, SsGeH3, SssGeH2, SsssGeH, SssssGe, SsAsH2, SssAsH, SsssAs, SdsssAs, SssssssAs, SsSeH, SdSe, SssSe, SaaSe, SdssSe, SddssSe, SsBr, SsSnH3, SssSnH2, SsssSnH, SssssSn, SsI, SsPbH3, SssPbH2, SsssPbH, SssssPb

Geometrical/Shape (3-D)[a]

kp0	Kappa zero
kp1-kp3	Kappa simple indices
ka1-ka3	Kappa alpha indices

QC

E_{HOMO}	Energy of the highest occupied molecular orbital
E_{HOMO-1}	Energy of the second highest occupied molecular orbital
E_{LUMO}	Energy of the lowest unoccupied molecular orbital
E_{LUMO+1}	Energy of the second lowest unoccupied molecular orbital
ΔH_f	Heat of formation
μ	Dipole moment

Source: Reprinted from Basak, S.C., Mills, D., *SAR QSAR Environ. Res.*, 12, 481, 2001. With permission.

[a] Since the time that these studies were performed, we have reclassified the kappa shape indices as TS and TC descriptors because, although they do represent molecular shape, they are calculated topologically.

energy of the highest occupied molecular orbital (E_{HOMO}) and energy of the lowest unoccupied molecular orbital (E_{LUMO}) calculated at semiempirical AM1 and various ab initio basis sets: STO-3G; 6-31G (d); 6-311G; 6-311G*; aug-cc-pVTZ; MP2; CCSD; and CCSD (T) levels.

31.3   HIERARCHICAL QSAR

Theoretical molecular descriptors may be classified into four hierarchical subsets based on the level of complexity and demand for computational resources:

$$TS < TC < 3\text{-}D < QC$$

At the low end of the hierarchy are the TS descriptors. This is the simplest of the four classes molecular structure is viewed only in terms of atom connectivity, not as a chemical entity, and thus no chemical information is encoded. Examples include path length descriptors [13], path or cluster connectivity indices [13,14], and number of circuits. The TC descriptors are more complex in that they encode chemical information, such as atom and bond type, in addition to encoding information about how the atoms are connected within the molecule. Examples of TC descriptors include neighborhood complexity indices [23], valence path connectivity indices [13], and electrotopological state indices [17]. The TS and TC are two-dimensional descriptors which are collectively referred to as TIs (Section 31.2.1). They are straightforward in their derivation, uncomplicated by conformational assumptions, and can be calculated very quickly and inexpensively. The 3-D descriptors encode 3-D aspects of molecular structure. At the upper end of the hierarchy are the QC descriptors, which encode electronic aspects of chemical structure. As was mentioned previously, QC descriptors may be obtained using either semiempirical or ab initio calculation methods. The latter can be prohibitive in terms of the time required for calculation, especially for large molecules.

The approach in hierarchical QSAR (HiQSAR) is to include the more complex and resource intensive descriptors only if they result in significant improvement in the predictive quality of the model. We begin by building a model using only the TS descriptors, followed by the creation of additional models based on the successive inclusion of the hierarchical descriptor classes. In comparing the resulting models, the contribution of each descriptor class is elucidated. In addition, the hierarchical approach enables us to determine whether or not the higher level descriptors are necessary to predict the property or activity under consideration. In situations where these complex descriptors are not useful, we can avoid spending the time required for their calculation.

In most cases, we have found that the TIs which can be calculated quickly and inexpensively, are sufficient for the prediction of various chemical properties, toxicities, and activities; the inclusion of 3-D and QC descriptors does not result in significant improvement in model quality [24–30].

31.4   STATISTICAL METHODS

Most of our recent QSAR modeling has been accomplished using three comparative regression methodologies, namely, ridge regression (RR) [31,32], principal component regression (PCR), [33] and partial least squares (PLS) [34]. Each of these methodologies makes use of all available descriptors, as opposed to subset regression, and is useful when the number of descriptors is large with respect to the number of chemicals available for modeling and when the descriptors are highly intercorrelated. Formal comparisons have consistently shown that using a small subset of available descriptors is less effective than using alternative regression methods that retain all available descriptors, such as RR, PCR, and PLS [35,36].

RR is similar to PCR in that the independent variables are transformed to their principal components (PCs). However, while PCR utilizes only a subset of the PCs, RR retains them all but downweighs them based on their eigenvalues. With PLS, a subset of the PCs is also used, but the PCs are selected by considering both the independent and dependent variables. Statistical theory suggests that RR is the best of the three methods, and this has been generally borne out in multiple comparative studies [30,36–38]. Thus, some of our published studies report RR results only.

The predictive quality of the models is judged according to the cross-validated R^2, known as q^2, obtained using the leave-one-out (LOO) approach, which is calculated as follows:

$$q^2 = 1-(\text{PRESS}/\text{SS}_{\text{Total}}) \tag{31.2}$$

where
  PRESS is the prediction sum of squares
  SSTotal is the total sum of squares

Unlike R^2 which tends to increase upon the addition of any descriptor, q^2 tends to decrease upon the addition of irrelevant descriptors and is a reliable measure of model predictability [39]. Also unlike R^2, q^2 may be negative, which is indicative of an extremely poor model.

RR, PCR, and PLS are appropriate methodologies when the number of descriptors exceeds the number of observations, and they are designed to utilize all available descriptors in order to produce an unbiased model whose predictive ability is accurately reflected by q^2, regardless of the number of independent variables in the model.

It is important to note that theoretic argument and empiric study have shown that the LOO cross-validation approach is preferred to the use of an "external test set" for small to moderate sized chemical databases [39]. The problems with holding out an external test set include: (1) structural features of the held out chemicals are not included in the modeling process, resulting in a loss of information, (2) predictions are made only on a subset of the available compounds, whereas LOO predicts the activity value for all compounds, and (3) personal bias can easily be introduced in selection of the external test set. The reader is referred to Hawkins et al. [39] and Kraker et al. [40] in addition to Section 31.6 for further discussion of proper model validation techniques.

The t value associated with each model descriptor, defined as the descriptor coefficient divided by its standard error, is a useful statistical metric. Descriptors with large $|t|$ values are important in the predictive model and, as such, can be examined in order to gain some understanding of the nature of the property or activity of interest. It should be stated, however, that the converse is not necessarily true, and thus no conclusions can be drawn with respect to descriptors with small $|t|$ values.

31.5   QSARs FOR PHYSICOCHEMICAL, ENVIRONMENTAL, PHARMACOLOGICAL, AND TOXICOLOGICAL PROPERTIES

This section provides specific examples of our QSAR studies involving the prediction of physicochemical, environmental, pharmacological, and toxicological properties. We have used software programs including POLLY v. 2.3, [18] Triplet, [15] Molconn-Z, [19] and Gaussian 03W [41] for the calculation of more than 350 molecular descriptors, each of which is derived solely from chemical structure without the need for any additional experimental data. Table 31.1 provides a list of the descriptors typically used by our research group for QSAR model development, along with brief descriptions and hierarchical classification. It should be noted that some of our HiQSAR studies have not included the QC descriptor class due to the amount of time required for the calculation of those indices. Where feasible, analyses of these data sets will continue with the addition of the QC descriptors. In addition to the examples described below, we have found our HiQSAR approach to be successful in predicting properties, activities, and toxicities including:

- Mutagenicity of aromatic and heteroaromatic amines [25]
- Complement-inhibitory activity of benzamidines [42]
- Partitioning of environmental pollutants [43]
- Boiling point of structurally heterogeneous chemicals [44]
- Acute toxicity of benzene derivatives [45]
- Biological partition coefficients [29,38,46]
- Dermal penetration of polycyclic aromatic hydrocarbons [24]
- Receptor binding affinity of dibenzofurans [30]
- Allergic contact dermatitis [47]

31.5.1   Vapor Pressure

The assessment of rate and distribution of environmental pollutants in various phases including air, water, and soil is important for the risk assessment of chemicals [48]. The partitioning of chemicals among different phases is usually assessed using a critical list of physical properties including vapor pressure (VP), aqueous solubility, air:water partition coefficient, and octanol–water partition coefficient.

Pollutants with high VP tend to concentrate more in the vapor phase as compared to soil or water. Therefore, VP is a key physicochemical property essential for the assessment of chemical distribution in the environment. This property is also used in the design of various chemical engineering processes [49]. Additionally, VP can be used for the estimation of other important physicochemical properties. For example, one can calculate Henry's law constant, soil sorption coefficient, and partition coefficient from VP and aqueous solubility. We were therefore interested to model this important physicochemical property using quantitative structure–property relationships (QSPRs) based on calculated molecular descriptors [27].

The set of 469 chemicals used in this study was obtained from the Assessment Tools for the Evaluation of Risk (ASTER) database [50] and it represents a subset of the Toxic Substances Control Act (TSCA) Inventory [51] for which VP was

TABLE 31.2

Chemical Class Composition of the VP Data Set

Compound Classification	No. of Compounds	Pure	Substituted
Total data set	469		
Hydrocarbons	253		
Nonhydrocarbons	216		
Nitrocompounds	4	3	1
Amines	20	17	3
Nitriles	5	4	1
Ketones	7	7	0
Halogens	97	92	5
Anhydrides	1	1	0
Esters	18	16	2
Carboxylic acids	2	2	0
Alcohols	10	6	4
Sulfides	38	37	1
Thiols	4	4	0
Imines	2	2	0
Epoxides	1	1	0
Aromatic compounds[a]	15	10	4
Fused-ring compounds[b]	1	1	0

Source: Reproduced from Basak, S.C. and Mills, D. *J. Chem. Inf. Comput. Sci.*, 41, 692, 2001. With permission.

[a] The 15 aromatic compounds are a mixture of 11 aromatic hydrocarbons and four aromatic halides.

[b] The only fused-ring compound was a polycyclic aromatic hydrocarbon.

measured at 25°C with a pressure range of approximately 3–10,000 mmHg. The molecular weights of the compounds in this data set range from 40 to 338, and the chemical diversity is described in Table 31.2. It should be noted that the six QC descriptors included in the study, namely, E_{HOMO}, E_{HOMO-1}, E_{LUMO}, E_{LUMO+1}, ΔH_f, and μ, were calculated for the AM1 semiempirical Hamiltonian using MOPAC v. 6.0 [52] in the Sybl interface [53].

Results in Table 31.3 indicate that the combination of TS and TC descriptors resulted in a highly predictive RR model ($q^2 = 0.895$); the addition of three-dimensional and QC indices to the set of independent variables did not result in significant improvement in model quality. It may be noted that we have observed such results for various other physicochemical and biological properties including mutagenicity [25,54], boiling point [55], blood:air partition coefficient [37], tissue:air partition coefficient [46], etc. [24,30,45,56]. Only in limited cases, e.g., halocarbon toxicity [12], the addition of QC indices after TS and TC parameters resulted in significant improvement in QSAR model quality.

It is interesting to note that of the three linear regression methods used, viz., RR, PCS, and PLS, RR outperformed the other two methods significantly. This is in line with our earlier observations with HiQSARs using the three methods [30,37,38,46].

TABLE 31.3

RR, PCR and PLS Regression VP Model Metrics

	RR		PCR		PLS	
Model Type	q^2	PRESS	q^2	PRESS	q^2	PRESS
TS	0.444	135	0.451	133	0.445	134
TS + TC	0.895	25.3	0.479	126	0.480	126
TS + TC + 3D	0.902	23.7	0.481	125	0.468	129
TS + TC + 3D + AM1	0.906	22.8	0.488	124	0.465	129
TS	0.444	135	0.451	133	0.445	134
TC	0.851	35.9	0.473	127	0.524	115
3D	0.552	108	0.453	132	0.556	107
AM1	0.201	193	0.189	196	0.203	193

Source: From Basak, S.C. and Mills, D., *ARKIVOC*, 2005(x), 308, 2005. With permission.

It is instructive to look at the top 10 molecular descriptors, based on t value, in the RR VP model derived from TS + TC indices (Table 31.4). They can be looked upon as representing the following features: (a) size (totop, DN^2Z_1), (b) hydrogen bonding (HB_1, SHHBa), (c) polarity (Qv), (d) heterogeneity of atom types (IC_0), and (e) presence of various types of heteroatoms and functional groups (SssO, SsF, $SsNH_2$, SaaO).

TABLE 31.4

Important Topological Descriptors for the Prediction of VP, Based on t Value, from the TS + TC RR Model

Descriptor Label	Description	$|t|$
SssO	Sum of the E-States for $-O-$	10.07
SsF	Sum of the E-States for $-F$	8.58
HB_1	General hydrogen bonding descriptor	7.76
$SsNH_2$	Sum of the E-states for $-NH_2$	6.83
IC_0	Mean information content or complexity of a hydrogen-filled graph based on the 0 order neighborhood of vertices	6.57
SaaO	Sum of the E-states for O:	6.56
SHHBa	Hydrogen bond acceptor index	6.21
DN^2Z_1	Triplet index from distance matrix, square of graph order (number of vertices), and atomic number	6.13
Qv	General polarity descriptor	6.06
totop	Total TI	5.87

Source: From Basak, S. C. and Mills, D., *ARKIVOC*, 2005(x), 308, 2005. With permission.

In the LSER approach, a combination of molecular size, hydrogen bonding, and polarity are used to estimate partitioning behavior of chemicals [5,57]. The presence of specific heteroatoms, functional groups, and different atom types, as encoded by information theoretic, triplet, and electrotopological indices, will probably be related to dipole–dipole interactions among the molecules and also specific regional interactions such as hydrogen bonding. Such factors have been found to be useful in predicting VP by Liang and Gallagher [58], Katritzky et al. [48], Engelhardt et al. [59], and Staikova et al. [60].

31.5.2 ESTROGEN RECEPTOR BINDING AFFINITY

An endocrine disrupting chemical is an exogenous substance that causes an adverse health effect in an intact organism, or its progeny, consequent to changes in endocrine function. The adverse effect can be produced through a variety of mechanisms including direct receptor binding with or without subsequent activation, change in the number of hormone receptors in a cell, and modified production or metabolism of natural hormones. Many of these substances have been associated with developmental, reproductive, and other health problems in wildlife and laboratory animals, and there is evidence to suggest that certain chemicals are producing endocrine disrupting effects on humans [61].

The most well-studied biological mechanism for producing endocrine disruption is the estrogenic response, examined through estrogen receptor binding. As such, we modeled calf estrogen receptor binding affinity using the HiQSAR approach for a set of 35 hydroxy-substituted 2-phenylindoles (Figure 31.1) [62]. Results indicated that RR generally outperforms PLS and PCR, and a very good predictive model is obtained using the TC descriptors alone, with a q^2 of 0.920 (Table 31.5) [8]. While good results are obtained using the TS and TC descriptors, addition of the 3-D descriptors does not result in improvement in model quality, and the model obtained using 3-D descriptors alone, is quite poor.

31.5.3 CELLULAR TOXICITY OF HALOCARBONS

Halocarbons are important industrial chemicals used worldwide as solvents and synthetic intermediates. Crebelli et al. [63–65] developed data for chromosomal

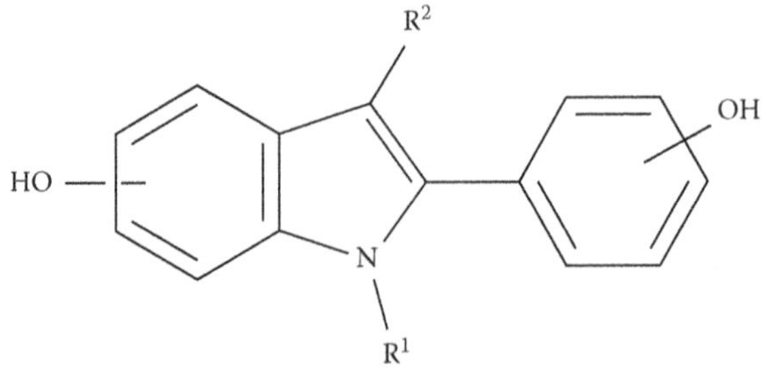

FIGURE 31.1 Chemical structure of hydroxy-substituted 2-phenylindoles.

TABLE 31.5
Summary Statistics for Predictive Estrogen Receptor Binding Affinity Models

Independent Variables	RR		PCR		PLS	
	q^2	*PRESS*	q^2	*PRESS*	q^2	*PRESS*
TS	0.880	17.9	0.363	95.1	0.822	26.6
TS + TC	0.877	18.3	−2.67	548	−2.01	450
TS + TC + 3-D	0.878	18.3	−2.69	551	−1.99	446
TS	0.880	17.9	0.363	95.1	0.822	26.6
TC	0.920	11.9	−1.32	346	−0.749	261
3-D	0.499	74.8	0.311	103	0.346	97.6

Source: Reprinted from Basak, S.C., Mills, D., and Gute, B.D., in *Advances in Quantum Chemistry,* Elsevier, in press. With permission.

malsegregation, lethality, and mitotic growth arrest in *Aspergillus nidulans* diploid strain P1. Crebelli's group conducted experimental analysis of these chemicals, as well as developed predictive QSAR models using a combination of physicochemical properties and QC indices calculated using the STO-3G basis set.

We have carried out HiQSAR development using TS, TC, and 3-D descriptors, in addition to AM1 semiempirical QC descriptors obtained using MOPAC 6.00 [52] and ab initio QC descriptors calculated with Gaussian 98W [66] using the STO-3G, 6-311G, and aug-cc-pVTZ basis sets, with results provided in Table 31.6.

The results show that, for the set of 55 halocarbons, a very high level of ab initio calculation was required before there was any significant improvement in model quality over and above the models derived from easily calculable TS and TC descriptors. When ranking the TS + TC + 3-D + cc-pVTZ model descriptors

TABLE 31.6
HiQSAR Model Results for Toxicity of the 55 Halocarbons to *A. nidulans*

Model	RR		PCR		PLS	
	q^2	*PRESS*	q^2	*PRESS*	q^2	*PRESS*
TS	0.290	90.00	0.240	96.38	0.285	90.64
TS + TC	0.770	29.13	0.426	72.84	0.644	45.13
TS + TC + 3-D	0.780	27.87	0.438	71.23	0.645	44.98
TS + TC + 3-D + AM1	0.775	28.49	0.492	64.37	0.753	21.29
TS + TC + 3-D + STO-3G	0.772	28.95	0.489	64.78	0.613	49.02
TS + TC + 3-D + 6–311G	0.777	28.26	0.510	62.14	0.631	46.75
TS + TC + 3-D + cc-pVTZ	0.838	20.59	0.507	62.49	0.821	22.67

Source: Reprinted from Basak, S.C., Mills, D., and Gute, B.D. in *Biological Concepts and Techniques in Toxicology: An Integrated Approach,* Taylor & Francis, New York, 2006, 61–82. With permission.

according to t values, it was found that indices related to reactivity make an important contribution, with vertical electron affinity (VEA) and energy of the lowest
unoccupied molecular orbital (E_{LUMO}) possessing the highest values.

VEA was obtained from the energy differences between the optimized structures
of neutral and charged species. The optimized geometry of neutral forms was used to
compute the energy of the corresponding anions to obtain the vertical EA values.

We have also formulated HiQSARs for toxicity of hepatocytes tested in vitro for
a subset of 20 of these chemicals [67].

31.6 PROPER VALIDATION OF QSARs IS CRITICAL

The literature of the past three decades has witnessed a tremendous explosion in the
use of computed descriptors in QSAR. But it is noteworthy that this has exacerbated
another problem: rank deficiency. This occurs when the number of independent
variables is larger than the number of observations. Stepwise regression and other
similar approaches, which are popularly used when there is a rank deficiency, often
result in overly optimistic and statistically incorrect predictive models. Such models
would fail in predicting the properties of future, untested cases similar to those used
to develop the model. It is essential that subset selection, if performed, be done
within the model validation step as opposed to outside of the model validation step,
thus providing an honest measure of the predictive ability of the model, i.e., the "true
q^2" [39,40,68,69]. Unfortunately, many published QSAR studies involve subset
selection *followed by* model validation, thus yielding a "naïve q^2," which inflates
the predictive ability of the model. The following steps outline the proper sequence
of events for descriptor thinning and LOO cross-validation, e.g.,

1. Leave out compound #1.
2. Use cross-validation to select the optimal descriptors without compound #1.
3. Fit the model to the selected descriptors and the remaining $n-1$ compounds.
4. Apply this model to compound #1 and obtain the prediction.
5. Repeat steps 1–4 for each of the remaining $n-1$ compounds.
6. Use predictors to calculate the q^2 value.

In order to show the inflation of q^2, which results from the use of improper
statistical methods, we have performed comparative studies involving stepwise
regression and RR [68,70]. In these studies, comparative models were developed
for the prediction of rat fat:air and human blood:air partitioning of chemicals.
For the former, proper statistical methods yielded a model with a q^2 value of
0.841, while the stepwise approach was associated with an inflated q^2 of 0.934.
Likewise, the rat fat:air model derived using proper methods had a q^2 value of 0.854,
while the stepwise approach yielded a model with an inflated q^2 of 0.955.

31.7 CHIRALITY INDICES AND MOLECULAR OVERLAY

Biological activities such as enzyme reactions and metabolic changes are highly
stereospecific, hence enantiomers and diastereomers may have entirely different

pharmacological activities and the extent to which they differ in their activities depends on the extent of interaction with the receptors. Ariens [71] pointed out that enantiomers might have positive activity, negative activity (inhibitory), or even synergistic biological activities. After critical reviews by Ariens [72,73] and the thalidomide episode, the U.S. Food and Drug Administration (FDA) changed its policy on stereoisomeric drugs and requires the pharmacological and toxicological profiling of both the enantiomers of a racemate before marketing a racemate [74]. Pharmaceutical companies have switched to single enantiomer (enantiopure) chiral drugs. This trend is indicated by the fact that in 2006, 80% of the small molecule drugs approved by FDA were chiral and 75% were single enantiomers. It is expected that 200 chiral compounds could enter the development process every year [75]. The use of enantiopure chemicals is not only increasing in the pharmaceutical industry but also in agricultural chemicals in order to avoid the unnecessary loading of the environment with the less active or inactive isomer. Hence, biological and toxicological profiling of enantiomers and diastereomers is becoming essential to market a racemic or chiral pure chemical compound as a drug or agrochemical. Moreover, successful QSAR models that include diastereomers and enantiomers are expected to play a significant role in computer assisted chiral synthesis.

QSAR modeling using simple computed molecular descriptors or physicochemical properties fails in handling compounds that exhibit polychiral diastereomerism. The reason for the limitation of the molecular descriptors derived from adjacency and distance matrices of molecular graphs in differentiating enantiomers is their identical scalar properties. In other words, enantiomers are isometric, i.e., for each distance between two given atoms in one isomer, there is a corresponding identical distance between a pair of atoms in the other. Thus, distance matrices for the enantiomers have identical entries and consequently the various TIs derived from the distance matrices cannot differentiate enantiomers. The same is true for 3-D distance matrices of enantiomers. There have been several attempts [76–82] to develop molecular descriptors for diastereomers using the graph theoretical approach. Continuous chirality measures (CCM) and continuous symmetry measure (CSM) were introduced by Zabrodsky et al. [83,84]. Many of the topology-based approaches tried to apply a correction to the commonly used TIs such as the connectivity indices. Hence, they are not stand-alone indices in the sense that they need the calculation of the corresponding TI before applying the chiral correction. Quantum chemistry–based methods have limitations with respect to large molecules and larger data sets containing thousands of molecules. Natarajan et al. [22] developed a simple approach to develop a new series of chirality indices called relative chirality indices (RCI). The new chirality indices are able to discriminate 3-D structural differences from 2-D oriented embedded graphs. A similar approach based on physicochemical properties was simultaneously reported by Zhang and Aires-de-Sousa [85].

31.7.1 CALCULATION OF RELATIVE CHIRALITY INDICES

According to the CIP rule, different degrees of priorities are assigned to the four chemical groups attached to the chiral carbon, "a" being given the highest priority, then "b," etc. Differences in the disposition of the groups a, b, c, and d around the asymmetric carbon is given below:

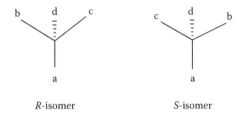

R-isomer   S-isomer

The least important chemical group (d) is placed at the rear, and the clockwise or the anticlockwise arrangement of the other three groups (a, b, c) is considered to assign the configuration as R or S.

The three groups of highest priority attached to a chiral center were viewed from a reference point to calculate the new chirality metric. The groups/atoms a, b, c and d are then assigned valence delta-values of atoms (δ^v) according to the method of Kier and Hall [13]. When the group has more than one atom, δ^v for the group a, b, or c is calculated considering the relative proximities of the atoms to the chiral center, and decreasing importance with increasing topological distance (through bond) was assigned while calculating the contribution of atoms other than hydrogen in a group. The group delta value for any group (δ_i^v) attached to a chiral carbon is calculated as:

$$\delta_i^v = \delta_{n_1}^v + \left(\delta_{n_2}^v/2\right) + \left(\delta_{n_3}^v/4\right) + \left(\delta_{n_4}^v/8\right) + \cdots \qquad (31.3)$$

where
   n_1 is the atom attached directly to the chiral center (nearest neighbor)
   n_2 is separated by one atom
   n_3 by two atoms, etc.

Relative chirality indices ($^V RCI$) for a pair of enantiomers are calculated as

$$^V RCI_R = \delta_a^v + \left(\delta_a^v + \delta_a^v\delta_b^v\right) + \left(\delta_a^v + \delta_a^v\delta_b^v + \delta_a^v\delta_b^v\delta_c^v\right) + \delta_a^v\delta_b^v\delta_c^v\delta_d^v \qquad (31.4)$$

$$^V RCI_S = \delta_a^v + \left(\delta_a^v + \delta_a^v\delta_c^v\right) + \left(\delta_a^v + \delta_a^v\delta_c^v + \delta_a^v\delta_b^v\delta_c^v\right) + \delta_a^v\delta_b^v\delta_c^v\delta_d^v \qquad (31.5)$$

To obtain $^V RCI$ for molecules containing more than one chiral center, the root mean square of $^V RCI$ for all the chiral centers is taken.

$$^V RCI = \sqrt{\frac{1}{N}\sum_{i=1}^{N}(^V RCI_i)^2} \qquad (31.6)$$

Information regarding the fourth group is encoded in the new index by the fourth term $\delta_a\delta_b\delta_c\delta_d$. When "d" is hydrogen, $\delta_a\delta_b\delta_c\delta_d$ becomes zero; otherwise it contributes to the RCI for the chiral center.

In addition to the valence connectivity, the formula weights of the groups and the electrotopological state of the various atoms and groups in a, b, c, and d were also used to calculate $^W RCI$ and $^I RCI$, respectively. In order to calculate $^I RCI$, one has to replace δ^v of the groups a, b, c, and d by their δ^I (electrotopological state) in the formula to calculate $^V RCI$. In the same way, $^W RCI$ for enantiomers and diastereomers can be calculated using the formula weights of a, b, c, and d. It is to be noted that simple connectivity and bond order connectivity of vertices (atoms) were not useful in generating RCI due to their inability to differentiate atom types, and this consequently gave degenerate values for several structures.

RCI were calculated for the chiral α-amino acids and were found to discriminate diastereomers and enantiomers very well. Singularity of *meso* compounds is also addressed and was illustrated using tartaric acid. All three scales of RCI have the same set of values for the *meso* isomers. Cysteine is the sulfur analog of serine and had the same set of $^V RCI$ values (*R*-isomer 194.25 and *S*-isomer 164.25). This shows that bioisosterism is addressed by the valence connectivity scale used for calculating $^V RCI$ values. Hence, we have gone from the qualitative two-way discrimination provided solely by the CIP rule to quantitative measures of chirality capable of ordering a set of diastereomers. We have introduced three different scales for RCI and any other new scale can be introduced based on a physicochemical property, electronic parameters, etc. Thus, a series of chirality measures that encode the distribution of groups based on different properties around the chiral center can be computed for modeling activities of diastereomeric compounds. The scale that gives the best correlation will help in interpreting the "handedness" (chirality) that the receptor is looking for.

We tried to use RCI to model the insect repellency of two diastereomeric repellants namely, 1-methylisopropyl 2-(2-hydroxyethyl)piperidine-1-carboxylate (Picaridin) and cyclohex-3-enyl 2-methylpiperidin-1-yl ketone (AI3-37220). The mosquito repellency (proportion not biting) of diastereomers of AI3-37220 was modeled well by $^V RCI$ but it failed in the case of Picaridin diastereomers. One can expect a correlation between the calculated indices and the biological activity if the proper ordering of a set of chiral molecules by calculated indices parallels the ordering of them by the receptor. Otherwise, the indices will only discriminate the structures without any necessary correlation with biological function. Hence, in the case of Picaridin, we were unsuccessful in selecting the proper measure of chirality that parallels what the receptor is looking for. Recently, we found that the $^V RCI$ is able to model the dopamine D2 receptor affinity for 3-(3-hydroxyphenyl) piperidine (unpublished results).

31.7.2 Hierarchical Molecular Overlay

It is important to emphasize that the degree of biological specificity varies widely across biological systems. Certain systems show a wide range of specificity and others show a very narrow range of selectivity. Modeling biological activities that have a wider range of specificity is easier, and calculated indices or properties might yield reasonable results. In the case of highly specific systems that do not allow even a small structural change such as a methyl ($-CH_3$) or a methylene (CH_2) group, it is very difficult to model biological responses by the conventional QSAR approach. The concept "structurally similar chemicals have similar biological activity" fails, and the reductionistic approach of using computed molecular descriptors might be unsuccessful. Moreover, if the biological receptor or binding mode is not known, indices based QSAR models will not work. For situations of this sort, we suggest the molecular overlay approach because it is holistic and 3-D geometries are considered. In the molecular overlay method, optimized geometries of each diastereomer is overlaid on the most active diastereomer and the root mean square distances (RMSD) between the specified pairs of atoms are measured to assess the degree of similarity (match). As the most active analog is taken as a template to rank the similarity of the other compounds, some information about the binding site is indirectly considered. However, the molecular overlay approach has two limitations namely, it is computer intensive and it cannot identify a chemical that is more active than the one used as the template. The geometry optimization routine can be selected based on the number of compounds to be screened.

Initial study [86] on molecular overlay was done for AI3-37220 and Picaridin using MM2 optimized geometries. In order to find the level of theory necessary to get reasonable predictions, the QC methods were used [87] in a graduated manner from semiempirical AM1, to Hartree Fock (STO3G, 3-21G, 6-31G, and 6-311G) to density functional theory (B3LYP/6-31G, B3LYP/6-311G). The output of a lower level geometry was used as the input for the next higher level of theory. Thus, a hierarchical geometry optimization scheme was followed for each diastereomer from AM1 to HF/STO3G to HF/3-21G to HF/6-31G to HF/6-311G to B3LYP/6-31G to B3LYP/6-311G. The freeware visual molecular dynamics (VMD) [88] was used to overlay the structures and compute RMSD between the pairs of atoms. We used this method to order the mosquito repellency of the diastereomers of AI3-37220 and Picaridin. The results obtained by using the hierarchical approach were far better than those obtained using the MM2 method where we could only differentiate the most active and the least active diastereomers clearly, but failed to rank the repellency with the degree of match/mismatch of structures. With the hierarchical overlay method, we were able to rank the repellency of the diastereomers using the RMSD and thus provided a novel structure-to-structure comparison tool. Unlike several of the commercially available modeling tools, we were able to use geometries optimized by any level of theory. The hierarchical approach indicated that in the case of molecules such as AI3-37220, which have relatively rigid cyclic groups, optimization at the HF 3-21G basis set level seems to be sufficient to give good results. For more flexible molecules such as Picaridin, however, HF 6-31G was found be the method of choice. In the case of both AI3-37220 and Picaridin, the advanced DFT methodology did not significantly improve the predictions made with lower levels of DFT theory (B3LYP/6-31G). We would also like to mention some of the advantages of the molecular overlay method over other 3-D QSAR approaches such as comparative molecular field analysis (CoMFA) [89]. CoMFA requires a prior knowledge of binding site, and a binding hypothesis proposed is based on the ligand–receptor complex. In the molecular overlay method, no prior knowledge of binding site is required and moreover molecular geometry optimized at any level of theory can be used.

31.8   CONCLUSION

In this chapter, we have shown how calculated molecular descriptors, viz., topological, geometrical, and QC indices, can be used in the development of predictive models for the estimation of physicochemical, environmental, biochemical, and toxicological properties of different congeneric as well as diverse groups of molecules. In the majority of cases, a combination of TS and topochemical descriptors gave the best models. But in the case of halocarbon toxicity, the addition of QC descriptors (aug cc-pVTZ level) made some significant improvement in model quality. Chirality descriptors have been developed and used in models involving stereoisomeric compounds. It is noteworthy that no matter what the number, class or quality of descriptors are, the use of proper statistical methods is critical for the assurance of predictive quality. We have particularly exemplified this point with blood:air and fat:air partition coefficient data [68,70] The three basic needs of successful QSARs are (a) good quality data, (b) a set of descriptors which capture the essence of the property, bioactivity, or toxicity to be modeled, and (c) proper statistical procedures which can develop robust models.

ACKNOWLEDGMENTS

This is contribution number 474 from the Center for Water and the Environment of the Natural Resources Research Institute. Research reported in this paper was supported in part by Grant F49620-02-1-0138 from the United States Air Force. The U.S. Government is authorized to reproduce and distribute reprints for Governmental purposes notwithstanding any copyright notation thereon. The views and conclusions contained herein are those of the authors and should not be interpreted as necessarily representing the official policies or endorsements, either expressed or implied, of the Air Force Office of Scientific Research or the U.S. Government.

REFERENCES

1. Hansch, C., Leo, A., Hoekman, D. *Exploring QSAR: Hydrophobic, Electronic, and Steric Constants,* American Chemical Society, Washington, D.C., 1995.
2. Hammett, L. P. *Physical Organic Chemistry,* McGraw-Hill, New York, 1940.
3. Taft, R. W. Jr. In *Steric Effects in Organic Chemistry,* Newman, M. S., Ed., Wiley, New York, 1956.
4. Hansch, C., Fujita, T. *p-α-π* Analysis. A method for the correlation of biological activity and chemical structure. *J. Am. Chem. Soc.* 1964, 86(8), 1616–1626.
5. Kamlet, M. J., Abboud, J. -L. M., Abraham, M. H., Taft, R. W. Linear solvation energy relationships. 23. A comprehensive collection of the solvatochromic parameters, π*, α and β, and some methods for simplifying the general solvatochromic equation. *J. Org. Chem.* 1983, 48, 2877–2887.
6. Debnath, A. K., Debnath, G., Shusterman, A. J., Hansch, C. A QSAR investigation of the role of hydrophobicity in regulating mutagenicity in the Ames test: 1. Mutagenicity of aromatic and heteroaromatic amines in *Salmonella typhimurium* TA98 and TA100. *Environ. Mol. Mutagen.* 1992, 19, 37–52.
7. Ferguson, J. The use of chemical potentials as indices of toxicity. *Proc. R. Soc. Lond. Ser. B Biol. Sci.* 1939, 127, 387–404.

8. Basak, S. C., Mills, D., Gute, B. D. Predicting bioactivity and toxicity of chemicals from mathematical descriptors: A chemical-cum-biochemical approach. In *Advances in Quantum Chemistry*, Klein, D. J., Brandas, E., Eds., Elsevier, Amsterdam, 2004, in press.

9. Dyguda, E., Grembecka, J., Sokalski, W. A., Leszczynski, J. Origins of the activity of PAL and LAP enzyme inhibitors: Toward ab initio binding affinity prediction. *J. Am. Chem. Soc.* 2005, *127*, 1658–1659.

10. Parthasarathi, R., Padmanabhan, J., Elango, M., Subramanian, V., Roy, D. R., Sarkar, U., Chattaraj, P. K. Application of quantum chemical descriptors in computational medicinal chemistry for chemoinformatics. *Indian J. Chem.* 2006, *45A*, 111.

11. Sarkar, U., Padmanabhan, J., Parthasarathi, R., Subramanian, V., Chattaraj, P. K. Toxicity analysis of polychlorinated dibenzofurans through global and local electrophilicities. *J. Mol. Struct. (Theochem)* 2006, *758*, 119–125.

12. Basak, S. C., Balasubramanian, K., Gute, B. D., Mills, D., Gorczynska, A., Roszak, S. Prediction of cellular toxicity of halocarbons from computed chemodescriptors: A hierarchical QSAR approach. *J. Chem. Inf. Comput. Sci.* 2003, *43*, 1103–1109.

13. Kier, L. B., Hall, L. H. *Molecular Connectivity in Structure–Activity Analysis*, Research Studies Press, Letchworth, Hertfordshire, UK, 1986.

14. Randic, M. On characterization of molecular branching. *J. Am. Chem. Soc.* 1975, *97*, 6609–6615.

15. Filip, P. A., Balaban, T. S., Balaban, A. T. A new approach for devising local graph invariants: Derived topological indices with low degeneracy and good correlational ability. *J. Math. Chem.* 1987, *1*, 61–83.

16. Basak, S. C., Roy, A. B., Ghosh, J. J. Study of the structure–function relationship of pharmacological and toxicological agents using information theory. Avula, X. J. R., Bellman, R., Luke, Y. L., Rigler, A. K., Eds., *Proceedings of the Second International Conference on Mathematical Modelling*, University of Missouri-Rolla: Rolla, Missouri, 1979, pp. 851–856.

17. Kier, L. B., Hall, L. H. *Molecular Structure Description: The Electrotopological State*, Academic Press, San Diego, CA, 1999.

18. POLLY v. 2.3, Copyright of the University of Minnesota, 1988.

19. Molconn-Z Version 3.5, Hall Associates Consulting, Quincy, MA, 2000.

20. Basak, S. C., Mills, D. Mathematical chemistry and chemobioinformatics: A holistic view involving optimism, intractability, and pragmatism. In Graovac, A., Gutman, I., Eds., Proceedings of the 22nd International Course and Conference on the Interfaces among Mathematics, Chemistry and Computer Sciences; Math/Chem/Comp 2007, 2008.

21. Basak, S. C., Mills, D., Mumtaz, M. M. Use of graph invariants in the protection of human and ecological health. In Basak, S. C., Balakrishnan, R., Eds., Lecture Notes of the First Indo-US Lecture Series on Discrete Mathematical Chemistry, 2007.

22. Natarajan, R., Basak, S. C., Neumann, T. A novel approach for the numerical characterization of molecular chirality. *J. Chem. Inf. Model.* 2007, *47*, 771–775.

23. Roy, A. B., Basak, S. C., Harriss, D. K., Magnuson, V. R. Neighborhood complexities and symmetry of chemical graphs and their biological applications. In *Mathematical Modelling Science and Technology*, Avula, X. J. R., Kalman, R. E., Liapis, A. I., Rodin, E. Y., Eds., Pergamon Press, New York, 1983, pp. 745–750.

24. Gute, B. D., Grunwald, G. D., Basak, S. C. Prediction of the dermal penetration of polycyclic aromatic hydrocarbons (PAHs): A hierarchical QSAR approach. *SAR QSAR Environ. Res.* 1999, *10*, 1–15.

25. Basak, S. C., Mills, D. Prediction of mutagenicity utilizing a hierarchical QSAR approach. *SAR QSAR Environ. Res.* 2001, *12*, 481–496.

26. Basak, S. C., Mills, D., Gute, B. D., Hawkins, D. M. Predicting mutagenicity of congeneric and diverse sets of chemicals using computed molecular descriptors:

A hierarchical approach. In *Quantitative Structure–Activity Relationship (QSAR) Models of Mutagens and Carcinogens*, Benigni, R., Ed., CRC Press, Boca Raton, FL, 2003, pp. 207–234.

27. Basak, S. C., Mills, D. Development of quantitative structure–activity relationship models for vapor pressure estimation using computed molecular descriptors. *ARKIVOC* 2005, *2005*(x), 308–320.

28. Basak, S. C., Natarajan, R., Mills, D. Structure–activity relationships for mosquito repellent aminoamides using the hierarchical QSAR method based on calculated molecular descriptors, 2005, pp. 958–963.

29. Basak, S. C., Mills, D., Gute, B. D. Prediction of tissue:air partition coefficients—theoretical vs experimental methods. *SAR QSAR Environ. Res.* 2006, *17*(5), 515–532.

30. Basak, S. C., Mills, D., Mumtaz, M. M., Balasubramanian, K. Use of topological indices in predicting aryl hydrocarbon receptor binding potency of dibenzofurans: A hierarchical QSAR approach. *Indian J. Chem.* 2003, *42A*, 1385–1391.

31. Hoerl, A. E., Kennard, R. W. Ridge regression: Biased estimation for nonorthogonal problems. *Technometrics* 1970, *12*, 55–67.

32. Hoerl, A. E., Kennard, R. W. Ridge regression: Applications to nonorthogonal problems. *Technometrics* 2005, *12*, 69–82.

33. Massy, W. F. Principal components regression in exploratory statistical research. *J. Am. Stat. Assoc.* 1965, *60*, 234–246.

34. Wold, S. Discussion: PLS in chemical practice. *Technometrics* 1993, *35*, 136–139.

35. Rencher, A. C., Pun, F. C. Inflation of R^2 in best subset regression. *Technometrics* 1980, *22*(1), 49–53.

36. Frank, I. E., Friedman, J. H. A statistical view of some chemometrics regression tools. *Technometrics* 1993, *35*(2), 109–135.

37. Basak, S. C., Mills, D., El-Masri, H. A., Mumtaz, M. M., Hawkins, D. M. Predicting blood:air partition coefficients using theoretical molecular descriptors. *Environ. Toxicol. Pharmacol.* 2004, *16*, 45–55.

38. Basak, S. C., Mills, D., Hawkins, D. M., El-Masri, H. Prediction of human blood: air partition coefficient: A comparison of structure-based and property-based methods. *Risk Anal.* 2003, *23*(6), 1173–1184.

39. Hawkins, D. M., Basak, S. C., Mills, D. Assessing model fit by cross-validation. *J. Chem. Inf. Comput. Sci.* 2003, *43*, 579–586.

40. Kraker, J. J., Hawkins, D. M., Basak, S. C., Natarajan, R., Mills, D. Quantitative structure–activity relationship (QSAR) modeling of juvenile hormone activity: Comparison of validation procedures. *Chemometr. Intell. Lab. Syst.* 2007, *87*, 33–42.

41. Gausssian 03W, Version 6.0 (Revision C.02), Gaussian, Inc.: Wallingford CT, 2004.

42. Basak, S. C., Gute, B. D., Lucic, B., Nikolic, S., Trinajstic, N. A comparative QSAR study of benzamidines complement-inhibitory activity and benzene derivatives acute toxicity. *Comput. Chem.* 2000, *24*, 181–191.

43. Basak, S. C., Mills, D. Prediction of partitioning properties for environmental pollutants using mathematical structural descriptors. *ARKIVOC* 2005, *2005*(ii), 60–76.

44. Basak, S. C., Mills, D. Use of mathematical structural invariants in the development of QSPR models. *MATCH (Commun. Math. Comput. Chem.)* 2001, *44*, 15–30.

45. Gute, B. D., Basak, S. C. Predicting acute toxicity of benzene derivatives using theoretical molecular descriptors: A hierarchical QSAR approach. *SAR QSAR Environ. Res.* 1997, *7*, 117–131.

46. Basak, S. C., Mills, D., Hawkins, D. M., El-Masri, H. A. Prediction of tissue:air partition coefficients: A comparison of structure-based and property-based methods. *SAR QSAR Environ. Res.* 2002, *13*, 649–665.

47. Basak, S. C., Mills, D., Hawkins, D. M. Predicting allergic contact dermatitis: A hierarchical structure–activity relationship (SAR) approach to chemical classification using topological and quantum chemical descriptors. *J. Comput. Aided Mol. Des.*, 2008, 22, 339–343.

48. Katritzky, A. R., Wang, Y., Sild, S., Tamm, T. QSPR studies on vapor pressure, aqueous solubility, and the prediction of water–air partition coefficients. *J. Chem. Inf. Comput. Sci.* 1998, *38*, 720–725.

49. Daubert, T. E., Jones, D. K. Project 821: Pure component liquid vapor pressure measurements. AIChE Symp. Ser., 86. 1990; pp. 29–39.

50. Russom, C. L., Anderson, E. B., Greenwood, B. E., Pilli, A. ASTER: An integration of the AQUIRE data base and the QSAR system for use in ecological risk assessments. *Sci. Total Environ.* 1991, *109/110*, 667–670.

51. United States Environmental Protection Agency, What is the TSCA Chemical Substance Inventory? http://www.epa.gov/opptintr/newchems/pubs/invntory.htm, 2006.

52. Stewart, J. J. P. 1990, MOPAC Version 6.00, QCPE #455, Frank J. Seiler Research Laboratory, US Air Force Academy, CO.

53. Tripos Associates, Inc., SYBYL v. 6.2, St. Louis, MO, 1995.

54. Basak, S. C., Gute, B. D., Grunwald, G. D. Relative effectiveness of topological, geometrical, and quantum chemical parameters in estimating mutagenicity of chemicals. In *Quantitative Structure–Activity Relationships in Environmental Sciences VII*, Chen, F., Schuurmann, G., Eds., SETAC Press, Pensacola, FL, 1998, pp. 245–261.

55. Basak, S. C., Gute, B. D., and Grunwald, G. D. A comparative study of topological and geometrical parameters in estimating normal boiling point and octanol–water partition coefficient. *J. Chem. Inf. Comput. Sci.* 1996, *36*, 1054–1060.

56. Basak, S. C., Gute, B. D., Grunwald, G. D. A hierarchical approach to the development of QSAR models using topological, geometrical, and quantum chemical parameters. In *Topological Indices and Related Descriptors in QSAR and QSPR*, Devillers, J., Balaban, A. T., Eds., Gordon and Breach Science Publishers, Amsterdam, 1999, pp. 675–696.

57. Kamlet, M. J., Doherty, R. M., Abraham, M. H., Marcus, Y., Taft, R. W. Linear solvation energy relationships. 46. An improved equation for correlation and prediction of octanol/water partition coefficients of organic nonelectrolytes (including strong hydrogen bond donor solutes). *J. Phys. Chem.* 1988, *92*, 5244–5255.

58. Liang, C., Gallagher, D. A. QSPR prediction of vapor pressure from solely theoretically-derived descriptors. *J. Chem. Inf. Comput. Sci.* 1998, *38*, 321–324.

59. Engelhardt, H., McClelland, H. E., Jurs, P. C. Quantitative structure–property relationships for the prediction of vapor pressures of organic compounds from molecular structures. *J. Chem. Inf. Comput. Sci.* 2000, *40*, 967–975.

60. Staikova, M., Wania, F., Donaldson, D. J. Molecular polarizability as a single-parameter predictor of vapour pressures and octanol–air partitioning coefficients of non-polar compounds: A priori approach and results. *Atmos. Environ.* 2004, *38*, 213–225.

61. United States Environmental Protection Agency Endocrine Disruptor Research Inititative. http://www.epa.gov/endocrine/ 2006.

62. von Angerer, E., Prekajac, J., Strohmeier, J. 2-Phenylindoles. Relationship between structure, estrogen receptor binding affinity, and mammary tumor inhibiting activity in the rat. *J. Med. Chem.* 1984, *27*, 1439–1447.

63. Benigni, R., Andreoli, C., Conti, L., Tafani, P., Cotta-Ramusino, M., Carere, A., Crebelli, R. Quantitative structure–activity relationship models correctly predict the toxic and aneuploidizing properties of halogenated methanes in *Aspergillus nidulans*. *Mutagenesis* 1993, *8*, 301–305.

64. Crebelli, R., Andreoli, C., Carere, A., Conti, G., Conti, L., Cotta-Ramusino, M., Benigni, R. The induction of mitotic chromosome malsegregation in *Aspergillus nidulans*. Quantitative structure activity relationship (QSAR) analysis with chlorinated aliphatic hydrocarbons. *Mutat. Res.* 1992, *266*, 117–134.

65. Crebelli, R., Andreoli, C., Carere, A., Conti, L., Crochi, B., Cotta-Ramusino, M., Benigni, R. Toxicology of halogenated aliphatic hydrocarbons: Structural and molecular determinants for the disturbance of chromosome segregation and the induction of lipid peroxidation. *Chem. Biol. Interact.* 1995, *98*, 113–129.

66. Gaussian 98W (Revision A.11.2), Gaussian, Inc.: Pittsburgh, PA, 1998.

67. Gute, B. D., Balasubramanian, K., Geiss, K., Basak, S. C. Prediction of halocarbon toxicity from structure: A hierarchical QSAR approach. *Environ. Toxicol. Pharmacol.* 2004, *16*, 121–129.

68. Basak, S. C., Mills, D., Hawkins, D. M., Kraker, J. J. Proper statistical modeling and validation in QSAR: A case study in the prediction of rat fat:air partitioning. In *Computation in Modern Science and Engineering, Proceedings of the International Conference on Computational Methods in Science and Engineering 2007 (ICCMSE 2007)*, Simos, T. E., Maroulis, G., Eds., American Institute of Physics, Melville, New York, 2007, pp. 548–551.

69. Basak, S. C., Natarajan, R., Mills, D., Hawkins, D. M., Kraker, J. J. Quantitative structure–activity relationship modeling of juvenile hormone mimetic compounds for *Culex pipiens* Larvae—with discussion of descriptor thinning methods. *J. Chem. Inf. Model.* 2006, *46*, 65–77.

70. Basak, S. C., Mills, D., Hawkins, D. M., Kraker, J. J. Quantitative structure–activity relationship (QSAR) modeling of human blood:air partitioning with proper statistical methods and validation. *Chem. Biodivers.*, accepted.

71. Ariens, E. J. Stereochemistry, a basis for sophisticated nonsense in pharmacokinetics and clinical pharmacology. *Eur. J. Clin. Pharmacol.* 1984, *26*, 663–668.

72. Ariens, E. J. Racemic therapeutics: Problems all along the line. In *Chirality in Drug Design and Synthesis*, Brown, C., Ed., Academic Press, New York, 1990, pp. 29–43.

73. Ariens, E. J. Stereospecificity of bioactive agents: General aspects of exemplified by pesticides and drugs. In *Stereoselectivity of Pesticides, Biological and Chemical Problems; Chemicals in Agriculture*, Vol. 1, Ariens, E. J., van Rensoen, J. J. S. W. W., Eds., Elsevier Science, New York, 1988.

74. Strong, M. FDA policy and regulation of stereoisomers: Paradigm shift and the future of safer, more effective drugs. *Food Drug Law J.* 1999, *54*, 463–487, and references therein.

75. Tahyer, A. M. C & EN August 6, 2007, www.cen-online.org.

76. Schultz, H. P., Schultz, E. B., Schultz, T. P. Topological organic chemistry. 9. Graph theory and molecular topological indices of stereoisomeric organic compounds. *J. Chem. Inf. Comput. Sci.* 1995, *35*, 864–870.

77. Golbraikh, A., Bonchev, D., Tropsha, A. Novel chirality descriptors derived from molecular topology. *J. Chem. Inf. Comput. Sci.* 2001, *41*, 147–158.

78. Randic, M. Graph theoretical descriptors of two-dimensional chirality with possible extension to three-dimensional chirality. *J. Chem. Inf. Comput. Sci.* 2001, *41*, 639–649.

79. Aires-de-Sousa, J., Gasteiger, J. New description of molecular chirality and its applications to the prediction of the preferred enantiomer in stereoselective reactions. *J. Chem. Inf. Comput. Sci.* 2001, *41*, 369–375.

80. Aires-de-Sousa, J., Gasteiger, J., Gutman, I., Vidovic, D. Chirality code and molecular structure. *J. Chem. Inf. Comput. Sci.* 2004, *44*, 831–836.

81. Dervarics, M., Otvos, F., Martinek, T. A. Development of a chirality-sensitive flexibility descriptor for $3 + 3D$-QSAR. *J. Chem. Inf. Comput. Sci.* 2006, *46*, 1431–1438.

82. Yang, C., Zhong, C. Chirality factors and their application to QSAR studies of chiral molecules. *QSAR Comb. Sci.* 2005, *24*, 1047–1055.
83. Zabrodsky, H., Peleg, S., Avnir, D. Continuous symmetry measures. *J. Am. Chem. Soc* 1992, *114*, 7843–7851.
84. Zabrodsky, H., Avnir, D. Continuous Symmetry Measures. 4. Chirality. *J. Am. Chem. Soc.* 1995, *117*, 462–473.
85. Zhang, Q., Aires-de-Sousa, J. Physicochemical stereodescriptors of atomic chiral centers. *J. Chem. Inf. Comput. Sci.* 2006, *46*, 2278–2287.
86. Natarajan, R., Basak, S. C., Balaban, A. T., Klun, J. A., Schmidt, W. F. Chirality index, molecular overlay and biological activity of diastereoisomeric mosquito repellents. *Pest. Manag. Sci.* 2005, *61*, 1193–1201.
87. Basak, S. C., Natarajan, R., Nowak, W., Miszta, P., Klun, J. A. Three dimensional structure-activity relationships (3D-QSAR) for insect repellency of diastereoisomeric compounds: A hierarchical molecular overlay approach. *SAR QSAR Environ. Res.* 2007, *18*, 237–250.
88. Humphrey, W., Dalke, A., Schulten, K. VMD-visual molecular dynamics. *J. Molec. Graphics* 1996, *14*, 33–38.
89. Cramer III, R. D., Patterson, D. E., Bunce, J. D. Comparative molecular field analysis (CoMFA). 1. Effect of shape on binding of steroids to carrier proteins. *J. Am. Chem. Soc.* 1988, *110*, 5959–5967.
90. Basak, S. C. and Mills, D. Quantitative structure-property relationships (QSPRs) for the estimation of vapor pressure: A hierarchical approach using mathematical structural descriptors. *J. Chem. Inf. Comput. Sci.* 2001, *41*, 692–701.

32 Chemical Reactivity: Industrial Application

Abhijit Chatterjee

CONTENTS

The aim of this chapter is to discuss chemical reactivity and its application in the real world. Chemical reactivity is an established methodology within the realm of density functional theory (DFT). It is an activity index to propose intra- and intermolecular reactivities in materials using DFT within the domain of hard soft acid base (HSAB) principle. This chapter will address the key features of reactivity index, the definition, a short background followed by the aspects, which were developed within the reactivity domain. Finally, some examples mainly to design new materials related to key industrial issues using chemical reactivity index will be described. I wish to show that a simple theory can be state of the art to design new futuristic materials of interest to satisfy industrial needs.

32.1   INTRODUCTION

Rapid advances are taking place in the application of DFT to describe complex chemical reactions. Researchers in different fields working in the domain of quantum chemistry tend to have different perspectives and to use different computational approaches. DFT owes its popularity to recent developments in predictive powers for physical and chemical properties and its ability to accurately treat large systems. Both theoretical content and computational methodology are developing at a pace, which offers scientists working in diverse fields of quantum chemistry, cluster science, and solid state physics.

   The HSAB principle classifies the interaction between acids and bases in terms of global softness. Pearson proposed the global HSAB principle [1,2]. The HSAB principles classify the interaction between acids and bases in terms of global softness and this has been validated further [3–11]. The global hardness was defined as the second derivative of energy with respect to the number of electrons at constant temperature and external potential, which includes the nuclear field. The global softness is the inverse of global hardness. Pearson also suggested a principle of maximum hardness (PMH) [12,13], which states that for a constant external potential the system with the maximum global hardness is most stable. PMH has also been studied extensively to further probe into both inter- and intramolecular interactions [14–19]. In recent days, DFT has gained widespread use in quantum chemistry [20–23]. For example, some DFT-based local properties, e.g., Fukui functions and local softness have already been successfully used for the reliable predictions in various types of electrophilic and nucleophilic reactions. On the other hand, the reactivity index finds its application in material designing. We proposed a reactivity index scale for heteroatomic interaction with zeolite framework [24]. The scale holds well for unisite interaction or in other way with one active site preset in the molecule, but does not offer any impressive results for the system with two or more active sites. The choice of template is an important criterion for the synthesis of zeolite. Reactivity index offers an important tool to derive the suitable template for the synthesis of zeolite like ZSM-5 via the investigation of a range of reactivity index using DFT of different representative template molecules along with zeolite framework. As a result, a priori rule was formulated to choose the best template for a particular zeolite synthesis. Moreover, the role of water during nucleation process can also be monitored efficiently in terms of solvation energy to rationalize the fundamental mechanism of crystal growth [25]. A range of reactivity index along with DFT was fruitful to determine the activity of nitrogen heterocyclics present in biomacromolecules and their suitable sorbent from the dioctahedral smectite family [26]. In another approach, a novel function λ [27] was introduced for quantitative description of weak adsorption cases, which was so far qualitative inside the domain of DFT. In addition to that, several calculations were performed to derive group softness [28] for inter- and intramolecular reactivities for nitroaromatics and their adsorption over clay matrices and contributed to the development of the methodology and its application in various systems [29–31].

There are different ways to describe the reactivity index, where the idea is to find donor/acceptor capability of an atom present in a molecule interacting with another molecule or the interaction is within itself. This is the main concept, and depending on the interaction that is taking place, one can look into local softness of the atom, which is approaching the other interacting species or the group of atoms together approaching the active site. It has also been mentioned that if one wishes, one can describe the interaction between atoms for an intermolecular interaction through the concept of an equilibrium using the idea of reactivity index. Hence the concept reactivity index tells you the activity of atom center and its capability to interact with other species in its localized/nonlocalized neighbor.

The aim here is to show the application of the simple theory of reactivity index along with some useful derivation of the theory in terms of resolving key issues. Some examples are provided to have an understanding of the applicability of this to industrial issues. Finally, a newest example of the application of reactivity index will performed to show the use of single-wall nanotube (SWNT) for gas sensing, and a rule was prescribed with a graphical representation of electrophilic and nucleophilc reactivity index.

32.2 THEORY

In DFT, hardness (η) is defined as [13]

$$\eta = 1/2(\delta^2 E/\delta N^2)v(r) = 1/2(\delta\mu/dN)_v \tag{32.1}$$

where
  E is the total energy
  N is the number of electrons of the chemical species
  μ is the chemical potential

The global softness, S, is defined as the inverse of the global hardness, η:

$$S = 1/2\eta = (\delta N/\delta\mu)_v \tag{32.2}$$

Using the finite difference approximation, S can be approximated as

$$S = 1/(IE - EA) \tag{32.3}$$

where IE and EA are the first ionization energy and electron affinity of the molecule, respectively.
  The Fukui function $f(r)$ is defined by Ref. [14]

$$f(r) = [\delta\mu/dv(r)]_N = [\delta\rho(r)/\delta N]_v \tag{32.4}$$

The function "f" is thus a local quantity, which has different values at different points in the species, N is the total number of electrons, μ is the chemical potential, and v is the potential acting on an electron due to all nuclei present. Since $\rho(r)$ as a function of N has slope discontinuities and Equation 32.1 provides the following three reaction indices [14]:

$$f^-(r) = [\delta\rho(r)/\delta N]_v^- \text{ (governing electrophilic attack)}$$

$$f^+(r) = [\delta\rho(r)/\delta N]_v^+ \text{ (governing nucleophilic attack)}$$

$$f^0(r) = 1/2[f^+(r) + f^-(r)] \text{ (for radial attack)}$$

In a finite difference approximation, the condensed Fukui function [14] of an atom, say x, in a molecule with N electrons is defined as

$$f_x^+ = [q_x(N+1) - q_x(N)] \quad \text{(for nucleophilic attack)}$$

$$f_x^- = [q_x(N) - q_x(N-1)] \quad \text{(for electrophilic attack)} \tag{32.5}$$

$$f_x^0 = [q_x(N+1) - q_x(N-1)]/2 \quad \text{(for radial attack)}$$

where q_x is the electronic population of atom x in a molecule.

The local softness $s(r)$ can be defined as

$$s(r) = (\delta\rho(r)/\delta\mu)_v \tag{32.6}$$

Equation 32.3 can also be written as

$$s(r) = [\delta\rho(r)/\delta N]v[\delta N/\delta\mu]_v = f(r)S \tag{32.7}$$

Thus, local softness contains the same information as the Fukui function $f(r)$ plus additional information about the total molecular softness, which is related to the global reactivity with respect to a reaction partner, as stated in HSAB principle. Thus, the Fukui function may be thought of as a normalized local softness. Atomic softness values can easily be calculated by using Equation 32.4, namely

$$s_x^+ = [q_x(N + 1) - q_x(N)]S$$
$$s_x^- = [q_x(N) - q_x(N - 1)]S \tag{32.8}$$
$$s_x^0 = S[q_x(N + 1) - q_x(N - 1)]/2$$

We have further explained the interaction energy scheme as follows. Let us consider a process where AB is the final product formed at equilibrium after combination of isolated A and B species present. With the existing knowledge, this is recognized that A and B interacts in two steps: (1) interaction will take place through the equalization of chemical potential at constant external potential and (2) A and B approach the equilibrium state through changes in the electron density of global system generated by making changes in the external potential at constant chemical potential. Thus, within DFT we can write

$$\Delta E_{inter} = E[\rho_{AB}] - E[\rho_A] - E[\rho_B] \tag{32.9}$$

where ρ_{AB}, ρ_A, ρ_B are the electron densities of the system AB at equilibrium and of the isolated systems A and B, respectively

For the application of the reactivity index to propose intra- and intermolecular reactivities, Equation 32.9 can be used.

In terms of the potentials, we can write

$$\Delta E_{inter} = \Delta E_v + \Delta E_\mu \tag{32.10}$$

where $\Delta E_v = -1/2[(\mu_A - \mu_B)^2/(S_A + S_B)](S_A S_B)$

$$\Delta E_\mu = -1/2 N_{AB}^2 k[1/(S_A + S_B)] \tag{32.11}$$

where N_{AB} is the total number of electrons, k is the proportionality constant between S_{AB} and $S_A + S_B$, product of N^2 and K is λ:

$$\Delta E_\mu = (-1/2)\lambda/(S_A + S_B) \tag{32.12}$$

If the interaction takes place through j site of A in AB complex

$$\lambda_{Aj} = q_{Aj}^{eq} - q_{Aj}^{0} \qquad (32.13)$$

where
   q_{Aj}^{eq} is the density of jth atom of A in complex AB
   q_{Aj}^{0} is the density in isolated system

In our all studies, all calculations with molecular clusters have been carried out with DFT [32] using DMol3 code of Accelrys. A gradient corrected functional BLYP [33] and DNP basis set [34] was used throughout the calculation. Single-point calculations of the clusters in their cationic and anionic forms, at the optimized geometry of the original neutral clusters were also carried out to evaluate Fukui functions and global and local softness. The condensed Fukui function and atomic softness were evaluated using electrostatic potential (ESP) driven charges.

32.3   DISCUSSION WITH EXISTING EXAMPLES

The above mentioned theory has indicated that the application domain of the theory will be related to chemical activity. We are mainly dealing with charge density for the purpose of prescribing a reaction followed by a process and eventually the way of designing a material. Various applications of reactivity index theory and its detailed description was recently published [35] in recent past. According to the literature, two main 5 issues are dealt with chemical reactivity index: (1) the chemical reactivity theory approach and its application for resolving chemical concern of importance within the helm of DFT and (2) application of DFT on resolving structure–property relationship in catalysis, reactions, and small molecules. With that background, this is time again to show how this theory can be applicable to address issues in industry, and our main concerned industries are chemical, pharmaceutical, drug, semiconductor, and also polymers where people want to design molecule or material for a specific inter- or intramolecular interaction. Following are the few examples where chemical reactivity index can efficiently apply to solve the industrial problems.

   1. *Scaling the activity of fluorophore molecules*: Anthracenes bearing aliphatic or aromatic amino substituents, which behave as molecular sensors, have shown their potential to act as photon-induced electron transfer (PET) systems. In this PET, the fluorophore moieties are responsible for electron release during protonation and deprotonation. The principle of HSAB deals with both intra- and intermolecular electron migrations. It is possible to calculate the localized properties in terms of Fukui functions in the realm of DFT and thus calculate and establish a numerical matchmaking procedure that will generate a priori rule for choosing the fluorophore in terms of its activity. We calculated the localized properties for neutral, anionic, and cationic systems to trace the course of the efficiency. A qualitative scale is proposed in terms of the feasibility of intramolecular hydrogen bonding. To investigate the effect of the environment of the nitrogen atom on

protonation going from mono- to diprotonated systems, the partial density of states has been calculated and the activity sequence has been with reactivity indices. The results show that location of the nitrogen atom in an aromatic ring does not influence the PET, but for aliphatic chains the effect is prominent. Furthermore, the protonation/deprotonation scenario has been explained. The results show that the reactivity indices can be used as a suitable property for scaling the activity of fluorophore molecules for the PET process [28].

2. *Adsorption of ozone-depleting chlorofluorocarbons (CFC)*: Adsorption of ozone-depleting CFC over zeolite is of major global environmental concern. To investigate the nature of CFCs including fluoro, chlorofluoro, and hydrofluoro/chloro carbons (CF_4, CF_3Cl, CF_2Cl_2, $CFCl_3$, CHF_3, $CHCl_3$) adsorption first-principle to calculation were performed on faujasite models [36,37]. Experimentally, it is observed that separation of halocarbons is possible using Na–Y, though the cause is unknown. Reactivity index within the realm of HSAB principle was used to monitor the activity of the interacting CFCs using DFT, to propose a qualitative order. The importance of both H-bonding and cation–F/Cl interactions in determining the low-energy sorption sites were monitored and rationalized. The host–guest interactions show a distinctive difference between the adsorption phenomenon between H–Y and Na–Y and as well for Cl and F. It is observed that Cl has more favorable interaction with hydrogen of H–Y compared to Na–Y and for F, the situation is reversed. To validate this trend, periodic optimization calculations were performed. The interaction energy as obtained matches well with the reactivity index order resulted from cluster calculations. This study is a combination of DFT and periodic calculation to rationalize the electronic phenomenon of the interaction process.

3. *Designing of stable clay nanocomposite*: Resorcinol forms a novel nanocomposite in the interlayer of montmorillonite. This resorcinol oligomer is stable inside the clay matrixes even above the boiling point of the monomer. A periodic ab initio calculation was performed with hydrated and nonhydrated montmorillonite, before and after intercalation of resorcinol [38]. For the most feasible dimer- and tetramer-shaped oligomers of resorcinol, the intramolecular and intermolecular hydrogen bonding feasibility has been tested using the DFT-BLYP approach and the DNP basis set, in the gas phase and in the presence of aqueous solvent. After locating the active site through Fukui functions within the realm of the HSAB principle, the relative nucleophilicity of the active cation sites in their hydrated state has been calculated. A novel quantitative scale in terms of the relative nucleophilicity and electrophilicity of the interacting resorcinol oligomers before and after solvation, is proposed. Besides that, a comparison with a hydration situation and also the evaluation of the strength of the hydrogen bridges have been performed using mainly the dimer- and cyclic tetramer-type oligomers of resorcinol. Using periodic ab initio calculations, the formation mechanisms were traced: (1) resorcinol molecules combine without any interaction with water, (2) resorcinol oligomerizes through water. Both the

mechanisms are compared and the effect of water on the process is elucidated. The results show that resorcinol molecules combine after hydration
only and hence they are stable at higher temperature. The fittings of the
oligomers were also tested as well by periodic calculation to compare the
stability of the oligomers inside the newly formed clay nanocomposite.

4. *Effect of dopants on Brönsted and Lewis acid site*: The influence of both
bivalent and trivalent metal substituents from a range of metal cation (Co,
Mn, Mg, Fe, and Cr) on the acidic property (both Brönsted and Lewis)
of metal substituted aluminum phosphate MeAlPOs is monitored [39].
The influence of the environment of the acid site is studied both by
localized cluster and periodic calculations to propose that the acidity of
AlPOs can be predictable with accuracy so that AlPO material with desired
acidity can be designed. A semiquantitative reactivity scale within the
domain of HSAB principle is proposed in terms of the metal substitutions
using DFT. It is observed that for the bivalent metal cations, Lewis
acidity linearly increases with ionic size, where as the Brönsted acidity is
solely dependent on the nearest oxygen environment. Intramolecular
and intermolecular interactions show that once the active site of the interacting species is identified, the influence of the environment can be prescribed. Mg(II)-doped AlPO-34 shows highest Brönsted acidity, whereas
Cr(III)-doped species shows lowest acidity. Fe(II)/Fe(III)-doped AlPO-34
show highest Lewis acidity, whereas Mn(III)-doped, Mg(II)-doped species
show lowest acidity.

5. *Effect of divalent cations on the swelling of clay*: We used both localized
and periodic calculations on a series of monovalent (Li^+, Na^+, K^+, Rb^+,
Cs^+) and divalent (Mg^{2+}, Ca^{2+}, Sr^{2+}, Ba^{2+}) cations to monitor their effect
on the swelling of clays [40,41]. The activity order obtained for the
exchangeable cations among all the monovalent and divalent series studied:
$Ca^{2+} > Sr^{2+} > Mg^{2+} > Rb^+ > Ba^{2+} > Na^+ > Li^+ > Cs^+ > K^+$. We have
shown that, in case of dioctahedral smectite, the hydroxyl groups play a
major role in their interaction with water and other polar molecules in the
presence of an interlayer cation. We studied both types of clays, with a
different surface structure and with/without water using a periodic calculation. Interlayer cations and charged 2:1 clay surfaces interact strongly with
polar solvents; when it is in an aqueous medium, the clay expands and the
phenomenon is known as crystalline swelling. The extent of swelling is
controlled by a balance between relatively strong swelling forces and
electrostatic forces of attraction between the negatively charged phyllosili-
cate layer and the positively charged interlayer cation. We have calculated
the solvation energy at the first hydration shell of an exchangeable cation,
but the results do not correspond directly to the experimental *d*-spacing
values. A novel quantitative scale is proposed with the numbers generated
by the relative nucleophilicity of the active cation sites in their hydrated
state through Fukui functions within the realm of the HSAB principle. The
solvation effect thus measured shows a perfect match with experiment,
which proposes that the reactivity index calculation with a first hydration

shell could rationalize the swelling mechanism for exchangeable cations. The conformers after electron donation or acceptance propose the swelling mechanism for monovalent and divalent cations.

6. *Effect of solvation on the interaction of chromophore*: Amino-functional silanol surface is mostly used for the immobilization of inorganic ions, molecules, organic, or biochemical molecules onto the mesopore surface. In analytical chemistry, the metal ion uptake was visualized through colorimetric sensors using chromophore molecules. One needs to know the structure–property correlation between the chromophore and silylating agent while choosing a chromophore, which is very important to design the sensors. We have used two chromophores representative of hydrophobic and hydrophilic type and used density functional calculation on all the interacting molecules in both the unsolvated phase and solvated medium within the domain of HSAB principle to look at the localized activity of the interacting atoms of these reacting molecules to formulate a rule to choose the best chromophore. The mechanism of interaction between chromophore and the silylating agent has also been postulated. The results were compared with experiment, and it is observed that solvation plays a detrimental role in the binding of chromophore with silylating agent. The results also show that the range of reactivity index can be used as a suitable property to scale the activity of chromophore molecules suitable for the sensing process. It is observed that the hydrophobic chromophore binds stronger with both the metal and the silylating agent whereas for the hydrophilic one, it binds only with the silylating agent when solvated, and in all cases the metal ion binding is weaker compared to that of the hydrophobic one [42].

7. *Prediction of interaction between metal clusters with oxide surface*: The HSAB principle classifies the interaction between acids and bases in terms of global softness. In the last few years, the reactivity index methodology was well established and had found its application in a wide variety of systems. This study deals with the viability of the reactivity index to monitor metal cluster interaction with oxide. Pure gold cluster of a size between 2 and 12 was chosen to interact with clean alumina (100) surface. A scale was derived in terms of intra- and intermolecular interactions of gold cluster with alumina surface to rationalize the role of reactivity index in material designing [43].

8. *Recent exemplary studies with reactivity indices for silica nanowires*: Structural and dynamic properties of the building block of silica nanowires, $(SiO_2)_6$, are investigated by quantum molecular dynamics simulations [44]. The energy component analysis shows that the lower electrostatic interaction differentiates the global minimum from the other structures. With the dominant electrostatic interaction, we further observe that the PMH can be employed to justify the molecular stability of this system. Time profiles of a few density functional reactivity indices exhibit correlations of dynamic fluctuations between HOMO and LUMO and between chemical potential and hardness. Electrophilicity, nucleaofugality, and electrofugality

indices [44] are found to change concurrently and significantly, indicating that the nanostructures sampled during the dynamic process are exceedingly reactive and rich in chemistry.

9. *A comparison of porphyrin and pincer activity rationalized through reactivity index*: Porphyrin and pincer complexes are both important categories of compounds in biological and catalytic systems. Structure, spectroscopy, and reactivity properties of porphyrin pincers are systematically studied for selection of divalent metal ions. It is reported that the porphyrin pincers are structurally and spectroscopically different from their precursors and are more reactive in electrophilic and nucleophilic reactions. These results are implicative in chemical modification of hemoproteins and understanding the chemical reactivity in heme-containing and other biologically important complexes and cofactors [45].

10. *Study on CDK2 Inhibitors Using Global Softness*: The reactivity index is well popular in pharmaceutical and drug applications. In particular, one problem of drug design is that one has to synthesize and screen thousands, sometimes, millions of candidate chemicals in developing one successful drug. There was a very successful study with reactivity index, long back, on human immunodeficiency virus (HIV) [46]. The cyclin-dependent kinases (CDKs) are a class of enzymes involved in the eukaryotic cell-cycle regulation. A recent theoretical study was on a series of CDK2 inhibitors using a set of global reactivity indices defined in terms of the density of states [47]. The related series were classified on the basis of the correlations obtained for the complete set of compounds and the sites targeted within the active site of CDK2. The comparison between the biological activity and the electronic chemical potential obtained through Fermi level yields poor results, thereby suggesting that the interaction between the hinge region of CDK2 and the ligands may have a marginal contribution from the charge transfer component. The comparison between the biological activity and global softness shows a better correlation, suggesting that polarization effects dominate over the CT contribution in the interaction between the so-called hinge region and the ligand. This result is very encouraging to show that the role of reactivity index in the intermolecular interaction, can be further extrapolated to the intermolecular region to study the occupied states.

11. *An exemplary application of reactivity in gas sensor with single-wall carbon nanotube (CWT)*: In this part, we wish to explore interatomic interaction as well as intramolecular interation through the center of activity. Since the discovery of the structure of CNTs or SWNT, much effort has been devoted to finding the uses of these structures in applications ranging from field-emission devices to other nanodevices [48,49]. Kong et al. [50] proposed for the first time the use of CNTs as gas sensors. Experimental data have shown that transport properties of SWNT change dramatically upon exposure to gas molecules at ambient temperature [51]. Main advantage of the open SWNT bundles is that they provide a larger number of adsorption sites. As a result, the adsorption capacity is significantly increased and

several new structures and phase transitions were observed [52]. A recent study of Andzelm et al. [53] indicates that the semiconducting SWNTs can serve as gas sensors for several gases like CO, NH_3, H_2, etc. However, NH_3 shows an intriguing behavior compared to the other gases. NH_3 molecule binds weakly with CNTs, yet can change the conductance significantly. This discrepancy was explained by assuming that the NH_3 binds at defects. For a semiconducting SWNT exposed to 200 ppm of NO_2, it was found that the electrical conductance can increase by three orders of magnitude in a few seconds. On the other hand, exposure to 2% NH_3 caused the conductance to decrease up to two orders of magnitude [54]. Sensors made from SWNT have high sensitivity and a fast response time at room temperature, which are important advantages for sensing applications.

We have studied the interaction of CNT with different gas molecules such as O_2, N_2, H_2, CO_2, and NO_2 to have an understanding of the adsorption behavior of the selected gases in defect-free CNT. We will as well focus on to figure out the effect of variation in the conductance with gas sorption, by applying external electric field. It is very difficult to obtain conductance by quantum mechanical calculation as it will be very much CPU intensive, but measurement of conductance is an utmost important parameter to prove the efficiency of the nanotubes as gas sensors, which is the experimental way of measuring sensors. Thus, a method was developed by calculating the change in the reactivity index before and after the application of the electric field. The reactivity index provides information about the activity of the gas molecules over SWNT, and if the activity changes then the sensing behavior will change. This is a simplistic approach, which is cost-effective to new material design for the sensor industry.

Here, we have first optimized all the molecules and half of the CNT using the same level of theory with DFT as mentioned earlier. We have then calculated the Fukui function for the individual molecules. The results for the global softness, Fukui function, and the local softness values were calculated for the nucleophilic and electrophilic behavior and are shown in Table 32.1. Figure 32.1a and b and Figure 32.2a and b represent the plot of electrophilic and nucleophilic Fukui function for the H_2 and CO_2 molecules, respectively. We have also calculated the same Fukui function and plotted the electrophilic function only for the CNT. A drastic change in the Fukui function values was observed for H_2 and CO_2. Hydrogen shows a lower softness values compared to that of CO_2. The order of activity for all the interacting molecules after adsorption over CNT is $CO_2 > NO_2 > N_2 > O_2 > H_2$. To validate this order, we have optimized half of the nanotube first and monitored the electrophilic Fukui function (Figure 32.3). This is followed by the adsorption of H_2 and CO_2 through a grand canonical Monte Carlo (GCMC) simulation methodology using the Sorption tools of Accelrys [55]. The configurations are sampled from a grand canonical ensemble. In the grand canonical ensemble, the fugacity of all components, as well as the temperature, are fixed as if the framework was in open contact with an infinite sorbate reservoir with a fixed temperature. The reservoir is completely described by the temperature plus the fugacity of all components and does not have to be simulated explicitly. The adsorption isotherm for the gas molecule has

TABLE 32.1

Global Softness, Fukui Function, and Local Softness of the Interacting Molecules

System	Global Softness	f_x^+	s_x^+	f_x^-	s_x^-
N_2 (free)	1.956	0.5	0.98	0.5	0.98
N_2 (adsorbed)	1.956	0.22	0.43	0.22	0.43
O_2 (free)	1.589	0.22	0.34	0.22	0.34
O_2 (adsorbed)	1.589	0.24	0.38	0.24	0.38
H_2 (free)	2.150	0.36	0.77	0.36	0.77
H_2 (adsorbed)	2.150	0.07	0.15	0.07	0.15
C of CO_2 (free)	2.666	0.22	0.59	0.45	1.20
C of CO_2 (adsorbed)	2.666	0.30	0.78	0.50	1.30
O of CO_2 (free)	2.666	0.27	0.72	0.39	1.04
O of CO_2 (adsorbed)	2.666	0.01	0.03	0.04	0.11
N of NO_2 (free)	2.680	0.21	0.56	0.21	0.56
N of NO_2 (adsorbed)	2.680	0.36	0.96	0.26	0.69
O of NO_2 (free)	2.680	0.14	0.37	0.14	0.37
O of NO_2 (adsorbed)	2.680	0.03	0.08	0.12	0.32

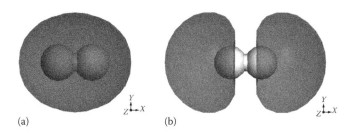

(a)   (b)

FIGURE 32.1   (a) The electrophilic Fukui function of hydrogen is plotted as an isosurface with a grid of 0.2 Å. (b) The nucleophilic Fukui function of hydrogen is plotted as an isosurface with a grid of 0.2 Å.

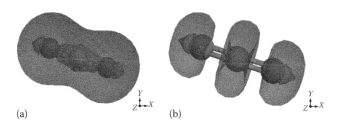

(a)   (b)

FIGURE 32.2   (a) The electrophilic Fukui function of carbon dioxide is plotted as an isosurface with a grid of 0.2 Å. (b) The nucleophilic Fukui function of carbon dioxide is plotted as an isosurface with a grid of 0.2 Å.

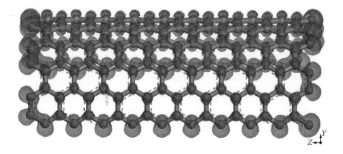

FIGURE 32.3 The electrophilic Fukui function of single-wall CNT is plotted as an isosurface with a grid of 0.2 Å.

also been calculated to compare with the experiment. A Langmuir-type isotherm for a fixed pressure of gas was observed. The minimum energy adsorption configuration is shown in Figures 32.4 and 32.5, respectively, for carbon dioxide and hydrogen. We have as well plotted the electrophilic Fukui function for the adsorption complex of carbon dioxide and hydrogen as oriented over the CNT, as shown in Figures 32.6 and 32.7, respectively. We have then taken the geometry of the local minima and optimized the geometry with DFT by using the same level of theory in which we have optimized the whole complex molecule. The next step is the calculation of binding energy in the presence and absence of the electric field. We have calculated the binding energy in presence of an electric field of 0.05 a.u. in the same direction with the CNT length. The results of binding energy are shown in Table 32.2. There is a variation in the order of energy, but the trend remains the same. The energy gap looks very narrow for the adsorption complex. Finally, we have performed

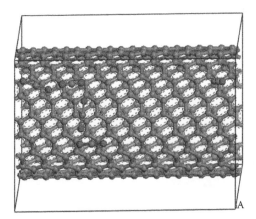

FIGURE 32.4 (See color insert following page 302.) The localized minima as obtained after the GCMS simulation with carbon dioxide adsorption over single-wall CNT with a fixed fugacity of 100 kPa.

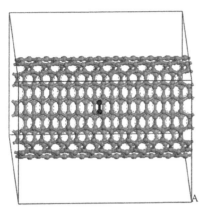

FIGURE 32.5 The localized minima as obtained after the GCMS simulation with hydrogen adsorption over single-wall CNT with a fixed fugacity of 100 kPa.

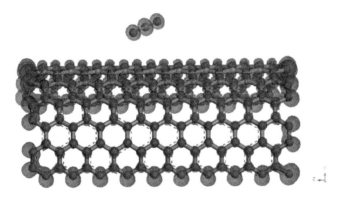

FIGURE 32.6 (See color insert following page 302.) The electrophilic Fukui function of carbon dioxide adsorbed over single-wall CNT is plotted as an isosurface with a grid of 0.2 Å.

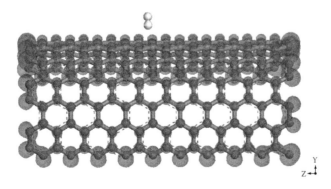

FIGURE 32.7 The electrophilic Fukui function of hydrogen adsorbed over single-wall CNT is plotted as an isosurface with a grid of 0.2 Å.

TABLE 32.2

Binding Energy and the Interaction Energy for the Interacting Molecule with SWNT

Molecule	Binding Energy (eV)		λ (eV)
	Normal	Presence of Electric Field 0.05 a.u. in the Tube Direction	
N_2	0.42	0.46	1.56
O_2	0.37	0.54	1.07
H_2	0.15	0.65	0.86
CO_2	0.78	0.18	2.91
NO_2	0.56	0.32	1.81

only the interaction energy calculation using λ. No change in the order of activity was found, but the numbers are more robust than the binding energy. The order obtained by binding energy calculations follow the same trend as obtained from Fukui function calculation. This shows that the method is robust and can be dependable for localized interactions. This is an example to design new material for an application with a recent need where experimentation is hard and designing is difficult.

32.4   CONCLUSION

In this chapter, we have presented an overview of the reactivity index theory from concept to industrial application. We have demonstrated that a theory within the DFT domain based on the theory of electronegativity and explored in the realm of electron affinity and ionization potential, is capable to deliver a simple correlation to predict the intermolecular and intramolecular interactions. If one can predict the localized interaction between interacting species carefully, then it will be possible to rationalize many chemical phenomena. The main issue of the industry is to reduce cost and to design novel material for a specific application, which is time consuming due to the trial and error process involved in this and as well expensive. They need a reliable as well as a faster way to screen the reactants and propose the products, which can be handled well by computer simulation technology with current reactivity index. We have tried to share with you its capability through the various application examples from the research of our group and as well some recent applications to show that reactivity is an emerging area for material designing from nanocluster through nanowire, nanotube to biomaterial applications. The effort will only be successful if one believes in this and tries to explore around to make it more robust and develop the way to apply this unique theory to all possible materials of choice and interest.

REFERENCES

1. Pearson, R. G. 1987. Recent advances in the concept of hard and soft acids and bases. *J. Chem. Edu.* 64: 561–567.
2. Parr, R. G. and Pearson, R. G. 1983. Absolute hardness: Companion parameter to absolute electronegativity. *J. Am. Chem. Soc.* 105: 7512–7516.
3. Geerlings, P. and De Proft, F. 2000. HSAB principle: Applications of its global and local forms in organic chemistry. *Int. J. Quantum Chem.* 80: 227–236.
4. Ayers, P. W. and Levy, M. 2000. Perspective on Density functional approach to the frontier-electron theory of chemical reactivity—Parr, R. G. and Yang, W. 1984. *J Am Chem Soc* 106: 4049–4050. *Theor. Chem. Acc.* 103: 353–360.
5. Chattaraj, P., Lee, K. H., and Parr, R. G. 1991. HSAB principle. *J. Am. Chem. Soc.* 113: 1855–1856.
6. Baeten, A., Tafazoli, M., Kirsch-Volders, M., and Geerlings, P. 1999. Use of the HSAB principle in quantitative structure–activity relationships in toxicological research: Application to the genotoxicity of chlorinated hydrocarbons. *Int. J. Quantum Chem.* 74: 351–355.
7. Mendez, F., Romero, M. d. L., De Proft, F., and Geerlings, P. 1998. The basicity of p-substituted phenolates and the elimination–substitution ratio in p-nitrophenethyl bromide: A HSAB theoretical study. *J. Org. Chem.* 63: 5774–5778.
8. Mendez, F., Tamariz, J., and Geerlings, P. 1998. 1,3-Dipolar cycloaddition reactions: A DFT and HSAB principle theoretical model. *J. Phys. Chem. A* 102: 6292–6296.
9. Ayers, P. W., Parr, R. G., and Pearson, R. G. 2006. Elucidating the hard/soft acid/base principle: A perspective based on half-reactions. *J. Chem. Phys.* 124: 194107.
10. Ayers, P. W. 2005. An elementary derivation of the hard/soft-acid/base principle. *J. Chem. Phys.* 122: 141102.
11. Ayers, P. W. 2007. The physical basis of the hard/soft acid/base principle. *Faraday Discuss.* 135: 161–190.
12. Parr, R. G. and Chattaraj, P. K. 1991. Principle of maximum hardness. *J. Am. Chem. Soc.* 113: 1854–1855.
13. Pearson, R. G. 2004. Theoretical and computational development. *Int. J. Quantum Chem.* 56: 211–215.
14. Chattaraj, P. K. 1996. The maximum hardness principle: An overview. *Proc. Indian Natn. Sci. Acad., Part A* 62: 513–519.
15. Chattaraj, P. K. and Parr, R. G. 1993. Density functional theory of chemical hardness, in *Chemical Hardness, Structure and Bonding*, Vol. 80, eds. K. D. Sen and D. M. P. Mingos, Springer-Verlag, Berlin, pp. 11–25.
16. Ayers, P. W. and Parr, R. G. 2000. Variational principles for describing chemical reactions: The Fukui function and chemical hardness revisited. *J. Am. Chem. Soc.* 122: 2010–2018.
17. Torrent-Sucarrat, M., Luis, J. M., Duran, M., and Sola, M. 2002. Are the maximum hardness and minimum polarizability principles always obeyed in nontotally symmetric vibrations? *J. Chem. Phys.* 117: 10561–10570.
18. Torrent-Sucarrat, M., Luis, J. M., Duran, M., and Sola, M. 2001. On the validity of the maximum hardness and minimum polarizability principles for nontotally symmetric vibrations. *J. Am. Chem. Soc.* 123: 7951–7952.
19. Parr, R. G. and Yang, W. T. 1995. Density functional theory of the electronic structure of the molecule. *Ann. Rev. Phys. Chem.* 46: 701–728.
20. Parr, R. G. and Yang, W. T. 1989. *Density-Functional Theory of Atoms and Molecules*. Oxford University Press, New York.
21. Geerlings, P., De Proft, F., and Langenaeker, W. 2003. Conceptual density functional theory. *Chem. Rev.* 103: 1793–1873.

22. Chermette, H. 1999. Chemical reactivity indexes in density functional theory. *J. Comp. Chem.* 20: 129–154.

23. Ayers, P. W., Anderson, J. S. M., and Bartolotti, L. J. 2005. Perturbative perspectives on the chemical reaction prediction problem. *Int. J. Quantum Chem.* 101: 520–534.

24. Chatterjee, A., Iwasaki, T., and Ebina, T. 1999. Reactivity index scale for interaction of heteroatomic molecules with zeolite framework. *J. Phys. Chem. A* 103: 2489–2494.

25. Chatterjee, A., Ebina, T., and Iwasaki, T. 2001. Best dioctahedral smectite for nitrogen heterocyclics adsorption—A reactivity index study. *J. Phys. Chem. A* 105: 10694–10701.

26. Chatterjee, A. and Iwasaki, T. 2001. A reactivity index study to choose the best template for a particular zeolite synthesis. *J. Phys. Chem. A* 105: 6187–6196.

27. Chatterjee, A., Iwasaki, T., and Ebina, T. 2002. 2:1 Dioctahedral smectites a selective sorbent for dioxins and furans—a reactivity index study. *J. Phys. Chem. A.* 106: 641–648.

28. Chatterjee, A., Iwasaki, T., Ebina, T., and Mizukami, F. 2003. Intermolecular reactivity study to scale adsorption property of para- and meta-substituted nitrobenzene over 2:1 dioctahedral smectite. *J. Chem. Phys.* 118: 10212–10220.

29. Chatterjee, A., Onodera, Y., Ebina, T., and Mizukami, F. 2003. 2,3,7,8-tetrachloro dibenzo-*p*-dioxin can be successfully decomposed over 2: 1 dioctahedral smectite—a reactivity index study. *J. Mol. Graphics Model.* 22: 93–104.

30. Chatterjee, A., Suzuki, T., Onodera, Y., and Tanaka, D. A. P. 2003. A density functional study to choose the best fluorophore for PET sensor. *Chemistry* 9: 3920–3929.

31. Chatterjee, A., Onodera, Y., Ebina, T., and Mizukami, F. 2004. Effect of exchangeable cation on the swelling property of 2:1 dioctahedral smectite—a periodic first principle study. *J. Chem. Phys.* 120: 3414–3424.

32. Delley, B. 1990. An all-electron numerical-method for solving the local density functional for polyatomic molecules. *J. Chem. Phys.* 92: 508–517.

33. Becke, A. D. 1988. A multicenter numerical-integration scheme for polyatomic molecules. *J. Chem. Phys.* 88: 2547–2553.

34. Lee, C. T., Yang, W. T., and Parr, R. G. 1988. Development of the Colle–Salvetti correlation-energy formula into a functional of the electron-density. *Phys. Rev. B.* 37: 785–789.

35. Chatterjee, A. 2002. Edited the Special Issue on "Application of Density Functional Theory in Chemical Reactions", *Int. J. Molec. Sci.*, Issue 4 (April) pp. 234–444. http://www.mdpi.org/ijms/list02.htm http://www.mdpi.org/ijms/index.htm

36. Chatterjee, A., Ebina, T., and Iwasaki, T. 2003. Adsorption structures and energetic of fluoro- and chlorofluorocarbons over faujasite—a first principle study. *Stud. Surf. Sci. Catal.* 145: 371–374.

37. Chatterjee, A., Ebina, T., Iwasaki, T., and Mizukami, F. 2003. Cholorofluorocarbons adsorption structures and energetic over faujasite type zeolites—a first principle study. *THEOCHEM* 630: 233–242.

38. Chatterjee, A., Ebina, T., and Mizukami, F. 2005. Effects of water on the structure and bonding of resorcinol in the interlayer of montmorillonite nanocomposite—a periodic first principle study. *J. Phys. Chem. B* 109: 7306–7313.

39. Chatterjee, A. 2006. A reactivity index study to rationalize the effect of dopants on Brönsted and Lewis acidity occurring in MeAlPOs. *J. Mol. Graphics Model.* 24: 262–270.

40. Chatterjee, A., Ebina, T., Onodera, Y., and Mizukami, F. 2004. Effect of exchangeable cation on the swelling property of 2:1 dioctahedral smectite—a periodic first principle study. *J. Chem. Phys.* 120: 3414–3424.

41. Chatterjee, A. 2005. Application of localized reactivity index in combination with periodic DFT calculation to rationalize the swelling mechanism of clay type inorganic material. *J. Chem. Sci.* 117: 533–539.

42. Chatterjee, A., Balaji, T., Matsunaga, H., and Mizukami, F. 2006. A reactivity index study to monitor the role of solvation on the interaction of the chromophores with amino-functional silanol surface for colorimetric sensors. *J. Mol. Graphics Model.* 25: 208–218.

43. Chatterjee, A. and Kawazoe, A. 2007. Application of the reactivity index to propose intra and intermolecular reactivity in metal cluster interaction over oxide surface. *Mater. Trans.* 48: 2152–2158.

44. Zhong, A., Rong, C., and Liu, S. 2007. Structural and dynamic properties of (SiO2)6 silica nanostructures: A quantum molecular dynamics study. *J. Phys. Chem. A* 111: 3132–3136.

45. Huang, Y., Zhong, A., Rong, C., Xiao, X., and Liu, S. 2008. Structure, spectroscopy, and reactivity properties of porphyrin pincers: A conceptual density functional theory and time-dependent density functional theory study. *J. Phys. Chem. A* 112: 305–311.

46. Maynard, A. T., Huang, M., Rice, W. G., and Covell, D. G. 1998. Reactivity of the HIV-1 nucleocapsid protein p7 zinc finger domains from the perspective of density-functional theory. *Proc. Natl. Acad. Sci. USA* 95: 11578–11583.

47. Renato, C., Alzate-Morales, J. H., William, T., Santos Juan C., and Ca'rdenas, C. 2007. Theoretical study on CDK2 inhibitors using a global softness obtained from the density of states. *J. Phys. Chem. B* 111(12): 3293–3297.

48. Ijima, S. 1991. Helical microtubules of graphitic carbon. *Nature* 354: 56–58.

49. De Heer, W. A., Chatelain, A., and Ugarte, D. 1995. A carbon nanotube field-emission electron source. *Science* 270: 1179–1180.

50. Kong, J., Franklin, N. R., Zhou, C. W., Chapline, M. G., Peng, S., Cho, K. J., and Dai, H. J. 2000. Nanotube molecular wires as chemical sensors. *Science* 287: 622–625.

51. Collins, P. G., Bradley, K., Ishigami, M., and Zettl, A. 2000. Extreme oxygen sensitivity of electronic properties of carbon nanotubes. *Science* 287: 1801–1804.

52. Jakubek, J. Z. and Simard, B. 2004. Two confined phases of argon adsorbed inside open single walled carbon nanotubes. *Langmuir* 20: 5940–5945.

53. Andzelm, J., Govind, N., and Maiti, A. 2006. Nanotube-based gas sensors—role of structural defects. *Chem. Phys. Lett.* 421: 58–62.

54. Thomas, W., Tombler, Chongwu Zhou, Jing Kong, and Hongjie Dai. 2000. Gating individual nanotubes and crosses with scanning probes. *Appl. Phys. Lett.* 76: 2412–2416.

55. Frenkel, D. and Smit, B. 2002. *Understanding Molecular Simulation: From Algorithms to Applications*, 2nd edn. Academic Press, San Diego.

33 Electronic Structure of Confined Atoms

Jorge Garza, Rubicelia Vargas, and K. D. Sen

CONTENTS

33.1   PARTICLE-IN-A-BOX AS AN EXAMPLE OF A CONFINED PARTICLE

Recently, there has been a renewed interest in studying the electronic structure and reactivity of the spatially confined atoms and molecules among the community of chemists and physicists. This is mainly due to the possible application of such studies in the development of new materials, specifically in the nanotechnology field. Though simple in appearance, the one-electron systems under confined potential exhibit several surprising characteristics not generally encountered in the free atoms and molecules. On purely theoretical grounds, understanding the behavior of the spatially confined systems is computationally challenging and conceptually enticing at the same time. In this chapter, we shall be concerned with the spatially confined atoms i.e., N-electronic system with the nuclear charge Z located at the center and spherically confined by potential walls located at a finite radius, R_c. Here, the wave function describing the N-electrons of the confined atom must be found through the Schrödinger equation including the new boundary conditions.

Before starting the discussion on confined atoms, we shall briefly describe the simplest standard confined quantum mechanical system in three dimensions (3-D), namely the particle-in-a-(spherical)-box (PIAB) model [1]. The analysis of this system is useful in order to understand the main characteristics of a confined system. Let us note that all other spherically confined systems with impenetrable walls located at a certain radius, R_c, transform into the PIAB model in the limit of $R_c \to 0$. For the sake of simplicity, we present the model in one-dimension (1-D). In atomic units (a.u.) ($m_e = 1$, $q_e = 1$, and $\hbar = 1$), the Schrödinger equation for an electron confined in one-dimensional box is

$$\left(-\frac{1}{2}\frac{d^2}{dx^2} + V(x)\right)\varphi_n(x) = \varepsilon_n\varphi_n(x) \tag{33.1}$$

with

$$V(x) = \begin{cases} 0 & 0 < x < R_c \\ \infty & x \le 0 \text{ and } x \ge R_c \end{cases} \tag{33.2}$$

In this case, the box length is represented by R_c. The solution of the Equation 33.1, in atomic units, is

$$\varepsilon_n = \frac{n^2\pi^2}{2R_c^2} \tag{33.3}$$

and

$$\varphi_n(x) = \left(\frac{2}{R_c}\right)^{1/2}\sin\left(\frac{n\pi x}{R_c}\right) \tag{33.4}$$

with $n = 1, 2, \ldots, \infty$. We note that this system satisfies the Dirichlet boundary conditions, $\varphi_n(0) = \varphi_n(R_c) = 0$. For the 3-D box, Equations 33.3 and 33.4 remain unchanged except that the variable x is to be replaced by r, the variable of radial distance from the origin.

From Equation 33.3 we obtain the behavior of the energy of each state as a function of the confinement length, R_c. It follows that if R_c is decreased (compression) then the energy will be increased, and vice versa. The wave function displays a similar behavior, i.e., its amplitude increases when the confinement length is decreased. We would like to draw attention to an important characteristic of these wave functions. In the Figure 33.1, we have plotted the ground state function for three values of R_c: 0.5, 1.0, and 2.0 a.u. It is clear that the three functions satisfy the boundary conditions, but the derivatives of these functions evaluated on the boundaries are different from 0.

From Equation 33.4 we can evaluate these derivatives at the boundaries,

$$\left.\frac{d\varphi_n}{dx}\right|_{x=0} = \frac{n\pi}{R_c}\left(\frac{2}{R_c}\right)^{1/2}, \tag{33.5}$$

$$\left.\frac{d\varphi_n}{dx}\right|_{x=R_c} = (-1)^n\left.\frac{d\varphi_n}{dx}\right|_{x=0}. \tag{33.6}$$

These equations suggest that a reduction in the confinement "length" results in the identical increase in the magnitude of the derivatives at the two conjugate boundaries. Clearly, we can associate these derivatives with the derivatives of the energy with respect of the confinement length as

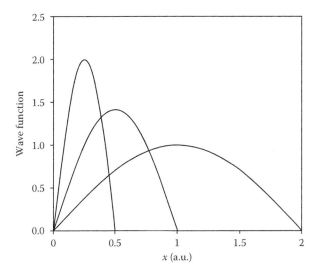

FIGURE 33.1   Ground state wave function for three values of R_c: 0.5, 1.0, and 2.0 a.u. See how the derivatives of the wave function, on the boundaries, are increased when R_c is reduced.

$$\frac{\partial \varepsilon_n}{\partial R_c} = -\frac{1}{2}\left(\frac{d\varphi_n}{dx}\Big|_{x=0}\right)^2 = -\frac{1}{2}\left(\frac{d\varphi_n}{dx}\Big|_{x=R_c}\right)^2. \tag{33.7}$$

This is an important result because the pressure p on the walls of a container induced by the electron cloud, may be estimated as

$$p = -\left(\frac{\partial E}{\partial V}\right)_T \tag{33.8}$$

for the systems obeying the Dirichlet boundary conditions.

In this equation, E represents the electronic system energy, V the container volume, and T the temperature. In our case we have $T=0$, $E=\varepsilon_n$, and the pressure is given by

$$p = -\frac{\partial \varepsilon_n}{\partial R_c} = \frac{n^2 \pi^2}{R_c^3}. \tag{33.9}$$

Thus, for the PIAB, the pressure is directly related to the slope of the wave function evaluated at the boundaries defining the length of confinement. As noted earlier, when the confinement length is reduced the pressure will be increased. However, depending on the quantum state (different principal quantum number, n), the pressure will be different, even if R_c is the same. From Equations 33.3 and 33.9 a relationship between the energy and the pressure can be established to obtain a plot similar to that presented in Figure 33.2.

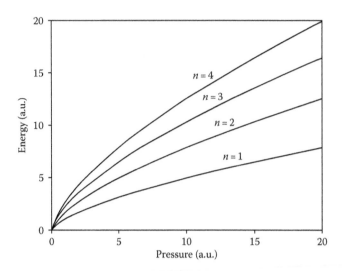

FIGURE 33.2 Energy as a function of the pressure for $n = 1, 2, 3$, and 4 of the PIAB.

In this plot, we can see that if we increase the pressure, the energy also will be increased but the rate of this increment will be different for each state. The results discussed for the PIAB model are particular situations of generalizations reported for systems confined with Dirichlet boundary conditions [2]. We must remember these results for further discussion through this chapter. Let us conclude this section with the remark that the state dependence of the effective pressure at the given value of R_c can be analogously understood in terms of the different electron densities and their derivatives at the boundaries. In most general case of atoms and molecules, scaled densities may have to be employed in order to include the excited states. In the next section, we present some basic results on such connections between wave function and electron density.

33.2 DERIVATIVES OF ELECTRON DENSITY AND THE WAVE FUNCTION

Consider a general one-dimensional system with Schrödinger equation

$$-\frac{1}{2}\psi'' = [E - V(x)]\psi. \tag{33.10}$$

For this case, the electron density $\rho = \Psi^2$. For simplicity we assume that Ψ is real, such that the first derivative

$$\rho' = 2\psi\psi' \tag{33.11}$$

and the second derivative

$$\rho'' = 2\Psi\Psi'' + 2(\Psi')^2. \tag{33.12}$$

Three distinct possibilities of the position variable are obtained from Equation 33.12, depending upon the sign of the first term $\Psi\Psi''$.

1. Over the classically *allowed* region where $E - V(x) > 0$, according to Equation 33.10, Ψ'' has the opposite sign of Ψ, i.e., $\Psi\Psi'' \leq 0$. So in this region, from Equation 33.12, ρ'' could also be negative.
2. In the classically forbidden region, on the other hand, Ψ'' has the same sign as Ψ and it follows from Equation 33.12 that ρ'' is positive.
3. At the classical turning point, $\Psi'' = 0$, which turns ρ'' positive.

Let us consider the locations of position space (values of x) where ρ has either a maximum $[\rho'' < 0]$, minimum $[\rho'' > 0]$, or a point of inflexion $[\rho'' = 0]$.

At its maxima, ρ must satisfy $\rho'' < 0$, which suggests that such locations must lie within the classically allowed region. The same applies for the existence of inflection points in ρ. At the nodes of Ψ, Equation 33.12 implies that $\rho'' > 0$. At such points, ρ in fact has a minimum. Finally, when the potential is reflection symmetric then at the origin $(x = 0)$, either $\Psi' = 0$ (even function) or $\Psi = 0$ (odd function). For even functions, $\rho'' < 0$ if the origin is within the classically allowed region and $\rho'' > 0$ if the potential is such that the origin is within the classically forbidden region. For odd functions, it is always true that $\rho'' = 0$ at the origin. These results are strictly valid for the one-dimensional case.

33.3   CONFINED HYDROGEN ATOM

Just 10 years after the free hydrogen atom problem was solved in quantum mechanics, Michels and coworkers [3] proposed the model of the spherically confined hydrogen atom and obtained the approximate solutions of the corresponding Schrödinger equation with impenetrable boundary. The idea was originally conceived in order to simulate the effect of pressure on the polarizability of hydrogen atom. In this model, the hydrogen atom is centrally confined inside a spherical cavity surrounded by impenetrable wall. The Schrödinger equation for a hydrogen-like atom (in a.u.) is written as

$$\left(-\frac{1}{2}\nabla^2 - \frac{Z}{r}\right)\varphi(\mathbf{r}) = \varepsilon\varphi(\mathbf{r}). \tag{33.13}$$

In this the three-dimensional can be transformed to work just with the radial coordinate, r. For this purpose, the wave function is written as a product of two functions. The first one depends only on the radial coordinate and the other on the angular coordinates denoted by $R(r)$ and $Y(\theta, \phi)$, respectively. In the solution of the hydrogen atom, it is found that $R(r)$ depends on the quantum number

$n = 1, 2, \ldots$ and Y depends on two quantum numbers, $l = 0, 1, 2, \ldots, n-1$ and $m = -l, -l+1, \ldots, 0, \ldots, l-1, l$ [1]. In this way, the wave function will depend on three quantum numbers

$$\varphi_{nlm}(r, \theta, \phi) = R_{nl}(r)Y_{lm}(\theta, \phi). \tag{33.14}$$

In addition to these new two functions, the Laplacian operator ∇^2 is written in spherical coordinates as

$$\nabla^2 = \frac{\partial^2}{\partial r^2} + \frac{2}{r}\frac{\partial}{\partial r} - \frac{1}{r^2}\hat{L}^2. \tag{33.15}$$

This is a convenient way to express the Laplacian operator because the Y function is an eigenfunction of the angular momentum operator $\hat{L}^2$

$$\hat{L}^2 Y_{lm}(\theta, \varphi) = l(l+1)Y_{lm}(\theta, \varphi). \tag{33.16}$$

By using Equations 33.15 and 33.16 within Equation 33.13 and eliminating the Y_{lm} function, an equation for the radial function is obtained as

$$\left(-\frac{1}{2}\frac{d^2}{dr^2} - \frac{1}{r}\frac{d}{dr} + \frac{l(l+1)}{2r^2} - \frac{Z}{r}\right)R_{nl}(r) = \varepsilon\, R_{nl}(r). \tag{33.17}$$

In the solution of this equation, the energy is a function just of the quantum number n. Thus, states with the same n but different l have the same energy. This is the so-called accidental degeneracy which implies that while the *effective* potential

$$\frac{l(l+1)}{2r^2} - \frac{Z}{r}$$

which yields the radial wave function changes as quantum number l changes for a given n, leading to different total wave functions, and their energy remains the same as it depends only on n. For the confined hydrogen-like system, we will define the potential energy operator as

$$V(r) = \begin{cases} -\frac{Z}{r} & r < R_c \\ \infty & r \geq R_c \end{cases}. \tag{33.18}$$

With this form of the operator, we preserve the spherical symmetry. In the solution of this equation, a new function is defined as

$$\psi_{nl}(r) = rR_{nl}(r). \tag{33.19}$$

Substitution of this equation in Equation 33.17 conduces to

$$\left(-\frac{1}{2}\frac{d^2}{dr^2} + \frac{l(l+1)}{2r^2} + V(r)\right)\psi_{nl}(r) = \varepsilon\psi_{nl}(r). \tag{33.20}$$

The boundary conditions on $\psi_{nl}(r)$ are determined by the boundary conditions of $R_{nl}(r)$. Because $R_{nl}(r)$ is finite in the origin, then $\psi_{nl}(0)=0$. Furthermore, as we have a potential wall of infinite height, similar to that found in the PIAB, the resulting wave function on the surface of this wall must vanish. Thus, we have the Dirichlet boundary conditions for this problem

$$\psi_{nl}(0) = \psi_{nl}(R_c) = 0. \tag{33.21}$$

This problem does not admit an analytic solution, but it has been solved with several techniques to obtain very accurate values [4–6]. In our case, we will use a simple technique to obtain many of the characteristics exhibited by the confined atoms.

In the first stage of our approach, the wave function will be represented by a set of functions. In order to decide which basis set we will use, the radial coordinate will be changed by the scaled variable x defined as

$$x = \frac{\pi}{R_c} r. \tag{33.22}$$

With this scaled variable, the Equation 33.20 becomes

$$\left(-\frac{1}{2}\frac{d^2}{dx^2} + \frac{l(l+1)}{2x^2} + V(x) \right)\psi_{nl}(x) = \lambda\psi_{nl}(x) \tag{33.23}$$

with $\lambda = (R_c/\pi)^2\varepsilon$, and

$$V(x) = \begin{cases} -\frac{R_c}{\pi}\frac{Z}{x} & x < \pi \\ \infty & x \geq \pi \end{cases}. \tag{33.24}$$

With x, the boundary conditions are

$$\psi_{nl}(0) = \psi_{nl}(\pi) = 0. \tag{33.25}$$

Here we will use K functions from the solution of the PIAB, to represent $\psi_{nl}(x)$ as

$$\psi(x) = \sum_{j=1}^{K} c_j f_j(x) \tag{33.26}$$

with

$$f_j(x) = \left(\frac{2}{\pi}\right)^{1/2} \sin(jx). \tag{33.27}$$

In proceeding procedure, we must determine $\{c_j\}$; for this purpose, Equation 33.26 is substituted into Equation 33.23. If the resulting equation is multiplied on the left side by the f_k function and then integrated over x, one obtains the matrix equation for the set $\{c_j\}$,

$$\mathbf{HC} = \lambda \mathbf{SC}. \tag{33.28}$$

In this equation, $\mathbf{C}$ represents a vector with the $\{c_j\}$ set as its components. $\mathbf{H}$ is a matrix with $K \times K$ elements, with H_{kj} given by

$$H_{kj} = \int_0^\pi dx f_k(x)\left(-\frac{1}{2}\frac{d^2}{dx^2}\right)f_j(x) + \int_0^\pi dx f_k(x)\left(\frac{l(l+1)}{2x^2}\right)f_j(x) + \int_0^\pi dx f_k(x)\left(-\frac{R_c}{\pi}\frac{Z}{x}\right)f_j(x).$$
$$\tag{33.29}$$

$\mathbf{S}$ represents the overlap matrix, with elements

$$S_{kj} = \int_0^\pi dx f_k(x) f_j(x). \tag{33.30}$$

The advantage of these functions is their orthogonality property, giving a diagonal matrix for $\mathbf{S}$ with 0 or 1 as elements. With this form of $\mathbf{S}$, we have a standard eigenvalue algebraic problem where $\mathbf{C}$ and λ must be determined.

The integrals involved in the matrix elements of Equation 33.29 can be evaluated by several techniques; in our case we will use Mathematica to compute the integrals and to solve the eigenvalue algebraic problem. The Mathematica instructions (for any release) to use are

```
(* Equation (27) *)
f[j_,x_]:=Sqrt[2/Pi]Sin[j*x];
(* First integral of the equation (29) *)
firsterm[k_,j_]:=Integrate[f[k,x]*j*j*Sin[j*x],
  {x,0,Pi}]/Sqrt[2*Pi];
(* Second integral of the equation (29) *)
secondterm[k_,j_]:=NIntegrate[f[k,x]*f[j,x]/(x*x),
  {x,0,Pi}];
(* Third integral of the equation (29) *)
thirdterm[k_,j_]:=NIntegrate[f[k,x]*f[j,x]/x,{x,0,Pi}];
(* Matrix H as a function of the angular momentum, l, the
confinement radius, Rc, and the number of functions, func *)
matrixh[l_,Rc_,func_]:=Table[firsterm[k,j]+l*(l+1)
*secondterm[k,j]/2-Rc*thirdterm[k,j]/Pi,{j,1,func},
  {k,1,func}];
```

With these five instructions, we can solve the confined hydrogen atom, $Z = 1$. For example, if we want to obtain the orbitals with $l = 0$ for $R_c = 1.0$ and $K = 25$, then the following instruction will give the answer

```
Rc = 1.0;
MatrixForm[Pi*Pi*Sort[Eigenvalues[math[0,Rc,25]]]/
   (Rc*Rc)]
```

By using this procedure, we have calculated the energies for the confined hydrogen atom for $l = 0, 1, 2, 3$ and $n = 1, 2, 3, 4$, corresponding to two confinement radii, $R_c = 1.0$ and 7.0 a.u. These results are presented in Table 33.1, where it is found that the energy depends on two quantum numbers n and l. Thus, there is a breaking of the degeneracy in l. For example, the 2s orbital has different energy than the 2p orbital. In addition to this effect, we observe the following energy ordering for $R_c = 7.0$ a.u.: $\varepsilon_{1s} < \varepsilon_{2p} < \varepsilon_{2s} < \varepsilon_{3d} < \varepsilon_{3p} < \varepsilon_{4f} < \varepsilon_{3s} < \varepsilon_{4s} < \varepsilon_{4d} < \varepsilon_{4p}$. Interestingly, the ordering presented by the shell M ($n = 3$), where the 3d orbital has the lowest energy, 3p the next, and 3s has the highest energy. The magnitude of this effect is more pronounced when the confinement radius is decreased. For example, when $R_c = 1.0$ a.u. the energy ordering is $\varepsilon_{1s} < \varepsilon_{2p} < \varepsilon_{3d} < \varepsilon_{2s} < \varepsilon_{4f} < \varepsilon_{3p} < \varepsilon_{4d} < \varepsilon_{3s} < \varepsilon_{4p} < \varepsilon_{4s}$. From these examples, we conclude that there are many crossings between the orbital energies depending on the confinement radii [7,8]. From the PIAB problem, it was concluded that the energy rate as a function of the confinement depends on the confinement radius and consequently on the pressure. From the confined hydrogen atom, a more subtle dependence on its parameters is observed; for a given value of n, there are crossings of the orbital energies with different angular momentum as R_c is varied.

TABLE 33.1

Energy Values for the Confined Hydrogen Atom

	Principal Quantum Number (*n*)	Angular Momentum (*l*)			
		0	1	2	3
R_c		s	p	d	f
7.0	1	−0.4988			
	2	−0.0509	−0.0874		
	3	0.3928	0.2580	0.0966	
	4	1.0582	1.8311	1.2397	0.2776
1.0	1	2.3740			
	2	16.5704	8.2234		
	3	40.8633	27.4760	14.9675	
	4	75.1308	56.7653	39.3153	22.8958

Note: Energies are in atomic units.

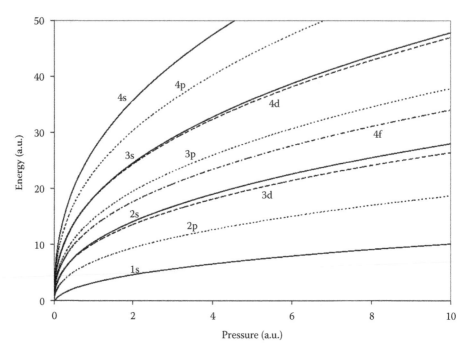

FIGURE 33.3 Energy as a function of the pressure $n = 1$, 2, 3, and 4 of the confined hydrogen atom. For different values of l, there are labels for the orbitals: s for $l = 0$, p for $l = 1$, d for $l = 2$, f for $l = 3$.

The dependence of the energy with respect to the pressure for the orbitals with $n = 1$, 2, 3, and 4 is displayed in Figure 33.3. This plot was generated by numerical evaluation of Equation 33.8 which, for the confined hydrogen atom, is given by

$$p = -\frac{1}{4\pi R_{\mathrm{c}}^2}\left(\frac{\partial \varepsilon_{nl}}{\partial R_{\mathrm{c}}}\right).\tag{33.31}$$

We evaluate numerically this derivative by using 3 values of R_{c}, each value is spaced in 0.05 a.u., beginning with $R_{\mathrm{c}} = 1.0$ a.u. and extending up to $R_{\mathrm{c}} = 7$ a.u., and the energy was computed for each l on these points.

This plot reveals the main characteristics of the confined atoms; depending on the angular momentum, each orbital behaves in a different way, because the energy is increased rapidly when the wave function contains several nodes. This behavior is related to the kinetic energy of each state; this quantity is increased when the orbital present several nodes and the atom is submitted to small confinement radii. We conclude this section with two observations. Firstly, the notation of (nl) defining the states of the confined hydrogen is retained although the integral values of n only refer to the free hydrogen atom. This is due to the fact that the nodal characteristics of the wave functions of the confined system remain the same as in free state with changed locations. Secondly, the fact that the higher l orbitals are relatively *less* destabilized under confinement, can be qualitatively understood from the relative sizes of the orbitals of the *free* hydrogen atom. The average radius $\langle r \rangle$ of the (nl) state of the free hydrogen atom is given by (in a.u.) $1/2[3n^2 - l(l+1)]$. Considering the example of 3s, 3p, and 3d orbitals, the average radius decreases with increasing l. If we choose R_c corresponding to the average size of 3d orbital, then in order to confine 3p and 3s orbitals to the same size, one has to compress both of them. However, 3s orbital must be compressed (confined) more than 3p orbital and no compression is needed for 3d orbital, which is already at R_c. This is why the relative destabilization under confinement is inversely proportional to l for a fixed n.

33.4 CONFINED MANY-ELECTRON ATOMS

We have discussed in the previous section the simplest confined atom, the hydrogen atom. Due to the different possibilities of the orbital energy level ordering under confinement, the electron configuration of a confined many-electron atom will have the possibility of obeying different Aufbau principle, which depends on the nature of confinement. Let us think in the potassium atom, when this atom is free (not confined) the electron configuration is $1s^2 2s^2 2p^6 3s^2 3p^6 4s^1$ (^1S). In the hypothetical case of noninteracting electrons in an atom, the electron configuration could be determined by the hydrogen atom. Thus, for the confined K atom we could obtain crossings between the orbital energies and consequently, the electron configuration will be different depending on the confinement imposed on the atom. However, we know that the electrons in an atom are interacting particles; consequently, we need a theory to describe the electron structure of a confined atom. In this chapter, we will use the Kohn–Sham model [9] to obtain the electron structure of the confined many-electron atoms.

The radial Kohn–Sham equations to solve are

$$\left(-\frac{1}{2}\frac{d^2}{dr^2} + \frac{l(l+1)}{2r^2} + v_J(r) + v_{xc}^\sigma(r) + V(r) \right)\psi_{nl}^\sigma(r) = \varepsilon\psi_{nl}^\sigma(r), \qquad (33.32)$$

where σ denotes the spin associated to each electron, α or β, and $V(r)$ is the same than that defined in Equation 33.18. In Equation 33.32, $v_J(r)$ represents the spherical Coulomb potential obtained from

$$v_J(r) = \frac{1}{4\pi}\int d\Omega\, \frac{\delta J}{\delta \rho^\sigma(r)} = \frac{1}{4\pi}\int d\Omega \int dr'\, \frac{\rho(r')}{|r - r'|}. \qquad (33.33)$$

In this equation $\rho^\sigma(r)$ is the electron density with σ spin, defined as

$$\rho^\sigma(\mathbf{r}) = \sum_{i=1}^{N^\sigma} \left|\varphi_i^\sigma(\mathbf{r})\right|^2 \qquad (33.34)$$

with N^σ as the number of electrons with σ spin, the sum of these components give the total electron density

$$\rho(\mathbf{r}) = \sum_{i=1}^{N^\sigma} \rho^\sigma(\mathbf{r}). \tag{33.35}$$

The exchange-correlation potential, $v_{xc}^\sigma(r)$, is given by

$$v_{xc}^\sigma(r) = \frac{1}{4\pi} \int d\Omega \frac{\delta E_{xc}}{\delta \rho^\sigma(\mathbf{r'})}. \tag{33.36}$$

In this chapter, we will use just the exchange contribution in the local density approximation,

$$v_{xc}^\sigma(r) = -\frac{4}{3} c_x \{\rho^\sigma(r)\}^{1/3} \tag{33.37}$$

with

$$c_x = \frac{3}{4} \left(\frac{3}{\pi}\right)^{1/3}.$$

Equation 33.32 can be solved by numerical techniques. For numerical details, we refer the reader to Refs. [10,11]. In the remaining part of this section, we present theoretical predictions derived from such calculations which can be compared with experimental findings. According to the Figure 33.3, we expect the 3d orbital energy to stay below the 4s orbital energy for small confinements.

We will now consider the orbital energies of the electrons with α spin, for the K atom confined inside a cavity with rigid walls. In Figure 33.4, we have focused on

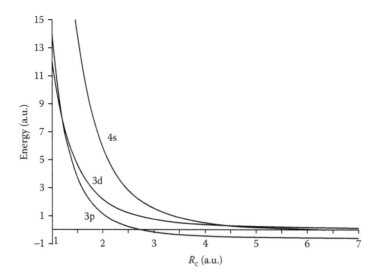

FIGURE 33.4 Orbital energies for the confined K atom. See how the ordering is changing when the confinement radius is changing.

the highest orbitals, occupied and unoccupied. We note that for $R_c > 5$ a.u., the 4s orbital energy is lower than the 3d orbital, and consequently the electron configuration corresponds to the ^1S state. However, for small R_c values, the 3d orbital energy is more stable than the 4s orbital energy. We can see, from the Figure 33.4, the crossing between ε_{4s} and ε_{3d}, around $R_c = 4.5$ a.u. This result suggests that the electron configuration $1s^2 2s^2 2p^6 3s^2 3p^6 3d^1$ can be more stable than that corresponding to the free atom. To verify this hypothesis, we must compute the total energy of the confined atom for the two-electron configurations and find the most stable configuration, depending on the total energy.

In Figure 33.5, we presented the energy for the two-electron configurations as a function of the confinement radius. It is clear from this figure that there is a crossing between the energies with different electron configuration. In fact, we see the crossing between the total energies in a similar R_c than that presented by the orbital energies (see Figure 33.4). Here is an important prediction of experimental significance coming from the confined atom model; the confinement induces changes on the electron configuration in an atom. Most interesting question is about the pressure at which such a transition in electron configuration actually takes place. From the confined hydrogen atom we found the difference between a energy versus R_c plot and a energy versus pressure plot. In the same way we numerically built the energy versus pressure plot in the hydrogen atom, we can apply such an approach for any atom. In the Figure 33.6, the energy versus pressure plot for the confined K atom is depicted for low pressure values. It is clear from this plot that the electron configuration transition is presented for small pressures.

We know that for same R_c, each orbital exhibits a different pressure. Thus, for the point where the pressure and the total energy coincide in both electron

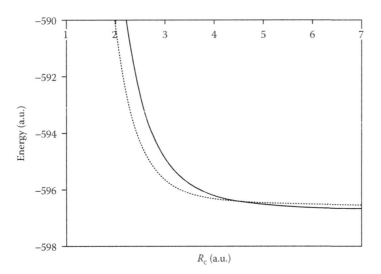

FIGURE 33.5 Total energy as a function of the confinement radius for the configurations ^1S (solid line) and ^1D (dotted line).

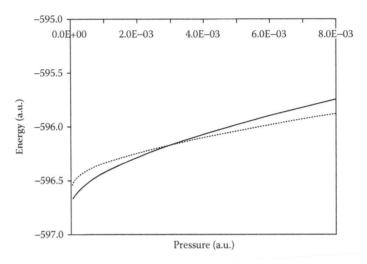

FIGURE 33.6 Total energy as a function of the pressure for the configurations ^1S (solid line) and ^1D (dotted line).

configurations, the R_c must be different. In Table 33.2 we are reporting some values for R_c, total energy and pressure for the two-electron configurations considered in our discussion. The numbers with bold font are those where there is an intersection between the total energy and the pressure for both configurations. These values are useful if we want to apply an interpolation method and find the intersection point to predict the pressure where the transition is presented. We want to remark that for the transition pressure, the R_c and consequently the volume, are not the same for both electron configurations. From the Table 33.2, it is clear that the ^1S configuration presents bigger volumes than the configuration ^1D for similar pressure

TABLE 33.2
R_c, Total Energy and Pressure for Two Electron Configurations of the Confined K Atom

Electron Configuration					
$1s^2 2s^2 2p^6 3s^2 3p^6 4s^1$			$1s^2 2s^2 2p^6 3s^2 3p^6 3d^1$		
R_c	Energy	Pressure	R_c	Energy	Pressure
3.952630	−596.161422	0.003051	3.573680	−596.156887	0.003108
3.973680	−596.173850	0.002933	3.594740	−596.167195	0.002962
3.994740	**−596.185930**	**0.002821**	**3.615790**	**−596.177137**	**0.002823**
4.015790	**−596.197672**	**0.002714**	**3.636840**	**−596.186722**	**0.002691**
4.036840	**−596.209084**	**0.002611**	**3.657900**	**−596.195972**	**0.002567**
4.057900	−596.220189	0.002514	3.678950	−596.204896	0.002449
4.078950	−596.230987	0.002419	3.700000	−596.213513	0.002338

Note: All quantities are in atomic units.

values. In a process where an atom is under pressure, the electron configuration transition and also a volume change will be presented. In this way, these electron configuration transitions can be visualized as first-order transitions.

From an experimental point of view, the electron configuration transitions are observed principally in alkali metals when they are submitted under high pressures [12]. When the atom is confined in relatively small confinement radii, we have noted for the K atom that its electron configuration is similar to that observed in transition metals as the d-shell is available now. Experimentally, K shows a transition metal behavior when it is under high pressures, and in these conditions this metal forms alloys with nickel [13]. Most of the alkali metals exhibit this interesting transition named s–d transition. It is impressive how a disarmingly simple model of the confined atom within rigid walls, can give a good understanding of such transitions.

As we have seen, an atom under pressure changes its electron structure drastically and consequently, its chemical reactivity is also modified. In this direction we can use the significant chemical concepts such as the electronegativity and hardness, which have foundations in the density functional theory [9]. The intuition tells us that the polarizability of an atom must be reduced when it is confined, because the electron density has less possibility to be extended. Furthermore, it is known that the polarizability is related directly with the softness of a system [14]. Thus, we expect atoms to be harder than usual when they are confined by rigid walls. Estimates of the electronegativity, χ, and the hardness, η, can be obtained from [9]

$$\chi \approx \frac{I + A}{2} \tag{33.38}$$

and

$$\eta \approx I - A. \tag{33.39}$$

In these equations, I represents the ionization potential and A denotes the electron affinity. It is important to mention that I and A are referred to the vertical process, and this means the evaluation of the cation and anion must be at the R_c of the neutral atom. In the Figure 33.7, we present the η behavior as a function of the confinement radius, for the Kr atom. It follows from this figure that our intuition gave us a good idea about the hardness behavior in a confined atom. However, we must take into account the transitions discussed above, in particular, when the atom is confined within smaller radii. Because the most stable orbitals exhibit smaller number of nodes than that corresponding to the free atom, they are less hard and consequently, the hardness of the system is reduced. Thus, a confined atom cannot be infinitely hard due to the electron transitions [15]. In terms of the chemical reactivity, the confined atoms are less electronegative than the free atoms and then the bonding is not preferred. For this reason, an atom can give the main characteristics of the electron structure of solids when they are under high pressure, as in the case of the s–d transition in alkaline metals.

In this chapter, we dealt with rigid walls, but there are reports where soft walls are used. However, the solution of the equations related to these models is more

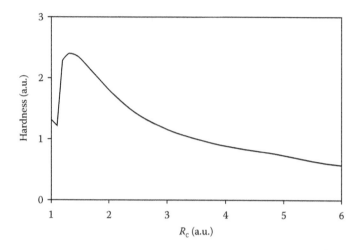

FIGURE 33.7   Hardness for the Kr atom as a function of the confinement radius.

complicated than that presented here. It is worthwhile to note that such models are
essential in order to include the interactions with the environment of the central atom,
a necessary requirement to simulate the experimental conditions. This is currently an
open field of research [16]. Finally, the same model can be applied to other model
system such as the D-dimensional isotropic harmonic oscillator with new and
interesting effects on the energy level ordering [17].

REFERENCES

1. Levine, I. N. 2000. *Quantum Chemistry*, 5th ed. New Jersey: Prentice-Hall.
2. Arteca, G. A., Fernández, F. M. and Castro, E. A. 1984. Approximate calculation
   of physical properties of enclosed central field quantum systems. *J. Chem. Phys.* 80:
   1569–75.
3. Michels, A., de Boer, J. and Bijl, A. 1937. Remarks concerning molecular interaction and
   their influence on the polarisability. *Physica* 4: 981–94.
4. Sommerfeld, A. and Welker, H. 1938. Artificial limiting conditions in the Kepler
   problem. *Ann. Phys.* 32: 56–65.
5. Ley-Koo, E. and Rubinstein, S. 1979. The hydrogen atom within spherical boxes with
   penetrable walls. *J. Chem. Phys.* 71: 351–57.
6. Aquino, N., Campoy, G. and Montgomery, H. E., Jr. 2007. Highly accurate solutions for
   the confined hydrogen atom. *Intl. J. Quantum Chem.* 107: 1548–58. Reports 100 decimal
   accuracy in eigenvalues.
7. Goldman, S. and Joslin, C. 1992. Spectroscopic properties of an isotropically compressed
   hydrogen atom. *J. Phys. Chem.* 96: 6021–27.
8. Connerade, J. P., Dolmatov, V. K. and Lakshmi, P. A. The filling of shells in compressed
   atoms. 2000. *J. Phys. B At. Mol. Opt. Phys.* 33: 251–64.
9. Parr, R. G. and Yang, W. 1989. *Density-Functional Theory of Atoms and Molecules*.
   New York: Oxford University Press.
10. Garza, J., Vargas, R. and Vela, A. 1988. Numerical self-consistent-field method to solve
    the Kohn–Sham equations in confined many-electron atoms. *Phys. Rev. E.* 58: 3949–54.

11. (a) Sen, K. D., Garza, J., Vargas, R. and Vela, A. 2000. Atomic ionization radii using Janak's theorem. *Chem. Phys. Lett.* 325: 29–32. (b) Garza, J., Vargas, R., Vela, A. and Sen, K. D. 2000. Shell structure in free and confined atoms using the density functional theory. *J. Mol. Structure Theochem.* 501–502: 183–88.

12. McMahan, A. K. 1984. Alkali-metal structures above the s–d transition. *Phys. Rev. B.* 29: 5982–85.

13. Parker, L. J., Atou, T. and Badding, J. V. 1996. Transition element-like chemistry for potassium under pressure. *Science* 273: 95–97.

14. Vela, A. and Gázquez, J. L. 1990. A relationship between the static dipole polarizability, the global softness, and the Fukui function. *J. Am. Chem. Soc.* 112: 1490–92.

15. Garza, J., Vargas, R., Aquino, N. and Sen, K. D. 2005. DFT reactivity indices in confined many-electron atoms. *J. Chem. Sci.* 117: 379–86.

16. (a) Gorecki, J. and Byers Brown, W. 1988. Padded-box model for the effect of pressure on helium. *J. Phys. B: At. Mol. Opt. Phys.* 21: 403–10. (b) Marín, J. L. and Cruz, S. A. 1992. Use of the direct variational method for the study of one- and two-electron atomic systems confined by spherical penetrable boxes. *J. Phys. B: At. Mol. Opt. Phys.* 25: 4365–71. (c) Bielińska-Wąż, D., Karwowski, J. and Diercksen, G. H. F. 2001. Spectra of confined two-electron atoms. *J. Phys. B: At. Mol. Opt. Phys.* 34: 1987–2000. (d) Connerade, J. P., Dolmatov, V. K., Lakshmi, P. A. and Manson, S. T. 1999. Electron structure of endohedrally confined atoms: Atomic hydrogen in an attractive shell. *J. Phys. B: At. Mol. Opt. Phys.* 32: L239–45. (e) Lo, J. H. M. and Klobukowski, M. 2007. Relativistic calculations on the ground and excited states of AgH and AuH in cylindrical harmonic confinement. *Theor. Chem. Acc.* 118: 607–22.

17. Sen, K. D., Montgomery, H. E. Jr. and Aquino, N. 2007. Degeneracy of confined D-dimensional harmonic oscillator. *Int. J. Quantum Chem.* 107: 798–806.

34 Computation of Reactivity Indices: Integer Discontinuity and Temporary Anions

Frank De Proft and David J. Tozer

CONTENTS

34.1 INTRODUCTION

Density functional theory (DFT) uses the electron density $\rho(\mathbf{r})$ as the basic source of information of an atomic or molecular system instead of the many-electron wave function Ψ [1–7]. The theory is based on the Hohenberg–Kohn theorems, which establish the one-to-one correspondence between the ground state electron density of the system and the external potential $v(\mathbf{r})$ (for an isolated system, this is the potential due to the nuclei) [6]. The electron density uniquely determines the number of electrons N of the system [6]. These theorems also provide a variational principle, stating that the exact ground state electron density minimizes the exact energy functional $E[\rho(\mathbf{r})]$.

When a molecule A is attacked by another molecule B, it will be perturbed in either its number of electrons N_A or its external potential $v_A(\mathbf{r})$. At the very early stages of the reaction, the total electronic energy of A, E_A can be expressed as a Taylor series expansion around the isolated system values N_A^0 and $v_A^0(\mathbf{r})$

$$E_A\left[N_A^0 + \Delta N_A, \; v_A^0(\mathbf{r}) + \Delta v_A(\mathbf{r})\right] = E_A\left[N_A^0, v_A^0(\mathbf{r})\right]$$

$$+ \left(\frac{\partial E_A}{\partial N_A}\right)_{v_A} \Delta N_A$$

$$+ \int \left[\frac{\delta E_A}{\delta v_A(\mathbf{r})}\right]_{N_A} \Delta v_A(\mathbf{r}) d\mathbf{r}$$

$$+ \frac{1}{2} \left(\frac{\partial^2 E_A}{\partial N_A^2}\right)_{v_A} (\Delta N_A)^2$$

$$+ \int \left[\frac{\delta \partial E_A}{\delta v_A(\mathbf{r}) \partial N_A}\right] \Delta N_A \Delta v_A(\mathbf{r}) d\mathbf{r}$$

$$+ \frac{1}{2} \iint \left[\frac{\delta^2 E_A}{\delta v_A(\mathbf{r}) \delta v_A(\mathbf{r}')}\right]_{N_A} \Delta v_A(\mathbf{r}) \Delta v_A(\mathbf{r}') d\mathbf{r} d\mathbf{r}'$$

$$+ \cdots \tag{34.1}$$

This equation is central in the so-called perturbational perspective to chemical reactivity [8] and introduces a number of response functions (also called charge sensitivities [9]). In conceptual DFT [3,4,8,10–12], it was realized that many chemical concepts, which were earlier often vaguely defined but readily used by chemists, can be identified with these derivatives. As can be seen, some involve the differentiation of the energy with respect to the number of electrons, and these will be mainly discussed in this chapter. The first order derivative of the energy E_A with respect to the number of electrons N_A at constant external potential was proven to be equal to the chemical potential μ, the Lagrange multiplier that is associated with the constraint that the electron density, at all times, should integrate to the total number of electrons when minimizing the energy functional $E[\rho(\mathbf{r})]$ with respect to the density [13]. Moreover, this quantity can be identified with the negative of the electronegativity χ

$$\mu_A = -\chi_A = \left(\frac{\partial E_A}{\partial N_A}\right)_{v_A} \tag{34.2}$$

This is a central quantity in chemistry [14], which was introduced by Pauling as the "power of an atom in a molecule to attract electrons to itself" and was quantified originally by thermochemical data [15]. The density functional definition of this quantity can be viewed as a generalization of the Mulliken definition [16] and is in accordance with earlier work of Iczkowski and Margrave [17].

Parr and Pearson have introduced the absolute hardness as the second derivative of the energy E_A with respect to the number of electrons N_A at constant external potential [18]

$$\eta_A = \frac{1}{2} \left(\frac{\partial^2 E_A}{\partial N_A^2}\right)_{v_A} \tag{34.3}$$

This concept was introduced qualitatively in the late 1950s and early 1960s by Pearson, in the framework of his classification of Lewis acids and bases, leading to the introduction of the hard and soft acids and bases (HSAB) principle [19–21]. This principle states that hard acids prefer to bond to hard bases and soft acids to soft bases. In many contributions, the factor of $1/2$ is omitted. The inverse of the hardness was introduced as the softness $S = 1/\eta$ [22]. A third quantity, which can be expressed as a derivative with respect to the number of electrons is the Fukui function, was introduced by Parr and Yang [23,24]:

$$f_A(\mathbf{r}) = \left(\frac{\delta \partial E_A}{\partial N_A \delta v_A(\mathbf{r})} \right) = \left[\frac{\delta \mu_A}{\delta v_A(\mathbf{r})} \right]_{N_A} = \left(\frac{\partial \rho_A(\mathbf{r})}{\partial N_A} \right)_{v_A} \tag{34.4}$$

This quantity can be viewed as a generalization of Fukui's frontier molecular orbital (MO) concept [25] and plays a key role in linking Frontier MO theory and the HSAB principle. It can be interpreted either as the sensitivity of a system's chemical potential to an external perturbation at a particular point $\mathbf{r}$, or as the change of the electron density $\rho(\mathbf{r})$ at each point $\mathbf{r}$ when the total number of electrons is changed. The former definition has recently been implemented to evaluate this function [26,27] but the derivative of the density with respect to the number of electrons remains by far the most widely used definition.

This chapter will be concerned with computing the three response functions discussed above—the chemical potential, the chemical hardness, and the Fukui function—as reliably as possible for a neutral molecule in the gas phase. This involves the evaluation of the derivative of the energy and electron density with respect to the number of electrons.

34.2 THEORETICAL BACKGROUND—QUANTITIES FOR FRACTIONAL NUMBER OF ELECTRONS

The dependence of the total electronic energy E on the number of electrons was established in a hallmark paper by Perdew et al. [28]. Using an ensemble treatment, these authors demonstrated that a plot of E versus the number of electrons comprises a series of straight line segments, with derivative discontinuities at the integer values of N. These manifest themselves as integer discontinuities in the exact exchange-correlation potential, and as a result, the exchange-correlation potentials on the electron-deficient and electron-abundant sides of the integer, denoted as v_{XC}^- and v_{XC}^+, respectively, will differ by some system-dependent positive constant Δ_{XC} at all points in space [28–30]:

$$v_{XC}^+ - v_{XC}^- = \Delta_{XC} \tag{34.5}$$

Using the fact that the energy is linear with respect to the number of electrons and Janak's theorem [31], the orbital energies of the $N - n$ and $N + n$ electron system become equal to the exact ground state vertical ionization energy and electron affinity, respectively:

$$\mu^- = \left.\frac{\partial E}{\partial N}\right|_{N-n} = \varepsilon_N(N-n) = -I^0 \tag{34.6}$$

$$\mu^+ = \left.\frac{\partial E}{\partial N}\right|_{N+n} = \varepsilon_{N+1}(N+n) = -A^0 \tag{34.7}$$

where $0 < n < 1$. In these equations, μ^- represents the chemical potential of the N-electron system evaluated from the electron deficient side; μ^+ is the same quantity from the electron-abundant side. The fact that derivative discontinuity occurs at the integers, implies that the derivative in Equation 34.2 does not exist formally for integer N, as the left and right side derivatives μ^- and μ^+ are different. In practical applications, one uses the average of both quantities as the estimate of the chemical potential of the system (we will now denote this quantity as μ^0):

$$\mu^0 = \frac{\mu^- + \mu^+}{2} = -\frac{I^0 + A^0}{2} \tag{34.8}$$

Since $\mu = -\chi$, this approximation for μ reduces to the Mulliken definition for the electronegativity [16]. Within the ensemble approach, the hardness η as defined in Equation 34.3 would be zero for noninteger N and undefined for integer values of the electron number. One obtains quantitative values for this concept for integer N, using a finite difference approximation of the chemical potentials μ^+ and μ^-, i.e.,

$$\eta^0 \equiv \frac{1}{2}(\mu^+ - \mu^-) = \frac{1}{2}(-A^0 + I^0) = \frac{I^0 - A^0}{2} \tag{34.9}$$

which is the original Parr and Pearson definition of the absolute hardness [18]. It is interesting to note that the estimates obtained in Equations 34.8 and 34.9 from the concepts defined in Equations 34.2 and 34.3 can also be obtained by assuming a quadratic relationship between E and N, for which certain qualitative arguments can be given [8].

Consider again Equations 34.6 and 34.7. When $n \rightarrow 0$, the orbital energies can be identified with the highest occupied molecular orbital (HOMO) and lowest unoccupied molecular orbital (LUMO) energies of the N-electron system, determined using v_{XC}^- and v_{XC}^+ respectively. The exact HOMO energy of the N-electron system determined with v_{XC}^- is thus exactly equal to the vertical ionization energy of the N-electron system.

$$\varepsilon_{HOMO}^- = -I^0 \tag{34.10}$$

whereas the exact LUMO energy determined with v_{XC}^+ is exactly equal to the vertical electron affinity of the N electron system:

$$\varepsilon_{LUMO}^+ = -A^0 \tag{34.11}$$

Finally, the expressions for μ^0 and η^0 become

$$\mu^0 = \frac{\varepsilon_{HOMO}^- + \varepsilon_{LUMO}^+}{2} \tag{34.12}$$

$$\eta^0 = \frac{\varepsilon_{LUMO}^+ - \varepsilon_{HOMO}^-}{2} \tag{34.13}$$

Not only the derivative of the energy shows integer discontinuities, the same is true for the derivative of the electron density with respect to particle number. This implies that the change in the electron density ρ_A due to an infinitesimal increase in the number of electrons N_A is different from the density change due to an infinitesimal decrease of N_A. The derivative associated with the first process is f_A^+, which is defined as

$$f_A^+(\mathbf{r}) \equiv \left(\frac{\partial \rho_A(\mathbf{r})}{\partial N_A}\right)_{v_A}^+ \tag{34.14}$$

where the superscript "$+$" on the derivative indicates that the derivative is taken on the electron-abundant side of the integer N_A. It can be anticipated that the molecule A will readily accept electrons into regions where this function is large and thus this function constitutes a reactivity index to probe the attack of a model nucleophile [23,24]. Next, consider the change in ρ_A upon loss of electrons N_A:

$$f_A^-(\mathbf{r}) \equiv \left(\frac{\partial \rho_A(\mathbf{r})}{\partial N_A}\right)_{v_A}^- \tag{34.15}$$

In this case, the superscript "$-$" on the derivative indicates that the derivative is taken on the electron-deficient side of the integer N_A; one can expect that the molecule will readily donate electrons from regions where $f_A^-(\mathbf{r})$ will be large and the derivative can thus be used to probe an electrophilic attack [23,24]. The average of these quantities was introduced as the Fukui function for a neutral (radical) attack [23,24]

$$f_A^0(\mathbf{r}) \equiv \frac{f_A^+(\mathbf{r}) + f_A^-(\mathbf{r})}{2} \tag{34.16}$$

Perdew et al. also showed that the electron density entering the definition of the energy functional for a non-integer number of electrons is also an ensemble sum [28]:

$$\rho_{N+n} = (1 - n)\rho_N + n\rho_{N+1} \tag{34.17}$$

again with $0 < n < 1$. When using this equation, i.e., in the grand canonical ensemble, zero temperature limit, it can be shown that the finite difference approximations to these equations are exact [24]:

$$f_A^+(\mathbf{r}) = \rho_{A,N+1}(\mathbf{r}) - \rho_{A,N}(\mathbf{r}) \tag{34.18}$$

and

$$f_A^-(\mathbf{r}) = \rho_{A,N}(\mathbf{r}) - \rho_{A,N-1}(\mathbf{r}) \tag{34.19}$$

where $\rho_{A,N+1}(\mathbf{r})$, $\rho_{A,N}(\mathbf{r})$, and $\rho_{A,N-1}(\mathbf{r})$ are all evaluated at the external potential of the N-electron system. These equations are traditionally used to evaluate this reactivity index, which can then be combined with the global softness of the system to yield the local softness [22], the descriptor of choice when evaluating orbital-controlled reactions [32–34].

The natural way to approximate the chemical potential and chemical hardness in DFT is to evaluate them directly from the calculated ionization energy and electron affinity

$$\mu = -\frac{I+A}{2} \tag{34.20}$$

and

$$\eta = \frac{I-A}{2} \tag{34.21}$$

where I and A are obtained from total electronic energy calculations on the $N-1$, N, and $N+1$ electron systems, at the geometry of the neutral molecule:

$$I = E_{N-1} - E_N \tag{34.22}$$

$$A = E_N - E_{N+1} \tag{34.23}$$

The experimental ionization energy can typically be reproduced within a few tenths of an electronvolt by standard DFT functionals (see, e.g., Ref. [5]). The same is true for positive electron affinities [5,35,36]; a positive experimental electron affinity indicates that the anion is stable with respect to electron loss. In such cases, the chemical potential and hardness can be calculated to a similar accuracy [35]. However, in many cases, the experimental electron affinity is negative rather than positive, as measured experimentally by the technique of electron transmission spectroscopy [37,38]. These anions pose a fundamental problem: they are unstable with respect to electron loss and so cannot be described by a standard DFT ground state total energy calculation. In practice, medium-sized basis set calculations on the anion do give energies above that of the neutral, and so reasonable estimates for the negative affinity can be obtained. However, this simply reflects an artificial binding of the electron by the finite basis set. The addition of diffuse functions allows the electron to leave the system, and so the electron affinity becomes near-zero. Consequently, the chemical potential and hardness become $\sim -I/2$ and $\sim I/2$, respectively. This basis set dependence makes Equations 34.20 and 34.21 a less attractive approach for calculating the chemical potential or hardness of a system with a significant negative experimental electron affinity. The same problems of course arise when one wishes to compute the electron densities associated with these metastable anions. In the gas phase, this is always necessary for the computation of f_A^+ and f_A^0. In the following sections, an approximate solution to these problems will be presented, enabling the qualitative estimate of chemical potentials, hardness values, and Fukui functions of systems with metastable anions.

34.3 APPROXIMATE EXPRESSIONS FOR LOCAL FUNCTIONALS

Local exchange-correlation functionals such as generalized gradient approximations (GGA) are continuum approximations, which can, at best, average over the discontinuity. In regions where the HOMO and LUMO are significant, they provide an approximate average description [39–41]:

$$v_{XC} \approx \frac{v_{XC}^+ + v_{XC}^-}{2} \tag{34.24}$$

From this, it follows that the eigenvalues of the LUMO and HOMO orbitals obtained by a local functional are approximately shifted from the exact values by

$$\varepsilon_{LUMO} \approx \varepsilon_{LUMO}^+ - \frac{\Delta_{XC}}{2} = \mu^+ - \frac{\Delta_{XC}}{2} \tag{34.25}$$

and

$$\varepsilon_{HOMO} \approx \varepsilon_{HOMO}^- + \frac{\Delta_{XC}}{2} = \mu^- + \frac{\Delta_{XC}}{2} \tag{34.26}$$

Numerical examples of this are available in Ref. [42]. Rearranging Equations 34.25 and 34.26 and substituting into Equations 34.12 and 34.13, gives

$$\mu^0 \approx \frac{\varepsilon_{HOMO} + \varepsilon_{LUMO}}{2} \tag{34.27}$$

and

$$\eta^0 \approx \frac{\varepsilon_{LUMO} - \varepsilon_{HOMO}}{2} + \frac{\Delta_{XC}}{2} \tag{34.28}$$

It therefore follows that a Koopmans-type expression [43] for the chemical potential in terms of GGA HOMO and LUMO energies:

$$\mu \approx \frac{\varepsilon_{HOMO} + \varepsilon_{LUMO}}{2} \tag{34.29}$$

will be reasonably accurate. However, the analogous expression for the hardness [43,44]

$$\eta \approx \frac{\varepsilon_{LUMO} - \varepsilon_{HOMO}}{2} \qquad (34.30)$$

underestimates the exact hardness by approximately half the integer discontinuity. We could gain an improved approximation to the hardness if Δ_{XC} was known. This quantity can be determined from high-quality electron densities, but a more appealing and more easily applicable route is to use the fact that $\Delta_{XC}/2$ is the exact asymptotic potential for a functional that averages over v_{XC}^{+} and v_{XC}^{-}. It is therefore approximately equal to the exact asymptotic potential $v_{XC}(\infty)$ of a local functional, which in turn can be well approximated by the generalized Koopmans' theorem

$$v_{XC}(\infty) \approx \varepsilon_{HOMO} + I \qquad (34.31)$$

where ε_{HOMO} is the local functional HOMO orbital energy and I is an approximate ionization energy determined from the neutral and cation total energies. Hence [45,46]

$$\frac{\Delta_{XC}}{2} \approx \varepsilon_{HOMO} + I \qquad (34.32)$$

which can be combined with Equation 34.28 to give an unconventional hardness expression

$$\eta = \frac{\varepsilon_{LUMO} - \varepsilon_{HOMO}}{2} + \varepsilon_{HOMO} + I \qquad (34.33)$$

Hence, the hardness can be approximated as half the HOMO–LUMO gap plus the near-exact asymptotic potential [47].

Equation 34.33 has a simple interpretation in terms of the chemical potential. One can rewrite it as

$$\eta = \frac{\varepsilon_{LUMO} + \varepsilon_{HOMO}}{2} + I = \mu + I \qquad (34.34)$$

corresponding to the addition of an approximate ionization energy to the chemical potential in Equation 34.29. The key point is that the chemical potential can be approximated using Equation 34.29, because taking the sum approximately cancels the contributions from Δ_{XC} in Equations 34.25 and 34.26. Finally, it can also be seen that the hardness expression can be rewritten as

$$\eta = \frac{I - (-\varepsilon_{LUMO} - \varepsilon_{HOMO} - I)}{2} \qquad (34.35)$$

which corresponds to evaluating Equation 34.21 with the conventional ionization energy but the unconventional electron affinity

$$A = -\varepsilon_{LUMO} - \varepsilon_{HOMO} - I \qquad (34.36)$$

This expression enables the computation of the electron affinity without the explicit computation of the energy of the anion, which will prove to be beneficial for the calculation of negative affinities [47].

34.4 COMPUTATION OF CHEMICAL POTENTIAL AND HARDNESS

We now assess the quality of the various chemical potential and hardness approximations [47]. We consider a series of representative, neutral closed-shell molecules containing first and second row atoms: F_2, Cl_2, H_2CO, C_2H_4, CO, PH_3, H_2S, HCN, HCl, CO_2, NH_3, HF, H_2O, and CH_4. The experimental electron affinity is positive for the first two molecules, but becomes increasingly negative across the series. Where possible, calculations were performed at near-experimental reference geometries taken from Ref. [30]; for the molecules without sufficiently accurate reference geometry, MP2/aug-cc-pVTZ geometries were used instead. All chemical potential and hardness calculations were performed using the aug-cc-pVTZ basis set [48] using the CADPAC [49] program. Experimental values for the chemical potentials and absolute hardness were determined using Equations 34.8 and 34.9, with vertical ionization energies and electron affinities from Ref. [50]. We confirmed that variations in the results due to the choice of functional were minimal and were significantly smaller than the variations due to the different chemical potential and hardness expressions; we use the Perdew–Burke–Ernzerhof (PBE) functional [51] throughout.

Table 34.1 presents the chemical potential and hardness values relative to the experiment. Mean and mean absolute errors, relative to the experiment, are denoted d and $|d|$, respectively. First consider the results from Equations 34.20 and 34.21, which use the calculated I and A values. The chemical potential and hardness values are fairly accurately reproduced for both F_2 and Cl_2, both of which have positive electron affinities. However, the results degrade as the table is descended, reflecting the increasingly negative experimental electron affinity. As you descend the table, the computed affinities degrade and they are not sufficiently negative. The chemical potential and hardness values in Table 34.1 significantly underestimate the experiment, with a mean absolute error of about 1.4 eV. The third and fourth columns of values in Table 34.1 list the chemical potential and hardness values determined using the HOMO–LUMO orbital energies (Equations 34.29 and 34.30). (For all systems, the HOMO and LUMO energies are negative.) In the case of the chemical potential, as anticipated, a significant improvement is encountered due to the cancellation of Δ_{XC}; the mean absolute error is just 0.6 eV. By contrast, the hardness values are significantly and uniformly underestimated, reflecting the absence of the discontinuity term in Equation 34.28. The mean absolute error is 4.7 eV. The values in the fifth column of Table 34.1 are determined from Equation 34.33, through simple correction of the HOMO–LUMO gap values. The uniform underestimation is eliminated (the mean error is near zero) and the mean absolute error is significantly reduced to 0.5 eV, which is now comparable to the error in the chemical potentials. Admittedly, this error is larger than that can be obtained from direct evaluation using Equation 34.21, on

TABLE 34.1

Chemical Potential and Hardness Values (in eV) Determined Using the Different Approaches with the PBE Functional and the aug-cc-pVTZ Basis Set, Compared to Experimental Values Determined Using the Data in Ref. [50]

Molecule	$\mu = -\dfrac{I+A}{2}$ Equation 34.20	$\eta = \dfrac{I-A}{2}$ Equation 34.21	$\mu = \dfrac{\varepsilon_{HOMO} + \varepsilon_{LUMO}}{2}$ Equation 34.29	$\eta = \dfrac{\varepsilon_{LUMO} - \varepsilon_{HOMO}}{2}$ Equation 34.30	$\eta = \dfrac{\varepsilon_{LUMO} - \varepsilon_{HOMO}}{2} + \varepsilon_{HOMO} + I$ Equation 34.33	$\mu^o = -\dfrac{I^o + A^o}{2}$ Equation 34.8	$\eta^o = \dfrac{I^o - A^o}{2}$ Equation 34.9
F_2	−7.98	7.36	−7.62	1.83	7.72	−8.5	7.2
Cl_2	−6.03	5.16	−5.76	1.54	5.42	−6.3	5.2
H_2CO	−5.09	5.66	−4.45	1.79	6.29	−4.7	6.2
C_2H_4	−5.08	5.58	−3.92	2.84	6.74	−4.4	6.2
CO	−6.42	7.44	−5.52	3.52	8.34	−6.1	7.9
PH_3	−5.08	5.44	−3.68	3.05	6.84	−4.3	6.2
H_2S	−5.02	5.38	−3.58	2.73	6.82	−4.2	6.3
HCN	−6.78	7.17	−5.07	3.97	8.88	−5.7	8.0
HCl	−6.17	6.56	−4.57	3.47	8.15	−4.7	8.0
CO_2	−6.47	7.20	−4.98	4.10	8.69	−5.0	8.8
NH_3	−5.26	5.70	−3.45	2.73	7.51	−2.6	8.2
HF	−7.93	8.39	−5.31	4.34	11.01	−5.1	11.1
H_2O	−6.19	6.61	−4.09	3.16	8.71	−3.1	9.5
CH_4	−6.75	7.22	−4.91	4.55	9.06	−2.9	10.7
d/eV	−1.33	−1.34	0.05	−4.71	0.04		
$|d|$/eV	1.44	1.36	0.63	4.71	0.51		
m	0.82	1.41	1.20	1.48	1.16		
c/eV	0.21	−1.3	0.90	3.21	−1.31		
R^2	0.28	0.64	0.73	0.64	0.87		

Note: d and $|d|$ denote mean and mean absolute errors, respectively, relative to the experiment. m, c, and R^2 denote the gradient, intercept, and correlation parameters, respectively, of the correlation plots, relative to the experiment.

systems with positive electron affinities, and this can be traced to the inherent approximations used in the derivation of Equation 34.33. However, hardness values determined using this expression do not degrade as you descend the table; it is as accurate for systems near the bottom of the table as it is for F_2 and Cl_2.

In applications, the chemical potentials and hardness are used to discuss trends in chemical reactivity. It is thus important that calculated values exhibit a good correlation with experimental values. To investigate this, Table 34.1 also presents the line parameters m (slope), c (intercept), and R^2 (square of the correlation coefficient), describing the correlation of the two chemical potentials and the three hardness expressions with the experimental values. The correlation plots are given in Figures 34.1 and 34.2. In the case of chemical potential, the correlation between the experimental values and the calculated values using computed I and A values is nonexistent. As can be seen, estimating this quantity using the frontier orbital energies largely improves the correlation, the R^2 value now increasing to 0.73 and the slope is somewhat higher than unity. When dropping the entry for the CH_4 molecule, the R^2 value improves to 0.87, which is comparable to the slope for the correlation with the new expression for chemical hardness. In the case of hardness, calculations using the conventional approaches (Equations 34.21 and 34.30) have similar slopes ($m = +1.41$ to $+1.48$) and correlation parameters ($R^2 = 0.64$) which are far from unity. The intercept is particulary large ($c = +3.21$ eV) when Equation 34.30 is used. The correlation is much better for the third approach, Equation 34.33, with improved slope ($m = +1.16$) comparable to the slope of the correlation curve for the chemical potential, and correlation parameter ($R^2 = 0.87$).

The focus until now has been on the evaluation of the chemical potential and the hardness, without explicit computation of the electron affinity. This has been achieved by implicitly approximating the affinity in terms of the Kohn–Sham eigenvalues and the ionization energy (Equation 34.36). Table 34.2 presents electron affinities determined using this expression, with the aug-cc-pVTZ basis set. For systems with significant negative experimental affinities, the results are a notable improvement over both the aug-cc-pVTZ and cc-pVTZ results from the conventional evaluation (Equation 34.23). The new results exhibit a correlation parameter of $R^2 = 0.76$, compared to values of 0.30 and 0.48 from the standard expression with the large and small basis sets, respectively. Next, we have investigated a much more elaborate set of molecules with known experimental negative electron affinities, as considered previously in Ref. [52], limiting ourselves to neutral closed-shell systems [53]. It should be remarked that the experimental ETS values are in a rather narrow range, i.e., from -0.07 to -3.80 eV. All geometries of the neutral molecules are now optimized at the B3LYP [54]/ 6-311+ G** [55] level. Table 34.3 compares the results from Equation 34.36 determined using the aug-cc-pVTZ basis set with conventional results from Equation 34.23, determined using both aug-cc-pVTZ and cc-pVTZ. Once again, the best results are obtained using Equation 34.36, with a mean absolute error of just 0.49 eV and a correlation coefficient of 0.92. Figure 34.3 presents the correlation plot for the latter expression for the affinity.

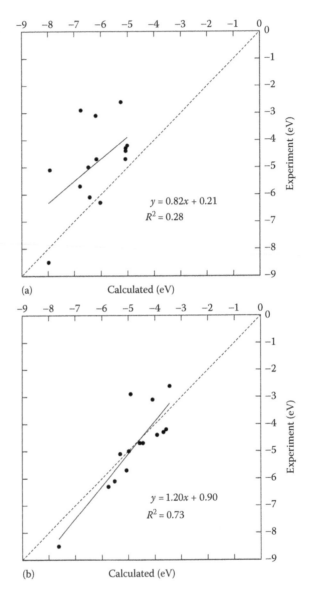

(a)   Calculated (eV)

(b)   Calculated (eV)

FIGURE 34.1 Correlation of the chemical potential, calculated using (a) Equation 34.20; (b) Equation 34.29 with experimental values. Data are taken from Table 34.1.

34.5 COMPUTATION OF THE FUKUI FUNCTION FOR A NUCLEOPHILIC ATTACK $F^+(\mathbf{r})$

Until now, we have been concerned with the computation of global properties. We now consider the Fukui function for nucleophilic attack (Equation 34.18), which is a local property that requires the calculation of the electron density of the anion, which

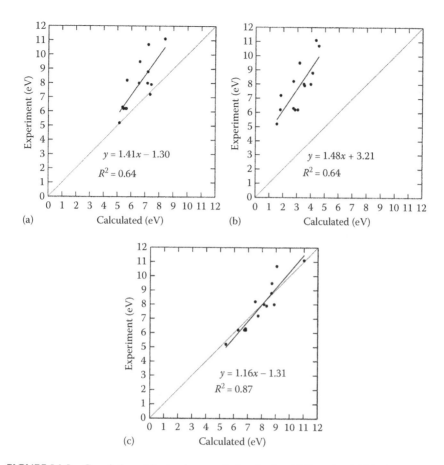

FIGURE 34.2 Correlation of the hardness, calculated using (a) Equation 34.21; (b) Equation 34.30; and (c) Equation 34.33, with experimental values. Data are taken from Table 34.1. (Reprinted from Tozer, D.J. and De Proft, F., *J. Phys. Chem. A*, 109, 8923, 2005. With permission.)

in many cases will be metastable. How can the findings we gathered in the previous section allow us to calculate electron densities of these systems? The key issue is to bind the excess electron, and we shall consider two approaches to do this [56]. The first is simply to choose an artificially compact basis set; the second is to apply a potential wall. The key feature of the calculations is that the degree of binding will be controlled by a knowledge of the negative affinity, estimated using the unconventional expression in Equation 34.36.

First, we summarize our potential wall approach which uses ideas from Ref. [45]. A sphere of radius R_X is constructed around each constituent atom X

$$R_X = \lambda B_X \tag{34.37}$$

TABLE 34.2

Electron Affinities Calculated Using Equation 34.23, with the aug-cc-pVTZ and cc-pVTZ Basis Sets, and Equation 34.36, with the aug-cc-pVTZ Basis, Compared to the Experimental Affinity, A^0

Molecule	A (Equation 34.23)[a]	A (Equation 34.23)[b]	A (Equation 34.36)[a]	A^0
F_2	+0.63	−0.43	−0.10	+1.24
Cl_2	+0.87	+0.47	+0.34	+1.02
H_2CO	−0.58	−1.51	−1.84	−1.5
C_2H_4	−0.50	−2.46	−2.82	−1.8
CO	−1.02	−2.34	−2.82	−1.8
PH_3	−0.37	−2.59	−3.16	−1.9
H_2S	−0.37	−2.30	−3.24	−2.1
HCN	−0.38	−3.47	−3.81	−2.3
HCl	−0.39	−2.21	−3.58	−3.3
CO_2	−0.72	−3.31	−3.70	−3.8
NH_3	−0.44	−2.65	−4.06	−5.6
HF	−0.47	−2.91	−5.70	−6.0
H_2O	−0.42	−2.71	−4.62	−6.4
CH_4	−0.47	−2.75	−4.15	−7.8

Source: Reprinted from Tozer, D.J. and De Proft, F., *J. Phys. Chem. A*, 109, 8923, 2005. With permission.

Note: All calculated quantities were obtained using the PBE functional. All quantities are in eV.

[a] Calculated using the aug-cc-pVTZ basis set.

[b] Calculated using the cc-pVTZ basis set.

TABLE 34.3

Electron Affinities Calculated Using Equation 34.23, with the aug-cc-pVTZ and cc-pVTZ Basis Sets (Equation 34.36), with the aug-cc-pVTZ Basis, Compared to the Experimental Affinity, A^0

Molecule	A Equation 34.23[a]	A Equation 34.23[b]	A Equation 34.36[b]	A^0
1,1-Dichloroethylene	−1.53	−0.36	−1.39	−0.75
1,3-Cyclohexadiene	−1.13	−1.67	−1.14	−0.80
Acetaldehyde	−1.84	−2.82	−2.04	−1.19
Adenine	−1.08	−0.09	−0.96	−0.64
Bromobenzene	−1.16	−0.22	−1.05	−0.70
Chlorobenzene	−1.23	−0.28	−1.15	−0.75
Chloroethylene	−1.98	−0.44	−1.98	−1.29
Chloromethane	−2.26	−0.38	−3.13	−3.45
cis-Dichloroethylene	−1.63	−0.35	−1.52	−1.12
Cytosine	−1.01	0.03	−0.84	−0.36
Ethylene	−2.52	−0.51	−2.82	−1.78

(*continued*)

TABLE 34.3

Electron Affinities Calculated Using Equation 34.23, with the aug-cc-pVTZ and cc-pVTZ Basis Sets (Equation 34.36), with the aug-cc-pVTZ Basis, Compared to the Experimental Affinity, A^0

Molecule	A Equation 34.23[a]	A Equation 34.23[b]	A Equation 34.36[b]	A^0
Fluorobenzene	−1.39	−0.33	−1.41	−0.87
Naphthalene	−0.37	−0.21	−0.34	−0.20
Norbornadiene	−1.40	−0.35	−1.54	−1.04
Pyrazine	−1.29	−0.20	−0.55	−0.07
Pyridazine[c]	−0.56	—	−0.87	−0.32
Pyrimidine	−0.76	−0.45	−0.88	−0.25
Styrene[c]	−0.52	—	−0.51	−0.25
Thiophene	−1.60	−0.39	−1.64	−1.17
trans-Dichloroethylene	−1.49	−2.28	−1.35	−0.82
Trichloroethylene	−1.19	−0.36	−1.01	−0.58
Uracil	−0.68	−0.31	−0.63	−0.21
1,2,4-Trimethylbenzene[c]	−1.62	—	−1.55	−1.07
Acetone	−1.79	−0.29	−2.14	−1.51
Aniline[c]	−0.25	—	−1.71	−1.13
Anisole	−1.61	−0.28	−1.55	−1.09
cis-Butene	−2.51	−0.38	−2.70	−2.22
Cyclohexene	−2.07	−0.34	−2.45	−2.07
Furan	−2.41	−0.41	−2.37	−1.76
m-Xylene	−1.60	−0.32	−1.57	−1.06
o-Xylene	−1.66	−0.32	−1.65	−1.12
Phenol	−1.63	−0.25	−1.59	−1.01
Propene	−2.62	−0.45	−2.74	−1.99
Pyrrole	−2.29	−0.29	−2.50	−2.38
trans-Butene	−2.71	−0.43	−2.77	−2.10
Trimethylethylene	−2.56	−0.36	−2.51	−2.24
CO_2	−3.48	−0.76	−3.71	−3.80
Guanine	−1.60	0.12	−1.17	−0.46
d (eV)	−0.44	0.77	−0.47	
|d| (eV)	0.53	1.02	0.49	
m	1.13	0.09[d]	1.03	
c (eV)	0.65	−1.22[d]	0.51	
R^2	0.80	0.00[d]	0.92	

Source: Reprinted partially from De Proft, F., Sablon, N., Tozer, D.J., and Geerlings, P., *Faraday Discuss.*, 135, 151, 2007. With permission.

Note: All calculated quantities were obtained using the PBE functional. All quantities are in eV.

[a] Calculated using the cc-pVTZ basis set.

[b] Calculated using the aug-cc-pVTZ basis set.

[c] Calculation of the electron affinity using Equation 34.23 could not be performed because the energy-calculation did not converge.

[d] These values are not representative because of the low value of R^2.

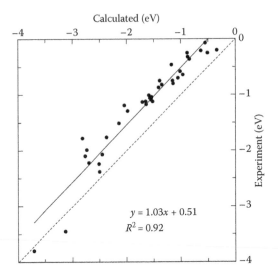

FIGURE 34.3 Correlation of the vertical electron affinity, calculated using Equation 34.36 at the PBE/aug-cc-pVTZ level, with experimental values. All values are in eV; data are taken from Table 34.3. (Reprinted from De Proft, F., Sablon, N., Tozer, D.J., and Geerlings, P., *Faraday Discuss.*, 135, 151, 2007. With permission.)

where

    λ is a dimensionless parameter to be determined

    B_X is the Bragg–Slater radius of atom X

Within any sphere, the exchange-correlation potential in the Kohn–Sham equation is defined to be the conventional functional derivative of E_{XC}:

$$v_{XC}^{\sigma}(\mathbf{r}) = \frac{\delta E_{XC}[\rho^{\alpha}, \rho^{\beta}]}{\delta \rho^{\sigma}(\mathbf{r})} \tag{34.38}$$

where σ denotes α or β spin. However, for points that lie outside all the spheres, the conventional potential is replaced by a constant, spin-dependent potential, in order to bind the excess electron

$$v_{XC}^{\sigma}(\mathbf{r}) = \mu^{\sigma} \tag{34.39}$$

A Kohn–Sham calculation is then performed on the anion using the potential defined according to Equations 34.38 and 34.39. The electronic energy of the anion is determined using the conventional Kohn–Sham energy expression with the regular E_{XC} term.

    The constants μ^{α} and μ^{β} determine the height of the potential wall and so control the asymptotic decay of the α and β anion orbitals. A natural choice is to demand that, in the limit of a complete basis set, the α HOMO of the anion decays as the LUMO of the neutral and the β HOMO of the anion decays as the HOMO of the neutral. Mathematically, this requires

$$\mu^{\alpha} = \varepsilon^{\alpha}_{HOMO,N+1} - \varepsilon_{LUMO,N}$$
$$\mu^{\beta} = \varepsilon^{\beta}_{HOMO,N+1} - \varepsilon_{HOMO,N} \tag{34.40}$$

where

$\varepsilon^{\alpha}_{HOMO,N+1}$ and $\varepsilon^{\beta}_{HOMO,N+1}$ are the α and β HOMO eigenvalues of the anion in the self-consistent-field procedure

$\varepsilon_{LUMO,N}$ and $\varepsilon_{HOMO,N}$ are the LUMO and HOMO eigenvalues of the neutral, obtained from a separate calculation

Next, we consider the only remaining quantity to be defined, λ, introduced in Equation 34.37. When λ is large, the effect of the wall will be negligible and the calculations will revert to standard Kohn–Sham calculations; if a diffuse basis set is used, the anion energy will be close to that of the neutral, yielding a near-zero electron affinity. As λ becomes smaller, however, the potential wall will artificially bind the excess electron, causing the anion energy to increase and the affinity to become increasingly negative. The wall therefore has the same effect as a compact basis set, without the loss of mathematical flexibility. In our approach, we now choose λ to be of value for which the electron affinity determined using Equation 34.23 equals that given by Equation 34.36. The Fukui function is then evaluated using the corresponding electron density. It should be noted that we do not apply the potential wall in the calculation of the electronic energy, eigenvalues, and ionization energy of the neutral, as it would break the spin symmetry.

We will consider the molecules H_2CO, C_2H_4, NH_3, and H_2O, which have increasingly negative electron affinities. Again, all calculations reported were performed using the PBE GGA functional [51], with the CADPAC program [49]. Affinities determined using Equation 34.36 with the aug-cc-pVTZ basis set [48] (as obtained in Section 34.4) are -1.84, -2.84, -4.06, and -4.62 eV, respectively. In order to maintain a fully theoretical analysis, we shall use the values from Equation 34.36 as our reference values, denoting them as A_{ref}. All molecules were oriented as to have their highest order axis coinciding with the z-axis; planar molecules were in the yz plane. The Fukui function was then plotted in the z-direction, 0.5 a.u. above this axis in the x-direction.

We first consider conventional Kohn–Sham calculations. Table 34.4 presents electron affinities determined using Equation 34.23 for a series of basis sets, compared with the A_{ref} values. The first row of numbers was determined using the extensive, diffuse aug-cc-pVTZ basis set. The affinities are significantly less negative than A_{ref}, reflecting the tendency for the excess electron to leave the system. The density of the anion is therefore rather close to the density of the neutral, and so the Fukui functions $f^+(\mathbf{r})$ which are presented with long-dashed lines in Figure 34.4a through d, have a relatively small magnitude. We regard these Fukui functions as unreliable. The remainder of Table 34.4 highlights the significant effect that the

TABLE 34.4

Electron affinities, A, Determined Using Equation 34.23 with Conventional DFT Calculations (No Potential Wall), for a Series of Basis Sets, Compared with Reference Values, A_{ref}, Determined Using Equation 34.36 with the aug-cc-pVTZ Basis Set

Basis set	H_2CO	C_2H_4	NH_3	H_2O
$A = E_N - E_{N+1}$ (Equation 34.23)				
aug-cc-pVTZ	−0.58	−0.50	−0.44	−0.42
3-21G	−2.75	−3.63	−5.87	−5.99
6-31G	−2.15	−3.40	−4.58	−4.46
6-31G*	−2.28	−3.38	−4.67	<u>−4.58</u>
6-311G	−1.78	−2.86	−3.09	−2.97
6-311G*	<u>−1.88</u>	<u>−2.83</u>	−3.15	−3.06
cc-pVDZ	−2.10	−2.98	<u>−3.90</u>	−3.97
cc-pVTZ	−1.51	−2.46	−2.65	−2.71
$A_{ref} = -(\varepsilon_{LUMO} + \varepsilon_{HOMO}) - I$ (Equation 34.36)				
aug-cc-pVTZ	−1.84	−2.82	−4.06	−4.62

Source: Reprinted from Tozer, D.J. and De Proft, F., *J. Chem. Phys.*, 127, 034108, 2007. With permission.

Note: All values are in eV. Underlined values are those for which A is in closest agreement with A_{ref}.

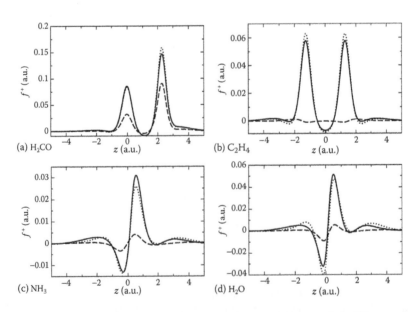

FIGURE 34.4 Fukui functions determined using Equation 34.18. Long-dashed lines are determined using conventional calculations (no potential wall) with the aug-cc-pVTZ basis set; dotted lines use conventional calculations with the optimally compact basis sets identified in Table 34.4; solid lines use the potential wall approach of Section 34.5, with the aug-cc-pVTZ basis. (Reprinted from Tozer, D.J. and De Proft, F., *J. Chem. Phys.*, 127, 034108, 2007. With permission.)

compact basis sets artificially have on the affinities determined using Equation 34.23. For each molecule, the affinity underlined is the value that best agrees with A_{ref}. For H_2CO and C_2H_4, which have relatively small magnitude negative affinities, the 6-311G and the 6-311G* basis sets give the best agreement. For NH_3 and H_2O, which have much more negative affinities, the more compact cc-pVDZ and 6-31G* basis sets are necessary. The dotted lines in Figure 34.4a through d present the Fukui functions determined using the optimal basis set identified in Table 34.4. As can be seen, the Fukui functions now exhibit significantly more features; they represent the best that can be obtained with the conventional approach. The solid lines in Figure 34.4a through d are the Fukui functions calculated by the potential wall approach,

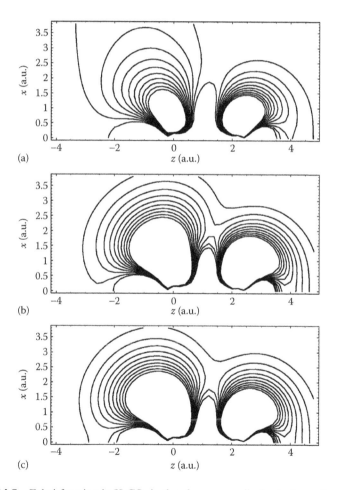

FIGURE 34.5 Fukui function in H_2CO, in the plane perpendicular to the molecular plane. The carbon atom is at $z=0$, $x=0$; the oxygen atom is at $z=2.27$ a.u., $x=0$. The increment between contour lines is $+0.001$ a.u.; the lowest contour line is at 0.0 a.u. See text for details. (Reprinted from Tozer, D.J. and De Proft, F., *J. Chem. Phys.*, 127, 034108, 2007. With permission.)

computed for each system using the optimal values of λ. They are very similar to those obtained using the optimally compact basis sets. In a later contribution [57], it was shown, for the large number of molecules discussed in Section 34.4, that the optimal value of λ is fairly constant over a whole range of molecules. It was also shown that this approach provides an interesting method to study the molecular orbitals and spin densities of metastable anions.

Finally, Figure 34.5 presents contour plots of $f^+(\mathbf{r})$ for H_2CO. Figure 34.5a uses the conventional DFT calculation with the diffuse, aug-cc-pVTZ basis set. The features around the C atom (located at $z = 0$, $x = 0$) are exaggerated; the excess electron, which is tending to leave the molecule, occupies an orbital with significant contributions from diffuse basis functions centered around this atom. Figure 34.5b uses conventional calculations with the optimally compact 6-311G* basis set; the tendency for the electron to leave has been removed. Figure 34.5c uses the approach of this section with the aug-cc-pVTZ basis; the plot is very similar to that in Figure 34.5b. As can be seen, all three figures clearly identify the C atom as the site of nucleophilic attack, as evidenced by the more extended contours around this atom.

34.6 CONCLUSIONS

In this chapter, we have focussed on the calculation of three important reactivity indices: the chemical potential, the chemical hardness, and the Fukui function $f^+(\mathbf{r})$ for a nucleophilic attack on neutral molecules in the gas phase. For the first two global molecular properties, it is shown that the conventional approach, based on which these quantities are evaluated directly from computed DFT ionization energies and electron affinities, works reasonably well for molecules with positive electron affinities. However, this direct approach is less appropriate for systems with a significant negative experimental electron affinity. Next, based on an analysis used in earlier studies [28,39–41], we have reiterated why an alternative approximation, based on Koopmans' theorem [43] using local functional eigenvalues, is a good approximation for the chemical potential but significantly underestimates the hardness. We have used this analysis to design a hardness approximation [47] that can be interpreted as a simple correction to the Koopmans' expression (see Equation 34.33); the correction eliminates the uniform underestimation. Based on the study on a set of typical systems, it was established that the inherent approximations in the derivation lead to relatively large errors of 0.5 eV on an average, but the results do not degrade as the electron affinity becomes more negative and the correlation with experiment is good. For systems with large negative experimental electron affinities, the results are an improvement over those of the conventional approaches. An interesting interpretation of these schemes is that they correspond to a regular evaluation using a conventional approximate ionization energy from Equation 34.22, but an unconventional electron affinity from Equation 34.36. This latter expression was subsequently proven to be useful in studies of negative affinities, overcoming the fundamental basis set breakdown of the standard approach [47,53]. Clearly, this is still a relatively open field of research and further work is needed in establishing a deeper understanding of the good performance of this equation.

Next, we investigated two approaches for modeling electron densities of temporary anions in DFT [56], among others used in the evaluation of the Fukui function. The calculations rely on an artificial binding of the excess electron in the anion, in one case by a compact basis set and in the other by a potential wall. The key feature of the calculations is that the degree of binding is controlled in both cases by a knowledge of the negative electron affinity of the corresponding neutral, approximated using the unconventional expression for the electron affinity in Equation 34.36. In the former case, the affinity is used to identify an optimally compact basis set. In the latter, it is used to choose the Bragg–Slater radius scaling parameter λ which determines the position of the potential wall; the height of the wall is independently obtained from a consideration of the asymptotic decay of the anion HOMO and LUMO. As an illustration, Fukui functions for nucleophilic attack have been determined for four molecules with increasingly negative affinities. They yield very similar results, which are notably different from those determined without artificial electron binding. Contour plots for H_2CO exhibit the same features. It turns out from a study on the large set of molecules possessing negative affinities used previously [57], that the optimal value of the parameter λ does not vary significantly for different molecular systems.

ACKNOWLEDGMENTS

The authors acknowledge many people for discussions and collaborations on many different aspects related to this chapter: N. Sablon, A. Borgoo, Prof. P. Geerlings, Dr. A.M. Teale, M.J.G. Peach, Dr. M. Torrent-Sucarrat, and Prof. P.W. Ayers. F.D.P. also wishes to thank Professor Pratim Chattaraj, with whom he has had many interesting discussions on conceptual DFT and related aspects, for his kind invitation to contribute to this volume.

NOTES AND REFERENCES

1. R. G. Parr and W. Yang, *Density Functional Theory of Atoms and Molecules*, Oxford University Press, New York, 1989.
2. R. M. Dreizler and E. K. U. Gross, *Density Functional Theory*, Springer-Verlag, Berlin Heidelberg, New York, 1990.
3. R. G. Parr and W. Yang, *Ann. Rev. Phys. Chem.* 46, 710, 1995.
4. W. Kohn, A. D. Becke, and R. G. Parr, *J. Phys. Chem.* 100, 978, 1996.
5. W. Koch and M. Holthausen, *A Chemist's Guide to Density Functional Theory*, Wiley-VCH, Weinheim, 2nd ed., 2001.
6. P. Hohenberg and W. Kohn, *Phys. Rev.* 136, B864, 1964.
7. W. Kohn and L. J. Sham, *Phys. Rev.* 140, 1133, 1965.
8. P. W. Ayers, J. S. M. Anderson and L. J. Bartolotti, *Int. J. Quantum Chem.* 101, 520 2005.
9. R. F. Nalewajski and J. Korchowiec, *Charge Sensitivity Approach to Electronic Structure and Chemical Reactivity*, World-Scientific, Singapore, 1997.
10. H. Chermette, *J. Comp. Chem.* 20, 129, 1999.
11. P. Geerlings, F. De Proft and W. Langenaeker, *Chem. Rev.* 103, 1793, 2003.
12. P. Geerlings and F. De Proft, *Phys. Chem. Chem. Phys.* 10, 3028, 2008.
13. R. G. Parr, R. A. Donnelly, M. Levy and W. E. Palke, *J. Chem. Phys.* 69, 3801, 1978.

14. For a detailed account on different electronegativity scales, see, e.g., J. Mullay in *Electronegativity* (*Structure and Bonding*, Vol. 66), K. D. Sen and C. K. Jørgenson editors, Springer-Verlag, Berlin-Heidelberg, 1987, p.1.

15. L. Pauling, *J. Am. Chem. Soc.* 54, 3570, 1932.

16. R. S. Mulliken, *J. Chem. Phys.* 2, 782, 1934.

17. R. P. Iczkowski and J. L. Margrave, *J. Am. Chem. Soc.* 83, 3547, 1961.

18. R. G. Parr and R.G. Pearson, *J. Am. Chem. Soc.* 105, 7512, 1983.

19. R. G. Pearson, *J. Am. Chem. Soc.* 85, 3533, 1963.

20. R. G. Pearson, *Science* 151, 172, 1966.

21. R. G. Pearson, *Chemical Hardness*, John Wiley and Sons, 1997.

22. W. Yang and R. G. Parr, *Proc. Natl. Acad. Sci.* 82, 6723, 1985.

23. R. G. Parr and W. Yang, *J. Am. Chem. Soc.* 106, 4049, 1984.

24. For a perspective on the orginal paper introducing this quantity, see P. W. Ayers and M. Levy, *Theor. Chem. Acc.* 103, 353, 2000.

25. K. Fukui, *Science* 218, 747, 1987.

26. P. W. Ayers, F. De Proft, A. Borgoo and P. Geerlings, *J. Chem. Phys.* 126, 224107, 2007.

27. N. Sablon, F. De Proft, P. W. Ayers and P. Geerlings, *J. Chem. Phys.* 126, 224108, 2007.

28. J. P. Perdew, R. G. Parr, M. Levy, and J. L. Balduz, *Phys. Rev. Lett.* 49, 1691, 1982.

29. For a perspective on the previous reference, see Y. Zhang and W. Yang, *Theor. Chem. Acc.* 103, 346, 2000.

30. N. C. Handy and D. J. Tozer, *Mol. Phys.* 94, 707, 1998.

31. J. F. Janak, *Phys. Rev. B* 18, 7165, 1978.

32. G. Klopman, *Chemical Reactivity and Reaction Paths*, Wiley, New York, 1974.

33. W. Langenaeker, F. De Proft and P. Geerlings, *J. Phys. Chem.* 99, 6424, 1995.

34. P. K. Chattaraj, *J. Phys. Chem. A 2001* 105, 511, 2001.

35. F. De Proft and P. Geerlings, *J. Chem. Phys.* 106, 3270 1997.

36. J. C. Rienstra-Kiracofe, G. S. Tschumper, H. F. Schaefer, III, S. Nandi and G. B. Ellison, *Chem. Rev.* 102, 231, 2002.

37. L. Sanche and G. J. Schulz, *Phys. Rev. A* 5, 1672, 1972.

38. K. D. Jordan and P. D. Burrow, *Chem. Rev.* 87, 557, 1987.

39. J. P. Perdew and M. Levy, *Phys. Rev. Lett.* 51, 1884, 1983.

40. J. P. Perdew and M. Levy, *Phys. Rev. B* 56, 16021, 1997.

41. D. J. Tozer, *J. Chem. Phys.* 119, 12697, 2003.

42. M. J. Allen and D. J. Tozer, *Mol. Phys.* 100, 433, 2002.

43. T. Koopmans, *Physica* 1, 104, 1934.

44. R. G. Pearson, *Proc. Natl. Acad. Sci. USA* 83, 8440, 1986.

45. D. J. Tozer and N. C. Handy, *J. Chem. Phys.* 109, 10180, 1998.

46. G. K.-L. Chan, *J. Chem. Phys.* 110, 4710, 1999.

47. D. J. Tozer and F. De Proft, *J. Phys. Chem. A* 109, 8923, 2005.

48. R. A. Kendall, T. H. Dunning Jr., and R. J. Harrison, *J. Chem. Phys.* 96, 6796, 1992.

49. R. D. Amos et al. CADPAC6.5, The Cambridge Analytic Derivatives Package, 1998.

50. Ionization energies: F_2 and Cl_2: J. M. Dyke, G. D. Josland, J. G. Snijders and P. M. Boerrigter, *Chem. Phys.* 91 1984, 419; H_2CO: W. von Niessen, G. Bieri and L. Asbrink, *J. Electron Spectrosc. Relat. Phenom.* 21 1980, 175; CH_4 and C_2H_4: G. Bieri, L. Asbrink. *J. Electron Spectrosc. Relat. Phenom.* 20 1980, 149; CO: A. W. Potts and T. A. Williams. *J. Electron Spectrosc. Relat. Phenom.* 3 1974, 3; PH_3: A. H. Cowley, R. A. Kemp, M. Lattman and M. L. McKee. *Inorg. Chem.* 21 1982, 85; H_2S: G. Bieri, L. Asbrink and W. Von Niessen. *J. Electron Spectrosc. Relat. Phenom.* 27 1982, 129; HCN: J. Kreile, A. Schweig and W. Thiel. *Chem. Phys. Lett.* 87 1982, 473; HCl: D. W. Turner, C. Baker, A. D. Baker and C. R. Brundle, *Molecular Photoelectron Spectroscopy*, Wiley-Interscience, London, 1970; CO_2: J. H. D. Eland and Berkowitz, *J. Chem. Phys.* 67 1977, 5034; NH_3: H. Baumgartel, H.-W. Jochims, E. Ruhl, H. Bock, R. Dammel, J. Minkwitz, and R. Nass. *Inorg. Chem.* 28 1989,

943; HF: M. S. Banna and D. A. Shirley. *J. Chem. Phys.* 63 1975, 4759; H_2O: K. Kimura, S. Katsumata, Y. Achiba, T. Yamazaki, and S. Iwata, *Handbook of HeI Photoelectron Spectra of Fundamental Organic Compounds*, Japan Scientific Society Press, Tokyo, 1981; All electron affinities: R.G. Pearson. *Inorg. Chem.* 27 1988, 734, except for F_2 and Cl_2 which are taken from J. A. Ayala, W. E. Wentworth and E. C. M. Chen, *J. Phys. Chem* 85, 768, 1981.

51. J. P. Perdew, K. Burke and M. Ernzerhof, *Phys. Rev. Lett.* 77, 3865, 1996.
52. D. M. A. Vera and A. B. Pierini, *Phys. Chem. Chem. Phys.* 6, 2899, 2004.
53. F. De Proft, N. Sablon, D. J. Tozer and P. Geerlings, *Faraday Discuss* 135, 151, 2007.
54. A. D. Becke, *J. Chem. Phys.* 98 1993, 5648; C. Lee, W. Yang and R. G. Parr, *Phys. Rev. B* 37 1988, 785; P. J. Stevens, F. J. Delvin, C. F. Chablaoski and M. J. Frisch, *J. Phys. Chem.* 98 1994, 11623.
55. For a detailed account on these types of basis sets, see, e.g., W. J. Hehre, L. Radom, P. v. R. Schleyer and J. A. Pople, *Ab Initio Molecular Orbital Theory*, Wiley, New York, 1986.
56. D. J. Tozer and F De Proft, *J. Chem. Phys.* 127, 034108, 2007.
57. N. Sablon, F. De Proft, P. Geerlings, and D. J. Tozer, *Phys. Chem. Chem. Phys.* 9, 5880, 2007.

Index

Printed and bound by CPI Group (UK) Ltd, Croydon, CR0 4YY

24/10/2024

01778302-0019